Climate Change and Global Health

Primary, secondary and tertiary effects

Climate Change and Global Health

Primary, secondary and tertiary effects

Edited by

Colin D. Butler

Australian National University,
Canberra, ACT, Australia

Kerryn Higgs

University of Tasmania,
Hobart, Tas, Australia

and

The Club of Rome
Winterthur, Switzerland

CABI

CABI is a trading name of CAB International

CABI
Nosworthy Way
Wallingford
Oxfordshire OX10 8DE
UK

Tel: +44 (0)1491 832111
E-mail: info@cabi.org
Website: www.cabi.org

CABI
200 Portland Street
Boston
MA 02114
USA

Tel: +1 (617)682-9015
E-mail: cabi-nao@cabi.org

A catalogue record for this book is available from the British Library, London, UK.

ISBN-13: 9781800620001 (hardback)
9781800620018 (ePDF)
9781800620025 (ePub)

DOI: 10.1079/9781800620025.0000

Commissioning Editor: David Hemming
Editorial Assistant: Emma McCann
Production Editor: James Bishop

Typeset by Straive, Pondicherry, India
Printed and bound in the UK by CPI Group (UK) Ltd, Croydon CR0 4YY

Contents

Dedication

Dedicated, with gratitude, to three of the most important people to inspire us: Drs Maurice King, A.J. (Tony) McMichael and Will Steffen.

Contributors

Dr Ågot Aakra: SINTEF Industry, Oslo, Norway. Email: agot.aakra@sintef.no

Adjunct Professor Khaled Abass: Environmental Health Sciences, University of Sharjah, UAE.; and Research Unit of Biomedicine and Internal Medicine, University of Oulu, Finland. Email: khaled.megahed@oulu.fi

Professor Robyn Alders: Development Policy Centre, *and* Institute for Climate, Energy and Disaster Solutions, The Australian National University, Canberra, ACT, Australia; Global Health Programme, Chatham House, London; Kyeema Foundation, Brisbane, Qld, Australia. Email: Robyn.Alders@anu.edu.au

Dr A. Kofi Amegah: Department of Biomedical Sciences, University of Cape Coast, Cape Coast, Ghana. Email: aamegah@ucc.edu.gh

Dr Janetrix Hellen Amuguni: Department of Infectious Diseases and Global Health, Tufts University, Medford/Somerville, Massachusetts, USA. Email: Janetrix.Amuguni@tufts.edu

Dr Gro Dehli Andersen: Division of Mental and Physical Health, Norwegian Institute of Public Health, Oslo, Norway. Email: gro.dehli.andersen@fhi.no

Dr Gulrez Shah Azhar: RAND Corporation, Santa Monica, California, USA. Email: gulrezdoc@gmail.com

Associate Professor Katherine Barraclough: School of Medicine, University of Melbourne, Melbourne, Vic, Australia; Department of Nephrology, Royal Melbourne Hospital, Melbourne, Vic, Australia. Email: Katherine.Barraclough@mh.org.au

Dr Jim Berner: Senior Director for Science, Division of Community Health, Alaska Native Tribal Health Consortium, Anchorage, Alaska, USA. Email: bernerb@aol.com

Dr Alexander J. Blum: Johns Hopkins University School of Medicine, Baltimore, Maryland, USA. Email: ablum16@jhmi.edu

Professor Justin Borevitz: Institute for Climate, Energy and Disaster Solutions *and* Research School of Biology, The Australian National University, Canberra, ACT, Australia. Email: justin.borevitz@anu.edu.au

Visiting Professor Menno Bouma: Instituto de Salud Global, Barcelona, Spain; independent scientist. Email: mennojbouma@gmail.com

Dr Devin C. Bowles: Faculty of Health, University of Canberra, Canberra, ACT, Australia. Email: devincombsbowles@yahoo.com

Assistant Professor Mark Braidwood: Department of Family Medicine, Queens University, Kingston, Ontario, Canada. Email: braidwood.mark@icloud.com

Dr Anne Lise Brantsæter: Division of Climate and Environmental Health, Norwegian Institute of Public Health, Oslo, Norway. Email: AnneLise.Brantsaeter@fhi.no

Honorary Professor Colin D. Butler: National Centre for Epidemiology and Population Health, College of Health & Medicine; *and* Institute for Climate, Energy and Disaster Solutions, The Australian National University, Canberra, ACT, Australia; *and* co-founder: Benevolent Organisation for Development, Health and Insight (BODHI Australia and BODHI US). Email: colin.butler@anu.edu.au

Dr Cyril Caminade: Abdus Salam International Centre for Theoretical Physics, Trieste, Italy; Department of Livestock and One Health, Institute of Infection, Veterinary and Ecological Sciences, University of Liverpool, UK. Email: ccaminad@ictp.it

Professor Katrina J. Charles: School of Geography and the Environment, University of Oxford, Oxford, UK. Email: katrina.charles@ouce.ox.ac.uk

Associate Professor Fiona Charlson: Queensland Centre for Mental Health Research, Queensland Health, Wacol, Qld, Australia; School of Public Health, University of Queensland, Brisbane, Qld, Australia. Email: f.charlson@uq.edu.au

Dr Moumita Sett Chatterjee: Ergonomics and Occupational Physiology Laboratory, Department of Physiology, University of Kalyani, West Bengal, India. Email: settm123@gmail.com

Professor Matthew Chersich: Wits Planetary Health Research, University of the Witwatersrand, Republic of South Africa; and Public Health and Primary Care, School of Medicine, Trinity College Dublin, Ireland. Email: matthew.chersich@witsphr.ac.za

Dr Rebecca Colvin: Crawford School of Public Policy, The Australian National University, Canberra, ACT, Australia. Email: rebecca.colvin@anu.edu.au

Dr Namukolo Covic: International Livestock Research Institute, Addis Ababa, Ethiopia. Email: N.Covic@cgiar.org

Professor Christopher B. Daniels: Green Adelaide Landscape Board, Department for Environment and Water, Adelaide, SA, Australia. Email: chris.daniels@sa.gov.au

Dr Richard Denniss: The Australia Institute, Canberra, ACT, Australia. Email: rd@australiainstitute.org.au

Dr Cybele Dey: Discipline of Psychiatry and Mental Health, School of Clinical Medicine Faculty of Medicine and Health, University of New South Wales, Sydney, NSW, Australia. Email: c.dey@unsw.edu.au

Dr Hubert Dirven: Division of Climate and Environmental Health, Norwegian Institute of Public Health, Oslo, Norway. Email: Hubert.Dirven@fhi.no

Professor Yuming Guo: Climate, Air Quality Research Unit, School of Public Health and Preventive Medicine, Monash University, Melbourne, Vic, Australia. Email: Yuming.Guo@monash.edu

Dr Kristine Bjerve Gutzkow: Division of Climate and Environmental Health, Norwegian Institute of Public Health, Oslo, Norway. Email: KristineBjerve.Gutzkow@fhi.no

Professor Tari Haahtela: Helsinki University Hospital, University of Helsinki, Finland. Email: tari.haahtela@haahtela.fi

Dr Ivan C. Hanigan: Faculty of Health Science and School of Population Health, Curtin University, Perth, WA, Australia. Email: ivan.hanigan@curtin.edu.au

Dr Andrew Harmer: Centre for Primary Care and Public Health, Queen Mary University of London, London, UK. Email: a.harmer@qmul.ac.uk

Professor Budi Haryanto: Research Centre for Climate Change, Universitas Indonesia, Jakarta, Java Indonesia. Email: bharyanto@ui.edu

Dr Kerryn Higgs: School of Social Sciences, University of Tasmania, Hobart, Tas, Australia; Associate Member, The Club of Rome, Winterthur, Switzerland. Email: kerryn1@smartchat.net.au

Dr Susanne Hyllestad: Division of Infection Control, Norwegian Institute of Public Health, Oslo, Norway. Email: Susanne.Hyllestad@fhi.no

Dr Christine Instanes: Division of Climate and Environmental Health, Norwegian Institute of Public Health, Oslo, Norway. Email: Christine.Instanes@fhi.no

Dr Ruth Irwin: Independent scholar, Australia and New Zealand. Email: ruth.irwin@gmail.com

Professor Ollie Jay: Faculty of Medicine & Health, University of Sydney, Sydney, Australia. Email: ollie.jay@sydney.edu.au

Dr Solveig Jore: Division of Infection Control, Norwegian Institute of Public Health, Oslo, Norway. Email: solveig.jore@fhi.no

Dr Ke Ju: Climate, Air Quality Research Unit, School of Public Health and Preventive Medicine, Monash University, Melbourne, Vic, Australia. Email: ke.ju@monash.edu

Professor Tord Kjellstrom: Health and Environment International Trust, Nelson, New Zealand. Email: kjellstromt@yahoo.com

Dr Marit Låg: Division of Climate and Environmental Health, Norwegian Institute of Public Health, Oslo, Norway. Email: Marit.Lag@fhi.no

Associate Professor Jason K.W. Lee: Heat Resilience and Performance Centre, Yong Loo Lin School of Medicine, National University of Singapore, Singapore. Email: phsjlkw@nus.edu.sg

Associate Professor Shanshan Li: Climate, Air Quality Research Unit, School of Public Health and Preventive Medicine, Monash University, Melbourne, Vic, Australia. Email: Shanshan.Li@monash.edu

Dr Irakli Loladze: Bryan College of Health Sciences, Lincoln, Nebraska, USA; School of Mathematical and Statistical Sciences, Arizona State University, Tempe, Arizona, USA. Email: loladze@gmail.com

Assistant Professor Rosemary A. McFarlane: Health Research Institute, University of Canberra, Canberra, ACT, Australia. Email: ro.mcfarlane@canberra.edu.au

Professor Martin McKee: London School of Hygiene and Tropical Medicine, London, UK. Email: Martin.McKee@lshtm.ac.uk

Dr Helle Margrete Meltzer: Division of Climate and Environmental Health, Norwegian Institute of Public Health, Oslo, Norway. Email: hellemargrete.meltzer@fhi.no

Professor Glen Mola: Department of Obstetrics, Gynaecology and Reproductive Health, School of Medicine and Health Sciences, The University of Papua New Guinea, Port Moresby, Papua New Guinea. Email: gmola4@gmail.com

Professor Andrew B. Morse: Department of Livestock and One Health, Institute of Infection, Veterinary and Ecological Sciences, University of Liverpool, UK. Email: a.p.morse@liverpool.ac.uk

Dr Juliet Nabyonga-Orem: World Health Organization, Africa Regional Office, Brazzaville, Congo. Email: nabyongaj@who.int

Dr Nick H. Ogden: Public Health Risk Sciences Division, National Microbiology Laboratory, Public Health Agency of Canada, Saint-Hyacinthe, Canada. Email: nicholas.ogden@canada.ca

Professor Johan Øvrevik: Division of Climate and Environmental Health, Norwegian Institute of Public Health; Department of Biosciences, University of Oslo, Oslo, Norway. Email: Johan.Ovrevik@fhi.no

Dr Rebecca Patrick: Melbourne School of Population and Global Health; Faculty of Medicine, Dentistry and Health Sciences, University of Melbourne, Melbourne, Vic, Australia. Email: rebecca.patrick@unimelb.edu.au

Dr Md Rezanur Rahaman: National Centre for Epidemiology and Population Health, College of Health & Medicine, The Australian National University, The Australian National University, Canberra, ACT, Australia. Email: MdRezanur.Rahaman@anu.edu.au

Professor Delia Randolph: University of Greenwich, London, UK; International Livestock Research Institute, Nairobi, Kenya. Email: D.C.Randolph@greenwich.ac.uk

Dr Shilpa Rao: Division of Climate and Environmental Health, Norwegian Institute of Public Health, Oslo, Norway. Email: Shilpa.Rao@fhi.no

Professor Arja Rautio: Arctic Health, University of Oulo, Oulo, Finland. Email: arja.rautio@oulu.fi

Mary Robinson: Former President of Ireland; Adjunct Professor of Climate Justice, Trinity College Dublin, Ireland.

Dr Tilman Ruff, AO: Immediate past Co-President, Board member, International Physicians for the Prevention of Nuclear War, Boston, Massachusetts, USA. International Campaign to Abolish

Nuclear Weapons Australia; Honorary Principal Fellow, School of Population and Global Health, University of Melbourne, Naarm/Melbourne, Vic, Australia. Email: tar@unimelb.edu.au

Associate Professor Subhashis Sahu: Ergonomics and Occupational Physiology Laboratory, Department of Physiology, University of Kalyani, West Bengal, India. Email: ssahu@klyuniv.ac.in

Professor Jonathan M. Samet: Colorado School of Public Health, Anschutz, Colorado, USA. Email: jon.samet@cuanschutz.edu

Professor Photini Sinnis: Bloomberg School of Public Health, School of Medicine, Johns Hopkins University, Baltimore, Maryland, United States. Email: psinnis1@jhu.edu

Associate Professor Julie P. Smith: National Centre for Epidemiology and Population Health, College of Health & Medicine, The Australian National University, Canberra, ACT, Email: julie.smith@anu.edu.au

Dr Jessica Stanhope: School of Allied Health Science and Practice, The University of Adelaide, Adelaide, SA, Australia. Email: jessica.stanhope@adelaide.edu.au

(The late) **Emeritus Professor Will Steffen:** Fenner School of Environment and Society, The Australian National University, Canberra, ACT, Australia.

Professor Peter Stoett: Faculty of Social Science and Humanities, Ontario Tech University, Oshawa, Ontario, Canada. Email: Peter.Stoett@ontariotechu.ca

Dr Stewart Sutherland: College of Health & Medicine, Australian National University, Canberra, ACT, Australia. Email: stewart.sutherland@anu.edu.au

Dr Cathrine Thomsen: Division of Climate and Environmental Health, Norwegian Institute of Public Health, Oslo, Norway. Email: cathrine.thomsen@fhi.no

Associate Professor Collin Tukuitonga: Division of Public Health, University of Auckland, New Zealand. Email: collin.tukuitonga@auckland.ac.nz

Dr May van Schalkwyk: London School of Hygiene and Tropical Medicine, London, UK. Email: May.VanSchalkwyk@lshtm.ac.uk

Dr Sue Wareham: Medical Association for the Prevention of War, Canberra, ACT, Australia. Email: sue.wareham@gmail.com

Professor Philip Weinstein: School of Public Health, The University of Adelaide, Adelaide, SA, Australia. Email: philip.weinstein@adelaide.edu.au

Dr Bo Wen: Climate, Air Quality Research Unit, School of Public Health and Preventive Medicine, Monash University, Melbourne, Vic, Australia. Email: bo.wen@monash.edu

Rob White: Emeritus Distinguished Professor of Criminology, School of Social Sciences, University of Tasmania, Hobart, Tas, Australia. Email: r.d.white@utas.edu.au

Dr Kirsty Wild: Department of Epidemiology and Biostatistics, School of Population Health, University of Auckland, New Zealand. Email: k.wild@auckland.ac.nz

Kate Wingett: Faculty of Medicine and Health, University of Sydney, Sydney, NSW, Australia. Email: kwin5980@uni.sydney.edu.au

Professor Alistair Woodward: Department of Epidemiology and Biostatistics, School of Population Health, University of Auckland, New Zealand. Email: a.woodward@auckland.ac.nz

Professor Caradee Y. Wright: South African Medical Research Council, South Africa; Department of Geography, Geoinformatics and Meteorology, University of Pretoria, Pretoria, Republic of South Africa; Department of Environmental Health, University of Johannesburg, Republic of South Africa. Email: Caradee.Wright@mrc.ac.za

Dr Yao Wu: Climate, Air Quality Research Unit, School of Public Health and Preventive Medicine, Monash University, Melbourne, Vic, Australia. Email: Yao.Wu@monash.edu

Dr Pei Yu: Climate, Air Quality Research Unit, School of Public Health and Preventive Medicine, Monash University, Melbourne, Vic, Australia. Email: pei.yu@monash.edu

Foreword

The ongoing and reckless emission of huge quantities of colourless greenhouse gases poses a great threat to humanity. Climate change interacts with many undesirable aspects of human behaviour, including inequality, racism and other manifestations of injustice. Climate change policies, as practised by most countries in the global North, not only interact with these long-standing forms of injustice, but also exemplify a new form, of startling magnitude. In the past, colonial empires (mostly ruled by governments of the North) acted with racism, contempt and disregard towards people of colour who were born into 'their' vast realms. But colonial policies rarely represented an existential threat for the survival of entire populations. Modern climate change policies, reflected by rising greenhouse gas concentrations (rather than promises of action made by rich countries) *do* represent such an existential threat. However, the people at risk are not only in the global South, but also in the North; such as those living in increasingly uninsurable flood zones, or those at risk of uncontrollable wildfire.

In the 20th century many industrialized locations in the global North acted, effectively, to reduce visible, localized air pollution, such as from smokestacks and car exhaust pipes. Adverse health effects were clear; affordable, if not always ethical, alternatives existed, such as the 'offshoring' of pollution-intensive manufacturing. If greenhouse gases were both visible and confined to either their origin or the site of fossil fuel extraction, these same populations would have acted, long ago, to curtail their production, even without proof of harm to their health.

In spite of insufficient political leadership, technologies and lifestyles with less environmental harm have emerged, though as yet they are not used at sufficient scale. While a major task of contemporary climate change activists is to persuade people in high-income settings to act altruistically to help the lives of those who live far away, a more reachable goal may be to promote 'win-wins', attainable from action to slow greenhouse gas pollution. Lifestyle changes to protect climate (e.g. less consumption of animal products, a switch to public transport, solar panels and electric cars) will not only help future generations in the global South, but also bring local, more immediate benefits, including to health; a topic which is increasingly understood, and which this book helps to reveal.

The evidence of climate change for which human activities are responsible is undeniable. Yet policy makers have for decades failed to address its causes and plan for the consequences. At times, I have felt a great frustration at the lack of urgency of my generation to address this most existential of threats. I have even had sleepless nights when I consider the state of the world my grandchildren may grow old in. However, I still have hope that the rising generation will not only fight for, but secure, climate justice for all.

Climate justice is a transformative concept – compelling a shift from a discourse on melting icecaps into a civil rights movement with the people and communities most vulnerable to climate impacts at its heart. Climate justice links human rights and development to achieve a human-centred approach, safeguarding the rights of the most vulnerable people and sharing the burdens and benefits of climate change and its impacts equitably and fairly.

This book provides invaluable evidence for public health professionals about the effects of climate, their vulnerability to climate change and how appropriate policies can enhance the resilience of communities to climatic change. It provides a comprehensive overview of a rapidly evolving field of research, which will be invaluable to those with responsibility for protecting populations from the effects of extreme events (e.g. heatwaves, floods, droughts and intense tropical cyclones) that are likely to increase in frequency with climate change. It will also be of interest to the wider public health and development communities, who need to be aware of how our changing climate could cause shifts in disease risk, including changes in the distribution of vector-borne and other climate-sensitive communicable diseases, and a range of other health outcomes.

We owe much to the public health pioneers who first pointed to the potential health impacts of climate change, more than 30 years ago, including Andy Haines, Martin Parry and the late Paul Epstein and Tony McMichael. Their leadership was crucial in raising the profile of this topic, increasingly accepted as a legitimate, indeed an increasingly vital, focus of concern for public health. This book demonstrates why it is imperative to move towards 'low-carbon' development, which achieves improvements in health and well-being without undermining the life-support systems on which civilization ultimately depends.

Mary Robinson
UN High Commissioner for Human Rights (1997–2002)
August, 2023

Introduction to the Second Edition

Welcome to the second edition. We share, with an increasing number of academics and activists, a concern that in the Anthropocene (the human-dominated era), civilization and thus population health is on a perilous trajectory. Humans traverse a narrow track, whose route is shaped by the struggle to meet the needs and wants of over eight billion (and growing) people, and the natural laws and other limits of the biosphere. While our resources are augmented by daily sunlight, they can easily be damaged, not only by time and geophysical forces (e.g. earthquakes) but also by conscious human decisions, such as missiles and machetes.

The warmth from that sunlight, on which we all depend, is increasingly being trapped within the Earth system, particularly by the greenhouse gases carbon dioxide, methane and nitrous oxide. Of the many risks and threats to human health that humanity must confront, the book focuses especially on those arising from climate change. However, these risks cannot be separated from other drivers of global health, including politics, power imbalances, human exploitation (both of other people and of nature) and the shadow of nuclear weapons; themes which recur throughout the book.

Necessarily, all chapters are time-bound synopses. In a book of this length, chapters cannot be comprehensive, nor can all topics be covered. Instead, we seek to identify key principles, with citations that introduce non-specialists and students to the vast and rich literature from which all contributors draw. We also believe that understanding the past helps us to understand and, to an extent, to forecast the future. Thus, several chapters explore historical aspects. Many also respect a strong 'ecological' theme.

Climate change and planetary health

A succinct definition of planetary health is the 'health of civilization and the state of the natural systems on which it depends' (Whitmee *et al.*, 2015). This definition is relevant to this book's scope, and some readers may wonder if we should have altered our title. There are several reasons we resisted. One is to reflect the substantial overlap of authors and subjects with the first edition. Another is that climate change, while a substantial component of planetary health, does not fully overlap with planetary health. Chapter 4 (the updated version of the titular chapter from the first edition) explores this topic in more detail, and is now called 'Climate Change, Global Health and Planetary Health'.

Changes to this edition

This edition has eight Sections, compared with seven in the first edition. Section 2, which has been added, has six chapters grouped around the theme of ecology and health – including 'human' ecology and human 'carrying capacity'. Another important change is the addition of a chapter on the risk of nuclear weapons, together with three mini-chapters on aspects of inequality (each of these is added to Section 1).

Finally, we thank the many authors, reviewers and others who have helped us to finalize this book. We especially thank Dr Kim van Daalen, whose input led to an expansion of the book's scope. We also thank the staff of CABI, especially David Hemming. All errors are our responsibility, of course.

Reference

Whitmee, S., Haines, A., Beyrer, C., Boltz, F., Capon, A.G. *et al.* (2015) Safeguarding human health in the Anthropocene epoch: report of The Rockefeller Foundation–Lancet Commission on Planetary Health. *The Lancet* 386, 1973–2028. DOI: 10.1016/S0140-6736(15)60901-1

Acronyms

AC: air-conditioning
ACT: artemisinin combination therapy
AFIDEP: African Institute for Development Policy
AGRA: Alliance for a Green Revolution in Africa
AIDS: acquired immunodeficiency syndrome
AMAP: Arctic Monitoring and Assessment Programme
AMOC: Atlantic meridional overturning circulation
AMR: anti-microbial resistance
AQI: air quality index
B: boron
BAU: business as usual
BCE: Before the Common (or Current or Christian) Era
BIPOC: Black, Indigenous and People of Colour
BMS: breastmilk substitutes
BP: Before Present
BPNI: Breastfeeding Promotion Network of India
BRIC: Brazil, Russia, India and China
BSE: bovine spongiform encephalopathy
Ca: calcium
CCAB: climate change-attributable burden
CD: Conference on Disarmament
CDC: Centers for Disease Control and Prevention
CDR: carbon dioxide removal
CE: Common Era, Current Era (= AD)
CFE: carbon fertilization effect
CFP: carbon footprint
CH_4: methane
C–H–O: carbon–hydrogen–oxygen
CI: confidence interval
CKD: chronic kidney disease
Cl: chlorine
CMA: China (or Chinese) Meteorological Administration
CMF: commercial milk formula

CO: carbon monoxide
CO_2: carbon dioxide
CO_2e: carbon dioxide equivalent
Co: cobalt
COP: Conference of the Parties
COVID-19: coronavirus disease 2019
Cr: chromium
Cu: copper
DALY: disability-adjusted life years
DARA: Development Assistance Research Associates
DDT: dichlorodiphenyltrichloroethane
DENV: dengue virus
DHM: donor human milk
DRC: Democratic Republic of the Congo
EEZ: Exclusive Economic Zone
EID: emerging infectious diseases
EIP: extrinsic incubation period
EMP: electromagnetic pulse
EPA: Environmental Protection Agency
EROEI: energy return on energy investment
EU: European Union
EV: electric vehicle
F: fluorine
FACE: free-air carbon dioxide enrichment
FAO: Food and Agriculture Organization of the United Nations
FAS: Federation of American Scientists
FBD: Foodborne disease
Fe: iron
FGM: female genital mutilation
FHI: Family Health International
FIFA: Fédération Internationale de Football Association
G20: Group of 20 (intergovernmental forum of 19 countries, EU and African Union (AU))
G77: Group of 77 (UN coalition of 135 developing countries)
GDP: gross domestic product
GEC: global environmental change
GHG: greenhouse gas
GHI: global health initiatives
GMC: General Medical Council
GNAFC: Global Network Against Food Crises
GT: gigaton(ne), i.e. 1 billion (1,000,000,000, or 10^9) tonnes
GtC: gigatonnes of carbon
GVP: global virome project
GWh: gigawatt hour
HAP: Heat Action Plan; household air pollution; household-associated pollution
HCC: human carrying capacity
HEU: highly enriched uranium
HIV: human immunodeficiency virus
HKH: Hindu Kush and Himalayan
HLTV1: human T-cell lymphotropic virus
HNW: high net worth
HPV: human papilloma virus
HRW: Human Rights Watch

I: iodine
IAEA: International Atomic Energy Agency
IBFAN: International Baby Foods Action Network
ICAN: International Campaign to Abolish Nuclear Weapons
ICN: International Council of Nurses
ICPD: International Conference on Population and Development
ICRC: International Committee of the Red Cross
IDB: Inter-American Development Bank
IDP: internally displaced person
IEA: International Energy Agency
IFAD: International Fund for Agricultural Development
IFMSA: International Federation of Medical Students Associations
IgE: immunoglobin (epsilon chain)
IIER: Institute for International Economic Research; International Institute of Engineers and Researchers
ILO: International Labour Organization (UN)
IMF: International Monetary Fund
IMR: infant mortality rate
IOM: International Organization for Migration
IPB: International Peace Bureau
IPBES: Intergovernmental Science-Policy Platform on Biodiversity and Ecosystem Services
IPC: Integrated Food Security Phase Classification
IPCC: Intergovernmental Panel on Climate Change
IPFM: International Panel on Fissile Materials
IPPNW: International Physicians for the Prevention of Nuclear War
IRC: International Rescue Committee
IRENA: International Renewable Energy Agency
IUCN: International Union for the Conservation of Nature
JE: Japanese encephalitis
K: potassium
kPA: kilopascal
kt: kilotonne (1 thousand tonnes, i.e where tonne = 1,000 kgs)
Li: lithium
LSHTM: London School of Hygiene and Tropical Medicine
LtG: limits to growth
MA: Millennium Assessment
MAD: mutually assured destruction
MARA: Mapping Malaria Risk in Africa
MEA: Millennium Ecosystem Assessment
MERS: Middle East respiratory syndrome
Mg: magnesium
MIT: Massachusetts Institute of Technology
MMR: maternal mortality ratio
Mn: manganese
Mo: molybdenum
Mt: megatonne (1 million tonnes, i.e where tonne = 1,000 kgs)
N: nitrogen
N_2O: nitrous oxide
Na: sodium
NASA: National Aeronautics and Space Administration
NCCHAP: National Climate Change and Health Action Plans
NCDs: non-communicable diseases

nd: no date
NDCs: nationally determined contributions
NGO: non-government organization
NH_3: ammonia
Ni: nickel
NIWA: National Institute of Water and Atmospheric Research
NOAA: National Oceanic and Atmospheric Administration
NOx: nitrogen oxides
NPP: net primary productivity
NRM: natural resource management
NSW: New South Wales
NTD: neglected tropical disease
O_3: ozone
O & G: obstetrics and gynaecology
OAS: Organization of American States
OECD: Organisation for Economic Co-operation and Development
OED: *Oxford English Dictionary*
OH: hydroxyl
OHPBL: One Health problem-based learning
OHHLEP: One Health High-Level Expert Panel
OIE: Office International des Epizooties
OPEC: Organization of the Petroleum Exporting Countries
P: phosphorus
PAT: population, affluence, technology
PB: planetary boundaries
PBDE: polybrominated diphenyl ethers
PBL: problem-based learning
PCB: polychlorinated biphenyl
PFAS: per- and poly-fluoroalkyl substances
PHC: primary healthcare
PICTs: Pacific Islands Countries and Territories
PM: particulate matter
PMGH: Port Moresby General Hospital
PMR: perinatal mortality rate
PNG: Papua New Guinea
POP: persistent organic pollutant
ppm: parts per million
PPP: purchasing power parity
PTSD: post-traumatic stress disorder
RCP: representative concentration pathway
S: sulfur
SADC: South African Development Community
SARS: severe acute respiratory syndrome
SDGs: Sustainable Development Goals
Se: selenium
SIDS: Small Islands Developing States
SIDS: sudden infant death syndrome
SIPRI: Stockholm International Peace Research Institute
Sn: tin
SO_x: sulfur oxide

SO_2: sulfur dioxide
SPCZ: South Pacific Convergence Zone
SRHR: sexual and reproductive health and rights
s.s.: *sensu stricto*
SSA: sub-Saharan Africa
START: Strategic Arms Reduction Treaty
STI: sexually transmitted infections
SV: simian virus
TB: tuberculosis
TC: tropical cyclone
TPNW: Treaty on the Prohibition of Nuclear Weapons
Tw: wet-bulb temperature
UCDP: Uppsala Conflict Data Program
UCS: Union of Concerned Scientists
UHC: universal health coverage
UHI: urban heat island(s)
UK: United Kingdom
UN: United Nations
UNCTAD: UN Conference on Trade and Development
UNDESA: UN Department of Economic and Social Affairs
UNDP: UN Development Programme
UNEP: UN Environment Programme
UNESCAP: UN Economic and Social Commission for Asia and the Pacific
UNESCO: UN Educational, Scientific and Cultural Organization
UNFCCC: UN Framework Convention on Climate Change
UNFPA: UN Population Fund (formerly UN Fund for Population Activities)
UNHCR: UN High Commission for Refugees
UNICEF: UN International Children's Fund
UNSNA: UN System of National Accounts
US: United States
USA: United States of America
UV: ultraviolet
V: vanadium
VC: vectorial capacity
vCJD: variant Creutzfeldt-Jakob disease
VOC: volatile organic compounds
WASH: water, sanitation and hygiene
WBGT: wet-bulb globe temperature
WCS: Wildlife Conservation Society
WFP: World Food Programme
WFPHA: World Federation of Public Health Associations
WHM: Women's Health Movement
WHO: World Health Organization
WIREs: Wiley Interdisciplinary Reviews
WM: Women's Movement
WMA: World Medical Association
WMO: World Meteorological Organization
WOAH: World Organisation for Animal Health
WPRO: Regional Office for the Western Pacific

WWF: World Wide Fund for Nature
WWII: World War II
XR: Extinction Rebellion
YOMPI: young, old, malnourished, pregnant or immunosuppressed
Zn: zinc
ZNiSO: *Zdorov'e Naseleniya i Streda Obitaniya* [*Public Health and Life Environment*]

Acknowledgements

Many people have helped with this book; some wish to remain anonymous. In particular, we thank Dr Akeem Ali, Emeritus Prof. Glenn Albrecht, Assist. Prof. Mark Braidwood, Prof. Tony Capon, Dr Cyril Caminade, Ken Cassman (Emeritus Prof. of Agronomy, Univ. of Nebraska, USA), Prof. Kristie Ebi, Jenny Goldie, Dr Richard Grossman, Dr Andrea Hansen, Dr Andrew Harmer, Dr Tracie Hendriks, Dr Simon Hodder, Prof. Tiina Ikäheimo, Assoc. Prof. Fay Johnston, the late Emeritus Prof. Joe Ladou, Dr Prasad Liyanage, Audrey Meltzer, Dr Murray May, Dr Monica Nitschke, Prof. John Potter (Research Centre for Hauora and Health, Massey University, Wellington, New Zealand and Prof. Emeritus of Epidemiology, Univ. of Washington, Seattle, Washington, USA), Prof. Mala Rao CBE., Assoc. Prof. Atanu Sarkar, Dr Ellen Tullo, Dr Kim van Daalen, Dr Sam Varvastian, Prof. Maxine Whittaker, Prof. Philip Weinstein, Prof. John Wiseman and Prof. Jakob Zinsstag.

We also thank all photographers and those who are in the photos; who, where named, have provided signed permission.

1 The Anthropocene: A Planet Under Pressure

Will Steffen
Fenner School of Environment and Society, The Australian National University, Canberra, Australia

Abstract

This chapter explores the vast, human-driven changes that are fundamentally changing our planetary home. Using the concept of the Earth as a single, integrated system – the Earth system – the chapter describes how the rapidly expanding impacts of human activities have changed the structure and functioning of the Earth System so much that we have entered a new geological epoch, the Anthropocene. The future trajectory of the Anthropocene is yet to be determined: can our societies achieve global sustainability or are we headed for collapse?

1.1 The Nature of the Earth System

The advent of climate change, and the broader issue of global change (Steffen *et al.*, 2004), has begun to alter humanity's perspective on our planetary home. To understand humanity's role in the functioning of the Earth system (the interacting physical, chemical and biological processes that link atmosphere, land, ocean and ice into a single planetary system (Steffen *et al.*, 2020)), a basic understanding of Earth history is essential.

For most of its 4.6 billion-year history, life has been a feature of the Earth system and has wrought vast and significant changes to the planet's environment, including the atmosphere. But, for contemporary issues, the past 60 million years or so are most relevant (Zalasiewicz *et al.*, 2012). During that period, the Earth has cooled slowly from a much warmer state, the continents have migrated into their current positions and the great polar ice sheets have appeared from the earlier ice-free world – first the Antarctic ice sheets about 34 million years ago and then the northern hemisphere ice sheets, only about 10–12 million years ago.

Fully modern humans – *Homo sapiens* – are a much more recent phenomenon; we are only about 200,000–250,000 years old as a species (Oppenheimer, 2004) and our hominid ancestors are only a few million years old. The period of the planet during which humans evolved is called the late Quaternary; it is characterized by a rich biodiversity dominated by mammals and a climate that oscillates in a strikingly regular pattern between long, cold ice ages and shorter, intervening warm periods (Petit *et al.*, 1999; EPICA Community Members, 2004).

The regular climatic pattern of the late Quaternary provides strong evidence that the Earth as a whole operates as a single system, albeit a highly complex one. Many specific features of the late Quaternary environment support this perspective (Scheffer, 2009).

 Climate Change and Global Health: Primary, Secondary and Tertiary Effects, 2nd Edition (eds C.D. Butler and K. Higgs)
DOI: 10.1079/9781800620025.0001

- The two states of the late Quaternary are well defined, with the ice ages apparently the more stable state.
- The oscillations between these states are exceptionally regular, with a periodicity of about 100,000 years through the past million years, and a periodicity of about 40,000 years before that. This is an example of 'phase locking', in which the internal dynamics of the Earth system are in synchronization with a regular, but weak, external forcing agent. In this case, regular variations in the Earth's orbit are the weak external forcing and the waxing and waning of the northern hemisphere ice sheets are probably the feature of the Earth's internal dynamics that locks into various modes of the Earth's orbit.
- Perhaps the most important feature of the Earth system is the 'limit cycles'; that is, as the Earth cycle oscillates between ice ages and warm periods, the variation in temperature, greenhouse gas (GHG) concentrations and other features of the environment are tightly constrained between well-defined limits.
- Feedbacks – both reinforcing (positive) and damping (negative) – are essential features of the complex Earth system in the late Quaternary. The two most important of these are: (i) the waxing and waning of the great northern hemisphere ice sheets, which change the reflectivity of the Earth's surface; and (ii) the concentration of atmospheric GHGs, the most important of which is carbon dioxide (CO_2). Both of these feedbacks significantly alter the energy balance at the Earth's surface and modulate the rather modest variations in incoming solar radiation.

1.2 The Expanding Human Enterprise

For nearly all of our existence, the relationship between humans and the Earth system has essentially been one way: the features of the Earth system described above, and many modes of natural variability on finer spatial and temporal scales, have had significant influences, sometimes devastating ones, on the viability of human societies. But humans have operated under the planetary radar screen in terms of their influences back onto the functioning of the Earth system. Our activities may have had environmental impacts at local and sometimes even regional levels, but we did not have any appreciable influence on the functioning of the Earth system on the global level (Steffen *et al.*, 2007, and references therein).

That benign relationship changed with the advent of the Industrial Revolution late in the 18th century, first in England and then elsewhere in Europe and the New World. The most striking feature of the Industrial Revolution was the widespread use of fossil fuels, which broke through a long-standing bottleneck in the development of the human enterprise. Until then, the extent and rate of activity of civilization was constrained tightly by the limits of animal muscle power; that is, the power of humans and their domesticated animals, augmented to a small extent by water power. Fossil fuels shattered that bottleneck.

The impacts of fossil fuel use were not confined to their direct impact on energy production and transport. Arguably more important were the flow-on effects that a cheap and plentiful new energy source facilitated, together with science, governance and popular acceptance. These effects included: (i) the capability to 'fix' unreactive nitrogen from the atmosphere and to make it available to support food production; (ii) the provision of fresh water in reliable ways to most human settlements; and (iii) the design and construction of much more effective sewage systems, leading to a great improvement in human health. These factors led to an explosion of the human population, and to their capability to produce and consume goods and services (McNeill, 2000; Hibbard *et al.*, 2006).

The vast changes in the human enterprise since the Industrial Revolution are presented in Fig. 1.1. The timescale begins at 1750, to capture the beginning of the Industrial Revolution, and continues to 2010. The graphs encompass indicators not only of population and economic activity but also of resource use, communication, transport and globalization. One feature stands out strongly in the figures: the year 1950. In each of the panels, the rate of increase rises sharply about that year, and there is no appreciable value at all for several of the indicators before 1950.

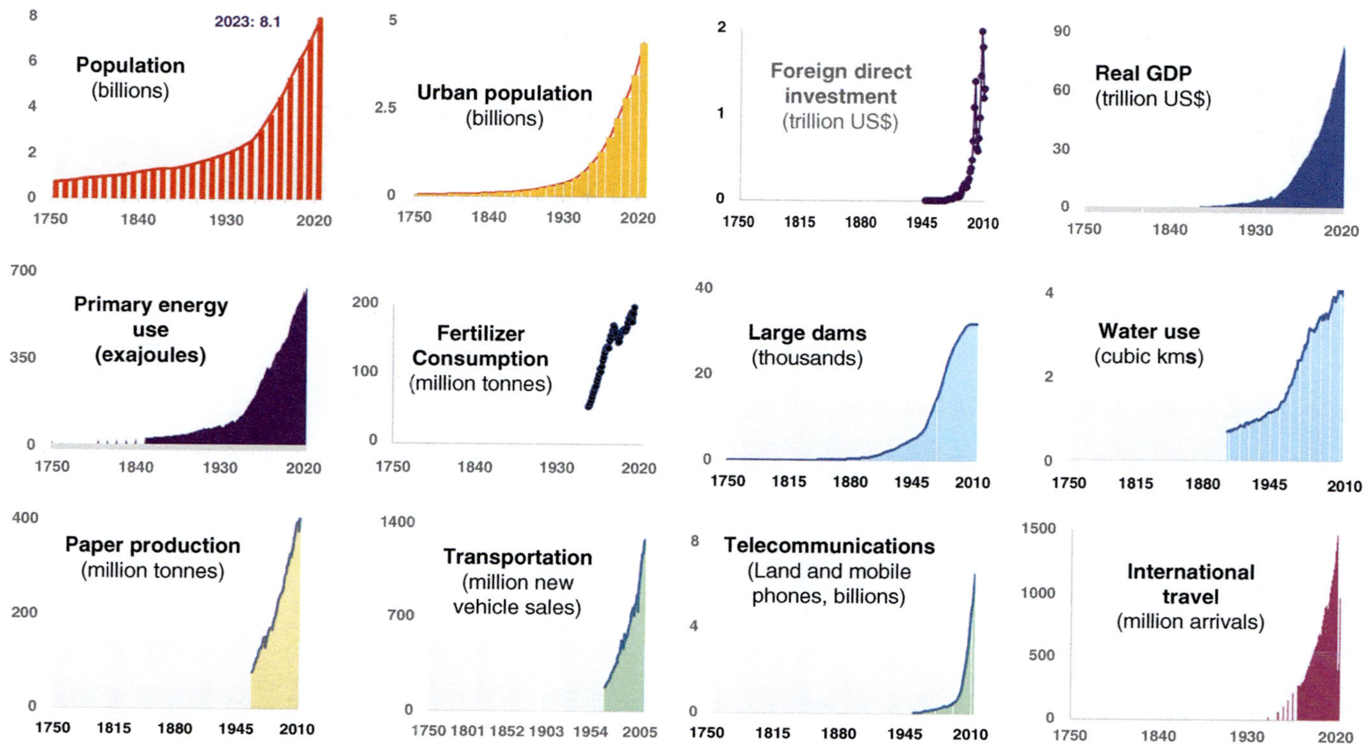

Fig. 1.1. Increasing rates of change in several indicators of human activity since the beginning of the Industrial Revolution. Significant increases in the rates of change of all indicators occur simultaneously around 1950, ushering in a period of dramatic and unprecedented change, often called the Great Acceleration. (adapted from Steffen *et al.*, 2015a). Figure prepared by Colin Butler, used with permission.[1]

This remarkable explosion of the human enterprise soon after the Second World War is sometimes called the 'Great Acceleration' (Hibbard *et al.*, 2006). Many excellent narratives have been created to explain this 20th century phenomenon (e.g. McNeill, 2000), and out of these, a few important drivers of the Great Acceleration can be distilled (Hibbard *et al.*, 2006).

- Global connectivity – people, finance and goods and services became much more connected around the globe as new technologies in communications and transport built the foundation for what is now known as globalization.
- Neoliberal economics – a free-market economic system became dominant and, by the end of the 20th century, had achieved virtually complete coverage of the entire global economic system. The Bretton Woods institutions, created following the Second World War, such as the World Bank and the International Monetary Fund, played a key role in this trend.
- Armies of scientists and engineers – the war effort generated a large cadre of scientists and engineers, and this significant technical capacity was tapped by the civilian economy soon after hostilities had ceased.
- The economic growth imperative – the need to rebuild economies after the devastation of the Depression and Second World War led to a period of rapid growth and the embedding of continuous, strong economic growth as a core value of late 20th century societies (see also Chapter 8, this volume).

The post-1950 Great Acceleration was supported by an abundance of natural resources, especially cheap fossil fuels such as petroleum and coal, and was driven almost entirely by the wealthy countries of the world, which represented no more than 20% of the human population. The first decade of the 21st century brought some changes to this pattern. The environmental contribution from wealthy countries is now amplified by several rapidly industrializing economies, such as those of Brazil, Russia, India and China (the so-called BRIC countries), as well as South Africa and Indonesia. Furthermore, the era of abundant natural resources, especially cheap fossil fuels, seems to be coming rapidly to an end, with the concept of 'peak oil' (e.g. Sorrell *et al.*, 2009) becoming the poster child for a growing array of resource constraints.

1.3 Global Change

The Great Acceleration is an important phenomenon in terms of human history, but it has perhaps even greater significance for the history of planet Earth. The imprint of the human enterprise on the environment can be seen at remarkably early times in our planetary existence, but almost always at local and regional levels only. As shown in Fig. 1.2, human activities have recently become so pervasive that their imprint on the environment at the global level is obvious. The intertwining planetary-level changes in human activities and the environment is now often called 'global change' (e.g. Steffen *et al.*, 2004).

Undoubtedly, the most well-known of the changes depicted in Fig. 1.2 is that of the energy balance at the Earth's surface, resulting in what is commonly known as 'climate change' or 'global warming'. The warming of the Earth's surface is unequivocal, and there are many other associated changes in the climate system, ranging from alteration in the patterns of precipitation to a slowly but inexorably rising sea level (IPCC, 2021). Although natural variability makes some aspects of climate change difficult to interpret, there is a high degree of scientific consensus that human activities, most notably the increase of GHGs in the atmosphere, are the primary cause of the temperature increase and associated climate changes we have experienced since the mid-20th century (IPCC, 2021). Perhaps most worrying is that the influence of human-driven climate change can already be seen in the characteristics of some extreme climatic events (IPCC, 2012), the aspect of climate that most directly and obviously affects human well-being.

While much of the media attention has been focused on climate change, largely missing from the public discourse is the fact that many other human-driven changes in the global environment are perhaps even more pronounced than climate change. A good example is the nitrogen cycle, which has been even more modified by human action than the carbon cycle (Galloway

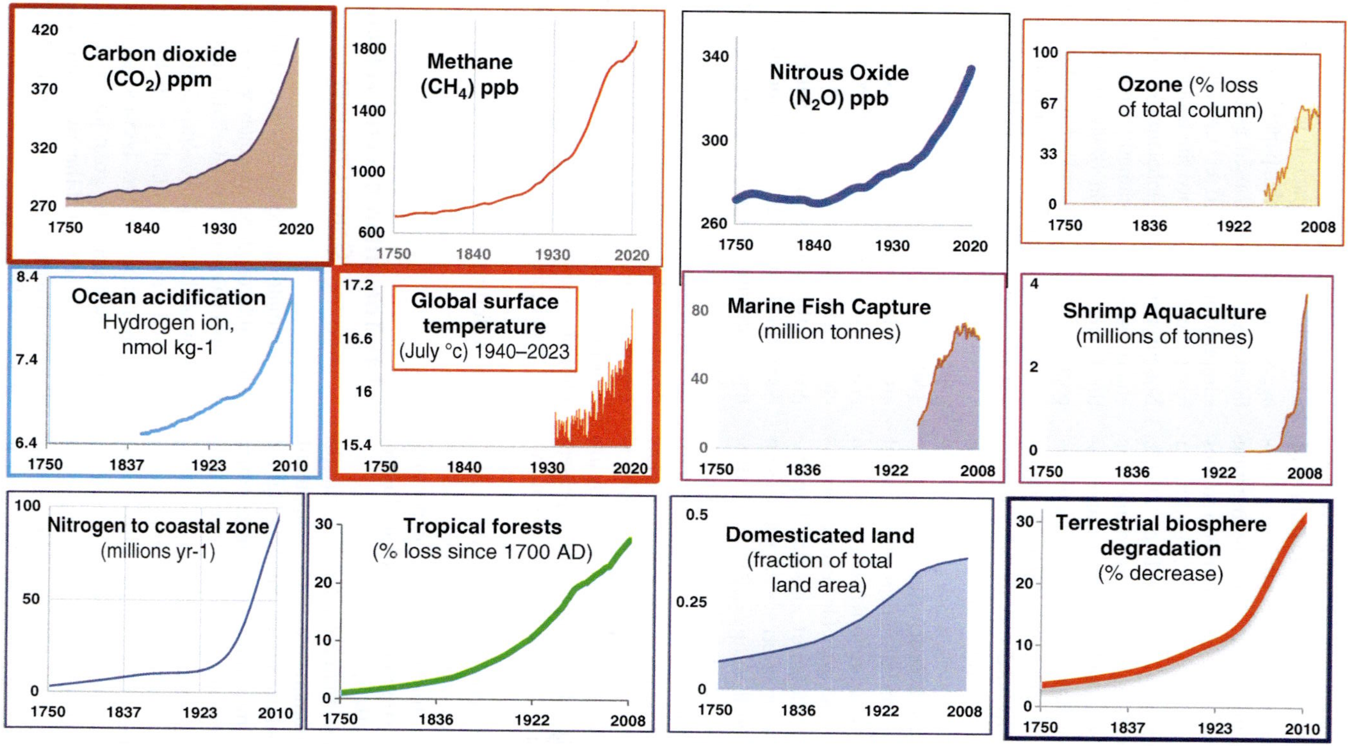

Fig. 1.2. Globally aggregated changes in the Earth System. These result from the dramatic increase in human activity shown in Fig. 1.1. Date ranges as shown in each panel; temperature data extend to July 2023. Adapted from Steffen *et al.*, 2015a, updated by Colin Butler, used with permission.[2]

and Cowling, 2002). Currently, human activities – the direct fixation of unreactive atmospheric nitrogen (N_2) into reactive forms used primarily in fertilizers and the planting of leguminous crops – fix more nitrogen from the atmosphere than all of the natural terrestrial fixation processes combined. In essence, human activities are inserting into the Earth system about 135 million tonnes (Mt) per annum of additional reactive nitrogen compounds (Galloway and Cowling, 2002). These then cascade through the environment as additional flows from the land to freshwater systems and the coastal seas, as a range of gaseous compounds that contribute to local air pollution and add to the GHG burden in the atmosphere, and as reactive compounds that slowly accumulate in the soil.

The loss of biodiversity (IPBES, 2019) is another change to the planetary environment that can be attributed unequivocally to human activities. This impact goes far beyond the loss of charismatic species, but also has implications for the provision of ecosystem services on which human well-being depends (IPBES, 2019). It is estimated that the rate of species extinctions, as measured by extinctions of mammal, bird and amphibian species, is now 100–1000 times greater than the natural background level, and is expected to rise by another factor of 10, at least, by 2100, if present trends continue (Millennium Ecosystem Assessment, 2005). It is even possible that there is a planetary-level tipping point beyond which biodiversity loss will become so rapid and uncontrollable that the Earth may experience the sixth great extinction event in its history (Barnosky *et al.*, 2011).

Nowhere more than in land systems has the direct impact of human activities been apparent, as shown in the bottom left two panels of Fig. 1.2. This is not surprising as humans, being terrestrial creatures, obtain most of their ecosystem services from terrestrial ecosystems. The loss of tropical forests and woodlands has been a dominant feature of land-system change in the 20th century and has continued through the first decades of the 21st century. Overall, the human imprint on the Earth's land surface is perhaps best summarized as the increase in 'domesticated land'; that is, in croplands and pastures, at the expense of forests and other more natural land covers (bottom middle panels of Fig. 1.2).

Methods have been developed to integrate the overall human impact on the global environment. One of them is the application of the I = PAT identity (Holdren and Ehrlich, 1974), in which the overall impact (I) is a function of the aggregation of the global population (P); their affluence (A), which is used as a proxy for consumption of goods and services; and the technology (T) used to produce the goods and services. Not surprisingly, the global impact of the human enterprise shows a dramatic rise since 1950, in the Great Acceleration (Kolbert, 2011). While population increase is, indeed, a factor in growing environmental impact, it is not nearly as important as the increase in consumption per capita.

A second method of estimating the aggregate global impact of humanity is the so-called 'global footprint analysis', which estimates the area of land and ocean per person required to provide all of the resources needed to support each person's consumption and to absorb and process the wastes that are generated (Global Footprint Network, 2021). The country-by-country results are aggregated to give a global total, which can be expressed as the number of planet Earths required to support the human enterprise. At present, that total is about 1.6 planets and rising, which implies that the human enterprise is now in overshoot, consuming more of the Earth's natural resources than can be replenished annually; that is, we are eating into the natural capital of the planet and thus reducing its capacity to support future generations.

1.4 The Anthropocene: A Planetary Game Changer

The ultimate consequence of the rapidly accelerating suite of global changes was summarized by Nobel Laureate, the late Paul Crutzen, in the concept of the Anthropocene (Crutzen, 2002), a proposed new geological epoch in which the human enterprise has become significant in comparison with the great forces of nature. In essence, the arrival of the Anthropocene implies that the Earth is leaving its current geological epoch, the Holocene.

Lasting now for over 10,000 years, the Holocene epoch is the most recent geological epoch in the Earth's history and has proven to be an exceptionally accommodating environment for human development. It was, after all, during the Holocene that we developed agriculture, villages and cities, and the more complex civilizations, including the contemporary civilization we now know.

The arrival of the Anthropocene implies a profound shift in the human–environment relationship at the global level, challenging the assumption that the Earth system will automatically continue to provide an accommodating environment for further human development. Our interaction with the Earth system has passed beyond a simple pressure–impact relationship into one governed by the more chaotic, unpredictable dynamics of complex systems.

We are learning rapidly about these complex system dynamics. An important advance in Earth System science over recent decades has been the elucidation of a number of 'tipping elements' in the Earth System – critical regions or processes where small changes in driving variables can trigger abrupt changes in critical processes, with significant implications for the functioning of the whole Earth system (Fig. 1.3) (Schellnhuber, 2002; Lenton *et al.*, 2008; Richardson *et al.*, 2011; Lenton *et al.*, 2019).

Tipping elements in the Earth system are basically of three types:

1. Those that involve the melting of ice, such as the loss of Arctic sea ice, the potential loss of most or all of the Greenland ice sheet and large emissions of methane from melting permafrost.
2. Changes in circulation, such as a shift in the behaviour of the South Asian monsoon, or a slowing or shutting down of the North Atlantic thermohaline circulation (i.e. the 'Gulf Stream').
3. Flips of major biomes from one state to another, such as the conversion of the Amazon basin from a tropical rainforest to a grassland or a savannah.

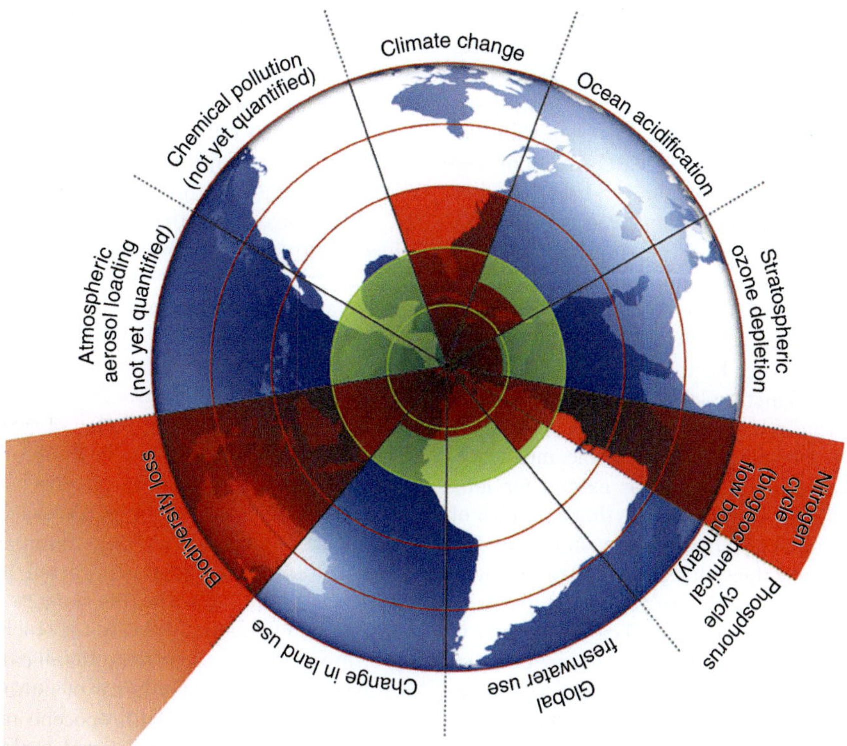

Fig. 1.3. Potential tipping elements in the Earth system. These are critical regions where small changes in driving variables can trigger abrupt changes in critical processes that have significant implications for the functioning of the Earth System as a whole (Schellnhuber, 2002). (Reprinted with kind permission of Springer.)

Although much uncertainty surrounds the nature of these tipping elements, we know from the palaeo-record that abrupt changes in all of them have occurred in the past and we are beginning to estimate where the tipping point might lie for some (Lenton *et al.*, 2008, 2019), where the term tipping point refers to a threshold along an important controlling variable (such as temperature) beyond which the tipping element changes into a fundamentally different state. A recent synthesis of tipping points in the Earth System shows evidence that many tipping points are already under way, and that we are already dangerously close to several cryosphere tipping points. The synthesis concludes that a global cascade of tipping points cannot be ruled out, presenting an existential threat to humanity (Lenton *et al.*, 2019).

Returning now to the complex systems analysis outlined in Section 1.1 above, what are the ultimate implications of the modifications of the global environment by human activities? That is, where is the Anthropocene going?

It is clear that the Earth as a whole operates as a single complex system, oscillating between two well-defined states in the late Quaternary. Up to now, the assumption has been that, depending on the amount of GHG emissions, the climate will stabilize at a temperature 2, 3, 4 or more degrees C above pre-industrial levels (IPCC, 2021); that is, the aggregate amount of emissions determines the final state of the climate system (Meinshausen *et al.*, 2009).

A complex systems perspective suggests that this may not be so. It is conceivable that there is a planetary-level threshold or tipping point beyond which the Earth system as a whole will move to a much warmer, stable, mostly ice-free state – a state that has existed for long periods in the past. This implies that a number of the individual tipping elements shown in Fig. 1.3, especially those that change the reflectivity (albedo) of the Earth's surface (mainly loss of ice), or those that add significant amounts of GHGs to the atmosphere (methane outburst from permafrost), act in concert once a temperature threshold (or series of closely placed temperature thresholds) is crossed and push the Earth system into a much warmer state. Once such a threshold (or thresholds) is crossed, there is nothing that humanity could do to prevent the transition (Lenton *et al.*, 2019).

If there is such a tipping cascade, it is unlikely to be initiated below a temperature rise of 1.5°C, but the risk of such a cascade rises rapidly beyond that level (Lenton *et al.*, 2019).

1.5 Challenges of the 21st Century

Achieving sustainability, however that concept is defined, is widely agreed to be the most fundamental challenge facing humanity in the 21st century, and was formalized in the post-2015 sustainable development goals (SDGs) (UN, 2015). While attempting to bring the bulk of the world's population out of poverty is at the core of the SDGs, there is a growing recognition that protecting the environment is not an optional extra but essential to support the ongoing human enterprise. This imperative is nowhere more profound than at the global level, where a destabilized and rapidly changing global environment would make sustainability at any other level very difficult or virtually impossible. Yet, this is the risk posed as humanity heads further into the Anthropocene, with no sign of slowing down.

There are two fundamentally different approaches to managing the human–environment relationship in the Anthropocene. One aims actively to manage the Earth system itself, steering it in ways perceived to be beneficial to humanity. *Geoengineering* is a term often used for this approach, and reflects the dominant belief of the Great Acceleration, namely that humanity can now shape, steer and transform the environment to suit its own purposes.

The most prominent example of geoengineering is the proposal to counteract GHG-induced warming by inserting aerosols into the stratosphere to scatter some of the incoming solar radiation and thus cool the planet (Royal Society, 2009). This approach, using simple cause–effect logic, ignores the complexity of Earth system dynamics. Injecting aerosols into the stratosphere would change rainfall patterns in ways that we cannot predict, would not counteract the rapid increase in ocean acidity caused by the dissolution of atmospheric CO_2 and would likely lead to other as yet unanticipated changes in Earth system dynamics.

The other approach is based on *planetary stewardship*. This is a fundamentally different philosophy aimed at reducing human pressure on the Earth system and allowing it to function with less interference. Stewardship is a basic human approach to respecting the integrity of the environment on which we depend, and is deeply rooted in our existence, found in all major religions and in indigenous cultures. It approaches humanity's relationship with the environment by managing human values and behaviour, rather than by attempting to control, manipulate, dominate and transform the environment to suit us.

The most prominent global-scale example of the stewardship philosophy is the planetary boundaries approach (Rockström *et al.*, 2009a, 2009b; Steffen *et al.*, 2015b; Richardson *et al.* 2023), which recognizes that the Holocene is the only state of the Earth system that we know for sure can support contemporary society. The approach aims to define a safe operating space for humanity, allowing Earth system functioning to continue in a way that we know can support our continued development and well-being. In short, the approach recognizes that the Earth system has intrinsic, hard-wired properties and processes (e.g. tipping elements, Section 1.4), many of which have thresholds beyond which abrupt and/or irreversible change, detrimental to human well-being, can occur. These features form the basis for boundaries.

The planetary boundaries approach is based on nine features and processes of the Earth system that are sufficient to define the state of the system. They are: climate change, ocean acidification, stratospheric ozone depletion, biogeochemical flows (the nitrogen and phosphorus cycles), global freshwater use, land-system change, biosphere integrity, atmospheric aerosol loading and novel entities (such as chemical pollution). The 'safe operating space' is defined as the set of boundaries that define a Holocene-like state of the Earth system.

We have already transgressed six of the boundaries: climate change, biosphere integrity, land-system change, the nitrogen cycle, freshwater change; and novel entities. The climate change issue is already well known. Much less known is the seriousness and urgency of the erosion of biosphere integrity. At its most fundamental level, the genetic code embodied in biodiversity acts as a giant control panel, complete with its set of in-built redundancies and back-up systems, that guides biosphere functioning, in turn providing the range of ecosystem services on which we depend (Millennium Ecosystem Assessment, 2005). In fact, without the biosphere, the composition of the atmosphere would be vastly different than it is today and the functioning of the Earth system as a whole would be significantly altered. The planet would be unrecognizable. Although there is, indeed, considerable redundancy and resilience in the Earth's biosphere, it too has its intrinsic limits and planetary-scale shifts are possible (Barnosky *et al.*, 2011).

Implementation of the planetary boundaries approach has already generated criticism (e.g. Lewis, 2012). Perhaps the most common is based on the long-standing tension between development and environment; that is, setting environmental limits will constrain development and constrain the task of developing countries to lift their populations from poverty (e.g. Kosoy *et al.*, 2012). However, one analysis argues that, in fact, there are significant synergies – not conflicts – between respecting the planetary boundaries and improving many aspects of social equity, including improving the material well-being of the poor in developing countries (Steffen and Stafford Smith, 2013). The synergies are especially apparent for the so-called aggregated planetary boundaries, such as those for the phosphorus and nitrogen cycles, land-use change and biodiversity loss. A transfer of resources or access to ecosystem services from those areas of overuse (i.e. the wealthy countries) to those that lack resources or services will simultaneously help keep Earth within the boundaries and improve social equity outcomes. Such analyses hold out hope for the future – that we can build an effective planetary stewardship regime while addressing the great inequities that still plague civilization.

1.6 Conclusion

Confronting the Anthropocene is the most complex, difficult and important challenge that humanity has yet faced in its existence on Earth. Are we headed for a new era of sustainability and enhanced human well-being, or are we headed for environmental degradation, conflict and collapse? The jury is out. As Maurice Strong, convenor of the Earth Summit, wrote: 'Where on Earth are we going?'

Notes

[1] Data (until 2010) and method for these panels are available at the supplement to Steffen *et al.*, 2015a. Additional data sources (after 2010): UN Population Division, US Dept. of Agriculture, and ourworldindata.org.

[2] Data (until 2010) and method for these panels are available at the supplement to Steffen *et al.*, 2015a. Additional data sources (after 2010): CSIRO, Copernicus.

References

Barnosky, A.D., Matzke, N., Tomiya, S., Wogan, G.O.U, Swartz, B. *et al.* (2011) Has the Earth's sixth mass extinction already arrived? *Nature* 471, 51–57. DOI: 10.1038/nature09678

Crutzen, P.J. (2002) Geology of mankind. The Anthropocene. *Nature* 415, 23. DOI: 10.1038/415023a

EPICA Community Members (2004) Eight glacial cycles from an Antarctic ice core. *Nature* 429, 623–628. DOI: 10.1038/nature02599

Galloway, J.N. and Cowling, E.B. (2002) Reactive nitrogen and the world: two hundred years of change. *Ambio* 31, 64–71. DOI: 10.1579/0044-7447-31.2.64

Global Footprint Network (2021) Our human development initiative. Available at: https://www.footprintnetwork.org (accessed 9 February 2021).

Hibbard, K., Crutzen, P., Lambin, E., Liverman, D., Mantua, N. *et al.* (eds) (2006) Integrated History and Future of People on Earth, Dahlem Workshop Report 96. MIT Press, Boston, Massachusetts, pp. 341–375.

Holdren, J.P. and Ehrlich, P.R. (1974) Human population and the global environment: population growth, rising per capita material consumption and disruptive technologies have made civilization a global ecological force. *American Scientist* 62, 282–292. Available at: https://www.jstor.org/stable/27844882.

IPBES (Intergovernmental Science-Policy Platform on Biodiversity and Ecosystem Services) (2019) Nature's Dangerous Decline. Available at: http://bit.ly/IPBESReport (accessed 10 February 2022).

IPCC (2012) Managing the Risks of Extreme Events and Disasters to Advance Climate Change Adaptation. A Special Report of Working Groups I and II of the Intergovernmental Panel on *Climate Change* (eds: Field, C.B., Barros, V., Stocker, T.F., Qin, D. *et al.*). Cambridge University Press, Cambridge, UK, and New York.

IPCC (2021) Summary for Policymakers. In: Masson-Delmotte, V., Zhai, P., Pirani, A., Conners, S.L., Péan, C. *et al.* (eds) *Climate Change 2021: The Physical Science Basis*. Working Group I Contribution to the Sixth Assessment Report of the Intergovernmental Panel on Climate Change. Cambridge University Press, Cambridge, UK, and New York.

Kolbert, E. (2011) Enter the Anthropocene: age of man. *National Geographic* 219, 60–77. Available at: https://education.nationalgeographic.org/resource/age-man-enter-anthropocene/ (accessed 9 July 2022).

Kosoy, N., Brown, P.G., Bosselmann, K., Duraiappah, A., Mackey, B. *et al.* (2012) Pillars for a flourishing Earth: planetary boundaries, economic growth delusion and green economy. *Current Opinion in Environmental Sustainability* 4, 74–79. DOI: 10.1016/j.cosust.2012.02.002,

Lenton, T.M., Held, H., Kriegler, E., Hall, J.W., Lucht, W. *et al.* (2008) Tipping elements in the Earth's climate system. *Proceedings of the National Academy of Sciences (USA)* 105, 1783–1785. DOI: 10.1073/pnas.0705414105

Lenton, T.M., Rockström, J., Gaffney, O., Rahmstorf, S., Richardson, K. *et al.* (2019) Climate tipping points – too risky to bet against. *Nature* 575, 593–596. DOI: 10.1038/d41586-019-03595-0

Lewis, S.L. (2012) We must set planetary boundaries wisely. *Nature* 485, 417. DOI: 10.1038/485417a

McNeill, J.R. (2000) *Something New Under the Sun: An Environmental History of the Twentieth-Century World*. W.W. Norton, New York.

Meinshausen, M., Meinshausen, N., Hare, W., Raper, S.C.B., Frieler, K. *et al.* (2009) Greenhouse-gas emission targets for limiting global warming to 2°C. *Nature* 458, 1158–1162. DOI: 10.1038/nature08017

Millennium Ecosystem Assessment (2005) *Ecosystems and Human Well-being. Synthesis*. Island Press, Washington, DC.

Oppenheimer, S. (2004) *Out of Eden. The Peopling of the World*. Constable and Robinson, London.

Petit, J., Jouzel, J., Raynaud, D., Barkov, N., Barnola, J.-M. *et al.* (1999) Climate and atmospheric history of the past 420,000 years from the Vostok ice core, Antarctica. *Nature* 399, 428–436. DOI: 10.1038/20859

Richardson, K., Steffen, W., Liverman, D., Barker, T., Jotzo, F. *et al.* (2011) *Climate Change: Global Risks, Challenges and Decisions*. Cambridge University Press, Cambridge, UK.

Richardson, K., Steffen, W., Lucht, W., Bendtsen, J., Cornell, S.E., *et al.* 2023. Earth beyond six of nine planetary boundaries. *Science Advances* 9, eadh2458, DOI: 10.1126/sciadv.adh2458

Rockström, J., Steffen, W., Noone, K., Persson, Å., Chapin, F.S. *et al.* (2009a) Planetary boundaries: exploring the safe operating space for humanity. *Ecology and Society* 14(2), 32. Available at: http://www.ecologyandsociety.org/vol14/iss2/art32/ (accessed 10 February 2022).

Rockström, J., Steffen, W., Noone, K., Persson, Å., Chapin, F.S. *et al.* (2009b) A safe operating space for humanity. *Nature* 461, 472–475. DOI: 10.1038/461472a

Royal Society (2009) *Geoengineering the Climate: Science, Governance and Uncertainty*. The Royal Society, London.

Scheffer, M. (2009) *Critical Transitions in Nature and Society*. Princeton University Press, Princeton, New Jersey.

Schellnhuber, H.J. (2002) Coping with Earth System complexity and irregularity. In: Steffen, W., Jäger, J., Carson, D. and Bradshaw, C. (eds) *Challenges of a Changing Earth: Proceedings of the Global Change Open Science Conference*. Amsterdam, The Netherlands, 10–13 July 2001. Springer-Verlag, Berlin, Heidelberg, Germany, and New York, pp. 151–156.

Sorrell, S., Speirs, J., Bentley, R., Brandt, A. and Miller, R. (2009) *An Assessment of the Evidence for a Near-term Peak in Global Oil Production*. UK Energy Research Centre, London.

Steffen, W. and Stafford Smith, M. (2013) Planetary boundaries, equity and global sustainability: why wealthy countries could benefit from more equity. *Current Opinion in Environmental Sustainability* 5, 403–408. DOI: 10.1016/j.cosust.2013.04.007

Steffen, W., Sanderson, A., Jäger, J., Tyson, P.D., Moore, B. III *et al.* (2004) *Global Change and the Earth System: A Planet Under Pressure*. The IGBP Book Series, Springer-Verlag, New York.

Steffen, W., Crutzen, P.J. and McNeill, J.R. (2007) The Anthropocene: are humans now overwhelming the great forces of nature? *Ambio* 38, 614–621. DOI: 10.1579/0044-7447(2007)36[614:TAAHNO]2.0.CO;2

Steffen, W., Broadgate, W., Deutsch, L., Gaffney, O. and Ludwig, C. (2015a) The trajectory of the Anthropocene: The Great Acceleration. *Anthropocene Review* 2, 81–98. DOI: 10.1177/2053019614564785

Steffen, W., Richardson, K., Rockström, J., Cornell, S.E., Fetzer, I. *et al.* (2015b) Planetary boundaries: guiding human development on a changing planet. *Science* 347(6223). DOI: 10.1126/science.1259855

Steffen, W., Richardson, K., Rockström, J., Schellnhuber, H.J., Dube, O.P. *et al.* (2020) The emergence and evolution of Earth System Science. *Nature Reviews: Earth and Environment* 1, 54–63. DOI: 10.1038/s43017-019-0005-6

UN (2015) *United Nations Secretary-General's High-level Panel on Global Sustainability. Resilient People, Resilient Planet: A Future Worth Choosing*. United Nations, New York.

Zalasiewicz, J., Crutzen, P. and Steffen, W. (2012) The Anthropocene. In: Gradstein, F.M., Ogg, J.G., Schmitz, M. and Ogg, G.M. (eds) *A Geological Time Scale 2012*. Elsevier, Amsterdam, Netherlands, pp. 1033–1040.

2.i Rising Inequality is Neither Inevitable Nor Necessary

Richard Denniss
The Australia Institute, Canberra, Australia

Abstract
This short chapter argues that gross inequality is conspicuous and growing and that gross domestic product (GDP) growth is not, in itself, the solution to this problem. When pursuing the well-being of citizens, it is the shape and content of economic activity that matters most, not the size. An economy that values and measures the provision of services, rather than the regular purchase and disposal of resource-intensive material objects for the symbolic purpose of displaying status, is called for.

2.i.1 Introduction

While billionaires battle each other to be the first to blast off into space, billions of people struggle to find the clean water, healthy food and safe housing that they need to meet their most basic needs. Nothing could make the disparity between the lives of those with the least and those with the most more stark than the fact that in the middle of a global pandemic, where over 5 million people have lost their lives prematurely and where billions of people have seen their incomes slashed, the world's wealthiest citizens had a bumper year (Hariharan, 2020).

Rising inequity is not inevitable, and nor is it essential to drive economic growth. According to Grudnoff (2020), the data clearly show that the most equal countries in the world are not only the happiest and healthiest, but also the wealthiest. Cuba and Costa Rica may be two of the few exceptions. The Nordic countries (Denmark, Finland, Iceland, Norway and Sweden) have the most even distribution of income in the world (measured by the Gini coefficient) and also have high levels of gross domestic product (GDP) per capita, high rates of GDP growth and high levels of productivity growth. Empirical evidence simply does not support the assertion that there is a strong, or indeed any, trade-off between the pursuit of equity and the pursuit of national income.

2.i.2 The Myth of Trickle-down Economics

It is important to distinguish between what political and business leaders say about the benefits of policies that cause inequality and what economic theory and evidence actually reveal about the effectiveness of these policies. For example, the idea that cutting taxes for high-income earners will stimulate economic activity and help the entire community (often referred to as 'trickle-down' economics) was proposed by economist Arthur Laffer (born 1940); the evidence for

DOI: 10.1079/9781800620025.0002i

such an approach was weak at the time and, in hindsight, it is clear that such policies have failed in their own terms (Estevez, 2021). Indeed, bodies like the Organisation for Economic Co-operation and Development (OECD), World Bank and International Monetary Fund (IMF) are now adamant that reducing inequality helps to boost GDP growth. For example, an IMF paper (Dabla-Norris *et al.*, 2015) argued that 'if the income share of the top 20 percent (the rich) increases, then GDP growth actually declines over the medium term, suggesting that the benefits do not trickle down. In contrast, an increase in the income share of the bottom 20 percent (the poor) is associated with higher GDP growth.'

While there is no strong, or indeed any, evidence that tolerance for inequality assists GDP growth, there is a very strong link between the pursuit of GDP growth and resultant inequity. Put simply, once policy makers decide that economic growth should be the primary national goal, then, by definition, goals such as social cohesion, strong concern with the development of other countries and environmental sustainability must be downgraded.

Despite the overwhelming attention paid to the rate of GDP growth in many countries, there is nothing in economic theory, and no empirical evidence, to support the assertion that maximizing the rate of GDP growth is the best way to boost the well-being or security of a nation's citizens. Such goals should not, therefore, be the primary goal of either economic policy or democratic decision making.

It is true that a country with a larger GDP may be able to spend more on education, health or defence for the betterment and protection of its people. However, it is also likely that countries that cut taxes or harm their natural environment (while attempting to increase their GDP) may end up with less capacity to invest in the well-being of their citizens, even if measured GDP rises.

In short, while the annual value of economic transactions within a country (measured as GDP) is widely discussed, what matters more for the well-being of citizens is the composition of those transactions, not their quantity. For example, while spending US$1 billion repairing cars that have been vandalized and spending US$1 billion to buy bicycles for low-income earners both contribute to GDP, the social, environmental and long-term economic consequences of each decision will be markedly different. When it comes to the well-being of citizens, it is the shape of economic activity that matters most, not the size (Denniss, 2017).

2.i.3 Economies That Focus on Services, Not Status

While it is true that it is not possible to have infinite economic growth on a finite planet, it is also true that there is almost no limit to the amount of education, aged care, preventive health, massages, haircuts and online music that a population can consume. A community that favours the regular purchase and disposal of resource-intensive material objects for the symbolic purpose of displaying status or being fashionable (such as cars, clothes, bathrooms and appliances) will impose far greater costs on the natural environment, and demand a far greater share of the world's scarce resources, than one that prioritizes the production and consumption of services, especially where publicly provided.

While the production of enormous quantities of 'stuff' in low-income countries for consumption in high-income countries is often depicted as a means to 'lift countries out of poverty' (UNCTAD, 2006), such an approach cannot deliver a significant reduction in global income disparities; nor is it environmentally or socially sustainable. Put simply, if those in low-income countries who produce the appliances, clothes and cars that are consumed in enormous quantities in the wealthiest countries could afford to purchase a similar amount of 'stuff' themselves, then the demand on the world's natural resources would need to grow far beyond the limits of the natural environment.

If billions of people in low-income countries were ever to consume the amount of natural resources used by the average citizen of the USA or Australia on a per capita basis, then the capacity of the natural environment to provide that amount of food, fibre, energy, clean water and mineral resources would far exceed the most optimistic estimates of 'sustainable limits'. However, the fact that those living in poverty cannot possibly consume the same bundle of material goods that those in rich countries take for granted today does not mean that their incomes and their standard of living cannot, and should not, grow steadily.

2.i.4 Conclusion

There are only two ways to close the gap separating the economic advantages of the most privileged from those of the average world citizen. One way is for those with the least to each consume dramatically more physical resources. A second is to dramatically reduce the material density of the average 'consumption bundle' consumed in high-income countries and high-income élites everywhere. Both routes might be possible.

To be clear, while a massive switch away from a culture of rapidly producing and disposing of enormous amounts of 'stuff' towards the provision of a much wider range of services would have an enormous impact on the shape of the world's economies, it need not have a significant impact on the size of the global economy. Inevitably, it will be much easier to distribute the provision of knowledge and services more equally than scarce natural resources and physical commodities.

References

Dabla-Norris, E., Kochhar, K., Suphaphiphat, N., Ricka, F. and Tsounta, E. (2015) *Causes and Consequences of Income Inequality: A Global Perspective*. IMF Discussion Paper SDN/15/13. Available at: https://www.imf.org/external/pubs/ft/sdn/2015/sdn1513.pdf (accessed 6 March 2023).

Denniss, R. (2017) *Curing Affluenza: How To Buy Less Stuff And Save The World*. Black Inc, Melbourne, Australia.

Estevez, E. (2021) Review of Amadeo, K.: Why trickle-down economics works in theory but not in fact: the shortcomings of supply-side economics. *The Balance*, 28 March. Available at: https://www.thebalance.com/trickle-down-economics-theory-effect-does-it-work-3305572 (accessed 6 March 2023).

Grudnoff, M. (2020) Tax and wellbeing. Available at https://australiainstitute.org.au/wp-content/uploads/2020/12/Tax-and-wellbeing-updated.pdf (accessed 4 October 2021).

Hariharan, A. (2020) Billionaires Bezos, Musk & Zuckerberg rode the coronavirus storm to get even richer. *Economic Times*, 13 November. Available at: https://economictimes.indiatimes.com/magazines/panache/billionaires-bezos-musk-zuckerberg-rode-the-coronavirus-storm-to-get-even-richer/articleshow/79193877 (accessed 4 October 2021).

UNCTAD (UN Conference on Trade and Development) (2006) Poverty reduction through production. Press release UNCTAD/PRESS/PR/2006/014. Available at: https://unctad.org/press-material/poverty-reduction-through-production (accessed 4 October 2021).

2.ii Climate Change and the Scourge of Carbon Inequality

Kerryn Higgs
School of Social Sciences, University of Tasmania, Hobart, Australia; Associate Member, The Club of Rome, Winterthur, Switzerland

Abstract

The great majority of the warming of the climate has resulted from emissions involved in the activities of globalized corporations, often owned in the global North and disproportionately serving Northern consumers. The decline in Earth system biodiversity and transgression of other planetary boundaries such as biogeochemical flows and land use change are also attributable to these activities. Despite decades of initiatives ostensibly aimed at 'lifting millions out of poverty', the world remains sharply divided between the rich (relatively few) and the rest, mainly in the global South, still experiencing the effects of centuries of colonial rule. Solutions to the climate emergency require solutions to the intrinsic inequality of this economic system, where rich world consumers monopolize the spoils of massive corporate extraction. As well as inequality in the usual sense of wealth and income, this involves major inequality in carbon footprints – 'carbon inequality'.

2.ii.1 Introduction: Whose Emissions?

The attribution of greenhouse gas (GHG) emissions to specific nations, regions or groups relies on various methods (Chancel, 2022). It is relatively simple to quantify the gross current emissions of each country, and this figure is often used, but it ignores cumulative emissions over the industrial era since emissions began. If justice or fairness is considered, cumulative tallies are more appropriate. China is currently the highest gross GHG emitter, but China has not overtaken the USA's cumulative contribution and was estimated to emit just over half of US cumulative emissions in 2022.[1]

In addition, the cumulative figure ignores per capita emissions, which will also give a fairer account of actual benefit to members of a given population. In gross terms China's emissions are gaining on those of the USA (UCS, 2022); however, China's population is more than four times that of the USA.

Another factor in determining who is responsible for emissions is the extent to which a country's industrial apparatus serves the needs of its local population or produces items for export to others. Methods of quantifying emissions centre on where the emissions arise, so the gross current emissions figure ignores the implications of trade across national boundaries. 'Fair shares' calculations (Hickel, 2020) attribute the emissions arising from production to the country where the products are actually used and puts these in the context of a fair division of emissions within a budget that is likely to avoid climate destruction. Thus, as William Lamb notes (quoted by Tandon, 2022), those who emphasize China's emissions as the main villain are being wilfully ignorant, at best.

DOI: 10.1079/9781800620025.0002ii

2.ii.2 The Role of Consumption and the 'Wasteful Wealthy'

Accumulated GHG emissions are associated with the wealth concentrated in the North today, built as it has been on fossil fuels, industrial agriculture and extraction (often unimpeded) throughout the world over several centuries.

The average carbon footprint of people living in sub-Saharan Africa is approximately 0.06 tonnes per year, while the average in the USA is 14.5 tonnes (Chancel, 2022). This disparity reflects the inequality in the levels of consumption enjoyed by people in different parts of the world and at different levels of income and wealth. In the case of countries of the lower-middle- and low-income sectors, both current and cumulative per capita emissions are very small indeed. Consumption is concentrated in the rich world. This is not to deny that élite minorities from the South are among the world's top consumers, and that some sectors in the North are impoverished – but these are outliers to the general pattern of gross inequality between nations and regions.

Consumption is particularly relevant in a globalized world, where trade transfers confuse who the true beneficiaries of production are. According to Kanemoto *et al.* (2013), 'up to 30% of global emissions are linked to production for export'. These emissions are routinely attributed to the country where the manufacture is located, even though the products are used somewhere else altogether.

Bruckner *et al.* (2022), when estimating national carbon emissions, started with direct energy use (e.g. heating, cooling, vehicle fuels) and then included the carbon emissions embodied in products that are imported. In this analysis, GHG emissions were assigned to the countries that use goods and services, rather than the countries that produce them, thus capturing the contribution of consumers in the North to emissions arising elsewhere.

Bruckner *et al.* (2022) also examined the emissions of the global rich (wherever they are based) and found that the top 1% of the world population emit more than the bottom 50%. A study by Otto *et al.* (2019) confirmed this claim, finding that a mere 0.54% of the wealthiest people on Earth are responsible for nearly 14% of all GHG emissions that flow from consumption choices. They are an extremely small number of people (approximately 40 million in all), and their 14% share of GHG emissions contrasts starkly with that of the poorest half of the world population (approximately 4 billion) who are estimated to contribute just 10% of emissions (Otto *et al.*, 2019). These figures are based on consumption or lifestyle choices only, and would increase significantly if emissions flowing from the investment choices of the same group were to be added. See Chancel (2022) for one attempt to do this.

Overall, carbon emissions flow primarily from the lifestyle and investment choices of the rich and, especially, the super-rich; efforts at carbon emissions reductions are likely to fail unless we modify the impact of this group, who are responsible for a disproportionately large contribution to global warming. Bruckner *et al.* (2022) noted that, at the very top of the world's high net worth (HNW) individuals, about 500,000 people each emit multiple hundreds of tonnes of carbon dioxide (CO_2) every year.

In Hickel's (2020) equality-based attribution of emissions (on a per capita basis and taking consumption into account), China's emissions remain far behind those of the USA and the North as a whole. In 2015, in this calculation, 'the USA was responsible for 40% of excess global CO_2 emissions.[2] The European Union (EU-28) was responsible for 29%. The G8 nations (the USA, EU-28, Russia, Japan and Canada) were together responsible for 85%.'

2.ii.3 What Emissions Levels Are Compatible with IPCC Targets?

Bruckner *et al.* (2022) note that '[t]o limit global warming to 1.5°C or 2.0°C, per capita carbon footprints need to be in a global target range of about 1.6–2.8 tonnes (t) CO_2 p.a.'[3] This is an extremely small level compared with the actual carbon footprint of nations in the North. In 2022, the spread of carbon footprints ranged from less than 0.01 t CO_2 p.a. for more than a million people in sub-Saharan countries to hundreds of tonnes of CO_2 for the 500,000 HNW individuals at the top of the 'global expenditure spectrum' (Bruckner *et al.*, 2022). Bruckner told Tandon (2022) at Carbon Brief: 'If we want to reduce our carbon emissions, we really need to do something about the consumption patterns of the super-rich.' This same message is tackled in detail by Wiedmann *et al.* (2020).

Crucially, more than half of the global population have carbon footprints below or within the target range (Bruckner *et al.*, 2022) and although it is clear that the bottom 50% will require some increase in emissions, to address unmet basic needs for food, water, shelter, sanitation, education and health care, this increase could be minimized if an energy transition to renewables is implemented. In any case, Bruckner *et al.* (2022) concluded that eradicating 'extreme poverty' would raise global GHG by only 1%, a very modest amount when compared with the excess carbon footprints of people in the North, especially those of the top 1% or even 10% of the global population. A majority of the citizens of the global North fall within that 10% of the global population.

2.ii.4 Barriers to Reductions

The vast majority of GHGs are emitted in the service of the high-carbon habits endemic in the North, based on a consumerist lifestyle that has been intentionally developed by the corporations that depend for their growth on inciting an appetite for endless products (Higgs, 2014). 'The harms have been borne disproportionately by low-income people, people of colour, and indigenous people whereas those who are wealthy, typically white, and more closely connected to government and corporate power have been able to avoid costs, while appropriating benefits' (Stoddard *et al.*, 2021).

Thus, the most pressing need is for an immediate reduction in the consumption habits of the rich and the ultra-rich – and even the comfortable middle classes of the North. Clearly, emissions that are well over the world average are a major impediment to achieving the necessary targets. As noted by Stoddard *et al.* (2021), there has been a 'pervasive failure in industrial, modern societies to imagine desirable ways of living that are neither wedded to the carbon economy nor dependent on narratives of progress reliant on perpetual economic growth'. This failure is reflected in the fact that, despite rapid growth in renewable energy, fossil energy also grew, so that the fossil fuel contribution to the world's final energy consumption in 2019 remained at the same level as it was in 2000, around 80% (Diesendorf, 2022).

As well as the inertia created by Northern consumers' attachment to their relentless high-carbon lifestyle, resistance to change is buttressed by the overtly obstructionist tactics of corporations, their lobbyists, the politicians they influence with donations, and the vast panoply of think tanks producing material designed to mislead the public and to shape national legal and regulatory systems in favour of a carbon-based growth economy (Higgs, 2014).

Ultimately, everyone will pay the price, as the heat, drought, fires and flooding in North America, Europe and Australia suggest. However, much of the horror is being borne by people who have contributed little or nothing to the GHG build up – the catastrophic 2022 floods in Pakistan being one of the most glaring examples (see Chapter 13, this volume).

2.ii.5 Equality and IPCC Process?

As noted by Stoddard *et al.* (2021), 'three decades of carefully crafted political statements, rumors of significant decoupling, and promises of technical breakthroughs have barely dented the apparently inexorable rise in emissions ... [which are] 60% higher than they were in 1990'. Instead, we have seen decades of promises, delays and muted incrementalism instead of the rapid change that is needed, and 'a pervasive failure to direct mitigation efforts toward high emitters and emission-intensive practices' (Stoddard *et al.*, 2021).

The basic assumptions of the IPCC framework privilege technical solutions over transforming personal habits, lifestyle or behaviour and prefer a market-based approach that ignores redistributive options. Inequality is also built into IPCC emissions reduction formulae: proposed emissions reductions are calculated as a percentage reduction below current emissions, a framework which accepts the status quo and divides the burden from that starting point rather than attempting 'a fair division of the burdens of the status quo itself' (Stoddard *et al.*, 2021). The status quo embodies more than two centuries of cumulative emissions that are not reflected at all.

Perhaps the most disappointing aspect of the IPCC negotiations process has been the failure of the rich countries to deliver help to the rest of the world, even where promises had been made. In 2001, the Least Developed Countries Fund was set up and the relevant countries asked to prepare adaptation plans to meet the most urgent threats; the estimated funding to implement these most urgent needs was US$2 billion. By 2009, less than 10% of this sum had been provided and Oxfam (2009) warned that the G8 needed to provide US$2 billion immediately for adaptation, US$50 billion a year in regular adaptation payments and US$100 billion a year to assist in cutting emissions.

At the Conference of the Parties (COP) 15 in Copenhagen that year, the rich nations agreed to contribute US$100 billion p.a. by 2020. The Organisation for Economic Co-operation and Development (OECD) (2021) figure for 2019 was still only US$80 billion – and there are questions about whether the OECD's estimate can be regarded as accurate. More than half of that money was in the form of repayable loans which were added to the recipients' debts; more than half was devoted to mitigation and not available for adaptation; some of it was existing development aid, rebranded as climate aid; and very little was given to the 'least developed countries' – one research team could only identify a total of US$5.9 billion between 2014 and 2018 for these (Timperley, 2021). As well as being under target, the contributions were not specifically designated to assist with impacts of the ever-increasing climate change disasters flowing from the cumulative GHG emissions of the rich countries.

Despite reluctance from the North, fearing they might be held legally responsible, it was finally agreed, as the COP27 negotiations (Egypt, 2022) dragged on through a second additional night, to establish a 'loss and damage' fund to compensate poor nations and help them cope with encroaching climate disasters. First raised by vulnerable nations, this issue had taken 30 years to get onto the COP agenda. Although agreed, it remained unclear whether richer developed nations would insist on China paying into the fund and whether the fund would assist only the most vulnerable countries or be more widely available. And given past performance, it is unclear whether the pledge will be met.

2.ii.6 Conclusion

Levels of consumption and levels of carbon emissions embodied in the items consumed reflect massive global inequalities in income and wealth. National emissions, as currently quantified, exclude relevant data: cumulative emissions over the industrial era, emissions that occur elsewhere to supply consumer products and emissions per capita.

To reverse or even contain the climate emergency, we must reduce per capita emissions to a range of 1.6–2.8 tonnes CO_2 p.a., a mere fraction of the current emission levels found in the global North. This is where reductions must bite – and especially, of course, with the ultra-rich who emit 14% of all GHGs though they are a tiny fraction of the world population.

Given the wealth, power and influence of the countries of the North, it is not surprising that they dominate the UN processes set up to solve the problem. These processes have done little so far to reduce annual emissions or to ameliorate the situations of the bottom half. The central need is for the rich to own up to their role – and contract.

Notes

1 Figures provided by the Union of Concerned Scientists (2022) put the USA's cumulative emissions at 416,738 Mt (24.5% of world emissions) and China's at 235,527 Mt (13.9%).

2 That is 'in excess of the planetary boundary of 350 ppm [parts per million] atmospheric CO_2 concentration. This approach is rooted in the principle of equal per capita access to atmospheric commons.'

3 Tonnes of carbon dioxide per year. Note that these figures do not include methane emissions, which also contribute to global warming, especially in the first 20 years after emission.

References

Bruckner, B., Hubacek, K., Shan, Y., Zhong, H. and Feng, K. (2022) Impacts of poverty alleviation on national and global carbon emissions. *Nature Sustainability* 5, 311–320. DOI: 10.1038/s41893-021-00842-z

Chancel, L. (2022) Global carbon inequality over 1990–2019. *Nature Sustainability* 5, 931–938. DOI: 10.1038/s41893-022-00955-z

Diesendorf, M. (2022) It'll be impossible to replace fossil fuels with renewables by 2050, unless we cut our energy consumption. *The Conversation*, 25 August 2022. Available at: https://theconversation.com/itll-be-impossible-to-replace-fossil-fuels-with-renewables-by-2050-unless-we-cut-our-energy-consumption-189131 (accessed 18 March 2023).

Hickel, J. (2020) Quantifying national responsibility for climate breakdown: an equality-based attribution approach for carbon dioxide emissions in excess of the planetary boundary. *The Lancet Planetary Health* 4(9), e399–e404. DOI: 10.1016/S2542-5196(20)30196-0

Higgs, K. (2014) *Collision Course: Endless Growth on a Finite planet*. MIT Press, Cambridge, Massachusetts.

Kanemoto, K., Moran, D., Lenzen, M. and Geschke, A. (2013) International trade undermines national emissions reduction targets; new evidence from air pollution. *Global Environmental Change* 24, 52–59. DOI: 10.1016/j.gloenvcha.2013.09.008

OECD (Organisation for Economic Co-operation and Development) (2021) Statement from OECD Secretary-General Mathias Cormann on climate finance in 2019. Available at: https://www.oecd.org/newsroom/statement-from-oecd-secretary-general-mathias-cormann-on-climate-finance-in-2019.htm (accessed 23 March 2023).

Otto, I., Kim, K., Dobrovsky, N., and Lucht, W. (2019) Shift the focus from the super-poor to the super-rich. *Nature Climate Change* 9, 82–84. DOI: 10.1038/s41558-019-0402-3

Oxfam (2009) Finance Ministers must find funds to secure climate deal. Available at: https://www.oxfam.org/en/press-releases/finance-ministers-must-find-funds-secure-climate-deal (accessed 18 March 2023).

Stoddard, I., Anderson, K., Capstick, S., Carton, W., Depledge, J. *et al.* (2021) Three decades of climate mitigation: why haven't we bent the global emissions curve? *Annual Review of Environment and Resources* 46, 653–689. DOI: 10.1146/annurev-environ-012220-011104

Tandon, A. (2022) Eradicating 'extreme poverty' would raise global emissions by less than 1%. *Carbon Brief*, 14 February 2022. Available at: https://www.carbonbrief.org/eradicating-extreme-poverty-would-raise-global-emissions-by-less-than-1/ (accessed 6 April 2023).

Timperley, J. (2021) How to fix the broken promises of climate finance. *Nature* 598, 400–402. DOI: 10.1038/d41586-021-02846-3

UCS (Union of Concerned Scientists) (2022) Each country's share of CO_2 emissions, published Jul 16, 2008, updated January 14, 2022. Available at: https://www.ucsusa.org/resources/each-countrys-share-co2-emissions# (accessed 22 March 2023).

Wiedmann, T., Lenzen, M., Keyßer, L.T. and Steinberger, J.K. (2020) Scientists' warning on affluence. *Nature Communications* 11, 3107. DOI: 10.1038/s41467-020-16941-y

2.iii Inequality Is Driving Us Over a Cliff

Colin D. Butler
National Centre for Epidemiology and Population Health, The Australian National University, Canberra, Australia

Abstract
This short chapter is a short introduction to the vast literature on inequality, seeking to tie it with the concept of 'environmental brinkmanship' – that is, that the elevated economic, political and agenda-setting position of a tiny minority of our species is so 'stratospheric' that they are willing to risk the entire human enterprise. The chapter argues that this recklessness is primarily made possible by the collective judgement of 'Davos people' that they can insulate themselves from what is coming – if our 'business as usual' trajectory persists for much longer. Enlightened global leadership is desperately needed, not only to consider ecological limits, but also to implement a massive redistribution of wealth and power. This challenge is daunting, but if it is not recognized and responded to, then a 'breakdown world', characterized by 'barbarianization', may have the same result, within this century.

2.iii.1 Introduction

This short chapter proposes that the extent of global inequality is a powerful, under-recognized factor that has long helped to delay recognition, by those in both the global upper and middle class, of the perils of 'planetary overload' (McMichael, 1993). These populations have, for over a generation, lived with the illusion that they are largely immune to the risk of adverse global environmental change, including to the climate. Instead, when risk has been recognized, there has been a widespread perception that it is people in the global South (the poorest global classes) who are not only the most vulnerable, but will act as a buffer, effectively insulating those who are currently better off. While this assumption may be valid in the short run, such thinking underestimates the linkages in the global system, and the longer-run risk to humanity, especially if global heating exceeds 1.5°C above pre-industrial levels. The assumption that those who are currently wealthier will long remain so is an important factor leading global civilization towards a precipice, beyond which lies the prospect of ruin, for populations of all layers of income and wealth.

2.iii.2 The Scope of Global Inequality

Stark, a book by satirist Ben Elton (1989), described 'a secret consortium with more money than God, and the social conscience of a dog on a croquet lawn'. This consortium has a plan; it is for a tiny number of rich humans to escape a dying Earth and start anew, in space (a 'star ark'). Such a proposition, absurd in 1989, is still ludicrous. Yet, Elon Musk, currently the world's richest man, appears to take the prospect of 'terraforming' Mars seriously (Gagnon, 2021).

 Climate Change and Global Health: Primary, Secondary and Tertiary Effects, 2nd Edition (eds C.D. Butler and K. Higgs)
DOI: 10.1079/9781800620025.0002iii

The human population is extremely unequal, with dimensions that are not only social and economic, but also nutritional, educational and aspirational. Inequalities embrace location, housing, employment, maternal health literacy, access to health care, opportunities and their lack. They extend to psychological assimilation of perceptions of superiority and inferiority that can remain pervasive and entrenched and be transmitted across many generations. Such mechanisms do not start at birth, nor even conception. They are neither exclusively social nor fully genetic, though genetic factors can play a role. For example, there is increasing recognition that the experience of disadvantage, including discrimination leading to stress, can itself trigger biological factors that are heritable, predisposing successive generations to disadvantage (Bohacek and Mansuy, 2013). This is in addition to the effects on behaviour, also transmitted across generations, that are caused by modelling, or copying the behaviour of role models and others. This phenomenon has also been called 'social epigenetics' and it exists in many settings (Martin *et al.*, 2022).

Stratification of many forms occurs among most, if not all, human populations, and many non-human animal populations as well (Strauss and Shizuka, 2022). Many egregious examples of national inequality exist, including the caste system in India, described as a form of apartheid by Indian Prime Minister Manmohan Singh in 2006 (Rajagopal, 2007), and (until recently) apartheid in South Africa, and before then, in the Belgian Congo (Zeilig, 2008). The Jim Crow laws in the USA are another example. The processes and legacies of colonization and imperialism, at their worst, illustrate not only forms of global inequality, but also indifference to, and even the 'invisibilization' of, many native and Indigenous peoples, akin to how some 'untouchables' have been treated in India (Thiyagarajan, 2021).

2.iii.3 Measuring global income inequality

Although there is wide understanding of the difference between 'developing' and 'developed' countries (also known as the global South and the global North) there is far less understanding that, if the world was to be considered as one economic entity, its income distribution would be almost as unequal, or perhaps even more unequal, than any currently existing nation state. This is true even using income adjusted for purchasing power parity (PPP), an indicator that inflates the income for populations in the global South.[1]

Using PPP data published by the World Bank, Milanovic (2022) estimated the global Gini coefficient (one of the most frequently used inequality measures) to have been 62% in 2013, slightly less than the unadjusted Gini coefficient for the most unequal nation he reported: the Republic of South Africa (66%). In comparison, Milanovic reported that the equivalent measure for the USA, in 2013, was only 41%. Although the PPP method has been extensively criticized (Butler, 2002; Reddy and Minoiu, 2007)[2], estimates of global inequality based on international traded currency (such as US dollars or pounds sterling) would be far more unequal.

It is implausible that global inequality distribution can be substantially fairer than Milanovic reported and it is likely that it is worse. One omission from such analysis is that PPP calculations ignore *international* purchasing power (Butler, 2002), such as to stay in a hotel in a country in the global North in order to participate in climate change negotiations. For example, Huq (2019) pointed out the 25th Conference of the Parties (COP) was the longest ever, having lasted two extra days (and nights) beyond the original plan. He decried this tendency, now 'standard practice', as 'deeply unfair to the most vulnerable developing countries whose delegates cannot stay on'.

2.iii.4 Global classes and their carbon footprints

At the top of the global class system are billionaires, including those with access to a private jet. One analysis has shown that fewer than one million people emit multiple hundreds of tonnes, per person, of the main greenhouse gas, carbon dioxide (CO_2) (Bruckner *et al.*, 2022). Robinson (2017) quoted extensively from a book called *Superclass* by David Rothkopf (2009), former manager of Kissinger Associates, in which its author estimated that as few as 6000–7000 people dominate and control some US$100 trillion in assets, setting the agendas for policy planning institutions including the Trilateral Commission, the Bilderberg Group, the G8 and G20, the World Trade Organization, the World Bank and the International Monetary Fund, the Bank of International Settlements and the World Economic Forum.

Sklair (2000) identified a 'transnational capitalist class' in his global system analysis. Its members consume 'luxury goods and services' and live in gated communities 'watched over by armed guards and electronic surveillance'. Well below this layer are approximately two billion people, chronically undernourished in micronutrients (Darnton-Hill, 2019). Of this population, an estimated 702–828 million endured hunger in 2021 (FAO *et al.*, 2022). Many of these people experience pervasive energy undernourishment, leaving them physically stunted and cognitively impaired, for life.

Kevin Bales (1999), in his book *Disposable People: New Slavery in the Global Economy*, estimated that Earth then had 27 million enslaved people. In a more recent analysis Jackson *et al.* (2019) estimated that this number had grown to over 40 million, including 15.4 million individuals in forced marriages.

The poorest fraction of the global population has a very low carbon footprint. Bruckner *et al.* (2022) estimated that the per capita CO_2 emissions of more than a million people in some sub-Saharan countries is as low as 0.01 tonnes; that is, at a minimum, lower by a multiple of 20,000 compared with the most carbon frugal of the richest class. The same authors calculated that, in order to limit global warming to 1.5°C or 2.0°C above pre-industrial levels, per person carbon footprints need to be between 1.6 and 2.8 tonnes CO_2.

Although the world's poor have a low carbon footprint, they can, collectively, still cause substantial local damage, such as through over-grazing or the over-harvesting of forest products, for subsistence. Some, obeying the orders of others, in order to pursue the consumerist lifestyles which more than a billion people already experience, may cause extensive environmental damage, for example by wielding chainsaws or driving bulldozers to clear rainforests, or by working in factory fishing trawlers. But the ultimate responsibility for such damage is higher in the chain.

2.iii.5 Regulating the Global Class System

Central to the globalization of the class system is the understanding that wealth and power are not found exclusively in the global North, just as poverty, vulnerability and servitude are not characteristics unique to the global South. Some powerful figures in the South, such as the late Mobutu Sese Soko, president of the resource-rich Democratic Republic of the Congo (DRC) for over three decades, used their influence to enrich themselves and their cronies, rather than their people. Effectively, President Mobutu was recruited to the most powerful class, remaining there as long as he retained the confidence of his peers, and his policies aligned with the imperatives of global corporations and the dictates of 'rules-based free trade' and globalization.

The chief enforcers for global inequality are economic, political and military. Wade (2005) identified how positive (strengthening) feedbacks, which he termed the Matthew effect ('to him that hath shall be given'), perpetuate and sometimes steepen inequalities, at the global scale. Economic enforcers that he described include the World Bank, the International Monetary Fund and other instruments of global and globalizing neoliberal capitalism.

The regulators of the global class system extend far beyond politicians.[3] They also include the military, police, civil servants, political advisers, the mainstream media, cultural figures and educators. Tacit allegiance to this system pervades the United Nations and many of its pledges, including the Sustainable Development Goals (SDGs), which exhibit forms of cognitive dissonance (e.g. Hickel, 2019), including the incoherence between their 'greening business as usual' approach and the biophysical realities now facing the planet, from ecological devastation to the climate crisis. Many conservative editors, publishers and broadcasters also participate by seeking to control the flow of information in ways that help to protect an intensely unfair global order, and present that order as both 'normal' and 'unchangeable'.

In another satire the political scientist and prominent critic of capitalism, Susan George, sketches how a working party, representing the group who could also be called 'Davos person'[4], describe, in meticulous detail, ways whereby populations from the poorest global stratum become effectively dispensable via means that appear entirely natural, such as famine (George, 1999).

Analyses of the global class system, though extensive, have limited impact in less specialized academic and scientific circles, where, for example, they are categorized as 'Marxist', or 'activist'. Such terms act as signals, with the effect of reducing genuine broad-scale engagement with the arguments. This is especially so in the global North, where scholarship and policies have, in recent decades, been dominated by neoliberalism. A minor example of this is the collapse of interest in learning and teaching economic history.

Many barriers inhibit close contact between the different classes. While most white-collar workers in the global North know some tradespeople and may form sincere friendships with them, few will venture into the dismal territories and circumstances described by Bales (1999) or seek to visit subterranean mines in the DRC, where child slaves still toil to obtain cobalt, essential for the 'clean' energy transition (Sovacool, 2021). Most middle-class visitors to the global South will stay in a hotel, rather than a hut. Even in countries in the global South, genuine engagement by middle-class people with those who are significantly poorer is uncommon. Of course, there are exceptions, including many individuals and some organizations from the North (and the South) who work to advance health, human rights and opportunities, in both the North and the South. Some people spend years or even devote their entire lives to such goals. Their endeavours are supported, more passively, by many people who sense that excessive inequality is not only unjust, but a cause of instability and thus danger, even to elites. Collectively, such actions do help counter inequality, but they fall far short of eradicating it, despite well-publicized promises (Hickel, 2016).

2.iii.6 Environmental Brinkmanship

Many writers blame capitalism for our peril (Feola *et al.*, 2021). However, the environmental record of non-capitalist societies, such as the Soviet Union and the communist countries of Eastern Europe, was no better than the capitalist nations at the time.[5] Could it be that an under-appreciated cause for humanity's ever-nearing approach to a critical tipping point is not capitalism (or socialism, or the failure, to date, of 'Green capitalism') but the extent of inequality, caused by the inability, in high-income settings, to restrain capitalism's excesses? From this perspective, the recent slight decline in global inequality (reported by Milanovic, 2022) is not enough, even if true and, thus, encouraging. This may be because the scale of global inequality long ago passed a threshold, triggering actions that would lead to the current peril. The scale of redistribution of political and economic entitlement now needs such a radical shift that it seems unlikely to occur in the absence of a massive war, which is itself highly undesirable.

George (1999) recognized how the actions of the global elite work to push the global ecology nearer to a tipping point, a process that Butler (2000) called 'environmental brinkmanship'. Glavovic *et al.* (2021) did not directly blame inequality or capitalism, but instead identified a broken 'science–society contract' as underpinning the failure to reduce greenhouse gas emissions. Cologna and Oreskes (2022) disagreed, instead identifying, as part of a large literature, 'powerful political and cultural forces' that obstruct the sustainability transition. They pointed out that science remains relatively well-funded. However, they did acknowledge that the assumption that policy makers will act appropriately, in response to valid scientific knowledge, is flawed. In this respect the contract is broken, if by contract we mean that information entails a social responsibility to respond. Why is there such a failure? Here, it is suggested that this failure is caused by the extent of inequality, and the consequent collective cognitive incapacity of hundreds of millions of educated citizens (the world's 'middle class') to realize that they are at risk. This collective incapacity means that there is insufficient pressure on decision makers to act with the required urgency.

Green and Healy (2022) also recognized that socio-economic inequalities weaken the social foundations of collective action. This argument is not identical to the better-known one that high-net-worth richest individuals are doing the greatest damage to the Earth system (Otto *et al.*, 2019). It is that the extent of inequality substantially conceals the dangers faced by the global middle class, because of regional and planetary overload, perhaps within a generation.

Population growth of the poor

Population growth among the poorest class reduces the prospect of 'developing' economies to meet basic needs, such as health, housing, jobs and education, helping to supply a large, renewable population of workers and potential workers, thus depressing wage growth (Butler and Higgs, 2018; Daly, 2022). Although the SDGs contain many targets which, if attained, will accelerate the demographic transition (i.e. in low-income settings), family planning programmes in such settings have been substantially underfunded for decades (Grollman *et al.*, 2018).

2.iii.7 Conclusion

Analyses such as this are uncommon, at least within public health. Here, it has been argued that an important cause for this lack is that most people who are aware of the peril from exceeding planetary boundaries belong to the more affluent classes. Although changing, too many in this group have either felt comparatively insulated from the numerous threats from planetary overload or have been oblivious. Some, such as Musk, may believe their rhetoric, while others, including climate negotiators who repeatedly make broken promises to compensate people of the global South, are following orders (Timperley, 2021). Mainstream media almost completely ignore the general indifference of the middle class to the fate of the poor, in the causation and continuation of global inequality. This is understandable; we do not want to feel guilty or responsible.

Philip Alston, former UN special rapporteur on extreme poverty and human rights is an exception. He was reported, in 2019, as stating that the world is increasingly at risk of 'climate apartheid', where 'the rich pay to escape heat and hunger caused by the escalating climate crisis while the rest of the world suffers'. He also predicted that climate change is 'likely to undermine not only basic rights to life, water, food, and housing for hundreds of millions of people, but also democracy and the rule of law' (Carrington, 2019). Soon after, Alston left the UN.

The fate of civilization cannot be left to tax-minimizing and other world-desiring billionaires, whose self-interests do not extend to reducing inequality and who do not support the fundamental changes to the global power structure that are needed (Butler, 2019). Inequalities will persist. However, if humanity is to avoid a catastrophic future, those in the second class (those with not only sufficient education and nutrition to perceive our predicament, but also with enough 'entitlement' to exert a discernible influence, should they choose to act) must comprehend the risk that we all face; it is not only the global poor who face desolation if 'business as usual' persists.

Prominent strategies that have been suggested to enable a global civilization with more longevity than now seems apparent include the 'Green New Deal' and the 'Just Transition' movement (Barry, 2021). During the 1930s, US President Franklin Roosevelt advocated for a fairer American system, to 'attempt to rescue liberal democracy via experimental intervention in the economy' (Hill-Andrews, 2019). The challenge faced by the world today, in its near-future, is likely to far exceed that of the Great Depression in the USA, and even World War II.

Enlightened global leadership is needed desperately, not only to consider ecological limits, but also to implement a massive redistribution of wealth and power. This challenge is daunting, but, if it is not recognized and responded to, then a 'breakdown world', characterized by 'barbarianization', may have the same result, within this century (Swart, 1996).

Notes

[1] PPP adjusted measures capture the partial reality that a unit of tradeable currency (e.g. a Euro) will buy more goods and services in a nation in the South than within the European Union. This means that the per capita PPP income of a country in the South will be substantially higher than in foreign exchange terms, for example by over three for India in 2021 (World Bank, nd).

[2] Particularly because China does not participate in the surveys used to calculate PPP data.
[3] The collaboration, generally tacit, between those with power in the North and South has been identified as an element of the global 'claste' system (Butler, 2000). This neologism attempts to enrich analysis of the global class system by asserting that global inequality has caste-like qualities, particularly of indifference to and invisibilization of its poorest stratum.
[4] Named after the Swiss town where many members of the transnational capitalist class meet annually for the annual World Economic Forum.
[5] Analysis here of the policies of China and Russia is necessarily restricted, but neither nation has a shortage of billionaires. Some Russian policy makers appear to consider that their country's vast size and northerly latitude is protective. Russian wealth is substantially based on the export of fossil fuels. Thus, while ideologically separate to the 'West', Russian policy concerning climate change is, in practice, little different. China is widely credited with accelerating its energy transition. In some ways China may lead the world with recognition of the severity of the climate change risk. However, China is still expanding its fossil fuel use.

References

Bales, K. (1999) *Disposable People: New Slavery in the Global Economy*. University of California Press, Berkeley, California.

Barry, J. (2021) Green republicanism and a 'Just Transition' from the tyranny of economic growth. *Critical Review of International Social and Political Philosophy* 24, 725–742. DOI: 10.1080/13698230.2019.1698134

Bohacek, J. and Mansuy, I.M. (2013) Epigenetic inheritance of disease and disease risk. *Neuropsychopharmacology* 38, 220–236. DOI: 10.1038/npp.2012.110

Bruckner, B., Hubacek, K., Shan, Y., Zhong, H. and Feng, K. (2022) Impacts of poverty alleviation on national and global carbon emissions. *Nature Sustainability* 5, 311–320. DOI: 10.1038/s41893-021-00842-z

Butler, C.D. (2000) Inequality, global change and the sustainability of civilisation. *Global Change and Human Health* 1, 156–172. DOI: 10.1023/A:1010029222095

Butler, C.D. (2002) Inequality and Sustainability. PhD thesis, Australian National University. Available at: https://openresearch-repository.anu.edu.au/handle/1885/46233 (accessed 3 September 2022).

Butler, C.D. (2019) Philanthrocapitalism: promoting global health but failing planetary health. *Challenges* 10, 24. DOI: 10.3390/challe10010024

Butler, C.D. and Higgs, K. (2018) Health, population, limits and the decline of nature. In: Marsden, T. (ed.) *The Sage Handbook of Nature*. Sage, London, pp. 1122–1149.

Carrington, D. (2019) 'Climate apartheid': UN expert says human rights may not survive. *The Guardian*, Available at: https://www.theguardian.com/environment/2019/jun/25/climate-apartheid-united-nations-expert-says-human-rights-may-not-survive-crisis (accessed 26 August, 2022).

Cologna, V. and Oreskes, N. (2022) Don't gloss over social science! A response to: Glavovic *et al.* (2021) 'the tragedy of climate change science'. *Climate and Development* 14(9), 839–841. DOI: 10.1080/17565529.2022.2076647

Daly, H. (2022) *Reflections on Population*. Contribution to GTI Forum 'The Population Debate Revisited'. Great Transition Initiative. Available at: https://greattransition.org/gti-forum/population-daly (accessed 28 August 2022).

Darnton-Hill, I. (2019) Public health aspects in the prevention and control of vitamin deficiencies. *Current Developments in Nutrition* 3, nzz075. DOI: 10.1093/cdn/nzz075

Elton, B. (1989) *Stark. An Earth Shattering Novel*. Sphere Books, London.

FAO, IFAD, UNICEF, WFP and WHO 2022 *The State of Food Security and Nutrition in the World 2022. Repurposing food and agricultural policies to make healthy diets more affordable*. Food and Agriculture Organization, Rome. DOI: 10.4060/cc0639en

Feola, G., Koretskaya, O. and Moore, D. (2021) (Un)making in sustainability transformation beyond capitalism. *Global Environmental Change* 69, 102290. DOI: 10.1016/j.gloenvcha.2021.102290

Gagnon, B. (2021) Elon Musk (Space X) has gone nuts. *Space Education and Strategic Applications Journal* 2, 15–20. DOI: 10.18278/sesa.2.1.3

George, S. (1999) *The Lugano Report: On Preserving Capitalism in the Twenty-First Century*. Pluto Press, London.

Glavovic, B.C., Smith, T.F. and White, I. (2021) The tragedy of climate change science. *Climate and Development* 14(9), 829–833. DOI: 10.1080/17565529.2021.2008855

Green, F. and Healy, N. (2022) How inequality fuels climate change: The climate case for a Green New Deal. *One Earth* 5, 635–649. DOI: 10.1016/j.oneear.2022.05.005

Grollman, C., Cavallaro, F.L., Duclos, D., Bakare, V., Martínez Álvarez, M. *et al.* (2018) Donor funding for family planning: levels and trends between 2003 and 2013. *Health Policy and Planning* 33, 574–582. DOI: 10.1093/heapol/czy006

Hickel, J. (2016) The true extent of global poverty and hunger: questioning the good news narrative of the Millennium Development Goals. *Third World Quarterly* 37, 749–767. DOI: 10.1080/01436597.2015.1109439

Hickel, J. (2019) The contradiction of the sustainable development goals: Growth versus ecology on a finite planet. *Sustainable Development* 27(5), 873–884. DOI: 10.1002/sd.1947

Hill-Andrews, O. (2019) 'A new and hopeful type of social organism': Julian Huxley, J.G. Crowther and Lancelot Hogben on Roosevelt's New Deal. *The British Journal for the History of Science* 52, 645–671. DOI: 10.1017/S0007087419000657

Huq, S. (2019) For the vulnerable, UN climate talks are no longer fit for purpose. *Climate Home News.* Available at: https://www.climatechangenews.com/2019/12/17/vulnerable-un-climate-talks-no-longer-fit-purpose/ (accessed 3 September, 2022).

Jackson, B., Bales, K., Owen, S., Wardlaw, J. and Boyd, D.S. (2019) Analysing slavery through satellite technology: how remote sensing could revolutionise data collection to help end modern slavery. *Journal of Modern Slavery* 4, 169–199. DOI: 10.22150/jms/URDJ6988

Martin, C.L., Ghastine, L., Lodge, E.K., Dhingra, R. and Ward-Caviness, C.K. (2022) Understanding health inequalities through the lens of social epigenetics. *Annual Review of Public Health* 43, 235–254. DOI: 10.1146/annurev-publhealth-052020-105613

McMichael, A.J. (1993) *Planetary Overload. Global Environmental Change and the Health of the Human Species*. Cambridge University Press, Cambridge, UK.

Milanovic, B. (2022) After the financial crisis: The evolution of the global income distribution between 2008 and 2013. *Review of Income and Wealth* 68, 43–73. DOI: 10.1111/roiw.12516

Otto, I.M., Kim, K.M., Dubrovsky, N. and Lucht, W. (2019) Shift the focus from the super-poor to the super-rich. *Nature Climate Change* 9, 82–84. DOI: 10.1038/s41558-019-0402-3

Rajagopal, B. (2007) The caste system – India's apartheid? *The Hindu*, 18 August. Available at: http://www.thehindu.com/2007/08/18/stories/2007081856301200.htm (accessed 3 September 2022).

Reddy, S.G. and Minoiu, C. (2007) Has world poverty really fallen? *Review of Income and Wealth* 53, 484–502. DOI: 10.1111/j.1475-4991.2007.00240.x

Robinson, W.I. (2017) Debate on the new global capitalism: Transnational capitalist class, transnational state apparatuses, and global crisis. *International Critical Thought* 7, 171–189. DOI: 10.1080/21598282.2017.1316512

Rothkopf, D. (2009) *Superclass: The Global Power Elite and the World They Are Making*. Farrar, Straus and Giroux, New York (cited by Robinson, 2017)

Sklair, L. (2000) The transnational capitalist class and the discourse of globalisation. *Cambridge Review of International Affairs* 14, 67–85. DOI: 10.1080/09557570008400329

Sovacool, B.K. (2021) When subterranean slavery supports sustainability transitions? Power, patriarchy, and child labor in artisanal Congolese cobalt mining. *The Extractive Industries and Society* 8, 271–293. DOI: 10.1016/j.exis.2020.11.018

Strauss, E.D. and Shizuka, D. (2022) The ecology of wealth inequality in animal societies. *Proceedings of the Royal Society B: Biological Sciences* 289, 20220500. DOI: 10.1098/rspb.2022.0500

Swart, R. (1996) Security risks of global environmental changes. *Global Environmental Change* 6, 187–192. DOI: 10.1016/0959-3780(96)00014-3

Thiyagarajan, S. (2021) The deplorable parallels between apartheid and caste. *The Globalist.* Available at: https://www.theglobalist.com/the-deplorable-parallels-between-apartheid-and-caste/ (accessed 2 September, 2022).

Timperley, J. (2021) The broken $100-billion promise of climate finance – and how to fix it. *Nature* 598, 400–402. Available at: https://media.nature.com/original/magazine-assets/d41586-021-02846-3/d41586-021-02846-3.pdf (accessed 22 June, 2022).

Wade, R.H. (2005) The causes of increasing world poverty and inequality; or, why the Matthew effect prevails. *International Journal of Health Services* 35, 631–653. DOI: 10.2190/5YLF-X5K4-TGFH-EM56

World Bank (nd) Indicators. Available at: https://data.worldbank.org/indicator/ (accessed 3 September 2022).

Zeilig, L. (2008) From 'whiteman in rags' to revolutionary nationalist: Patrice Lumumba 1925–1960. *Social Dynamics* 34, 227–258. DOI: 10.1080/02533950802280147

3 Nuclear Weapons, Climate Disruption and Planetary Health

Tilman A. Ruff

International Physicians for the Prevention of Nuclear War, Boston, Massachusetts, USA; International Campaign to Abolish Nuclear Weapons, Australia; School of Population and Global Health, University of Melbourne, Australia

Abstract

This chapter explores the acute existential threat that nuclear weapons pose to humankind and the biosphere, including the global climate. If not eliminated first, they will destroy civilization, catastrophically damage the biosphere and risk extinction for humans and many other species. Fewer than 2% of the devices in the current global nuclear arsenal, if exploded on cities, would cause abrupt ice age conditions, putting over 2 billion people at risk of starvation. The 2017 UN Treaty on the Prohibition of Nuclear Weapons provides the best available pathway for all nations to fulfil their obligation to eliminate nuclear weapons. Scientists and health professionals have vital roles and responsibilities in work to end nuclear weapons.

3.1 Context: Existential Threats

Collisions of sizeable celestial bodies with planet Earth have been responsible for most major extinction events, such as that of the dinosaurs 65 million years ago. We may now be able to anticipate and avoid many such catastrophic collisions and should work collectively to improve these capabilities. In billions of years we can expect our Sun to expand into a red giant star and render the Earth uninhabitable.

However, there are two other kinds of unprecedented existential threats which have emerged recently and are of human origin. Though they are upon us, it is not yet too late to take the urgent decisive preventive action they demand. The first is the subject of much of this book: environmental disruption, degradation and depletion of vital resources and ecosystems. Poorly controlled global heating poses the greatest of these interrelated challenges. The second, more acute and potentially less reversible, is the danger of nuclear war.

The World Health Assembly, the pre-eminent annual meeting of the World Health Organization (WHO), identified nuclear weapons as the greatest immediate threat to human health and welfare (WHO, 1983). The modest, often minimal, attention generally given to this issue by health professionals, including in their training and organizations, as for many other professions, is starkly at odds with this reality. Preventing use of nuclear weapons, by accident or design, necessitates their eradication – an urgent and feasible precondition for securing and sustaining planetary and human survival and health. Two senior climate scientists recently wrote: 'We must solve the problem of nuclear weapons so that we have the luxury of devoting our time to solving the climate crisis' (Robock and Prager, 2021).

 Climate Change and Global Health: Primary, Secondary and Tertiary Effects, 2nd Edition (eds C.D. Butler and K. Higgs)
DOI: 10.1079/9781800620025.0003

3.2 What Nuclear Weapons Do

Evidence of the true extent of the effects of nuclear weapons has frequently not been collected, or has been covered up, misrepresented or disregarded by governments, to reinforce the myths that nuclear weapons are weapons like any other, only bigger, and can be used to serve legitimate military purposes and enhance security (Ruff, 2013). Public acceptance of the massive investments and risks involved to develop, produce, deploy and threaten to use nuclear weapons depends on an understanding of their effects that is not firmly based on evidence of what nuclear weapons actually do. No meaningful health or humanitarian response, reconciliation or recovery is possible after the catastrophic indiscriminate radioactive violence of a nuclear war. The concept of 'winners' would be meaningless; there would only be losers.

The bomb that destroyed Hiroshima and killed 140,000 people in the first five months, and is still killing people 78 years later, was by today's standards a small, tactical-size weapon, 15 kilotonnes (15 kt) of high-explosive equivalent. The average explosive power of the weapons in today's arsenal is almost 11 times larger (162 kt); the biggest nuclear weapon deployed currently at 5 megatonnes (5 Mt) is more than 330 times larger. A single nuclear weapon (in Russia in 1961) has been detonated with more than four times the destructive power of all explosives used in all wars throughout human history.

Acute effects

The effects of nuclear weapons differ both quantitatively and qualitatively from those of any other weapon (Table 3.1). Nuclear weapons produce an enormous blast wave that causes trauma both directly (such as lung trauma and eardrum rupture), and indirectly through powerful winds that can turn objects, including people, into missiles. An intense flash of light and initial ionizing radiation are produced together with intense heat, which causes direct vaporization, incineration and burns, and ignites anything flammable over a large area.

Apart from their essentially limitless destructive power, a defining feature of nuclear weapons is the release of huge amounts of radioactivity in the initial pulse as well as radioactive fallout containing hundreds of different radioisotopes with half-lives ranging from fractions of a second to millions of years. Fallout is dispersed by wind and water over great distances, eventually worldwide. Ionizing radiation causes acute toxicity to many organs (acute radiation sickness) at high doses; and in the long term, at any level, causes dose-related and heritable genetic damage and subsequent lifelong increased risk of cancer and chronic disease.

An intense burst of radiowaves produced by a nuclear explosion (electromagnetic pulse) (EMP) would cause extensive disruption of electrical equipment. The EMP from a nuclear explosion high in the atmosphere would cover a continental-size area with voltage a million times greater than lightning. Some military equipment is specifically hardened against such an effect; virtually all civilian equipment, including electricity grids, is not. An EMP would disrupt the vast array of electrical and electronic equipment on which the infrastructure of modern societies is increasingly dependent – including electricity and water supply, telecommunications, computer systems, transport networks, medical equipment, the internet, cars and trucks, traffic lights, banking, appliances, and most commerce and trade.

The largest nuclear weapons currently deployed are 5 Mt – the equivalent amount of high explosive would fill a freight train 2414 km long. The energy released by such a bomb could turn 5 Mt of ice to steam. Within a thousandth of a second, conditions akin to the centre of the Sun would be produced: 100×10^6 °C and 100×10^6 atmospheres of pressure in a fireball which would rapidly expand to 1.8 km across. Within 4.7 km in every direction, winds of 750 km/h and a blast wave over 140 kilopascals (kPa) would crush, collapse, or explode all buildings, including those of steel and reinforced concrete, and turn the debris into missiles with lethal velocity. Glass and steel would melt; concrete would explode. Wherever they were, all living things would die almost immediately – vaporized, crushed, charred, irradiated.

Out to about 7.5 km in every direction, winds of 460 km/h and blast pressures of 80 kPa would break apart concrete and steel buildings and sweep out their walls, floors, and ceilings. Aluminium would be vaporized. Adults would

Table 3.1. Effects of a nuclear explosion

Effect		Typical energy distribution	Notes
Blast	Crater (for groundburst[a]) Blast overpressure Blast dynamic pressure	50%	Important in military planning
Thermal radiation	Heat flash Fire initiation	35%	Dominant in humanitarian effects
Radiation	Initial nuclear radiation	5%	
	Local fallout (greater for groundburst than airburst[a]) Regional, global fallout	10%	
Electromagnetic pulse	Damages electrical/electronic equipment potentially on a continental scale		

[a]In a groundburst the nuclear fireball touches the Earth's surface; in an airburst it does not

be hurled over 100 m at high speed. Essentially everyone would be killed or seriously injured, including by crush injuries, penetrating wounds, ruptured lungs and other internal organs, transected spinal cords and nerves, severe haemorrhage and deep burns.

As far as 12.3 km in every direction, winds of 260 km/h and blast pressures of 35 kPa would crush wooden and brick buildings, including houses, schools, shops and many factories. People would be hurled 7 m. Asphalt would melt. Windows would be fragmented into more than 4000 projectile glass shards per square metre. Many people would be deaf from ruptured eardrums. In less than 10 seconds a city would be completely devastated.

Stretching 22.6 km in every direction, over an area of 1605 km^2, everything flammable would ignite – wood, paper, cloth, vegetation, plastics, petrol, and oil from ruptured tanks and cars; further fuelled by ruptured gas pipes, downed electricity lines and leaking chemicals. Within half an hour, thousands upon thousands of fires would coalesce into a giant firestorm 45 km across, with temperatures of more than 800°C, sucking in air and creating winds of more than 320 km/h, consuming available oxygen. Within this area, every living thing would die and shelters would become crematoria.

Still further out, windows would be shattered, buildings damaged, broken debris turned into projectiles. Streets would be impassable. There would be no functioning ambulances, paramedics, fire engines or police, no power or communications or functioning hospitals. The vast majority of injured and burnt people would die alone without so much as a hand or voice to comfort them and without any relief for their agonizing pain.

Smaller weapons in the range of tens to hundreds of kilotonnes are about 100 times more efficient per kilotonne of explosive force at killing people and igniting fires compared with megatonne-range weapons (Toon *et al.*, 2007).

WHO reports

In 1983, the World Health Assembly considered the first report of an international committee of experts on the effects of nuclear war on health and health services, coordinated by WHO. It endorsed the committee's conclusion 'that it is impossible to prepare health services to deal in any systematic way with the catastrophe resulting from nuclear warfare, and that nuclear weapons constitute the greatest immediate threat to the health and welfare of mankind' (WHO, 1983). The committee's report stated: 'It is obvious that no health service in any area of the world would be capable of dealing adequately with the hundreds of thousands of people seriously injured by blast, heat or radiation from even a single 1-megatonne bomb.' The committee concluded that: 'the only approach to the treatment of the health effects of nuclear explosions is primary prevention of such explosions, that is, the primary prevention of atomic war' (WHO, 1983, pp. 5–6).

WHO issued an updated report in 1987, including new research on urban superfires caused by nuclear explosions, concluding that deaths caused by such fires could be 3–4 times greater than those caused by blast (WHO, 1987). The report also considered the emerging evidence of global climate impact – nuclear winter – that would follow nuclear war even involving only a small fraction of the global arsenal detonated over large fuel-dense modern cities. It concluded: 'After a major nuclear war famine and diseases would be widespread and social, communication and economic systems around the world would be disrupted' (WHO, 1987, p. 5).

3.3 New Evidence of Climate Impacts

In the decades since, we have learned much about the multiplicity of impacts of nuclear explosions and war. The evidence grows ever clearer as to the catastrophic effects. With striking consistency, the more we learn, the worse are the assessed impacts of nuclear war. New evidence on the health impacts of ionizing radiation demonstrates greater and more diverse impacts than previously understood; it confirms the lack of a threshold dose for adverse health effects, shows that a dose rate reduction factor should not be applied, and confirms increased risk at all doses for a range of non-communicable diseases (NCDs), particularly cardiovascular disease (Ruff, 2017; Little *et al.*, 2023). Field studies of a wide range of plants and animals in the contaminated regions around Chernobyl and Fukushima demonstrate extremely pervasive effects, from genetic and individual developmental, phenotypic, fitness and fertility impacts, to population, species and ecosystem effects, far greater than those expected based on *in vitro* studies (Mousseau, 2021).

Major humanitarian organizations, including the Red Cross Red Crescent movement and UN agencies, have re-examined capacities nationally and globally to respond to nuclear detonations, and concluded, as WHO did in the 1980s, that no capacity exists or is feasible to respond meaningfully to humanitarian needs in the aftermath of even a single nuclear detonation in a city.

However, the most important new evidence relates to climate impacts. Nuclear weapons are extremely efficient at igniting vast numbers of simultaneous fires over large areas. These would consume all flammable materials and coalesce into massive confluent fires within which no one could survive the > 800°C heat, intense smoke and oxygen depletion. Atmospheric scientists estimated that even the relatively small tactical-size nuclear weapon exploded on Hiroshima (15 kt of high-explosive equivalent) released about 1000 times as much energy in the fires it ignited as in the explosion itself (Toon *et al.*, 2007). In Hiroshima approximately 13 km^2 of the city burned completely. As described above, detonation of the largest currently deployed nuclear weapon, up to 5 Mt in size, would engulf a city in a confluent megafire more than 45 km in diameter, 1600 km^2 in area (Ruff, 2013).

Atmospheric and climate effects of regional nuclear war: the example of India–Pakistan

The scenario for regional nuclear war most often studied by atmospheric scientists is a war between India and Pakistan. This possibility is all too realistic, as the two nations have waged war four times since their independence in 1949, and mobilized up to 1 million troops on two further occasions. Violence erupts across their disputed border in Kashmir almost daily. They possess two of the world's six growing nuclear arsenals. Both have policies that create high risks of nuclear escalation in a war between them. The most recently updated scenario involves use of 250 nuclear weapons of 15, 50, or 100 kt in size, half the combined arsenal they were estimated on current trajectory to possess in 2025 (Toon *et al.*, 2019). These constitute 2% of the number of nuclear weapons worldwide and amount to less than 1% of their explosive yield, because the average size of the current 12,119 nuclear weapons is 162 kt (Kristensen *et al.*, 2024).

Such a war would produce between 83 and 183 million acute casualties in cities across both nations, including 52–127 million deaths, depending on the size of the weapons used. Radioactive contamination, severe social and economic disruption, and people attempting to flee on an unprecedented scale would extend across South Asia and beyond. Such a war would also produce between 16 and 37 Mt of black carbon in sooty smoke from burning cities (Toon *et al.*, 2019). This smoke would loft quickly into the upper stratosphere and mesosphere, beyond the reach of clouds and precipitation in the lower atmosphere (troposphere). The sun would heat the rising smoke by 50–80°C, lofting and keeping it high in the atmosphere, where it would blanket the Earth for over a decade. It would reduce global average surface temperatures by 3–6°C, well within the range of minimum temperatures during the peak of the last ice age 20,000 years ago, 3–8°C colder than present. Unevenly distributed temperature declines of 8–15°C would cover much of the large North American and Eurasian land masses where most of the world's grain is grown.

Global precipitation would also decline by up to 40%, with particular disruption of the South Asian monsoon on which food production for 1.5 billion people critically depends. Incident sunlight would decline by up to 30%. Scientists expect these darker, drier conditions and colder temperatures to be associated with cold spells and shorter frost-free growing seasons in temperate regions. An unprecedented increase in ultraviolet flux (30–100% increases during summer outside the tropics) would exacerbate these changes (Mills *et al.*, 2015). Stratospheric ozone would be extensively depleted in the hot smoky stratosphere, with harmful effects of increased ultraviolet (UV) radiation on plant and animal development and health in both aquatic and terrestrial environments. Even a smaller India–Pakistan nuclear war releasing 5 Mt of soot would produce peak global ozone loss of 25%, and up to 55% loss at higher latitudes, leading to a peak increase of 40% in the UVB wavelengths associated with DNA damage, with recovery taking 12 years (Bardeen *et al.*, 2021). Most agricultural production would abruptly cease in higher-latitude regions, including Canada, northern areas of Europe, Russia, China, Korea and Japan (Jägermeyr *et al.*, 2020).

Radioactive fallout and toxic chemical contamination from damaged pipelines and industrial and storage sites would affect large areas of agricultural land. Social, economic, transport and trade turmoil would disrupt global distribution of fertilizer, fuel, machinery and equipment, parts, seeds, pesticides, food storage facilities, and transport on which modern agriculture, food storage and distribution depend. The direct climatic changes alone would cause a decline in net primary productivity (NPP) of between 10% and 20% in the oceans and between 15% and 40% on land over multiple years (Toon *et al.*, 2019). NPP is the net amount of carbon per square metre per year converted into plant matter after accounting for what plants use for their own respiration. This loss would be comparable to the total current annual human use of food and fibre. Agricultural losses could not be offset by the world's fisheries, especially with pre-existing widespread overfishing as occurs currently (Scherrer *et al.*, 2020).

Scientists continue to discover new effects that would exacerbate the harm. Recent findings indicate that various nuclear war scenarios could induce an El Niño-like pattern of unprecedented magnitude across the Pacific, with associated reductions in equatorial Pacific phytoplankton productivity of about 40% (Coupe *et al.*, 2021). Researchers recently identified large and abrupt exacerbations in global ocean acidification as a consequence of nuclear conflict, including potential inability for marine calcifying organisms like shellfish and corals to maintain their shells or skeletons in a corrosive environment (Lovenduski *et al.*, 2020). Even the smallest nuclear war would devastate ocean systems, leading to sharp declines in fish stocks, expansion of ice sheets into coastal communities and changes in ocean currents that would take decades or longer to reverse (Harrison *et al.*, 2022). Sea temperature impacts from nuclear war are particularly extreme in coastal regions where the majority of fish catch and marine ecosystem services are provided. Entirely different conditions for nutrients vital to plankton at the base of food chain would be widespread through the world's oceans. Some nuclear little ice age changes are likely to last hundreds of years or longer (Harrison *et al.*, 2022).

Nutritional impact

The world is ill prepared to withstand any sustained and substantial decline in food production. In July 2022, the Food and Agriculture Organization (FAO) of the United Nations (UN) estimated between 702 and 828 million people faced hunger in 2021, 150 million more than in 2019 (FAO *et al.*, 2022). Worsened by the COVID-19 pandemic and the war in Ukraine, FAO estimated the number of people experiencing moderate or severe food insecurity in 2021 at 2.31 billion. The Global Report on Food Crises 2023 estimated that the number of people facing high levels of food insecurity has increased from 105 million people in 48 countries in 2016, to 258 million people in 58 countries in 2022 (FSIN and GNAFC, 2023). The FAO forecast for the year 2022/23 for global cereal stocks is equivalent to about 112 days of consumption (FAO, 2023).

It has been known for decades that extrapolation of the physical effects of nuclear detonations cannot begin to characterize the reality of the world after nuclear war, and that the climatic effects of nuclear war and disruption of global food support systems could result in the deaths of multiple billion humans (Harwell *et al.*, 1989; Helfand *et al.*, 2016). Recent work has provided more detailed global and country-specific estimates of the impacts of regional and global nuclear war on both terrestrial and aquatic food production and nutrition (Xia *et al.*, 2022). Soot injections greater than 5 Mt, as would be produced by burning cities ignited by 100 Hiroshima size nuclear explosions, would lead to mass food shortages. In the three scenarios described above for an India–Pakistan war utilizing 250 nuclear weapons 15, 50 or 100 kt in yield, with 16–37 Mt of smoke, the number of people estimated to die of starvation by the end of the second year after such a war would range between 926 million and 2.081 billion – more than 16 times the number of direct fatalities. Food stores, livestock and aquatic food production would be unable to compensate for persistently reduced crop output. Adaptation measures such as killing livestock and reducing food waste would have quite limited impact. Crop reduction and starvation would be most severe in mid and high northern latitudes. For example, assuming livestock has been slaughtered and animal feed diverted to human consumption, an India–Pakistan war producing 37 Mt of smoke would result, after two years, in the starvation of over a billion people in China, 125 million in Russia, 131 million in the USA and 39 million in the UK.

These salutory estimates are nevertheless quite conservative, as they used an FAO database reflecting a global population of 6.7 rather than the current 8 billion; they do not account for direct or radiation-related deaths, increased solar ultraviolet radiation, disruption of agricultural inputs, food storage and distribution, displacement, societal collapse or conflict, electromagnetic pulse effects, or contamination of productive land by radiation and chemical pollution.

Epidemics of various infectious diseases would inevitably accompany famine of such unprecedented magnitude, as well as conflict within and between nations over unequal, inadequate and rapidly diminishing food reserves. The combination would likely substantially exacerbate the human toll. Human civilization would not be expected to survive a shock of this magnitude.

Effects of a Russia–USA nuclear war

Nuclear war utilizing a substantial portion of the deployed Russian and US nuclear arsenals is conservatively estimated to produce 150 Mt of black carbon in smoke from burning cities. Over the course of a week, this would plummet global surface temperatures to more than 9°C colder than present, with temperatures in extensive areas of the Eurasian and North American land masses dropping by 20–35°C, and locations such as Ukraine and Iowa barely experiencing days with minimum temperatures above freezing for two years (Coupe *et al.*, 2019). Such a severe nuclear winter would put the survival of humans and many other species in severe jeopardy. For a war between Russia and the USA using 4400 nuclear weapons 100 kt in size, less than half their weapons stockpile, over 5 billion people would be expected to succumb to famine by the end of year 2. In the most favourable livestock and food waste scenario for human survival, by the end of year 2 more than 99% of the population of China, Canada and northern Europe would be starving, along with more than 75% of the population of Russia and the USA (Xia *et al.*, 2022).

Policy implications of nuclear famine and other health effects of nuclear war

Immediate localized destruction would cause catastrophic local health impacts. Widespread direct health impacts would be caused by dispersed radioactive fallout and potentially an electromagnetic pulse from a high-altitude nuclear explosion that would incapacitate all civilian electrical and electronic infrastructure on a continental scale. However, the indirect effects of disruption to essential infrastructure and services, the 'health hardware' needed to sustain life and health in an era of unprecedented urbanization and global interdependence, would likely be greater. Disruption would extend far beyond areas of direct impact and affect shelter, heating, transport and communications, trade and the monetary system, supply and distribution of food, water, clothing, medicines, electricity, fuel, health care, and disposal of the dead, waste and sewage. Even without use of nuclear weapons, modern conflicts with extensive bombing of cities and infrastructure – such as in Vietnam, Iraq, Syria, Yemen and Ukraine – have generally caused far more civilian than military deaths, and far more deaths from indirect effects than direct violence. Nuclear war would cause widespread disruption of human societal functioning, and of ecosystem functioning, on an unprecedented scale.

The major cause of casualties worldwide from a nuclear war would be from the abrupt onset of nuclear ice age conditions and resultant mass starvation. Human civilization is unlikely to survive such severe global disruption. The findings of ice age conditions inducing mass global starvation from even a localized regional nuclear war contradict the commonly claimed theoretical basis for 'nuclear deterrence of mutually assured destruction'. Instead, they characterize nuclear weapons as risking 'self-assured destruction' from what amount to global suicide bombs. Nuclear weapons overwhelmingly endanger the security of all peoples – especially in the states that wield and are most likely to be directly targeted by them – and render meaningless any concept of winning a nuclear war (Robock and Toon, 2012). Given their catastrophic, indiscriminate, inhumane and pervasive transgenerational impacts, nuclear weapons can have no legitimate, logical or legal military purpose. Humanity has made significant progress in stigmatizing and controlling most other types of indiscriminate and inhumane weapons that violate international humanitarian law (the laws of war): from expanding bullets and blinding laser weapons, to biological and toxin weapons, chemical weapons, anti-personnel landmines and cluster munitions. Nuclear weapons are far more indiscriminately destructive than any of these.

3.4 The Current Risk of Nuclear War

The world has no experience, no proven pathways, and extremely limited mechanisms to curtail escalation once the threshold of use of nuclear weapons in war is crossed. While precise plans for nuclear war are generally not public, the consistent assessment of experts and military and other leaders with direct experience of managing nuclear weapons and nuclear war exercises is that once nuclear war erupts, rapid escalation to involve most of the nuclear weapons deployed in the arsenals of the involved nations would be highly probable (Hyten, 2018; Toon *et al.*, 2019; Perry and Collina, 2020).

Risks of nuclear war are growing (UN, 2021). The Science and Security Board of the Bulletin of the Atomic Scientists in 2023 moved the hands of the Doomsday Clock 'forward, largely ... because of the mounting dangers of the war in Ukraine. The Clock now stands at 90 seconds to midnight – the closest to global catastrophe it has ever been' (Science and Security Board, 2023). No nuclear armed state is currently disarming, or engaged in nuclear disarmament negotiations. First the USA, followed by Russia, abrogated hard-won treaties negotiated between them which were fruits of the end of the first Cold War, and which constrained nuclear weapons numbers and types. Together these two countries hold 89% of all nuclear weapons (Kristensen *et al.*, 2023a). These treaties include the Anti-Ballistic Missile Treaty (which limited missile defences, reducing the incentive for relentless proliferation), the Intermediate Nuclear Forces Treaty (which eliminated short- and medium-range nuclear missiles from the Soviet – later Russian – and US arsenals), the Open Skies Treaty (which increased nuclear transparency), and the more recent Joint Comprehensive Plan of Action (Iran nuclear deal that provided effective constraints on Iran's nuclear programme until the US Trump administration abrogated it). Were it not for the incoming Biden administration's quick agreement to extend the New START (Strategic Arms Reduction Treaty) just two days before it would otherwise have expired in February 2021, there would have been no treaty restraints in force in and beyond 2021 on US and Russian nuclear weapons, despite a resurgent Cold War between them. In early 2023, Russia suspended its participation in the New START. This treaty is due to expire in February 2026; no negotiations are in sight regarding a successor. Should this last remaining treaty constraining Russian and US nuclear weapons expire, it is assessed that these two nations could approximately double their deployed nuclear weapons within a matter of months (Korda, 2023). China's growing nuclear arsenal, though no more than 8% the size of that of Russia or the USA, is subject to no specific treaty constraints; this is also the case for those of France, India, Israel, North Korea, Pakistan and the UK. For the first time in three decades, the number of warheads in military stockpiles (assigned to operational forces) is again increasing. Of the 12,119 estimated total nuclear weapons in early 2024, approximately 2000 British, French, Russian and US warheads are on high alert, ready to be launched within minutes of a launch order (Kristensen *et al.*, 2023a).

Russia's 2022 invasion of Ukraine has dramatically raised nuclear dangers at the same time as it has eroded international cooperation required to address a variety of global risks. The 2023 Doomsday Clock Statement continued: 'worst of all, Russia's thinly veiled threats to use nuclear weapons remind the world that escalation of the conflict – by accident, intention, or miscalculation – is a terrible risk. The possibility that the conflict could spin out of anyone's control remains high' (Science and Security Board, 2023). The war in Ukraine has brought NATO states and Russia into increasingly direct conflict. Between them, these states hold 93% of the world's nuclear weapons, many of them on high alert, and all have policies of first use of nuclear weapons. Apart from deliberate escalation, the potential for military, including nuclear, escalation is ever-present: via direct force encounter, error, inadvertence or misunderstanding, potentially compounded by disinformation and cyberwarfare.

These dangers are exacerbated by Russia's attacks, occupation and disruption of the Chernobyl and Zaporizhzhya nuclear power plants, for the first time weaponizing and holding hostage such massive repositories of long-lived radioactive materials in a war zone. These civilian nuclear facilities are effectively serving as military bases, weapons stores and large, pre-positioned, potential radiological weapons, with the potential to greatly magnify the transboundary, long-lasting health consequences of the war. Uncontrolled radioactive release could occur from nuclear reactors or the associated spent fuel pools, which are without engineered containment and usually contain greater amounts of long-lived fission products than do reactors. Both reactors and spent fuel pools require continuous cooling through electrically pumped circulating water. Core meltdown can occur within a few minutes of loss of coolant (Rotblat, 1981), as the 2011 Fukushima nuclear disaster demonstrated. Fallout extent and long-term radiation dose caused by damaged nuclear facilities could be several times greater than would occur from nuclear weapons (Harwell *et al.*, 1989). At time of writing in April 2024, all efforts by the International Atomic Energy Agency to demilitarize and secure the plants have been rebuffed. The deliberate destruction of the Nova Kakhovka dam in June 2023 not only caused extensive downstream flooding, loss of drinking and irrigation water supplies, dispersal of landmines, oil and toxic chemicals; it has also put in jeopardy the ultimate source of cooling water for the Zaporizhzhya nuclear power plant, Europe's largest (IAEA, 2023).

Modernizing and expanding nuclear arsenals at enormous and escalating cost

All nine nuclear armed states are investing massively in modernizing, and a number in expanding, their nuclear arsenals (Norwegian People's Aid, 2023). Modernization means new, faster, stealthier and longer range, with more flexible and accurate capacities. Some delivery vehicles, including most of Russia's 1558 non-strategic nuclear weapons, can be armed with either conventional or nuclear warheads, indistinguishable until point of impact (Kristensen *et al.*, 2024). These changes lower the overall threshold for use of nuclear weapons (Pytlak and Acheson, 2022). Military stockpiles of nuclear weapons, i.e. those assigned to operational forces and excluding those that are retired and awaiting dismantlement, are again increasing, as are the number on high alert – ready to be launched within minutes. China, India, North Korea, Pakistan, the UK and possibly Russia are all increasing their stockpiles (Kristensen *et al.*, 2023a). All the nuclear-armed states apart from Israel and the UK test-launched nuclear capable missiles in 2022; North Korea launched at least 37 missiles (Norwegian People's Aid, 2023).

Both Russia and the USA, owning between them 89% of all nuclear weapons, are comprehensively replacing and modernizing their warheads, missiles and launch platforms. They are also increasing the role of nuclear weapons in their military policies, and the range of circumstances in which they might be used, including against conventional and cyber-attacks (Kristensen *et al.*, 2023b; Kristensen and Korda, 2023). Russia is testing and deploying entirely new types of nuclear weapons, including nuclear-powered cruise missiles, hypersonic glide delivery vehicles atop ballistic missiles (also being fielded by China and the USA), and very long-range nuclear-powered torpedoes designed to explode in waters close to cities; as well as a new dual-capable ground-launched cruise missile, and a new long-range dual-capable air-launched ballistic missile (Kristensen *et al.*, 2023b). The USA is producing new nuclear warheads for the first time in three decades, modernizing all types of nuclear weapons – ballistic and cruise missiles, bombs delivered by aircraft, and the submarines, ships and aircraft that carry them. It is also upgrading the nuclear weapons it provides to the UK and the nuclear bombs it stations in Belgium, Germany, Italy, Netherlands and Turkey for delivery by the air forces of those countries (Kristensen and Korda, 2023).

Current estimates of global spending on development and production of nuclear weapons reached US$82.9 billion in 2022, a real increase of US$6.9 billion from 2020, despite constraints of the COVID-19 pandemic (ICAN, 2023). The total cost of nuclear weapons programmes, including environmental clean-up and legacy costs, is far greater. The USA spends the most on military and nuclear weapons: in financial year 2023, nuclear weapons-related costs reached US$90.34 billion (Dodge, 2023). Military spending, 39% of the global total in 2022, consumes half of all discretionary US government spending. In the USA, nuclear warhead spending is currently at an all-time record high, with projected expenditures over the next three decades of over US$2 trillion to comprehensively refurbish the nuclear arsenal and the facilities that produce nuclear weapons (Pytlak and Acheson, 2022). While Russia's military spending in 2022 (US$86.4 billion) was estimated to be only 10% of that of the USA (US$877 billion) (SIPRI, 2023), the proportion it spends on nuclear weapons is more than double that of the USA (ICAN, 2023).

Opportunity costs: weapons versus United Nations and related programmes

Such vast expenditures on weapons that create a hazardous legacy even in their production have enormous social, environmental and public health opportunity costs. Estimates by the Sustainable Development Solutions Network (SDSN) place average total annual investment required between 2019 and 2030 to fully finance achievement of the Sustainable Development Goals agreed by all nations at US$1,011 billion (SDSN, 2019). That amounts to less than half of annual military expenditures, which reached a record high of US$2,240 billion in 2022 – 3.7% higher than in 2021 (SIPRI, 2023). That increase occurred despite the COVID-19 pandemic and the associated severe economic downturn, increase in poverty and food insecurity. The increase from the previous year, US$127 billion, is greater than the entire UN annual climate finance goal (SIPRI, 2023). In 2019 the combined annual budgets of the WHO, UNICEF, the UN itself, the UN High Commissioner for Refugees (UNHCR), the International Committee of the Red Cross (ICRC) and the UN Office of Disarmament Affairs amounted to less than 30% of direct spending on nuclear weapons, and around 1% of overall military expenditure (IPB, 2020). Operating an F-35 nuclear-capable combat aircraft for one hour costs as much as a nurse earns in a year (Organisation for Economic Co-operation and Development average); the cost of one Virginia Class nuclear submarine could fund 9180 fully equipped ambulances; the cost of one Trident II nuclear missile could buy 17 million facemasks (IPB, 2020). By September 2021, at least one dose of COVID-19 vaccine had reached fewer than 3% of people in low-income countries and WHO fell short by US$900 million in funds they needed to cover the period until March 2022 for their essential role in ending the acute phase of the pandemic – equivalent to 1.2% of annual direct nuclear weapons spending (WHO, 2021).

Doomsday Clock reflects growing insecurity

Leaders of all nuclear-armed states have, in recent years, issued specific nuclear threats, with military leaders confirming their active planning to fight nuclear war (Richard, 2021; Norwegian People's Aid, 2023). In 2020, the Bulletin of the Atomic Scientists moved its authoritative Doomsday Clock to 100 seconds to midnight, further forward than it had ever been before, explaining that 'the international security situation is now more dangerous than it has ever been, even at the height of the Cold War' (Science and Security Board, 2020). In 2021 and in 2022, the Clock hands remained in the same position, as 'the potential to stumble into nuclear war – ever present – has grown' (Science and Security Board, 2021) and the world was assessed as loitering at 'the doorstep of doom' (Science and Security Board, 2022). The unprecedented moving forward of the Clock hands in 2023 was mirrored in the assessment of the UN Secretary-General that 'humanity is just one misunderstanding, one miscalculation away from nuclear annihilation' and 'a time of nuclear danger not seen since the height of the Cold War' (UN, 2022).

In 2019, the US intelligence community's annual assessment to Congress of worldwide threats warned that the effects of climate change and environmental degradation increase stress on communities around the world and intensify global instability and the likelihood of conflict, increasing the danger of nuclear war (US Congress House of Representatives, 2019). Over the past decade, according to the Uppsala Conflict Data Programme (UCDP), the number of armed conflicts has steadily grown, particularly the number of 'internationalized intrastate' conflicts – within a state but involving at least one nation (disproportionately nuclear-armed nations) outside the state in the conflict (UCDP, 2022). Each of such conflicts thus poses a risk of nuclear escalation.

Corruption of the information ecosystem, disinformation and resultant exacerbation and manipulation of societal divisions undermine respect for evidence and effective governance, including of nuclear weapons. Insurrectionists at the US Capitol on 6 January 2021 came close to capturing then Vice-President Mike Pence and the 'nuclear football' that accompanies the vice-president as the back-up for US nuclear launch commands.

Cyberwarfare increases vulnerability of nuclear weapons systems

Another major area of increasing risk of use of nuclear weapons is growing use of cyberwarfare by both states and non-state actors. Attacks on civilian and military nuclear facilities included extensive hacking in December 2020 of the US National Nuclear Security Administration, which maintains US nuclear weapons (Bertrand and Wolff, 2020). Complex global systems of early warning, command, control, communications and intelligence are connected to nuclear weapons. They are complex, dispersed and interlinked – and vulnerable to cyberattack. As General James Cartwright, former head of US Strategic Command, stated, it 'might be possible for terrorists to hack into Russian or American command and control systems and launch nuclear missiles, with a high probability of triggering a wider nuclear conflict' (Cirincione, 2015).

British, French, Russian and US authorities between them keep 2000 nuclear warheads on high alert, all mounted on delivery vehicles and ready for use within minutes of a launch order (Kristensen *et al.*, 2024). These warheads are particularly vulnerable to digital sabotage and inadvertent or unauthorized launch, especially where authority to launch nuclear weapons rests entirely with one person, such as the US president. Many states, including China, Iran, Israel, North Korea, Russia and the USA, engage in offensive cyber operations (Perlroth, 2021).

Buyers of cybertools for offensive use may include governments, government proxies and terrorist organizations. For terrorist organizations, cyberwarfare creates the prospect of exploding nuclear weapons without themselves needing to acquire or deliver them. Frequently buyers find tools in a lucrative global black and grey market offering hacking tools, especially 'Zero-day exploits'. These tools exploit software or hardware flaws and vulnerabilities for which no corrective patch yet exists (Perlroth, 2021). Government staff as part of their work, moonlighting staff, or government contractors can develop offensive digital tools. Individual or organized hackers and cybercriminals or private for-profit companies can also produce them almost anywhere. Targets of hacking and digital sabotage to date include banking and health systems, Sony Corporation, electricity grids, water treatment facilities, airports, electoral systems, oil company and oil pipeline computer systems, government departments and services, uranium enrichment centrifuges and nuclear power plants. Increasing digital sophistication of nuclear weapons and delivery systems may increase their vulnerability to digital sabotage (Perlroth, 2021).

Source materials for nuclear weapons lack adequate control

Vast stocks of fissile materials, the highly enriched uranium (HEU) and plutonium from which nuclear weapons can be built, persist in civilian and military stockpiles in tens of countries. There are no effective international constraints on the production of these materials. Every state with a civilian nuclear industry is also capable of producing fissile materials; and any state that can enrich uranium to reactor grade (currently 14) can enrich it to weapons grade. The operation of nuclear reactors inevitably converts some of the uranium in the fuel into plutonium. The global fissile material stockpile in May 2022 was estimated by the International Panel on Fissile Materials (IPFM) to contain 550 t of separated plutonium and 1250 t of HEU (IPFM, 2022a, 2022b). Assuming 3 kg of military (weapon-grade) plutonium and 5 kg of civilian (reactor-grade) plutonium is a nuclear weapon equivalent, the global stock of separated plutonium in 2020 corresponded to about 140,000 weapon equivalents. For HEU, assuming 25 kg per warhead (less material is sufficient), the current global stockpile is enough for more than 50,000 nuclear weapons (IPFM, 2022c).

Apart from removal of relatively modest quantities of HEU from civilian stockpiles in 34 countries plus Taiwan (IPFM, 2022b), the challenges of ceasing production of these materials, eliminating them where possible, and keeping the remaining quantities in consolidated storage in the safest possible form at the highest possible levels of security, remain largely unaddressed. The main UN forum specifically tasked with negotiating disarmament treaties, the Geneva-based Conference on Disarmament (CD), has discussed a treaty limiting production of fissile materials for decades. However, because of its requirement for consensus decisions and the persistent objections of one state that feels it would be disadvantaged relative to its principal regional rival by such a treaty, the CD has been unable to progress such negotiations, indeed it has been unable to agree on an agenda of work since 1996.

3.5 The Role of Scientists and Health Professionals Towards the Eradication of Nuclear Weapons

The stable and hospitable climate needed for the survival and health of humans and the biosphere requires protection from rampant global heating and from an abrupt nuclear ice age. Planetary health depends on both, and addressing these twin existential dangers requires approaches that do not risk exacerbating either. An increasingly climate-stressed world is an even more dangerous and unsustainable place to keep arsenals of doomsday weapons.

Evidence-based advocacy by expert custodians of scientific and health evidence has been key to much of the progress towards constraining the nuclear arms race, and is discussed in more detail in Chapter 39. The most recent major example of such work is found in the humanitarian initiative on nuclear weapons, a collaboration since 2010 between many of the world's governments, civil society and international organizations, the world's largest humanitarian organization, the Red Cross Red Crescent movement being prominent among them. To date the main fruit of this initiative is the UN Treaty on the Prohibition of Nuclear Weapons (TPNW), adopted through the UN General Assembly in New York in 2017, which entered into legal force on 22 January 2021. By

April 2024, this treaty had 70 states parties and 23 additional signatory states.

This historic treaty is the first to categorically and comprehensively prohibit nuclear weapons. This treaty is helping to drive accelerating divestment by financial institutions from companies that profit from manufacturing the worst, and now illegal, weapons of mass destruction (Snyder, 2022). It also contains the first treaty-based obligations for states to assist victims of nuclear weapons use and testing; and it obligates them to assist in remediation of environments so contaminated. The treaty also contains the only internationally agreed and codified framework providing for the elimination of nuclear weapons. The treaty provides flexible pathways for nuclear-armed states to disarm either before or after joining the treaty. It specifies plans for time-bound dismantling of the weapons and the facilities that produce and maintain them, subject to verification by a competent international authority. Thus, the treaty provides the most promising available pathway for all states to fulfil their obligations to negotiate in good faith to achieve nuclear disarmament.

Experience with treaties prohibiting other types of inhumane and indiscriminate weapons shows that they often influence even states that oppose and have not signed the treaty. Biological and chemical weapons, landmines and cluster munitions are all now less often justified, produced, traded, deployed and used after their prohibition – even by states not having joined the respective bans (Ruff, 2018).

It was an organization of health professionals, International Physicians for the Prevention of Nuclear War (IPPNW), that founded the International Campaign to Abolish Nuclear Weapons (ICAN). This campaign coalition became the major civil society partner with governments in development of the TPNW, resulting in award of the Nobel Peace Prize to ICAN in 2017 'for its work to draw attention to the catastrophic humanitarian consequences of any use of nuclear weapons and for its ground-breaking efforts to achieve a treaty-based prohibition of such weapons' (Norwegian Nobel Committee, 2017). Many health and humanitarian organizations, including the World Federation of Public Health Associations (WFPHA), the Red Cross and Red Crescent movement, the World Medical Association (WMA), the International Council of Nurses (ICN), and the International Federation of Medical Student Associations (IFMSA), have joined with IPPNW to welcome the entry into force of the treaty and urge all nations to join and faithfully implement it (ICRC *et al.*, 2021).

As the risks of nuclear war are not only continuing but growing, effective and coordinated evidence-based advocacy by scientists and health professionals on the health imperative to eliminate nuclear weapons has never been needed more.

References

Bardeen, C.G., Kinnison, D.E., Toon, O.B., Mills, M.J., Vitt, F. *et al.* (2021) Extreme ozone loss following nuclear war results in enhanced surface UV radiation. *Journal of Geophysical Research: Atmospheres* 126, e2021JD035079. DOI: 10.1029/2021JD035079

Bertrand, N. and Wolff, E. (2020) Nuclear weapons agency breached amid massive cyber onslaught. *Politico.* Available at: https://www.politico.com/news/2020/12/17/nuclear-agency-hacked-officials-inform-congress-447855 (accessed 8 March 2023).

Cirincione, J. (2015) What happens when our nuclear arsenal is hacked? *San Francisco Chronicle.* Available at: https://www.sfchronicle.com/opinion/openforum/article/What-happens-when-our-nuclear-arsenal-is-hacked-6333739.php (accessed 8 March 2023).

Coupe, J., Bardeen., C.G., Robock, A. and Toon, O.B. (2019) Nuclear winter responses to global nuclear war in the whole atmosphere community climate model version 4 and the Goddard Institute for Space Studies modelE. *Journal of Geophysical Research: Atmospheres* 124, 8522–8543. DOI: 10.1029/2019JD030509

Coupe, J., Stevenson, S., Lovenduski, N.S., Rohr, T., Harrison, C.S. *et al.* (2021) Nuclear niño response observed in simulations of nuclear war scenarios. *Communications Earth and Environment* 2, 18. DOI: 10.1038/s43247-020-00088-1

Dodge, R. (2023) *How much do you pay for nuclear weapons?* Physicians for Social Responsibility, Los Angeles. Available at: https://www.psr-la.org/nuclear-costs (accessed 28 May 2023).

FAO (2023) *FAO cereal supply and demand brief*. Food and Agriculture Organization of the UN, Rome. Available at: https://www.fao.org/worldfoodsituation/csdb/en/ (accessed 27 May 2023).

FAO, IFAD, UNICEF, WFP and WHO (2022) *The State of Food Security and Nutrition in the World 2022*. Food and Agriculture Organization of the UN, Rome.

Federation of American Scientists (nd) https://fas.org/initiative/status-world-nuclear-forces/ (accessed 5 April 2024).

FSIN and GNAFC (2023) *Global Report on Food Crises 2023*. Food Security Information Network and Global Network Against Food Crises, Rome.

Harrison, C.S., Rohr, T., DuVivier, A., Maroon, E.A., Bachman, S., Bardeen, C.G. *et al.* (2022) A new ocean state after nuclear war. *AGU Advances* 3, e2021AV000610. DOI: 10.1029/2021AV000610

Harwell, M.A., Hutchinson, T.C., Cropper, W.P., Harwell, C.C., Grover, H.D. *et al.* (1989) *Environmental Consequences of Nuclear War: SCOPE 28. Volume II: Ecological and Agricultural Effects*, 2nd edn. Scientific Committee on Problems of the Environment (SCOPE) of the International Council of Scientific Unions. Wiley & Sons, Chichester, UK.

Helfand, I., Haines, A., Ruff, T., Kristensen, H., Lewis, P. *et al.* (2016) The growing threat of nuclear war and the role of the health community. *World Medical Journal* 62, 86–94. Available at: https://ippnw.ch/wp-content/uploads/2020/11/WMJ-Nuclear_plus-28.10.16.pdf (accessed 10 March 2023).

Hyten, J. (2018) Speech. The Mitchell Institute Triad Conference. 17 July 2018. Available at: https://www.stratcom.mil/Media/Speeches/Article/1577239/the-mitchell-institute-triad-conference/ (accessed 27 May 2023).

IAEA (2023) *Update 167 – IAEA Director General Statement on Situation in Ukraine. 21 June 2023*. International Atomic Energy Agency, Vienna. Available at: https://www.iaea.org/newscenter/pressreleases/update-167-iaea-director-general-statement-on-situation-in-ukraine (accessed 25 June 2023).

ICAN (2023) *Wasted: 2022 global nuclear weapons spending*. International Campaign to Abolish Nuclear Weapons, Geneva. Available at: https://www.icanw.org/wasted_2022_global_nuclear_weapons_spending (accessed 25 June 2023).

ICRC, ICN, IFMSA, IPPNW, WFPHA *et al.* (2021) *International health and humanitarian organisations welcome the entry into force of the Treaty on the Prohibition of Nuclear Weapons*. International Committee of the Red Cross, International Council of Nurses, International Federation of Medical Students Associations, International Physicians for the Prevention of Nuclear War, World Federation of Public Health Associations *et al.* Available at: https://www.ippnw.org/wp-content/uploads/2021/01/FINAL-English-Global-Health-Statement.docx.pdf (accessed 8 March 2023)

IPB (2020) *Global campaign on military spending. Infographics and materials*. International Peace Bureau, Berlin. Available at: https://demilitarize.org/resources/gdams-healthcare-not-warfare-infographic/ (accessed 28 May 2023).

IPFM (2022a) *Materials: Plutonium*. International Panel on Fissile Materials. Available at: https://fissilematerials.org/materials/plutonium.html (accessed 28 May 2023).

IPFM (2022b) *Materials: Highly enriched uranium*. International Panel on Fissile Materials. Available at: http://fissilematerials.org/materials/heu.html (accessed 28 May 2023).

IPFM (2022c) *Global Fissile material Report 2022*. International Panel on Fissile Materials. Available at: https://fissilematerials.org/publications/2022/07/global_fissile_material_r.html (accessed 28 May 2023).

Jägermeyr, J., Robock, A., Elliott, J., Müller, C., Xia, L. *et al.* (2020) A regional nuclear conflict would compromise global food security. *Proceedings of the National Academy of Science (USA)* 117, 7071–7081. DOI: 10.1073/pnas.1919049117

Korda, M. (2023) *If arms control collapses, US and Russian strategic nuclear arsenals could double in size*. Federation of American Scientists (FAS), Washington, DC. Available at: https://fas.org/publication/if-arms-control-collapses-us-and-russian-strategic-nuclear-arsenals-could-double-in-size/ (accessed 27 May 2023).

Kristensen, H.M. and Korda, M. (2023) United States nuclear weapons, 2023. *Bulletin of the Atomic Scientists* 79(1), 28–52. DOI: 10.1080/00963402.2022.2156686

Kristensen, H.M., Korda, M., Johns E., Kohn, K. (2023a) Status of world nuclear forces. Federation of American Scientists (FAS), Washington, DC. Available at: http://fas.org/issues/nuclear-weapons/status-world-nuclear-forces/ (accessed 27 May 2023)

Kristensen, H.M., Korda, M., Reynolds, E. (2023b) Russian nuclear weapons, 2023. *Bulletin of the Atomic Scientists* 79(3), 174–199. DOI: 10.1080/00963402.2023.2202542

Kristensen, H.M., Korda, M., Johns, E., Knight, M. and Kohn, K. (2024) Status of world nuclear forces. Federation of American Scientists. Available at: http://fas.org/issues/nuclear-weapons/status-world-nuclear-forces/ (accessed 5 April 2024).

Little, M.P., Azizova, T.V., Richardson, D.B., Tapio, S., Bernier, M-O. *et al.* (2023) Ionising radiation and cardiovascular disease: systematic review and meta-analysis. *BMJ* 380, e072924. DOI: 10.1136/ bmj-2022-072924

Lovenduski, N.S., Harrison, C.S., Olivarez, H., Bardeen, C.G., Toon, O.B. *et al.* (2020) The potential impact of nuclear conflict on ocean acidification. *Geophysical Research Letters* 47, e2019GL086246. DOI: 10.1029/2019GL086246

Mills, M.J., Toon, O.B., Lee-Taylor, J. and Robock, A. (2015) Multi-decadal global cooling and unprecedented ozone loss following a regional nuclear conflict. *Earth's Future* 2, 161–176. DOI: 10.1002/2013EF000205/full

Mousseau, T.A. (2021) The biology of Chernobyl. *Annual Review of Ecology, Evolution, and Systematics* 52, 87–109. DOI: 10.1146/annurev-ecolsys-110218-024827

Norwegian Nobel Committee (2017) Press release. Available at: https://www.nobelprize.org/prizes/peace/2017/press-release/ (accessed 9 March 2023).

Norwegian People's Aid (2023) Nuclear weapons ban monitor 2022. Available at: https://banmonitor.org (accessed 27 May 2023).

Perlroth, N. (2021) *This Is How They Tell Me the World Ends. The Cyber Weapons Arms Race*. Bloomsbury Publishing, London.

Perry, W.J. and Collina, T.Z. (2020) *The Button*. BenBella Books, Dallas, Texas.

Pytlak, A. and Acheson, R. (2022) *Assuring Destruction Forever: 2022 Edition*. Women's International League for Peace and Freedom, New York. Available at: https://www.reachingcriticalwill.org/resources/publications-and-research/publications/15669-assuring-destruction-forever-2022-edition (accessed 28 May 2023)

Richard, C.A. (2021) Forging 21st-century strategic deterrence. *Proceedings* 147, 1416. US Naval Institute (USNI), Annapolis, Maryland. Available at: https://www.usni.org/magazines/proceedings/2021/february/forging-21st-century-strategic-deterrence (accessed 29 May 2023)

Robock, A. and Prager, S.C. (2021) Geoscientists can help reduce the threat of nuclear weapons. *Eos* 102, 27–29. DOI: 10.1029/2021EO210633

Robock, A. and Toon, O.B. (2012) Self-assured destruction: the climate impacts of nuclear war. *Bulletin of the Atomic Scientists* 68, 66–74. DOI: 10.1177/0096340212459127

Rotblat, J. (1981) *Nuclear Radiation in Warfare*. Stockholm International Peace Research Institute, Sweden. Taylor & Francis, London.

Ruff, T.A. (2013) The health consequences of nuclear explosions. In: Fihn, B. (ed.) *Unspeakable Suffering – the Humanitarian Impact of Nuclear Weapons.* Women's International League for Peace and Freedom, Geneva, and New York, pp. 14–27.

Ruff, T.A. (2017) Health implications of ionising radiation. In: Van Ness, P. and Gurtov, M. (eds) *Learning from Fukushima. Nuclear Power in East Asia.* ANU Press, Canberra, Australia, pp. 221–260.

Ruff, T. (2018) Negotiating the UN treaty on the prohibition of nuclear weapons and the role of ICAN. *Global Change, Peace & Security* 30, 233–241. DOI: 10.1080/14781158.2018.1465908

Scherrer, K.J.N., Harrison, C.S., Heneghan, R.F., Galbraith, E., Bardeen, C.G. *et al.* (2020) Marine wild-capture fisheries after nuclear war. *Proceedings of the National Academy of Science (USA)* 117, 29748–29758. DOI: 10.1073/pnas.200825611

Science and Security Board (2020) Closer than ever. It is 100 seconds to midnight. 2020 Doomsday Clock Statement. *Bulletin of the Atomic Scientists*, Chicago. Available at: https://thebulletin.org/doomsday-clock/2020-doomsday-clock-statement/ (accessed 8 March 2023).

Science and Security Board (2021) This is your COVID wake-up call: It is 100 seconds to midnight. 2021 Doomsday Clock Statement. 27 January 2021. *Bulletin of the Atomic Scientists*, Chicago. Available at: https://thebulletin.org/doomsday-clock/current-time/ (accessed 9 March 2023)

Science and Security Board (2022) At doom's doorstep: It is 100 seconds to midnight. *Bulletin of the Atomic Scientists*, Chicago. Available at: https://thebulletin.org/doomsday-clock/2022-doomsday-clock-statement (accessed 28 January 2024)

Science and Security Board (2023) A time of unprecedented danger: It is 90 seconds to midnight. 2023 Doomsday Clock Statement. *Bulletin of the Atomic Scientists*, Chicago. Available at: https://thebulletin.org/doomsday-clock/current-time/ (accessed 8 March, 2023).

SDSN (2019) *SDG costing and financing for low-income developing countries*. Sustainable Development Solutions Network, Paris, New York and Kuala Lumpur. Available at: https://resources.unsdsn.org/sdg-costing-financing-for-low-income-developing-countries (accessed 8 March 2023).

SIPRI (2023) *Trends in World Military Expenditure, 2023.* SIPRI Fact Sheet. Stockholm International Peace Research Institute, Solna, Sweden. Available at: https://sipri.org/publications/2023/sipri-fact-sheets/trends-world-military-expenditure-2022 (accessed 29 January 2024).

Snyder, S. (2022) *Rejecting Risk*. PAX and International Campaign to Abolish Nuclear Weapons, Utrecht, The Netherlands. Available at: https://www.dontbankonthebomb.com/policy-analysis-report-rejecting-risk/ (accessed 28 May 2023)

Toon, O., Turco, R., Robock, A., Bardeen, C., Oman, L. *et al.* (2007) Atmospheric effects and societal consequences of regional scale nuclear conflicts and acts of individual nuclear terrorism. *Atmospheric Chemistry and Physics* 7, 1973–2002. DOI: 10.5194/acp-7-1973-2007

Toon, O.B., Bardeen, C.G., Robock, A., Xia, L., Kristensen, H. *et al.* (2019) Rapidly expanding nuclear arsenals in Pakistan and India portend regional and global catastrophe. *Science Advances* 5, eaay5478. DOI: 10.1126/sciadv.aay547

UN (2021) *Our Common Agenda – Report of the Secretary General*. United Nations, New York.

UN (2022) *Secretary-General's remarks to the Tenth Review Conference of the Parties to the Treaty on the Prohibition of Nuclear Weapons*. 1 August, 2022. United Nations, New York. Available at: https://www.un.org/sg/en/content/sg/statement/2022-08-01/secretary-generals-remarks-the-tenth-review-conference-of-the-parties-the-treaty-the-non-proliferation-of-nuclear-weapons-bilingual-delivered-follows-scroll-down-for (accessed 28 May 2023)

Uppsala Conflict Data Program (UCDP) (2022). State-based: Armed conflict by type and year (1946–2021). Uppsala. Available at: https://ucdp.uu.se/downloads/charts/ (accessed 28 May 2023)

US Congress House of Representatives (2019) H.Res.302 Embracing the goals and provisions of the Treaty on the Prohibition of Nuclear Weapons. *116th Congress (2019–2020)*. Introduced 10 April 2019. Available at: https://www.congress.gov/bill/116th-congress/house-resolution/302/text (accessed 8 March 2023)

WHO (1983) *Resolution WHA36.28: The role of physicians and other health workers in the preservation of peace as the most significant factor for the attainment of health for all*. World Health Organization, Geneva.

WHO (1987) *Effects of Nuclear War on Health and Health Services*. World Health Organization, Geneva.

WHO (2021) WHO COVID-19 SPRP: Updated Appeal September 2021 – March 2022. September update: an urgent call to fund the COVID-19 emergency response. World Health Organization, Geneva. Available at: https://www.who.int/publications/m/item/who-preparedness-and-response-progress-september-update (accessed 23 January 2024).

Xia, L., Robock, A., Scherrer, K., Harrison, C.S., Bodirsky, B.L. *et al.* (2022) Global food insecurity and famine from reduced crop, marine fishery and livestock production due to climate disruption from nuclear war soot injection. *Nature Food* 3, 586–596. DOI: 10.1038/s43016-022-00573-0

4 Climate Change, Global Health and Planetary Health

Colin D. Butler
National Centre for Epidemiology and Population Health, The Australian National University, Canberra, Australia

Abstract
This chapter explores the contested meanings of global health and its relationship to healthcare delivery in low-income settings, inequality of multiple kinds, colonialism and eco-climatic change. Planetary health is discussed, as is the re-invigorated movement to 'decolonize' global health. Adapted from the titular chapter in this book's first edition (written by the late Professor Tony McMichael), the chapter argues that climate change and other forms of planetary overload are pushing 'life support mechanisms' towards a highly undesirable tipping point. The chapter suggests that neocolonialism underpins disparities in global health outcomes, provision and application, and the existence of an associated assumption that poor populations will bear the vast share of the burden of adverse planetary change. It finds that the related assumption that climate change and other forms of planetary overload can thus continue indefinitely – at least from the perspective of the global North – is a profound error.

> 'As medicine and global health struggle over decolonisation, we need to undertake a more realistic and rigorous analysis of where power lies politically, economically, militarily, and culturally' (Horton, 2021).

4.1 Introduction

The phrase 'global health', like 'sustainability', is widespread in the research and policy arenas. Both terms have diverse and contested meanings. Each can be misused. As Kelley Lee has commented, 'the term "global health" has generated a considerable amount of interest in recent years; so much so that many policy makers, development specialists, health professionals, and even academics have attempted to jump on the bandwagon claiming global health gains' (Lee *et al.*, 2002). More recently, a movement seeking to 'decolonize' global health has strengthened (Abimbola and Pai, 2020). Its supporters build on the idea, perhaps first expressed by Frantz Fanon (Zeilig, 2016), that medicine has been 'co-opted as a tool for colonial control' (Horton, 2018).[1] However, Horton (2021) also suggested that 'the idea that one can purge global health of colonialism is a comforting but deceptive myth'. Inequalities in socio-economic development and other forms of power, at scales from local to planetary, cannot be divorced from human health. These forms of inequality extend far beyond the nationality of authorial teams and even how funds for health delivery and research are allocated (Erondu *et al.*, 2021). Inequality is also embedded in the conceptual tools

 Climate Change and Global Health: Primary, Secondary and Tertiary Effects, 2nd Edition (eds C.D. Butler and K. Higgs)
DOI: 10.1079/9781800620025.0004

that are considered acceptable ways to improve and sustain health. However, even though substantial inequalities are likely to endure, their substantial amelioration is necessary if civilization is to survive. There are, therefore, self-interested as well as altruistic reasons for those with position and means to work towards a fairer world.

This goal is difficult, given the inequality of the starting point. Those with advantages virtually always use their power to maintain, or even to increase, their leverage over others with less assets and influence (Wade, 2004). How can the health of the world's population become more equal – and at a desirable level? This would involve improving health but that necessitates improvement in its social, political and commercial determinants (Labonté *et al.*, 2005). It is also uncomfortable territory for many epidemiologists; Rothman *et al.* (1998) argued that 'epidemiologists cannot be expected to solve every problem, especially not those beyond our expertise. However well motivated, epidemiologists cannot rid the world of poverty.'

While true, this does not mean that public health cannot contribute, with other disciplines, towards this goal. And social epidemiologists certainly have a role. Better global health for all must include a better diet (starting, where possible, with exclusive breastfeeding for six months), sanitation and other physical, nutritional and environmental factors that enable improved literacy, numeracy and reasoning. Motivated and empowered teachers are needed, with resources far beyond classrooms. It needs the rule of law and identification and resistance to gross judicial corruption. A free and vigorous press will help. Above all, greater fairness requires a willingness by existing power holders to concede some (perhaps many) of their privileges, and to pay their share of tax (Hickel, 2019). These challenges are formidable, particularly as those who most benefit from global inequality insist on 'growing the pie' with increased 'trickle-down', rather than fundamental reform.

The related phrase 'globalization of health' is also relevant (Schrecker, 2012). It recognizes that the processes of globalization – and the alleged international commitment to socio-economic development in lower-income countries, such as that expressed in the Sustainable Development Goals (Higgs, 2022) – have facilitated: (i) much more complete and standardized health metrics around patterns of health and disease in all countries; and (ii) the opportunity to address and coordinate at global level issues such as infectious-disease surveillance and control, rising antimicrobial resistance and the continuing rise of non-communicable diseases as populations urbanize and seek to join the pervasive consumer culture dominant in the global North (Beaglehole *et al.*, 2011).

It is here argued that global health is far more than a rebranding of international (or tropical) health activities. Although healthcare delivery in low-income settings is a component, global health can be considered an attempt to extend our conceptual frame and research-and-policy agenda. Some will argue that it also stresses a larger-scale, more *ecological* dimension of influence on the health risks to communities and populations. This dimension distinguishes much of its content from that of the great body of existing, generally beneficial, international health activity (McMichael, 2013).

Although the dividing line is unclear, international health comprises activities of the following sorts: (i) vaccine trials for those infectious diseases that cause high population burdens of disease and premature deaths; (ii) large multi-country trials of polypills, aspirin, etc. to reduce the incidence of cardiovascular disease; and (iii) the widening evaluation and uptake of pesticide-impregnated bed-nets to fend off malaria (especially in sub-Saharan Africa). International health also deals with cross-border regulations relating to infection control, hygiene, safety, food trade, etc. These are bedrock international health activities, essential to the improvement and maintenance of health around the world.

The crucial connotation of the word 'global' is that, today, there is also a new category of large, escalating and unfamiliar risks within population health. Increasingly, populations are being influenced by transnational forces, as the size and intensity of the human enterprise expands and as the aggregated human impact on the natural environment's life-supporting systems increases (McMichael and Butler, 2011). In consequence, we are putting unprecedented demands and stresses on much of nature, and thus on the stability and resilience of cultures, societies and communities. This new era of an intensively interconnected growth-oriented world with an increasingly human-dominated biosphere is now often referred to as the Anthropocene (Crutzen, 2002).

The resultant emerging challenges to population health include:

- health and environmental impacts (positive and negative) of the liberalization of regional and global trade, capital mobility, and the resultant changes in the conditions of the production workforce and in consumer choices and behaviour;
- spread of Western culture and consumerism, the power of modern transnational corporations to influence both production and national governmental policies;
- health consequences of urbanization (largely unplanned in relation to actual human biological and psychological needs);
- many and diverse health risks from human-induced global climate change, escalating and broadening over time;
- fundamental risks to population health sustainability because of disruptions of larger biophysical and ecological systems and processes in nature (e.g. food shortages, water scarcity and a destabilized microbial world); and
- worldwide increase in human mobility (migrant and refugee flows).

Some of today's more systemic environmental changes are genuinely 'global' in scale (e.g. stratospheric ozone depletion, human-induced climate change, ocean acidification). Some are greatly amplified and globally generalized versions of previously more localized stressors (e.g. soil exhaustion and erosion, freshwater depletion, biodiversity losses).

These global changes present a different type of challenge to health systems from those posed by responses to more localized health hazards or health inequalities, such as: (i) remediating local environmental pollution caused by toxic chemicals or radiation leaks; (ii) strengthening basic primary care and public health services; (iii) educating girls and women in less literate societies (hygiene, family planning, nutrition, etc.); and (iv) evaluating the efficacy of specific interventions (e.g. pesticide-impregnated bed-nets against malaria and DOTS[2] therapy for tuberculosis). Those more locally focused intervention programmes are central to the continuing business of international health, with action primarily at national or local level. It is therefore unhelpful for this long-standing international health agenda and the additional agenda of global change and health to be lumped together as 'global health'. That impedes understanding of the qualitative shift in scale, complexity and time-frame of the contemporary risks to health arising from large-scale environmental change, including the future sustainability of human population health. Nevertheless, such lumping together does occur; global health will continue to have subtly different meanings to different actors.

4.2 Planetary Health

These concepts are so different from *narrow* interpretations of global health (i.e. as the descendent of tropical or international health) that a new term might seem warranted. Yet such a suggestion is unlikely to be widely adopted and a new candidate term is likely to compound confusion. However, since the first edition of this book, an old term, 'planetary health' (dating at least from the 1980s), has been revived. Though related to 'ecological public health', 'ecosystem health' and 'ecohealth' (Butler, 2020), its most succinct definition differs from these precursors by specifying that the 'health of civilization' as well as the 'state of the natural systems on which it depends' are key (Whitmee *et al.*, 2015). Climate change is a substantial contributor to these natural systems, as are other threatened (or exceeded) planetary boundaries, such as those of biodiversity and freshwater (Wang-Erlandsson *et al.*, 2022). However, the 'health of civilization' also depends on many social factors, such as inequality, health services, inclusion of minorities (including Indigenous populations) and the global exchange of ideas and wealth. Although a broad interpretation of planetary health could embrace these concepts, in practice such interpretations are rare. Thus an expansion of the definition of planetary health to specify social factors, for example the 'state of the *eco-social* systems on which civilization depends' (i.e. rather than natural), seems appropriate, though probably unlikely.

This leads to another foundational problem. A divide between the natural science and social science communities has long been recognized (Snow, 1959). The crisis we now face, as a result of climate change and other aspects of 'planetary overload' (McMichael, 1993), has arisen, in no small part, because social scientists, especially economists, have largely disregarded the increasingly anguished warnings of the natural science community, particularly from ecologists (Ripple *et al.*, 2017). These warnings have also been sidelined because of the belief that the populations at greatest risk from planetary overload are in the majority, recently decolonized world. This belief may be flawed: a risk exists of an accelerating decline in health for all of us, at local, regional and global scale, due to positive (reinforcing) feedbacks that could unravel both ecological conditions and human capacity to respond to interacting crises.

Supporters of renewable energy argue that improvement in its capacity and dissemination can rapidly ameliorate the challenge of climate change (Fthenakis *et al.*, 2021). Others point to broader and deeper issues that renewable energy – which as yet remains far from affordable at large scale – will not solve (Seibert and Rees, 2021). Whatever the truth of this, time is running out: until atmospheric levels of greenhouse gases level off, renewable energy remains a theoretical cure and for only one aspect of planetary ill health.

4.3 Responding to the Challenge

These transboundary influences on population health are of a kind and scale that has no satisfactory local response. The impacts of global climate change cannot be avoided merely by local adaptive action. Pandemics also cannot be shut out indefinitely. Averting or controlling these complex larger-scale forces requires: (i) a much greater supranational understanding and sharing of information; (ii) the engagement of regional and world-level agencies; (iii) disinterested aid commitments by high-income countries; (iv) appropriate international laws and policies; and (v) a broad-based integrative research effort (Fidler *et al.*, 2009).

MacFarlane and colleagues (2008) wrote:

> If global health is about the improvement of health worldwide, the reduction of disparities, and protection of societies against global threats that disregard national borders, it is essential that academic institutions reach across geographic, cultural, economic, gender, and linguistic boundaries to develop mutual understanding of the scope of global health and to create collaborative education and research programs.

Much of this may sound unrealistic in a fragmented and self-interested nation state-based world. However, this is a moment of great and now unavoidable transition for the world community. A huge shift is occurring in human ecology and in our relationships with the environment on which it depends absolutely. We can continue to think myopically and sleepwalk into ever-greater problems – or we can contrive an extraordinary coordinated global response, powered either by new understanding and foresight or by a cascade of crises. Or, probably, both.

4.4 The Profile of Global Change-related Health Risks

Consider, as examples, what might be required to prevent the risks to global and regional population health from three sources:

- accelerated generation of new avian-derived influenza strains in South-East and East Asia as commercial food production practices evolve and wild-bird migration patterns respond to changes in climate, winds, food sources and surface water;
- decline in seafood due to warming, acidification and deoxygenation of the world's coastal seas; and
- increasing numbers of people displaced by crowding, conflict, food shortages and loss of low-lying land.

Obviously, effective prevention will require an understanding – across government, private sector and general public – of the underlying drivers of these new and growing stressors. How are human actions causing these seismic shifts in demographic profile, settlement patterns, wealth distribution and the viability and functioning of environmental systems at large?

Equally, there is need for the health sector to be enlisted, soon and deliberately, with other sectors of government and community to change or transform the ways that humans live, build, move, produce, consume and share. Emerging systemic global health threats can be lessened or eliminated via such integrated strategies.

In public health terms, both secondary and tertiary prevention will also be required. The complexity of these large-scale changes, our unfamiliarity with their modes of impact (especially when acting in combination) and current shortcomings in our research and response capacities mean that some changes in the patterns of health risk and health outcome will inevitably occur because of these global changes. One example is the climate change-related weather disasters (such as the Pakistani flood crisis of 2022), some with unprecedented damage to human health and survival.

The 'global' dimension of 'global health' refers primarily, as already discussed, to the scale, complexity and systemic nature of these transboundary influences. Their advent has great implications for professional training, research concepts and methods, and for the types of interventions needed (over a range of timescales) to avert or minimize the risks to human health. This will require: (i) coming to terms with some unfamiliar and less exact concepts of causality and of public health intervention; (ii) a shared understanding of the inherent complexity and related uncertainties that characterize many of these risk sources, pathways and outcomes; and (iii) the development of new interdisciplinary research methods and collaborative networks.

4.5 Global Changes – Systemic Risks to Health

Most of the worldwide socio-demographic changes are new only in the scale, complexion and magnitude of their social–political impacts, not in their existence. For example, rural populations have been drifting into cities for millennia but the rapidity of that movement in the past several decades is new – as is the fact that, worldwide, we are now an urban species (more than 50%).

Population itself is often sidestepped in global-change discussions, despite its unprecedented size and absolute growth rate. The UN projects that today's 8 billion people will exceed 9 billion by 2050 (UNDESA Population Division, 2019). This warrants much more attention and policy response.

In partial contrast to the current socio-demographic changes, the various global environmental changes are unprecedented, not only in their scale and their system-disrupting nature but also because of their recent debut as critical environmental thresholds have been passed for the first time in our species' existence. Their significance can be understood better if viewed within a long historical perspective.

Since settled agrarian living first emerged around 11,000 years ago, human societies have sought to make life more secure and more comfortable. Over time, that has entailed a slow increase in the extent to which humans have re-engineered local environments, mostly for food production. Over the past two centuries, some human societies have succeeded, both in material terms across a broader production front and in terms of life expectancy, beyond the dreams of our ancestors. However, adverse consequences for the natural environment and for human health are now accruing rapidly. The success was due largely to the extraordinary subsidy of high-density, portable energy extracted from fossil fuels (Fig. 4.1). Consequently, the global human environmental footprint has become huge and damaging (McMichael and Butler, 2011). We are currently placing a demand on Earth's capacity to supply, replenish and absorb that is 40–50% greater than it can bear for a prolonged time. Hence, we are experiencing global environmental depletion and disruption.

Major environmental systems are now deteriorating, particularly the climate system, the global nitrogen cycle and biodiversity (Steffen *et al.*, 2015). These systems underpin life support for all species. The sustained health of human populations everywhere depends on them absolutely.

Such disruptions of entire biogeophysical systems, occurring at global scale, have no precedent in human history. Climate change makes that point clearly. Before the mid-20th century, it was not possible for humans, in aggregate, to generate enough greenhouse gas (GHG) emissions to cause a change in the composition of the lower atmosphere that would induce geologically rapid and continuing warming. Now, alongside the accelerating clearance of forests (carbon sinks), it is.

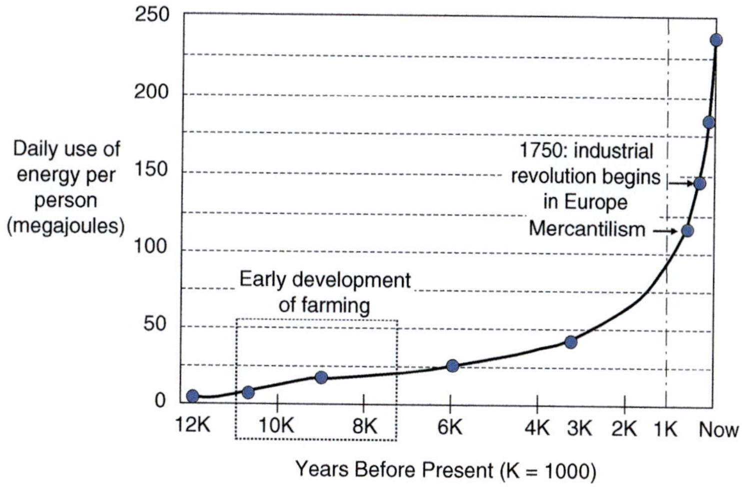

Fig. 4.1. Changes in globally averaged daily energy use per person during the Holocene, entailing a 16-fold increase since the early farming period through to the present. (Source: A.J. McMichael. Based on data in Smil, 1994; Boyden, 2004; Ponting, 2007.)

4.6 Climate Change: Overview of Recent Science

Human actions are causing a rise in GHG concentrations in the troposphere, predominantly by mobilizing geologically stored carbon as carbon dioxide (CO_2). Other industrial and agricultural emissions, particularly methane, nitrous oxide and elemental black carbon, are also significant and very potent GHGs, but with much shorter atmospheric lifetimes than CO_2.

This build-up of GHGs is increasing the radiative-forcing property of the lower atmosphere, increasing the absorption of outward-bound infrared radiation from Earth's solar-heated surface. That radiation energy manifests as heat. Most of the unusually rapid global warming since the mid-1970s (0.7°C global average increase) is attributable to this human-caused increase in GHG concentration. The logical expectation, both from theory and from the 'back-casting' empirical testing of global climate models against the actual observed record of temperature change, is that substantial human-caused warming will continue this century. However, if all net human-generated GHG emissions were terminated today (i.e. we achieved net zero emissions), warming might slow or cease[3] (MacDougall *et al.*, 2020) – provided that the Earth system itself did not still trigger additional releases, such as from carbon stores in the tundra, or through dieback of the Amazon forest (Lovejoy and Nobre, 2018).

In essence, global climate change is a consequence of massive human intervention in the planet's carbon cycle. The perennial cycling of mobile carbon between soil, vegetation, the atmosphere and the ocean has been occurring within the frame of a more-or-less steady dynamic equilibrium throughout the post-glacial, or Holocene, epoch, i.e. the past 11,000 years. Now, human activity is withdrawing carbon stocks that have long been 'on deposit' (for several hundred million years) and adding them to the already circulating life-supporting carbon currency. Concentrations of CO_2 in our atmosphere are now higher than at any time in the past million years and are probably higher than at any time in the past 20 million years. This glut of circulating carbon has tipped the balance: the system is now out of equilibrium and a cascade of (mostly) adverse consequences is gathering momentum – for environmental conditions, ecosystem viability and the health and survival of almost all species.

4.7 Climate Change and Human Health

Epidemiologists and other health researchers have had insufficient interest, over the past half-century, in studying the risks to health from variations in weather and climate. Those variations, viewed as natural and unalterable, were interpreted as offering little opportunity for interesting or useful scientific discovery. Since the early 1990s, that situation has changed markedly. Today, there is a surge of research interest in understanding how climate change affects human health and survival, predicted as it is to increase and extend for many decades into this century and possibly much longer.

Climate change is not, however, a one-dimensional risk factor of the ilk of asbestos exposure or cigarette smoking. Indeed, the spectrum of risks to human health from climate change is not just wide but also seemingly unbounded. So, too, are the social and political ramifications.

The various regional manifestations of climate change will bear differentially on communities. These manifestations will include: (i) extremes of warming or drying; (ii) the shift of monsoonal rains and seasons; (iii) heightening of particular extreme weather events (e.g. cyclones, wildfires, flooding and heatwaves); (iv) shortages of freshwater as river flows decrease and evaporation increases; and (v) changes in the geographical range and rate of infectious disease agent transmission. Populations living in diverse social, economic and physical conditions will be affected in different ways by changes in their regional climate. Poorer and remote populations will be particularly vulnerable via exposures to physical hazards, undernutrition, diarrhoeal diseases, various vector-borne infectious diseases and the adverse consequences of displacement and relocation. In the Arctic region, communities are already experiencing profound declines in the health and number of native animal populations, especially caribou, with resultant negative health effects (Jaakkola *et al.*, 2018).

This presents the world with a huge moral and political problem of heightened inequity in population health and safety, a topic covered in greater detail in other chapters. Suffice to say, here, that there is no intrinsic incompatibility between two great and urgent global objectives: (i) developing a path to successful low-carbon, low-environmental-intensity socio-economic development; and (ii) taking action to rein in fossil fuel use, destruction of natural habitat, overharvesting of many species, overuse of synthetic nitrogen-based fertilizers and so on. The crises of the present are not in competition with the looming crises of the near future. We can, and must, find a way to bring these two global agendas into synergistic synchronization with each other (Friel *et al.*, 2008).

The broad scientific consensus assessment is that climate change will have predominantly adverse impacts on human health and that these will become more severe and generalized as climate change gathers force. Meanwhile, there will also be several benefits to human health from climate change – in at least some regions of the world and at least during the earlier stages of the process. Examples noted in the literature include: the likely reduction in wintertime death from extremes of cold if winters become milder in temperate countries; gains in local food yields in some mid- to higher-latitude regions; and the contraction of transmission zones for some vector-borne infections if conditions become too hot and dry for vector survival.

Categories of human health impacts

Overall, there are three broad categories of causal pathways by which changes in climate affect human health: (i) direct-acting climatic environmental exposures to excessive heat, physical hazards, etc.; (ii) health risks resulting from disruptions and other changes to ecosystems and interspecies relations; and (iii) risks from social and economic disruptions and the more ominous risks from tensions and conflict that are partly driven by climatic changes. This classification of primary, secondary and tertiary impacts has been revived by Butler and Harley (2010) (see also Fig. 4.2).

A different category of risk to health arises as governments, commercial interests and communities implement adaptive measures to lessen the impacts of climate change and its diffuse consequences. Such adaptations may inadvertently be detrimental, in particular to powerless, remote, or marginal groups. For example, the diversion of agriculture into biofuel production is now a substantial contributor to some forms of food shortage and to rising food prices. Two-fifths of the US maize crop is being directed to biofuel production, underpinned by federal tax incentives. As the world approaches 'peak oil', this helps to drive food shortages and price rises.

Starting at the least complex end of this three-way classification, there are various relatively direct health impacts of higher temperatures and

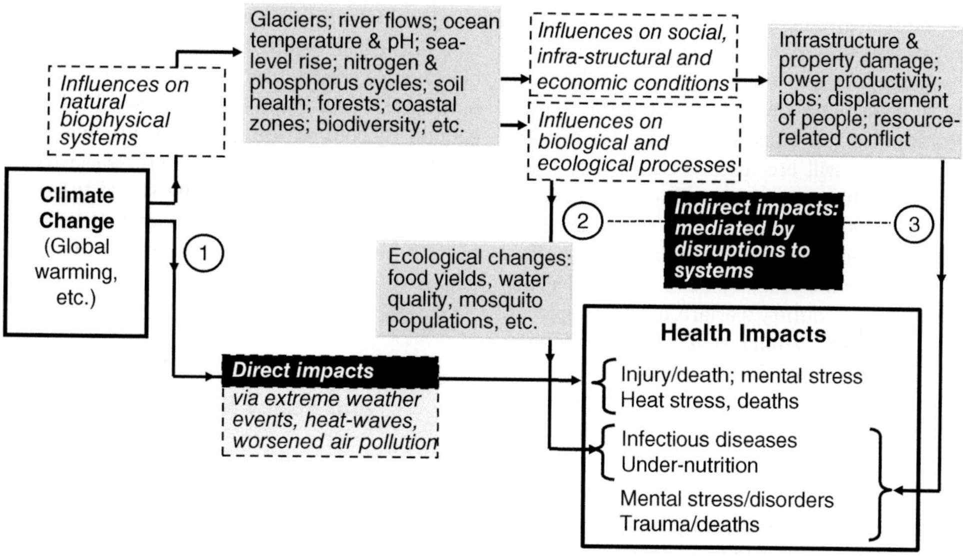

Fig. 4.2. Range of paths by which climate change can influence the health and survival of human populations/communities. The circled numbers refer to (1) primary, (2) secondary and (3) tertiary effects.

the immediate and physical risks from exposure to more frequent and more extreme weather events. Data on injuries, deaths and manifest disease such as gastroenteritis are often available; rates and risks can be estimated. This primary category of direct-acting influences on health has now begun to attract a wider range of epidemiological research. Questions abound. Will warmer summers in temperate countries affect patterns of clothing and outdoor behaviour and thus contribute to increased risks of skin cancer? Does the increase in coastal groundwater salinity due to ground subsidence (Steckler *et al.*, 2022) and rising sea level (Khan *et al.*, 2020) pose a risk of hypertension in poorer rural populations dependent on well water for drinking (e.g. in Bangladesh) and might that increase the risk of hypertensive eclampsia of pregnancy?

The secondary health impacts entail more complex processes entailing changes in climatic conditions affecting various geophysical phenomena, biotic processes and ecological relationships, all with consequences for human health. Examples include impacts on patterns of occurrence of vector-borne infectious diseases, changes to allergic conditions and interactions with air pollution, including via increased forest fires.

The tertiary impacts span a range of systemic risks to society embedded in feedback loops. Although primary and secondary effects are likely to impact behaviour patterns and mental and emotional health as well as individual and community morale (for example, the death of a child from malaria or a heatwave associated with violence), tertiary effects are likely to cause the greatest physical and psychological loss. These can occur: (i) as part of the complex aftermath of extreme weather disasters; (ii) from loss of livelihoods and security in rural settings as climatic adversity affects agricultural yields and supportive services; and (iii) from displacement or outmigration of groups or whole communities in response to actual or perceived threats from climate change.

For example, as physical environments change and animal and plant species recede or shift, so the traditional sources of food for various indigenous populations decline. In northern Scandinavia, thinner ice sheets on rivers and lakes and shorter cold seasons endanger the management of reindeer herds as they are moved across the landscape by Saami herders. In northern Canada, supplies of seal, fish and caribou dwindle as temperatures rise, snow cover declines and ice floes diminish. Consequently, there is greater reliance on imported, processed, less healthy foods (Evengård and McMichael, 2011).

There are similarly fundamental threats to social stability and geopolitical security from climate-related changes in river flows, regional food yields and access to habitable land (for example, sea level rise will impinge on small island states, delta populations and low-lying coastal communities). Shortages in food and fundamental resources will breed tensions, as they have throughout history. Severe cold weather in early 17th century Europe resulted in food shortages, tripling in staple food prices, a doubling in the frequency of famine years, epidemic outbreaks and a surge in intercountry warfare, and the displacement of people (Zhang *et al.*, 2011). Climate change, acting as a risk multiplier, is a recognized promoter of conflict and warfare (German Advisory Council on Global Change, 2008).

Detecting health (or directly health-related) impacts attributable to climate change

There should, by now, be many emerging opportunities to assess the extent to which climate change is now affecting human health. All such empirical evidence will strengthen the message to communities and policy makers about the potential seriousness of unabated climate change for humans themselves. However, attributing changes in health outcomes to climate change is complex. Most health outcomes have a range of plausible contributory causal factors: environmental, social, behavioural, biomedical and genetic. By comparison, glaciologists explaining acceleration in the melting of Arctic sea ice have an easier task. In general, the signal-to-noise ratio in assessing climatic contributions to human health outcomes is lower than that for glaciers (Fig. 4.3).

4.8 Focus and Purpose of Research and Policy Response

The preferred strategy is always primary prevention: elimination of the risk at source. The 1987 Montreal Protocol for the elimination of stratospheric ozone-destroying gases is an example. The risks of increased exposure to ultraviolet radiation to biological targets, including the human skin and eye, were recognized as real, serious and preventable. But the underlying cause of that particular environmental health hazard was almost eliminated. The same thinking must become paramount in our response to climate change. A change in the world's climatic

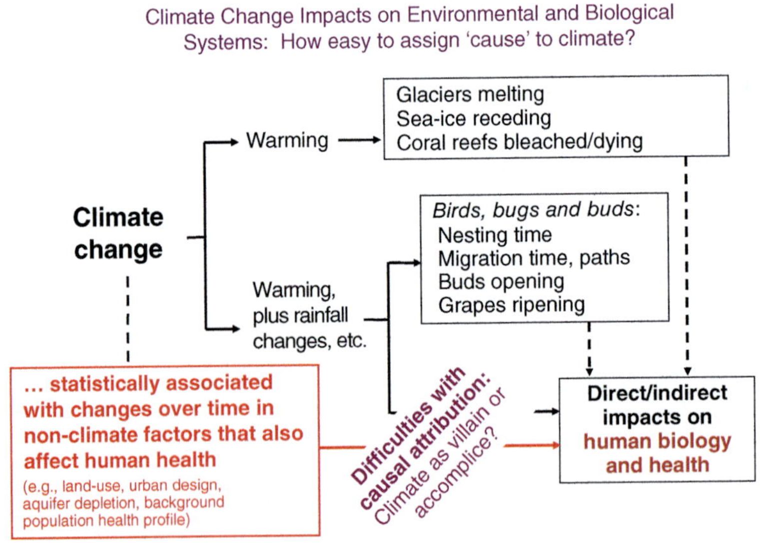

Fig. 4.3. The detection of climate change impacts on qualitatively differing systems: geophysical processes, natural ecological relationships and human well-being and health. The process of attribution is much more complex and difficult for that third human-impact category – especially since many other aspects of human culture, circumstances and behaviour bear on human health, and often change over time.

conditions, associated with now-plausible temperature increases in this century of 4–5°C, would almost certainly cause major surprise step changes as critical thresholds are passed. Feedback processes that reinforce these changes are most likely and, therefore, the prospect of runaway climate change now needs to be contemplated (Schellnhuber, 2008).

Adaptive strategies to ameliorate risks from human-induced climate change should not distract from the great first-order need to actually avert climate change. In pursuing such mitigation – now urgently needed before the necessary international schedule of annual percentage emissions reduction becomes unachievable – health researchers and public health practitioners have at least two important messages to convey.

The first is profound. If our collective actions have reached the level whereby they are disrupting and weakening life-support processes necessary for our civilization, then that is surely a signal that we have overplayed our hand and that we must therefore curb our current climate-disrupting economic activities and heighten our attempts to lower fertility rates further (especially in poorer countries, where unwanted pregnancies persist at high rates).

The second message is positive. Most mainstream actions taken to reduce GHG emissions and atmospheric concentrations will, at the same time, confer local and often immediate health co-benefits on the community or population undertaking that action. Although that may seem to offer most health benefits to richer countries – where private car usage should be constrained, physical activity encouraged, houses made more energy efficient and urban air pollution reduced – many of those benefits would also flow to lower-income countries, where urbanization is proceeding rapidly (Haines *et al.*, 2009). Further, there would be wider-reaching benefits to crop yields, via the multilateral reduction of black carbon and methane emissions, with some consequent reduction in the amount of ground-level ozone in air (Shindell *et al.*, 2012).

4.9 Conclusion

Humankind has entered an accelerated and expanded phase of globalization. The scale and connectivity of human activity and associated demographic, social, economic and environmental global changes are exerting increasing influences on human well-being, health and survival.

Those influences, whether positive or negative, occur via diverse pathways. For example, on the positive side, they facilitate many aspects of international healthcare and mainstream public health activities and rapid sharing of global electronic information (including the surveillance of infectious diseases); on the negative side, they drive the spread of urban middle-class consumer culture and its health-related behaviours, and the health risks to communities and populations from disruptions to the climate and the biosphere's life-support capacity. All those examples transcend national, even regional, boundaries; all reflect aspects of recent globalization; all impinge on whole groups or communities of people and thus reflect disruptions, distortions, or re-alignments in human ecology. They entail new and unfamiliar shifts in our natural and social environments.

The climate system is fundamental to the biological and psychological health of human populations, via food yields, water flows, constraints on infectious disease agents, and physical buffering against natural disasters, such as reefs, mangroves, forests, wetlands, etc. Hence, a fuller understanding of the risks to health and survival from climate change should help focus public concern on the long-term and fundamental consequences of climate change and spur its mitigation by governments.

The advent of these transboundary social, demographic and environmental influences on human health – especially those due to systemic environmental disruption and depletion – broadens the role and responsibilities of the health sector. We have an urgent collective responsibility to lessen and alleviate these risks – including those to future generations, who are in danger of inheriting an impoverished, disrupted and less habitable world.

Acknowledgement

This chapter is influenced by the chapter 'Climate Change and Global Health', written by the late A.J. (Tony) McMichael, published in the first edition of this book.

Notes

[1] These observations are not recent in the field now called global health. Maurice King's edited book, widely regarded as the foundational textbook for primary health care, was subtitled 'A Primer on the Medicine of Poverty' (King, 1966). Published at the dawn of the 'warm decade for social justice', Brazilian educator Paolo Freire's influential book *The Pedagogy of the Oppressed* (Freire, 1970) called for the empowerment of the illiterate so that they might analyse and improve their social situation. Freire, who was born into the Brazilian middle class, experienced hunger as a child, following the loss of his family's economic base. This enabled a sense of 'solidarity with the children from the poor outskirts of town' (Macedo, 2000, p. 13).

[2] Directly observed treatment, short course.

[3] There is uncertainty; some models continue to warm for decades to millennia; others cool substantially.

References

Abimbola, S. and Pai, M. (2020) Will global health survive its decolonisation? *The Lancet* 396, 1627–1628. DOI: 10.1016/S0140-6736(20)32417-X

Beaglehole, R., Bonita, R., Horton, R., Adams, C., Alleyne, G. *et al.* (2011) Priority actions for the non-communicable disease crisis. *The Lancet* 377, 1438–1447. DOI: 10.1016/S0140-6736(11)60393-0

Boyden, S.V. (2004) *The Biology of Civilisation*. UNSW Press, Sydney, Australia.

Butler, C.D. (2020) Planetary health. In: Kobayashi, A. (ed.) *International Encyclopedia of Human Geography*, 2nd edn. Elsevier, Amsterdam, pp. 153–157. DOI: 10.1016/B978-0-08-102295-5.10413-5.

Butler, C.D. and Harley, D. (2010) Primary, secondary and tertiary effects of eco-climatic change: the medical response. *Postgraduate Medical Journal* 86, 230–234. DOI: 10.1136/pgmj.2009.082727

Crutzen, P.J. (2002) Geology of mankind. *Nature* 415, 23. DOI: 10.1038/415023a

Erondu, N.A., Aniebo, I., Kyobutungi, C., Midega, J., Okiro, E. *et al.* (2021) Open letter to international funders of science and development in Africa. *Nature Medicine* 27, 742–744. DOI: 10.1038/s41591-021-01307-8

Evengård, B. and McMichael, A.J. (2011) Vulnerable populations in the Arctic. *Global Health Action* 4. DOI: 10.3402/gha.v4i0.11210.

Fidler, D.P., Drager, N. and Lee, K. (2009) Managing the pursuit of health and wealth: the key challenges. *The Lancet* 373, 325–331. DOI: 10.1016/S0140-6736(08)61775-4

Freire, P. (1970) *Pedagogy of the Oppressed*. The Continuum International Publishing Group Inc, New York and London.

Friel, S., Marmot, M., McMichael, A.J., Kjellstrom, T. and Vågerö, D. (2008) Global health equity and climate stabilisation: a common agenda. *The Lancet* 372, 1677–1683. DOI: 10.1016/S0140-6736(08)61692-X

Fthenakis, V., Raugei, M., Breyer, C., Bhattacharya, S., Carbajales-Dale, M. *et al.* (2021) Comment on Seibert, M.K.; Rees, W.E. Through the Eye of a Needle: An Eco-Heterodox Perspective on the Renewable Energy Transition. *Energies* 14, 4508. DOI: 10.3390/en15030971

German Advisory Council on Global Change (2008) *Climate Change as a Security Risk*. Earthscan, London.

Haines, A., McMichael, A.J., Smith, K.R., Roberts, I., Woodcock, J. *et al.* and on behalf of the Task Force on Climate Change Mitigation and Public Health (2009) Public health effects of strategies to reduce greenhouse-gas emissions: overview and implications for policy makers. *The Lancet* 374, 2104–2114. DOI: 10.1016/S0140-6736(09)61759-1

Hickel, J. (2019) A letter to Steven Pinker (and Bill Gates, for that matter) about global poverty. *Class, Race and Corporate Power* 7, Article 3. Available at: https://digitalcommons.fiu.edu/classracecorporatepower/vol7/iss1/3 (accessed 24 February 2023)

Higgs, K. (2022) How sustainable are the UN Sustainable Development Goals? In: Williams, S.J. and Taylor, R. (eds) *Sustainability and the New Economics. Synthesising Ecological Economics and Modern Monetary Theory.* Springer Nature, Cham, Switzerland, pp. 69–88.

Horton, R. (2018) Offline: Frantz Fanon and the origins of global health. *The Lancet* 392, 720. DOI: 10.1016/S0140-6736(18)32041-5

Horton, R. (2021) Offline: The myth of 'decolonising global health'. *The Lancet* 398, 1673. DOI: 10.1016/S0140-6736(21)02428-4

Jaakkola, J.J.K., Juntunen, S. and Näkkäläjärvi, K. (2018) The holistic effects of climate change on the culture, well-being, and health of the Saami, the only indigenous people in the European Union. *Current Environmental Health Reports* 5, 401–417. DOI: 10.1007/s40572-018-0211-2

Khan, J.R., Awan, N., Jahan, A.R., Sultana, N. and Muurlink, O. (2020) The association between drinking water salinity and hypertension in coastal Bangladesh. *Global Health Journal* 4, 153–158. DOI: 10.1016/j.glohj.2020.11.001

King, M. (ed.) (1966) *Medical Care in Developing Countries. A Primer on the Medicine of Poverty and a Symposium from Makerere*. Oxford University Press, Nairobi, Kenya.

Labonté, R., Schrecker, T. and Gupta, A.S. (2005) *Health for Some? Death, Disease and Disparity in a Globalizing Era*. Centre for Social Justice, Toronto, Canada.

Lee, K., Buse, K. and Fustukian, S. (2002) An introduction to global health policy. In: Lee, K., Buse, K. and Fustukian, S. (eds) *Health Policy in a Globalising World*. Cambridge University Press, Cambridge, UK, pp. 4–5.

Lovejoy, T.L. and Nobre, C. (2018) Amazon tipping point. *Science Advances* 4, eaat2340. DOI: 10.1126/sciadv.aat2340

MacDougall, A.H., Frölicher, T.L., Jones, C.D., Rogelj, J., Matthews, H.D. *et al.* (2020) Is there warming in the pipeline? A multi-model analysis of the zero emissions commitment from CO_2. *Biogeosciences* 17, 2987–3016. DOI: 10.5194/bg-17-2987-2020

Macedo, D. (2000) *Introduction to Freire, Pedagogy of the Oppressed*. Continuum International Publishing Group Inc, New York and London, pp. 11–26.

MacFarlane, S.B., Jacobs, M. and Kaaya, E.E. (2008) In the name of global health: trends in academic institutions. *Journal of Public Health Policy* 29, 383–401. DOI: 10.1057/jphp.2008.25

McMichael, A.J. (1993) *Planetary Overload. Global Environmental Change and the Health of the Human Species*. Cambridge University Press, Cambridge, UK.

McMichael, A.J. (2013) Globalization, climate change and health. *New England Journal of Medicine* 368, 1335–1343. DOI: 10.1056/NEJMra1109341

McMichael, A.J. and Butler, C.D. (2011) Promoting global population health while constraining the environmental footprint. *Annual Review of Public Health* 32, 179–197. DOI: 10.1146/annurev-publhealth-031210-101203

Ponting, C. (2007) *A New Green History of the World*. Vintage Books, London.

Ripple, W.J., Wolf, C., Newsome, T.M., Galetti, M., Alamgir, M. *et al.* (2017) World scientists' warning to humanity: a second notice. *BioScience* 57, 1026–1028. DOI: 10.1093/biosci/bix125

Rothman, K.J., Adami, H.-O. and Trichopoulos, D. (1998) Should the mission of epidemiology include the eradication of poverty? *The Lancet* 352, 810–814. DOI: 10.1016/S0140-6736(98)01327-0

Schellnhuber, H.J. (2008) Global warming: stop worrying, start panicking? *Proceedings of the National Academy of Sciences (USA)* 105, 14239–14240. DOI: 10.1073/pnas.0807331105

Schrecker, T. (ed.) (2012) *The Ashgate Research Companion to the Globalization of Health*. Ashgate Publishing, Farnham, UK.

Seibert, M.K. and Rees, W.E. (2021) Through the eye of a needle: An eco-heterodox perspective on the renewable energy transition. *Energies* 14, 4508. DOI: 10.3390/en14154508

Shindell, D., Kuylenstierna, J.C.I., Vignati, E., van Dingenen, R., Amann, M. *et al.* (2012) Simultaneously mitigating near-term climate change and improving human health and food security. *Science* 335, 183–188. DOI: 10.1126/science.1210026

Smil, V. (1994) *Energy in World History*. Westfield Publishing, Boulder, Colorado.

Snow, C.P. (1959) *The Two Cultures and the Scientific Revolution*. Cambridge University Press, Cambridge, UK.

Steckler, M.S., Oryan, B., Wilson, C.A., Grall, C., Nooner, S.L. *et al.* (2022) Synthesis of the distribution of subsidence of the lower Ganges-Brahmaputra delta, Bangladesh. *Earth-Science Reviews* 224, 103887. DOI: 10.1016/j.earscirev.2021.103887

Steffen, W., Richardson, K., Rockström, J., Cornell, S.E., Fetzer, I. *et al.* (2015) Planetary boundaries: Guiding human development on a changing planet. *Science* 347. DOI: 10.1126/science.1259855

UNDESA Population Division (2019) *World Population Prospects, 2019 Revision*. United Nations Department of Economic and Social Affairs Population Division, New York.

Wade, R.H. (2004) On the causes of increasing world poverty and inequality, or why the Matthew effect prevails. *New Political Economy* 9, 163–188. DOI: 10.1080/1356346042000218050

Wang-Erlandsson, L., Tobian, A., van der Ent, R.J., Fetzer, I., te Wierik, S. *et al.* (2022) A planetary boundary for green water. *Nature Reviews Earth & Environment*. DOI: 10.1038/s43017-022-00287-8

Whitmee, S., Haines, A., Beyrer, C., Boltz, F., Capon, A.G. *et al.* (2015) Safeguarding human health in the Anthropocene epoch: report of The Rockefeller Foundation–Lancet Commission on planetary health. *The Lancet* 386, 1973–2028. DOI: 10.1016/S0140-6736(15)60901-1

Zeilig, L. (2016) *Frantz Fanon: Militant Philosopher of the Third World Revolution*. IB Tauris Bloomsbury Publishing, London.

Zhang, D.D., Lee, H.F., Wang, C., Li, B., Pei, Q. *et al.* (2011) The causality analysis of climate change and large-scale human crisis. *Proceedings of the National Academy of Sciences (USA)* 108, 17296–17301. DOI: 10.1073/pnas.1104268108

5 One Health: From Rinderpest to the Threat of a Four-Degree World

Colin D. Butler[1] **and Rosemary A. McFarlane**[2]
[1]*National Centre for Epidemiology and Population Health, The Australian National University, Canberra, Australia;* [2]*Health Research Institute, University of Canberra, Canberra, Australia*

Abstract

This chapter summarizes some of the key issues related to One Health, including its history, named pioneers, some of its definitions, the milieu which gave it impetus and its relationship with politics, ecology and reductionism. Some of the links between human, animal and environmental health are discussed, including a focus on the now extinct cattle disease, rinderpest. Unusually for a chapter on One Health, it also describes some of the proven and speculative links between infectious diseases and laboratory processes, including vaccine development and trials. The chapter outlines some of the challenges that face One Health practitioners, identifying not only a requirement for more 'joined-up', systemic thinking, but also a need for courage, sometimes in the face of lethal, very powerful opposition. The chapter concludes by suggesting that the greatest challenge for One Health may be to shift its support from rhetoric to reality.

5.1 Introduction

The One Health movement seeks a safer, better world. It aims to enhance the health and well-being of three connected elements: humans, animals and the environment (Evans and Leighton, 2014). One Health has many definitions. Amuasi *et al.* (2020) defined it as 'the complex interconnectedness and interdependence of all living species and the environment'. Zinsstag *et al.* (2021) defined One Health as 'any added value in terms of health and humans and animals, financial savings, social resilience and environmental sustainability achievable by the cooperation of human and veterinary medicine and other disciplines when compared to the two medicines and other disciplines working separately'. The added value this refers to is a call for holistic (systemic) thinking, and a rejection of reductionism and the conceit that humans can attain mastery over nature. Recognizing this diversity, the One Health high-level expert panel (OHHLEP *et al.*, 2022) recently published its own definition and an accompanying set of principles:

> One Health is an integrated, unifying approach that aims to sustainably balance and optimize the health of people, animals, and ecosystems. It recognizes the health of humans, domestic and wild animals, plants, and the wider environment (including ecosystems) are closely linked and interdependent. The approach mobilizes multiple sectors, disciplines, and communities at varying levels of society to work together to foster well-being and tackle threats to health and ecosystems, while addressing the collective need for healthy food, water, energy, and air, taking action on climate change and contributing to sustainable development.

 Climate Change and Global Health: Primary, Secondary and Tertiary Effects, 2nd Edition (eds C.D. Butler and K. Higgs)
DOI: 10.1079/9781800620025.0005

One Health has ancient roots. Many Indigenous peoples express reverence for 'country' (homeland) and respect, even love, for other species, even if they are hunted and killed to provide food and other materials necessary for human survival. Some Indigenous people demonstrate knowledge and practices that conserve biodiversity, in order to maintain sufficient ecosystem services for human well-being (Gadgil *et al.*, 1993). Some principles of One Health, including a recognition of zoonotic infections (infectious diseases that originate in non-human animals), can be traced to the civilizations of Mesopotamia, Egypt, Greece, India and China (Zinsstag *et al.*, 2011). For example, almost 4000 years ago, heavy penalties were applied in Babylonia to the owners of rabid dogs responsible for the deaths of humans (if the killed person was a slave, then a discount of over 60% was applied) (Tarantola, 2017). Central Asian nomads have traditionally recognized an association between diseased rodents and human sickness; leading to the cultural shunning of sick animals and moving camp. This disease is now known as the plague, caused by *Yersinia pestis* (Lynteris, 2013).

The formal history of One Health has had many pioneers, including veterinarians and social epidemiologists (Table 5.1) (Woods *et al.*, 2017). Zoologist, veterinarian and philosopher[1] Calvin Schwabe, whose book *Veterinary Medicine and Human Health* was first published in 1964 (Schwabe, 1964), is probably the most important single individual in the framing of One Health. Schwabe is credited with coining the term 'One Medicine' (Zinsstag *et al.*, 2011).[2] One of many tributes to Schwabe described him as seeing the world 'as an ecosystem of planetary proportions constituted of interdependent civilizations and cultures, in which human history and human progress (are) inexorably linked with the co-evolution of the animal kingdom' (Kass *et al.*, 2006).

This chapter summarizes some of the key issues: the history of One Health, the milieu that

Table 5.1 Some of the key figures in the movement increasingly called One Health, from the 17th to the 20th centuries.

Name	Lifespan	Major One Health contribution(s)
Giovanni Lancisi	1654–1720	Developed 'stamping out' response to rinderpest epidemics (requested by Pope Clement XI), proposed swamp reduction to lower malaria risk (Italian)
Claude Bourgelat	1712–1779	Founded first European veterinary faculty (Lyons, France). (Veterinary medicine was a discipline in China in Zhou dynasty (11th–13th centuries CE))
Edward Jenner	1749–1823	English physician whose recognition that cowpox infection protected against smallpox is a classic example of One Health
Louis-René Villermé	1782–1863	French social epidemiologist who made important contributions challenging the miasma theory, showing poverty to be a more powerful explanation for diseases formerly attributed to miasma
Claude Bernard	1813–1878	Promoted experiments on animals as a key route to improve understanding of human physiology (French)
Rudolf Virchow	1821–1902	German physician and pathologist, who made many contributions, including to social medicine, credited with coining the word 'zoonosis'
Louis Pasteur	1822–1895	French microbiologist and vaccine developer known for his discoveries of the principles of microbial fermentation, pasteurization and vaccination
William Osler	1849–1919	The leading English-speaking physician of his day, studied under Virchow; helped promote principles later called 'one medicine'
James Steele	1913–2013	Helped motivate WHO to establish a veterinary public health unit, was also instrumental in establishing the US Centers for Disease Control (American)
Calvin Schwabe	1927–2006	Veterinary epidemiologist who coined the term 'one medicine' (American)

gives it impetus and its relationship with politics, ecology and reductionism. There is a partial discussion of the links between human and animal health. Other issues include antimicrobial resistance (AMR), intensive livestock farming, the microbiome, parasitology, food safety and climate change.

5.2 The 'Milieu' that Created the One Health Movement

Recent events have impelled the emergence of One Health as a strong movement, in concert with other reforming attempts. One Health, today, is arguably the best-known term for expressing concerns specific to human and animal health, and increasingly the health of the environment, including ecological change or social inequalities (Table 5.2).

The pressures that have generated this emergence can be considered 'signs' of the Anthropocene, stimulating 'symptoms'; together these have awakened an increasing number of scholars, practitioners and the public to concerns about limits to growth (Meadows *et al.*, 1972) and the sustainability of civilization. In the closing decades of the 20th century, the link between these challenges and human health was recognized by many individuals and several groups, with names such as planetary health (first used in the 1980s), conservation medicine and ecosystem health.[3] In 2004, the first issue of *EcoHealth*, then associated with the International Association of Ecology and Health, was published. In 2005, the final reports of the Millennium Ecosystem Assessment were released, representing an influential global consortium of over 1000 scientists, consolidating and publicizing knowledge of the time, about not only the state of the global ecosystem but also their relationship with health.

The Wildlife Conservation Society (WCS), founded in 1895, with headquarters at the Bronx Zoo in New York, was also important. From the late 1990s, a small group of veterinarians working there developed collaborations with scientists studying emerging infectious diseases (animal, including human) as well as with veterinarians involved with global livestock health. In 2003 the WCS and a working group of the International Union for the Conservation of Nature (IUCN) co-organized a workshop in Durban, South Africa, which used the words 'One Health' (Woods *et al.*, 2017).

In 2004, a meeting was held called 'One World, One Health'. This involved the World Health Organization (WHO), the UN Food and Agriculture Organization (FAO), the WCS and the International Union for the Conservation of Nature (IUCN).[4] Attendees framed 12 'Manhattan principles', described as 'priorities for establishing

Table 5.2. Characteristics and time of emergence of some of the key interests and sub-disciplines relevant to One Health and associated reform movements.

Term	Origin	Inequality / social justice	Zoonoses	NTDs parasitology	Animal health	AMR	Ecology/ wildlife	GEC
Conservation Medicine	1996	?	++	?	?	?	+++	+
EcoHealth	2004	++	++	+++	+++	+	++	+
Ecological Public Health	1970s	++	?	++	?	?	++	++
One Medicine	1960s	?	+++	+++	+++	+++	+	+
One Health	2000s	+	+++	+++	+++	+++	+	+
Planetary Health	1980s	++	+	+	?	+	++	+++
Political ecology of health	Late 20th c	+++	?	?	?	?	++?	?
Social medicine	19th c	+++						

AMR, antimicrobial resistance; c, century; GEC, global environmental change; NTDs, neglected tropical diseases. Question marks indicate uncertainty; the scales are subjective, based on the judgement of the authors. Blank squares should not be considered as criticism; no branch in this (or any) field can be comprehensive.

a more holistic approach to preventing epidemic/epizootic disease and for maintaining ecosystem integrity for the benefit of humans, their domesticated animals, and the foundational biodiversity that supports us all' (Cook *et al.*, 2004). This meeting also set out to establish an 'international, interdisciplinary approach for combating threats to the health of life on Earth'.

However, perhaps inevitably, the issues that One Health was and still is chiefly associated with are a subset of these wider threats. Dimensions such as poverty, inequality and nuclear war are recognized by some One Health practitioners as relevant, but the range of skills needed to address these issues is too vast for a single discipline. All movements need collaborators.

There have been many amplifiers of Schwabe's work, particularly in the context of livestock and the health of pastoralists in Africa (Zinsstag *et al.*, 2005, 2011). Many others have contributed; today there is a flourishing journal called *One Health*.

The most obvious impetus for the formation and endurance of the One Health movement is the spate of newly recognized infectious diseases, some with epidemic and pandemic potential (Table 5.3). Several of these infections have penetrated into high-income settings, with high economic costs. The best-recognized examples of these include HIV/AIDS, severe acute respiratory syndrome (SARS), variant Creutzfeld Jacob Disease (vCJD) and, most dramatically, COVID-19. Most, if not all, of these conditions have a zoonotic origin.

5.3 Politics and One Health

All 'movements' are political, including those that claim not to be.[5] But movements that directly

Table 5.3 Some of the most important infectious diseases recognized in the decades leading to the emergence of One Health.

Infectious disease	Time first recognized [location]	Animal link/epidemiology
SV40[a]	1960s [global]	Primate virus, spread via contaminated polio vaccines
Marburg	1967 [Germany/ Yugoslavia]	Laboratory workers infected, via use of Ugandan green monkeys (whose kidneys were used for polio vaccine production)
Lassa fever[a]	1969 [Nigeria]	Via rodent urine
Ebola[a]	1976 [Zaire (now DRC)]	Bats to bushmeat to humans (unproven)
H1N1	1977 [global]	Live vaccine trial escape? (Rozo and Gronvall, 2015)
HIV/AIDS	1980s [Belgian Congo (now DRC)]	Chimpanzees to humans probably directly; via polio vaccine contamination considered by some as also possible
Bovine spongiform encephalopathy	1986 [UK]	Prion disease spread by feeding livestock remains to (vegetarian) cattle.
vCJD	1996 [UK]	From eating prion-positive beef, from asymptomatic cows.
H5N1 influenza[a]	1997 (humans) [China, South-East Asia]	Associated with intensive chicken farming.
Nipah	1998 [Malaysia, Singapore]	Bats to pigs to pig farmers and abattoir workers.
Nipah[a]	2004 [Bangladesh, West Bengal, India]	Bats to human, via fomite and bat faecal contact.
Nipah	2015 [Philippines]	Link to horse slaughtering and horsemeat ingestion
SARS	2003 [China]	Thought to be from bats to humans via intermediate 'novel' species.

DRC, Democratic Republic of the Congo; SV, simian virus; [a]indicates multiple animal-to-human transmission, as broadly defined (for example, as soon as the cause of BSE and vCJD was understood, further transmission was interrupted, but until then, multiple transmissions occurred).

challenge the status quo, including the pre-existing distribution of wealth, power and disease, are inevitably identified as political, even subversive. Rudolf Virchow (see Table 5.1) is widely recognized as a champion of social justice – even as the founder of social medicine (Mackenbach, 2009).[6]

Contemporary practitioners of One Health, who seek to reduce the burden of human and animal disease and to protect ecosystems, are challenging a powerful paradigm of human health and food security at any cost. To do this successfully will require effort and courage, including by challenging the views and power of mono-disciplinary, reductionist gatekeepers, who still dominate many senior positions in industry, governments, academia and journals.

5.4 Links Between Human and Animal Health

The links between human and animal health are based on two key insights. The first is that humans are animals, physiologically and emotionally, though not cognitively. The second is that humans and animals sometimes share infections, including parasites, which can cross the species barrier.[7]

Humans, like bonobos and chimpanzees, our closest living evolutionary relatives, have evolved to hunt other animal species and to eat meat, at least occasionally (Stanford, 2000). Starting perhaps 36,000 years ago, with dogs (Perri, 2016), humans have domesticated many species, for food (and other materials, such as sinews and hides), protection, labour and companionship. As human populations expanded in size, some infectious diseases not only passed from animals to humans, but have then been able to persist in human populations. Wolfe *et al.* (2007) cited studies which conclude that human populations of at least several hundred thousand people are needed to sustain measles, rubella and pertussis. In other cases, significant infectious diseases in humans have been transmitted to animal populations, including tuberculosis (Smith *et al.*, 2009). These are sometimes called reverse zoonoses.

One of the most important domesticated animal species is cattle – vital in Africa, Europe and many parts of Asia. Cattle provide food, milk, labour, hides and, for some populations, blood. They can provide food and other forms of security for pastoral populations, where rain is variable, mobility essential and crop cultivation difficult or impossible. Cattle domestication (from aurochs) is thought to have started about 10,000 years ago in the Near East (Götherström *et al.*, 2005). Cattle are still prone to the devastating infectious diseases (epizootic) of foot-and-mouth disease, but rinderpest ('cattle plague') has been recently declared eradicated. It is thought that both diseases were recognized by Aristotle, and repeated outbreaks have been described for over 2000 years, some with devastating effects for human health (Pastoret *et al.*, 2006). In its classical, virulent form, rinderpest could kill as many as 80–90% of infected cattle, as well as buffalo, yaks and wildlife ungulate species.

The Great European famine (1315–1317) was compounded by a devastating epizootic in cattle disease (probably rinderpest), which soon after (1319–1320) killed over half the cows in England. This created a scarcity of milk that lasted until at least 1332 (DeWitte and Slavin, 2013). In the late 19th century (1883–1902) a conjunction of events in eastern Africa was so devastating that it was called by the Maasai *emutai* – 'to wipe out' This crisis was caused by a combination of an outbreak of smallpox in humans, drought and devastating cattle diseases, probably rinderpest and bovine pleuropneumonia. Social breakdown followed and famine was severe (Gillson, 2006). Some families were so desperate that children were pawned (Tiki and Oba, 2009).

Many diseases have crossed from cattle into humans. Among the most important are smallpox (probably), anthrax and measles, which is thought to have diverged from rinderpest about 2600 years ago (Düx *et al.*, 2020). Measles is not yet extinct, but smallpox (like rinderpest) is now completely eradicated, apart from in laboratories (Butler, 2019).

Much of this is familiar and accepted by One Health practitioners, as are many parasitic and bacterial infections (e.g. brucellosis, echinococcus, Q fever, salmonella and trypanosomiasis). More controversial, and less well-known to One Health workers, are the links between the oncogenic (cancer-promoting) simian virus 40 (SV40), an accidental contaminant of both injected and oral polio vaccines, given to millions of children (Bookchin and Schumacher, 2004).[8] More speculatively, the spread of HIV/AIDS has been hypothesized as also being linked with polio vaccines, accidentally contaminated through the use of chimpanzee kidney cells to grow polio virus in order to make vaccines at Kisangani (then Stanleyville) in the Belgian Congo (now the Democratic Republic of the Congo) in the 1950s (Jain, 2020). Although this theory is generally considered to have been disproven by the dating of the first cases of HIV to the first half of the 20th century, this assumes the reliability and accuracy of the 'molecular clock' used to date the evolution of viruses (Warnock *et al.*, 2015).

Other medical practices[9] have proven and theoretical links relevant to disease transmission and One Health. These practices including xenotransplantation (Weiss, 2001), the accidental release of pathogens from laboratories and even 'gain of function' studies, in which the properties of pathogens are deliberately modified, or where the range of hosts is deliberately expanded, in attempts to gain fundamental understanding, and also to better promote new treatments and vaccines (Wain-Hobson, 2013).

Others have warned that smallpox may not only escape from a laboratory, but be deliberately synthesized by rogue actors (Noyce and Evans, 2018). The cause of SARS-Co-V-2, which has extraordinary characteristics, remains unknown, but a laboratory origin for it remains plausible (UNEP, 2022; Butler, 2022).

5.5 Ecology and One Health

The discipline of ecology, a term coined in 1866 by Ernst Haeckel (1834–1919), is a significant step away from 'reductionism' – the division of discovery into smaller and smaller sub-disciplines.[10] Ecology is derived from the Greek word for house (*oikos*), or home, as well as study (*logos*). Ecology concerns the study of relationships between components of nature, both living and non-living. Animals and humans, are, of course, a part of nature, but ecology is not primarily concerned with health, be it animal, human, plant or of any other form. There are, however, overlaps.

One Health is, similarly, a part of a wider movement to reduce scientific reductionism. Specialization evolved because the explosive growth of knowledge in recent centuries prevented mastery of any one scientific field. Reductionism allowed some individuals to keep more or less completely up to date in a limited field of study, in turn enabling progress. But this has a price. Its most obvious cost is of separation and rivalry among disciplines, leading at times to mutual incomprehension. This division is particularly apparent in the context of the unfolding crisis of civilization due to the interaction of numerous factors outlined in this book; one that will dominate this century. While a few individuals, collaborations and research groups regard this risk as of utmost gravity (Ripple *et al.*, 2017), most scientific disciplines, even in 2023, still consider the prospect of collapse far-fetched.

There is accruing evidence that other animals have cognition and emotion (Sekar and Shiller, 2020). While trees are generally not considered to be sentient, they nurture their young and cooperate (Gorzelak *et al.*, 2015). Sacred groves, which help conserve biodiversity (Baviskar, 1999), not only provide 'cultural' ecosystem services, but may also beneficially influence the microbiome of humans who spend time in them.

The Earth system, the entirety of life and non-life on Earth, has also been hypothesized to interact in ways that maintain homeostasis, suitable for life, but not necessarily for civilization (Xu *et al.*, 2020). Adding to this wonder, our solar system is one of hundreds now known to exist, in a galaxy which is itself one of countless others.

5.6 One Health in Action

The greatest challenge to One Health, or to any holistic approach to quadripartite health challenges (see Table 5.2), is to move beyond the conceptualization of interdependencies to action. Importantly WHO, FAO, the World Organisation for Animal Health (WOAH, formerly the Office International des Epizooties (OIE)) and the United Nations Environment Programme (UNEP) are committed to a fourfold collaboration to address health threats at the human–animal–ecosystem interface. While this coalition has had a strong focus on zoonotic diseases it is also concerned with significant livestock and plant diseases that have consequences for food security and trade, AMR, food-borne diseases and other food safety-related issues. The quadripartite has developed capacity and resources to replicate its collaborative approach in other countries, at multiple levels of governance. In addition to health, and the food and agricultural sector, the interface with the environment extends at least as far as wildlife hosts of zoonotic diseases and water and waste disposal (particularly where this attracts and feeds stray, pest and wild animals). How successfully this has been translated to country-level policies and action varies, and is still largely driven by individual diseases or concerns.

Rabies is a disease that showcases the benefits of a One Health approach – one where synergies of surveillance, data sharing, resources and coordinated programmes have been demonstrated, particularly in Asian countries (Acharya *et al.*, 2020). While much of the high-profile discussion of One Health has concerned emerging zoonoses of pandemic potential, it is worth noting that rabies is also classified as a neglected tropical disease (NTD). Importantly, most NTDs are diseases of the human–animal–ecosystem interface, occurring in low-income settings. Combining financial, logistic and service delivery of vaccination for livestock and pastoralists has improved outcomes for both (Schelling *et al.*, 2007). Institutional and educational capacity building for national One Health strategies is exemplified by Rwanda (Nyatanyi *et al.*, 2017). Elsewhere in the East African community, training of One Health capacity has expanded (see Chapter 5.i). In this region, endemic infectious diseases (including NTDs), diseases related to poor food safety (see Chapter 5.ii) and reverse zoonoses such as those that pose risk to great apes in conservation reserves are relevant. So too are zoonoses with epidemic and possibly pandemic potential such as Ebola.[11] Remote area logistics and cold storage are also relevant to the production and distribution of vaccines, including for yellow fever.

SARS-CoV-2 also highlights the importance of a One Health response, coordinated across sectors. Understanding this pandemic's emergence is part of this. The early identified association between the Wuhan seafood market and the virus (either because the market was the original source of animal-to-human transmission or a venue for human-to-human spread, with emergence elsewhere) brought closure of that and some other so called 'wet markets' across China. It also resulted in the destruction of farmed and legally and illegally obtained wildlife.[12] The livelihoods of market store-holders, farmers engaged in state-endorsed wildlife farming and remote communities where new biodiversity laws led to fines for previously remunerated activities were all impacted. Whether this drives wildlife trade further underground or accelerates its stronger regulation, or even its end, is uncertain. In many low-income settings, consumers prefer to buy ingredients in open (wet) markets rather than supermarkets where processed and discretionary foods are prevalent, and that may increase diet- and lifestyle-related illnesses such as type II diabetes. Food safety, food security, biodiversity conservation and communicable and non-communicable diseases are intertwined in this example, highlighting the importance of One Health responses.

One Health is an important example of successful multisector collaboration; considerable effort has been invested to continue to develop systems, networks and policy embedded in normal processes as well as the development of a 'surge' or emergency capacity, such as to respond to zoonotic disease outbreaks. Even within its current scope, climate change is highly relevant to the concerns of One Health. The habitable zones of many plants, animals (including those of agricultural importance), vectors and diseases are sensitive to changing weather patterns and sea level rise. This leads to changes in the intensity of disease transmission, geographical range and seasonality (Carlson *et al.*, 2022). Changes in food availability and habitat for bats may not only bring bats and other species into closer contact with humans, but also contribute to viral 'spillover' (Eby *et al.*, 2022). More speculatively, so may heatwaves.

Changing patterns of animal parasites and other pathogens were noted over a decade ago in the Arctic, often associated with wildlife declines (Bradley *et al.*, 2005). Compounded by changes in ice and snow cover, many northern-latitude populations who are highly dependent on reindeer, caribou and wildlife have been at the frontier of climate impacts on food security, human health and culture (Davidson *et al.*, 2011). The latter include zoonoses and vector-borne diseases such as tick-borne encephalitis, Lyme disease, tularaemia and environmental pathogens such as vibriosis and botulism (Waits *et al.*, 2018). The interdependency of human and animal health, food safety and security in the Arctic is such that the circumpolar Arctic Council has been operating a One Arctic–One Health project since 2015 (Abass *et al.*, 2019).

In addition to the institutional capacities described earlier, projects such as this support participatory community-based approaches for identifying and responding to health and well-being issues in communities. This includes information from traditional and local knowledge. Similarly, Small Island Developing States (SIDS) in the Pacific are addressing the triple burden of infectious and non-infectious diseases magnified by climate change and its impacts on marine and terrestrial ecosystems (McFarlane *et al.*, 2019). As the impacts of climate change are felt more broadly, the One Health approach will continue to be invaluable.

Although links between wildlife and human disease are important and irrefutable, some authors (e.g. Holmes *et al.*, 2018) have warned of exaggerated benefits that will accrue from exploration of spaces which the medical anthropologist Paul Farmer called 'geographies of blame' (Jephcott, 2023). Such an excursion may be exhibited by the global virome project (GVP), which seeks to identify between 631,000 and 827,000 unknown viruses which it claims have 'zoonotic potential' (Carroll *et al.*, 2018). This may be overreach, given that a comprehensive literature review identified 1415 unique, potentially infectious microorganisms and prions of which only 217 were viruses and prions (Taylor *et al.*, 2001). Although new viruses have since been discovered (including SARS-Co-V-2), this number is three orders of magnitude less than that which supporters of the GVP apprehend. Koch and Caceres-Escobar (2022) have also cast doubt on the infectious disease risk posed by wildlife in the absence of modern livestock production. Likewise, it is a simplistic aspiration that increased land area under conservation in line with the 2030 Global Biodiversity Framework (UN Convention of Biological Diversity) will by itself reduce pandemic risk.

5.7 Antimicrobial Resistance

Robinson *et al.* (2016) reported that the Review on AMR attributed 700,000 deaths annually to drug-resistant pathogens[13] and that this could increase more than tenfold by 2050. AMR is a One Health issue for several reasons. One is the overuse of antibiotics in intensive livestock production, including aquaculture. This practice is claimed to be essential as a growth promoter. However, the evidence for this is flimsy; decades of opposition have led to more judicious use in some settings (Collignon, 2015). Intensive livestock production, which exposes animals to sub-therapeutic doses, for prolonged periods, creates ideal conditions for bacteria to acquire genes that confer resistance, which can then be transmitted to human-adapted pathogens or to human gut microbiota via people, contaminated food or the environment (Robinson *et al.*, 2016).

5.8 Conclusion: Triumphs and Challenges

One of the first papers to use the words One Health claims that a 'one health' approach to disease can be 'much more effective and inexpensive than the traditional "quarantine and stamping out" strategy for fighting an illness after an outbreak has already begun' (Karesh and Cook, 2005). To date, however, this promise is unfulfilled; mass culling of animals as diverse as chickens, pigs, mink and bees continues. Some critics of One Health also have pointed out that its focus on zoonotic diseases have led to a relative neglect of its third pillar, the environment (Essack, 2018). But probably the greatest challenge for One Health is to shift its support from rhetoric to reality. A genuinely sustainable world with the 'tri-incidence' of a healthy environment, animal health and human health requires many fundamental changes, which challenge many existing power holders. Thinking in academic silos also needs reform. Resources do exist to greatly improve all three elements of the One Health triad, but far too many resources are currently spent on the world's military, on overconsumption of luxuries, and the failure of 'trickle down' economics to lift hundreds of millions from poverty.

One Health has, however, had many successes. Rinderpest, like smallpox, is extinct, its fire quenched not only by a vaccine, but also by the coordinated efforts of thousands of vaccine givers and administrators. Many other vaccines are under development. The pressure for wildlife preservation and the re-wilding of places is increasing.

One of Charles Darwin's books focused on the emotions of animals, but his insight was suppressed for over a century (de Waal, 2020). There is increased appreciation, including within the conservation movement, of the sentience and emotional lives of other species (Düx *et al.*, 2020). The insight that non-human animals have (at least to an extent) inner lives may help not only human health, but also the living and working conditions for other creatures. With the inclusion of UNEP into the quadripartite of intergovernmental organizations promoting One Health, there is international support to bring the health of the environment more prominently into the human–animal–environmental health nexus. However, climate change will bring change to all of these relationships – not just in the ecology of diseases but also in the determinants of individual, species and integrated 'one health'. Cities typify this challenge as both people and wildlife seek refuge, with increased proximity, within greener, climate-adaptation urban landscapes.[14] A diversity of other challenges that require a One Health approach keep arising also, including monkeypox and deepening food insecurity. Finally, we note the importance for transparency, accountability and cooperation, including by laboratory researchers and funders (Relman, 2020).

Notes

[1] One of whose books was called Science, Spirit, Wholeness: A Quaker Scientist's Sense of God.

[2] Schwabe's keynote address, given in 2002 at a conference to celebrate his own career, was subtitled 'importance of an integrating mindset'. In it, Schwabe declined to speculate on the future of 'One Medicine'.

[3] The journal Ecosystem Health was published from 1995 until the early 2000s.

[4] Other representatives included: the US Centers for Disease Control and Prevention; the Canadian Cooperative Wildlife Health Centre; the Laboratoire Nationale de Sante Publique of Brazzaville (Republic of Congo); and the US Department of Agriculture.

[5] Many individuals and movements claim to be apolitical. Such claims are often accepted uncritically, especially where the claimant unquestioningly accepts not only the status quo, the starting conditions, but the generally acceptable method of improving conditions, even if such methods appear tepid or even futile to others.

[6] For example, his rashness led to the German chancellor of the time, Otto von Bismarck, challenging him to a duel. As the challenged party, Virchow was entitled to choose the weapons. He selected two externally identical cooked sausages, one of which was contaminated with deadly trichinae. Von Bismarck refused to accept the duel, preserving Virchow's honour and the life of Germany's greatest then living scientist.

[7] Some of these are (universally) called zoonoses (e.g. SARS). Others are not always (e.g. HIV), as there are varying definitions of the term.

[8] SV40 was an inadvertent contamination of some of the primate kidney tissue used to grow polio virus, which was then either killed or attenuated (for the Salk and Sabin vaccines, respectively). Although it was thought that the processes to prepare these vaccines would similarly kill or weaken SV40 (and any other contaminant), this proved incorrect.

[9] Malaria experiments, in which humans were inoculated with chimpanzee blood, were undertaken at the Institute of Tropical Medicine in Antwerp from the late 1930s to 1955 (cited by Weiss, 1999).

[10] Schwabe frequently called for a non-reductionist, 'integrative' mindset (Schwabe, 2004) and argued that reductionism and integration rely on different parts of the brain.

[11] Butler argued that the risk of a pandemic from Ebola is lower than commonly appreciated, due to its limited 'stealth' transmission (UNEP, 2022), and also assuming a conservative definition of pandemic (see Chapter 7).

[12] The extent of this crackdown is uncertain, possibly deliberately.

[13] Including fungi and parasites. AMR is also greatly contributed to by over-the-counter medications, fake medications and inappropriate prescribing. These factors result, in part, from low health literacy, inappropriate financial incentives, corruption and inadequate regulation. In turn, these factors, while more prevalent in low-income settings, also occur in wealthy countries.

[14] Achieving universal access to greenspace, 30% canopy cover from urban forest for shade and reduced 'heat island' and other green infrastructure as recommended (e.g. by the UN Sustainable Development Goals) to improve human liveability of cities under climate change potentially benefits wildlife, particularly as ex-urban habitats deteriorate. However, in some circumstances proximity between human and wildlife may lead to increased conflict, injury and disease, for both humans and animals.

References

Abass, K., Waits, A., Emelyanova, A., Miettinen, I.T., Lavikainen, A. *et al.* (2019) *One Arctic – One Health*. Finnish Food Authority Research Reports 3/2019. Helsinki, Finland.

Acharya, K.P., Acharya, N., Phuyal, S., Upadhyaya, M. and Lasee, S. (2020) One-health approach: A best possible way to control rabies. *One Health* 10, 100161. DOI: 10.1016/j.onehlt.2020.100161

Amuasi, J.H., Lucas, T., Horton, R. and Winkler, A.S. (2020) Reconnecting for our future: The Lancet One Health Commission. *The Lancet* 395, 1469–1471. DOI: 10.1016/S0140-6736(20)31027-8

Baviskar, A. (1999) Vanishing forests, sacred trees: a Hindu perspective on eco-consciousness. *Asian Geographer* 18, 21–31. DOI: 10.1080/10225706.1999.9684045

Bookchin, D. and Schumacher, J. (2004) *The Virus and the Vaccine*. St Martins Griffin, New York.

Bradley, M.J., Kutz, S.J., Jenkins, E. and O'Hara, T.M. (2005) The potential impact of climate change on infectious diseases of Arctic fauna. *International Journal of Circumpolar Health* 64, 468–477. DOI: 10.3402/ijch.v64i5.18028

Butler, C.D. (2022) Comparing and contrasting two United Nations Environment Programme reports on COVID-19. *Science in One Health,* 1, 100003. DOI: 10.1016/j.soh.2022.100003

Butler, D. (2019) Sequence and destroy: the quest to eliminate the last stocks of deadly rinderpest virus. *Nature* 572(7767), 18. DOI: 10.1038/d41586-019-02098-2

Carlson, C.J., Albery, G.F., Merow, C., Trisos, C.H., Zipfel, C.M. *et al.* (2022) Climate change increases cross-species viral transmission risk. *Nature* 607, 555–562. DOI: 10.1038/s41586-022-04788-w

Carroll, D., Daszak, P., Wolfe, N.D., Gao, G.F., Morel, C.M. *et al.* (2018) The global virome project. *Science* 359, 872–874. DOI: 10.1126/science.aap7463

Collignon, P. (2015) Antibiotic resistance: are we all doomed? *Internal Medicine Journal* 45, 1109–1115. DOI: 10.1111/imj.12902

Cook, R.A., Karesh, W.B. and Osofsky, S.A. (2004) *The Manhattan Principles on 'One World One Health'*. Wildlife Conservation Society, New York.

Davidson, R., Simard, M., Kutz, S.J., Kapel, C.M.O., Hamnes, I.S. *et al.* (2011) Arctic parasitology: why should we care? *Trends in Parasitology* 27, 239–245. DOI: 10.1016/j.pt.2011.02.001

de Waal, F. (2020) *Mama's Last Hug: Animal Emotions and What They Tell Us About Ourselves*. Granta, London.

DeWitte, S. and Slavin, P. (2013) Between famine and death: England on the eve of the Black Death – Evidence from paleoepidemiology and manorial accounts. *Journal of Interdisciplinary History* 44, 37–60. DOI: 10.1162/JINH_a_00500

Düx, A., Lequime, S., Patrono, L.V., Vrancken, B., Boral, S. *et al.* (2020) Measles virus and rinderpest virus divergence dated to the sixth century BCE. *Science* 368, 1367–1370. DOI: 10.1126/science.aba9411

Eby, P., Peel, A.J., Hoegh, A., Madden, W., Giles, J.R. *et al.* (2022) Pathogen spillover driven by rapid changes in bat ecology. *Nature* 613, 340–344. DOI: 10.1038/s41586-022-05506-2

Essack, S.Y. (2018) Environment: the neglected component of the One Health triad. *The Lancet Planetary Health* 2, e238–e239. DOI: 10.1016/S2542-5196(18)30124-4

Evans, B.R. and Leighton, F.A. (2014) A history of One Health. *Scientific and Technical Review of the Office International des Epizooties* 33, 413–420. DOI: 10.20506/rst.33.2.2298

Gadgil, M., Berkes, F. and Folke, C. (1993) Indigenous knowledge for biodiversity conservation. *Ambio* 22, 151–156. http://www.jstor.com/stable/4314060

Gillson, L. (2006) A 'large infrequent disturbance' in an East African savanna. *African Journal of Ecology* 44, 458–467. DOI: 10.1111/j.1365-2028.2006.00662.x

Gorzelak, M.A., Asay, A.K., Pickles, B.J. and Simard, S.W. (2015) Inter-plant communication through mycorrhizal networks mediates complex adaptive behaviour in plant communities. *AoB PLANTS* 7, plv050. DOI: 10.1093/aobpla/plv050

Götherström, A., Anderung, C., Hellborg, L., Elburg, R., Smith, C. *et al.* (2005) Cattle domestication in the Near East was followed by hybridization with aurochs bulls in Europe. *Proceedings of the Royal Society B: Biological Sciences* 272, 2345–2351. DOI: 10.1098/rspb.2005.3243

Holmes, E.C., Rambaut, A. and Andersen, K.G. (2018) Pandemics: spend on surveillance, not prediction. *Nature* 558, 180–182. DOI: 10.1038/d41586-018-05373-w

Jain, S.L. (2020) The WetNet: What the oral polio vaccine hypothesis exposes about globalized interspecies fluid bonds. *Medical Anthropology Quarterly* 34, 504–524. DOI: 10.1111/maq.12587

Jephcott, F.L. (2023) Propagating visions of a forest reservoir. *Medical Anthropology* 42, 383–396. DOI: 10.1080/01459740.2023.2166411

Karesh, W.B. and Cook, R.A. (2005) The human–animal link. *Foreign Affairs* 84, 38–50. Available at: https://www.foreignaffairs.com/articles/2005-07-01/human-animal-link (accessed 15 November 2020).

Kass, P.H., McCapes, R.H. and Pritchard, W.R. (2006) In Memorium. Calvin W. Schwabe. Available at: https://senate.universityofcalifornia.edu/_files/inmemoriam/html/calvinwschwabe.htm, (accessed 20 October 2020).

Kock, R. and Caceres-Escobar, H. (2022) *Situation analysis on the roles and risks of wildlife in the emergence of human infectious diseases*. IUCN, Gland, Switzerland.

Lynteris, C. (2013) Skilled natives, inept coolies: marmot hunting and the Great Manchurian pneumonic plague (1910–1911). *History and Anthropology* 24, 303–321. DOI: 10.1080/02757206.2012.697063

Mackenbach, J.P. (2009) Politics is nothing but medicine at a larger scale: reflections on public health's biggest idea. *Journal of Epidemiology and Community Health* 63, 181–184. DOI: 10.1136/jech.2008.077032

McFarlane, R.A., Horwitz, P., Arabena, K., Capon, A., Jenkins, A. *et al.* (2019) Ecosystem services for human health in Oceania. *Ecosystem Services* 39, 100976. DOI: 10.1016/j.ecoser.2019.100976

Meadows, D., Meadows, D., Randers, J. and Behrens, W. III (1972) *The Limits to Growth*. Universe Books, New York.

Noyce, R.S. and Evans, D.H. (2018) Synthetic horsepox viruses and the continuing debate about dual use research. *PLOS Pathogens* 14, e1007025. DOI: 10.1371/journal.ppat.1007025

Nyatanyi, T., Wilkes, M., McDermott, H., Nzietchueng, S., Gafarasi, I. *et al.* (2017) Implementing One Health as an integrated approach to health in Rwanda. *BMJ Global Health* 2, e000121. DOI: 10.1136/bmjgh-2016-000121

OHHLEP (One Health High-Level Expert Panel), Adisasmito, W.B., Almuhairi, S., Behravesh, C.B., Bilivogui, P. *et al.* (2022) One Health: A new definition for a sustainable and healthy future. *PLOS Pathogens* 18, e1010537. DOI: 10.1371/journal.ppat.10105

Pastoret, P.-P., Yamanouchi, K., Mueller-Doblies, U., Rweyemamu, M.M., Horzinek, M. *et al.* (2006) Rinderpest — an old and worldwide story: history to c. 1902. In: Barrett, T., Pastoret, P.-P. and Taylor, W.P. (eds) *Rinderpest and Peste des Petits Ruminants*. Academic Press, Oxford, UK, pp. 86–104.

Perri, A. (2016) A wolf in dog's clothing: Initial dog domestication and Pleistocene wolf variation. *Journal of Archaeological Science* 68, e4. DOI: 10.1016/j.jas.2016.02.003

Relman, D.A. (2020) Opinion: To stop the next pandemic, we need to unravel the origins of COVID-19. *Proceedings of the National Academy of Sciences (USA)* 117, 29246–29248. DOI: 10.1073/pnas.2021133117

Ripple, W.J., Wolf, C., Newsome, T.M., Galetti, M., Alamgir, M. *et al.* (2017) World scientists' warning to humanity: a second notice. *BioScience* 57, 1026–1028. DOI: 10.1093/biosci/bix125

Robinson, T.P., Bu, D.P., Carrique-Mas, J., Fèvre, E.M., Gilbert, M. *et al.* (2016) Antibiotic resistance is the quintessential One Health issue. *Transactions of The Royal Society of Tropical Medicine and Hygiene* 110, 377–380. DOI: 10.1093/trstmh/trw048

Rozo, M. and Gronvall, G.K. (2015) The reemergent 1977 H1N1 strain and the gain-of-function debate. *mBio* 6, e01013-15. DOI: 10.1128/mBio.01013-15

Schelling, E., Bechir, M., Ahmed, M.A., Wyss, K., Randolph, T.F. *et al.* (2007) Human and animal vaccination delivery to remote nomadic families, Chad. *Emerging Infectious Diseases* 13, 373–379. DOI: 10.3201/eid1303.060391

Schwabe, C.W. (1964) *Veterinary Medicine and Human Health*. Williams & Wilkins Co., Baltimore, Maryland.

Schwabe, C. (2004) Keynote address: The calculus of disease-importance of an integrating mindset. *Preventive Veterinary Medicine* 62, 193–205. https://europepmc.org/article/med/19434958

Sekar, N. and Shiller, D. (2020) Engage with animal welfare in conservation. *Science* 369, 629–630. DOI: 10.1126/science.aba7271

Smith, N.H., Hewinson, R.G., Kremer, K., Brosch, R. and Gordon, S.V. (2009) Myths and misconceptions: the origin and evolution of Mycobacterium tuberculosis. *Nature Reviews Microbiology* 7, 537–544. DOI: 10.1038/nrmicro2165

Stanford, C.B. (2000) The brutal ape vs. the sexy ape? *American Scientist* 88, 110–111. https://www.jstor.org/stable/27857987

Tarantola, A. (2017) Four thousand years of concepts relating to rabies in animals and humans. Its prevention and its cure. *Tropical Medicine and Infectious Disease* 2, 5. DOI: 10.3390/tropicalmed2020005

Taylor, L., Latham, S. and Woolhouse, M. (2001) Risk factors for human disease emergence. *Philosophical Transactions of the Royal Society B* 356, 983–939. DOI: 10.1098/rstb.2001.0888

Tiki, W. and Oba, G. (2009) Ciinna – the Borana Oromo narration of the 1890s Great Rinderpest epizootic in North Eastern Africa. *Journal of Eastern African Studies* 3, 479–508. DOI: 10.1080/17531050903273750

UNEP (2022) *COVID-19: A Warning. Addressing Environmental Threats and the Risk of Future Pandemics in Asia and the Pacific*. United Nations Environment Programme, Bangkok, Thailand. Available at: https://wedocs.unep.org/20.500.11822/40871 (accessed 21 October 2022).

Waits, A., Emelyanova, A., Oksanen, A., Abass, K. and Rautio, A. (2018) Human infectious diseases and the changing climate in the Arctic. *Environment International* 121, 703–713. DOI: 10.1016/j.envint.2018.09.042

Wain-Hobson, S. (2013) Pandemic influenza viruses: time to recognize our inability to predict the unpredictable and stop dangerous gain-of-function experiments. *EMBO Molecular Medicine* 5, 1637–1641. DOI: 10.1002/emmm.201303475

Warnock, R.C.M., Parham, J.F., Joyce, W.G., Lyson, T.R. and Donoghue, P.C.J. (2015) Calibration uncertainty in molecular dating analyses: there is no substitute for the prior evaluation of time priors. *Proceedings of the Royal Society B: Biological Sciences* 282, 20141013. DOI: 10.1098/rspb.2014.1013

Weiss, R.A. (1999) Is AIDS man-made? *Science* 286, 1305. DOI: 10.1126/science.286.5443.1305

Weiss, R.A. (2001) The Leeuwenhoek Lecture 2001. Animal origins of human infectious disease. *Philosophical Transactions: Biological Sciences* 356, 957–977. DOI: 10.1098/rstb.2001.0838

Wolfe, N.D., Dunavan, C.P. and Diamond, J. (2007) Origins of major human infectious diseases. *Nature* 447, 279–283. DOI: 10.1038/nature05775

Woods, A., Bresalier, M., Cassidy, A. and Dentinger, R.M. (eds) (2017) *Humans, Other Animals and 'One Health' in the Early Twenty-First Century*. Palgrave MacMillan, London.

Xu, C., Kohler, T.A., Lenton, T.M., Svenning, J.-C. and Scheffer, M. (2020) Future of the human climate niche. *Proceedings of the National Academy of Science (USA)* 117, 11350–11355. DOI: 10.1073/pnas.1910114117

Zinsstag, J., Schelling, E., Wyss, K. and Mahamat, M.B. (2005) Potential of cooperation between human and animal health to strengthen health systems. *The Lancet* 366, 2142–2145. DOI: 10.1016/S0140-6736(05)

Zinsstag, J., Schelling, E., Waltner-Toews, D. and Tanner, M. (2011) From 'one medicine' to 'one health' and systemic approaches to health and well-being. *Preventive Veterinary Medicine* 101, 148–156. DOI: 10.1016/j.prevetmed.2010.07.003

Zinsstag, J., Waltner-Toews, D. and Tanner, M. (2021) Why One Health? In: Zinsstag, J., Schelling, E., Crump, L., Whittaker, M., Tanner, M. and Stephen, C. (eds) *The Theory and Practice of Integrated Health Approaches*, 2nd edn. CABI, Wallingford, UK, pp. 15–24.

5.i A Practical, Integrated Way to Build a One Health Workforce – Using One Health Problem-based Learning Cases for In-Service Training Programmes in Africa

Janetrix Hellen Amuguni
Department of Infectious Diseases and Global Health, Tufts University, Medford/Somerville, Massachusetts, USA

Abstract
The One Health approach calls for a paradigm shift to develop, implement and sustain strengthened training, to build collaborations and partnerships and to redesign educational programmes. While other pedagogical approaches are possible, a promising method is by bringing a wide range of professionals together to engage in problem-based learning (PBL). This short chapter synthesizes our experience and learnings gleaned from experience at a number of workshops recently held in parts of Africa.

5.i.1 Introduction: The Challenge to Integrate Education for One Health

The health security of nations is threatened by new infectious, environmental and behavioural risks and rapid epidemiological and demographic changes (Bouskill and Smith, 2019). These events are driven by climate change, increasing human populations, amplified interactions between humans and animals and surging global travel (Amuguni *et al.*, 2018). As shown during the coronavirus pandemic, many national health systems are struggling to keep up. Controlling infections has becomes more complex and costly, placing additional demands on the public health workforce. Acknowledging the interactions and the risks that zoonotic diseases represent to public health has led to appeals for more horizontal collaboration. This is among not only the sector agencies, but also departments and ministries and disciplines responsible for public health: health professions, veterinary services and those who protect the environment (World Bank, 2012).

A key approach to build global health security is to develop a multidisciplinary public health workforce that is well trained and equipped to respond to these recurring public health threats. This requires the redesign of the educational system to impart a more integrative, dynamic understanding in the public health workforce (Amuguni *et al.*, 2018). Professional competencies are too siloed; the requirements of an increasingly multidisciplinary, complex world make revamping public health education imperative. The Commission on Education for Health Professionals for the 21st Century, launched in 2010, is one instrument for this (Bhutta *et al.*, 2010). Unfortunately, many professional training programmes are still producing graduates who are ill equipped to handle current challenges. The premise of engaging

 Climate Change and Global Health: Primary, Secondary and Tertiary Effects, 2nd Edition (eds C.D. Butler and K. Higgs)
DOI: 10.1079/9781800620025.0005i

multiple disciplines to address shared problems that span the human–animal–environmental interface is an attempt to lower existing disciplinary barriers (American Veterinary Medical Association, 2008; Barrett and Osofsky, 2013). The One Health approach calls for a paradigm shift to develop, implement and sustain strengthened training, to build collaborations and partnerships and to redesign educational programmes. One way it can do this is by bringing a wide range of professionals together to engage in problem-based learning (PBL). Of course, other pedagogical approaches are also possible.

PBL is an interactive, learner-focused method of teaching. Its advocates claim that it is well suited to develop the intellectual inquiry, teamwork and communication skills that are necessary for a modern public health workforce. PBL has been used with remarkable results to educate professions as diverse as nursing, medicine, veterinary medicine, agriculture, engineering and economics, and is increasingly being adopted in many African health science institutions (Sayyah *et al.*, 2017). PBL uses actual events, often discussed in small groups, to develop critical thinking skills for investigating 'real world' problems. It seeks to provide a nurturing environment in which all curriculum elements are systematically aligned to help learning (Downing, 2013).

Examples of One Health problem-based experiences from real life can be as simple, yet important, as helping animal owners understand the importance of rational antibiotic use, in order to reduce the emergence of antimicrobial resistance, in both humans and animals, or as complex as employing a large team to track and contain a zoonotic disease outbreak. PBL is promoted as the ideal forum for teaching and coaching One Health approaches (Kilroy, 2004). PBL has been lauded as promoting not just knowledge of facts, but also the development of 'integrative thought processes, problem-solving, creativity, interpersonal communication, and resourcefulness' (McLoda, 2003) – essential competencies for effective One Health programme implementation.

Developing One Health PBL (OHPBL) cases that can be used across different disciplines and sectors is a holistic way to integrate One Health concepts and principles into the health sciences curricula and in-service training programmes, including short workshops. These cases analyse problems that span the human–animal–environmental domains, taking into consideration One Health competencies such as communication, leadership, risk analysis, collaboration and partnership, gender, and behaviour change. Participants are encouraged to think critically across the three domains; solving complex health problems that go beyond their discipline; extending their capacity to think critically and collaborate with different disciplines; and recognizing the value of interdisciplinary partnerships. In one such development activity, potential OHPBL cases were

Box 5.i.1

This PBL case made me think about public health issues for the first time, i.e. what happens after I call the health department to report a notifiable disease. That was interesting. I liked how it made me explore the impact that we can have on Public Health. Usually I never think about this – Vet, Kenya

I learned more about the veterinary side of medicine, zoonotic diseases, and environmental issues – I previously just focus on the patients who walk into my clinic. It was eye opening. When we discussed that Rift Valley fever may be circulating in animals before it gets to humans, I realized that I need to collaborate more with vets, so am just not waiting for sick patients to walk into my office – I guess that is what One Health is about – Physician, Uganda

I always think that no one recognizes the role of the environment. This case highlighted how an earthquake can release dormant spores and cause an outbreak. There are other examples. We need to collaborate with different disciplines, including engineers and entomologists. This was a very practical example of what happens when we just work in our own silos – Environmental scientist, Rwanda

I'm glad I was included to help develop and train my department on gender issues in One Health using this antimicrobial stewardship case. Many scientists ignore the role of gender and culture in disease outbreaks and even other issues. In my country many mothers are responsible for purchasing antibiotics for their children, but no one thinks of them as a target audience – Gender specialist, Cameroon.

grouped under infectious diseases transmissible across the human–wildlife–domestic-animal interface – diseases whose epidemiology is evolving as a result of climatic and other environmental changes and others such as toxins and antimicrobial resistance. Representative student evaluations are reported in Box 5.i.1.

OHPBL cases provide extra knowledge, skills and behaviours needed for the public health workforce to effectively collaborate with other disciplines. They can help identify gaps in existing health sciences training programmes and as frameworks for training-programme evaluation. Combining One Health with PBL is an opportunity, within public health, to facilitate increased understandings, necessary to create a stronger and more efficient integrated health system, with inputs from multiple stakeholders.

References

American Veterinary Medical Association (2008) *One Health Initiative Task Force. One Health: A New Professional Imperative*. Available at: https://www.avma.org/KB/Resources/Reports/Documents/onehealth_final.pdf (accessed 14 July, 2021).

Amuguni, H., Bikaako, W., Naigaga, I. and Bazeyo, W. (2018) Building a framework for the design and implementation of One Health curricula in East and Central Africa: OHCEAs One Health training modules development process. *One Health* 7, 100073. DOI: 10.1016/j.onehlt.2018.08.002

Barrett, M.A. and Osofsky, S.A. (2013) One Health: Interdependence of people, other species, and the planet. In: Katz, D.L., Elmore, J.G., Wild, D.M.G. and Lucan, S.C. (eds) *Jekel's Epidemiology, Biostatistics, Preventive Medicine, and Public Health*, 4th edn. Elsevier/Saunders, Philadelphia, Pennsylvania, pp. 364–377.

Bhutta, Z.A., Chen, L., Cohen, J., Crisp, N., Evans, T. *et al.* (2010) Education of health professionals for the 21st century: a global independent Commission. *The Lancet* 375, 1137–1138. DOI: 10.1016/S0140-6736(10)60450-3

Bouskill, K.E. and Smith, E. (2019) *Global Health and Security: Threats and Opportunities*. Rand Corporation, Santa Monica, California. Available at: https://www.rand.org/pubs/perspectives/PE332.html (accessed 14 July 2021).

Downing, K.E. (2013) Using problem-based learning to facilitate student learning. In: *Proceedings of the Association of College and Research Libraries*, 16th National Conference 'Imagine, Innovate, Inspire'. ACRL, Chicago, pp. 621–624. Available at: https://www.ala.org/acrl/sites/ala.org.acrl/files/content/conferences/confsandpreconfs/2013/papers/Downing_Using.pdf (accessed 26 July 2022).

Kilroy, D.A. (2004) Problem based learning. *Emergency Medicine Journal* 21, 411–413. DOI: 0.1136/emj.2003.012435

McLoda, T. (2003) Problem-based learning in allied health and medicine. *The Internet Journal of Allied Health Sciences and Practice* 1, Article 6. DOI: 10.46743/1540-580X/2003.1005

Sayyah, M., Shirbandi, K., Saki-Malehi, A. and Rahim, F. (2017) Use of a problem-based learning teaching model for undergraduate medical and nursing education: a systematic review and meta-analysis. *Advances in Medical Education and Practice* 8, 691–700. DOI: 10.2147/AMEP.S143694

World Bank (2012) *People, Pathogens and our Planet: The Economics of One Health*. World Bank, Washington, DC. Available at: https://openknowledge.worldbank.org/handle/10986/11892 (accessed 14 July 2021).

5.ii Food Systems, Food Safety and One Health

Delia Randolph
Natural Resources Institute, University of Greenwich, Chatham Maritime, Kent, UK; International Livestock Research Institute, Nairobi, Kenya

Abstract
This short chapter focuses on foodborne diseases, the global health burden of which rivals that of malaria or tuberculosis. Most of this burden is from the consumption of animal-source foods and fresh produce sold in the informal markets of low- and middle-income countries. The chapter considers the causes of foodborne diseases, their One Health characteristics, their health and economic burden, their trends, their associated food matrices, their associated myths and, finally, their management. Considerable progress has occurred with the management of these conditions in high-income settings; some of the lessons learned there can also be applied in low- and middle-income countries.

5.ii.1 Introduction

Food is fundamental to life; many human diseases have their origin in, or are linked to, agro-food systems – that is, processes, activities, infrastructure and environment that encompass the primary production, processing, distribution, waste disposal, preparing and consuming of food (Fig 5.ii.1). Food systems have always been dynamic, but the pace of change accelerated rapidly in the second half of the last century. This was driven by dramatic changes in demographics, land use, urbanization, consumption habits and climate as well as technological innovation. This rapid transformation has been called the Green Revolution or third agricultural revolution. Like the two preceding agricultural revolutions (sedentarization and animal domestication) this has also been accompanied by epidemiological transitions.

Sedentarization brought crowd diseases, while animal domestication (the second agricultural revolution) was associated with massive disease spillover from livestock (zoonoses and emerging diseases). However, this third revolution is especially associated with diseases linked to over- and undernutrition, a new wave of zoonoses associated with livestock intensification and diseases caused by pathogens that have acquired resistance to antimicrobials.

Food systems transitions are also associated with emergence and change in foodborne diseases (FBDs) (Wolfe *et al*., 2007; Grace, 2015ab). FBDs may be defined as sicknesses caused by hazards commonly transmitted through ingested food and beverages. They comprise a broad group of illnesses, caused by microbial pathogens, parasites, chemical contaminants and biotoxins. This section on FBDs covers their causes, their One Health characteristics, their health and economic burden, the food matrices associated with FBDs, myths around FBDs and, finally, their management.

DOI: 10.1079/9781800620025.0005ii

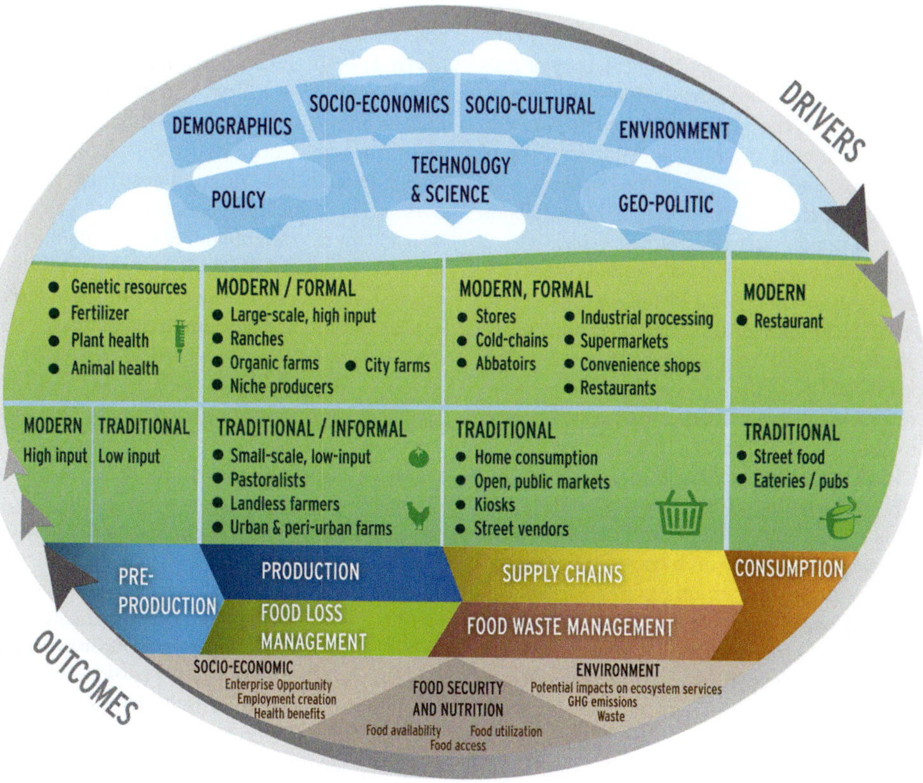

Fig. 5.ii.1. Interpretation of the international food and agriculture system and its drivers, activities, actors and outcomes. All elements of growing, harvesting, storing, processing, distributing, consuming and managing the food and agriculture system are encompassed in the UN's Sustainable Development Goals. GHG, greenhouse gas. Source: Authors.

5.ii.2 Foodborne Diseases as a One Health Issue

FBDs are a quintessential One Health problem. First, they occur at the interface of human health, animal health and ecosystems. Secondly, the most important FBDs are zoonotic, managed by multiple sectors (typically including health, veterinary, trade and tourism). Lastly, many FBDs are associated with the types of food whose production has profound impacts on ecosystems and biodiversity (livestock and cereals). The Quadripartite,[1] which has been spearheading international One Health, considers FBDs as one of its four priority areas, along with emerging zoonoses, neglected zoonoses and tropical diseases, and antimicrobial resistance (FAO *et al.*, 2022). These four key One Health disease clusters are closely interlinked. Examples of zoonotic FBDs include campylobacteriosis and cysticercosis. FBDs can also be plant associated (e.g. aflatoxicosis, konzo) and emerging (hepatitis E). Animal-source food production is also strongly associated with antimicrobial resistance. Finally, many emerging zoonoses have been linked to the exploitation of wild animals for food (e.g. HIV/AIDs and Ebola) or with land use change for food production (e.g. vector-borne diseases associated with irrigation) or with food storage (e.g. Lassa fever with grain storage).

5.ii.3 Causes and Impacts of Foodborne Diseases

Thirty-five of the most important FBDs are together responsible for a health burden of more than 40 million lost disability-adjusted life years

(DALYs) annually, similar to that caused by malaria or tuberculosis (Havelaar *et al.*, 2015; Gibb *et al.*, 2019). The most important category of hazards is microbial pathogens, which are responsible for 54% of the burden, followed by parasites and heavy metals (each responsible for 22% of the burden) and toxins (2% of the burden).

Most of the health burden of FBDs falls on low- and middle-income countries (LMICs), accompanied by an annual economic burden estimated at more than US$110 billion (Jaffee *et al.*, 2019). And while most infectious diseases tend to decline as GDP increases, FBDs tend to first worsen before eventually declining. This is because as LMICs get richer and more urbanized, consumption of the most high-risk foods (fresh produce and animal-source food) increases. Secondly, in response to increased demand, food chains are becoming longer and more complex, increasing the spread of hazards. Mature and well-governed value chains control foodborne disease by insisting on high standards along the value chain. However, in LMICs the expansion of value chains is happening in advance of effective governance, thus increasing risk (Grace, 2015ab). This implies that, for LMICs, the FBD burden will continue to increase for decades to come (Kristkova *et al.*, 2017).

Certain groups of people are more vulnerable to FBDs; these have been summarized as YOMPIs or 'young, old, malnourished, pregnant, or immunosuppressed'. Although children younger than five years are only 9% of the global population, they bear more than 40% of the global health burden of FBDs (Havelaar *et al.*, 2015).

Only in the past decade has good evidence on the burden of FBD started to emerge, and attribution to different food sources is even less clear. However, multiple lines of evidence suggest that animal-source foods and fresh vegetables are the riskiest products (Hoffmann *et al.*, 2017; Sapp *et al.*, 2022). Consumers of street foods, ready-to-eat foods, foods eaten raw and complementary foods (foods given to infants as a complement to breastmilk or formula) are also especially vulnerable to acquiring FBD (Grace *et al.*, 2018).

5.ii.4 Common Myths Around Foodborne Diseases

Despite the importance of FBDs, myths about them abound, especially in LMICs (where evidence is weakest). Some of the most prevalent among policy makers include the following.

Myth 1: *FBDs are mild and transitory*. Although gastroenteritis is a common manifestation of FBD and most such cases are mild, the greatest health burden from FBD is the result of more severe, though less common gastrointestinal illnesses. These include septicaemia (a complication of many microbial FBDs), Guillain-Barre syndrome (campylobacteriosis), epilepsy (cysticercosis), blindness in newborns (from toxoplasmosis) and mental retardation from lead poisoning (Havelaar *et al.*, 2015; Gibb *et al.*, 2019).

Myth 2: *The most serious FBDs are those caused by manmade chemicals such as pesticides and fertilizers*. Current best evidence on FBD suggests that biological hazards are more important than chemicals. However, it is more difficult to measure the health impacts of diseases caused by chemicals (Grace, 2015ab).

Myth 3: *Food produced using modern, intensive agriculture is safer than food from small-scale agriculture*. While some aspects of modern agriculture reduce health risk, others increase it. For example, intensive (high-input–high-output) farmers are more likely to use antimicrobials than smallholders. As a result, 'formal sector' milk has been found to have higher levels of antimicrobial-resistant bacteria (Sharma *et al.*, 2022).

Myth 4: *Food sold in modern retail outlets (supermarkets, convenience stores) is safer than food sold in traditional retail outlets (open markets, small stores)*. While food sold through modern retail outlets is generally safe, the same is not always true in LMICs where there are added challenges in maintaining cold-chains and accountability is less established. For example, one study found that pork sold from supermarkets in Hanoi was less safe than that sold in the traditional 'wet market' (Fahrion *et al.*, 2013).

Food safety detection and management is also hampered by surprisingly widespread myths and misunderstandings common among lay people (Grace, 2015ab). These include the following.

- Food that is unsafe looks, tastes and smells bad.
- Thoroughly cooking food makes it safe.
- If you have FBD, it was caused by the last thing you ate.
- If someone gets sick from bad food, everyone who eats it gets sick.
- Drinking alcohol along with bad food will protect you.

Some cultural beliefs protect the most vulnerable from FBD, such as the Vietnamese taboo for pregnant women to avoid eating dogs (Hoa *et al.*, 2014). But others increase risk, such as the Ugandan belief that eating raw eggs is therapeutic (Cole *et al.*, 2008). In Vietnam, eating raw fish is the main risk factor for acquiring fish-borne trematode infections (Tran *et al.*, 2019), which, unless treated, can be lifelong.

5.ii.5 Management of Foodborne Diseases

Although there is considerable concern over FBDs in high-income countries (HICs), most evidence suggests that the food sold there is relatively safe; much safer than it was historically. This can be attributed to better technologies, systems, governance and incentives. Much of this improvement was initially driven by outrage from a public that was becoming increasingly informed and empowered, following exposés of filthy conditions and fraudulent practices in the rapidly industrializing food industries. A free press and investigative journalists were key partners in this reform.

Despite the enormous health and financial burden of foodborne disease in LMICs, FBD management has been relatively neglected by donors. Worse still, most of the relevant investments of the past decade have been directed towards areas with little prospect of cost efficacy (for example, promotion of export trade and the control of FBDs of relatively low health burden such as those caused by toxins) (GFSP, 2019). A recent review of FBD interventions in LMICs categorized them according to nodes along the 'farm-to-fork' value chain they addressed. This found that most interventions were at points where the impact on food safety was hardest to measure (infrastructure provision). Even when interventions were appropriately targeted, there was little evidence of long-term continuation (sustainability) or scalability (Grace *et al.*, 2018).

5.ii.6 Conclusion: Challenges and Pathways

There are numerous challenges to assuring food safety in LMICs. Government systems to support food safety are often emerging, but consumer choice is limited by income and lack of information. The result is that the most important incentives for safe food production in developing countries (private demand and effective private or public regulation) are lacking. In this context, new approaches to food safety that support and are supported by a range of incentives – social, market, or regulatory based – need to be developed to encourage farmers and other value-chain actors to produce quality products that are safe. Because of the high level of involvement of the poor and women in producing for, and selling in, informal markets, agricultural research and development interventions that aim to improve livelihoods of women and the poor should engage with informal markets (Chenevix Trench *et al.*, 2011; Grace *et al.*, 2008; Vorley, 2013).

Fortunately, alternative, market-based approaches to improving food safety in informal markets are emerging (Grace *et al.*, 2020). These light-touch interventions change practice through capacity building, rewards (incentives) and provision of an enabling policy environment; they are especially suited to contexts where demand for food safety is high. They often target informal market agents because actors who aggregate product from many producers (for example, traders, processors, slaughterhouses) play a key role in maintaining and improving the quality of food, and they may also be easier to reach since there are fewer of them compared with either producers or consumers; and because they are consumer-facing they have more incentives to operate hygienically (Leahy *et al.*, 2022).

Note

[i] Consisting of the World Health Organization (WHO), the Food and Agriculture Organization of the United Nations (FAO), the United Nations Environment Programme (UNEP) and the World Organisation for Animal Health (WOAH).

References

Chenevix Trench, P., Narrod, C., Roy, D. and Tiongco, M. (2011) *Responding to Health Risks along the Value Chain*. IFPRI 2020 Conference Paper 5, International Food Policy Research Institute (IFPRI), Washington, DC.

Cole, D., Grace, D. and Diamond, M. (2008) Researcher approaches to evidence on urban agriculture and human health. In: Cole, D., Lee-Smith, D. and Nasinyama, G. (eds) *Healthy City Harvests: Generating Evidence to Guide Policy on Urban Agriculture*. International Potato Center (CIP) and Makerere University Press, Lima, Peru, pp. 31–47.

Fahrion, A.S., Lapar, M.L., Nguyen N.T., Do N.T. and Grace D. (2013) Food-borne hazards in a transforming pork value chain in Hanoi: Basis for future risk assessments. *Vietnamese Journal of Preventive Medicine* 23(4), 18–25. Available at: https://cgspace.cgiar.org/handle/10568/33815 (accessed 13 May 2023).

FAO, UNEP, WHO and WOAH (2022) *One Health Joint Plan of Action (2022–2026). Working Together for the Health of Humans, Animals, Plants and the Environment*. Food and Agriculture Organization, Rome. DOI: 10.4060/cc2289en

GFSP (2019) *Food Safety in Africa: Past Endeavors and Future Directions*. Global Food Security Programme, World Bank, Washington, DC.

Gibb, H.J., Barchowsky, A., Bellinger, D., Bolger P.M., Carrington, C. *et al.* (2019) Estimates of the 2015 global and regional disease burden from four foodborne metals – arsenic, cadmium, lead and methylmercury. *Environmental Research* 174,188–194. DOI: 10.1016/j.envres.2018.12.062

Grace, D., Omore, A., Randolph, T., Kang'ethe, E., Nasinyama, G.W. *et al.* (2008) Risk assessment for *Escherichia coli* O157:H7 in marketed unpasteurized milk in selected East African countries. *Journal of Food Protection* 71, 257–263. DOI: 10.4315/0362-028X-71.2.257

Grace, D. (2015a) Food safety in low and middle income countries. *International Journal of Environmental Research and Public Health* 12, 10490–10507. DOI: 10.3390/ijerph120910490

Grace, D. (2015b) *Food Safety in Developing Countries: An Overview,* Evidence on Demand, Hemel Hempstead, UK.

Grace, D., Alonso, S., Mutua, F., Roesel, K., Lindahl, J. *et al.* (2018) *Food Safety Investment Expert Advice*. International Livestock Research Institute (ILRI), Nairobi, Kenya.

Grace, D., Alonso, S., Bett, B., Cook, E., Lee, H.S. *et al.* (2020) Zoonoses. In: Mcintire, J. and Grace, D. (eds) *The Impact of the International Livestock Research Institute*. CABI, Wallingford, UK, pp. 302–337.

Havelaar, A.H., Kirk, M.D., Torgerson, P.R., Gibb, H.J., Hald, T. *et al.* (2015) World Health Organization global estimates and regional comparisons of the burden of foodborne disease in 2010. *PLoS Medicine* 12(12), e1001923. DOI: 10.1371/journal.pmed.1001923

Hoa, N.T.T., Nguyen-Viet, H., Sinh, D.X., Phuc, P.D., Thanh, N.T. *et al.* (2014) Rapid integrated assessment on food safety related to pork consumers in Hung Yen and Nghe An provinces. *Journal of Practical Medicine* 5, 230–233. Available at https://hdl.handle.net/10568/56584 (accessed 13 May 2023).

Hoffmann, S., Devleesschauwer, B., Aspinall, W., Cooke, R., Corrigan, T. *et al.* (2017) Attribution of global foodborne disease to specific foods: Findings from a World Health Organization structured expert elicitation. *PLoS ONE* 12, e0183641. DOI: 10.1371/journal.pone.0183641

Jaffee, S., Henson, S., Unnevehr, L., Grace, D. and Cassou, E. (2019) *The Safe Food Imperative: Accelerating Progress in Low- and Middle-Income Countries*. Agriculture and Food Series, World Bank, Washington, DC. Available at: https://openknowledge.worldbank.org/handle/10986/30568 (accessed 24 June 2023)

Kristkova, Z.S., Grace, D. and Kuiper, M. (2017) *The Economics of Food Safety in India – A Rapid Assessment*. Wageningen University & Research, Amsterdam, The Netherlands.

Leahy, E., Mutua, F., Grace, D., Lambertini, E. and Thomas, L.F. (2022) Foodborne zoonoses control in low- and middle-income countries: Identifying aspects of interventions relevant to traditional markets which act as hurdles when mitigating disease transmission. *Frontiers in Sustainable Food Systems* 6, 913560. DOI: 10.3389/fsufs.2022.913560

Sapp, A.C, Amaya, M.P., Havelaar, A.H. and Nane, G. (2022) Attribution of country level foodborne disease to food group and food types in three African countries: Conclusions from a structured expert judgment study. *PLoS Neglected Tropical Diseases* 16, e0010663. DOI: 10.1371/journal.pntd.0010663

Sharma, G., Leahy, E., Deka, R.P., Shome, B.R., Bandyopadhyay, S. *et al.* (2022) Antibiotic use, knowledge, and practices of milk vendors in India's informal dairy value chain. *Frontiers in Sustainable Food Systems* 6, 1058384. DOI: 10.3389/fsufs.2022.1058384

Tran, A.K.T., Doan, H.T., Do, A.N., Nguyen, V.T., Hoang, S.X. *et al.* (2019) Prevalence, species distribution, and related factors of fish-borne trematode infection in Ninh Binh Province, Vietnam. *BioMed Research International* 2019, 8581379. DOI: 10.1155/2019/8581379

Vorley, B. (2013) *Meeting Small-scale Farmers in Their Markets: Understanding and Improving the Institutions and Governance of Informal Agrifood Trade*. International Institute for Environment and Development/Hivos/Mainumby, London; The Hague, The Netherlands; La Paz, Bolivia.

Wolfe, N.D., Dunavan, C.P. and Diamond, J. (2007) Origins of major human infectious diseases. *Nature* 447, 279–283. DOI: 10.1038/nature05775

6 Land Use, Biodiversity Loss and Health

Jessica Stanhope,[1] Christopher B. Daniels[2] and Philip Weinstein[3]
[1]School of Allied Health Science and Practice, The University of Adelaide, Australia; [2]Green Adelaide Landscape Board, Department for Environment and Water, Adelaide, Australia; [3]School of Public Health, The University of Adelaide, Australia

Abstract

Provisioning, regulating and cultural ecosystem services link land use change directly to human health outcomes. We use a case study of land use change – conversion of natural ecosystems into agroecosystems – to illustrate how land use change can alter these services, with potentially adverse health outcomes for humans. In a second case study, we discuss how some of the adverse effects of loss of contact with nature in urban settings can be reversed by the provision of urban green spaces and their associated ecosystem services. To ensure that our future land use takes account of both environmental (biodiversity) and human (health) needs requires multisectoral approaches that are ultimately dependent on partnerships. Natural resource management (NRM) is built on such partnerships, takes the diversity of needs of different sectors into account and builds resilience so that each system can better cope with dramatic perturbations such as climate change.

6.1 Introduction

Healthy populations are directly reliant on healthy ecosystems. For the purposes of discussing human health outcomes in this chapter, we define a healthy ecosystem as one capable of providing adequate ecosystem services to sustain human health. Ecosystem services refer to the functions of a healthy ecosystem that provide benefits to human wellbeing (MEA, 2003) – an anthropocentric perspective, but one that facilitates the allocation of resources to land management (e.g. conservation and restoration) and human health concurrently. The Millennium Ecosystem Assessment (MEA, also known as the Millennium Assessment (MA)) divides these services into four basic but not necessarily mutually exclusive categories: provisioning, regulating, cultural and supporting (Table 6.1). Supporting services do not directly benefit humans, but are required for the other three services to occur. This conceptual framework places the relationship between ecosystem services and human well-being (including health) in a broader context that includes drivers that are both direct (e.g. food availability) and indirect (e.g. economic and socio-political factors, climate change, land use), and considers different scales that may be temporal (e.g. short and long term) and/or spatial (e.g. local, regional and global) (MEA, 2003).

This approach has been criticized as not taking sufficient account of local community perspectives, but remains the most useful framework for establishing the mechanistic linkage mechanisms that can be targeted by restoration ecologists aiming to improve human health (Weinstein, 2005; Speldewinde *et al.*, 2014). It is ecosystem services that provide the critical link

 Climate Change and Global Health: Primary, Secondary and Tertiary Effects, 2nd Edition (eds C.D. Butler and K. Higgs)
DOI: 10.1079/9781800620025.0006

Table 6.1. Examples of ecosystem services. Based on the examples provided by the Millennium (Ecosystem) Assessment (MEA, 2003).

Supporting	Provisioning	Regulating	Cultural
Nutrient cycling	Food	Climate	Spiritual values
Primary production	Fresh water	Air quality	Religious values
Soil formation	Fibre	Water	Recreation
	Genetic resources	Diseases	Ecotourism
	Biochemicals	Pests	Education
	Natural medicines	Natural hazards	Aesthetic
	Pharmaceuticals	Erosion	

between understanding environmental change (including both climate change and land use change) and human health – two traditionally separate disciplinary areas that have only recently been pushed closer together in the face of otherwise intractable global problems. Foremost among these intersectoral problems are: mortality from 'water-washed' diseases (e.g. trachoma and diarrhoeal diseases) and crop failures when land use and climate change create desertification; starvation from depletion of fish stocks; and biodiversity loss, with disruption of food webs, ecosystems and ecosystem services. Both climate change and land use change are driven by increased population pressure – an ever-increasing population size as well as growing per capita consumption; the global population has doubled over the past half century, from just under 4 billion people to almost 8 billion (UN, 2019). Despite some uncertainty about the future of this trajectory, there is no question that population pressure and overutilization of resources are already pushing biodiversity loss and climate change close to the point of no return (Whitmee *et al.*, 2015).

To support these larger populations and increased consumption, land use change occurs, not only for housing and transport corridors but also to increase provisioning ecosystem services via agriculture. Agriculture has facilitated the growth of large populations but requires significant landscape change and forest clearing. Similarly, water provision to support larger populations may require manipulation of waterways, e.g. reservoirs. Both are changes to land use that allow for exploitation of provisioning services but at the cost of biodiversity and worsening climate change. Climate change mitigation cannot be entertained without considering this other elephant in the room: land use change (Weinstein and Daszak, 2020).

In this chapter, we explore in more detail the three major types of ecosystem services that demonstrably link land use change to human health outcomes: provisioning, regulating and cultural services. We use two case studies to illustrate how land use change may alter these services, and thus human wellbeing (including health). The first case study examines the conversion of natural ecosystems into agroecosystems; and the second examines the development of biodiverse urban green spaces. We finish by discussing the possible cross-disciplinary interventions that may prevent and/or reverse the adverse effects of inappropriate landuse – interventions that would ideally target the health of the environment and human health concurrently.

6.2 Case 1: Conversion of Natural Ecosystems to Agroecosystems

To support increasing human populations, agriculture is required for the provision of food. This example illustrates how the exploitation of an ecosystem to increase a specific provisioning service can adversely affect ecosystem health, and thereby other ecosystem services, as well as human health. The conversion of natural ecosystems into agroecosystems is an important example of the trade-offs of different ecosystem services that occur with land use change.

Provisioning ecosystem services

Provisioning services are those where healthy ecosystems provide resources crucial to human existence, including the provision of water, food, fibre and fuel. In the case of converting natural ecosystems to agroecosystems, food production – an agroecosystem service – is increased, but at the expense of other provisioning services. For example, agriculture changes water consumption and distribution patterns in both time and space, thereby potentially reducing water availability in other places and/or at other times. Such changes in water availability may lead to potential reductions in food production capacity elsewhere (entailing possible conflict and famine) and also reduce the biodiversity upon which ecosystem function and the sustainability of ecosystem services depend. Trees that have not already been felled to make way for agriculture may not survive changes in local hydrology, and their loss results in an undersupply of construction materials and fuel (both of which are used by humans to protect against climate extremes). The conversion of natural ecosystems to agroecosystems therefore provides an increase in the provisioning service of food but at the expense of the provision of other services like water, fuel and fibre.

Regulating ecosystem services

Regulating services are those that buffer an ecosystem against change, and provide resilience to change in a system that might otherwise destabilize. The classic example of a regulating service is that of climate control, and most graduates of sciences will be aware of the critical role that the Amazon forest plays in stabilizing regional and global climate and producing oxygen; cutting down this rainforest exacerbates the effects of climate change, including biodiversity loss and the erosion of ecosystem services associated therewith (Peng *et al.*, 2020). Thus, a healthy Amazon buffers against climatic excursions that might otherwise produce weather extremes with both direct (e.g. heat stroke) and indirect (e.g. crop failure) consequences for human health. Conversion of important natural ecosystems, like the Amazon, into agroecosystems may therefore have an impact on the regulation of air quality and climate.

Another good example is natural hazard regulation, afforded by wetlands and mangroves in coastal areas; such ecosystems buffer against storm surges and tidal waves, again providing protection against both direct (e.g. drowning) and indirect (e.g. saline incursion spoiling potable water) threats to human health. Recent work in Australia showed that such coastal wetlands saved the economy from $27 billion dollars in storm damage over the past 50 years (Mulder *et al.*, 2020). Despite agroecosystems increasing food provisioning services, this land use change comes at the potentially significant cost of natural hazard regulation.

Perhaps one of the highest-profile regulating services – writing in a time of the coronavirus disease 2019 (COVID-19) pandemic – is that of the regulation of emerging infectious diseases (EIDs). EIDs are those that are caused either by previously unknown pathogens (e.g. human immunodeficiency virus), or increases in distribution or severity (e.g. dengue virus), an increased number of cases (e.g. tuberculosis), or some combination of the above, for example severe acute respiratory syndrome 2 (SARS-CoV-2), the virus that causes COVID-19. Many EIDS have only human-to-human transmission and are therefore driven largely by sociopolitical factors. However, approximately 60% are zoonotic (Jones *et al.*, 2008), and these are driven predominantly by the ecological factors that influence both the animal reservoirs of the pathogens and the arthropod vectors that often transmit them. For example, vector-borne EIDs have lost the biodiversity-dependent regulating mechanisms of competition between and predation upon vectors, and dilution of the pathogens that they carry. 'Competition' and 'predation' are fairly straightforward concepts that limit mosquito populations by providing competition for the limited food resources available in the freshwater bodies in which mosquito larvae breed, and reducing the number of both larval and adult mosquitoes by increasing predation thereupon (O'Sullivan *et al.*, 2008). In healthy, biodiverse ecosystems, there is a greater variety and a higher abundance of competitors and predators to provide these regulating services (O'Sullivan *et al.*, 2008). 'Dilution' is a more complex concept, best illustrated by the emergence of Lyme disease in the USA. In ecosystems along the East Coast with high species richness and evenness, people being bitten by ticks are less likely to get infected by the causative pathogen *Borrelia burgdorferi* (Ostfeld and Keesing, 2000). The reason is that the mammals that ticks bite differ in their capacity as pathogen reservoirs, and if there are many poor resevoirs available for ticks to feed on, ticks in those ecosystems are less likely, on average, to carry the pathogen. Thus, people living in close proximity to tick-infested areas, as well as city-based hikers, tourists and forest workers entering those areas, are at lower risk of Lyme disease in healthy, biodiverse ecosystems where regulating services are intact. As agroecosystems typically have very low biodiversity, the disease risk is typically higher than it would be in natural ecosystems. The combination of ecological drivers with the globalization of travel and trade account for approximately two-thirds of EID problems (Semenza *et al.*, 2016). Surprisingly then, very little attention has been paid to the possibility of reducing the global disease burden from EID by investing in the conservation and restoration of healthy ecosystems that provide the regulating services that buffer against EIDs and reduce the probability of epidemics. It would appear that the drive to support larger population sizes has led to an increased need for agroecosystems to deliver the food provisioning service, but with impacts on the conservation of natural ecosystems, resulting in the loss of ecosystem services that include pathogen regulation.

Cultural ecosystem services

Cultural ecosystem services are those that enhance spirituality, provide a sense of place, or offer opportunities for recreation and education. People gain a sense of well-being and self-worth from these ecosystem services. From visiting sites of traditional cultural significance in a natural environment, to the social capital gained by sharing a hike in the wilderness, human quality of life remains inextricably linked to the natural environment. With any change in land use from conservation to agriculture (or to urban development, as discussed below), cultural ecosystem services are lost. Such loss puts us at risk of developing a type of nature deprivation syndrome – solastalgia (Albrecht, 2019), with associated risks of deteriorating mental health. This phenomenon has also been highlighted by the COVID-19 pandemic, where confinement to a non-natural environment has perhaps increased our awareness of how important cultural ecosystem services are to well-being generally and to mental health in particular (Stanhope and Weinstein, 2022). With the increasing conversion of natural ecosystems to agroecosystems, cultural ecosystem services are lost.

Summary

Natural ecosystems provide services that support the health and well-being of all human populations, in the form of provisioning services, regulating services and cultural services. When land use is altered to increase food production, for example, other ecosystem services can be negatively affected and adverse health outcomes can result at a population level.

6.3 Case 2: Urban Green Spaces

Urbanization has led to significant change in land use, often from natural to agricultural and now urban landscapes. With this change in land use, humans have experienced a decrease in their exposure to nature, with resulting health (Kondo *et al.*, 2018; Twohig-Bennett and Jones, 2018) and social consequences (Bogar and Beyer, 2016). For example, evidence suggests that less residential exposure to green spaces is associated with a range of adverse health outcomes, such as increased salivary cortisol, heart rate, blood pressure, diabetes, cancer, respiratory symptoms and disease, kidney disease, cancer, and cardiovascular and all-cause mortality, as well as poorer mood (Kondo *et al.*, 2018; Twohig-Bennett and Jones, 2018). Furthermore, living in urban areas with less residential green space exposure has also been associated with several other negative outcomes, including violence, and poorer attention (Kondo *et al.*, 2018). The development of urban green spaces could reverse some of these adverse effects, and may also provide an opportunity to reduce health inequity in urban areas. For example, a recent systematic review revealed that improvements in health outcomes associated with urban green space exposure were greatest for the children of mothers with lower levels of education and socioeconomic status (Islam *et al.*, 2020). The mechanisms linking exposure to green space and human health are not currently well understood, but may include exposure to microbes, biogenic volatile organic compounds (e.g. phytoncides), negative air ions, the sights and sounds of nature, reduced heat, reduced air pollution (Kuo, 2015) and increased physical activity (Keskinen, 2018) and social integration (Jennings and Bamkole, 2019). In light of this evidence, governments are now beginning to increase the number, size and quality of urban green spaces. As another type of land use change, here we explore how the provision of urban greenspace may enhance provisioning, regulating and cultural ecosystem services.

Provisioning ecosystem services

Urban green spaces may produce several provisioning services, depending on the particular features of the green space. For example, green spaces that feature community gardens and beehives may provide food. Wetlands within urban green spaces may also provide a water source, as the water is being held back. These are important provisioning services that have long been associated with urban green spaces, but emerging evidence suggests there are additional, novel provisioning services that have not previously been considered.

Different environmental microbiota (communities of microorganisms) are associated with different land use types and vegetation diversity. For example, a recent study identified that soil microbiome diversity increased with a restoration gradient (Liddicoat *et al.*, 2019), while vegetation type in urban areas is associated with differences in the diversity of the aerobiome (i.e. the microbiome of the air) (Mhuireach *et al.*, 2020). We are now beginning to understand the potential role that exposure to environmental microbiota may play in the development and maintenance of animal (including human) microbiota (Grönroos *et al.*, 2019; Liddicoat *et al.*, 2020; Stanhope *et al.*, 2022a). The gut microbiome in particular has received a lot of recent attention for its association with a wide range of health outcomes, including autism spectrum disorder, Alzheimer's disease, stroke, Parkinson's disease, multiple sclerosis (Cryan *et al.*, 2020), pain (Guo *et al.*, 2019), ulcerative colitis (Imdad *et al.*, 2018), and depression and anxiety (Simpson *et al.*, 2021). Urban green spaces might therefore provide humans with microbiota that can benefit their health – a new type of provisioning service.

With our increasing understanding of the role of urban green spaces in health, exposure to biogenic volatile organic compounds and negative air ions may also be added to the list of potential provisioning services. Biogenic volatile organic compounds are produced by plants and microbes, and may influence human health both directly and indirectly (via the environmental microbiota) (Stanhope *et al.*, 2020), thus reflecting another provisioning service. Similarly, negative air ions are produced by energy sources, including plants, shear forces of water, radiant or cosmic rays in the atmosphere, and sunlight (Jiang *et al.*, 2018). There is emerging evidence that their production may influence human health (Jiang *et al.*, 2018) and, if so, could also be considered a provisioning ecosystem service from urban green spaces.

Regulating ecosystem services

The introduction and optimization of urban green spaces may also enhance regulating ecosystem services. For example, vegetation improves air quality by removing carbon dioxide from the atmosphere, collecting pollutants, and by altering air flow to keep air pollution away from populated areas (Kuo, 2015; Kumar *et al.*, 2019). Similarly, green spaces can regulate water quality; for example, riparian vegetation around urban wetlands or waterways will improve water quality through natural filtration (Riis *et al.*, 2020). Diverse microbiota, as seen in well-designed green spaces, can also regulate pathogens through competition between microorganisms (Liddicoat *et al.*, 2018).

Importantly, urban green spaces may also mitigate the effects of climate change. The urban heat island effect can be reduced through the provision of green spaces (Kuo, 2015); specifically, trees planted over grass (Rahman *et al.*, 2020). Urban green spaces may also play a (small) role in reducing climate change through increased carbon sequestration in woody vegetation. All of these examples reflect important regulating ecosystem services, associated with urban green spaces.

Cultural ecosystem services

Cultural ecosystem services are perhaps the main reason green spaces have been integrated into urban areas over the past 150 years or so. Green spaces provide a space for socializing and exercising, as well as education and recreation, including the aforementioned community gardens and beekeeping. Furthermore, increasingly, urban green spaces have also been a site for sharing culture; through ecotourism, education boards, artwork, performances and outdoor cinemas. The inclusion of well-designed urban green spaces may enhance cultural ecosystem services, and therefore human wellbeing.

Summary

Urban land use change may include the development of urban green spaces, which deliver provisioning, regulating and cultural ecosystem services. Through these ecosystem services, green space benefits the well-being of communities and their presence constitutes an important part of public health infrastructure.

6.4 Partnerships for a Healthier Future

Clearly, land use change can impact biodiversity, ecosystem services and human health. If we are to halt and reverse the loss of ecosystem services that are critical to population health, intervention is required that targets population growth, conservation and restoration concurrently. A plethora of models, frameworks and recommendations for such interventions abounds, with the best major synthesis perhaps represented by the MEA (2003), and the most recent comprehensive plan put forward in the United Nations Sustainable Development Goals (UN, 2015). Unfortunately, all such plans are dependent on socio-political and economic drivers that, in some parts of the world, have changed very little in the past few decades. The recent COVID-19 pandemic has been presented, in this context, as a timely reminder that we need to put more time and energy into looking after and restoring nature, or she will bite back (Dobson *et al.*, 2020; Lawler *et al.*, 2021; Reaser *et al.*, 2021; Robinson *et al.*, 2022) – in this case in the form of a pandemic resulting from a perfect storm of ecosystem disruptions, huge human population concentrations and massive and rapid networks of travel and trade.

Decision making regarding land use change and the impact this change has on different ecosystem services is almost always a matter of optimization. The conversion of natural ecosystems into agroecosystems, as described in Case 1, provides a good example of trade-offs. Agroecosystems are required to support the food needs of increasing populations, hence we exploit this provisioning service; however, this comes at the cost of other ecosystem services. Land use change therefore has to be carefully considered. Even when introducing green spaces into urban areas, there may be trade-offs in terms of what communities themselves deem fit for purpose, and broad stakeholder consultation is required to ensure that such spaces are developed optimally, considering both environmental and human health needs. New modelling approaches are being developed to assist in the necessarily complex decision-making process required (Stanhope *et al.*, 2022b).

As discussed above, it is clear we rely on the ecosystem services that depend on the biodiversity of the land, soil, plants and animals that make up our landscape. In Australia, for example, we derive a significant proportion of our economic wealth from environmental assets, including through agriculture, mining and tourism. However, environmental assets and ecosystem services are very expensive to maintain and often impossible to replace. In addition, our natural resources are not finite but are both limited and susceptible to damage. Such limitation and damage can occur over short, medium or even very long periods of time. We call our environmental resources 'natural resources' and their management 'natural resource management' (NRM). NRM is required to ensure sustainable use and also to provide resilience so that each system can cope with dramatic change (such as climate change). NRM is, therefore, the integrated management of the land, water, soil, plants and animals, and how people use these resources for the betterment of themselves and their communities (Faull *et al.*, 2021). Good management focuses not only on the environment, but on how that management affects the quality of life for both future and present generations.

Natural resources also play a leading role in the education, recreation and health of both local residents and visitors. As we discuss in this chapter, human well-being is dependent on the availability of accessible and healthy environments that provide opportunities to spend time in contact with nature.

We now understand and accept that to best protect and manage our natural resources in a way that optimizes the delivery of ecosystem services as discussed in this chapter, we require strong and diverse partnerships. Partnerships represent the most effective and efficient mechanism to connect people with nature, the land and ecosystems. Many different forms of partnerships are now a growing and vibrant part of our culture. NRM is conducted through partnerships with local communities and local, state and federal governments. In Australia, the past few years have seen the emergence of complex multi-partnered organizations (involving natural resource management boards) such as Nature Play SA, Healthy Parks–Healthy People, Education for Sustainability and a range of other organizations and movements designed to engage people with biodiversity in our environment. These partnerships deliver demonstrably positive outcomes for the health and well-being of individuals, communities and populations – and the environment. Most recently, we have seen growing recognition of the role and importance of environmentally aware organizations (many of them volunteer) that also respond to environmental disasters to mitigate their effects on nature and ourselves. Fire-fighters, fauna rescue groups, wildlife parks and even individuals acting for the collective good delivered a huge amount of the recovery effort in response to natural disasters such as the terrible Australian bush fires of January 2020. In future we will see NRM as the mechanism to deliver integrated health and ecosystems outcomes. However, as NRM grows and develops it will continue to face complex organizational and socio-political challenges that will only be met through supporting cross-sectoral partnerships at all levels. Supporting such partnerships will be one of society's greatest challenges – and ultimately its major achievement – in combating both land use challenges and climate change.

6.5 Conclusion

Human health is reliant on healthy ecosystems. With increasing populations and changes in the way we live and consume goods, the ensuing land use changes are of a scale and intensity that is not sustainable. Because land use change generally results in biodiversity loss, it is most often also associated with a loss of ecosystem services and potentially therefore human health. We can, however, use land use change as a means of improving human health as well, for instance through the development and optimization of urban green spaces. Decisions regarding future land use must take into account both environmental (biodiversity) and human (health) needs, necessitating multi-sectoral approaches that are ultimately dependent on strong partnerships and collaborations. Natural resource management is built on such relationships, takes the diversity of needs of different sectors into account, and builds resilience so each system can better cope with dramatic perturbations such as climate change. By better managing land use, we can benefit biodiversity conservation and human health concurrently.

References

Albrecht, G.A. (2019) *Solastalgia. The Homesickness You Have at Home. Earth Emotions*. Cornell University Press, Ithaca, New York, pp. 27–62.

Bogar, S. and Beyer, K.M. (2016) Green space, violence, and crime: a systematic review. *Trauma Violence Abuse* 17, 160–171. DOI: 10.1177/1524838015576412

Cryan, J., O'Riordan, K., Sandhu, K., Peterson, V. and Dinan, T. (2020) The gut microbiome in neurological disorders. *The Lancet Neurology* 19, 179–194. DOI: 10.1016/S1474-4422(19)30356-4

Dobson, A.P., Pimm, S.L., Hannah, L., Kaufman, L., Ahumada, J.A. *et al.* (2020) Ecology and economics for pandemic prevention. *Science* 369, 379–381. DOI: 10.1126/science.abc3189

Faull, K., Sharrad, K., Daniels, C., Simpson, B. and Agius, P. (2021) *Our Country: Managing and Protecting Natural Resources*. Axiom Press, Adelaide, Australia.

Grönroos, M., Parajuli, A., Laitinen, O.H., Roslund, M.I., Vari, H.K. *et al.* (2019) Short-term direct contact with soil and plant materials leads to an immediate increase in diversity of skin microbiota. *Microbiology Open* 8, e00645. DOI: 10.1002/mbo3.645

Guo, R., Chen, L.-H., Xing, C. and Liu, T. (2019) Pain regulation by gut microbiota: molecular mechanisms and therapeutic potential. *British Journal of Anaesthesia* 123, 637–654. DOI: 10.1016/j.bja.2019.07.026

Imdad, A., Nicholson, M., Tanner-Smith, E., Zackular, J., Gomez-Duarte, O. *et al.* (2018) Fecal transplantation for treatment of inflammatory bowel disease. *Cochrane Database of Systematic Reviews* 11, CD012774. DOI: 10.1002/14651858.CD012774.pub2

Islam, M.Z., Johnston, J. and Sly, P.D. (2020) Green space and early childhood development: a systematic review. *Reviews on Environmental Health* 35, 189–200. DOI: 10.1515/reveh-2019-0046

Jennings, V. and Bamkole, O. (2019) The relationship between social cohesion and urban green space: an avenue for health promotion. *International Journal of Environmental Research and Public Health* 16, 452. DOI: 10.3390/ijerph16030452

Jiang, S., Ma, A. and Ramachandran, S. (2018) Negative air ions and their effects on human health and air quality improvement. *International Journal of Molecular Sciences* 19, 2966. DOI: 10.3390/ijms19102966

Jones, K.E., Patel, N.G., Levy, M.A., Storeygard, A., Balk, D. *et al.* (2008) Global trends in emerging infectious diseases. *Nature* 451, 990–993. DOI: 10.1038/nature06536

Keskinen, K. (2018) Nature as a facilitator for physical activity: defining relationships between the objective and perceived environment and physical activity among community-dwelling older people. *Health Place* 49, 111–119. DOI: 10.1016/j.healthplace.2017.12.003

Kondo, M.C., Fluehr, J.M., McKeon, T. and Branas, C.C. (2018) Urban green space and its impact on human health. *International Journal of Environmental Research and Public Health* 15, 445. DOI: 10.3390/ijerph15030445

Kumar, P., Druckman, A., Gallagher, J., Gatersleben, B., Allison, S. *et al.* (2019) The nexus between air pollution, green infrastructure and human health. *Environment International* 133, 105181. DOI: 10.1016/j.envint.2019.105181

Kuo, M. (2015) How might contact with nature promote human health? Promising mechanisms and a possible central pathway. *Frontiers in Psychology* 6, 1093. DOI: 10.3389/fpsyg.2015.01093

Lawler, O.K., Allan, H.L., Baxter, P.W.J., Castagnino, R., Corella Tor, M. *et al.* (2021) The COVID-19 pandemic is intricately linked to biodiversity loss and ecosystem health. *The Lancet Planetary Health* 5, 840–850. DOI: 10.1016/S2542-5196(21)00258-8

Liddicoat, C., Bi, P., Waycott, M., Glover, J., Breed, M. *et al.* (2018) Ambient soil cation exchange capacity inversely associates with infectious and parasitic disease risk in regional Australia. *Science of the Total Environment* 626, 117–125. DOI: 10.1016/j.scitotenv.2018.01.077

Liddicoat, C., Weinstein, P., Bissett, A., Gellie, N.J.C., Mills, J.G. *et al.* (2019) Can bacterial indicators of a grassy woodland restoration inform ecosystem assessment and microbiota-mediated human health? *Environment International* 129, 105–117. DOI: 10.1016/j.envint.2019.05.011

Liddicoat, C., Sydnor, H., Cando-Dumancela, C., Dresken, R., Liu, J. *et al.* (2020) Naturally-diverse airborne environmental microbial exposures modulate the gut microbiome and may provide anxiolytic benefits in mice. *Science of the Total Environment* 701, 134684. DOI: 10.1016/j.scitotenv.2019.134684

MEA (2003) *Ecosystems and Human Well-being. A Framework for Assessment*. Millennium Ecosystem Assessment: conceptual framework report. Island Press, Washington, DC.

Mhuireach, G., Wilson, H. and Johnson, B. (2020) Urban aerobiomes are influences by season, vegetation, and individual site characteristics. *EcoHealth* 18, 331–344. DOI: 10.1007/s10393-020-01493-w

Mulder, O.J., Mulder, K.P., Kubiszewski, I., Anderson, S.J., Costanza, R. *et al.* (2020) The value of coastal wetlands for storm protection in Australia. *Ecosystem Services* 46, 101205. DOI: 10.1016/j.ecoser.2020.101205

O'Sullivan, L., Jardine, A., Cook, A. and Weinstein, P. (2008) Deforestation, mosquitoes, and ancient Rome: lessons for today. *BioScience* 58, 756–760. DOI: 10.1641/B580812

Ostfeld, R.S. and Keesing, F. (2000) Biodiversity and disease risk: the case of Lyme disease. *Conservation Biology* 14, 722–728. DOI: 10.1046/j.1523-1739.2000.99014.x

Peng, W., Sonne, C., Lam, S.S., Ok, Y.S. and Alstrup, A.K.O. (2020) The ongoing cut-down of the Amazon rainforest threatens the climate and requires global tree planting projects: A short review. *Environmental Research* 181, 108887. DOI: 10.1016/j.envres.2019.108887

Rahman, M.A., Stratopoulos, L.M.F., Moser-Reischl, A., Zolch, T., Haberle, K.-H. *et al.* (2020) Traits of trees for cooling urban heat islands: a meta-analysis. *Building and Environment* 170, 106606. DOI: 10.1016/j.buildenv.2019.106606

Reaser, J.K., Witt, A., Tabor, G.M., Hudson, P.J. and Plowright, R.K. (2021) Ecological countermeasures for preventing zoonotic disease outbreaks: when ecological restoration is a human health imperative. *Restoration Ecology* 29, 13357. DOI: 10.1111/rec.13357

Riis, T., Kelly-Quinn, M., Aguiar, F.C., Manolaki, P., Bruno, D. *et al.* (2020) Global overview of ecosystem services provided by riparian vegetation. *BioScience* 70, 501–514. DOI: 10.1093/biosci/biaa041

Robinson, J.M., Aronson, J., Daniels, C.B., Goodwin, N., Liddicoat, C. *et al.* (2022) Ecosystem restoration is integral to humanity's recovery from COVID-19. *Lancet Planetary Health* 6, e769–e773. DOI: 10.1016/S2542-5196(22)00171-1

Semenza, J.C., Lindgren, E., Balkanyi, L., Espinosa, L., Almqvist, M.S. *et al.* (2016) Determinants and drivers of infectious disease threat events in Europe. *Emerging Infectious Diseases* 22, 581. DOI: 10.3201/eid2204

Simpson, C., Diaz-Arteche, C., Eliby, D., Schwartz, O., Simmons, J. *et al.* (2021) The gut microbiota in anxiety and depression – a systematic review. *Clinical Psychology Review* 83, 101943. DOI: 10.1016/j.cpr.2020.101943

Speldewinde, P., Slaney, D. and Weinstein, P. (2014) Is restoring an ecosystem good for your health? *The Science of the Total Environment* 502, 276–279. DOI: 10.1016/j.scitotenv.2014.09.028

Stanhope, J. and Weinstein, P. (2022) Public health lessons from the COVID-19 pandemic: the importance of green spaces for vulnerable populations. *Perspectives in Public Health* 142, 145–146. DOI: 10.1177/17579139211057594

Stanhope, J., Breed, M. and Weinstein, P. (2020) Exposure to greenspaces could reduce the high global burden of pain. *Environmental Research* 187, 109641. DOI: 10.1016/j.envres.2020.109641

Stanhope, J., Breed, M. and Weinstein, P. (2022a) Biodiversity, microbiomes, and human health. In: Rook, G. and Lowry, C. (eds) *Evolution, Biodiversity and a Reassessment of the Hygiene Hypothesis*. Progress in Inflammation Research, vol 89. Springer, Cham, Swizerland, pp. 67–104. DOI: 10.1007/978-3-030-91051-8_3

Stanhope, J., Mayfield, H.J., Guillaume, J.H.A., Sahin, O., Weinstein, P. *et al.* (2022b) Synergising decision making and interventions across human health and environment: concepts for designing a model for infectious diseases. *Socio-Environmental Systems Modelling* 3, 18126. DOI: 10.18174/sesmo.18126

Twohig-Bennett, C. and Jones, A. (2018) The health benefits of the great outdoors: A systematic review and meta-analysis of greenspace exposure and health outcomes. *Environmental Research* 166, 628-637. DOI: 10.1016/j.envres.2018.06.030

UN (2015) *Transforming Our World: The 2030 Agenda for Sustainable Development*. United Nations, Rome. Available at: https://sdgs.un.org/sites/default/files/publications/21252030%20Agenda%20for%20Sustainable%20Development%20web.pdf (accessed 20 July, 2022)

UN (2018) *The World's Cities in 2018*. United Nations, Rome. Available at: https://www.un.org/development/desa/pd/sites/www.un.org.development.desa.pd/files/files/documents/2020/Jan/un_2018_worldcities_databooklet.pdf (accessed 20 July 2022).

UN (2019) *World Population Prospects 2019: Highlights*. Available at: https://population.un.org/wpp/publications/files/wpp2019_highlights.pdf (accessed 22 July 2022).

Weinstein, P. (2005) Human health is harmed by ecosystem degradation, but does intervention improve it? A research challenge from the Millenium Ecosystem Assessment. *EcoHealth* 2, 228–230. DOI: 10.1007/s10393-005-6329-3

Weinstein, P. and Daszak, P. (2020) Failing efforts to mitigate climate change are a futile band-aid that will not stop other elephants filling the room. *EcoHealth* 17, 421–423. DOI: 10.1007/s10393-020-01512-w

Whitmee, S., Haines, A., Beyrer, C., Boltz, F., Capon, A.G. *et al.* (2015) Safeguarding human health in the Anthropocene epoch: report of The Rockefeller Foundation–Lancet Commission on planetary health. *The Lancet* 386, 1973–2028. DOI: 10.1016/S0140-6736(15)60901-1

6.i The Biodiversity Hypothesis for Health Emerged from a Natural Experiment in the Finnish and Russian Karelia

Tari Haahtela
Helsinki University Hospital, University of Helsinki, Finland

Abstract
In Finland, the 'allergy epidemic' became visible in the 1970s. But was the increasing prevalence a result of improved awareness and diagnostics rather than a true change? Comparing Finnish and Russian Karelia in the 2000s gave an indisputable answer: the increase is true and apparent in all urbanizing societies. The allergy epidemic was mainly caused by changes in living environment and lifestyle affecting the richness and diversity of the human microbiome, the main driver of the immune regulative circuits. Disconnection from natural environments causes a loss of internal nature, as well as externally. Non-communicable, chronic diseases of modern times seem to share common features: microbial imbalance, immune dysfunction and low-grade inflammation. The biodiversity hypothesis originates from these observations in Karelia. The nationwide Finnish Allergy Programme (2008–2018), which consciously increases human contact with nature, has attempted to improve health and immune tolerance; results are favourable.

6.i.1 Introduction

Allergies are rare where butterflies flourish in a biodiverse environment – that was the title of an invited commentary published in *Allergy* (the official journal of the European Academy of Allergy and Clinical Immunology) in 2009 (Haahtela, 2009). This unconventional paper resulted from the Karelia Allergy Study, which started in the late 1990s in Finnish and Russian Karelia. These are adjacent areas in Northern Europe, similar geo-climatically, but strictly divided by the 'Iron Curtain' drawn after World War II. In more than 70 years of separation, inhabitants acquired different lifestyles, living standards and exposure to the natural environment. People on the Russian side continued their small-scale agricultural life, while urbanization and technological development accelerated in Finnish Karelia. These circumstances have generated a natural experiment, allowing comparison of health determinants and health outcomes of two formerly more or less similar populations (Haahtela *et al.*, 2015).

Allergic diseases are rare in the Russian Karelia

Allergic diseases and asthma are, today, far more common in the Finnish children and adults than in their Russian counterparts. The contrast is astonishing; according to skin prick tests, 5% of the Russian children are sensitized to grass pollen compared with 28% of the Finnish children (von Hertzen *et al.*, 2006). In Russian Karelia, symptoms of pollen and food allergy are rare; peanut allergy is non-existent. In adults, after

©2024 CAB International. *Climate Change and Global Health: Primary, Secondary and Tertiary Effects*, 2nd Edition (eds C.D. Butler and K. Higgs)
DOI: 10.1079/9781800620025.0006i

the War, one divergence was also striking. Finnish and Russian adults born in the 1940s share the same level of sensitization to common allergens, but Finns born in the 1980s have a sensitization rate far higher than their Russian Karelian cousins (Laatikainen *et al.*, 2011). What explains such a different development in a relatively short period of time?

A genome-wide association study showed that genotypical differences between the Finnish and Russian populations did not explain the allergy gap (Ruokolainen *et al.*, 2020). Whether or not subjects become sensitized or develop clinical allergic manifestations seems to depend less on genetic factors and more on environmental factors, especially through epigenetic regulatory mechanisms (Song *et al.*, 2017). Overall, the research of gene–environment interaction is still at an early stage.

Microbes and mites in everyday life

Attempts to explain this puzzle first explored air pollution and common environmental chemicals, but without success. What about the drinking water? Indeed, a marked microbial contrast was found between the tap water in Finnish and Russian schools. The Russian water, mostly from the surface of Lake Ladoga (only occasionally purified with chlorine), had an allergy-protective effect, when potential confounders were taken into account (von Hertzen *et al.*, 2007). Later, more or less the same result was obtained from the house dust comparisons: the microbe-rich Russian dust was associated with allergy protection (von Hertzen *et al.*, 2010).

It seemed that contact with the environmental microbes in everyday life (from drinking, eating, breathing and touch) was modulating immunological pathways and balancing responses against foreign proteins to which humans are exposed via pollens, animal dander and food. Interestingly, the response of immunoglobulin E (IgE) to house dust mites (*Dermaphagoides* sp.) did not differ between the populations. However, the mite-positive Russians were mostly sensitized to mites only, and free of allergic symptoms, while the Finns were polysensitized and with atopic symptoms (von Hertzen and Haahtela, 2009). Many mites were found in the Russian home dust; the Finnish dust was almost mite-free. Mites are ectodermal parasites; much of the biological function of IgE is to protect humans from parasite invasion. Indeed, the tropical storage[1] mite, *Blomia tropicalis*, even invades the skin.

Biodiverse macro- and micro-environment

With the help of ecologists, a survey of plant richness and land use was conducted around the home environment of the Finnish schoolchildren. The children's skin microbiome was analysed, together with indicators of immune function. The result was surprisingly straightforward. A diverse natural environment was associated with a rich skin microbiome, balanced innate immunity and less atopic (IgE) sensitization (Hanski *et al.*, 2012). Furthermore, comparison of the gene expression of blood mononuclear cells of Finnish and Russian children identified nearly 300 differentially expressed genes. This contrast paralleled the skin and nasal epithelial microbiota disparity and was linked to the functional pathways of innate immunity. Altogether, the cross-talk between the human and environmental microbiome seemed to guide human gene activation. This, in turn, regulated the immune system to better distinguish signals of danger or non-danger and self or non-self (Ruokolainen *et al.*, 2020).

The biodiversity hypothesis was born: contact with natural environments enriches the human microbiome, promotes immune balance and protects from allergy and inflammatory disorders (von Hertzen *et al.*, 2011; Haahtela, 2019) (Fig. 6.i.1).

6.i.2 Direct Evidence of the Biodiversity Effect

In 2020, another Finnish group found direct proof of the importance of biodiversity (Roslund *et al.*, 2020). In an intervention lasting 28 days, researchers manipulated urban day-care outdoor

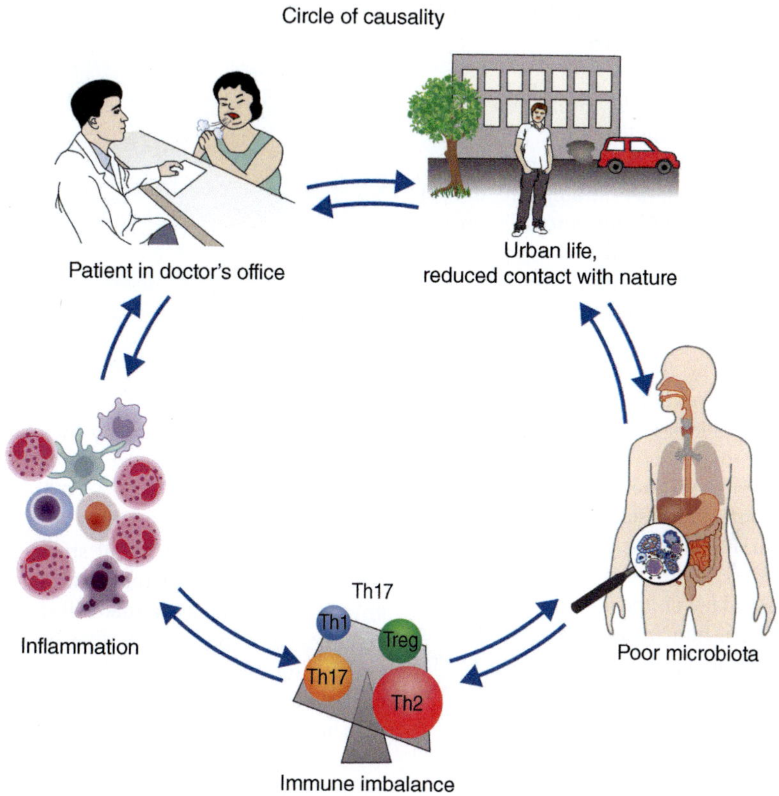

Fig. 6.i.1. The circle of suggested causality. Megatrends, including urbanization, increase non-communicable (inflammatory) diseases, including allergy (clockwise). Or, counterclockwise, from symptoms to background factors. Reprinted, with permission, from Haahtela (2019).

environments and analysed changes in the children's skin and gut microbiome as well as serum immune markers. Children in standard urban day care were included as controls. The intervention group showed greater diversity of environmental and skin bacterial communities, which, in turn, were associated with changes in serum immune regulatory markers. The study showed that favourable effects can be obtained with relatively short interventions, which is good news for disease prevention.

Biogenic chemicals in forest and green environments

The air in some forested environments contains biogenic chemicals, including volatile organic compounds (VOCs) such as limone, pinene and terpenes. These affect cellular autophagy, in which human cells remove unnecessary or dysfunctional material. Inhaled VOCs can be anti-inflammatory, antioxidant and anxiolytic (Antonelli *et al.*, 2020). Many studies indicate beneficial psychological and physiological effects from visiting forested and other natural environments. One Finnish study exposed volunteers to three different types of urban environments, finding that brief visits to 'green' environments were associated with favourable changes in pulse rate and blood pressure (Lanki *et al.*, 2017). Inhaling VOCs suggests a plausible 'pharmaceutical' effect, in addition to psychological pathways that may have also contributed to these favourable results. If so, this may be related to the Japanese practice of *Shinrin-yoku* or 'forest bathing' (Park *et al.*, 2009).

6.i.3 The Finnish Allergy Programme 2008–2018

These new ideas have also been tested in a 10-year Finnish nationwide public health programme, which resulted in a change of the traditional allergen avoidance strategy to one of tolerance/resilience, with an emphasis on nature relatedness. Revisiting the allergy paradigm and systematic education has mitigated the burden of allergic diseases in Finnish society (Haahtela *et al.*, 2017).

Encouraged by the results of the allergy programme, prevention of other chronic non-communicable diseases (e.g. diabetes, obesity and inflammatory bowel diseases) has been discussed in Finland. The contrast observed between Finnish and Russian Karelia for allergy has also been shown for type 1 diabetes (Kondrashova *et al.*, 2013).

Understanding of the interdependent relationship between humans and the natural environment is central to global and planetary health. Many insights and health gains from the Finnish experience are likely to apply at a larger scale. The biodiversity hypothesis has major and beneficial implications for society, including for city and mobility planning, food and energy production as well as for nature conservation (Halonen *et al.*, 2020).

Back to the butterfly effect

As part of the chaos theory, mathematician and meteorologist Edward Lorenz (1963) argued that 'slightly differing initial states can evolve into considerably different states'. Extending this, in 1972, he gave a talk called 'Predictability: does the flap of a butterfly's wings in Brazil set off a tornado in Texas?' The disappearance of butterflies indicates not only a loss of biodiversity, but also a loss of immuno-protective factors, a major determinant of human health, survival and well-being. Let us take care of the flying flowers, the silent pollinators (Fig. 6.i 2).

Fig. 6.i.2. A young tiger butterfly (*Ideopsis juventa*) flaps its wings; Bali, Indonesia, January 2001. Photographer: Tari Haahtela (used with permission).

Acknowledgements

Special thanks to my Australian collaborators and friends Peter LeSouëf, Guicheng Zhang and Susan Prescott.

Note

[1] There are two basic categories of mites: (i) house dust mites; and (ii) 'storage' mites, which feed on substances such as grain, flour, hay and straw and are also found in house dust.

References

Antonelli, M., Donelli, D., Barbieri, G., Valussi, M., Maggini, V. *et al.* (2020) Forest volatile organic compounds and their effects on human health: a state-of-the-art review. *International Journal of Environmental Research and Public Health* 17, 6506. DOI: 10.3390/ijerph17186506

Haahtela, T. (2009) Allergy is rare where butterflies flourish in a biodiverse environment. *Allergy* 64, 1799–1803. DOI: 10.1111/j.1398-9995.2009.02246.x

Haahtela, T. (2019) A biodiversity hypothesis. *Allergy* 74, 1445–1456. DOI: 10.1111/all.13763

Haahtela, T., Laatikainen, T., Alenius, H., Auvinen, P., Fyhrquist, N. *et al.* (2015) Hunt for the origin of allergy – comparing the Finnish and Russian Karelia. *Clinical & Experimental Allergy* 45, 891–901. DOI: 10.1111/cea.12527

Haahtela, T., Valovirta, E., Bousquet, J. and Mäkelä, M. (2017) The Finnish Allergy Programme 2008–2018 works. *European Respiratory Journal* 49, 1700470. DOI: 10.1183/13993003.00470-2017

Halonen, J.I., Erhola, M., Furman, E., Haahtela, T., Jousilahti, P. *et al.* (2020) The Helsinki Declaration 2020: Europe that protects. *The Lancet Planetary Health* 4, e503–e505. DOI: 10.1016/S2542-5196(20)30242-4

Hanski, I., von Hertzen, L., Fyhrquist, N., Koskinen, K., Torppa, K. *et al.* (2012) Environmental biodiversity, human microbiota, and allergy are interrelated. *Proceedings of the National Academy of Sciences (USA)* 109, 8334–8339. DOI: 10.1073/pnas.1205624109

Kondrashova, A., Seiskari, T., Ilonen, J., Knip, M. and Hyöty, H. (2013) The 'Hygiene hypothesis' and the sharp gradient in the incidence of autoimmune and allergic diseases between Russian Karelia and Finland. *Apmis* 121, 478–493. DOI: 10.1111/apm.12023

Laatikainen, T., von Hertzen, L., Koskinen, J.-P., Mäkelä, M.J., Jousilahti, P. *et al.* (2011) Allergy gap between Finnish and Russian Karelia on increase. *Allergy* 66, 886–892. DOI: 10.1111/j.1398-9995.2010.02533.x

Lanki, T., Siponen, T., Ojala, A., Korpela, K., Pennanen, A. *et al.* (2017) Acute effects of visits to urban green environments on cardiovascular physiology in women: A field experiment. *Environmental Research* 159, 176–185. DOI: 10.1016/j.envres.2017.07.039

Lorenz, E.N. (1963) Deterministic nonperiodic flow. *Journal of Atmospheric Science* 20, 130–141. DOI: 10.1175/1520-0469(1963)020<0130:DNF>2.0.CO;2

Park, B.J., Tsunetsugu, Y., Kasetani, T., Kagawa, T. and Miyazaki, Y. (2009) The physiological effects of Shinrin-yoku (taking in the forest atmosphere or forest bathing): evidence from field experiments in 24 forests across Japan. *Environmental Health and Preventive Medicine* 15, 18. DOI: 10.1007/s12199-009-0086-9

Roslund, M.I., Puhakka, R., Grönroos, M., Nurminen, N., Oikarinen, S. *et al.* (2020) Biodiversity intervention enhances immune regulation and health-associated commensal microbiota among daycare children. *Science Advances* 6, eaba2578. DOI: 10.1126/sciadv.aba2578

Ruokolainen, L., Fyhrquist, N., Laatikainen, T., Auvinen, P., Fortino, V. *et al.* (2020) Immune-microbiota interaction in Finnish and Russian Karelia young people with high and low allergy prevalence. *Clinical & Experimental Allergy* 50, 1148–1158. DOI: 10.1111/cea.13728

Song, Y., Khoo, S.-K., Lee, K.H., Mäkelä, M., Haahtela, T. *et al.* (2017) Dual responses of CD14 methylation to distinct environments: a role in asthma and allergy. *European Respiratory Journal*, 50, 1701228. DOI: 10.1183/13993003.01228-2017

von Hertzen, L. and Haahtela, T. (2009) Con: House dust mites in atopic diseases. *American Journal of Respiratory and Critical Care Medicine* 180, 113–119. DOI: 10.1164/rccm.200807-1001CO

von Hertzen, L., Mäkelä, M.J., Petäys, T., Jousilahti, P., Kosunen, T.U. *et al.* (2006) Growing disparities in atopy between the Finns and the Russians: A comparison of 2 generations. *Journal of Allergy and Clinical Immunology* 117, 151–157. DOI: 10.1016/j.jaci.2005.07.028

von Hertzen, L., Laatikainen, T., Pitkänen, T., Vlasoff, T., Mäkelä, M.J. *et al.* (2007) Microbial content of drinking water in Finnish and Russian Karelia – implications for atopy prevalence. *Allergy* 62, 288–292. DOI: 10.1111/j.1398-9995.2006.01281.x

von Hertzen, L., Hyvärinen, A., Laatikainen, T., Mäkelä, M., Nevalainen, A. *et al.* (2010) Risk of atopy associated with microbial components in house dust. *Annals of Allergy, Asthma & Immunology* 104, 269–270. DOI: 10.1016/j.anai.2010.01.008

von Hertzen, L., Hanski, I. and Haahtela, T. (2011) Natural immunity. Biodiversity loss and inflammatory diseases are two global megatrends that might be related. *EMBO Reports* 12, 1089–1093. DOI: 10.1038/embor.2011.195

7 Pandemics and Their Co-factors: a Short History

Colin D. Butler
National Centre for Epidemiology and Population Health, The Australian National University, Canberra, Australia

Abstract
This chapter reviews the history of pandemics from antiquity to the present. Recognized pandemics include plague, cholera, smallpox, influenza and, more recently, HIV/AIDS and COVID-19. Also discussed are the thousands of years of recognition, much of it vague, of the germ theory and the possibility that novel epidemics and even pandemics may emerge as the result of changes in the Anthropocene, including biodiversity loss, the farming of familiar and novel species, and emerging laboratory techniques.

7.1 Introduction: A Continuum Between Pandemics, Epidemics and Disease Outbreaks

The pandemic of COVID-19, starting in 2019, was caused by a newly identified beta coronavirus, whose distal origin is believed to have been a horseshoe bat. It has reinvigorated use of the word 'pandemic', and been a major trigger for the inclusion of this chapter.

The term pandemic was in occasional but ill-defined use between at least 1666 (Morens *et al.*, 2009) and 1918. In 1831, the creep of 'Asiatic' cholera towards Europe (it also reached Canada, the USA and Mexico) was carefully watched; its extensive mapping was later described as helping to launch the idea of global health (Koch, 2014). Indeed, geographical observation of the spread of disease seems a characteristic of many older uses of the term. However, the word 'pandemic' appears to have been used rarely (if at all) by contemporary observers in association with this extensive cholera epidemic. Today it is widely recognized that there have been seven cholera pandemics, with the first in 1817 (Lacey, 1995).

Smart (1876) defined 'pandemic' as more extensive than 'epidemic'. He described, as pandemic, a connected series of cases of dengue extending to the Eastern and Western hemispheres, from Calcutta [*sic*] to the USA. Smart provided detailed notes of the geographical spread of several such dengue pandemics. Clemow (1894) described the gradual progression of another pandemic disease, influenza (c. 1889–1892), using the word eight times in one paper, including in the title. However, he made no attempt to distinguish 'pandemic' from 'epidemic', a word he used over 20 times in the same article.

Starting near the end of the world's first 'total' war (1918) the world was beset by a truly devastating infectious disease, which killed as many as 100 million people, across almost all nations. This calamity, 'Spanish' influenza (described in more detail below), quashed the then growing confidence that the triumph of microbiology

 Climate Change and Global Health: Primary, Secondary and Tertiary Effects, 2nd Edition (eds C.D. Butler and K. Higgs)
DOI: 10.1079/9781800620025.0007

might see infectious diseases tamed (Tognotti, 2003). For this catastrophe, 'epidemic' seemed (and seems) too mild; as a consequence the word 'pandemic' became temporarily commonplace (Morens *et al.*, 2009), as it is, again, today.[1]

In 2014 the *Dictionary of Epidemiology* defined pandemic as an 'epidemic occurring over a very wide area, crossing international boundaries, and usually affecting a large number of people'. It pointed out that the infectious agent must be able to infect humans and to spread easily from human to human (Porta *et al.*, 2014). Tedros Ghebreyesus, director of the World Health Organization (WHO), speaking in March 2020, said that 'pandemic is not a word to use lightly or carelessly. It is a word that, if misused, can cause unreasonable fear, or unjustified acceptance that the fight is over, leading to unnecessary suffering and death' (WHO, 2020).

Morens *et al.* (2009) identified several pandemic characteristics, in addition to traceable disease movement over a large region (mentioned above). These included: (i) high attack rates and 'explosive' spread; (ii) transmissible by an infectious agent (rather than a spreading set of social determinants, thus excluding conditions like obesity); and (iii) comparative novelty. Although waves of plague, cholera, influenza and dengue have occurred repeatedly, their high crests of disease are, or were, at sufficient intervals (often a generation or longer) to qualify as novel, unlike more perennial conditions such as tuberculosis. Morens *et al.* (2009) also recognized, as Ghebreyesus seemed to, that the term pandemic should be reserved for the most severe conditions, with a potential for extraordinary mortality. Today, few people would call dengue 'pandemic', nor conjunctivitis (Kono *et al.*, 1972).

In summary, it seems reasonable to consider 'pandemic' as located at the end of a continuum. Excessively generous interpretations of this term may risk 'pandemic fatigue'. In 2009 WHO declared as 'pandemic' a global outbreak of H1N1 influenza, a decision which, in retrospect, looks excessive, as its global death toll, fewer than 400,000, was similar to that of a relatively mild seasonal influenza outbreak (Fineberg, 2014). This declaration led Godlee (2010), then editor of the *BMJ*, to call for greater transparency within WHO. Godlee pointed out that several relevant WHO advisers had conflicts of interest, especially with links to pharmaceutical companies, some of which made windfall profits through the sales of antivirals and vaccines, whose purchase was triggered by the pandemic declaration.

A premature pandemic declaration may not only cause unnecessary alarm, but may also divert funds from the prevention and treatment of conditions with a higher burden of disease. However, some conditions that do not fulfil strict pandemic criteria, such as severe acute respiratory syndrome (SARS), have had high economic costs, though a comparatively low burden of disease (Keogh-Brown and Smith, 2008).

7.2 The Triumph of the Germ Theory

To try to better convey an understanding of the history and future of pandemics, we next discuss the evolution of 'germ theory'. We challenge, as overly simplistic, the perception that, until John Snow's pioneering epidemiological studies of cholera in London in the 1850s, most infectious diseases were considered as spread by miasmas (bad air), even if in conjunction with social determinants. Smith (2002), for example, pointed out that John Snow observed that 'poverty and overcrowding forced people into situations where evacuations containing the cholera agent were involuntarily transmitted'.

In parts of the USA, cholera was considered contagious before Snow's work, by both the public and a minority of physicians (Smith, 2002). For example, wealthy New Yorkers left the city to avoid cholera, or bought in water from outside the city.

A more nuanced view of the evolution of the germ theory may be that there has long been considerable understanding that *some* maladies are contagious, not only from animals to humans but also from human to human. The observation by microbiologist and philosopher of science Ludwig Fleck is pertinent: '... many very solidly established scientific facts are undeniably linked, in their development, to pre-scientific, somewhat hazy, related proto-ideas or pre-ideas' (Brorson, 2006).[2]

Religious texts in India, some from 4000 years ago, included laws prohibiting social contact with unfortunate people with skin diseases, generally understood to be leprosy (Jacob and Franco-Paredes, 2008). Such stigma reduced the risk of disease transmission more than three millennia before 1377, when the Rector of the seaport of Ragusa (then part of the Venetian Republic) decreed a 30-day isolation period for ships, which evolved to quarantine (40 days) for land travellers (Gensini *et al.*, 2004). This interference with personal and economic freedom evolved in response to the Black Death, now widely considered as the beginning of the second pandemic of *Yersinia pestis*, recurring in many parts of Eurasia for hundreds of years, with a third pandemic originating in China in the 19th century (Bramanti *et al.*, 2019).

At the close of the 15th century a previously unknown disease, then called the 'Great Pox', spread through Europe and Asia, soon after the return of Christopher Columbus and his crew from the New World in 1493. Though long believed to have crossed from the Americas, this is now controversial. However, the sexual transmission of this terrifying malady, soon to be called syphilis,[3] was quickly recognized (Sefton, 2001).

The violent conquest of the so-called 'New World' by Europeans was assisted by a complex of diseases, including measles, malaria, smallpox and yellow fever, introduced to an immunologically naïve population, sometimes called part of the 'Columbian exchange' (Nunn and Qian, 2010). Smallpox was repeatedly spread to some American tribes by 'gifts' of contaminated blankets and clothing (Mayor, 1995). Szinicz (2005) documented numerous additional examples of biological warfare, including the alleged (Field, 2023) catapulting of corpses of plague victims in the Crimea in 1346–7, in an attempt by Mongols to penetrate deeper into Europe. Biological warfare suggests an understanding of contagion and is difficult to reconcile with the miasma theory, though this had considerable support and plausibility, at least for malaria, with its long-understood association with swamps and marshy areas, for example by the Roman agricultural theorist Varro (116–27 BCE) (Brorson, 2006).

The contagious nature of smallpox has also long been understood, at least by some. Prior to the introduction and then general use of vaccination to control it, initially in England (c. 1800), its spread was reduced by social distancing measures, including isolation of sufferers in pest houses and private homes and the cancellation of markets and other public gatherings (Davenport *et al.*, 2018). Sweating sickness, a still mysterious, often fatal condition which periodically affected England (1485 to mid-16th century), was also widely understood by contemporaries to be contagious (Flood, 2003).[4]

However, the understanding of communicable infectious diseases was far from complete. Pernick (2002) stated that 'many medieval European writers recognised eight or fewer contagious diseases, often including epilepsy but excluding smallpox'. Scientific progress in this field required numerous advances, including in microscopy, epidemiology and statistics. Bacteria, the name for the single celled microorganisms that are neither plant nor animal, were first observed by Antony van Leeuwenhoek, a discovery confirmed by Robert Hooke, the 'Leonardo of London', curator of the Royal Society in the late 17th century (Gest, 2004). The brilliant advances in microscopy made by van Leeuwenhoek permitted magnification of at least 275-fold (easily visualizing bacteria), but the secret of his superb technique died with him (Porter, 1976), slowing progress in the field. Although van Leeuwenhoek's discoveries did stimulate theories that linked his 'animalcules' to human disease, such ideas remained marginal for more than another 150 years (Fig. 7.1).

Snow, today so well-recognized for his mapping studies and other epidemiological work (with others) to show that cholera was associated with polluted water, died in 1858, with his theories still considered marginal (Halliday, 2001). Within another 25 years the miasma explanation had faded; the German microbiologist Robert Koch not only identified the causal organism of cholera (the curved rod-shaped *Vibrio cholera*) but was also developing his postulates (Gillies, 2016).[5] Koch's work helped cement the germ theory, although one of the postulates failed for cholera, as he was unable to identify a susceptible animal model. Also important were the French scientists and rivals, Claude Bernard and Louis Pasteur. Pasteur is sometimes reported (probably apocryphally) as acclaiming, at the end of his life, the value of Bernard's insight concerning the role of the *milieu intérieur* in infection and health as important in determining the severity and pattern of any disease, at both the individual and population scale (Lahaie and Watier, 2017).[6] By extension, the external milieu is also important, including

Fig. 7.1. In 1850 the educated public in London had some understanding that water contained 'animalcules' (now called bacteria), as suggested by this cartoon published in *Punch*, a popular satirical magazine. In 1883 Koch identified the causal bacterium for cholera, helping seal the end of the 'miasmatic' explanation for communicable diseases. Source: https://commons.wikimedia.org/wiki/File:A_Drop_of_Thames_Water,_by_Punch,_1850.jpg (public domain).

the context of social, nutritional, climatic and other environmental determinants.

Virological advances

In the 1890s investigators demonstrated that organisms even smaller than bacteria could cause disease, by showing the agents responsible for tobacco mosaic disease and (soon after) foot-and-mouth disease could penetrate a filter (or sieve) sufficiently fine to block the smallest bacteria (Tognotti, 2003; van Helvoort and Sankaran, 2019). A bacterium called *Haemophilus influenza* had been isolated by one of Koch's students, from biological material drawn from influenza sufferers, and pronounced, by its discoverer (Richard Pfeiffer), in 1892 as the cause. By 1905 this view was challenged by a few, but generally dominated until the close of the Great War (Tognotti, 2003). Nevertheless, the viral cause of influenza remained uncertain for over another decade, until a series of experiments in the early 1930s clearly showed that a filterable virus was the cause. It is now well recognized that both bacteria and viruses are capable of causing not only epidemics but also, in rare cases, pandemics.

7.3 Major Infectious Disease Outbreaks in Antiquity

In antiquity (including the time of ancient Greece and Rome), plague was a generic term for large-scale disease outbreaks we now definitely know to be communicable (infectious). Ancient religious texts mention ten plagues in Egypt. Although it is unknown if any of these were caused by *Yersinia pestis*, this organism is now thought to have fairly commonly infected humans in diverse areas of Eurasia during the Bronze Age, as early as 2900 BCE. There is also evidence that a more virulent form of plague evolved, by 1686 BCE, able to spread not only via fleas living with rodents but also from person to person, in its bubonic, pneumonic and septicaemic forms (Rasmussen *et al.*, 2015). It has also been postulated that these earlier outbreaks may have contributed to substantial migrations, including into Europe, in the Bronze Age, as well as population declines around 1000 BCE.

Four episodes of large-scale pestilence are well documented in antiquity in the peri-Mediterranean area, some with spread to northern Europe (Table 7.1). The first (described by the historian and general, Thucydides) occurred in Athens in the 5th century BCE. The later three were more widespread, affecting parts of the Roman Empire. These were the Antonine plague (also called the plague of Galen, c. 165 CE), the Cyprian plague (249–270 CE)[7] and, most famously, the Justinian plague, also often called the first pandemic of (bubonic) plague (starting c. 541 CE).

With others, the historian Kyle Harper (2017), credited a series of enormous volcanic eruptions (536 CE, 540 CE (Ilopongo, in modern El Salvador) and 547 CE) with triggering an 'antique little Ice Age' and suggesting that this extremely cold period helped to spread the Justinian plague, which in turn further depleted the already waning vitality of the Roman Empire (Büntgen *et al.*, 2016). This plague also helped the spread of Islam. Sabbatani *et al.* (2012) noted that nomadic Arab tribes, moving on desert or semi-desert territories, 'succeeded in escaping the contagion more easily'. Estimates of its death toll range from 25 million to 100 million people, especially in the Mediterranean region, but extending as far as Ireland.

7.4 The Black Death

In the calamitous mid-14th century (1348–1350), the Black Death killed over one-third of the population of Western Europe (Fig. 7.2) and up to 100 million globally, including in Africa and China (Alfani and Murphy, 2017). Its mortality may also have had an important environmental co-factor. It was preceded by decades of widespread undernutrition, as the climate cooled following the end of the Medieval Warm Period, in turn probably triggering the Great European Famine (1315–1317) and a devastating disease of cattle, probably rinderpest, which spread from central Europe to Britain. This pestilence killed over half the cows in England and created a scarcity of milk that lasted until at least 1332 (DeWitte and Slavin, 2013). Resultant undernutrition would have reduced the immunity of survivors, leaving them more physiologically vulnerable to infectious diseases.

In 1894 (during the third pandemic), plague was recognized as caused by the bacterium *Yersinia pestis* (Zietz and Dunkelberg, 2004)[8] and in 2011 this organism was isolated from the skeletal material of victims of the Black Death. The draft genome sequence was little different to modern *Y. pestis* strains (Bos *et al.*, 2011).

7.5 The Depopulation of the 'New World', Caused Primarily by Infectious Diseases

The death toll from infections accompanying the European invasions of the Americas was lower than from the Black Death, but was proportionately even more devastating. As many as 80–95% of the Indigenous peoples of the Americas died within 150 years of Columbus's landing, soon followed, as is well-known, by its exploration, conquest and pillage by European powers (Nunn and Qian, 2010). Introduced diseases, for which

Table 7.1. The four best-known major epidemics from antiquity, showing their location, origin, cause and potential predisposing factor.

Name	Location / [origin] / *pathogen*	Predisposing factor(s)
Athenian	Athens / [Ethiopia?] / Typhoid fever[a]	Siege
Antonine	Peri-Mediterranean (including North Africa) / [Ethiopia] / (Smallpox?)	Unknown
Cyprian	(Filovirus? Smallpox? Measles?)	Unknown
Justinian	Byzantine empire / Europe / (*Yersinia pestis*[a])	Antique ice age

[a]Indicates high confidence in the aetiology, for example from analysis of DNA extracted from dental pulp.

Fig. 7.2. The people of Tournai (in modern Belgium) bury victims of the Black Death, c. 1353. Miniature by Pierart dou Tielt illustrating the Tractatus quartus bu Gilles li Muisit. http://balat.kikirpa.be/photo.php?path=X004175&objnr=20049662 (public domain).

Indigenous Americans had no immunity, included smallpox, measles, typhus, cholera, malaria and yellow fever, the latter two of which were transported with African slaves (McNeill, 1976). The demographic and social impact of these and other infectious diseases was compounded by serfdom, slavery and violence. There were also barriers to reproduction of subjugated people, particularly via the use of Indigenous women and girls as domestic workers and concubines for the Spanish *encomenderos* – conquistadores and colonists to whom the land and the natives were granted (Livi-Bacci, 2006).

Elsewhere in the New World, an outbreak of smallpox[9] occurred in Australia in 1789, within two years of European invasion. Centred around Sydney, the site of first European settlement, this appears to have been deliberately spread by infectious material that had been brought from England for purposes of inoculation, about a decade before Jenner introduced vaccination (Mear, 2008). Smallpox rapidly spread among Australia's Indigenous peoples, who, similarly to those in the Americas, had no immunity.

7.6 Spanish Influenza

Epidemics of influenza (the name is derived from the Latin *influentia*, 'to flow into') have been retrospectively identified as occurring for well over 500 years. Morens *et al.* (2010) identified the first influenza 'pandemic' (which was thought to have spread from Asia) as occurring in 1510. They documented an additional 14 pandemics, until 2010.

Of these, Spanish influenza (called this as it was first reported by the uncensored Spanish media) had by far the largest impact. This pandemic, caused by influenza A, was the direct or indirect cause of at least 25 million and perhaps 100 million premature deaths, in almost all countries (Barry, 2004). Terrifyingly, it especially affected young adults, many of whom died within 2 days of the first symptom. An early focus of the disease is widely considered to have been the outbreak in Fort Funston, in Kansas, a US state with a very large population of hogs and chickens, many of which, in the absence of refrigeration, would have been kept near the troops (Humphreys, 2018). Influenza crossed the Atlantic from military camps in the USA in troopships, dispersing rapidly among troops, on both Allied and German sides of the front. However, the death toll from this pandemic was orders of magnitude more than any previous influenza outbreak, for reasons that remain poorly understood.

Oxford *et al.* (2002) posited that unique factors associated with World War I magnified the severity of the pandemic, not only by the global dispersal of troops but also via undernutrition and lung damage to troops, caused by the use of mutagenic poison gas by both armies. These writers also suggested that extreme crowding, including in makeshift hospitals, might have facilitated intense serial passage of the virus between people, favouring the emergence and dominance of strains with high infectivity and rapid and high mortality. In this way, crowding may have inadvertently acted to 'incubate' a highly pathogenic strain of influenza A. That is, viruses that killed their host too rapidly may not have suffered the penalty (i.e. of running out of additional hosts) that they would have in less crowded conditions. They may even have acquired an evolutionary edge over their slower-killing competitors. A similar theory has been advanced for the emergence of highly pathogenic influenza in intensely crowded chicken farms (Lebarbenchon *et al.*, 2010).

7.7 HIV/AIDS

The pandemic of HIV/AIDS has now killed over 30 million people. For many years the disease had a fatality rate of very close to 100%, although death was not always rapid. Today, strict adherence to antiviral treatment, if affordable, can lead to an almost normal life expectancy. Although there are two main strains of human immunodeficiency virus (HIV), which is the causative organism of AIDS (acquired immunodeficiency syndrome), only one strain (HIV-1) is widespread[10] and universally accepted as pandemic.

HIV-1 is generally thought to have first penetrated the human population, perhaps in the Democratic Republic of the Congo (DRC) before the 1950s, when one case was retrospectively identified (Zhu *et al.*, 1998). The dominant hypothesis for its first human transmission is contact between a hunter and the blood of a chimpanzee, infected with a simian immunodeficiency virus, probably in the early 20th century.

Irrespective of its origin, there is consensus that the prevalence of HIV/AIDS was greatly magnified, at first in sub-Saharan African populations, by factors such as the rarity of male circumcision, and the frequency of 'concurrent' heterosexual networks (Wilson and Halperin, 2008). Medical procedures, including blood transfusions, the unethical harvesting of plasma (Hayes, 2005) and the use of unsterilized needles (Pépin, 2013) also magnified its spread, as did male-to-male sexual contact in many countries. An alternative hypothesis, pursued especially by the investigative journalist Edward Hooper, postulates that HIV may have accidentally entered the human population via contaminated polio vaccines, cultured in kidney cells removed from chimpanzees that had been hunted, captured and housed near Kisangani, then called Stanleyville, in the DRC (Jain, 2020).[11] It is possible that both hypotheses are valid, though not exclusively.

7.8 COVID-19

COVID-19, transmitted by the newly identified coronavirus, SARS-Co-V-2, emerged in Wuhan, China, late in 2019. Many facts about its origin remain in dispute, including the month of the first case, and the timing of the first cases out of China. It is likely that the Chinese government perceives its self-interest is best served by obfuscating facts concerning its origin (Wood, 2023). Equally indisputably, COVID-19 is a pandemic, with a death toll, both direct and indirect (such as caused by 'lockdowns' and reductions in ordinary medical care), that may eventually approach that of HIV/AIDS.

Its cause is intensely disputed. Although the overwhelming number of articles in journals with high impact factors contend that it emerged from an as yet unidentified mammal bred and traded in a market (Worobey *et al.*, 2022), there are many counter-arguments suggesting that it could have evolved via serial passage, or even genetic engineering in a laboratory (Courtier-Orgogozo and de Ribera, 2022; Segreto *et al.*, 2021; Liu *et al.*, 2023). One curious theory about SARS-Co-V-2 is that its unexpectedly low death toll in Africa may have been because of protective genetic factors there, associated with the low penetration of Neanderthal genes (Zeberg and Pääbo, 2020).

7.9 Conclusion: Pandemics in the Anthropocene: a Two-Way Relationship?

Many definitions of the Anthropocene date its origin to about 1750, when the Industrial Revolution greatly magnified the burning of fossil fuels. However, Ruddiman *et al.* (2020) have long argued for a much earlier date, perhaps 7000 years ago. This entire period has been marked by agriculture and animal domestication of increasing scale, as well as deforestation. Although some zoonoses may have emerged before 7000 years BP (for example from dogs), most of the major infectious diseases that afflict humans have occurred in this period, including the epidemics and pandemics described in this chapter. Of these, all but cholera[12] have a zoonotic origin (see Chapter 5, this volume).

The number of disease episodes now widely identified as pandemic in the past 500 years has increased considerably (Fig. 7.3). While part of the explanation lies with human crowding, another is improved records and diagnosis. Terminology is also important. For example, Morens *et al.* (2010) identified 15 influenza 'pandemics' since 1510. However, the death toll of several (especially the last two, 1977 and 2009) has been minor.

Nonetheless, there are growing concerns that deforestation, climate change (Carlson *et al.*, 2022), intensive agriculture and the smuggling and farming of 'novel' species, such as raccoon dogs (*Nyctereutes procyonoides*) and civet cats (*Paradoxurus hermaphroditus*), are generating new zoonoses, and even the risk of new pandemics, perhaps on an unprecedented scale (Morse *et al.*, 2012). Such concerns appear justified, to many, by the immense suffering wrought not only by HIV/AIDS, but also by COVID-19.

There are many other factors associated with the Anthropocene, some of which greatly reduce the burden of infectious diseases, especially of bacterial conditions such as plague and cholera. Even the threat of influenza is dampened by antivirals, vaccines and a better

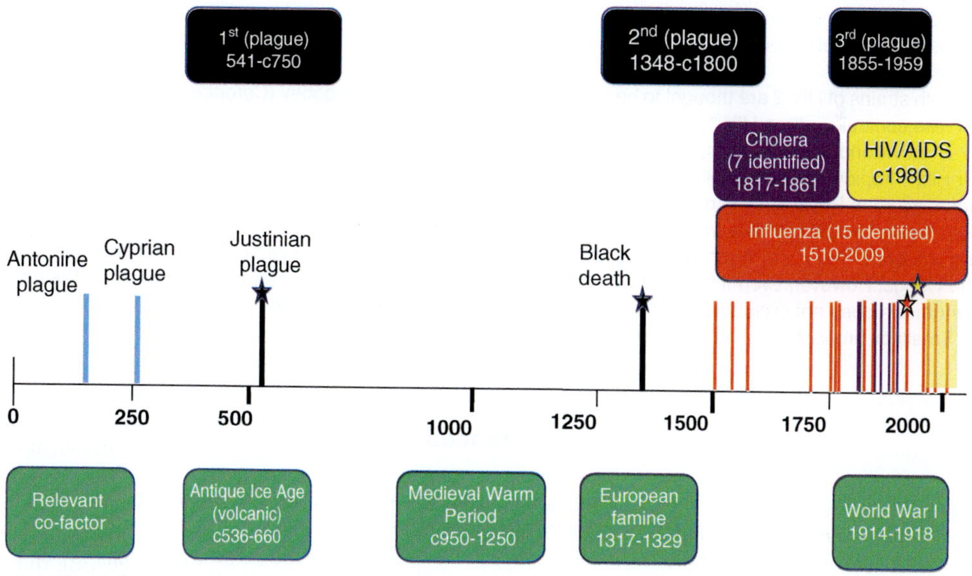

Fig. 7.3. Major plagues and pandemics until almost the present, over the past 2015 years. The durations of the plague pandemics are approximate and vary with the source. Three events have had an especially high and compressed mortality; each is marked with a star. The Medieval warm period was followed by the Little Ice Age (not shown), which was probably a factor in the great European famine (and following milk scarcity, which extended to 1329).

understanding of social distancing and other public health measures.

At the same time, medical and laboratory procedures may increase the risk of some infectious diseases, perhaps on a sufficient scale to cause a pandemic. One mechanism is the documented escape or deliberate release of pathogens from laboratories (Klotz and Sylvester, 2014). Another is via 'gain of function' research, such as the deliberate creation of 'chimeric' pathogens, and their serial passage using chimeric laboratory animals, such as 'humanized' mice (Segreto *et al.*, 2021). Butler (2022) proposed that such synthetic organisms should be classified as 'novel entities', one of the planetary boundaries, and thus relevant to planetary health.

Notes

[1] A search using Google Scholar identifies several papers, published between 1918 and 1922, with the word 'pandemic' in their title, referring to influenza.

[2] Traces of the germ theory are detectable in the writings of the Greek philosopher Anaxagorus as well as several Roman scholars, who speculated on the existence of minute 'seeds', from which diseases could grow.

[3] Named by Fracastoro, widely considered as a pioneering supporter of the germ theory (Howard-Jones, 1977).

[4] The aetiology of sweating sickness is generally thought to be viral, perhaps a flavivirus. One recently suggested possibility is tick-borne louping-ill virus, which has an epidemiology similar to Lyme disease (Wollert, 2017).

[5] The microorganisms: (i) must be shown to be present in all cases of the disease; (ii) must be in such numbers and distribution that all the symptoms of the disease can be explained; (iii) must be isolated and grown in pure culture; and (iv) it must be possible to reproduce the disease by introducing this pure culture into animals.

[6] Antoine Béchamp, Pasteur's most bitter rival, is also sometimes credited with having this insight, especially by practitioners of 'alternative' medicine. Such claims are probably exaggerated, although Béchamp did make important discoveries (Manchester, 2007).

[7] Named for its best-known chronicler, the Bishop of Carthage.

[8] Independently by Alexandre Yersin (1863–1943) and soon after by Shibasaburo Kitasato (1856–1931).

[9] The afflicted were reportedly shunned by the survivors, hinting at an innate aversion which could also be protective.

[10] Both strains of HIV-2 are thought to have originated from a sooty mangabey (*Cercocebus atys*). They are far less contagious than HIV-1, for reasons that are poorly understood.

[11] The evidence against Hooper's hypothesis rested on two key elements; one was the assurance of Dr Hilary Koprowski and his colleagues that chimpanzee kidneys were not used to grow polio virus. Seventeen years after the Royal Society meeting that discredited Hooper, its chair, Professor Robin Weiss, confided to Jain that he 'would not have trusted Koprowski more than I could throw him'. The second element discrediting Hooper's idea is phylogenetic dating of the first cases of HIV/AIDS to well before Koprowski's work in the Congo. However, even if those studies are correct, both theories could be true, i.e. the 'natural origin' hypothesis does not exclude an additional, inadvertent, anthropogenic source.

[12] Smallpox probably originated in cattle, but this is uncertain.

References

Alfani, G. and Murphy, T. (2017) Plague and lethal epidemics in the pre-industrial world. *The Journal of Economic History* 77, 314–343. DOI: 10.1017/S0022050717000092

Barry, J.M. (2004) *The Great Influenza: the Epic Story of the Deadliest Plague in History*. Viking, New York.

Bos, K.I., Schuenemann, V.J., Golding, G.B., Burbano, H.A., Waglechner, N. *et al.* (2011) A draft genome of *Yersinia pestis* from victims of the Black Death. *Nature* 478, 506–510. DOI: 10.1038/nature10549.

Bramanti, B., Dean, K.R., Walløe, L. and Stenseth, N.C. (2019) The third plague pandemic in Europe. *Proceedings of the Royal Society B: Biological Sciences* 286, 20182429. DOI: 10.1098/rspb.2018.2429

Brorson, S. (2006) The seeds and the worms: Ludwik Fleck and the early history of germ theories. *Perspectives in Biology and Medicine* 49, 64–76. DOI: 10.1353/pbm.2006.0004

Büntgen, U., Myglan, V.S., Ljungqvist, F.C., McCormick, M., Di Cosmo, N. *et al.* (2016) Cooling and societal change during the Late Antique Little Ice Age from 536 to around 660 AD. *Nature Geoscience* 9, 231–236. DOI: 10.1038/ngeo2652

Butler, C.D. (2022) Comparing and contrasting two United Nations Environment Programme reports on COVID-19. *Science in One Health* 1, 100003. DOI: 10.1016/j.soh.2022.100003

Carlson, C.J., Albery, G.F., Merow, C., Trisos, C.H., Zipfel, C.M. *et al.* (2022) Climate change increases cross-species viral transmission risk. *Nature* 607, 555–562. DOI: 10.1038/s41586-022-04788-w

Clemow, F. (1894) The recent pandemic of influenza: its place of origin and mode of spread. *The Lancet* 1, 139–143, 329–331. DOI: 10.1016/S0140-6736(01)66317-7

Courtier-Orgogozo, V. and de Ribera, F.A. (2022) SARS-CoV-2 infection at the Huanan seafood market. *Environmental Research* 214, 113702. DOI: 10.1016/j.envres.2022.113702

Davenport, R.J., Satchell, M. and Shaw-Taylor, L.M.W. (2018) The geography of smallpox in England before vaccination: A conundrum resolved. *Social Science & Medicine* 206, 75–85. DOI: 10.1016/j.socscimed.2018.04.019

DeWitte, S. and Slavin, P. (2013) Between famine and death: England on the eve of the Black Death – evidence from paleoepidemiology and manorial accounts. *The Journal of Interdisciplinary History* 44, 37–60. DOI: 10.1162/JINH_a_00500

Field, M. (2023) Catapulting corpses? A famous case of medieval biological warfare probably never happened. *Bulletin of the Atomic Scientists*. Available at: https://thebulletin.org/2023/08/catapulting-corpses-a-famous-case-of-medieval-biological-warfare-probably-never-happened/?utm_source=Newsletter&utm_medium=Email&utm_campaign=ThursdayNewsletter08102023&utm_content=DisruptiveTechnologiesBio_CatapultingCorpses_08102023 (accessed 11 August 2023).

Fineberg, H.V. (2014) Pandemic preparedness and response – lessons from the H1N1 influenza of 2009. *New England Journal of Medicine* 370, 1335–1342. DOI: 10.1056/NEJMra1208802

Flood, J.L. (2003) 'Safer on the battlefield than in the city': England, the 'sweating sickness', and the continent. *Renaissance Studies* 17, 147–176. DOI: 10.1111/1477-4658.00015

Gensini, G.F., Yacoub, M.H. and Conti, A.A. (2004) The concept of quarantine in history: from plague to SARS. *Journal of Infection* 49, 257–261. DOI: 10.1016/j.jinf.2004.03.002

Gest, H. (2004) The discovery of microorganisms by Robert Hooke and Antoni van Leeuwenhoek, Fellows of the Royal Society. *Notes and Records. The Royal Society Journal of the History of Science* 58, 187–201. DOI: 10.1098/rsnr.2004.0055

Gillies, D.A. (2016) Establishing causality in medicine and Koch's postulates. *International Journal of History and Philosophy of Medicine* 6, 10603. DOI: 10.18550/ijhpm.0603

Godlee, F. (2010) Conflicts of interest and pandemic flu. *BMJ* 340, c2947. DOI: 10.1136/bmj.c2947

Halliday, S. (2001) Death and miasma in Victorian London: an obstinate belief. *BMJ* 323, 1469–1471. DOI: 10.1136/bmj.323.7327.1469

Harper, K. (2017) *The Fate of Rome*. Princeton University Press, Princeton, New Jersey.

Hayes, A. (2005) AIDS, bloodheads & cover-ups: The 'ABC' of Henan's AIDS epidemic. *AQ: Australian Quarterly* 77, 12–40. www.jstor.org/stable/20638337

Howard-Jones, N. (1977) Fracastoro and Henle: a re-appraisal of their contribution to the concept of communicable diseases. *Medical History* 21, 61–68. DOI: 10.1017/s0025727300037170

Humphreys, M. (2018) The influenza of 1918: evolutionary perspectives in a historical context. *Evolution, Medicine, and Public Health* 2018, 219–229. DOI: 10.1093/emph/eoy024

Jacob, J.T. and Franco-Paredes, C. (2008) The stigmatization of leprosy in India and its impact on future approaches to elimination and control. *PLoS Neglected Tropical Diseases* 2, e113. DOI: 10.1371/journal.pntd.0000113

Jain, S.L. (2020) The WetNet: what the oral polio vaccine hypothesis exposes about globalized interspecies fluid bonds. *Medical Anthropology Quarterly* 34, 504–524. DOI: 10.1111/maq.12587

Keogh-Brown, M.R. and Smith, R.D. (2008) The economic impact of SARS: How does the reality match the predictions? *Health Policy* 88, 110–120. DOI: 10.1016/j.healthpol.2008.03.003

Klotz, L.C. and Sylvester, E.J. (2014) The consequences of a lab escape of a potential pandemic pathogen. *Frontiers in Public Health* 2, 116. DOI: 10.3389/fpubh.2014.00116

Koch, T. (2014) 1831: the map that launched the idea of global health. *International Journal of Epidemiology* 43, 1014–1020. DOI: 10.1093/ije/dyu099

Kono, R., Sasagawa, A., Ishii, K., Ochi, M., Sugiura, S. *et al.* (1972) Pandemic of new type of conjunctivitis. *The Lancet* 299, 1191–1194. DOI: 10.1016/S0140-6736(72)90921-X

Lacey, S.W. (1995) Cholera: calamitous past, ominous future. *Clinical Infectious Diseases* 20, 1409–1419. DOI: 10.1093/clinids/20.5.1409

Lahaie, Y.-M. and Watier, H. (2017) Contribution of physiologists to the identification of the humoral component of immunity in the 19th century. *MABS* 9, 774–780. DOI: 10.1080/19420862.2017.1325051

Lebarbenchon, C., Feare, C.J., Renaud, F., Thomas, F. and Gauthier-Clerc, M. (2010) Persistence of highly pathogenic avian influenza viruses in natural ecosystems. *Emerging Infectious Diseases* 16, 1057–1062. DOI: 10.3201/eid1607.090389

Liu, W.J., Liu, P., Lei, W., Jia, Z., He, X. *et al.* (2023) Surveillance of SARS-CoV-2 at the Huanan Seafood Market. *Nature*. DOI: 10.1038/s41586-023-06043-2

Livi-Bacci, M. (2006) The depopulation of Hispanic America after the conquest. *Population and Development Review* 32, 199–232. DOI: 10.1111/j.1728-4457.2006.00116.x

Manchester, K.L. (2007) Louis Pasteur, fermentation, and a rival. *South African Journal of Science* 103, 377–380. DOI: 10.10520/EJC96719

Mayor, A. (1995) The Nessus shirt in the New World: smallpox blankets in history and legend. *The Journal of American Folklore* 108, 54–77. DOI: 10.2307/541734

McNeill, W.H. (1976) *Plagues and Peoples*. Anchor Press, Garden City, New York.

Mear, C. (2008) The origin of the smallpox outbreak in Sydney in 1789. *Journal of the Royal Australian Historical Society* 94, 1–22.

Morens, D.M., Folkers, G.K. and Fauci, A.S. (2009) What Is a pandemic? *The Journal of Infectious Diseases* 200, 1018–1021. DOI: 10.1086/644537

Morens, D.M., Taubenberger, J.K., Folkers, G.K. and Fauci, A.S. (2010) Pandemic influenza's 500th anniversary. *Clinical Infectious Diseases* 51, 1442–1444. DOI: 10.1086/657429

Morse, S.S., Mazet, J.A.K., Woolhouse, M., Parrish, C.R., Carroll, D. *et al.* (2012) Prediction and prevention of the next pandemic zoonosis. *The Lancet* 380, 1956–1965. DOI: 10.1016/S0140-6736(12)61684-5

Nunn, N. and Qian, N. (2010) The Columbian exchange: a history of disease, food, and ideas. *Journal of Economic Perspectives* 24, 163–188. DOI: 10.1257/jep.24.2.163

Oxford, J.S., Sefton, A., Jackson, R., Innes, W., Daniels, R. *et al.* (2002) World War I may have allowed the emergence of 'Spanish' influenza. *The Lancet Infectious Diseases* 2, 111–114. DOI: 10.1016/s1473-3099(02)00185-8

Pépin, J. (2013) The origins of AIDS: from patient zero to ground zero. *Journal of Epidemiology and Community Health* 67, 473–475. DOI: 10.1136/jech-2012-201423

Pernick, M.S. (2002) Contagion and culture. *American Literary History* 14, 858–865. https://www.jstor.org/stable/3568029

Porta, M., Last, J.M., Greenland, S., Hernán, M.A. and dos Santos Silva, I. (eds) (2014) *A Dictionary of Epidemiology*. Oxford University Press, Oxford, UK.

Porter, J.R. (1976) Antony van Leeuwenhoek: tercentenary of his discovery of bacteria. *Bacteriological Reviews* 40, 260–269. DOI: 10.1128/br.40.2.260-269.1976

Rasmussen, S., Allentoft, M.E., Nielsen, K., Orlando, L., Sikora, M. *et al.* (2015) Early divergent strains of *Yersinia Pestis* in Eurasia 5,000 years ago. *Cell* 163, 571–582. DOI: 10.1016/j.cell.2015.10.009

Ruddiman, W.F., He, F., Vavrus, S.J. and Kutzbach, J.E. (2020) The early Anthropogenic hypothesis: a review. *Quaternary Science Reviews* 240, 106386. DOI: 10.1016/j.quascirev.2020.106386

Sabbatani, S., Manfredi, R. and Fiorino, S. (2012) The Justinian plague (part two). *Le Infezioni in Medicina* 20, 217–232. https://pubmed.ncbi.nlm.nih.gov/22992565/

Sefton, A.M. (2001) The Great Pox that was … syphilis. *Journal of Applied Microbiology* 91, 592–596. DOI: 10.1046/j.1365-2672.2001.01494.x

Segreto, R., Deigin, Y., McCairn, K., Sousa, A., Sirotkin, D. *et al.* (2021) Should we discount the laboratory origin of COVID-19? *Environmental Chemistry Letters* 19, 2743–2757. DOI: 10.1007/s10311-021-01211-0

Smart, W. (1876) On dengue, or dandy fever. *Transactions of the Epidemiological Society of London* 4, 58–71. https://pubmed.ncbi.nlm.nih.gov/29418979/

Smith, G.D. (2002) Behind the Broad Street pump: aetiology, epidemiology and prevention of cholera in mid-19th century Britain. *International Journal of Epidemiology* 31, 920–932. DOI: 10.1093/ije/31.5.920

Szinicz, L. (2005) History of chemical and biological warfare agents. *Toxicology* 214, 167–181. DOI: 10.1016/j.tox.2005.06.011

Tognotti, E. (2003) Scientific triumphalism and learning from facts: bacteriology and the 'Spanish flu' challenge of 1918. *Social History of Medicine* 16, 97–110. DOI: 10.1093/shm/16.1.97

van Helvoort, T. and Sankaran, N. (2019) How seeing became knowing: the role of the electron microscope in shaping the modern definition of viruses. *Journal of the History of Biology* 52, 125–160. DOI: 10.1007/s10739-018-9530-2

WHO (2020) WHO Director-General's opening remarks at the media briefing on COVID-19 – 11 March 2020. Available at: https://www.who.int/director-general/speeches/detail/who-director-general-s-opening-remarks-at-the-media-briefing-on-covid-19---11-march-2020 (accessed 15 December 2020).

Wilson, D. and Halperin, D.T. (2008) 'Know your epidemic, know your response': a useful approach, if we get it right. *The Lancet* 372, 423–426. DOI: 10.1016/S0140-6736(08)60883-1

Wollert, E. (2017) An Assessment of the 'Sweating Sickness' Affecting England During the Tudor Dynasty. PhD, Oregon State University, Portland, Oregon.

Wood, S. (2023) The Chinese Communist Party and the COVID-19 pandemic: Face loss, status anxiety, resentment. *Global Society* 37, 245–265. DOI: 10.1080/13600826.2022.2098701

Worobey, M., Levy, J.I., Serrano, L.M.M., Crits-Christoph, A., Pekar, J.E. *et al.* (2022) The Huanan market was the epicenter of SARS-CoV-2 emergence. *Science* 377, 951–959. DOI: 10.1126/science.abp8715

Zeberg, H. and Pääbo, S. (2020) The major genetic risk factor for severe COVID-19 is inherited from Neanderthals. *Nature* 587, 610–612. DOI: 10.1038/s41586-020-2818-3

Zhu, T., Korber, B.T., Nahmias, A.J., Hooper, E., Sharp, P.M. *et al.* (1998) An African HIV-1 sequence from 1959 and implications for the origin of the epidemic. *Nature* 391, 594–597. DOI: 10.1038/35400

Zietz, B.P. and Dunkelberg, H. (2004) The history of the plague and the research on the causative agent *Yersinia pestis*. *International Journal of Hygiene and Environmental Health* 207, 165–178. DOI: 10.1078/1438-4639-00259

8 Limits to Growth

Kerryn Higgs
School of Social Sciences, University of Tasmania, Hobart, Australia; Associate Member, Club of Rome, Winterthur, Switzerland

Abstract
All forms of what is known as 'development', whether in the global North or South, have been dependent on the continuing extraction of inputs from Nature. Until the middle of the 20th century, the natural world could accommodate human demands. But from that time, as populations grew, and affluence became a universal dream multiplying per capita impact, the growth of the human enterprise began to undermine the Earth system, ultimately threatening human health and civilization itself.

8.1 Background to Awareness: *The Limits to Growth,* 1972

The idea that there are physical limits on the human economy's expansion entered mainstream thinking in 1972, when *The Limits to Growth* (Meadows *et al.*, 1972) was published.[1] Written by a research team at the Massachusetts Institute of Technology (MIT), it identified five major aspects of future limitation: accelerating industrialization, rapid population growth, extensive undernutrition, depletion of non-renewable resources and environmental decline. The team wanted to determine whether, in the long run, growing populations pursuing expanded industrialization could avoid crippling environmental degradation and depletion of indispensable resources.

To address this question, they developed World3, a computer program based on one developed by the pioneering systems analyst Jay Forrester. World3 processed extensive data on interacting aspects of the economy and the environment in various combinations. The standard run, which assumed that industrialization, resource usage and population growth would continue along the same 'business as usual' (BAU) trajectory, led to collapse in the middle of the 21st century. Vast technological progress was found to postpone this outcome, but not prevent it. To avoid collapse, it was necessary to stabilize population and cut the flows of material extraction and waste. These findings were derided by most economists, but researchers who have analysed the data for the decades since 1972 have found that outcomes in the real world correspond surprisingly well with the BAU scenario or standard run (Hall and Day, 2009; Turner, 2014; Randers, 2018).[2]

Fifty years ago, the specific manifestations of the proposed limits were less obvious than those we face today. In 1972, global warming was less advanced, though the Meadows team recognized that carbon dioxide (CO_2) concentrations were rising exponentially. Indicators of environmental decline were restricted to pollutants that were then well-measured, including nuclear

 Climate Change and Global Health: Primary, Secondary and Tertiary Effects, 2nd Edition (eds C.D. Butler and K. Higgs)
DOI: 10.1079/9781800620025.0008

waste, global dichlorodiphenyltrichloroethane (DDT) production, lead found in the Greenland icecap and US mercury consumption. The reduction of dissolved oxygen in the oceans and declining catches of wild fish were also noted, as was waterway eutrophication. Like climate, land system change and biodiversity loss were less obvious problems in 1970 than they have become. Overall, the MIT team stressed that, at that time, we were still largely ignorant of where the planetary limits would lie.

Alongside environmental damage, the team also pointed to the depletion of key resources. This was widely misinterpreted by economists, who misread the claims and ignored the team's caution that precise dates and precise estimates of exhaustion were not being put forward. Rather, Meadows and colleagues sought to characterize the overall trend in the relationship between the Earth system and human expansion and to indicate that resource depletion would ultimately threaten industrial expansion (Higgs, 2014, pp. 257–270). Although there was little evidence of imminent short supply in 1970, the observed exponential growth in resource use pointed to a future crisis. Today, many resources are indeed less plentiful (phosphate rock, for example) and almost all are more difficult and energy-intensive to extract (Bardi, 2014; Michaux, 2017).

8.2 Key Elements of the Crisis

Among the recent elaborations of the 'limits to growth' (LtG) ideas is the planetary boundaries (PB) work of the team led by Will Steffen and Johan Rockström, which investigates the nature and extent of environmental limits in the 21st century (Steffen *et al.*, 2015a). The PB project aims to establish the boundaries beyond which human activities fatally undermine the various elements of ecological integrity on which all life, including ours, depends. Their most recent PB paper (Richardson *et al.*, 2023) argued that six of the nine PBs have been breached: biodiversity loss; biogeochemical disruption (especially that of the nitrogen (N) and phosphorus (P) cycles); land system change; climate change; freshwater change; and novel entities (previously characterized as pollution). These are interconnected threats, with land use change for example affecting climate, biodiversity and the expansion of N and P extraction. The PB researchers are representative of scientists more broadly. Two major warning messages to humanity have been published, each signed by hundreds of scientists, including many Nobel Laureates (Union of Concerned Scientists, 1992; Ripple *et al.*, 2017).

Biodiversity loss

The 2020 Living Planet Report (WWF, 2020) revealed a 68% reduction in numbers of mammals, birds, fish, amphibians and reptiles worldwide in the period between 1970 and 2016. The previous Living Planet Report (WWF, 2016) estimated a continuing annual reduction of 2% and correctly predicted the reduction identified in 2020. These studies count individual animals and do not chart extinction of species. However, entire species are also sliding to extinction at a rate of hundreds or thousands of times that of the background level in the geological past (Ceballos *et al.*, 2020). A comprehensive assessment by the Intergovernmental Science-Policy Platform on Biodiversity and Ecosystem Services (IPBES) found that a quarter of all species (approximately one million) face early extinction, many within decades (Díaz *et al.*, 2019). These authors stated that 'the world is increasingly managed to maximize the flow of material contributions from nature to keep up with rising demands for food, energy, timber, and more, with global trade increasing the geographic separation between supply and demand. This unparalleled appropriation of nature is causing the fabric of life on which humanity depends to fray and unravel.'

Disruption of biogeochemical cycles, nitrogen and phosphorous

As a result of the expansion of industrial farming over the past century (and especially since 1950), N and P are coursing through waterways, aquifers and continental shelves, leading to toxic groundwater, algal blooms, dead zones and fish kills. Humans 'now convert more N_2 from the atmosphere into reactive forms than all of the Earth's terrestrial processes combined', more than half through the manufacture of ammonia-based fertilizer (Rockström *et al.*, 2009).

Phosphorus was salvaged from animal manure and nightsoil in East Asian traditional farming (Paull, 2011). As European populations grew, graveyards and battlefields were ransacked for bones; eventually, ancient guano deposits were tapped, yielding both N and P. The superphosphate process, which made rock phosphate soluble, was achieved in 1842, and this mineral replaced the waning guano supply and became the main source of P (Smil, 2000). Production of P, largely sourced from only three countries, is likely to be effectively exhausted by 2100 (Cordell *et al.*, 2009). Phosphorus, an element, is indispensable to life, and no substitute is plausible, though it can be recovered or recycled from human waste (Cordell *et al.*, 2009). Both N and P are discharged to the ocean, whether as sewerage or agricultural runoff; there they are lost to human use (Butler and Higgs, 2018).

Climate

In October 2018, the Intergovernmental Panel on Climate Change (IPCC) warned that we had only 12 years (until 2030) to reduce carbon emissions by half, if global warming was to be limited to 1.5°C above preindustrial levels, the aim adopted in Paris (IPCC, 2018). At the 27th Conference of the Parties (COP27) held in Sharm Al Shaikh, Egypt, in 2022, the ambition of the nationally determined contributions (NDCs) remained inadequate to the task, pointing to an increment of at least 2.5°C. Reductions of as much as 17% in greenhouse gas emissions during the first year of pandemic crisis (2020) were fleeting as economic growth[3] was restored; public debate in government circles worldwide embraced the restoration of gross domestic product (GDP) growth as the central objective of post-pandemic policy. See section 8.4 below for discussion of the tight correlation between GDP growth and material footprint.

Many climate scientists warn that, unless emissions are drastically and rapidly cut, we will speed towards what they call 'hothouse Earth' (Steffen *et al.*, 2018). In this scenario, we will soon arrive in a climate similar to that of the mid-Pliocene, three million years ago (+2°C–3°C, CO_2e 400–450 ppm)[4] and, if drastic action continues to be brushed aside, the mid-Miocene, 17 million years ago (+5°C, as much as 500 ppm). While there is a slight chance that humans might adapt to the former, we have little hope of adapting to the latter (Schellnhuber, 2018).

Another key aspect of the climate threat is its impact on sea levels. We already observe beachside dwellings collapsing into the ocean on the east coasts of Australia and the US and high tides washing into American city centres. More concerning, Kulp and Strauss (2019) established new elevation data which show that sea level rise may be triple that predicted so far; they map likely effects on key international ports and cities (Pearl River Delta, China; Bangladesh; Jakarta; Bangkok) which show inundation that will jeopardize and ultimately halt industry and habitation for millions (M): '630M people live on land below projected annual flood levels for 2100, and up to 340M for mid-century, versus roughly 250M at present. We estimate one billion people now occupy land less than 10m above current high tide lines, including 230M below 1m.'[5] As populations increase, sea level refugees will have little hope of finding empty country in which to settle.

Land system change (clearing, conversion for agriculture)

The main driver of loss of biological diversity is land use conversion, classified as an 'increasing risk' according to Steffen *et al.* (2015b); this process is driven by both human population growth and the pursuit of greater affluence. The clearing of forests and woodlands and the draining of wetlands for human purposes continues, primarily for agriculture, but also due to urbanization and industrial expansion. Examples include the draining and burning of Indonesian peat forests, mainly for oil palm and pulpwood (Page and Hooijer, 2016), the 'reclamation' of east Asian mudflats (BirdLife Australia, nd) and the cutting of forests for pasture and soy across Latin America (FAO, 2016). According to the Food and Agriculture Organization's 2016 State of the World's Forests (FAO, 2016): 'in the tropics and subtropics large-scale commercial agriculture and local subsistence agriculture are responsible for about 40 percent and 33 percent of forest conversion, respectively, and the remaining 27 percent of deforestation happens due to urban growth, infrastructure expansion and mining'.

Land use changes impact on adjacent populations of humans and other species. These changes are also implicated in the increased risk of diseases crossing the species barrier to humans. As forests and savannahs are converted, humans and wild animals are pushed closer together. When habitat is destroyed, animals must seek food in and around human settlements and farms, as flying foxes do in Australia. Bats pose a particular risk, due to their carriage of multiple viruses, but many other animal species can act as reservoirs or carriers of infectious disease. Few zoonotic diseases become pandemic, but over 75% of emerging infectious diseases are zoonotic. Examples include avian influenza, Middle East respiratory syndrome or MERS (bats via camels), severe acute respiratory syndrome or SARS (bats, possibly via civet cat), Hendra and lyssavirus (fruit bats, via horses), Nipah virus (fruit bats), and Ebola (bats, via primates). Factory farming of chickens and pigs, in particular, also creates a situation ripe for the development of contagious diseases and is associated with some outbreaks of bird and swine flu (UNEP and ILRI, 2020).

Other aspects

In their most recent paper (Richardson *et al.*, 2023), the PB team argues that ozone depletion, ocean acidification and atmospheric aerosol loading remain within the "safe operating space", although ocean acidification is close to the zone of increasing risk. They have, however, added freshwater change and "novel entities" to the boundaries that have been breached. Novel entities includes pollutants covered in the first PB paper (Rockström *et al.*, 2009). Pollutants such as persistent organic pollutants (POPs), plastics, endocrine disruptors, heavy metals and nuclear wastes constitute a threat to human health and well-being and are poorly regulated in many cases (Butler and Higgs, 2018).

8.3 Social Implications

Fostering economic growth, as measured by GDP, formed the foundation of institutional development thinking in the post-war period. Initially, state intervention and planning were considered suitable tools for 'making the cake bigger' but from the 1980s, neoliberal ideology began to capture the policy debate and market-driven 'globalization' was soon deemed to be the only appropriate method. In both phases, the objective was to spur economic growth. The neoliberal phase embraced the market as the sole appropriate path; planning and regulation were rejected and replaced by the notion that market forces will automatically benefit everyone in the long run. In this approach, taxes should be cut, government programmes reduced and public enterprises privatized. Free trade was designated as the means to 'make economic growth and environmental protection mutually supportive' (Higgs, 2014, pp. 128–132, 239–254).

Neoliberalism favours the interests of those who benefit most from market transactions. By presenting GDP growth as 'the common good', it has created the illusion that furthering the interests of the people who benefit from the expansion of market transactions is actually 'beneficial for all' (Schmelzer, 2016). Retreat from regulation has facilitated all forms of extraction, including logging forests for timber and clearing them for agriculture. At the same time, the retreat from adequate taxation and government welfare programmes abandons the provision of services to the mercy of markets – which support business that realizes a profit. Many aspects of human health hinge on the provision of a clean water supply and sanitation, neither of which are likely to reach the poorest if profit is the benchmark. Many of the most impressive improvements in poverty in the early part of the second decade of the 21st century were achieved in Ghana, Vietnam and Brazil, where government programmes provided benefits such as cash for mothers, support for smallholders, and land grants (Lappé *et al.*, 2013). Such programmes are not favoured in the neoliberal agenda. In its response to the COVID-19 pandemic, Indonesia launched similar programmes to help poor families with food aid and young people with education (ILO, 2020).

Contrary to the neoliberal canon, the rewards of economic growth have fallen to those who were already rich far more than to the least advantaged (Alvaredo *et al.*, 2018; Oxfam International, 2018). This awkward fact justifies scepticism about growth as a development strategy and free market ideology as the best route to genuine and sustainable development.[6]

8.4 Material Extraction and Decoupling

The central focus of the LtG critique of GDP growth is the expansion of material extraction or human footprint. On the global scale, GDP growth has been correlated with extraction and waste-disposal for a century and, especially since 1950, with escalating carbon emissions, depletion of resources and ongoing transformation of land and marine environments (Steffen *et al.*, 2015b) (Fig. 8.1). The situation is not improving: in the second decade of the 21st century, we were using more materials per unit of GDP than in 2000 (UNEP, 2016).[7]

Economic growth has nevertheless been favoured as a panacea for social and environmental problems since 1950 and continues to be advanced by politicians, business organizations and the institutions involved in global sustainability policy (see Higgs, 2014, Part II). The pursuit of a 'jobs and growth' policy is familiar in Australia – and in most nations. The coal industry in Australia illustrates the flaws in this supposed correlation. Despite colossal claims from the industry, few jobs are created, as the coal industry increasingly shifts to high-tech, highly automated processes (Australia Institute, 2014).

Given that growth is crucial to the operation of the capitalist market system, it is unsurprising that it is widely claimed that growth can be decoupled from extraction (OECD, 2011; World Bank, 2012; UNEP, 2016, 2019; McAfee, 2019). However, there is no hard evidence of absolute decoupling, where GDP increases while material extraction declines. Rather, the empirical evidence shows global extraction increasing steadily since 1970 and accelerating during the past few decades (Fig. 8.2).

The United Nations Environment Programme (UNEP) estimated the volume of material extraction in 2015 at 87–88 billion tonnes a year, a figure that had tripled over the previous 50 years and is still growing (UNEP, 2019). UNEP's calculations include metals, non-metallic minerals (such as sand), fossil fuels, biomass, water and land conversion.[8]

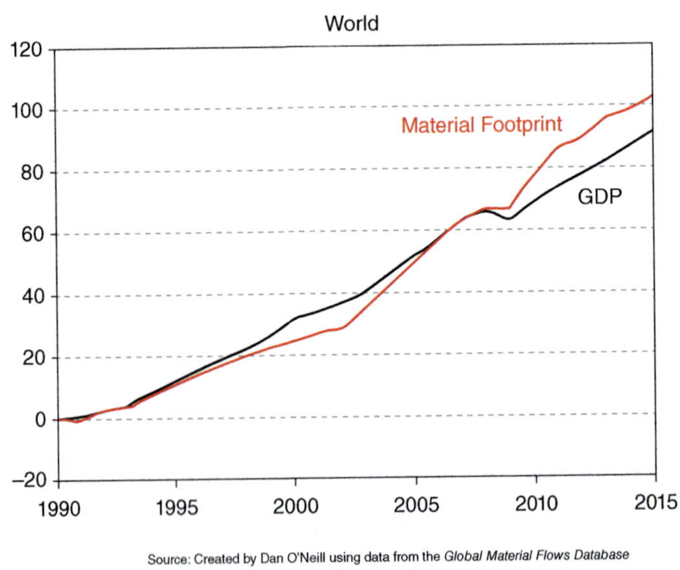

Fig. 8.1. Material footprint and GDP, 1990–2015. Credit: Dan O'Neill, data from Material Flows database: https://www.resourcepanel.org/global-material-flows-database (used with permission).

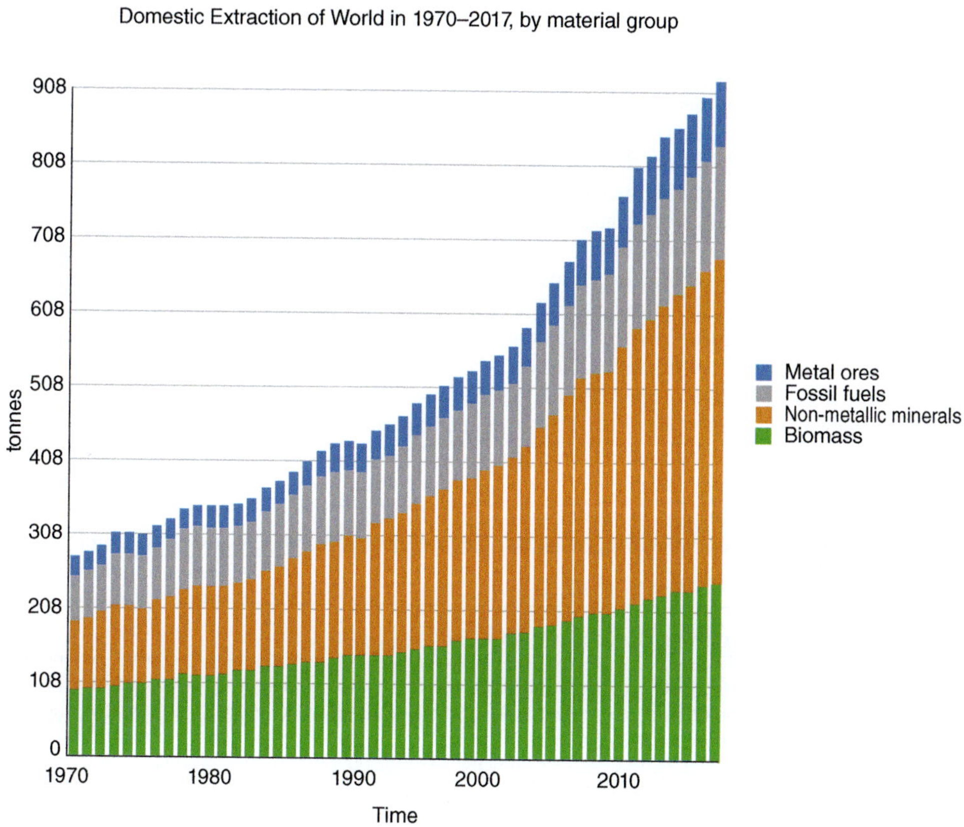

Fig. 8.2. Material Extraction, 1970–2017. Data from MaterialFlows.net. The material flow analysis portal (http://www.materialflows.net/visualisation-centre/data-visualisations/).

Estimates of a feasible target for sustainable resource extraction cluster around 50 billion tonnes a year (Hoekstra and Wiedmann, 2014; UNEP, 2014; Bringezu, 2015; Hickel, 2019). Even if such a figure is acceptable, we are already more than 50% over this target – and still heading in the wrong direction. According to UNEP (2019), if we continue along the BAU trajectory, gross annual resource use will increase drastically by 2060 to something like 190 billion tonnes. At the same time, greenhouse gas emissions will rise by 43%, forests will shrink by more than 20% and other habitats such as grassland and savannahs by a similar amount. If population growth exceeds the UN's medium projection, these figures will be greater (O'Neill *et al.*, 2018).

UNEP (2014, 2016, 2019) has argued for some years that material extraction must not only be decoupled from environmental impact to ensure future human well-being, but also that this task is feasible. Despite this optimism, UNEP concedes that greater efficiencies could cause a 'rebound effect', triggering accelerated economic growth, thus undermining or at least attenuating any success in lowering the physical throughput of resources. It suggests that tax measures could counteract this trend.

UNEP is aware that BAU will make the task impossible. It promotes worldwide implementation of a 'Towards Sustainability' pathway, involving 'an urgent and systemic transformation of how we use and manage natural resources' (UNEP, 2019). To reduce the volume of material extraction, UNEP advocates resource efficiency, technological advance and circular production involving maximum reuse and recycling. Such measures are welcome and indispensable, but are unlikely to be sufficient. The trend over the past decade runs unambiguously the other way – escalating material flows, more extraction, more consumption and more waste.

The pursuit of decoupling may be a mistaken goal in any case. Lenzen *et al.* (2016) noted that 'it is illogical to expect technological progress to stretch to unprecedented limits in order to let populations enjoy unchecked growth in numbers and affluence'. Chasing technical fixes alone appears to rely on magical thinking. Tackling 'unchecked growth in numbers and affluence' appears much harder but may be the only realistic pathway. Some decoupling has occurred but not on the scale needed. While we delay, the problems deepen.

In its 2019 assessment of progress made towards realizing Sustainable Development Goal 12 (sustainable production and consumption), the UN (2019) noted that 'worldwide material consumption has expanded rapidly, as has material footprint per capita'. It warned that 'urgent action is needed to ensure that current material needs do not lead to the over-extraction of resources or to the degradation of environmental resources'. Again, the UN could only offer the same solutions: 'improve resource efficiency, reduce waste, and mainstream sustainability practices across all sectors of the economy'.

Furthermore, much of the decoupling that has been claimed in the rich world relies on its transitions to 'service economies'. Such accounts ignore the outsourcing of manufacturing and extractive industries to the rest of world, with associated environmental impacts (Lenzen *et al.*, 2012).

There is also considerable debate about the concept and measurement of net energy, sometimes called 'energy return on energy invested' or 'energy return on investment' (Box 8.1).

Box 8.1. Energy return on energy investment (EROEI)

Fossil fuels (principally coal, oil and gas) remain the major sources of energy for purposes of transport, heating and electricity production (over 60% globally) and they also serve as a chemical stock for manufacture of plastics and fertilizer. But fossil fuels must be extracted from the environment, whether by drilling, manual digging or robotic shovels, all processes that require energy. Energy once obtained must then be distributed via an infrastructure that must be constructed and maintained. Coal needs road or rail transport networks; gas requires pipelines or tankers and electrical energy needs poles, wires and regulation, whether sourced from fossil fuel or renewables. Manufacture of wind turbines and solar panels also requires energy, as does the mining infrastructure described above. Life-cycle assessment allows a full quantification of the energy invested in any form of energy extracted, the basis of the concept of EROEI.

The EROEI of oil and coal was high originally – some analysts report an average EROEI of over 100 for oil in the late 19th century. In contrast, one review (Louwen *et al.*, 2016) found that, until 2017, solar panels may not have produced any energy globally beyond the amount expended in their manufacture and installation. Under this least optimistic scenario, solar panels were a sink for energy until very recently.

The EROEI for solar is argued to have increased considerably since this assessment, perhaps to 30 or 35, especially in sunny locations. Significantly, the climate impact of solar is less than that of any fossil fuel and will continue to decline as the efficiency of panels improves. The EROEI for wind is generally agreed to be even higher than for solar. Thus, these sources have promise as key substitutes for fossil fuel energy.

Researchers still examine the question of whether renewables will yield energy adequate to power the current consumption-oriented economy. Even if this could occur, critics such as Bardi and Sgouridis (2017) find that the window for successful transition is narrow: it needs a very large investment of available, mainly fossil-derived energy, while still providing enough energy to continue existing services. Moreover, the world's market-based economic system may fail to allocate the requisite resources in the available time-frame – by 2050 in this analysis, though more recent estimates suggest that the time-frame is shorter. In any case, Bardi and Sgouridis were sceptical that market forces can effect this transition. They calculated that, as of 2017, capital investment was only about 10% of what is necessary and energy investment was also inadequate. Though not impossible, the transition to renewables will be challenging and involves a substantially greater rate of energy and capital investment than is currently allocated.

8.5 The Problems of Population and Affluence

Although the rate of population growth has been slowing since the 1960s, the huge base created then still generates much the same increment as was seen at that time – we still add more than 80 million extra people every year. Population growth in wealthy countries such as Australia contributes significantly to impacts on land, oceans and climate. Levels of per capita consumption are much greater than those of poorer countries; and trade has allowed rich countries to reduce their environmental impact by exporting it. In Malaysia, Madagascar, Honduras and Papua New Guinea, 50–60% of species loss is directly attributable to activities conducted on behalf of the global North, for example clearing for crops such as coffee, sisal, cocoa and palm oil, or for timber. Average species decline attributable to such activities in the global South was found to be 30% (Lenzen *et al.*, 2012; Schandl *et al.*, 2016; Wiedemann *et al.*, 2020). Wherever it is found, whether in the rich world or amongst elite minorities in the South, affluence is a key driver of unsustainable material flows. The leading ecological economist Paul Ekins pointed out three decades ago that 'universal opulence' was simply not an option (Ekins, 1991). To provide even 20% of Northern affluence across the whole world population, he calculated, would require a total and immediate freeze on Northern consumption. That challenge has since been amplified by rising inequality (see Chapter 2, this volume) (Alvaredo *et al.*, 2018). Oxfam International (2018) estimated that 82% of all wealth created globally in 2017 flowed to the top 1%, with no increase whatsoever for the bottom 50%.[9] After many decades of extraordinary growth, affluence is still only available to a privileged minority (Higgs, 2014; Hickel, 2017).

UNEP also concedes that 'the level of well-being achieved in wealthy industrial countries cannot be generalized globally based on the same system of production and consumption", not without encroaching further on environmental thresholds that are already at risk (UNEP, 2016, pp. 29–30). Here, UNEP accepts that a different system of production and consumption will be necessary if well-being is to be available to all.

8.6 The Health of Human Populations

The disruption and tragedy of COVID-19 is a foretaste of the chaos that awaits us if 'business as usual' continues to be pursued. The dominant global discourse, both economically and academically, has for several decades been captured by excessive optimism, while voices calling for more fairness and warning of imminent limits to growth have been largely ignored or suppressed.

We are on a trajectory towards even greater catastrophes than those of the 20th century, unless there is a global revolution in how society is managed, especially our economic systems. This will require great leadership, truth-telling, courage and a whole-of-civilization transformation.

Daunting as this sounds, many ingredients do exist, including clean energy technologies, family planning strategies and the principles of taxation and redistribution. More enlightened leadership (both business and political), more widespread awareness of our interdependence with Nature and less blatant hypocrisy (e.g. by 'Davos man' – see Chapter 2) could foster a myriad of encouraging shoots, as billions of ordinary people seek to create a decent future for themselves, and therefore for all of us.

Any real solution must reduce material flows, conserve and restore land, ocean and climate while eliminating poverty and hunger, providing clean water and sanitation and producing health and well-being for all people. This will only be feasible in a reformed system of production and consumption, and if drastic redistribution is implemented between and within countries.

Notes

[1] Thomas Malthus theorized, at the end of the 18th century, that food production would not keep up with population growth, but did not deal with broad issues of the sustainability of modern industrial production, not least because he pre-dated it (see Higgs, 2014, pp. 40–43).

[2] Randers is in broad agreement but argues that the 'pollution crisis' run, a variant of BAU, more accurately reflects the trend.
[3] Economic growth refers to GDP growth throughout this chapter. GDP is measured conventionally, ignoring 'externalities'.
[4] CO_2e refers to carbon dioxide equivalent and takes account of methane and other GHGs. Atmospheric concentration is expressed in parts per million (ppm).
[5] These authors use M to denote million and m to denote metres.
[6] There is debate about whether inequality has been reduced. See Chapter 2 in this volume and Chapter 9 in Higgs (2014) for discussion of this issue and the associated problems of using PPP and the Gini coefficient as tools.
[7] See Ward *et al*. (2016) on the underlying physical realities that govern non-substitutable resources such as land, water, raw materials and energy.
[8] Cooper *et al*. (2018) included all forms of human sediment production in the mining and construction industries, i.e. not just materials that are used but all forms of material shifted from its original location (such as overburden, waste rock etc.); if these are taken into account, total shifting of sediment was 316 billion tonnes in 2015, much greater than UNEP's estimate where waste is excluded (2019).
[9] See also Chancel and Piketty (2017). Milanovic (2022) argued, using data to 2013, that inequality declined. This was primarily due to the reduction of poverty in China.

References

Alvaredo, F., Chancel, L., Piketty, T., Saez, E. and Zucman, G. (2018) *World Inequality Report*. Available at: http://wir2018.wid.world/ (accessed 31 January 2022).

Australia Institute (2014) *How many coal jobs? Check the facts*. The Australia Institute, Canberra. Available at: https://www.tai.org.au/content/how-many-coal-jobs-check-facts (accessed 31 January 2022).

Bardi, H. (2014) *Extracted: How the Quest for Mineral Wealth Is Plundering the Planet*. Chelsea Green, White River Junction, Vermont.

Bardi, U. and Sgouridis, S. (2017) In support of a physics-based energy transition planning: Sowing our future energy needs. *Biophysical Economics and Resource Quality* 2, 14. DOI: 10.1007/s41247-017-0031-2

BirdLife Australia (nd) *East Asia/Australasia Flyway*. Available at: http://datazone.birdlife.org/userfiles/file/sowb/flyways/8_East_Asia_Australasia_Factsheet.pdf (accessed 5 June 2023).

Bringezu, S. (2015) Possible target corridor for sustainable use of global material resources. *Resources* 4, 25–54. DOI: 10.3390/resources4010025

Butler, C.D. and Higgs, K. (2018) Health, population, limits and the decline of nature. In: Marsden, T. (ed.) *The Sage Handbook of Nature*. Sage, London, pp. 1122–1149.

Ceballos, G., Ehrlich, P.R. and Raven, P.H. (2020) Vertebrates on the brink as indicators of biological annihilation and the sixth mass extinction. *Proceedings of the National Academy of Sciences (USA)* 117, 13596–13602. DOI: 10.1073/pnas.1922686117

Chancel, L. and Piketty, T. (2017) *Indian income inequality, 1922–2014: From British Raj to Billionaire Raj?* WID.world Working Paper Series 2017/11. Available at: http://wid.world/document/chancelpiketty-2017widworld/ (accessed 31 January 2022).

Cooper, A.H., Brown, T.J., Price, S.J., Ford, J.R. and Waters, C.N. (2018) Humans are the most significant global geomorphological driving force of the 21st century. *The Anthropocene Review* 5, 222–229. DOI: 10.1177/2053019618800234

Cordell, D., Drangert, J.-O. and White, S. (2009) The story of phosphorus: global food security and food for thought. *Global Environmental Change* 19, 292–305, DOI: 10.1016/j.gloenvcha.2008.10.009

Díaz, S., Settele, J., Brondízio, E.S., Ngo, H.T., Agard, J. *et al.* (2019) Pervasive human-driven decline of life on Earth points to the need for transformative change. *Science* 366, eaax3100. DOI: 10.1126/science.aax3100

Ekins, P. (1991) The sustainable consumer society: a contradiction in terms? *International Environmental Affairs* 3, 243–258.

FAO (2016) *State of the World's Forests*. Food and Agriculture Organization, Rome. Available at: http://www.fao.org/3/a-i5588e.pdf (accessed 31 January 2022).

Hall, C.A. and Day, J.W. Jr (2009) Revisiting the limits to growth after peak oil. *American Scientist* 97, 230–237. DOI: 10.1511/2009.78.230

Hickel, J. (2017) *The Divide. A Brief Guide to Global Inequality and its Solutions*. Penguin, New York.

Hickel, J. (2019) The contradiction of the sustainable development goals: Growth versus ecology on a finite planet. *Sustainable Development* 27, 873–884. DOI: 10.1002/sd.1947

Higgs, K. (2014) *Collision Course: Endless Growth on a Finite Planet*. MIT Press, Cambridge, Massachusetts.

Hoekstra, A. and Wiedmann, T. (2014) Humanity's unsustainable environmental footprint. *Science* 344, 1114–1117. DOI: 10.1126/science.1248365

ILO (2020) Social protection responses to the Covid-19 crisis. Briefing note. International Labour Organization of the UN, Geneva. Available at: https://www.ilo.org/asia/publications/issue-briefs/WCMS_739587/lang--en/index.htm (accessed 31 January 2022).

IPCC (2018) *Global Warming of 1.5°C. Summary for Policy Makers*. Intergovernmental Panel on Climate Change, Geneva. Available at: https://www.ipcc.ch/sr15/chapter/spm/ (accessed 31 January, 2022).

Kulp, S.A. and Strauss, B.H. (2019) New elevation data triple estimates of global vulnerability to sea-level rise and coastal flooding. *Nature Communications* 10, Article no: 4844. DOI: 10.1038/s41467-019-12808-z

Lappé, F.M., Clapp, J., Anderson, M., Lockwood, R., Forster, T. *et al.* (2013) *Framing Hunger: A Response to The State of Food Insecurity in the World 2012*. Small Planet Institute, Cambridge, Massachusetts. Available at: https://www.iatp.org/sites/default/files/Framing%20Hunger.pdf (accessed 1 February 2022).

Lenzen, M., Moran, D., Kanemoto, K., Foran, B., Lobefaro, L. and Geschke, A. (2012) International trade drives biodiversity threats in developing nations. *Nature* 486, 109–112. DOI: 10.1038/nature11145

Lenzen, M., Malik, A. and Foran, B. (2016) Letter (reply to Schandl *et al.*, 2016). *Journal of Cleaner Production* 139, 796–798. DOI: 10.1016/j.jclepro.2016.08.037

Louwen, A., van Sark, W.G., Faaij, A. and Schropp, R. (2016) Re-assessment of net energy production and greenhouse gas emissions avoidance after 40 years of photovoltaics development. *Nature Communications* 7, 13728. DOI: 10.1038/ncomms13728

McAfee, A. (2019) *More From Less*. Simon and Schuster, London and New York.

Meadows, D., Meadows, D., Randers., J. and Behrens, W. III (1972) *The Limits to Growth*. Universe Books, New York.

Michaux, S. (2017) *Peak Industrial Output and the Limits to Growth as a Consequence of Depleting Natural Resources*. Data Consulting Dynamix. Available at: https://tuxdoc.com/embed/peak-industrial-output-and-the-limits-to-growth-as-a-consequence-of-depleting-natural-resources-and-appendices-acme-v2_pdf.html (accessed 27 January 2022).

Milanovic, B. (2022) After the financial crisis: The evolution of the global income distribution between 2008 and 2013. *Review of Income and Wealth* 68, 43–73. DOI: 10.1111/roiw.12516

O'Neill, D.W., Fanning, A.L., Lamb, W.F. and Steinberger, J.K. (2018) A good life for all within planetary boundaries. *Nature Sustainability* 1, 88–95. DOI: 10.1038/s41893-018-0021-4

OECD (2011) *Towards green growth*. Organisation for Economic Co-operation and Development, Paris. Available at: https://www.oecd.org/greengrowth/48012345.pdf (accessed 31 January 2022).

Oxfam International (2018) *Reward work, not wealth*. Oxfam International, Nairobi. Available at: https://www.oxfam.org/en/research/reward-work-not-wealth (accessed 31 January 2022).

Page, S.E. and Hooijer, A. (2016) In the line of fire: the peatlands of Southeast Asia. *Philosophical Transactions of the Royal Society B: Biological Sciences* 371, Article no. 20150176. DOI: 10.1098/rstb.2015.0176

Paull, J. (2011) The making of an agricultural classic: farmers of forty centuries or permanent agriculture in China, Korea and Japan, 1911–2011. *Agricultural Science* 2, 175–180. DOI: 10.4236/as.2011.23024

Persson L., Carney Almroth, B., Collins C.D., Cornell, S., de Wit, C.A. *et al.* (2022) Outside the safe operating space of the planetary boundary for novel entities. *Environmental Science and Technology* 56, 1510–1521. DOI: 10.1021/acs.est.1c04158

Randers, J. (2018) *The Limits to Growth (1972) in a 50-year perspective*. 50th Anniversary Conference of the Club of Rome, Oct 17, Rome. Available at: https://www.youtube.com/watch?v=ykFXvUDnw9Y (accessed 31 January 2022).

Richardson, K., Steffen, W., Lucht, W., Bendsten, J., Cornell, S. *et al.* (2023) Earth beyond six of nine planetary boundaries. *Science Advances* 9, eadh2458. DOI: 10.1126/sciadv.adh2458

Ripple, W.J., Wolf, C., Newsome, T.M., Galetti, M., Alamgir, M. *et al.* (2017) World scientists' warning to humanity: a second notice. *BioScience* 57, 1026–1028. DOI: 10.1093/biosci/bix125

Rockström, J., Steffen, W., Noone, K., Persson, Å., Chapin, F.S. *et al.* (2009) A safe operating space for humanity. *Nature* 461, 472–475. DOI: 10.1038/461472a

Schandl, H., Hatfield-Dodds, S., Wiedmann, T., Geschke, A., Cai, Y. *et al.* (2016) Decoupling global environmental pressure and economic growth: scenarios for energy use, materials use and carbon emissions. *Journal of Cleaner Production* 132, 45–56. DOI: 10.1016/j.jclepro.2015.06.100

Schellnhuber, H.J. (2018) *2018 Aurelio Peccei Lecture: 'Climate, Complexity, Conversion'*. 50th Anniversary Summit of the Club of Rome, October 17, Rome. Available at: https://www.youtube.com/embed/Rtg5QJlb484 (accessed 31 January 2022).

Schmelzer, M. (2016) *The Hegemony of Growth: The OECD and the Making of the Economic Growth Paradigm*. Cambridge University Press, Cambridge, UK.

Smil, V. (2000) Phosphorus in the environment: natural flows and human interferences. *Annual Review of Energy and the Environment* 25, 53–88. DOI: 10.1146/annurev.energy.25.1.53

Steffen, W., Richardson, K., Rockström, J., Cornell, S.E., Fetzer, I. *et al.* (2015a) Planetary boundaries: Guiding human development on a changing planet. *Science* 347, 736–746. DOI: 10.1126/science.1259855

Steffen, W., Broadgate, W., Deutsch, L., Gaffney, O. and Ludwig, C. (2015b) The trajectory of the Anthropocene: the Great Acceleration. *Anthropocene Review* 2, 81–98. DOI: 10.1177/2053019614564785

Steffen, W., Rockström, J., Richardson, K., Lenton, T.M., Folke, C. *et al.* (2018) Trajectories of the Earth System in the Anthropocene. *Proceedings of the National Academy of Sciences (USA)* 115, 8252–8259. DOI: 10.1073/pnas.1810141115

Turner, G.M. (2014) *Is Global Collapse Imminent? An Updated Comparison of The Limits to Growth with Historical Data*. Melbourne Sustainable Society Institute, The University of Melbourne, Melbourne, Australia.

UN (2019) *Report of the Secretary-General on SDG Progress 2019: Special Edition*. United Nations, Rome. Available at: https://sdgs.un.org/documents/report-secretary-general-sdg-progress-20-25571 (accessed 31 January 2022).

UNEP (2014) *Managing and Conserving the Natural Resource Base for Sustained Economic and Social Development*. United Nations Environment Programme, Nairobi, Kenya. Available at: irp_think_piece_contributing_to_the_sdgs_process.pdf (accessed 31 January 2022).

UNEP (2016) *Global material flows and resource productivity. Summary report*. United Nations Environment Programme, Nairobi, Kenya. Available at: https://www.resourcepanel.org/global-material-flows-and-resource-productivity-summary-english (accessed 31 January 2022).

UNEP (2019) *Global Resources Outlook 2019*. United Nations Environment Programme, Nairobi, Kenya. Available at: https://www.resourcepanel.org/reports/global-resources-outlook (accessed 31 January 2022).

UNEP, ILRI (2020) *Preventing the next pandemic: Zoonotic diseases and how to break the chain of transmission*. United Nations Environment Programme, Nairobi, Kenya; International Livestock Research Institute, Nairobi. Available at: https://www.unenvironment.org/resources/report/preventing-future-zoonotic-disease-outbreaks-protecting-environment-animals-and (accessed 31 January 2022).

Union of Concerned Scientists (1992) *World Scientists' Warning to Humanity*. Union of Concerned Scientists, Cambridge, Massachusetts.

Ward, J., Sutton, P., Werner, A., Costanza, R., Mohr, S. and Simmons, C. (2016) Is decoupling GDP growth from environmental impact possible? *PLoS ONE* 11, e0164733. DOI: 10.1371/journal.pone.0164733

Wiedmann, T., Lenzen, M., Keyßer, L.T. and Steinberger, J.K. (2020) Scientists' warning on affluence. *Nature Communications* 11, 3107. DOI: 10.1038/s41467-020-16941-y

World Bank (2012) Inclusive Green Growth – the Pathway to Sustainable Development. World Bank, Washington, DC. Available at: https://openknowledge.worldbank.org/handle/10986/6058 (accessed 31 January 2022).

WWF (2016) *Living Planet Report*. World Wide Fund for Nature, Gland, Switzerland. Available at: https://www.worldwildlife.org/pages/living-planet-report-2016 (accessed 31 January 2022).

WWF (2020) *Living Planet Report*. World Wide Fund for Nature, Gland, Switzerland. Available at: https://www.wwf.org.au/knowledge-centre/living-planet-report (accessed 28 January 2022).

9 Population, Neoliberalism and 'Human Carrying Capacity'

Colin D. Butler
National Centre for Epidemiology and Population Health, the Australian National University, Canberra, Australia

Abstract
This chapter reviews the history of human population size and policy over the past century, as an issue relevant for global health and development. It focuses on the role of neoliberalism (the ideology that claims free-market forces will maximize wealth for all) and the Roman Catholic church as central to the recent neglect and avoidance of this topic. This gap is evident in the teaching of not only global public health but also many other scientific disciplines. The chapter also analyses the concept of human carrying capacity, using a 'five capitals approach', i.e. natural, social, human, built and financial resources. Using this conceptual framework, the chapter discusses the conflict in Tigray, Ethiopia, as a case study. The following chapter (Sexual and Reproductive Health and Rights: The Relevance of Family Planning) complements this one.

> The Malthusian specter of overpopulation looms large in pervasive images of imminent ecological disaster in ways that are profoundly depoliticizing and that serve projects of militarization, misogyny, and racism (Ojeda *et al.*, 2019).

9.1 First Things First

The predominant causes for climate change are found in the lifestyles of the people who live and who have lived in the 'global North', as part of the richest one billion people on the planet – people like the author. An analysis in 2007 by the Global Carbon Project is instructive: it found that 'developing' and 'least-developed' economies (then forming approximately 80% of the world's population) accounted for only 23% of *cumulative* global carbon dioxide (CO_2) emissions from the mid-18th century until 2004 (Raupach *et al.*, 2007). An update, for the period 1850–2020, found that CO_2 emissions from the USA and European Union were 43% of the total, with China responsible for 14%. Although a fraction of China's cumulative emissions, especially recently, are to produce goods consumed in the West, the recent destruction of dozens of unoccupied, unfinished high-rise towers in different parts of China (He, 2022) is evidence that China, too, is a profligate producer of CO_2 emissions.

This chapter does not seek to attribute climate change (or any other approaching or exceeded planetary boundary) primarily to the actions of people or size of population in the majority world (the global South);[1] however, there are occasional claims, both in the academic literature and from some journalists[2], to the contrary. For example, Ojeda *et al.* (2019) claimed that the blame for climate change is 'often placed' on 'marginalized populations who have done little to cause the problem in the first place'. This chapter argues that such claims are best regarded as 'strawmen' –

 Climate Change and Global Health: Primary, Secondary and Tertiary Effects, 2nd Edition (eds C.D. Butler and K. Higgs)
DOI: 10.1079/9781800620025.0009

allegations that are barely, if at all, supported by the scholarly literature.

Nevertheless, human population size (both in absolute numbers and regional growth rates) does matter, not only for future climate change (and its consequences), but also for many other aspects of human well-being and global health. This is, for example, recognized by Project Drawdown, subtitled 'the most comprehensive plan ever proposed to reverse global warming'. It lists educating girls and family planning as among its top ten priorities, delivering the greatest payoff for future climate change (Hawken, 2017). Several of the Sustainable Development Goals (SDGs) promote improving determinants of slower population growth, in low-income settings, including female education and family planning (Target 3.7).[3] This and the following chapter seek to explore these complicated, challenging and often subtle ideas.

There are nuances. If technology could completely or even substantially 'decouple'[4] environmental damage from the consumption of goods and services, levels of atmospheric greenhouse gases (GHGs) would flatten and then decline. To date, despite claims to the contrary, there is scant evidence for decoupling (Vadén *et al.*, 2020). Another complication is that a substantial and growing fraction of GHG emissions released in low-income settings serve consumption in the global North. If a cotton shirt is made in a factory in Dhaka, Bangladesh, for sale in a Berlin boutique, then the emissions involved in growing the cotton and operating the factory should largely be attributed to the German wearer. This is not the case.

Another twist is that a few overconsumers live in 'developing' countries. The late President of the Democratic Republic of the Congo (DRC), Mobutu Sese Seko, illustrates this. Mobutu is reported to have occasionally flown Paul Bocuse and other top European chefs by Concorde to Gbadolite, his 'African Versailles', over 1000 km from the DRC capital, Kinshasa (Smith, 2015). There is also a significant number of people with comparatively low ecological footprints who live in high-income settings, including a substantial underclass of refugees and asylum seekers who seek greater opportunity but whose aspirations are often thwarted. Their GHG emissions are modest by global standards.

Despite these exceptions, it is clear that most responsibility for GHG emissions lies with those with the highest incomes, who also have the greatest political power (Otto *et al.*, 2019). This chapter will also argue that considerable responsibility for persistent poverty in low-income settings should be attributed to the same group: those in the world with the greatest political power. One consequence of their exercise of power (at least in part) is high population growth in most countries of the South, including for many sub-populations who are the poorest and most vulnerable (Finkle and Crane, 1985). The rest of this chapter will not discuss climate change but, instead, focus on other aspects of global health and population.

Venturing into the swamp of population, security and development

The topic of population and its relationship with security and development is treacherous: quicksand avoided not only by environmentalist journalists (Roberts, 2018) but also by many courses that teach public and global health. This chapter will seek to explain how topics that were once mainstream have been tainted and suppressed in a process that has been called the 'Hardinian[5] taboo' (King and Elliot, 1997). This chapter also proposes that lasting solutions for human development, in the context of the global environmental crisis, require courageous thinking, including a willingness to contemplate the existence of limits – not on well-being but on human transgressions against nature.

9.2 Human 'Carrying Capacity'

The recognition by humans that they modify, and sometimes degrade, the landscapes in which they live, including their resources such as water and wild animals, is ancient[6] (Catton, 1980), understood by most Indigenous peoples. However, for many thousands of years, humans could spread out, gradually colonizing the entire Earth (not yet Antarctica) in order to alleviate these stresses. In some cases, such migration was to territory not previously occupied by other human beings but in many cases, even long ago, migration did bring humans together, often in competition, and sometimes with other hominin species, such as Neanderthals.

Some ancient peoples appear to have developed ways to restrict their human population growth, for example the Indigenous people of Australia, especially in the arid interior, where options for farming[7] were restricted. The methods used may have included sub-incision, a surgical means to reduce male fertility, as well as rules and customs (Gebbie, 1981, pp. 337–358; Butler, 2012). Australia's Indigenous people still foster a deep respect for 'country' (the land to which they belong) and there is evidence for similar views in many other Indigenous belief systems. Respect for country extends to animal, plant and other natural resources.

At the same time, modern and ancient humans frequently modified their landscapes, by farming and in other ways that regulated species; in this way, they sought (and seek) to improve their living standards and, in many cases, to increase their own numbers. Humans are not unique in doing this, whether consciously or unconsciously.[8]

The geographer Sayre (2008) traced the term 'carrying capacity' to the 1840s,[9] when it was used in the context of international shipping to establish a safe load or limit. Sayre also explained that, by the late 19th century, the evolving field of ecology had begun to apply the term to natural systems, such as the carrying capacity of a grassland for a herd or flock.[10] The fact that the maximum size of a group of farmed animals is limited for any set of resources is understood by every successful farmer seeking to improve pasture, provide adequate water, or supply supplementary feed to their flock. Stocking density (and quality) is governed by resources (e.g. food, shelter, climate), as well as predators, pathogens and stressors, from heat to snow.

The idea that there may be a finite planetary human carrying capacity (HCC) is extremely controversial among social scientists. Joel E. Cohen (Professor of Populations at the Rockefeller University and Columbia University in New York City, USA) described HCC as a 'specious metaphor' that is 'superficially plausible but actually wrong'. He stated that 'Earth's capacity to support people is determined both by natural constraints, which some will emphasize, and by human choices, which others will emphasize' (Cohen, 2017). As in his lengthy book called *How Many People Can the Earth Support*, Cohen (1995) sidestepped the quagmire, in part by avoiding the issue of conflict among rival human groups. Demographer and editor Paul Demeny (1988) was also critical of the term; while accepting that human population size is limited, he also speculated that this limit might be far in the future. A team[11] led by Economics Nobel Laureate Kenneth Arrow was less reluctant to discuss the term 'carrying capacity'. Arrow *et al.* (1995) also suggested a rational basis for determining the level of 'fullness'. They cited research finding that humans appropriated, for their own use, an estimated 40% of the global photosynthetic product (Vitousek *et al.*, 1986), implying that limits to the human impact on our planet must be approaching, and was unlikely to be 'far in the future'. Demeny was either unaware of this finding or chose to overlook it. Arrow *et al.* (1995) also recognized that HCC is contingent on 'technology, preferences, and the structure of production and consumption'. The key word here is 'preference'. Some humans may prefer to live within their means; others to seek to predate on others.

In more recent years the term HCC has largely disappeared from interdisciplinary journals, although it resurfaced in 2020 in *The Lancet* when Abubakar (2020) commented on a major report on the likely evolution of the size of the human population in this century. Neither HCC nor population is mentioned in the 'planetary boundaries' literature which instead uses apparent proxies such as 'a safe operating space for humanity'. Phrases such as this suggest that living conditions for many, or perhaps most, people on Earth will deteriorate beyond the 'safe' zone. This does not necessarily mean that human numbers will decline in the unsafe zone, but it is possible.

It seems plausible that the contributors to the planetary boundaries papers wish to avoid a perception of being too definite, pessimistic, or controversial. However, one contributor, the physicist Hans Schellnhuber, the founding and recent director of the Potsdam Institute for Climate, used the term carrying capacity in his keynote address at the scientific meeting in association with the 2009 Conference of the Parties of the UN Framework Convention on Climate Change (not recently in print). In 2018, Schellnhuber wrote that 'climate change is now reaching the end-game, where very soon humanity must choose between taking unprecedented action, or accepting that it has been left too late and bear the consequences'.

9.3 Towards a Theory of Human Carrying Capacity

Estimating HCC for Earth is far more difficult than estimating how much food the planet can grow and allocating the resultant total number of calories to a theoretical population at a given distribution (the main approach adopted by Cohen (1995). This is primarily because humans do not yet cooperate sufficiently at either a regional or global scale. An additional complicating factor is that global environmental change, whether human-made or natural[12], could make Earth less fertile, or in other ways less conducive to supporting over 8 billion people.

Relevant to this discussion is the formula I = fn(PAT), where environmental impact (I) is a function[13] of population (P), affluence (A) and technology (T) (Ehrlich and Holdren, 1971). This formula implies that increased human population will cause additional environmental impact (harm), for any given level of affluence and technology, but does not directly lead to an estimate of human carrying capacity (Figs 9.1 and 9.2).

Recognizing these complexities, McMichael and Butler (2002) argued that HCC can be conceptualized as an emergent property of five kinds of capital, or wealth. These are: (i) 'natural' capital (e.g. a stable climate, compatible with global civilization, fertile soil, and adequate energy that can be harnessed, including fossil fuels); (ii) 'human' capital (e.g. skills, health, talents, grudges, allegiances); (iii) 'built' capital (e.g. cities, roads, technology, fences); (iv) 'financial' capital (a form of lubricant, allowing partial exchange of the other forms of capital); and (v) 'social' capital. Social capital (related to social cohesion) is consciously named last here, because discussing it veers towards the quagmire.

Humans are social animals, who cooperate well (generally) in groups such as families, football team supporters and tribes. Some groups cooperate at international scales, united by ethnicity, language, religion, or shared cultural belief. But

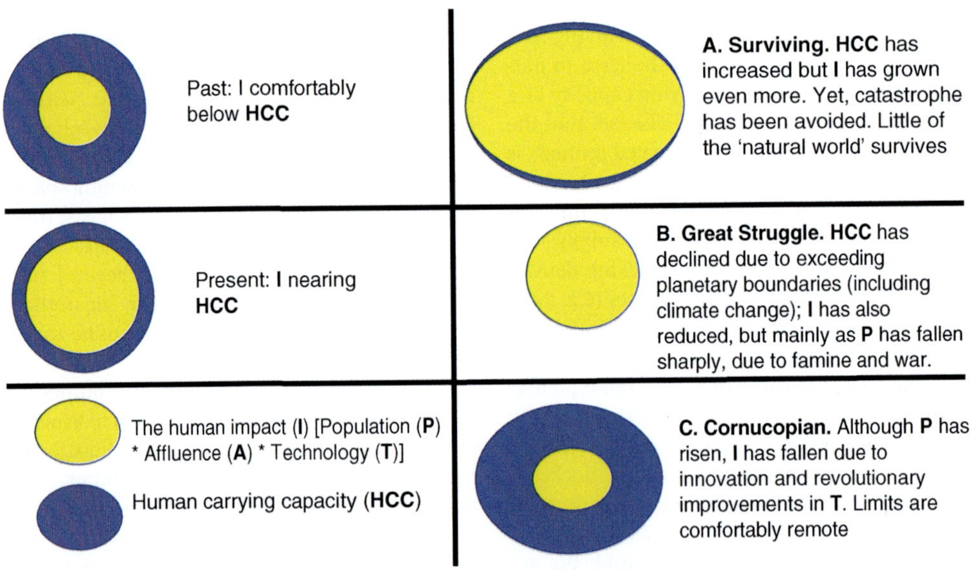

Fig 9.1. The relationship between population (P), affluence (A), technology (T), human impact on the environment (I) and HCC and three futures (A, B and C). As P rises, so does I, except in the 'cornucopian' scenario. A reduced P is avoided in scenario A, primarily because of the increased scale of cooperation, avoiding war and by distributing food surpluses to those in deficit. In scenario B, limits are passed, P declines in response (also reducing I but this should be no consolation). Optimists argue that the blue area is still large, but few natural scientists agree. HCC is a product of natural, human, social, built and financial capital.

A. Simple model: Environmental impact (I) and human carrying capacity (HCC)

Total **I** (yellow circle) is the sum of the impacts of numerous identifiable human groups, each with their own level of population (**P**), affluence (**A**) and technology (**T**). Total **I** is less than **HCC** (blue circle)

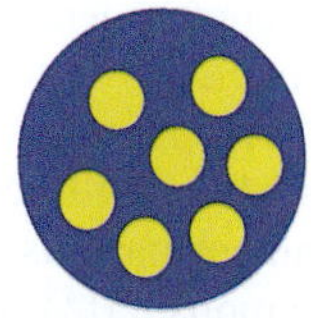

B. More realistic model: I is the sum of the impacts of numerous identifiable human groups (yellow circles). Conflict between groups is avoided (or at least minimized) if there is no perceived overlap between the groups. In reality, high greenhouse gas emissions by some groups does affect the living standards of other groups; today, no human group lives completely free from interference of some groups. Blue space is potential **HCC**, which could be used for either more humans, or left for other species.

C. Highly cooperative model: I has expanded (increased small yellow circles) but with minimal conflict, because of extraordinary co-operation between groups. If cooperation can be maximized then (perhaps) all the blue area could be occupied. Neither ancient nor contemporary history suggest this is likely, nor do the principles of evolution (which in part build on the insights of Thomas Malthus).

Fig 9.2. Three models of the relationship between environmental impact and HCC at the planetary scale. All are simplified, but models that consider total global resources and total population, without also considering distribution, cooperation and conflict, are the least insightful.

the scale of this cooperation is not yet global, and in many cases not even national, particularly in countries whose boundaries were determined by colonial powers, such as in Africa, Asia and the Americas (Kokole, 1994). The political scientist Christian Leuprecht (2010) coined the term 'demographic security dilemma' to explain why, in some cases, minority populations have comparatively high birth rates. He wrote that 'in instances where the dominant ethnic group enjoys overwhelming political, economic, and ethnic group military control, the sole remaining retaliatory "weapon" at the minority's disposal is demographic: fertility'.

One of the greatest difficulties of discussing human population size is the question of which population should be reduced. Whereas it may not be controversial to agree that the global population size is now (or soon will be) too high, and that it can be gradually lowered, it is definitely controversial to identify any single subgroup, or even a large number of sub-groups, whose population should decline. In fact, it is so controversial that its shadow inhibits almost all sustainability scholars from discussing the broader issue.

Consideration of the civil war in Ethiopia, focused mainly in the northern border state of Tigray, may help crystallize the generally abstract ideas presented in this chapter. (At the time of writing, peace talks look encouraging.) The causes of this war in the Horn of Africa, where subsistence farming is the main source of food, are complex and contested. A well-recognized factor is the resentment of many Ethiopians of non-Tigray ethnicity towards the Tigrayans. This reflects a comparatively low social capital within Ethiopia more broadly – although not among the people of Tigrayan ethnicity (Ibreck and de Waal, 2021).

In 2019/20 a severe outbreak of the desert locust (*Schistocerca gregaria*) occurred, which Salih *et al.* (2020) linked to climate change. It is here suggested that the crop losses caused by these locusts may be an under-appreciated factor in the initiation of this war, reducing the region's natural capital, and thus, at least temporarily, its HCC. Of course, given sufficient global (or even regional) social capital, any temporary loss of food caused by locusts or any other factor could easily have been replaced, either by donated food relief or by using financial capital to import food, possible in a world with ample and cheap food (see Chapter 22). But the use of financial capital by the primary victims of the war, the people of Tigray, is limited, as they have a low credit rating. Those with greater power are reluctant to lend, as they fear a loss of their investment.

There are also claims – albeit strongly contested – that forces allied to the Ethiopian government have used starvation as a weapon (Anonymous, 2021; Garrison, 2023). The war has also resulted in considerable damage to infrastructure which, in the case of banks, reduced the capacity to transfer funds into Tigray, for example from supporters and Tigrayan expatriates. Human capital is also reduced, not only as skilled people flee or are killed, but also as physical and emotional damage accrues in survivors, including by the widespread use of rape and other forms of sexual violence, some of it extreme and degrading (Mazurana *et al.*, 2021). As a result of war, the HCC of Tigray is clearly reduced. More speculatively, the war may have been initiated in part because the HCC was nearing a critical threshold.

9.4 1920–1980: Concerns about Planetary 'Overpopulation'

Recognition of the scale and distribution of the global human population as a major problem emerged after World War I (1914–1918), with, for example, the publication of books such as *The Population Problem* (Carr-Saunders, 1922) and *Standing Room Only?* (Ross, 1928). In 1927 the pioneering birth control advocate, Margaret Sanger, convened the World Population Conference. In its later years (the 1930s), the League of Nations, the United Nations (UN) predecessor, was strongly concerned with the issue of overpopulation (Bashford, 2008).

In 1944, Kirk noted that 'a generation ago ... the writings of demography were filled with the dangers of overpopulation' (Kirk, 1944). He also commented that:

> The day is rapidly passing when a handful of Europeans, equipped with superior weapons and a complacent and somehow contagious faith in white supremacy, can expect indefinitely to dominate the half of the world that is occupied by the coloured peoples.

After World War II, concerns about overpopulation continued at the highest levels for over four decades. The historian Bashford identified six organizations that arose from the newly formed UN and were concerned with aspects of family planning and world population policy, including the World Health Organization and UNESCO (UN Educational, Scientific and Cultural Organization). UNESCO was initially led by the eugenicist Julian Huxley,[14] who had attended the 1927 World Population Conference.[15] Concerns about population peaked in the late 1960s, as the rate of global population growth crested at about 2.1% per annum. In 1969 the UN Fund for Population Activities (UNFPA) was created, renamed in 1987 as the UN Population Fund (Bashford, 2008). More books appeared warning of the growing imbalance between population and resources, such as *Famine 1975! America's Decision: Who Will Survive?* (Paddock and Paddock, 1967) and, most famous of all, *The Population Bomb* (Ehrlich, 1968).

In 1974, in Bucharest, Romania, the UN held the first of three major decadal global conferences, to discuss population and development, following smaller meetings in 1954 and 1965. Richer countries, including the USA at the time, were strong promoters of family planning, but poorer countries were reluctant to be given orders by 'developed' countries. Resenting the unfairness of the international system, poorer countries instead called for a 'New International Economic Order'. Illustrating the mood of the meeting was the call by Indian representative Dr Karan Singh, the Indian Minister for Health, who asserted that 'development is the best contraceptive'[16] (Sinding, 2000). The meeting adopted a world population plan of action, which explicitly authorized each sovereign nation to make its own decisions about its population trajectory (Finkle and Crane, 1985). This is discussed in more detail in the following chapter.

9.5 The 1980s: The Rise of Neoliberalism and Its Influence on US Population Policy

When US President Ronald Reagan took office in 1981, he was initially expected to follow his predecessors and be sympathetic to the issue of family planning. But this did not occur. By 1984 Reagan responded to a question about the 'population explosion' by stating that it has been 'vastly exaggerated – over-exaggerated' (Reagan and Mondale, 1984; Finkle and Crane, 1985). A key factor in this approach was the ascendancy of neoliberalism and 'cornucopian' thinking.

Early in Reagan's presidency Julian Simon published his book *The Ultimate Resource* (Simon, 1981). This combined a typical pro-natalist view with cornucopianism. Simon considered that additional people could contribute to solutions, at least as much as they might lead to problems (as Mao Zedong the first chairman of the Chinese Communist Party, once also believed).[17] Simon and his supporters also argued that, provided markets can flourish unencumbered by regulations, the Limits to Growth – if they exist at all – are many generations away.

Funded in large part by billionaires such as the Koch Brothers,[18] the influence of neoliberalism grew rapidly in the USA (Higgs, 2014). Initially, the Population Association of America, representing US demographers and population specialists, were irate about the change in the official US position. This revised position between population growth and development was announced in the lead-up to the second major post-war global population conference, held in Mexico City in 1984. The Association released a statement commenting that the authors of the draft US report, were 'either unaware of 50 years of demographic research, or deliberately ignored it' (Finkle and Crane, 1985). In the USA, over 200 editorials were published challenging the government position, but without success. Cementing the influence of the cornucopians, the US National Academy of Science held a second enquiry into population and development, releasing a report in 1986 that was substantially influenced by Simon (National Research Council, 1986).[19]

Although not explicitly confirmed in any Western government archive, nor via an insider's leak, high population growth in low-income countries is convenient for capitalism, because large populations in a globalized world (which neoliberalism promoted) depress wages on a global scale. Marx and his followers also recognized the role of excess population.

> The greater this reserve army in proportion to the active labour-army, the greater is the mass of a consolidated surplus population, whose misery is in inverse ratio to the amount of torture it has to undergo in the form of labour. The more extensive, finally, the pauperized sections of the working class and the industrial reserve army, the greater is official pauperism. This is the absolute general law of capitalist accumulation. (Foster *et al.*, 2011).

This assertion is at the heart of my argument. Lack of explicit evidence may be seen as weakness, but it is here proposed that the acceptability of Simon's arguments by conservative (neoliberal) policy makers in support of *laissez faire* population growth in low-income settings was, in part, fostered by shared thinking about this issue, even if this cannot be documented. An alternative interpretation is that Simon's supporters sincerely believed that market forces would generate a 'win-win' for everyone, rich and poor. That view, which can also only be inferred, is naïve from the point of view of George Kennan (1948), the US elder statesman, who stated in a then top-secret government report:

> ... we have about 50% of the world's wealth but only 6.3% of its population. This disparity is particularly great as between ourselves and the peoples of Asia. In this situation, we cannot fail to be the object of envy and resentment. Our real task in the coming period is to devise a pattern of relationships which will permit us to maintain this position of disparity without positive detriment to our national security. To do so, we will have to dispense with all sentimentality and day-dreaming; and our attention will have to be concentrated everywhere on our immediate national objectives. We need not deceive ourselves that we can afford today the luxury of altruism and world-benefaction.'

9.6 The Catholic Church: Undermining Family Planning

The Catholic Church has also played a major role in suppressing discussion of human overpopulation, as did (possibly inadvertently) some in the women's movement (see next chapter) and evangelical protestants. In 1992, Milton P. Siegel, a key figure in the founding of WHO, claimed that the Vatican had successfully blocked the incorporation of family planning and birth control into official WHO policy.[20]

In the 1960s, Pope Paul VI commissioned an enquiry into Population and Birth Control, which met from 1964 until 1966. It was established to assess the legitimacy of 'artificial' forms of birth control, such as condoms and the oral contraceptive pill, the latter of which was then newly invented. The Commission had two layers, a group of 15 cardinals and bishops and a second tier of 64 lay (Catholic) experts representing several academic disciplines. Mumford (1986) reported that, according to commission member Thomas Burch, Pope Paul VI 'assigned them the task of finding a way to change the Church's position on birth control without destroying papal authority'. The Commission strongly supported a change in position, voting 69 to 10 in favour. However, a minority of six clerics submitted a dissenting report to the Pope. Co-author of this report was the young Archbishop of Cracow, Karol Wojtyla, later elected as Pope John Paul II, following the unexplained death, after one month in office, of Pope John Paul I. Curiously, the investigative reporter Yallop (1984) claimed that Pope John Paul I was a supporter of greater access to family planning. As the influence of neoliberalism grew, not only in the global South but via its instruments such as the World Bank (George, 1999), the Catholic Church lobbied extensively, especially in the USA, to reposition US family planning policy away from its earlier stance of encouraging greater access to contraception (Mumford, 1986).

The causes of the Hardinian taboo

Since the 1980s, in most high- and low-income settings (the People's Republic of China being the most important exception), discussion of the relationship between population growth and human development has been largely suppressed, forgotten or self-censored. While there have been many attempts to revive the debate (Kelley, 2001; Butler, 2004; Campbell *et al.*, 2007; Crist *et al.*, 2017) it remains largely underground, to this day. Of the many factors for this it is suggested that the taint of eugenics and the associated fear of being labelled racist are the most powerful.

Some past advocates of population 'control' supported eugenics[21] and forms of coercion such as forced sterilization and forced abortion (Hartmann, 1987). Others are likely to have supported attempts to lower fertility in low-income settings due to an apprehension that consequent turmoil could produce a security risk (Coole, 2021).

Coole wrote that 'Americans worried that rapid population growth was a destabilising factor encouraging Communist insurgencies'. Former US President Richard Nixon, an ardent supporter of family planning, wrote in his book *Seize the Moment: America's Challenge in a One-Superpower World*:

> At current rates the underdeveloped world's population will increase 23% by the year 2000 and 77% by the year 2025. Countries such as Mozambique, Ethiopia, Tanzania and Somalia will need to maintain real economic growth rates greater than 3% just to keep their per capita incomes from dropping. Unchecked population growth will put them on an ever-accelerating treadmill that will outpace any potential economic performance. (Nixon, 1992, pp. 263–4).

To interpret Nixon's motivation as entirely humanitarian would be naïve. The Green Revolution (discussed in Chapter 22) was called Green because, in part, it hoped to eliminate the chance of a Red (Communist) revolution. It was widely perceived that rapid population growth in the global South would ferment not only hunger but also other revolutionary tendencies.

9.7 Conclusion

Support for greater access to family planning among the poor (by anyone, including those who are privileged) can be motivated by altruism: a genuine desire to reduce poverty, improve health and lower the risk of famine and civil war. However, such support may conceal covert discrimination against the poor, exercised by those with more economic and political power. Historically, some supporters of family planning appear to have considered that a major reason for poverty is genetic, thus discounting social factors including a history of colonization, discrimination, undernutrition and inadequate education, which combine to thwart opportunity (Selden, 2005).

However, there is little if any evidence that modern support for international family planning programmes and awareness reflects disguised discrimination, eugenics, or prejudice, despite occasional protests to the contrary (Monbiot, 2020). Nor is there yet much evidence, by those concerned with climate change, of calls to reduce population growth in the South. This should be called for – especially in countries with high fertility rates – but not primarily to slow climate change. Far more pressing reasons and needs exist.

Some active supporters of family planning are from the Global South. They strive to highlight the benefits for health from increased child spacing, and improved human development enabled by the 'demographic dividend' and investment in human capital, including education (Starrs *et al.*, 2018). The topics of population, population growth, family planning and HCC warrant revival in education courses concerning global health, development, human rights and the global environment.

Notes

[1] Although much (probably most) of the responsibility for global biodiversity loss lies with the population of the global North, the size of the population in the global South is also a substantial contributor (Crist *et al.*, 2017).

[2] A prominent journalist who claims this is George Monbiot. His statement that he 'naively imagined that people in rich nations would at last stop blaming all the world's environmental problems on population growth. I was wrong. If anything, it appears to have got worse' is characteristic (Monbiot, 2020).

[3] Target 3.7 of the Sustainable Development Goals (SDGs) calls on countries 'by 2030, to ensure universal access to sexual and reproductive health-care services, including for family planning, information and education, and the integration of reproductive health into national strategies and programmes'.

[4] Decoupling, at its extreme, is a form of magic; the idea that humans can consume limitless goods and service with no environmental cost, not only to the climate but other aspects of nature. Partial decoupling is possible, but is often defeated by Jevons' paradox, i.e. increased use of a more environmentally benign technology.

[5] Garrett Hardin was a Californian ecologist, most famous for his article 'The Tragedy of the Commons' (published in *Science* in 1968).

[6] *Ecclesiastes* (5:11) has been translated as meaning 'when goods increase, they are increased that eat them'.

[7] Pascoe's book *Dark Emu* documents farming, both of fish and plants, in some parts of Australia prior to its invasion by Europeans in 1788. Gammage's book *The Biggest Estate on Earth: How Aborigines Made Australia* argues that much of Australia was managed by fire to make hunting easier. Gammage (p. 150) also points out that Indigenous Australians restricted their fertility.

[8] Termites farm fungi, beavers dam streams, birds create nests, and so on.

[9] The *Oxford English Dictionary* (*OED*) gives 1836 as the earliest use (in the *Boston Courier* 29 February: 'a fleet, with a carrying capacity of more than 1,500,000 tons, navigated by 40,000 seamen, with an investment of 50 millions of dollars in fixed capital'.

[10] The *OED* gives the first reference of this sort as 1857 (in the Melbourne Argus, June 3: 'it has to be remembered that, in adopting the system of assessment [of land values] according to carrying capacity, Government has not arrived at a simple and self-adjusting scheme'.

[11] Including Bert Bolin, inaugural chair of the Intergovernmental Panel on Climate Change.

[12] For example, a series of catastrophic volcanic eruptions occurred in the years 536–547, affecting all or almost all of the inhabited parts of the planet (Gibbons, 2018).

[13] The precise mathematical relationship between P, A and T is not defined.

[14] Huxley may not have supported the most extreme form of eugenics, as practised by the Nazis. For example, Huxley co-authored the anti-Nazi publication *We Europeans* (Huxley and Haddon, 1935). See also Weindling (2012).

[15] Not all, or even any, of the UN concerns for overpopulation were primarily motivated by racism. See Butler (2022); also see Chapter 10.

[16] In other words, if you (high-income countries) tolerate or even help to establish a more equal world system then fertility in poor countries (then generally called the Third World) will decline. The problem will be solved. In 1992 Singh revised his conclusion (Sinding, 2000), in 1993 personally confirming this by telling Potts, 'I have learnt that contraception is the best development' (Potts, 2015).

[17] As early as the 1950s Mao was struggling with his earlier confidence (Zhang, 2017).

[18] Simon represented the Cato Foundation, originally named after one of its founders, Charles Koch.

[19] Twelve of the 217 references were either to Simon, as first or sole author, or as chapters in his co-edited book *The Resourceful Earth: A Response to Global 2000* (Simon and Kahn, 1984). The next most cited authors were A.J Coale and A.C. Kelley (each with three citations). (Analysis by author.)

[20] The transcript of an interview in which Siegel asserted this is available at http://churchandstate.org.uk/2012/02/vatican-control-of-world-health-organization-policy/ (accessed 4 June, 2023).
[21] An extreme form of eugenics was practised by the Nazis. Less extreme versions were practised elsewhere, including in the US, UK, Australia and the USSR. Israel, today, is a leader in eugenics, albeit solely pre-natal (Glad, 2006).

References

Abubakar, I. (2020) The future of migration, human populations, and global health in the Anthropocene. *The Lancet* 396, 1133–1134. DOI: 10.1016/S0140-6736(20)31523-3

Anonymous (2021) Hunger as a weapon. *The Economist* 438 (229), 10.

Arrow, K., Bolin, B., Costanza, R., Dasgupta, P., Folke, C. *et al.* (1995) Economic growth, carrying capacity, and the environment. *Science* 268, 520–521. DOI: 10.1016/0921-8009(95)00059-3

Bashford, A. (2008) Population, geopolitics, and international organizations in the mid twentieth century. *Journal of World History* 19, 327–347. https://www.jstor.org/stable/40542618

Butler, C.D. (2004) Human carrying capacity and human health. *Public Library of Science Medicine* 1, 192–194. DOI: 10.1371/journal.pmed.0010055

Butler, C.D. (2012) Population trends and the environment. In*:* Friis, R.H. (ed.) *Praeger Handbook of Environmental Health*. Praegar, Westport, Connecticut, pp. 215–231.

Butler, C.D. (2022) Overpopulation: a century of debate that deserves re-examining. In: Saeverot, H. (ed.) *Meeting the Challenges of Existential Threats through Educational Innovation: A Proposal for an Expanded Curriculum*. Routledge, Abingdon, UK, pp. 67–84. DOI: 10.4324/9781003019480-5

Campbell, M., Cleland, J., Ezeh, A. and Prata, N. (2007) Return of the population growth factor. *Science* 315, 1501–1502. DOI: 10.1126/science.1140057

Carr-Saunders, A.M. (1922) *The Population Problem: A Study in Human Evolution*. Clarendon Press, Oxford, UK.

Catton, W.R. (1980) *Overshoot: The Ecological Basis of Revolutionary Change*. University of Illinois Press, Urbana, Illinois.

Cohen, J.E. (1995) *How Many People Can the Earth Support?* W.W. Norton and Co., New York.

Cohen, J.E. (2017) How many people can the Earth support? *The Journal of Population and Sustainability* 2, 37–42. Available at: https://whp-journals.co.uk/JPS/article/download/618/374 (accessed 4 April 2022).

Coole, D. (2021) The toxification of population discourse. A genealogical study. *The Journal of Development Studies* 57, 1454–1469. DOI: 10.1080/00220388.2021.1915479

Crist, E., Mora, C. and Engelman, R. (2017) The interaction of human population, food production, and biodiversity protection. *Science* 356, 260–264. DOI: 10.1126/science.aal2011

Demeny, P. (1988) Demography and the limits to growth. *Population and Development Review* 14, 213–244. https://www.jstor.org/stable/2808097

Ehrlich, P.R. (1968) *The Population Bomb*. Ballantyne, London.

Ehrlich, P.R. and Holdren, J.P. (1971) Impact of population growth. *Science* 171, 1212–1217. DOI: 10.1126/science.171.3977.1212

Finkle, J.L. and Crane, B. (1985) Ideology and politics at Mexico City: The United States at the 1984 International Conference on Population. *Population and Development Review* 11, 1–28. DOI: 10.2307/1973376

Foster, J.B., McChesney, R.W. and Jonna, R.J. (2011) The global reserve army of labor and the new imperialism. *Monthly Review*. Available at: https://monthlyreview.org/2011/11/01/the-global-reserve-army-of-labor-and-the-new-imperialism/ (accessed 28 November 2020).

Gammage, B. (2011) *The Biggest Estate on Earth. How Aborigines made Australia*. Allen and Unwin, Sydney, Australia.

Garrison, A. (2023) Ethiopia: The Tigray famine narrative was a total fabrication. *Black Agenda Report*. Available at: https://blackagendareport.com/ethiopia-tigray-famine-narrative-was-total-fabrication (accessed 3 March 2023).

Gebbie, D.A.M. (1981) *Reproductive Anthropology – Descent Through Woman*. John Wiley & Sons, Chichester, UK, and New York.

George, S. (1999) *The Lugano Report: On Preserving Capitalism in the Twenty-First Century*. Pluto Press, London.

Gibbons, A. (2018) Why 536 was 'the worst year to be alive'. *Science* 362, 733–734. DOI: 10.1126/science.aaw0632

Glad, J. (2006) *Future Human Evolution: Eugenics in the Twenty-first Century*. Hermitage Publishers, Schuylkill Haven, Pennsylvania.

Hardin, G. (1968) The Tragedy of the Commons. *Science* 162, 1243–1248. DOI: 10.1126/science.162.3859.12

Hartmann, B. (1987) *Reproductive Rights and Wrongs; the Global Politics of Population Control*. South End Press, Boston, Massachusetts.

Hawken, P. (2017) *Drawdown: The Most Comprehensive Plan Ever Proposed to Reverse Global Warming*. Penguin Books, New York.

He, L. (2022) Evergrande has been ordered to demolish 39 buildings in Chinese resort. *CNN Business*. Available at: https://edition.cnn.com/2022/01/04/investing/evergrande-stock-gain-resume-trading-intl-hnk/index.html (accessed 4 September 2022).

Higgs, K. (2014) *Collision Course: Endless Growth on a Finite Planet*. MIT Press, Cambridge, Massachusetts.

Huxley, J.S. and Haddon, A.C. (1935) *We Europeans. A Survey of 'Racial' Problems*. Jonathan Cape, London.

Ibreck, R. and de Waal, A (2021) Introduction: Situating Ethiopia in genocide debates. *Journal of Genocide Research* 24, 1–15. DOI: 10.1080/14623528.2021.1992920

Kelley, A.C. (2001) The population debate in historical perspective: Revisionism revised. In: Birdsall, N., Kelley, A.C. and Sinding, S.W. (eds) *Population Matters: Demographic Change, Economic Growth, and Poverty in the Developing World*. Oxford University Press, New York, pp. 24–54.

Kennan, G.F. (1948) *Report by the Policy Planning Staff. PPS No. 23: Review of Current Trends, US Foreign Policy*. Foreign Relations of the United States series, vol. 1, part 2. US Department of State. Government Printing Office, Washington, DC. Available at: https://history.state.gov/historicaldocuments/frus1948v01p/d4 (accessed 3 March 2023).

King, M. and Elliott, C. (1997) To the point of farce: a Martian view of the Hardinian taboo – the silence that surrounds population control. *BMJ* 315, 1441–1443. DOI: 10.1136/bmj.315.7120.1441

Kirk, D. (1944) Population changes and the postwar world. *American Sociological Review* 9, 28–35. Available at: https://www.jstor.org/stable/2086020

Kokole, O.H. (1994) The politics of fertility in Africa. *Population and Development Review* 20, 73–88. DOI: 10.2307/2807940

Leuprecht, C. (2010) The demographic security dilemma. *Yale Journal of International Affairs* 5, 60–74. Available at: https://heinonline.org/HOL/LandingPage?handle=hein.journals/yaljoina5&div=35 (accessed 4 April 2022).

Mazurana, D., Mekonen, H.K., Conley, B., de Waal, A. and Burns, D. (2021) *What 'Rape as a Weapon of War' in Tigray really means*. Available at: https://sites.tufts.edu/reinventingpeace/2021/08/10/what-rape-as-a-weapon-of-war-in-tigray-really-means/ (accessed 5 March 2022).

McMichael, A.J. and Butler, C.D. (2002) Global health trends, evidence for and against sustainable progress. In: Shkolnikov, V. (ed.) *Emerging Health Threats: determinants of diverging trends in mortality in a changing world*. IUSSP workshop 2002, Rostock, Germany, pp. 1465–1466. Available at: http://www.demogr.mpg.de›020619_paper25.pdf *(accessed 4 April 2022).*

Monbiot, G. (2020) Population panic lets rich people off the hook for the climate crisis they are fuelling. *The Guardian*. Available at: https://www.theguardian.com/commentisfree/2020/aug/26/panic-overpopulation-climate-crisis-consumption-environment (accessed 13 October 2020).

Mumford, S. (1986) *The Pope and the New Apocalypse*. Center for Research on Population, Research Triangle Park, North Carolina.

National Research Council (1986) *Population Growth and Economic Development: Policy Questions*. National Academy of Sciences Press, Washington, DC.

Nixon, R.M. (1992) *Seize the Moment: America's Challenge in a One-Superpower World*. Simon and Schuster, New York, pp. 263–264.

Ojeda, D., Sasser, J.S. and Lunstrum, E. (2019) Malthus's specter and the Anthropocene. *Gender, Place & Culture. A Journal of Feminist Geography* 27, 316–332. DOI: 10.1080/0966369X.2018.1553858

Otto, I.M., Kim, K.M., Dubrovsky, N. and Lucht, W. (2019) Shift the focus from the super-poor to the super-rich. *Nature Climate Change* 9, 82–84. DOI: 10.1038/s41558-019-0402-3

Paddock, W. and Paddock, P. (1967) *Famine 1975! America's Decision: Who Will Survive?* Little Brown, Boston, Massachusetts.

Pascoe, B. (2018) *Dark Emu. Aboriginal Australia and the Birth of Agriculture*. Magabala Books Aboriginal Corporation, Broome, Australia.

Potts, M. (2015) The road not taken: How the migrant crisis in Europe could have been ameliorated. Available at: https://blogs.berkeley.edu/2015/08/15/the-road-not-taken-how-the-migrant-crisis-in-europe-could-have-been-ameliorated/ (accessed 4 June, 2023).

Raupach, M.R., Marland, G., Ciais, P., Quéré, C.L., Canadell, J.G. *et al.* (2007) Global and regional drivers of accelerating CO_2 emissions. *Proceedings of the National Academy of Science (USA)* 104, 10288–10293. DOI: 10.1073/pnas.0700609104

Reagan, R. and Mondale, W. (1984) Debate on Foreign Policy. *The New York Times,* October 24, B4–B6.

Roberts, D. (2018) I'm an environmental journalist, but I never write about overpopulation. Here's why. *Vox.* Available at: https://www.vox.com/energy-and-environment/2017/9/26/16356524/the-population-question (accessed 13 October 2020).

Ross, E.A. (1928) *Standing Room Only?* Chapman and Hall, London.

Salih, A.A.M., Baraibar, M., Mwangi, K.K. and Artan, G. (2020) Climate change and locust outbreak in East Africa. *Nature Climate Change* 10, 584–585. DOI: 10.1038/s41558-020-0835-8

Sayre, N.F. (2008) The genesis, history, and limits of carrying capacity. *Annals of the Association of American Geographers* 98, 120–134. DOI: 0.1080/00045600701734356

Schellnhuber, H.J. (2018) Foreword. In: Spratt, D. and Dunlop, I. (eds) *What Lies Beneath.The Understatement of Existential Climate Risk.* Breakthrough National Institute for Climate Restoration, Melbourne, Australia, pp. 2–3.

Selden, S. (2005) Transforming better babies into fitter families: Archival resources and the history of the American eugenics movement, 1908–1930. *Proceedings of the American Philosophical Society* 149, 199–225. Available at: https://www.jstor.org/stable/4598925

Simon, J. (1981) *The Ultimate Resource*. Martin Robertson, Oxford, UK.

Simon, J.L. and Kahn, H. (1984) *The Resourceful Earth: A Response to Global 2000*. Blackwell, Oxford, UK.

Sinding, S.W. (2000) The great population debates: how relevant are they for the 21st century? *American Journal of Public Health* 90, 1841–1845. DOI: 10.2105/ajph.90.12.1841

Smith, D. (2015) Where Concorde once flew: the story of President Mobutu's 'African Versailles'. *The Guardian*. Available at: http://www.theguardian.com/cities/2015/feb/10/where-concorde-once-flew-the-story-of-president-mobutus-african-versailles (accessed 13 October 2020).

Starrs, A.M., Ezeh, A.C., Barker, G., Basu, A., Bertrand, J.T. *et al.* (2018) Accelerate progress – sexual and reproductive health and rights for all: report of the Guttmacher–Lancet Commission. *The Lancet* 391, 2642–2692. DOI: 10.1016/S0140-6736(18)30293-9

Vadén, T., Lähde, V., Majava, A., Järvensivu, P., Toivanen, T. *et al.* (2020) Decoupling for ecological sustainability: A categorisation and review of research literature. *Environmental Science & Policy* 112, 236–244. DOI: 10.1016/j.envsci.2020.06.016

Vitousek, P.M., Ehrlich, P.R., Ehrlich, A.H. and Mateson, P.A. (1986) Human appropriation of the products of photosynthesis. *Bioscience* 34, 368–373. DOI: 10.2307/1310258

Weindling, P. (2012) Julian Huxley and the continuity of eugenics in twentieth-century Britain. *Journal of Modern European History* 10, 480–499. DOI: 10.17104/1611-8944_2012_4_480

Yallop, D.A. (1984) *In God's Name (An Investigation into the Murder of Pope John Paul I)*. Corgi Books, London.

Zhang, J. (2017) The evolution of China's one-child policy and its effects on family outcomes. *Journal of Economic Perspectives* 31, 141–160. DOI: 10.1257/jep.31.1.141

10 Sexual and Reproductive Health and Rights: The Relevance of Family Planning

Colin D. Butler[1]

[1]*National Centre for Epidemiology and Population Health, The Australian National University, Canberra, Australia*

Abstract

Global female reproductive health remains unacceptably compromised. After reviewing important reproductive health conditions, the chapter suggests that an underlying reason for this dismal situation is that fertility rates in low-income settings remain too high; thus slowing fertility will not only bring many individual benefits (including to the health of mothers and children) but can also help precipitate beneficial social changes, generating helpful feedback loops, which further improve health. Controversially, it is also argued that (complementing neoliberalism and Catholicism; see previous chapter), a reason for persisting high fertility in some settings may be an inadvertent consequence of an 'overreach' led by elements of the women's movement, evident at the 1994 Cairo International Conference on Population and Development. The chapter also appeals for respectful dialogue between those who perceive sinister 'neo-Malthusian' motivations as underpinning concerns about high population growth and those who argue that high fertility is a factor in underdevelopment.

> There can be no prosperity without family planning (Karan Singh, 1992).

10.1 Introduction

This chapter concerns several aspects of sexual and reproductive health and rights (SRHR), which are vital for human health and well-being (Starrs *et al.*, 2018). A key domain of SRHR is gender-based violence across the age spectrum, including for the unborn girl child. Young men and boys can also experience forms of gender-based violence. 'Honour' killings are discussed, as are childhood marriage and sexually transmitted infections. Other topics relevant to SRHR, discussed here, include family planning, unintended pregnancy, unsafe abortion, and complications of pregnancy and childbirth. The relationship between breast feeding and fertility is briefly explored in Chapter 28.

Despite progress, the global health status of women and girls remains unacceptably compromised, particularly in terms of the absolute number who are affected. Some underlying reasons for this are presented, including a proposed relationship between population policy, demography, development and SRHR. The evolution and relationship of 'neo-Malthusianism' in the context of the three great global population conferences held between 1974 and 1994 is discussed. Readers are asked to imagine a counterfactual world that might exist today, had the idea that 'contraception is the best development' been allowed to flourish, instead of being suppressed. The chapter appeals for deeper, mutually respectful listening, by actors who would appear to share similar goals, yet regard with

 Climate Change and Global Health: Primary, Secondary and Tertiary Effects, 2nd Edition (eds C.D. Butler and K. Higgs)
DOI: 10.1079/9781800620025.0010

suspicion, even hostility, others who might be allies collectively pursuing a world with greater justice and prospects for sustainability. The chapter concludes by analysing case studies that illustrate issues relevant to the accessibility and uptake of family planning, including in Gaza, Guatemala and Myanmar.

10.2 Gender-based Violence

A 2013 study found that, at some point in their lives, nearly one in three women, globally, experience sexual violence (WHO *et al.*, 2013). Many forms of gender-based violence are recognized, including physical, sexual, financial, spiritual and psychological. Harmful practices include child marriage, sex-selective abortion, female genital mutilation (FGM), 'honour' killings and dowry-related violence as well as rape, sexual assault and harassment. In 2013 about 800,000 people, of whom 80% were women and girls, were estimated as being trafficked, each year, for prostitution, another form of gender-based violence (Deshpande and Nour, 2013). Such violence is also used as a weapon of war (Banwell, 2014).

Although most gender-based violence is inflicted on women and girls, men and boys can also be affected, particularly during childhood, during periods of conflict, and when they do not conform to social norms regarding sexual orientation and gender identity. Most perpetrators of gender-based violence are intimate partners, defined as past or current sexual partners, although they may also be family members, friends, acquaintances, strangers, colleagues, military and police officers. Religious practitioners in positions of trust with children can also use their position of authority to inflict unwanted sexual attention, sometimes including rape, as can some teachers and employers.

Female genital mutilation

FGM is an ancient practice, pre-dating Christianity and Islam, that seems primarily intended to reduce female sexual pleasure[1] and thus, it is believed, to reduce female sexual desire and hence the risk of marital infidelity (Berg and Denison, 2012). Acute complications of FGM include severe bleeding and perineal infections. More intrusive forms of FGM are associated with numerous complications in addition to effects that are clearly intended, yet have widespread acceptance in communities where FGM is common. In addition to psychological distress, these unintended complications include urogenital pain, anatomical disruption from perineal scarring, and additional risks during childbirth (Elneil, 2016). In cultures that follow this custom, women and girls are not considered as fit for marriage unless they have endured forms of FGM. Data and research on FGM are relatively scarce and recent, but at least 200 million girls and women in 30 countries are estimated to have been subjected to the practice of involuntary FGM in Africa and the Middle East (UNICEF, 2016). Even though most of these countries have made the practice of FGM illegal, it still continues.

Less destructive forms of FGM have been documented in at least six countries of South and South-East Asia, predominantly among Muslims (Cappa *et al.*, 2019). FGM has also been reported among migrant populations in Western countries. In the 19th century, FGM was also practised by some gynaecological surgeons in Europe and North America, in attempts to treat epilepsy and 'hysteria' (Elneil, 2016). Female genital mutilation is thus an extreme form of discrimination against women and a violation of their human rights, i.e. their right to health, security and physical integrity, and to be free from torture and cruel, inhuman, or degrading treatment.

Honour killings

'Honour' killings occur when the death of a woman is perpetrated by one or more family members, who are motivated by disapproval of her social behaviour in general or her sexual behaviour in particular. Women (and sometimes their partners) are not just punished, but murdered, for the 'crime' of family 'disgrace'; examples include: refusing a forced marriage; divorce; adultery; and even being the victim of rape. The social construction of 'honour' as a value system, norm, or tradition is the main justification for the perpetration of such violence (Pirnia *et al.*, 2020). In 2000, a report for the UN estimated that 5000 women are killed each year in the name of 'honour' (UNFPA, 2000). Chesler (2010) argued that this is a gross underestimate, although the scale appears modest in comparison with FGM. Like FGM, the practice of 'honour' killings has been observed among immigrant communities in several countries in the global North. Although Sikhs and Hindus do sometimes commit such murders, honour killings, both worldwide and in the West, are mainly Muslim-on-Muslim crimes.

10.3 Childhood Marriage and Motherhood

Although the practice of girl-child marriage (defined as younger than 18) has decreased substantially since 1990, it remains common in some regions, particularly in South Asia and sub-Saharan Africa (UNICEF, 2014). Surveys have found an estimated 7% of girls in 'developing' regions marry before the age of 15 years, while another 21% marry when they are younger than 18. The Universal Declaration of Human Rights (signed in 1948) states that individuals must enter marriage freely with full consent and must be at full age. As with the legal prohibition of gender-based violence, the outlawing of childhood marriage is often unenforced (Nour, 2009).

Children (especially first-born) of adolescent mothers face greater health risks than those born to older mothers, in part because young mothers usually have worse health and educational status than older mothers. The first-born children of mothers younger than 18 years have a higher risk of neonatal mortality, preterm birth and infant mortality. Even subsequent babies of adolescents face a very high risk of these outcomes, as well as of stunting and anaemia. Adolescent mothers are also more likely than adult women to experience pregnancy complications, including pregnancy-induced hypertension and spontaneous abortion. Globally, about 300,000 women per annum still die of complications of childbirth (WHO, 2019). The vast majority occur in low-resource settings and most could be prevented. Adolescent and nulliparous mothers are at a higher risk, although there is even greater risk among women older than 35 (Restrepo-Méndez and Victora, 2014).

Obstetric fistula: 'life in perpetual darkness'

Obstetric fistula is an extremely disabling condition, caused by prolonged obstructed labour, a risk greatly magnified by inaccessible obstetric services (Fig. 10.1). Such fistulae, which usually result in urinary or (less often) faecal incontinence, lead to lives of stigma, shunning and social isolation, likened by one sufferer as akin to living in perpetual darkness (Animut *et al.*, 2019) (Fig. 10.1). The World Health Organization (WHO) has estimated that at least 50,000 to 100,000 cases of obstetric fistulae occur every year and that over two million women with such fistulae remain untreated, almost exclusively in low- and middle-income countries (LMICs), often called the 'global South', despite the condition being both preventable and treatable (WHO, 2018).

10.4 Sexually Transmitted Infections and Infertility

There are more than 30 identified pathogens (bacteria, viruses and parasites) spread through sexual contact, of which eight are the most common. Of these sexually transmitted infections (STIs), four are curable by antibiotics if treated early: syphilis,[2] gonorrhoea, chlamydia and trichomoniasis. Over 350 million new infections involving these four conditions are estimated to occur globally each year (WHO, 2016). Two of these four (gonorrhoea and chlamydia) are major causes of pelvic inflammatory disease and female infertility. The other four of these eight common STIs — HIV, hepatitis B, herpes simplex virus and human papillomavirus (HPV) — are viral infections and are currently incurable, although their symptoms can be managed and their disease progression altered. Young people are especially susceptible to STIs but sexually active people can be at risk of infection at any age. The means to control each of these viral infections is different for each. For instance, HIV can be suppressed and even prevented by the use of medication. Mother-to-child transmission of hepatitis B is very common in certain areas, and can be prevented by the use of passive–active immunoprophylaxis[3] (Huang *et al.*, 2020). Herpes

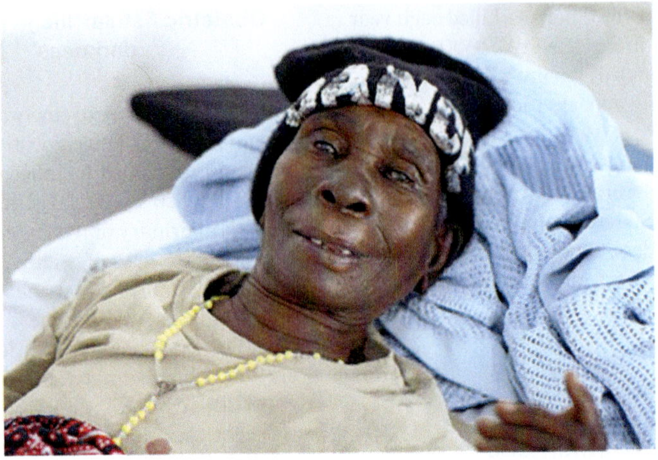

Fig. 10.1. 'Katrina (identified with permission) thinks she is about 75 and has been suffering from a fistula for over 50 years. She had a long four-day labour at home, delivered a stillborn child and was left leaking urine day and night from her fistula. Her husband left her, as did everyone else. She went to different hospitals over the years and had five operations but was no better. We operated on her in November 2021 and she was totally cured and overjoyed, 50 years of shame ended. However, she fell getting down off the bus when she went home after being cured and broke her hip. She struggled at home alone for 3 months with her broken hip, having to keep moving to look after herself so the hip never healed. Our outreach found her during their follow-up and brought her back a few weeks ago. She is now in traction and slowly getting better.' Photo credit: Dr Andrew Browning Fistula Surgeon, Tanzania.

simplex can be suppressed with medication, although the infection is not curable. Infection with HPV, which can cause cancer, can be prevented by immunization, particularly systematic programmes among young people (Simms *et al.*, 2019). All of these infections can be prevented by the use of condoms.[4]

Where concurrent sexual relationships are common, young women have a higher risk of acquiring HIV than do young men, particularly where male circumcision is also infrequent (Halperin and Epstein, 2004). Any STI that leads to genital erosion can increase the risk of acquiring HIV. These include herpes, syphilis and gonorrhoea. Several STIs, including HIV and syphilis, can be transmitted from mother to child.

Infertility (defined as no conception during a year of unprotected intercourse) affects as many as 10% of the world's women. The majority of these women, generally as couples, live in LMICs. Infertility is very stigmatizing in some cultures, for both women and men (Rouchou, 2013). Treatment is expensive and support groups for infertile couples in most low-income settings are rare.

10.5 The Unmet Need for Contraception

This chapter now reaches the most controversial and least understood topic relevant to sexual and reproductive health and rights. It expands on some of the ideas discussed in the previous chapter, on 'human carrying capacity'. This topic does not concern contraception or family planning as isolated factors but the relationship of these elements to wider human development, to human rights, and the 'rights' of nature, of non-human species, and even of natural features.

Betsy Hartmann, author of *Reproductive Rights and Wrongs: The Global Politics of Population Control* (1987), writes tellingly of her own experience with family planning in newly independent Bangladesh in 1975. Soon after her arrival in the village of Katni that year, Hartmann experienced first-hand what the population establishment calls the 'unmet need' for contraception. The village women were extremely curious about why Hartmann had no children 'at the ripe old age of twenty-four', and she told them about her use of birth control. The villagers seemed receptive and Hartmann was able to arrange a visit to Katni by family planning workers, women she described as wearing fine saris, with unfamiliar accents. For a while, many women in the village took the contraceptive pills which the visitors left. However, this supply quickly ran out. On a return visit to Katni seven years later, Hartmann learned that the family planning team had never returned.

A vast unmet need for contraception still exists. A recent UN report estimated that, in 2019, among the 1.9 billion women of reproductive age (nominally 15–49 years) about 10% (190 million) who want to avoid pregnancy do not use any contraceptive method. It was also reported that this number had increased by over 30 million since 2000. Furthermore, of the 922 million women estimated to use contraceptives, 80 million follow traditional methods[5] of family planning (UNDESA, 2019).

The context of the three great population conferences

Hartmann first visited Katni in the year following the first of three major decennial conferences (1974–1994) held by member countries of the United Nations (UN). These meetings (in Bucharest, Romania; Mexico City, Mexico; and Cairo, Egypt, respectively) were held to discuss the relationship between population, development and family planning. In the decades and centuries leading to the first of these conferences, fertility (though not necessarily population growth) was far higher than today, in almost all parts of the world.[6] However, by 1970, fertility was rapidly declining in most high-income countries apart from high-income oil-exporting countries such as Saudi Arabia. In the global South, population increase remained comparatively low, particularly before the end of World War II (WWII) in 1945, despite generally high fertility. Following the end of that war, advances in healthcare and sanitation and the willingness and capacity to disseminate knowledge of more modern forms of knowledge – including to improve health – led to a rapid reduction in infant and child mortality within low-income settings.

Demographic apprehensions: two reasons

A 1971 UN report estimated that the annual population increase in low-income countries, between 1850 and 1900, was about 3 million per annum (about 0.3% per annum). By the 1960s, this had increased by a multiple of 18 to about 54 million per year, over 20% per decade (UNDESA, 1971). This extraordinary increase in population growth in the South occurred because of the significant fall in infant and child mortality, due to interventions such as widely distributed vaccinations, antibiotics and the insecticide dichlorodiphenyltrichloroethane (DDT), which greatly reduced malaria mortality, together with the persistence of high fertility. Promoters of family planning held the view that, although high fertility was completely justifiable in the context of high mortality, it was problematic if it was not adjusted to take account of declining mortality. Population also grew in the global North after WWII, but the rise did not persist for as long.

The rapidly increasing population in the South, especially in LMICs in Asia, caused great political unease, primarily in high-income settings and among the leaders of the UN, who were mainly from Western countries at that time. This unease, which pre-dated WWII, became more marked after that war (Bashford, 2008; Butler, 2022). It is here argued that this concern about emerging 'overpopulation' in the South had two broad causes, the consequences of which have permeated the following decades, right to the present. Only one of these concerns was broadly humanitarian and altruistic.

Fear of 'revenge by Empire', or colonialism by other means

It is here proposed that, by the time of the Bucharest meeting (1974), much of the North's concern about global overpopulation was motivated not only by fear of famine in the South, but also by apprehension that the balance of global power, if crudely measured by total population size (an approach sometimes called 'mercantilist'), was shifting too far from Europe and North America. Although the South, in the 1970s, was already home to most of the global population, leaders in the North grew increasingly apprehensive of the rising demographic imbalance. Closely related was the concern that instability in the South (worsened by rapid population growth) could generate Leftist uprisings, preventable not only by more abundant food but also by more family planning (Gaud, 1968; Perkins, 1990; Stone, 2019). Emeritus Professor (of political and social theory) Diana Coole (2021) appeared to support this broad argument, for example by writing that 'the perception of Third World countries[7] was that the US favoured family planning as a quick technological fix in lieu of addressing world poverty and systemic inequality'.

The decades leading to and following WWII were a time when many formerly colonized peoples were agitating for, and in most cases achieving, political independence. But decolonization was often violent, as several European powers stubbornly clung to 'their' territories as well as the racist and brutal habits that had characterized their empires, acquired and exploited over previous centuries. It was also a time when the comparative advantage of Europe had been weakened by two catastrophic wars. Thus, a 'fear of the other', an apprehension of 'revenge by empire', is likely to have existed, at least subconsciously, among political leaders, intellectuals and populations, especially in Europe and its offshoots, such as the USA and Australia.

This speculation is difficult to prove by government documents, but can be inferred, such as in the writing of the so-called environmental determinists Paul Bracken[8] and Robert Kaplan (Kaplan, 2009). Traces still remain (i.e. fear of being 'swamped' by immigrants from the South) among many European populations, in a phenomenon that Gökariksel *et al.* (2019) called 'demographic fever dreams'. In the context of the Cold War (between the Western Bloc and Eastern Bloc, 1947–1991) and then recent decolonization, the more palatable humanitarian motivation to care about population growth, in what was then starting to becoming known as the South (Butler, 2007), was tainted; it was partly propaganda. It is here proposed that not only did this taint influence all three population conferences, but also that it survives to the present.

The humanitarian imperative

Above, it has been argued that concern about high population growth in the South was influenced by neocolonialism. However, it is surely an exaggeration to claim that all, or perhaps even most, advocates for enhanced family planning in the South were Eurocentric (North-centric), principally concerned with retaining or enhancing the grossly disproportionate distribution of the determinants of global health that existed in the 1970s.

The end of WWII and the birth of the United Nations had, in at least a non-trivial number of people, liberated a genuine wish for a fairer, safer world (Glendon, 2001). The 1960s were designated as the (first) 'United Nations development decade' (Esman, 1971). Idealism, even if accompanied by a wish for professional recognition, was a powerful motivator for many of the agricultural scientists who created the Green Revolution, including Norman Borlaug, its 'father'.

Another idealist was Lester Pearson, who noted in his Nobel Peace Prize lecture (1957) that more than 30 million people are added to our 'crowded planet' each year. This could be interpreted as code for 'the Third World is a threat – their population growth must be reduced'. But this, in my opinion, would be very narrow minded. In 1969, Pearson (by then Canadian Prime Minister) chaired a major report arguing that countries in the North should donate 0.7% of their gross domestic product towards developing the South. That target has rarely been met, but its endorsement suggests genuine goodwill from *some* Northern populations towards the South – as do the Sustainable Development Goals.

Ahlberg (2007) described the 'Food for Peace' Program in India, organized by the administration of US President Johnson, as 'Machiavelli with a heart'. The US Peace Corps, founded in 1961, has been described as a powerful propagator of 'American values' (Hall, 2007). It is well recognized, at least outside the USA, that American values have led to actions that have greatly harmed people in the global South, such as the use of the herbicide Agent Orange[9] in the Vietnam War and the alleged, attempted and actual assassinations of left-leaning leaders in the South, from Patrice Lumumba, the promising leader of the Democratic Republic of the Congo, to Chile's Dr Salvatore Allende (Falcoff, 1991). Yet, some Peace Corps volunteers became not only deeply exposed to poverty and inequality but also agents attempting to change the world for the better (Hall, 2007).

Not everyone in the North is racist; nor is every initiative by a Northern government purely for propaganda. Malcolm Potts, the first Medical Director of the International Planned Parenthood Federation, is a prominent humanitarian and a strong advocate for family planning. As mentioned in Chapter 9 (this volume), in 1993 Potts reported a conversation with the Indian politician and philosopher, Karan Singh, who agreed with him that 'contraception is the best development', thus reversing his famous position from almost two decades earlier (Potts, 2015). Singh had publicly revised his view at the 40th Anniversary Congress of the International Planned Parenthood Federation (1992), when he stated: 'There can be no prosperity without family planning.'

The economic argument that high rates of population growth in low-income settings can surpass the capacity to increase needed infrastructure (in quality as well as quantity) for development, such as schools, teachers and health workers, has long been made (Leibenstein, 1954; Nelson, 1956; Coale and Hoover, 1958). It is here proposed that this argument is valid, irrespective of alleged hidden motivation. This theory, sometimes summarized as harvesting the 'demographic dividend', proposes that economic and social 'take-off' in low-income settings can be assisted by slower population growth, wider child spacing and better schooling. Written negatively, theorists argue that failure to slow fertility can lead to demographic 'entrapment' (Brown and Jacobson, 1986; King, 1990). It is this argument (in essence) that critics contend was sidelined at Cairo, as will be discussed.

Arguably, both statements by Singh are correct, rather than contradictory. The tragedy is that, in the absence of substantial and genuinely intended aid from the North (a prospect that seems as elusive today as ever, despite the hope of Pearson and the rhetoric of US Presidents Truman and Kennedy), development without contraception is likely either to remain elusive or to progress at a snail's pace. In contrast, nations such as China, Taiwan and South Korea (all of whose societies and governments strongly endorsed family planning) have taken substantial steps out of poverty; countries such as the Philippines, Niger and South Sudan (with limited receptivity to family planning) have not.

Beneath the rhetoric, very few holders of significant economic and political power in the North genuinely seek a significantly more equal global distribution of wealth and political power. Development may be an excellent contraceptive but the South has largely to find its own path towards it. This truism held not only at each of the three great population conferences, but also at the 28 climate change "Conferences of the Parties" so far conducted.

Setbacks for family planning: from Bucharest to Cairo

The Bucharest conference was an unexpected setback for the family planning movement. It is best remembered for the phrase 'development is the best contraceptive', credited to Karan Singh, Indian Minister of Health who represented Prime Minister Indira Gandhi. In other words, the South seemed to be saying at Bucharest, give us 'development' (a fair share of Earth's wealth) and the North will get your wish: you will see fertility in the South decline.

The following two population conferences were even more disappointing for family planning advocates. While it was argued (see Chapter 9) that the main reasons for the loss of support for family planning were neoliberalism and the influence of Catholicism, some have suggested that some elements of feminism may also have contributed. For example, Maurice King, who attended the Cairo International Conference on Population and Development (ICPD), identified 'some aspects of the feminist agenda' as one of 29 'population demons' that he claimed were slowing the global demographic transition (King, 2003). King also suggested that the 'feminist agenda' had a powerful effect at the ICPD, by diverting priorities from population control and family planning towards the more nebulous, less demographically relevant notion of 'reproductive health'.

King was far from alone in this perception. Martha Campbell (2007) identified the ICPD as the 'turning point in removing the population subject from policy discourse':

> The important difference between [Cairo] and the previous decadal UN population conferences was its emphasis on drawing attention to the needs of women around the world. In the run-up ... and following ... Cairo, talking about population became politically incorrect in many circles. Drawing attention to any connection between population and the environment became taboo – again, because it was viewed, or promoted, as disadvantageous to women. It became inappropriate to say that slowing population growth will make it more possible to preserve the environment for future generations. 'Malthusian' and even 'demographic' became derogatory terms describing anyone still expressing interest in, or concerned about, population growth.

Campbell (2007) also argued that 'the strategy adopted at Cairo for addressing issues about pregnancy and childbearing was to combine family planning with all aspects of health particular to women, calling it reproductive health'. They also 'promoted the idea that all family planning efforts before Cairo were coercive, labelling them "population control" – and (thus) the term itself became purely derogatory'.

The difficulty of discussing family planning in positive terms lingers in the language in some contemporary feminist publications. There is evidence of dismissal, in this literature, of *any* humanitarian reasons to slow population growth in the South (e.g. Ojeda *et al.*, 2019; Bhatia *et al.*, 2020). Coole (2021) also traced some of this contemporary hostility to family planning to what she characterized as 'the International Women's Movement' active at Cairo.

10.7 Feminism, Family Planning and Two Pleas

Many elements contribute to the difficulties encountered when attempting to discuss population. These include neoliberalism and abuses of human rights associated with advocates for family planning, including forced and unhygienic sterilizations and abortions (Hartmann, 1987). The Catholic Church and the gradual ascendancy of 'cornucopian' thinking in much of the academic literature has also contributed. However, it is here argued that some in the women's movement may have also played a role in what Coole calls 'toxification', albeit inadvertently.

Dodgson (2000) argued that some reformist programmes were 'co-opted' by neoliberal elements influencing the UN. These included the call for a New International Economic Order (to improve the position of the global South) and the position promoted by the environmental movement at the 1992 Rio 'Earth Summit'. He discussed critics who suggested that some of the women who were active in Cairo (especially a group that he calls the 'Women's Health Movement' (WHM)) may have been similarly influenced. He wrote that the UN – which he characterized as essentially neoliberal – believed that the Cairo conference and the support of the WHM 'could be used to widen, enhance and legitimise its own model of sustainable development and the neoliberal principles on which it was based'. He added that 'critics of the WHM's influence upon the ICPD have argued that it was the WHM that co-opted the other actors involved in the ICPD process'. In doing so, 'the group of women who tried to steer the ICPD shifted it away from a discussion of over-population and population control towards a debate on women's reproductive rights, health and empowerment'.

This summary of Dodgson's paper does not prove that either he or the authors he cited were correct, but it is additional evidence of a long-standing debate about the success of the women's movement at Cairo. Dodgson is also clear that the WHM was not the sole representative of women's views at the ICPD; for example, he mentioned 'the decision not to foreground the issue of sustainable human development led some members of the WHM to criticise the tactics of the Women's Caucus at the Cairo conference'.

Preceding and paraphrasing Martha Campbell, Rosalind Petchesky (1995) argued that the Cairo Programme of Action 'enshrines an almost feminist vision of reproductive rights and gender equality in place of the old population control discourse'. C. Alison McIntosh and Jason Finkle (1995) stated that 'a new definition of population policy was advanced (at Cairo), giving prominence to reproductive health and the empowerment of women, while downplaying the demographic rationale for population policy'. Charles Westoff (1995) was immediately sceptical. Criticizing Cairo's 'platitudinous recommendations', he speculated a reason as:

> ... perhaps it was the effect of the women's well-organized efforts to substitute their own agenda, which enshrined the 'empowerment of women' and reproductive health concerns. Demographic targets were dismissed as potentially coercive. In a spirit of elevating the level of discourse, the population subject per se was submerged in the rhetoric of reproductive rights and 'sustainable development'. In fact, it became almost mandatory to explicitly denigrate the whole subject of population growth and subordinate it to other issues.

Westoff (1995) added:

> This reluctance to deal directly with the issue of high fertility and the tendency to bury the problem in other related subjects is dramatically evident from one calculation that an earlier draft produced an estimated 1,170 recommendations (a staggering number), of which only a dozen directly addressed population growth.

Westoff also noted that 'the U.S. delegation claimed a major victory in the consensus reached by the representatives of the 180 governments in attendance'. In Chapter 9 the author speculated that US-led neoliberal forces perceived economic benefits from the maintenance of high population growth rates in the global South. If this speculation is valid, then it provides a reason for the USA to be pleased with the outcomes of the Cairo meeting.

More recently, Coole (2021, p. 13) noted that 'the equation postcolonial feminists made between neo-Malthusian policies, racist eugenics and women's coercion was reprised by feminists from the (global) South. The toxification of population policies was greatly enhanced, as it had been in the US, by this integration of gender and race.' Karin Kuhlemann (2019, p. 84) identified 'social justice activists' (including what she called 'women's initiatives' communities) as 'one of the three ideational groupings that brought about population denialism and population taboo', the others being growthism and religious conservatism.

Steven Sinding, later Director-General of the International Planned Parenthood Federation, traced a much longer debate, between a narrower 'family planning approach' and advocates of a 'reproductive health approach' (Sinding, 2000). He stated that the Bucharest and Mexico City conferences were 'full of recommendations to take a holistic developmental approach'. He added that 'nearly all participants in the debate agree that a broader range of services is better than a narrower one: nothing is sacrificed and much is gained'. However, Sinding warned that the higher cost of the wider approach could cripple its implementation, requiring a commitment that 'neither developing countries nor donors are prepared to make'.

Coole (2021, p.14) was sympathetic to this argument, warning that:

> ... a narrow focus on reproductive rights and maternal health would stifle debate (references listed). [W]ithout a demographic driver, support for family planning programmes would further atrophy: not only causing high fertility levels to stagnate or rise again, but also undermining both women's rights to birth control and their life chances within degraded environments. With family planning no longer ringfenced or having an independent funding-stream, a danger was that inadequate health budgets would be redirected into other areas (as indeed occurred).

A more conciliatory interpretation for the failure to adequately fund family planning following the ICPD is simply that funds were insufficient. Yet funding for militarization and private jets was ample; perhaps at its heart, political will in the North, seeking to improve conditions in the South, was as inadequate then as it has been since.

Even if the representatives of the women's movement who were active in Cairo are completely blameless for the shortfall in family planning funding that followed the ICPD (i.e. the writers whose work I have briefly discussed above are completely misguided), some writers publishing in feminist journals today appear to equate words such as 'neo-Malthusian' with malevolence (e.g. Bhatia *et al.*, 2020). They also seem to deny any humanitarian motive to those who promote family planning (e.g. John Guillebaud,[10] King, Potts, Sinding and Singh), instead aggregating all supporters of rights-based family planning as part of a 'population control' strategy, defined as 'top-down, coercive interventions (often rooted in imperialist foreign policy) that have an explicit goal of lowering birth rates', based in 'ideologies shaped by eugenicist and neo-Malthusian thinking' (Bhatia *et al.*, 2020, p. 335).

Of course, not all feminist scholars dismissed the realities of limits. Coole discussed how the radical feminist Shulamith Firestone (in 1971) explicitly denied that reducing fertility was a racist objective; she asserted its ecological merits. But Firestone wrote this well before the ascendancy of neoliberalism had largely killed the debate. While there are other exceptions, including the book *Making Kin Not Population* (Clarke and Haraway, 2018), Firestone's opinion, among feminists, appears somewhat isolated. Illustrating this, two contributing female authors to this book, each from Taiwan, a country whose economic take-off is in part due to the adoption of family planning, pleaded (in the journal *Feminist Studies*) that 'engaging with population and environment does not have to repeat neo-Malthusian rationales' (Strathern *et al.*, 2019).

This chapter makes two pleas. The first is for humanitarians to consider more fairly exploring the health and other benefits that can flow to mothers and children, as well as more broadly, if birth rates are lowered in high-fertility settings, by both the shaping of norms such as better education for women and girls, and an increased use of voluntary contraception, in the context of less patriarchy. The second plea is for 'deep and respectful' listening among all parties who seek to genuinely advance human rights, especially for women and girls. The aphorism popularized at the Bucharest conference is correct; development will eventually reduce fertility. However, that will depend on sufficient resources and that civilization (including in the global South) does not self-destruct in the meantime. More importantly, a lowering of fertility, if provided by far-sighted leadership in LMICs, will hasten development. Such actions would, in a sense, be subversive, in that they will reduce global inequality. This reduction in fertility could be greatly assisted by the vigorous promotion of this idea by people in the North who care about human development, including those who care about climate change (Bryant *et al.*, 2009).

Some of these issues are discussed elsewhere in this book. Much of recent history (e.g. since 1970) can and often is interpreted as supporting the idea that Malthusian and neo-Malthusian arguments have been disproven. This belief can and needs to be contested. Supporters of this view are especially common among neoliberals and other followers of the New Right. Moreover, the demographic determinants of counter-examples, such as the 1994 genocide in Rwanda, have largely been ignored (rather than analysed, with postulated links carefully dismantled), for example in the journal *Population and Development Review*. A rare counter-example was the paper 'Land relations under unbearable stress: Rwanda caught in the Malthusian trap' (André and Platteau, 1998), published in an economic journal.

Today, the fears of ever-declining biodiversity held and expressed by many ecologists (Bradshaw *et al.*, 2021) and those of internecine, unending tribal conflicts expressed by some humanitarians, appear, broadly, valid. In April 2022, the International Committee of the Red Cross reported that 'one quarter' of the population of Africa faced a 'food security crisis'. Reasons listed were conflict, drought, rising food and fuel prices, and the lingering effects of the COVID-19 pandemic (Anonymous, 2022). Population pressure was not mentioned; nor was Malthus, nor was neoliberal-fuelled optimism, dominant as the 20th century closed, providing the background for the ICPD. Meadows *et al.* (1972), who wrote *The Limits to Growth*, did not predict collapse by 2000, and not even by 2025. It is premature to dismiss their grim prognosis.

10.8 Barriers to Family Planning in the South

This chapter has argued that mistrust between the global North and South has substantially inhibited family-planning funding, advocacy and uptake. But mistrust between groups in the global South can also operate to maintain relatively high rates of population growth in several of its parts. Of relevance is what the political scientist Christian Leuprecht (2010) has identified as a 'demographic security dilemma'. He described a cyclical relationship of interactions between a politically dominant low-fertility population and a large, youthful minority. Conflict arises when the dominant majority perceives that a 'youthful' minority population will either eventually overtake and displace the majority, or at least become of sufficient size to be perceived as a nuisance. The perception of an emerging demographic threat (within-country, rather than North vs South) leads to political suppression of the minority, in some cases, probably slowing the demographic transition[11] that would otherwise likely unfold for it.

The Rohingya

A postulated example of this phenomenon involves the Rohingya, the main Muslim minority in Buddhist-dominated Myanmar, who live mostly in Rakhine state, or in exile in Bangladesh (Blomquist, 2015).[12] Aung San Suu Kyi, the again-imprisoned Burmese leader who has expressed hostility to the Rohingya, is reported to have told a cautionary tale about how Muslims have come to dominate Indonesia, after centuries in the minority. Suu Kyi is alleged to have said that Myanmar wants to avoid a similar situation (Mandhana and Hookway, 2017). The Rakhine politician Shwe Maung is reported as telling *The Economist* that 'they are trying to Islamize us through their terrible birth rate'. Rakhine State government spokesman Win Myaing claimed to India's *The Hindu* that the Rohingya population growth was ten times that of native Buddhists (Winn, 2013). These examples imply that politically dominant actors in Myanmar would like to lower the fertility rate of the Rohingya, but do not want to foster the 'development' that might lead in this direction; instead, the perceived 'excess' population is periodically terrorized, with large numbers driven into exile. Living in exile however, Rohingya in Bangladesh are reported as consciously desiring high fertility, in part to "increase Muslim soldiers", so that might recover the land they perceive as ancestral (Hossain and Hossain, 2023).

India, Gaza and Israel

Such dynamics also apply in India, contributing to mistrust of Muslims (and sometimes Christians) by Hindus, even though Muslims remain a small minority in all Indian states except Kashmir. In China, there are credible accusations of active suppression of the fertility of the Uighurs and other Muslims, who constitute about 45% of the population of Xinjiang (Waller and Albornoz, 2021).

Mistrust and discrimination have also been suggested as contributing to the very high fertility rate of the Palestinians crowded into the Gaza strip, who were reported, in the 1990s, as having the world's highest fertility rate (7.8). Fertility is also high among ultra-orthodox Israeli Jews (as high as 7.6 in the mid-1990s). Fargues (2000) noted: 'For both sides, it seems that belligerence has produced excess fertility.'

Guatemala and N.W. Nigeria

Guatemala is the second most populous nation in central America. One of every 170 pregnant women dies giving birth; however, the rate of maternal death among its Indigenous (Mayan) population (40–50% of the total) is three times higher than in its more westernized ('Ladino') people. Contraceptive use among the Mayans is lower, as are literacy, fluency in Spanish and reproductive health knowledge. Festa and Shirazian (2018) identified this as due to a range of 'social, cultural, and economic barriers that limit Mayan use and adherence to family planning methods'. Guatemala has suffered an enduring conflict, causing hundreds of thousands of deaths, between its Mayan and Ladino combatants. The situation in Guatemala is different from that of China and Myanmar: there is no obvious evidence that the higher fertility of the Mayans is perceived as a threat by Ladinos; instead, discrimination[13] and poverty trap them into having larger families that then face additional difficulties.

There is agreement among demographers and economists that the demographic transition is 'exogenous' (caused by external factors). Supporters of family planning claim that lower population growth then triggers feedbacks that benefit society. While there is debate concerning the extent and mechanisms of this process (Canning and Schultz, 2012; Bongaarts, 2014)[14] it is clear that, in many parts of the world, changes to the desired number of children lagged, and in some cases still lag, in improvement in the life expectancy of children, leading to comparatively rapid population growth. One such region is north-western Nigeria, where surveys from the late 2000s reported a total fertility rate of

7.3, similar to the reported preferred number of children (eight) (Connolly *et al.*, 2019). There are many interlinked factors that drive this, including patriarchy, maternal education, lack of women's empowerment and other cultural values. A mechanism that has been used in several countries, including in this part of Nigeria, to try to lower desired family size is the creative use of entertaining serial dramas, broadcast in the local language by radio. These feature characters that model how to discuss having smaller families, birth spacing, contraceptive use and maternal and child health (Connolly *et al.*, 2019).

10.9 Conclusion

Many aspects of the global status of sexual and reproductive health and rights have improved, yet remain grossly deficient, especially in absolute terms. Even though nature appears to be boundless, ingenuity cannot fully compensate for its degradation; the world cannot support endless growth in human population size. Even if global population is now on course to stabilize at ten billion, that number is far too high. In addition, nature has intrinsic value; not all of it should be converted to goods and services to satisfy human wants and needs, which are influenced by human numbers as well as by human greed and income.

The use (or rejection) of family planning is a human right that warrants support by all who seek to improve the health status of women and girls (adults of the future). A fresh coalition between feminists and the promoters of reproductive health is needed. It is likely that many promoters of population 'control' (as defined by Bhatia *et al.*, 2020) held views inconsistent with global social justice. But not all did or do, even if there are counter-examples such as in Myanmar and China.

Target 3.7 of the Sustainable Development Goals (SDGs) aims, by 2030, 'to ensure universal access to sexual and reproductive health-care services, including for family planning, information and education,[15] and to integrate reproductive health into national strategies and programmes'. This chapter calls on all its readers to take this target seriously, as its achievement will do much to advance many other targets and goals. Other SDGs should also be pursued; many, such as universal education and equal rights for women and girls, will also accelerate the demographic transition in low-income settings, such as in many parts of sub-Saharan Africa. However, without acknowledging the reality of limits, most SDGs will remain well out of reach (Hickel, 2019; Higgs, 2022).

The populist leader of Turkey, President Erdoğan, has been described as 'locating Turkey's future strength in a young and large population and stoking fears of population decline' (Gökarıksel *et al.*, 2019). This mercantilist view of population remains too common, but can only be defused by good will and transparency. Finally, therefore, I appeal for the framers of the goals that will replace the SDGs to work with courage and new thinking, in order to further help dismantle the many barriers that impede global health, including its sexual and reproductive aspects.

Disclaimer

by Kerryn Higgs (co-editor)

Despite supporting the central idea that high fertility jeopardizes development and inhibits tackling poverty, I could not endorse this chapter. The claim that women's advocates who intervened in Cairo were 'led by elements of the women's movement', or 'represented' the Women's Movement (WM), is misleading. The WM consists of many diverse and often contradictory ideological strands with no central organization. The chapter does not establish who the advocates were (Americans? anti-imperialist?), or who influenced them.

It seems likely, given language such as 'overreach' and 'toxify', that hostility at Cairo flowed as much from the neo-Malthusians as from those focused on broader women's 'empowerment' (rather than population). Moreover, the programmes the women pushed (health, education) would have improved women's health and encouraged non-coercive family planning – which is supported by most feminists, unlike the Vatican. If these programmes were too costly for the UN, this should be blamed on rich-world failure, rather than 'feminist overreach'.

Notes

[1] By removing all or part of the clitoris, FGM prevents or inhibits the chance of a female climax, as comparatively few women experience vaginal orgasms. This also reduces the pleasure of partners.
[2] Tertiary syphilis is still untreatable.
[3] The administration of hepatitis B immunoglobulin and birth-dose hepatitis B vaccine in infants within 12 or 24 hours after birth.
[4] At least for penetrative genital-to-genital contact.
[5] Traditional methods of contraception include rhythm (e.g. fertility awareness-based methods, periodic abstinence), withdrawal, prolonged breastfeeding, taboos such as on marrying widows in some settings, and practices such as polyandry and monasticism (Abernethy, 1993).
[6] Gebbie (1981) argued that Indigenous fertility was comparatively low in arid Australia (pre-European invasion), because of cultural practices including ritualized sub-incision, a form of surgery for adult men.
[7] I.e. around the time of the Bucharest meeting.
[8] Kaplan, using ideas clearly linked to the Limits of Growth, writes of Yale University professor Paul Bracken, who declared a 'crisis of room'. In the past, sparsely populated geography acted as a safety mechanism. Yet this is no longer the case, Bracken argues, 'for as empty space increasingly disappears, the very "finite size of the earth" becomes a force for instability'.
[9] The term 'ecocide' was first used in relation to the use of this herbicide; subsequently calls have been repeatedly made to make ecocide a punishable international crime (Falk, 1973).
[10] Born in Burundi when it was still a colony, Guillebaud became the world's first professor of family planning.
[11] The demographic transition is the change from high mortality and high fertility (with a slow growth in population) to low mortality and low fertility (with slow growth or even a contraction of population size).
[12] Also known as "Forcibly Displaced Myanmar Nationals".
[13] A study showed that many Ladino providers hold biased views of Indigenous women; they feel that they are less intelligent and are less able to understand and properly use contraceptive methods. 'Many Ladino providers admitted to withholding information about certain contraceptive methods from indigenous women' (Festa and Shirazian, 2018, p. 643).
[14] 'On one side of the debate are those who accept "demand theories," which are based on conventional demographic transition theory. This view argues that fertility declines are driven by socioeconomic development. In this view, family planning programs are often seen as having only a marginal impact.' The alternative view is that family planning is important (Bongaarts, 2014).
[15] Unfortunately, the SDGs are silent concerning the *quality* and effectiveness of education; these omissions are a huge problem.

References

Abernethy, V.D. (1993) *Population Politics. The Choices That Shape Our Future*. Plenum Press, New York.

Ahlberg, K.L. (2007) 'Machiavelli with a heart': The Johnson administration's Food for Peace Program in India, 1965–1966. *Diplomatic History* 31, 665–701. DOI: 10.1111/j.1467-7709.2007.00642.x

André, C. and Platteau, J.-P. (1998) Land relations under unbearable stress: Rwanda caught in the Malthusian trap. *Journal of Economic Behavior and Organization* 34, 1–47. DOI: 10.1016/s0167-2681(97)00045-0

Animut, M., Mamo, A., Abebe, L., Berhe, M.A., Asfaw, S. *et al.* (2019) 'The sun keeps rising but darkness surrounds us': a qualitative exploration of the lived experiences of women with obstetric fistula in Ethiopia. *BMC Women's Health* 19, 37. DOI: 10.1186/s12905-019-0732-3

Anonymous (2022) ICRC: One quarter of Africans face 'food security crisis'. VOA News. Available at: https://www.voanews.com/a/icrc-one-quarter-of-africans-face-food-security-crisis/6516128.html (accessed 17 April 2022).

Banwell, S. (2014) Rape and sexual violence in the Democratic Republic of Congo: a case study of gender-based violence. *Journal of Gender Studies* 23, 45–58. DOI: 10.1080/09589236.2012.726603

Bashford, A. (2008) Population, geopolitics, and international organizations in the mid-twentieth century. *Journal of World History* 19, 327–347. https://www.jstor.org/stable/40542618

Berg, R.C. and Denison, E. (2012) Does female genital mutilation/cutting (FGM/C) affect women's sexual functioning? A systematic review of the sexual consequences of FGM/C. *Sexuality Research and Social Policy* 9, 41–56. DOI: 10.1007/s13178-011-0048-z

Bhatia, R., Sasser, J.S., Ojeda, D., Hendrixson, A., Nadimpally, S. *et al.* (2020) A feminist exploration of 'populationism': engaging contemporary forms of population control. *Gender, Place & Culture* 27, 333–350. DOI: 10.1080/0966369X.2018.1553859

Blomquist, R. (2015) Ethno-demographic dynamics of the Rohingya-Buddhist conflict. Georgetown Security Studies Review, Special edition: The Changing Calculus of Security and Violence, pp. 5–33. Available at: https://georgetownsecuritystudiesreview.org/wp-content/uploads/2016/02/Final-CCS-Conference-Journal.pdf (accessed 14 July 2021).

Bongaarts, J. (2014) The impact of family planning programs on unmet need and demand for contraception. *Studies in Family Planning* 45, 247–262. DOI: 10.1111/j.1728-4465.2014.00387.x

Bradshaw, C.J.A., Ehrlich, P.R., Beattie, A., Ceballos, G., Crist, E. *et al.* (2021) Underestimating the challenges of avoiding a ghastly future. *Frontiers in Conservation Science* 1, 615419. DOI: 10.3389/fcosc.2020.615419.

Brown, L.R. and Jacobson, J.L. (1986) Our Demographically Divided World. Worldwatch Paper 74. Worldwatch Institute, Washington, DC. Available at: https://files.eric.ed.gov/fulltext/ED277640.pdf (accessed 28 June 2023).

Bryant, L., Carver, L., Butler, C.D. and Anage, A. (2009) Climate change and family planning: least developed countries define the agenda. *Bulletin of the World Health Organization* 87, 852–857. DOI: 10.2471/BLT.08.0625

Butler, C.D. (2007) The North and South. In: Hedblad, A. (ed.) *International Encyclopedia of the Social Sciences*. MacMillan Reference, New York, pp. 542–544.

Butler, C.D. (2022) Overpopulation: a century of debate that deserves re-examining. In: Saeverot, H. (ed.) *Meeting the Challenges of Existential Threats through Educational Innovation: A Proposal for an Expanded Curriculum*. Routledge, Abingdon, UK, pp. 67–84. DOI: 10.4324/9781003019480-5

Campbell, M. (2007) Why the silence on population? *Population and Environment* 28, 237–246. DOI: 10.1007/s11111-007-0054-5

Canning, D. and Schultz, T.P. (2012) The economic consequences of reproductive health and family planning. *The Lancet* 380, 165–171. DOI: 10.1016/S0140-6736(12)60827-7

Cappa, C., Van Baelen, L. and Leye, E. (2019) The practice of female genital mutilation across the world: data availability and approaches to measurement. *Global Public Health* 14, 1139–1152. DOI: 10.1080/17441692.2019.1571091

Chesler, P. (2010) *Worldwide trends in honor killings*. Middle East Quarterly, Spring, 3–11. Available at: https://www.journals.uchicago.edu/doi/full/10.1086/701154?mobileUi=0 (accessed 23 June 2023).

Clarke, A.E. and Haraway, D. (eds) 2018 *Making Kin Not Population*. Prickly Paradigm Press, Chicago, Illinois.

Coale, A.J. and Hoover, E.M. (1958) *Population Growth and Economic Development in Low Income Countries*. Princeton University Press, Princeton, New Jersey.

Connolly, S.W., Jah, F., Barker, K., Ryerson, W., Okon, E. *et al.* (2019) Reproductive health and family planning behaviors in north west Nigeria. *Effects of a Hausa language radio drama. The Journal of Development Communication* 30, 30–44. Available at: http://jdc.journals.unisel.edu.my/ojs/index.php/jdc/article/view/153 (accessed 8 March 2022).

Coole, D. (2021) The toxification of population discourse. A genealogical study. *The Journal of Development Studies* 57, 1454–1469. DOI: 10.1080/00220388.2021.1915479

Deshpande, N.A. and Nour, N.M. (2013) Sex trafficking of women and girls. *Reviews in Obstetrics & Gynecology* 6, e22–e27. DOI: 10.3909/riog0214

Dodgson, R. (2000) Contesting neoliberal globalisation at UN global conferences: The Women's Health Movement, United Nations and the International Conference on Population and Development. *Global Society* 14, 443–463. DOI: 10.1080/13600820050085796

Elneil, S. (2016) Female sexual dysfunction in female genital mutilation. *Tropical Doctor* 46, 2–11. DOI: 10.1177/0049475515621644

Esman, M.J. (1971) Foreign aid: Not by bread alone (review). *Public Administration Review* 31, 92–100. https://www.jstor.org/stable/974963

Falcoff, M. (1991) Head-hunting: Assassination as a policy. *The National Interest* 24, 103–105. Available at: https://www.jstor.org/stable/42894754

Falk, R.A. (1973) Environmental warfare and ecocide – facts, appraisal, and proposals. *Bulletin of Peace Proposals* 4, 80–96. Available at: https://www.jstor.org/stable/4480206

Fargues, P. (2000) Protracted national conflict and fertility change: Palestinians and Israelis in the twentieth century. *Population and Development Review* 26, 441–482. DOI: 10.1111/j.1728-4457.2000.00441.x

Festa, J. and Shirazian, T. (2018) Family planning methods among Mayan women in Guatemala and the challenges facing the Indigenous population. An anthropological, epidemiological, and biomedical approach. In: Schwartz, D.A. (ed.) *Maternal Death and Pregnancy-Related Morbidity Among Indigenous Women of Mexico and Central America*. Springer, Cham, Switzerland, pp. 630–650.

Firestone, S. (1971) *The Dialectics of Sex*. The Women's Press, London.

Gaud, W.S. (1968) The Green Revolution: Accomplishments and Apprehensions. *Address to the Society for International Development*. Available at: http://www.agbioworld.org/biotech-info/topics/borlaug/borlaug-green.html (accessed 23 November 2022).

Gebbie, D.A.M. (1981) *Reproductive Anthropology – Descent Through Woman*. John Wiley & Sons, Chichester, UK, and New York.

Glendon, M.A. (2001) *A World Made New: Eleanor Roosevelt and the Universal Declaration of Human Rights*. Random House, New York.

Gökarıksel, B., Neubert, C. and Smith, S. (2019) Demographic fever dreams: Fragile masculinity and population politics in the rise of the global right. *Signs: Journal of Women in Culture and Society* 44, 561–587. Available at: https://www.journals.uchicago.edu/doi/full/10.1086/701154?mobileUi=0 (accessed 8 March 2022).

Hall, M.R. (2007) The impact of the US Peace Corps at home and abroad. *Journal of Third World Studies* 24, 53–58. Available at: http://www.jstor.org/stable/45194351

Halperin, D.T. and Epstein, H. (2004) Concurrent sexual partnerships help to explain Africa's high HIV prevalence: implications for prevention. *The Lancet* 364, 4–6. DOI: 10.1016/S0140-6736(04)16606-3

Hartmann, B. (1987) *Reproductive Rights and Wrongs; the Global Politics of Population Control*. South End Press, Boston, Massachusetts.

Hickel, J. (2019) The contradiction of the sustainable development goals: Growth versus ecology on a finite planet. *Sustainable Development* 27(5), 873–884. DOI: 10.1002/sd.1947

Higgs, K. (2022) How sustainable are the UN Sustainable Development Goals? In: Williams, S.J. and Taylor, R. (eds) *Sustainability and the New Economics*. Synthesising Ecological Economics and Modern Monetary Theory. Springer, Cham, Switzerland, pp. 69–88.

Hossain, M.A. and Hossain, M.B. (2023) Understanding fertility behavior of the Forcibly Displaced Myanmar Nationals in Bangladesh: A qualitative study. *PLoS ONE* 18, e0285675. DOI: 10.1371/journal.pone.0285675lands

Huang, H., Xu, C., Liu, L., Chen, L., Zhu, X. *et al.* (2020) Increased protection of earlier use of immunoprophylaxis in preventing perinatal transmission of Hepatitis B virus. *Clinical Infectious Diseases* 73, e3317–e3323. DOI: 10.1093/cid/ciaa898

Kaplan, R.D. (2009) The revenge of geography. *Foreign Policy* 172, 96–105. https://www.jstor.org/stable/20684874

King, M. (1990) Health is a sustainable state. *The Lancet* 336, 664–667. DOI: 10.1016/0140-6736(90)92156-C

King, M. (2003) The population demons. In: Mola, G., Thornton, J., Breen, M., Bullough, C., Philpott, H. *et al.* (eds) *Primary Mother Care and Population*. Spiegl, Stamford, UK, pp. 29.1–29.11.

Kuhlemann, K. (2019) The elephant in the room: the role of interest groups in creating and sustaining the population taboo. In: Almiron, N. and Xifra, J. (eds) *Climate Change Denial and Public Relations*. Strategic Communication and Interest Groups in Climate Inaction. Routledge, London, pp. 74–99.

Leibenstein, H. (1954) *A Theory of Economic-Demographic Development*. Princeton University Press, Princeton, New Jersey.

Leuprecht, C. (2010) The demographic security dilemma. *Yale Journal of International Affairs* 5, 60–74. Available at: https://researchoutput.csu.edu.au/en/publications/the-demographic-security-dilemma (accessed 8 March 2022).

Mandhana, N. and Hookway, J. (2017) Behind the silence of Myanmar's Aung San Suu Kyi. *Wall St Journal.* Available at: https://www.wsj.com/articles/behind-the-silence-of-myanmars-aung-san-suu-kyi-1507306551 (accessed 12 July 2021).

McIntosh, C.A. and Finkle, J.L. (1995) The Cairo conference on population and development: a new paradigm. *Population and Development Review* 21, 223–260. DOI: 10.2307/2137493

Meadows, D., Meadows, D., Randers., J. and Behrens III, W. (1972) *The Limits to Growth*. Universe Books, New York.

Nelson, R.R. (1956) A theory of the low-level equilibrium trap in underdeveloped economies. *The American Economic Review* 46, 894–908. https://www.jstor.org/stable/1811910

Nour, N.M. (2009) Child marriage: a silent health and human rights issue. *Reviews in Obstetrics & Gynecology* 2, 51–56. Available at: https://www.ncbi.nlm.nih.gov/labs/pmc/articles/PMC2672998/ (accessed 8 March 2022).

Ojeda, D., Sasser, J.S. and Lunstrum, E. (2019) Malthus's specter and the Anthropocene. Gender, Place & Culture. *A Journal of Feminist Geography* 27, 316–332. DOI: 10.1080/0966369X.2018.1553858

Pearson, L.B. (1957) *The Four Faces of Peace and the International Outlook*. McClelland and Stewart, Toronto. Available at: https://www.nobelprize.org/prizes/peace/1957/pearson/lecture/ (accessed 17 April 2022)

Perkins, J.H. (1990) The Rockefeller Foundation and the Green Revolution, 1941–1956. *Agriculture and Human Values* 7, 6–18. DOI: 10.1007/BF01557305

Petchesky, R.P. (1995) From population control to reproductive rights: feminist fault lines. *Reproductive Health Matters* 3, 152–161. DOI: 10.1016/0968-8080(95)90172-8

Pirnia, B., Pirnia, F. and Pirnia, K. (2020) Honour killings and violence against women in Iran during the COVID-19 pandemic. *The Lancet Psychiatry* 7, e60. DOI: 10.1016/S2215-0366(20)30359-X

Potts, M. (2015) The Road Not Taken: How the Migrant Crisis in Europe Could Have Been Ameliorated. Available at: https://blogs.berkeley.edu/2015/08/15/the-road-not-taken-how-the-migrant-crisis-in-europe-could-have-been-ameliorated/ (accessed 1 November 2020).

Restrepo-Méndez, M.C. and Victora, C.G. (2014) Maternal mortality by age: who is most at risk? *The Lancet Global Health* 2, e120–e121. DOI: 10.1016/S2214-109X(14)70007-5

Rouchou, B. (2013) Consequences of infertility in developing countries. *Perspectives in Public Health* 133, 174–179. DOI: 10.1177/1757913912472415

Simms, K.T., Steinberg, J., Caruana, M., Smith, M.A., Lew, J.B. *et al.* (2019) Impact of scaled up human papillomavirus vaccination and cervical screening and the potential for global elimination of cervical cancer in 181 countries, 2020-99: a modelling study. *The Lancet Oncology* 20, 394–407. DOI: 10.1016/S1470-2045(18)30836-2

Sinding, S.W. (2000) The great population debates: how relevant are they for the 21st century? *American Journal of Public Health* 90, 1841–1845. DOI: 10.2105/ajph.90.12.1841

Singh, K. (1992) Closing address: challenge for the future. In: Senanayake, P. and Kleinman, R.L. (eds) *Family Planning: Meeting Challenges, Promoting Choices. Parthenon*, Carnforth, UK, p. 726 [cited by Sinding, 2000].

Starrs, A.M., Ezeh, A.C., Barker, G., Basu, A., Bertrand, J.T. *et al.* (2018) Accelerate progress - sexual and reproductive health and rights for all: report of the Guttmacher – Lancet Commission. *The Lancet* 391, 2642–2692. DOI: 10.1016/S0140-6736(18)30293-9

Stone, G.D. (2019) Commentary: new histories of the Indian Green Revolution. *The Geographical Journal* 185, 243–250. DOI: 10.1111/geoj.12297

Strathern, M., Sasser, J.S., Clarke, A., Benjamin, R., TallBear, K. *et al.* (2019) Forum on Making Kin Not Population: Reconceiving generations. *Feminist Studies* 45, 159–172. https://www.jstor.org/stable/10.15767/feministstudies.45.1.0159

UNDESA (1971) The World Population Situation in 1970. United Nations Department of Economic and Social Affairs. Available at: https://www.un.org/development/desa/pd/sites/www.un.org.development.desa.pd/files/files/documents/2020/Jan/un_1970_the_world_population_situation_in_1970_0.pdf (accessed 12 July 2021)

UNDESA (2019) Family Planning and the 2030 Agenda for Sustainable Development. United Nations Department of Economic and Social Affairs. Available at: https://www.un.org/en/development/desa/population/publications/pdf/family/familyPlanning_DataBooklet_2019.pdf (accessed 12 July 2021).

UNFPA (2000) *The State of World Population 2000*. United Nations Fund for Population Activities, New York.

UNICEF (2014) *Ending Child Marriage: Progress and Prospects*. United Nations International Children's Emergency Fund, New York.

UNICEF (2016) *Female Genital Mutilation/Cutting: A Global Concern*. United Nations International Children's Emergency Fund, New York.

Waller, J. and Albornoz, M.S. (2021) Crime and no punishment? China's abuses against the Uyghurs. *Georgetown Journal of International Affairs* 22, 100–111. DOI: 10.1353/gia.2021.0000

Westoff, C.F. (1995) *International population policy. Society* 32, 11–15. DOI: 10.1007/BF02693318

WHO (2016) *Report on Global Sexually Transmitted Infection Surveillance 2015*. World Health Organization, Geneva.

WHO (2018) *Obstetric Fistula*. World Health Organization, Geneva. Available at: https://www.who.int/news-room/facts-in-pictures/detail/10-facts-on-obstetric-fistula (accessed 17 May 2021).

WHO (2019) *Maternal Mortality*. World Health Organization, Geneva. Available at: https://www.who.int/news-room/fact-sheets/detail/maternal-mortality (accessed 22 July 2021).

WHO, London School of Hygiene Medicine and South African Medical Research Council (2013) *Global and Regional Estimates of Violence Against Women: Prevalence and Health Effects of Intimate Partner Violence and Non-Partner Sexual Violence*. World Health Organization, Geneva.

Winn, P. (2013) Do 'rapidly breeding' Rohingya Muslims really threaten Myanmar's Buddhist identity? Global Post. Available at: https://www.pri.org/stories/2013-10-14/do-rapidly-breeding-rohingya-muslims-really-threaten-myanmars-buddhist-identity (accessed 12 July 2021).

10.i Reproductive Health in Papua New Guinea: A Vignette

Glen D.L. Mola

Department of Obstetrics, Gynaecology and Reproductive Health, School of Medicine and Health Sciences, The University of Papua New Guinea, Port Moresby, Papua New Guinea

Abstract

This short chapter distils insights from over 50 years of working to improve health in the diverse, populous, resource rich island of Papua New Guinea (PNG). Partly inspired by Maurice King, my career turned towards public health, and then to obstetrics, gynaecology and other aspects of reproductive health. Today, the health statistics for 'booked' mothers at Port Moresby General Hospital are excellent, compared with other low- and middle-income countries, and is achieved with possibly the lowest intervention rates in the world, such as for caesarean sections (5%). On the other hand, the supervised birth proportion for the rest of PNG is shamefully low; under 40%, a level static for more than 20 years. In the long run, improved survival of mothers and children, especially if accompanied by further reduction in the birth rate, should help trigger 'virtuous' feedbacks that improve well-being and prosperity in PNG.

10.i.1 Introduction

Now 52 years along the track of a medical career in Papua New Guinea (PNG), and having never written anything about my work (in spite of 100+ peer reviewed publications), I am going to tell a few stories about how I got to where I am, and what I plan to do with the last few years of my professional life. I trust this might be helpful to some young ones starting out.

10.i.2 Encountering the Ideas of Maurice King

After completing a medical degree (MBBS) at Melbourne University (Australia) in 1970, I came home (Fig 10.i.1) for the standard 24 months PNG residency – rotations through all the major clinical specialties (and a 3-month block in a rural district hospital). In this time I encountered (in the Arawa Hospital library) a copy of Maurice King's book *Medical Care in Developing Countries* (King, 1966). My introduction to public health in medical school had focused on epidemic control, sanitation, water supplies and other health inspection-type activities. King's book opened my eyes to the concept of community diagnosis and community medicine – and particularly in relation to what we called in those days 'developing country' healthcare. I can remember one night going home and constructing a population projection graph for Bougainville (where I was then working), and was amazed to see that, with a rate of increase of 3% p.a., the population would double every 25 years or so.[1] I have now been around long enough professionally to see this actually unfold in Bougainville, and indeed across the whole of PNG. Since I was a schoolboy in the 1960s, PNG's population has

 Climate Change and Global Health: Primary, Secondary and Tertiary Effects, 2nd Edition (eds C.D. Butler and K. Higgs)
DOI: 10.1079/9781800620025.0010i

Fig. 10.i.1. The author on the right (2022) with adopted son Gadaphy from the Yuri tribe of Gumine district in Simbu province of Papua New Guinea. Photographer Mr Kale Gari (used with permission).

increased fivefold, from about 2.5 million to as many as 12 million today.[2]

Maurice King's ideas were also responsible for my first real career direction: to paramedical education. One of the axioms of his book was that when health resources and 'man' power are scarce, healthcare should be provided by the lowest-level cadre who can competently do it. Clearly, in PNG, that meant a need to focus on quality training for health extension officers (the PNG version of medical assistants) and health inspectors.[3] Hence my decision to work at the Community Health Practice Centre in Kainantu, Eastern Highlands Province (Fig. 10.i.2), then as Dean of the College of Allied Health Sciences in Madang, and, in between, further study in public health at Sydney University in 1974.

For a number of personal and professional reasons I again switched career paths in 1979, undertaking clinical specialist training in obstetrics and gynaecology, and qualifying with UK and Australian fellowships. This helped me better understand reproductive public health from the population perspective and, more recently, its relevance to climate change and planetary health. In 1998 I joined with Maurice King in writing and editing *Primary Mothercare and Population*, a book which Tony McMichael (2003) described as a 'frontal attack, an exposé' (of the view that discussion of population is politically incorrect), and which 'articulates clearly the dual problem of, firstly overpopulation, and secondly, overconsumption – and their disparate distribution between the world's rich and poor'.

Primary Mothercare was intended to be the last in the primary care series – of *Primary Surgery*, *Primary Anaesthesia*, *Primary Childcare* and *Primary Laboratory Care*. However, its forthright, politically incorrect placement of population – especially 'demographic entrapment' – at the heart of the climate change and planetary survival debate meant that Oxford University Press (the publisher of all the previous primary care books) balked at publishing a version of the book that included a cartoon accusing demographers and UN agencies of being corrupt because they were 'supressing debate on population issues'.

There are hundreds of text books on obstetrics and gynaecology (O&G) and reproductive public health. *Primary Mothercare and Population* is the only comprehensive text book on obstetrics, gynaecology, family planning and reproductive public health that is written expressly for health workers in low- and middle-income countries (LMICs). The third edition of the book (Mola and King, 2023) has been published by Health Action Publishing (the successor to Teaching Aids at Low Cost) at a subsidized price (US$6.00).

10.i.3 Improving PNG's Reproductive Health

At a local level, my O&G unit at the Port Moresby General Hospital (PMGH) recorded an increase in its numbers of annual births of about 10%

Fig. 10.i.2. The author checking village people for tuberculosis on a medical patrol near Tinputz in Bougainville, PNG, in 1972. Photographer: the late Sr Janice SM. Reprinted with kind permission of the Papua New Guinea Society for Rural and Remote Health (Williams, 2018).

between 1978 and 2016 (Mola and Unger, 2018). Then, in 2016 we were offered some reproductive health assistance by the DAK Foundation, an Australian philanthropic trust (https://dak.org.au/). My professional colleagues (midwifery and medical) and I decided that the best way we could use this support was to set up an immediate post-partum family planning Implant[4] programme. Since 2017 the numbers of women presenting to the PMGH labour ward for supervised birthing care has plateaued at around 15,000. If it had kept on rising by 10% per annum (as occurred between 1970 and 2016), we would by now be overwhelmed by more than 25,000 women per annum seeking supervised birthing care. (There is no other public maternity facility in the national capital, Port Moresby; perhaps 2000 women give birth in private hospitals.) PMGH O&G division also offers immediate post-partum tubal ligation and vasectomy services, and about 1800 couples avail themselves of this family planning-completion strategy, annually. Along with the 5000 implants we provide, this means that almost half of women birthing at PMGH return home with an effective method of contraception. We estimate that less than 10% of the remainder subsequently attend city family planning clinics to start a modern method of contraception.

More than 90% of women who go on to deliver at PMGH receive regular antenatal care in Port Moresby's clinics – including 5000 p.a. who attend the public hospital clinic. These women receive some of the best maternity care in the world – despite many human and resource constraints. The perinatal mortality rate (PMR) among 'booked' mothers at PMGH maternity is generally less than half that of similar sized public maternity units in other LMICs. The maternal mortality ratio (MMR) is similarly low. The booked PMR for PMGH is less than 20 per 1000 births, and the MMR about 50 per 100,000 live births (Mola and Unger, 2018). This is achieved with possibly the lowest intervention rates in the world, for example caesarean section rate 5%, assisted vaginal delivery rate 5% and induction of labour rate of 5% (Litorp, 2015; Litorp *et al.*, 2015; Sobhy *et al.*, 2019).

On the other hand, the supervised birth proportion for the rest of PNG is shamefully low, at less than 40%, and in fact static at this level for more than 20 years. PMGH delivers ~4% of all the births in PNG but ~12% of all the supervised births in PNG. Women who are unable to get a supervised birth have a tenfold risk of dying in childbirth. Their babies have a threefold risk of death, and also a higher rate of non-fatal birth trauma, often with life-long consequences. See Table 10i.1.

In order to improve this, it is critical to increase the proportion of women choosing to deliver at rural health facilities, rather than remaining in their rural village home. Of 360,000 births annually in PNG, 30% are currently supervised in the 22 provincial hospitals that today exist, together with PMGH. Ten per cent give birth in 800 rural health facilities – but this means fewer than one birth per week in most of them. The provincial hospitals are overcrowded, and the rural health facilities are very under-utilized.

About 10 years ago I decided that my retirement would be more meaningful if we could achieve better birthing services for the rural majority of women (80%) who are currently at risk of morbid outcomes for both themselves and their newborns. To get women to come to their rural health facility there needed to be:

- well-trained health workers to provide quality maternal and newborn care; and
- rural facilities with the capacity to provide 24 hours per day quality care for birthing women.

To this end, with the help of the late Dr Lahui Geita (maternal health adviser to the National Department of Health) and the chief obstetrician for PNG (Dr Ligo Augerea) in 2012 we set up a 6-month in-service training programme for our basic enrolled nurses (called Community Health Workers (CHWs) in PNG). This training has now been carried out in 14 provincial hospitals; over 600 upskilled CHWs have been trained in groups of six to eight. And, over the past 20 years, many rural health facilities have had new birthing units built on their premises.

Table 10.i.1. Mortality rates in PNG, by place of delivery.

Facility	MMR/100,000	PMR/1000
Provincial Hospital	100	30–40
Rural Health Centre	200–300	40–50
At home, with no ability to refer	800–900	100

These trainings and rural health centre upgrades have been made possible with funding from various support agencies, including church agency health services and NGOs, for example DAK and Soroptomists International (https://www.soroptimistinternational.org/) and the Mola Foundation (a philanthropic trust directed by Perpetual Ltd, Melbourne, Australia). Government and UN agencies have also helped, including AusAid, UNFPA and UNICEF. Local provincial health authorities in PNG have also contributed significantly, such as by helping to select suitable CHWs for training and by paying the salaries of the trainees during their 6 months of in-service.

Initially, we hoped that decent facilities and well-trained staff would be sufficient to alter the behaviour of women and their families; to swap their traditional view of home births for a journey to their local rural health facility for a safer supervised birth. To begin with this did not happen. Most rural health facilities remained underused. Thus, in 2019 I decided that we should try an additional motivational strategy, building on an idea pioneered in Finland and Cambodia. The latter country managed to increase the supervised birth proportion from about 35% to 75% over 5 years in the late 1990s by incentivizing supervised birth in rural areas. Also, a pilot incentivization programme under way since 2009 in PNG's Milne Bay province had doubled supervised birth numbers in 15 participating rural health facilities.

So in 2019 I initiated a 'baby box' programme, complemented with food supplies, to twelve rural health facilities in Simbu province in PNG's highlands. The baby box contains everything that a mother and baby might need (including nappies, pins, baby blanket, booties, vest, towel, and, for mother, sanitary pads, soap, umbrella, new sarong, and for dad a file to sharpen his axe and a new spade – all wrapped up in a large plastic baby bath) (Fig. 10.i.3). Also supplied is brown rice, tinned tuna and coins for the nurse to purchase greens in the local market. Without food to eat, pregnant women are unable to stay around the Mama waiting houses in the rural health facilities for the onset of labour. This programme has now been running for 4 years, and in this time most of the participating rural health facilities have seen a more than doubling in the numbers of women attending for supervised birth.

Fig. 10.i.3. A baby box, a happy mother (Elsie Titus), a happy nurse (Maria Kaupa of the Gogmei health centre) and the happy author. Photographer Dr David Sinowai (used with permission).

10.i.4 Conclusion

Epidemiologically one can estimate that one woman's life is saved for every additional 150 women who come for supervised birth, and one newborn is saved for every 20 additional women who come for supervised birth in their local rural health centre. Over the 4 years of this programme's life, it is estimated that incentivization for this improved level of delivery care has averted 26 maternal deaths and 180 newborns lives have been saved. These benefits do not include the improvements in morbidity, nor the increased numbers of women who accept immediate post-partum contraceptive implants or referral to the provincial hospital for post-partum tubal ligation.

The cost of this programme is modest: about US$3000 per rural health facility p.a. This enables an additional 300 women p.a. to have a facility-supervised birth, by a health professional. I am hoping that we can extend this work to other provinces of PNG soon.

In the long run, improved survival of mothers and children, especially if accompanied by a further reduction in the birth rate, should help trigger 'virtuous' feedbacks that improve well-being and prosperity in PNG. These are benefits that Maurice King and his forebears and supporters have long alluded to (King, 1990).

Notes

[1] A total fertility rate of five approximates to an annual population increase of 3%.

[2] The population of PNG has been estimated for many years; censuses are difficult (Bourke and Allan, 2021).

[3] Health inspectors and medical assistants are high-school graduates who do a diploma or degree course for 3 years followed by an apprenticeship (or intern training) for a year.

[4] Contraceptive implants are nowadays a standard family planning technology (e.g. Implanon, which lasts 3 years) and Jadelle, which is supplied by the UNFPA to LMICs; these contain a depot of the hormone progestin in a silastic rod, and are implanted sub-dermally.

References

Bourke, R.M. and Allen, B. (2021) *Estimating the Population of Papua New Guinea in 2020.* Development Policy Centre Discussion Paper No. 90. Crawford School of Public Policy, Australian National University, Canberra, Australia. [Paper amended 2023 to include 2011 population numbers for Aitape and Rabaul.] DOI: 10.2139/ssrn.3770356

King, M. (ed.) (1966) *Medical Care in Developing Countries. A Primer on the Medicine of Poverty and a Symposium from Makerere*. Oxford University Press, Nairobi, Kenya.

King, M. (1990) Health is a sustainable state. *The Lancet* 336, 664–667. DOI: 10.1016/0140-6736(90)92156-C

Litorp, H. (2015) 'What about the mother?' Rising caesarean section rates and their association with maternal near-miss morbidity and death in a low-resource setting. PhD thesis, Uppsala University, Sweden. Available at: https://www.diva-portal.org/smash/get/diva2:772232/FULLTEXT01.pdf (accessed 27 June 2023).

Litorp, H., Mgaya, A., Kidanto, H.L., Johnsdotter, S. and Essén, B. (2015) 'What about the mother?' Women's and caregivers' perspectives on caesarean birth in a low-resource setting with rising caesarean section rates. *Midwifery* 31, 713–720. DOI: 10.1016/j.midw.2015.03.008

McMichael, T. (A.J.) (2003) Primary Mothercare and Population (book review). *BMJ* 327, 626. DOI: 10.1136/bmj.327.7415.626

Mola, G.D.L. and King, M. (2023) *Primary Mother Care and Population*, 3rd edn. Health Action Publishing, Rugby, UK.

Mola, G.D.L. and Unger, H. (2018) Strategies to reduce and maintain low perinatal mortality in resource-poor settings – Findings from a four-decade observational study of birth records from a large public maternity hospital in Papua New Guinea. *Australian and New Zealand Journal of Obstetrics and Gynaecology* 59, 1–9. DOI: 10.1111/ajo.12876

Mola, G.D.L., Thornton, J., Breen, M., Bullough, C., Philpott, H. *et al.* (eds) (2003) *Primary Mother Care and Population*. Oxford University Press, Oxford, UK.

Sobhy, S., Arroyo-Manzano, D., Murugesu, N., Karthikeyan, G., Kumar, V. *et al.* (2019) Maternal and perinatal mortality and complications associated with caesarean section in low-income and middle-income countries: a systematic review and meta-analysis. *The Lancet* 393, 1973–1982. DOI: 10.1016/S0140-6736(18)32386-9

Williams, R. (ed.) (2018) *The Last Doctor. A Compilation of Stories by the Papua New Guinea Society of Rural and Remote Health*. Papua New Guinea Society for Rural and Remote Health, Mt Hagen, Western Highlands Province, Papua New Guinea.

11 Heat Impacts, Adaptations and Inequities

Matthew Chersich[1,2]
[1]*Wits Planetary Health Research, University of the Witwatersrand, South Africa;*
[2]*Public Health and Primary Care, School of Medicine, Trinity College, Dublin, Ireland*

Abstract
Rising temperatures, especially prolonged temperature spikes, pose major threats to health. Vulnerable groups such as the elderly, pregnant women and infants are at particularly high risk. This chapter examines the impacts of heat exposure on human health, together with the range of adaptation options that can be considered. Interactions between heat effects on the natural and human world are highlighted. Globally, heat responses are marked by major inequities, with most affected areas least able to respond. Much, however, can be done, even with limited resources. Interventions may involve a combination of space cooling, adaptations to the built environment and nature-based solutions, depending on the specific risk groups, patterns of heat exposure and available financing in a particular setting.

11.1 Introduction

The primary manifestation of climate change is a progressive escalation in land and sea temperatures across the globe, accompanied by occasional multi-day periods of extreme heat or heatwaves, affecting people's health, well-being, livelihoods, food security and their surrounding ecosystem (Xu *et al.*, 2020; Ebi *et al.*, 2021). Poor and least-developed communities and countries are most affected by – yet least responsible for – climate change and the consequent escalation in temperatures. The world has already warmed by over 1.2°C, with record temperatures in most parts of the world. In 2023 the average global temperature was 1.45±0.12°C above the pre-industrial average; this is the hottest year on record (WMO, 2024). The reasons for this recent increase in heat, which is persisting in early 2024 (Carbon Brief, 2024)[1] are poorly understood; hypotheses to explain this include El Niño and reduced particulate air pollution (Schmidt, 2024).

In the period 2013–2022, children younger than 1 year collectively experienced 108% more days of heatwaves per year than in the period 1986–2005 (Romanello *et al.*, 2023). At the other age extreme, in 2023, heat-related mortality of people older than 65 years increased by an estimated 85% compared to mortality rates in the period 1990–2000. Heat-related mortality of people older than 65 years increased by an estimated 68% between 2000–2004 and 2017–2021 in the decade to 2022. Heatwaves, set against increasing background temperatures, are becoming stronger, more frequent, longer in duration and more complex. Heat extremes receive most media and policy attention, but their absolute effect on mortality and morbidity is less than that of more frequent exposures to moderately high temperatures. If unchecked, heat levels will exceed the capacity of the human and natural systems to successfully adapt in many areas, even with large-scale, transformational change. Some areas are set to become 'beyond adaptive capacity' and will

 Climate Change and Global Health: Primary, Secondary and Tertiary Effects, 2nd Edition (eds C.D. Butler and K. Higgs)
DOI: 10.1079/9781800620025.0011

no longer be viable for sustaining livelihoods, liveability and even survivability (Im *et al.*, 2017; Raymond *et al.*, 2020; Timofte, 2021; Tuholske *et al.*, 2021). Most areas of greatest concern lie in low- to mid-latitude regions, and include places with humidity-enhanced heat stress, as well as those characterized by extreme dry heat. These areas require less warming to exceed 'dangerous' thresholds than higher-latitude countries.

Humans, like all animals, operate in a specific thermal niche, with a set of physiological thresholds (hard biophysiological limits) that correspond to thermal comfort, heat stress, organ system compromise and death. 'Deadly heat' arises when conditions of air temperature and humidity surpass the physiological threshold for human adaptability. Core body temperatures may reach lethal levels under sustained periods of wet-bulb[2] temperatures of 35°C or more (or at lower temperatures if there is significant exertion or comorbidities) (see Chapter 12).

Heat vulnerability is complex, hindering and complicating assessments of risk and comparisons across settings. Simplistic measures such as maximum daily ambient temperatures are unable to capture this complexity. High diurnal heat fluxes and night-time temperatures are better measures of heat-health risks, as are extreme temperature peaks, especially in transitional seasons such as spring. Moreover, there are complex connections between environmental, infrastructural and socio-economic systems that generate cascading impacts and amplify existing risks or create new ones. In compounding events, multiple hazards occur simultaneously, such as heatwaves during periods of droughts or coinciding with wildfires. Further, infrastructure systems are interdependent and impacts on one system (e.g. electricity supply disruption) affect other systems (e.g. delivery of safe water). Additionally, some factors can amplify heat hazards, such as urban heat islands, air pollution and poverty. Moreover, and perhaps most importantly, one cannot consider heat impacts on humans separately from heat effects on the rest of the natural environment. Heat-related soil drying, losses in wildlife and agriculture and the progressive undermining of ecological resilience have major short- and long-run implications for human health, and indeed for the world as a whole. Clearly, the above layers of complexity need to be taken into account in risk assessments, but, equally, to reach higher levels of adaptation, it is necessary to combine multifaceted adaptation options into broad policy and climate financing portfolios, aiming to amplify effects across multiple sectors and ecosystems.

11.2 Places and Population at Risk

Risk from the built environment and workplace

Extreme heatwaves superimposed on background warming trends can together push heat-vulnerable people and whole geographical areas beyond physiological, social and mortality thresholds. Heat-health impacts, however, show marked variability between settings. This variability is sometimes referred to as 'thermal inequity'; it is a distributive justice concern (Mitchell and Chakraborty, 2015; Bonds and Martin, 2016). A disproportionate burden falls on individuals of lower socio-economic status, those living in substandard housing or homeless, and people in low-paid occupational groups, especially outdoor construction or agriculture workers. Heat impacts increasingly reflect entrenched racial disparities, including among the incarcerated (Colucci *et al.*, 2023). Ultimately, the degree of risk one faces is contingent on the opportunities one has to prepare for and respond to heat hazards. Increasingly, access to air conditioning is the pre-eminent determinant of heat outcomes, but universal air-conditioning access is neither feasible, nor without major drawbacks (Osunmuyiwa *et al.*, 2020). The pervasive use of air-conditioning in high-income countries and relative absence in resource-constrained settings echoes the nationalism that marked access to COVID-19 vaccines.

Extreme heat exposure is especially dangerous in urban areas, also referred to as urban heat islands (UHIs). In UHIs, increased heat capacity of urban structures or physical changes to surface-energy balances are compounded by reduced evapotranspiration and waste heat emissions from anthropogenic sources, such as air-conditioning, mechanical ventilation systems, transportation and biological metabolism. The phenomenon is driven by unrelenting expansion of the built environment, poor urban planning, population growth and urbanization. The urban heat problem is progressively worsening as temperatures gradually escalate, but takes full expression during heatwaves, with major consequences for health, vegetation and infrastructure, such as melting asphalt roads and buckled rail lines. Risks are highest in cities in urban tropical or subtropical parts of low- and middle-income countries. Women may experience higher impacts of extreme heat in areas where culturally they have limited mobility outside the home, which restricts their ability to access cool areas.

Overheating risk depends on the density of buildings and population, building characteristics, such as insulation, thermal resistance, presence of solar shading, thermal mass, ventilation, orientation and geographical location. Additionally, thermal circulation wind systems can advect heat to areas downwind of the city and microscale wind conditions within urban canyons can undermine thermal comfort at pedestrian level. People living in overcrowded subdivided apartments without temperature controls – typically in Asian countries – are at especially high risk. Risks are also high in residents of informal settlements or slums in Africa and South America, where housing is often made of corrugated iron sheeting, or even more flimsy material (Fig 11.1).

Thermal stress symptoms are also common in schools, health facilities and prisons that are poorly adapted to heat and lack temperature regulation. Temperatures inside these structures can exceed outdoor figures by 2–4°C and have large diurnal fluctuations (Naicker *et al.*, 2017).

The impact of heat in rural areas must not be ignored, particularly for farmers and outdoor labourers. Many rural areas have ageing populations, high poverty levels and few, if any, cooling spaces.

Within occupational settings, high temperatures undermine labour productivity, while increasing absenteeism and traumatic occupational injuries, likely from fatigue and decreased psychomotor performance (Spector *et al.*, 2019).

Fig. 11.1. Housing in informal settlement, Pune, India. Photographer: Colin D. Butler

Strenuous outdoor work, such as in agriculture, construction and mining (both in open-cast and underground mines) carries the highest risks, but indoor manufacturing and service sectors are also affected where cooling is not available. Reductions in work performance, increased health service use during heatwaves and insurance pay outs for occupational injuries generate significant economic costs (Foster *et al.*, 2020; Romanello *et al.*, 2023).

Risks from extreme heat in the natural environment

Within the health sector, discourse on heat exposure has largely been restricted to human health, with little attention given to heat impacts on the natural world and on human–nature interactions. Heatwaves desiccate vegetation and have significant negative impacts on crop production and thus food security. Warming also increases growth of pests, plant and animal pathogens, and weeds. Heat stress also diminishes livestock productivity, animal growth and fertility, complicates long-distance transport of animals and undermines cold chains for the safe passage of meat and dairy products (Gonzalez-Rivas *et al.*, 2020). Longer growing seasons for crops can potentially compensate for some of these negative agricultural effects.

High temperatures increase evapotranspiration with consequent desiccation of vegetation, fuel aridity and soil drying. These conditions set the stage for large, severe wildfires and prolonged fire seasons. Wildfires result in trauma injuries, respiratory distress from smoke, exposure to heavy particulate pollution, asthma, cardiovascular–respiratory conditions, degraded water quality and heightened anxiety levels, including post-traumatic stress disorder.

Warming also impacts the physiology, phenology and ecology of plant and animal species, changing the timing of development, distribution, migration and synchronization of species interactions. As the world warms, many plant and animal species are moving pole-wards or to higher altitudes. As with humans, death occurs when heat exposure exceeds physiological thresholds. Aberrations cascade and cumulate across ecosystems and food webs, and are of particular concern at the boundaries of ecological regions. Marine and lake heatwave events, which can last for weeks at a time, cause declines in fish populations during and following these events, threatening fisheries and aquaculture. Mass mortalities of fish, bird and mammal species can accompany marine heatwaves. Warming of sea and lake surface temperatures reduces mixing and water stratifies into hot and cold layers. Fish have little refuge in these circumstances as upper layers become too hot for survival, while oxygen concentrations in deeper waters are too low as they are not replenished though mixing. Heat-related bleaching of coral, with the destruction of these unique ecosystems, portends our future.

All the above heat-related changes in the natural environment undermine food security and safety, and have significant economic costs with associated knock-on effects on health and well-being. Moreover, we should not underestimate the negative mental health, cultural and livelihood implications of shifts in biomes, loss of biodiversity, compromised agricultural productivity and wildfires. Indeed, heat-related crop failures, with concomitant income losses and food insecurity, are key indirect drivers of migration, with all the marked health and social risks that follow. Heat is also a direct driver of migration through prolonged periods of high temperatures or repeated temperature spikes around the threshold of humans' physical tolerance.

Other environmental sequelae of increased temperatures, such as permafrost thawing and wildfires, increase entry of mercury into the food chain. Mercury is neurotoxic and nephrotoxic, and bioaccumulates across the food chain. Rising sea surface temperatures also accentuate rates of ciguatera poisoning. Ozone levels are augmented by oxidants formed in heat-sensitive photochemical reactions.

11.3 Health Impacts of Escalating Temperatures

Excess deaths during extreme heat events can cause considerable mortality, with deaths occurring predominantly in older individuals and from cardiorespiratory conditions. Myocardial infarction and stroke, in particular, display temperature sensitivity. Higher temperatures, especially marked heat fluxes, increase asthma, pneumonia episodes and pneumonia-related mortality and may compound the impact of air pollution. Renal conditions among heat-exposed outdoor workers are also frequent, including chronic kidney disease of unknown aetiology, which have been documented at especially high levels in certain rural populations around the world (see Chapter 11.ii). People with diabetes are at particular risk because autonomic nerve damage (a common complication of diabetes) impairs heat loss capacity by reducing dermal blood flow and by hindering thermoregulatory sweating, impairing the body's ability to transfer heat from its core to the environment.

Anatomical and physiological changes during pregnancy increase vulnerability of women and foetuses to heat exposure (Ravanelli *et al.*, 2019). High temperatures can overwhelm thermoregulation in pregnant women, cause dehydration and endocrine dysfunction, and affect placental function. Clinical sequelae include pre-term birth, pre-eclampsia, stillbirth and prolonged labour (Chersich *et al.*, 2020). In children, heat exposure and dehydration are linked with electrolyte imbalances, fever, and renal and respiratory disease. Less severe heat-exposure outcomes such as lethargy, headaches, rashes and cramps negatively affect children in school and play environments. Under extreme heat conditions, trauma centre visits and hospitalizations will increase for fluid replacement, renal failure, urinary tract infections, septicaemia and heat stroke, as well as unintentional injuries. Sports and military training during heatwaves are especially hazardous for exertional heat stroke, including in otherwise healthy young individuals.

Heatwaves amplify risks for most food-, water- and vector-borne infections (Semenza and Paz, 2021). This occurs through increased pathogen survival and proliferation, expansion of affected geographical areas, extended transmission seasons for pathogens, the development of new ecological niches, and enhanced pathogen loads in animal reservoirs, e.g. chickens. Indirect mechanisms raising infection risks include behavioural changes such as increased outdoor activities for example picnics or hiking, where food spoilage is more common, and contact with environmental pathogens more frequent.

On the other hand, escalating temperatures may reduce cold-related mortality in winter and transmission of some respiratory infections, including tuberculosis.

Impacts on mental health and behaviours

Heat-related psychopathology includes fluctuations in mood and sentiments, greater experiences of anxiety and acute stress, resulting in increased psychiatric hospital admissions for mood, schizophrenia and neurotic disorders. Additionally, heat profoundly affects quality of life (Haynes *et al.*, 2022). High summer temperatures are linked with decreased happiness and ratings of well-being and sleep. Heat impairs human capabilities through negative effects on concentration, cognitive performance, personal comfort and social interactions, affecting both children and adults (Park *et al.*, 2020). Dehydration heightens these impacts. Ecoanxiety, underscored by deep concerns around environmental destruction, is increasingly common, especially among youth (Pihkala, 2020). Heat is also associated with suicide and with interpersonal and collective aggression. Both physiological and behavioural pathways affect the likelihood of violence. Increased irritability and aggressive thoughts, and reduced positive emotions illustrate the physiological pathways, and people congregating in uncomfortably hot environments the behavioural ones. Men's use of negative coping mechanisms, such as alcohol, may be exacerbated by heatwaves and may promote gender-based violence (Chersich *et al.*, 2019), which has also been reported as increasing during extreme heat (van Daalen *et al.*, 2022).

11.4 Adapting to a Warming World

Adaptation encompasses measures taken by individuals, or collective efforts involving communities, local and national government, or industry and non-governmental partners. Many experts and international organizations promote a 'systems approach' to adaptation, in which interventions target multiple interlinked areas, aiming to shore up all parts of the 'system' and overall resilience. The World Health Organization (WHO), for example, advances an approach to adaptation that centres around the six WHO Health System Building Blocks (WHO, 2015).

The alternative, a more targeted approach, where a package of selected interventions is implemented, may be a more effective option, and more attractive for policy makers and funding agencies (WHO, 2015). A targeted approach may include, for example, coating roofs in informal settlements with white reflective paint and building a few designated 'cooling centres'. These debates echo previous public health discussions around comprehensive versus selective primary healthcare, or integrated versus standalone services for HIV treatment. Regardless of overall approach, clearly a broad range of interventions is required. These may be characterized by incremental steps or transformative shifts, a focus on engineering advances and technological fixes as opposed to behavioural change strategies, and may or may not include nature-based solutions. The optimum package of adaptation measures varies by setting and climate zone. For example, natural ventilation is an effective, low-cost and highly feasible measure for many areas in northern Europe, but is less effective in southern Europe.

Overall, adaptation measures to reduce heat impacts on health hold tremendous potential, though they have achieved very low coverage to date. Heat-related excess mortality has declined over time in a range of countries, closely tied to a progressive increase in air-conditioning coverage (Sera *et al.*, 2020). However, while there is evidence of higher heat tolerance over time, complete acclimatization is likely not possible. Indeed, significant relative risks of mortality persist at high temperatures and there is some indication of a slowing of progress in recent decades. There are also important uncertainties about the precise factors driving mortality declines and whether adaptation successes can be extended to low- and middle-income countries (Sera *et al.*, 2020). While very few robust evaluations have been done on the relative effectiveness of different interventions (Conti *et al.*, 2022), declines have mostly been attributed to a general strengthening of health systems, heightened preparedness and an increased prevalence of residential air-conditioning. Guidelines are needed on the temperature thresholds at which schools, outdoor work, sporting and other outdoor events need to be cancelled.

Migration is increasingly viewed as an adaptive measure. For those with means, this can take the form of short-term moves during hot periods, where people go from cities to rural areas or escape to mountainous regions. Over the longer term, permanent or semi-permanent migration within or between countries, in people exposed to lengthy temperature perturbations, is likely to become increasingly common (Balsari *et al.*, 2020).

Adaptive capacities vary markedly across settings. Overall effectiveness rests on extending and strengthening basic public health functions to meet the additional demand from climate-sensitive conditions. Local government leadership is critical. Effectiveness also hinges on levels of community resilience, participation and cohesion. There are, however, considerable challenges to adaptation success: leadership and governance are poor in many countries, and changes to the built environment may face competing interests of local government stakeholders and community groups. Some measures may inadvertently worsen inequities. Measuring the 'success' of complex adaptive interventions is difficult and poorly functioning health systems may limit the success of initiatives. Additionally, while water bodies and moist vegetated surfaces are effective cooling strategies, they may provide suitable habitats for vectors and pests - reducing the acceptability of this strategy for local populations. Finally, in many places buildings and infrastructure are crumbling and poorly suited to heat proofing interventions.

Heat action plans for local and national governments

Effective solutions to escalating temperatures need to extend beyond air-conditioning and changes to the built environment. A coordinated 'heat action plan' (HAP) helps to prepare communities for heat exposure, and then to protect exposed communities during periods of high temperatures. Evidence supports the effectiveness of HAPs, but there is little certainty about the relative effectiveness of individual components of these plans (Hasan *et al.*, 2021).

Preparing for heatwaves is an important first step. This involves creating national or regional early warning and information systems, heatwave plans and guidelines, and raising public awareness through campaigns.

Trusted media that inform people about what actions they need to take is critical. Priority behaviours to be covered in communication include: keeping hydrated by drinking sufficient fluids and avoiding alcohol and caffeine; accessing cool areas or seeking shade; using water sprays or other forms of evaporative cooling; altering the pattern of work; avoiding outdoors during the hottest times of the day; wearing protective and appropriate clothing, ideally cotton; and self-monitoring for symptoms of heat illness, especially dizziness, heavy sweating, fatigue, and cool moist skin with goosebumps.

A 'culture of heat' needs to be generated, where the general public is highly aware of heat risks and the range of adaptation measures available. People who have experienced climate shocks are more likely to implement risk management measures. Being aware of linkages between climate change and higher temperatures may also raise heat preparedness. Appreciation of anthropogenic climate change is by no means universal. Indeed, many countries are marked by denial and obfuscation, with major consequences for current and future generations.

Key actions to be considered in HAPs include: providing dedicated public cooling shelters; securing the availability of clean water and simple water purification systems in low-income settings; surveillance of heat-related diseases; outreach to vulnerable populations; and extended hours for public pools or other water bodies. Interventions need to focus on actively supporting the elderly, pregnant women, infants and other key risk groups during heatwaves. Homelessness and social isolation increase heatwave-related mortality.

Early warning systems provide warnings some days in advance of a heatwave, based on meteorological forecasts. Once the predefined temperature threshold is exceeded, warnings trigger a tiered set of actions intended to lower exposure to harmful temperatures and to assist people to take protective actions. Heatwave definitions are needed which, when met, trigger the warnings. Given differences between climate zones and variations in the built environment, not surprisingly there is a multitude of heatwave definitions, each suited to a different cluster of circumstances. The cut-off thresholds selected for defining a heatwave in each setting need to balance the specificity and sensitivity levels at which heat risks become apparent (WMO and WHO, 2015). Thresholds with poor specificity can cause needless anxiety and waste of resources, while poor sensitivity places lives in danger. Heatwave definitions and thus warning systems can be tiered to match the specific risk levels in different settings and population subgroups, contingent on key risk factors such as age, co-morbidities and dwelling type, for example. This would require identification of specific heat vulnerability 'hotspots' within urban areas, allowing for targeting of resources to higher risk populations in these areas. Warnings provided through cell phone apps, with levels set based on an individual's location, risk profile and physiological responses such as heart rate, hold much promise. The ClimApp tool is one such example (Kingma *et al.*, 2021).

Heat action plans or even government regulations for the workplace on their own may be insufficient for full protection; workers require knowledge about the risks of heat exposure and agency to slow down or take breaks on their own initiative. Changing working hours to cooler times of the day is a frequently proposed measure, as is increased drinking supplies for outdoor workers. Use of wearable sensors to analyse how warming is affecting a workers' physiological parameters and their productivity may be useful for tracking level of risk in workers and for testing new interventions.

Risk assessments, monitoring and evaluation are key steps in optimizing the effectiveness of HAPs. These activities can include: a baseline and serial assessments of population vulnerability, response capacity and projected risks; selection of priority actions and activities; identification of important indicators to track; ongoing collection of data on selected indicators; and periodic consideration of new information, as well as ongoing progress towards pre-specified goals. Once a heatwave has ended, it is important to evaluate the response. Key questions in evaluations include whether the early warning system provided timely, useful information, whether responses of different actors was adequately coordinated and what was the burden of heat-related conditions. Comparisons of outcomes over time allow for estimates of shifts in morbidity markers, such as hospital admissions. Cross-city or cross-region learning can be invaluable. While indicators assessing climate change and health have been developed, including by the

Lancet Countdown Report (Romanello *et al.*, 2023), more work is needed on local- or district-level measures of effectiveness of heatwave responses, especially in low- and middle-income countries. Monitoring of indoor temperatures, which are a better reflection of actual exposure than outdoor temperatures, in sentinel sites could provide useful information. Lastly, calculating the burden of disease attributable to heat is key to making valid estimates of the climate burden, a vital step towards developing a commensurate response to this unfolding crisis and the efficient targeting of available resources.

Space cooling

Air-conditioning reduces mortality in heatwaves by up to 80% (Broome and Smith, 2012), but the ever-increasing reliance on air conditioning carries multiple drawbacks, including high levels of energy use, with risks for power outages, and considerable noise pollution and heat generation. Moreover, ubiquitous AC in high-income countries has become a stark symbol of widening 'thermal inequity' or 'air-conditioning nationalism', as it provides very high levels of protection for people in wealthier countries, those largely responsible for climate change, while people in poorer countries are worst affected and have low or no access to cool areas, or to adaptation interventions anywhere near the effectiveness of AC. Indeed, many types of dwellings in informal settlements, but also in parts of southern and eastern Europe, cannot be cooled to comfortable levels during the summer without AC. Energy-poor households often live in thermally inefficient homes and the costs of indoor space cooling can be prohibitive. This can exacerbate 'energy poverty', as occurs with heating energy costs in the winter.

From a public health perspective, AC needs to be used judiciously and powered by renewable energy sources, to be more energy efficient to minimize GHG emissions and to employ refrigerants with lower global warming potential. Many governments may be unwilling to enforce such imperatives but may encourage behaviour change via incentives (e.g. rebates on efficient cooling equipment) or disincentives (e.g. more onerous permit approvals for non-compliant systems). Mechanical ventilation, such as fans, offers an alternative to AC systems but is less effective at higher temperatures.

Passive cooling is a design-based, widely used strategy to reduce thermal stress. Generally, passive cooling is achieved by modifying interactions between a building envelope and the natural elements, through shading, heat sinks and natural ventilation. Simple facade fixes such as overhangs, louvres and insulated walls are effective at shading buildings from solar radiation, while more complex modifications such as texture walls, diode roofs and heat sinks (e.g. roof ponds) are effective at minimizing heat gains from solar radiation and ambient heat. In addition, natural ventilation strategies such as solar chimneys and air vents can facilitate cool-air circulation within buildings and help dissipate heat. The effectiveness of passive cooling varies widely, depending, for example, on the location of the sun and wind direction.

Transformation of the built environment

There are limits to how much incremental low-cost adaptations can achieve through behaviour changes or evaporative cooling; transformative 'heat-proofing' actions are required to avoid bio-physiological 'heat stress limits'. This includes both changes in heat resistance of buildings and nature-based solutions. Long-term planning is needed for climate-sensitive urban designs, aiming to transform urban spaces towards resilience and sustainability. System transformations have long lead times, therefore requiring timely start of implementation and long-run planning, including in regions that are not yet experiencing high heat stress.

Building construction measures can include: spatial planning; use of cool materials in asphalt; increasing reflectivity of surfaces, especially rooftop albedo through reflective white paint; more stringent building-code regulations, including for housing, workplaces, schools and health facilities; a cool-roofs policy; and localized solutions such as awnings, louvres, mist sprays and evaporative materials.

Incorporating Indigenous knowledge into building design and urban planning can increase resilience to heat (Makondo and Thomas, 2018). For example, traditional earthen housing construction materials commonly used in

Africa provide thermal comfort advantages through thermal conductivity and diffusivity, sometimes incorporating 'solar chimneys' (Osman and Sevinc, 2019). Moreover, earthen materials are recyclable and environmentally 'cleaner' than concrete structures.

Nature-based solutions

In cities, green-blue spaces such as gardens, parks and water bodies (e.g. lakes, ponds and rivers) can reduce temperature by several degrees (Lin *et al.*, 2020). Nature-based solutions include street trees, urban forestry, green roofs and green walls (Hayes *et al.*, 2022). Shading from tree canopies reduces exposure to solar radiation, a key determinant of outdoor thermal comfort. Tree shade also can protect crops and livestock in agricultural settings and is used increasingly in agroforestry. Trees and other vegetation also have a cooling effect as a result of water being lost from their leaves through evaporation and transpiration (loss of water through pores in the leaves).

Planning and strategies for 'urban ecological infrastructure' is an important component of city climate change planning, and can include an urban forest strategy (Marando *et al.*, 2022). Tree planting and other urban greening initiatives need to target areas where heat vulnerability is largest, such as communities that currently lack urban tree canopy, or have low penetration of air-conditioning. It is important to choose tree species that provide the greatest amount of shading and evapotranspiration. Indigenous trees are often the most heat- and drought-resistant, and thus minimize water demand and costs.

In addition to reducing urban heat, greening has several environmental and health co-benefits and it reduces noise and captures stormwater, lowering risks of flooding. Greening improves social well-being through social benefits of shade and outdoor recreation, reducing the mental health impacts of heat. It also forms part of mitigation strategies in urban areas and improves air quality. Gardens in the community and backyards increase urban green spaces and can provide access to fruit and vegetables, connections with nature and enhanced social cohesion. These and other greening initiatives are good examples of 'bottom-up' community-led adaptation.

11.5 Conclusion

In many settings, existing adaptive capacity is progressively being challenged due to escalating heat levels; and more robust and intensive measures are urgently required. It is a public health imperative to clearly define the package of protective and health promotional interventions to reduce the burden of heat-related disease. Incremental and targeted adaptation measures have garnered most attention to date. However, given the considerable and escalating health risks posed by rising temperatures and heatwaves, complex systems-based transformational interventions and processes are required. Overall, measures to adapt to extreme heat have not received sufficient attention by researchers, policy makers and funders. Current patterns in heat adaptation are disturbing: ubiquitous air-conditioning in high-income countries and negligible, poorly effective protections and the potential for mass mortality elsewhere. An alternative future remains possible. Indeed, heatwaves provide a key opportunity for catalysing action for communities and policy makers, hastening equitable climate response activities and the adoption of more rigorous mitigation initiatives.

Note

[1] Hausfather (2024) states that the average global temperature was around 1.6°C above pre-industrial levels in early 2024.

[2] Wet-bulb globe temperatures (WBGTs) were developed as a convenient substitute for a measure sometimes called the 'effective temperature including the radiation component'. It combined readings from three thermometers: a natural wet-bulb, a globe and sometimes a dry-bulb. WBGT responds to all four elements of the thermal environment (radiant heat, air temperature, humidity and wind) – but not necessarily in the same way as the human body, because of differences in size and shape (Budd, 2008).

References

Balsari, S., Dresser, C. and Leaning, J. (2020) Climate change, migration, and civil strife. *Current Environmental Health Reports* 7, 404–414. DOI: 10.1007/s40572-020-00291-4

Bonds, E. and Martin, L. (2016) Treating people like pollution: Homelessness and environmental injustice. *Environmental Justice* 9, 137–141. DOI: 10.1089/env.2016.0021

Broome, R.A. and Smith, W.T. (2012) The definite health risks from cutting power outweigh possible bushfire prevention benefits. *Medical Journal of Australia* 197, 440–441. DOI: 10.5694/mja12.10218

Budd, G.M. (2008) Wet-bulb globe temperature (WBGT) – its history and its limitations. *Journal of Science and Medicine in Sport* 11, 20–32. DOI: 10.1016/j.jsams.2007.07.003

Chersich, M.F., Swift, C.P., Edelstein, I., Breetzke, G., Scorgie, F. *et al.* (2019) Violence in hot weather: Will climate change exacerbate rates of violence in South Africa? *South African Medical Journal* 109, 447–449. DOI: 10.7196/SAMJ.2019.v109i7.14134

Chersich, M.F., Duc Pham, M., Areal, A., Haghighi, M., Manyuchi, A. *et al.* (2020) Associations between high temperatures in pregnancy and risk of preterm birth, low birth weight, and stillbirths: systematic review and meta-analysis. *BMJ* 371, m3811. DOI: 10.1136/bmj.m3811

Colucci, A.R., Vecellio, D.J. and Allen, M.J. (2023) Thermal (In)equity and incarceration: A necessary nexus for geographers. *Environment and Planning E: Nature and Space* 6, 638–657. DOI: 10.1177/25148486211063488

Conti, A., Valente, M., Paganini, M., Farsoni, M., Ragazzoni, L. *et al.* (2022) Knowledge gaps and research priorities on the health effects of heatwaves: a systematic review of reviews. *International Journal of Environmental Research and Public Health* 19, 5887. DOI: 10.3390/ijerph19105887

Ebi, K.L., Capon, A., Berry, P., Broderick, C., de Dear, R. *et al.* (2021) Hot weather and heat extremes: health risks. *The Lancet* 398, 698–708. DOI: 10.1016/S0140-6736(21)01208-3

Foster, J., Hodder, S.G., Lloyd, A.B. and Havenith, G. (2020) Individual responses to heat stress: Implications for hyperthermia and physical work capacity. *Frontiers in Physiology* 11, 541483. DOI: 10.3389/fphys.2020.541483

Gonzalez-Rivas, P.A., Chauhan, S.S., Ha, M., Fegan, N., Dunshea, F.R. *et al.* (2020) Effects of heat stress on animal physiology, metabolism, and meat quality: A review. *Meat Science* 162, 108025. DOI: 10.1016/j.meatsci.2019.108025

Hasan, F., Marsia, S., Patel, K., Agrawal, P. and Razzak, J.A. (2021) Effective community-based interventions for the prevention and management of heat-related illnesses: a scoping review. *International Journal of Environmental Research and Public Health* 18. DOI: 10.3390/ijerph18168362

Hausfather, Z. (2024) State of the climate: 2024 off to a record-warm start. Carbon Brief. https://www.carbonbrief.org/state-of-the-climate-2024-off-to-a-record-warm-start/ (accessed 24 April, 2024).

Hayes, A.T., Jandaghian, Z., Lacasse, M.A., Gaur, A., Lu, H. *et al.* (2022) Nature-based solutions (NBSs) to mitigate urban heat island (UHI) effects in Canadian cities. *Buildings* 12, 925. DOI: 10.3390/buildings12070925

Haynes, V., Calgaro, E. and Dominey-Howes, D. (2022) The future of our suburbs: Analyses of heatwave vulnerability in a planned estate. *Geographical Research* 60, 152–167. DOI: 10.1111/1745-5871.12498

Im, E.-S., Pal, J.S. and Eltahir, E.A.B. (2017) Deadly heat waves projected in the densely populated agricultural regions of South Asia. *Science Advances* 3, e1603322. DOI: 10.1126/sciadv.1603322

Kingma, B.R.M., Steenhoff, H., Toftum, J. Daanen, H.A.M., Folkerts, M.A. *et al.* (2021) ClimApp – Integrating personal factors with weather forecasts for individualised warning and guidance on thermal stress. *International Journal of Environmental Research and Public Health* 18, 11317. DOI: 10.3390/ijerph182111317

Lin, Y., Wang, Z., Jim, C.Y., Li, J., Deng, J. *et al.* (2020) Water as an urban heat sink: Blue infrastructure alleviates urban heat island effect in mega-city agglomeration. *Journal of Cleaner Production* 262, 121411. DOI: 10.1016/j.jclepro.2020.121411

Makondo, C.C. and Thomas, D.S.G. (2018) Climate change adaptation: Linking indigenous knowledge with western science for effective adaptation. *Environmental Science & Policy* 88, 83–91. DOI: 10.1016/j.envsci.2018.06.014

Marando, F., Heris, M.P., Zulian, G., Udías, A., Mentaschi, L. *et al.* (2022) Urban heat island mitigation by green infrastructure in European Functional Urban Areas. *Sustainable Cities and Society* 77, 103564. DOI: 10.1016/j.scs.2021.103564

Mitchell, B.C. and Chakraborty, J. (2015) Landscapes of thermal inequity: disproportionate exposure to urban heat in the three largest US cities. *Environmental Research Letters* 10, 115005. DOI: 10.1088/1748-9326/10/11/115005

Naicker, N., Teare, J., Balakrishna, Y., Wright, C.Y. and Mathee, A. (2017) Indoor temperatures in low cost housing in Johannesburg, South Africa. *International Journal of Environmental Research and Public Health* 14, 1410. DOI: 10.3390/ijerph14111410

Osman, M.M. and Sevinc, H. (2019) Adaptation of climate-responsive building design strategies and resilience to climate change in the hot/arid region of Khartoum, Sudan. *Sustainable Cities and Society* 47, 101429. DOI: 10.1016/j.scs.2019.101429

Osunmuyiwa, O.O., Payne, S.R., Vigneswara Ilavarasan, P., Peacock, A.D. and Jenkins, D.P. (2020) I cannot live without air conditioning! The role of identity, values and situational factors on cooling consumption patterns in India. *Energy Research & Social Science* 69, 101634. DOI: 10.1016/j.erss.2020.101634

Park, R.J., Goodman, J., Hurwitz, M. and Smith, J. (2020) Heat and learning. *American Economic Journal: Economic Policy* 12, 306–339. DOI: 10.1257/pol.20180612

Pihkala, P. (2020) Anxiety and the ecological crisis: an analysis of eco-anxiety and climate anxiety. *Sustainability* 12, 7836. DOI: 10.3390/su12197836

Ravanelli, N., Casasola, W., English, T., Edwards, K.M. and Jay, O. (2019) Heat stress and fetal risk. Environmental limits for exercise and passive heat stress during pregnancy: a systematic review with best evidence synthesis. *British Journal of Sports Medicine* 53, 799–805. DOI: 10.1136/bjsports-2017-097914

Raymond, C., Matthews, T. and Horton, R.M. (2020) The emergence of heat and humidity too severe for human tolerance. *Science Advances* 6, eaaw1838. DOI: 10.1126/sciadv.aaw1838

Romanello, M., di Napoli, C., Green, C., Kennard, H., Lampard, P. *et al.* (2023) The 2023 report of the Lancet Countdown on health and climate change: the imperative for a health-centred response in a world facing irreversible harms. *Lancet* 402, 2346–2394. DOI: 10.1016/S0140-6736(23)01859-7

Schmidt, G. (2024) Climate models can't explain 2023's huge heat anomaly — we could be in uncharted territory. *Nature* 627, 467. DOI: 10.1038/d41586-024-00816-z

Semenza, J.C. and Paz, S. (2021) Climate change and infectious disease in Europe: Impact, projection and adaptation. *The Lancet Regional Health – Europe* 9, 100230. DOI: 10.1016/j.lanepe.2021.100230

Sera, F., Hashizume, M., Honda, Y., Lavigne, E., Schwartz, J. *et al.* (2020) Air conditioning and heat-related mortality: a multi-country longitudinal study. *Epidemiology* 31, 779–787. DOI: 10.1097/EDE.0000000000001241

Spector, J.T., Masuda, Y.J., Wolff, N.H., Calkins, M. and Seixas, N. (2019) Heat exposure and occupational injuries: review of the literature and implications. *Current Environmental Health Reports* 6, 286–296. DOI: 10.1007/s40572-019-00250-8

Timofte, C. (2021) *Scorching fate: the impact of the climate crisis on liveability in the Middle East and North Africa*. Institut Nowej Europy. Available at: https://ine.org.pl/wp-content/uploads/2022/01/Scorching-Fate-The-Impact-of-the-Climate-Crisis-on-Liveability-in-the-Middle-East-and-North-Africa.pdf (accessed 9 October, 2022).

Tuholske, C., Caylor, K., Funk, C., Verdin, A., Sweeney, S. *et al.* (2021) Global urban population exposure to extreme heat. *Proceedings of the National Academy of Sciences (USA)* 118, e2024792118. DOI: 10.1073/pnas.2024792118

van Daalen, K.R., Kallesøe, S.S., Davey, F., Dada, S., Jung, L. *et al.* (2022) Extreme events and gender-based violence: a mixed-methods systematic review. *The Lancet Planetary Health* 6, e504–e523. DOI: 10.1016/S2542-5196(22)00088-2

WHO (2015) *Operational framework for building climate resilient health systems*. World Health Organization. Available at: https://www.who.int/publications/i/item/9789241565073 (accessed 27 May, 2023).

WMO and WHO (2015) *Heatwaves and Health: guidance on warning-system development*. World Meteorological Organization and World Health Organization. Available at: https://climate-adapt.eea.europa.eu/en/metadata/guidances/heatwaves-and-health-guidance-on-warning-system-development (accessed 7 February, 2023).

WMO (2024) State of the Global Climate 2023. Available at: https://wmo.int/publication-series/state-of-global-climate-2023 (accessed 24 April, 2024).

Xu, C., Kohler, T.A., Lenton, T.M., Svenning, J.-C. and Scheffer, M. (2020) Future of the human climate niche. *Proceedings of the National Academy of Science (USA)* 117, 11350–11355. DOI: 10.1073/pnas.1910114117

11.i Thermoregulation: Risks and Protection

Ollie Jay
Faculty of Medicine & Health, University of Sydney, Sydney, Australia

Abstract
This short chapter provides an overview of the basic physiology of human hyperthermia and conditions including heat exhaustion and heatstroke, which can be fatal, and occur suddenly even in healthy young individuals. Ageing, diseases (including type 2 diabetes, cardiovascular and renal impairment) and some medications impart additional risk of heat-related health problems. Finally, the topic of heat acclimatization is discussed; this is possible but is likely limited by widespread use of mechanical cooling.

11.i.1 Basic Physiology and Risk Factors

Human hyperthermia ultimately arises from an accumulation of heat energy inside the human body, which itself is a function of a temporal mismatch between rates of internal heat generation and heat dissipation at the skin surface to the surrounding environment (Cramer and Jay, 2016). Internal metabolic heat production is determined by the absolute rate at which oxygen is consumed to facilitate cellular metabolism, which rises dramatically with the contraction of skeletal muscles during physical activity. Other modifiers of heat production include the substrates used to fuel these metabolic processes (i.e. carbohydrates or fats) and the efficiency with which metabolic energy can be converted into external work (Astrand and Saltin, 1961).

Upon exposure to a hot environment or following the commencement of physical activity, initial rises in body temperature trigger thermophysiological mechanisms that serve to moderate body heat storage. Cutaneous vasodilation redirects warm blood away from the body core towards the skin surface to support heat dissipation, while sweat secretion from eccrine glands, distributed across the body, greatly enhances the potential for evaporative heat loss. Hot ambient temperatures narrow, or reverse, the temperature gradient between skin and air and thus limit the ability to shed body heat via the dry heat transfer processes of convection and radiation. High ambient humidity blunts sweat evaporation (Gagge and Gonzalez, 1996). It therefore follows that in hot and humid conditions, because of the limited ability to lose body heat, the risk of critically high body temperatures and the development of heat-related illnesses rises greatly. Conditions at risk of occurring include heat exhaustion or heatstroke, even for healthy people free of thermoregulatory impairments (Kenny and Jay, 2013).

 Climate Change and Global Health: Primary, Secondary and Tertiary Effects, 2nd Edition (eds C.D. Butler and K. Higgs)
DOI: 10.1079/9781800620025.0011i

Other environmental modifiers of human heat stress risk include wind speed and cloud cover. Higher air flow across the skin influences convective heat loss if there is a temperature difference between the skin and air (Gagge and Gonzalez, 1996), and evaporative heat loss if all moisture on the skin surface does not completely evaporate in still air (Morris *et al.*, 2021). People in an outdoor environment under direct solar radiation in the middle of a sunny summer day will be exposed to black-globe temperatures (which help represent the additional effects of thermal radiation) that are ~10–15°C higher than the ambient air temperature measured in the shade.

Additional heat stress risk factors: ageing, disease and some medications

Disruptions to physiological heat loss mechanisms, particularly sweating, secondary to primary ageing, disease, or conditions such as dehydration, can render a person more physiologically vulnerable to overheating during heat exposure. Age-related impairments of sweating occur beyond the age of 60 and can deteriorate steeply thereafter (Kenney and Munce, 2003) but some of these ageing effects on thermoregulation can be tempered by maintaining regular levels of physical activity and thus aerobic fitness (Kenney and Munce, 2003). Relatively common co-morbidities such as type 2 diabetes have also been reported to affect sweating capacity (Kenny *et al.*, 2013). Some prescription medications, such as antidepressants that as a side-effect supress the action of the neurotransmitter responsible for sweat production, may also reduce the ability to regulate body temperature (Ebi *et al.*, 2021). Dehydration due to the inadequate replacement of body fluids lost through sweating (i.e. greater than 1% total body weight loss) causes changes in blood osmolality and blunts sweating, potentially exacerbating rises in core temperature (Horstman and Horvath, 1972). If left to ingest water *ad libitum*, young healthy people seem to drink enough to prevent these negative effects of dehydration on thermoregulation (Graham *et al.*, 2020). Older people have been reported to have blunted thirst sensations, indicating that they may be more susceptible to voluntary dehydration (Kenney, 1997).

Heat exhaustion and heatstroke: a potentially life-threatening condition

Most humans can safely operate without ill-effect at core temperatures up to at least 38.5–39°C. Above this approximate threshold, the risk of heat exhaustion, which is characterized by signs and symptoms such as nausea, dizziness, vomiting and ataxia, is elevated (Parsons, 2003). At core temperatures higher than ~40°C, heatstroke – a potentially life-threatening condition – may develop. The onset of this condition can be rapid and quickly evolve into a medical emergency if not recognized. Immediate treatment in the form of rapid whole-body cooling (preferably through water immersion) is essential and could prove to make the difference between life and death (Hosokawa *et al.*, 2021).

The pathophysiology of heatstroke is thought to be associated with very high temperatures of deep tissues coupled with low oxygen delivery to the splanchnic region, secondary to reduced blood flow during extreme heat exposure. These combined responses lead to increased gut permeability, resulting in endotoxins entering the circulation and setting off a cascade of events culminating in widespread clotting, multiple organ failure and ultimately death (Bouchama *et al.*, 2022).

Cardiovascular and renal disease: risk factors for heat-related morbidity and mortality

People with cardiovascular disease are among those at the highest risk of heat-related morbidity and mortality (Semenza *et al.*, 1996). This vulnerability is probably not due to a risk of overheating per se. The most likely mechanistic pathway is a low capacity to physiologically compensate the elevated cardiovascular strain that evolves during heat exposure – due to the rise in cardiac output that is required to maintain blood pressure in the face of a cutaneous vasodilation – leading to cardiovascular collapse rather than impaired heat loss and the development of heatstroke (Ebi *et al.*, 2021). This cardiovascular risk is further aggravated by an accompanying dehydration, as a lower blood volume leads to a reduced stroke volume and therefore an increased heart rate requirement for a given cardiac output. Similarly, people with pre-existing kidney disease are at a much higher risk of renal failure during heat extremes, again not necessarily due to thermoregulatory insufficiency. A reduced blood perfusion of the kidneys secondary to the heat-induced redistribution of blood flow to the skin may lead to low oxygen delivery and subsequently acute injury to parts of the kidneys most sensitive to hypoxia (Basile *et al.*, 2012). Repeated (chronic) dehydration can also cause kidney fibrosis and chronic kidney disease.

11.i.2 Physiological heat acclimatization

Physiological resilience to extreme heat exposure can be acquired through repeated exposure to hot environments. Seven to 14 consecutive days of 1–2 hours of deliberate exposure to extreme heat (> 40°C with high humidity) coupled with at least moderate-intensity exercise (i.e. > 60% of aerobic capacity) in a climate chamber (i.e. heat 'acclimation') induce an expansion of plasma volume in the blood and a marked elevation in sweating capacity (Gisolfi and Robinson, 1969). Collectively, these physiological adjustments serve to lower core temperature, at rest and during exertion, and improve cardiovascular stability (and reduce strain) and thermal comfort (Periard *et al.*, 2015). Given the prevalence of mechanical cooling in living and working spaces in many countries, it is currently unclear whether the natural free-living exposure to heat stress of most people (i.e. heat 'acclimatization'), even during the hottest summer months, is adequate to elicit complete adaptation and thus physiological protection.

References

Astrand, P.O. and Saltin, B. (1961) Oxygen uptake during the first minutes of heavy muscular exercise. *Journal of Applied Physiology* 16, 971–976. DOI: 10.1152/jappl.1961.16.6.971

Basile, D.P., Anderson, M.D. and Sutton, T.A. (2012) Pathophysiology of acute kidney injury. *Comprehensive Physiology* 2, 1303–1353. DOI: 10.1002/cphy.c110041

Bouchama, A., Abuyassin, B., Lehe, C., Laitano, O., Jay, O. *et al.* (2022) Classic and exertional heatstroke. *Nature Reviews Disease Primers* 8, 8. DOI: 10.1038/s41572-021-00334-6

Cramer, M.N. and Jay, O. (2016) Biophysical aspects of human thermoregulation during heat stress. *Autonomic Neuroscience* 196, 3–13. DOI: 10.1016/j.autneu.2016.03.001

Ebi, K.L., Capon, A., Berry, P., Broderick, C., de Dear, R. *et al.* (2021) Hot weather and heat extremes: health risks. *The Lancet* 398, 698–708. DOI: 10.1016/S0140-6736(21)01208-3

Gagge, A.P. and Gonzalez, R.R. (1996) Mechanisms of heat exchange. In: Fregly, M.J. and Blatteis, C.M. (eds) *Handbook of Physiology. Environmental Physiology.* American Physiological Society, Bethesda, Maryland, pp. 45–84.

Gisolfi, C. and Robinson, S. (1969) Relations between physical training, acclimatization, and heat tolerance. *Journal of Applied Physiology* 26, 530–534. DOI: 10.1152/jappl.1969.26.5.530

Graham, C., Morris, N.B., Harwood, A.E. and Jay, O. (2020) *Ad libitum* water consumption off-sets the thermal and cardiovascular strain exacerbated by dehydration during a 3-h simulated heatwave. *European Journal of Applied Physiology* 120, 391–399. DOI: 10.1007/s00421-019-04283-7

Horstman, D.H. and Horvath, S.M. (1972) Cardiovascular and temperature regulatory changes during progressive dehydration and euhydration. *Journal of Applied Physiology* 33, 446–450. DOI: 10.1152/jappl.1972.33.4.446

Hosokawa, Y., Racinais, S., Akama, T., Zideman, D., Budgett, R. *et al.* (2021) Prehospital management of exertional heat stroke at sports competitions: International Olympic Committee Adverse Weather Impact Expert Working Group for the Olympic Games Tokyo 2020. *British Journal of Sports Medicine* 55, 1405–1410. 10.1136/bjsports-2020-103854

Kenney, W.L. (1997) Thermoregulation at rest and during exercise in healthy older adults. *Exercise and Sport Sciences Reviews* 25, 41–76. DOI: 10.1249/00003677-199700250-00005

Kenney, W.L. and Munce, T.A. (2003) Invited review: aging and human temperature regulation. *Journal of Applied Physiology* 95, 2598–2603. DOI: 10.1152/japplphysiol.00202.2003

Kenny, G.P. and Jay, O. (2013) Thermometry, calorimetry, and mean body temperature during heat stress. *Comprehensive Physiology* 3, 1689–1719. DOI: 10.1002/cphy.c130011

Kenny, G.P., Stapleton, J.M., Yardley, J.E., Boulay, P. and Sigal, R.J. (2013) Older adults with type 2 diabetes store more heat during exercise. *Medicine & Science in Sports & Exercise* 45, 1906–1914. DOI: 10.1249/MSS.0b013e3182940836

Morris, N.B., Chaseling, G.K., English, T., Gruss, F., Maideen, M.F.B. *et al.* (2021) Electric fan use for cooling during hot weather: a biophysical modelling study. *Lancet Planetary Health* 5, e368–e377. DOI: 10.1016/S2542-5196(21)00136-4

Parsons, K.C. (2003) *Human Thermal Environments*. Taylor & Francis, London.

Periard, J.D., Racinais, S. and Sawka, M.N. (2015) Adaptations and mechanisms of human heat acclimation: applications for competitive athletes and sports. *Scandinavian Journal of Medicine & Science in Sports* 25, 20–38. DOI: 10.1111/sms.12408

Semenza, J.C., Rubin, C.H., Falter, K.H., Selanikio, J.D., Flanders, W.D. *et al.* (1996) Heat-related deaths during the July 1995 heat wave in Chicago. *New England Journal of Medicine* 335, 84–90. DOI: 10.1056/NEJM199607113350203

11.ii Heat and Kidney Disease

Katherine A. Barraclough
School of Medicine, University of Melbourne, Melbourne, Australia;
Department of Nephrology, Royal Melbourne Hospital, Melbourne, Australia

Abstract
This mini-chapter focuses on the relationship between heat, hydration and kidney diseases. Clear links exist between increased extreme heat and the development of acute kidney injury and kidney stone disease. Recent years have also seen the emergence of chronic kidney disease epidemics in some of the hottest regions of the world, primarily affecting young adult men from low-income agricultural communities who are required to perform strenuous manual labour under very hot conditions. As climate change drives further temperature rise over the coming decades, it is expected that the incidence and severity of all the above-mentioned kidney diseases will increase. To preserve kidney health, adaptation measures must include outreach and education to raise awareness of the risks of heat and to promote protective behavioural changes.

11.ii.1 Heat and Kidney Disease: Basic Science

The kidneys play a unique role in protecting individuals against heat and dehydration through their ability to concentrate the urine, minimizing fluid loss from the body. Simultaneously, the kidneys are highly susceptible to heat-related injury.

Extreme heat days, which are predicted to become more frequent and severe over the coming decades, increase loss of body water and salt through sweating. If fluid intake is insufficient and the ability of the kidneys to compensate for the loss is exceeded, this can lead to total body fluid deficit, constriction of blood vessels to maintain blood pressure, reduced blood flow to the kidneys and subsequent acute kidney injury (Johnson *et al.*, 2019a). Numerous studies have shown increases in hospital admissions from acute kidney injury during heat waves (Hansen *et al.*, 2008). In the severe 2003 European heat wave, kidney failure was identified as a major cause of excess mortality in the elderly (Conti *et al.*, 2007).

Extreme heat can also lead to heatstroke, whereby the body is unable to dissipate the heat load and core body temperature rises, typically above 40°C. Acute kidney injury is a common manifestation of heatstroke. It was, for example, identified in over 50% of those presenting with heatstroke during the 1995 heat wave in Chicago (Dematte *et al.*, 1998). The kidney injury in heatstroke can be caused by any combination of reduced blood flow to the kidney, direct thermal injury or kidney toxicity from muscle-cell contents that leak into the circulation following muscle damage by the heat. In turn, heatstroke-induced acute kidney injury has been associated with

 Climate Change and Global Health: Primary, Secondary and Tertiary Effects, 2nd Edition (eds C.D. Butler and K. Higgs)
DOI: 10.1079/9781800620025.0011ii

increased rates of chronic kidney disease and long-term kidney failure (Lee *et al.*, 2016).

Furthermore, heat exposure promotes kidney stone disease. Here, the primary mechanism is believed to be reduced total body water as a consequence of sweating leading to a compensatory increase in urine concentration to preserve total body water (Johnson *et al.*, 2019a). As a result, urinary supersaturation with stone-forming salts can arise, and from there, kidney-stone formation. Multiple studies have shown increased rates of kidney-stone disease at higher ambient temperatures. For instance, a study that examined the prevalence of stone disease across five different metropolitan areas in the USA demonstrated a 36–39% excess risk[1] of kidney stones with a mean daily temperature of 30°C relative to 10°C (Tasian *et al.*, 2014). Similarly, in Seoul, South Korea, the relative risk[2] of developing kidney-stone disease was reported to be 2.5 times higher with a mean daily temperature of 29°C compared with 13°C. Another study has suggested that the USA is likely to see an extra 1.6–2.2 million lifetime cases of renal stone disease by 2050 due to climate change-induced temperature rise. This would mean an additional cost to the healthcare system of US$0.9–1.3 billion annually (Brikowski *et al.*, 2008).

11.ii.2 Kidney Disease, Climate Change, and Adaptation

It is noteworthy that recent years have seen the emergence of chronic kidney disease epidemics in some of the hottest regions of the world, including in parts of Central America, Sri Lanka, India, the Middle East, Africa and North and South America (Sorensen and Garcia-Trabanino, 2019). Those affected are primarily young adult men from low-income agricultural communities required to perform strenuous manual labour under very hot conditions. However, cases have also been seen in women and children (without exposure to strenuous labour). Exposure to extreme heat is a unifying finding in affected individuals and there is typically an absence of known conditions that are risk factors for kidney disease, such as diabetes, hypertension and glomerular disease.

At the time of diagnosis, kidney disease is commonly advanced. Combined with the limited or no treatment options in most of the affected regions, the death toll is high. By 2012, this disease was believed to be responsible for in excess of 20,000 fatalities across Central America (Johnson *et al.*, 2019b), while in both Nicaragua and El Salvador chronic kidney disease is the second leading cause of death (Sorensen and Garcia-Trabanino, 2019).

The definitive underlying cause of chronic kidney disease in these cases remains unproven (Johnson *et al.*, 2019b). However, there is general agreement that heat is likely to be a key contributor, possibly acting alone, or exacerbating damage from an as yet unidentified infectious or toxic agent. The mechanism is thought to be repeated subclinical acute kidney injury from heat stress and dehydration cumulatively causing chronic damage.

As climate change drives further temperature rise over the coming decades, it is expected that the incidence and severity of all the above-mentioned kidney diseases will increase. As is the case for most climate-sensitive diseases, socially and economically disadvantaged populations from low- and middle-income countries will be disproportionately affected, due to their limited capacity for climate adaptation and lack of access to high-quality healthcare. Within these populations, particularly vulnerable groups include the elderly, children, those with pre-existing medical conditions (particularly kidney disease) and certain occupational groups (such as manual labourers).

Government-led climate mitigation strategies must be central in the response to minimize the impacts of climate change on kidney diseases and broader health, based on understanding that the impacts of unchecked climate change are likely to be catastrophic. Targeted adaptation measures are also imperative, given that many of the consequences of ongoing temperature rise are already unavoidable.

To preserve kidney health, adaptation measures must include outreach and education to raise awareness of the risks of heat and to promote protective behavioural changes. Preventive policies and actions need to focus on reducing heat and sun exposure, improving hydration with safe water and electrolyte replacement and ensuring safe working practices and living environments. Furthermore, there is a need for comprehensive oversight in at-risk populations, particularly certain occupational groups.

International collaborative research and funding will also be required to facilitate the development and implementation of appropriate preventive and treatment strategies in vulnerable regions, particularly resource-poor countries with little or no access to kidney replacement therapy.

Notes

[1] Excess risk: the difference in rate between a population group exposed to a putative causal factor (in this case a mean daily temperature of 30°C) and a population group not exposed to the factor (in this case a mean daily temperature of 10°C).

[2] Relative risk generally means 'rate ratio', that is, the ratio of two risks, usually of exposed and not exposed (Porta *et al*., 2014).

References

Brikowski, T.H., Lotan, Y. and Pearle, M.S. (2008) Climate-related increase in the prevalence of urolithiasis in the United States. *Proceedings of the National Academy of Sciences (USA)* 105, 9841–9846. DOI: 10.1073/pnas.0709652105

Conti, S., Masocco, M., Meli, P., Minelli, G., Palummeri, E. *et al*. (2007) General and specific mortality among the elderly during the 2003 heat wave in Genoa (Italy). *Environmental Research* 103, 267–274. DOI: 10.1016/j.envres.2006.06.003

Dematte, J.E., O'Mara, K., Buescher, J., Whitney, C.G., Forsythe, S. *et al*. (1998) Near-fatal heat stroke during the 1995 heat wave in Chicago. *Annals of Internal Medicine* 129, 173–181. DOI: 10.7326/0003-4819-129-3-199808010-00001

Hansen, A.L., Bi, P., Ryan, P., Nitschke, M., Pisaniello, D. *et al*. (2008) The effect of heat waves on hospital admissions for renal disease in a temperate city of Australia. *International Journal of Epidemiology* 37, 1359–1365. DOI: 10.1093/ije/dyn165

Johnson, R.J., Sánchez-Lozada, L.G., Newman, L.S., Lanaspa, M.A., Diaz, H.F. *et al*. (2019a) Climate change and the kidney. *Annals of Nutrition and Metabolism* 74 (suppl. 3), 38–44. DOI: 10.1159/000500344

Johnson, R.J., Wesseling, C. and Newman, L.S. (2019b) Chronic kidney disease of unknown cause in agricultural communities. *The New England Journal of Medicine* 380, 1843–1852. DOI: 10.1056/nejmra1813869

Lee, S., Kim, M.-S., Kim, J.H., Kwon, J.K., Chi, B.H. *et al*. (2016) Daily mean temperature affects urolithiasis presentation in Seoul: a time-series analysis. *Journal of Korean Medical Science* 31, 750–756. DOI: 10.3346/jkms.2016.31.5.750

Porta, M., Last, J.M., Greenland, S., Hernán, M.A. and dos Santos Silva, I. (eds.) (2014) *The IEA Dictionary of Epidemiology,* Oxford University Press, Oxford, UK.

Sorensen, C. and Garcia-Trabanino, R. (2019) A new era of climate medicine – addressing heat-triggered renal disease. *New England Journal of Medicine* 381, 693–696. DOI: 10.1056/NEJMp1907859

Tasian, G.E., Pulido, J.E., Gasparrini, A., Saigal, C.S., Horton, B.P. *et al*. (2014) Daily mean temperature and clinical kidney stone presentation in five US metropolitan areas: a time-series analysis. *Environmental Health Perspectives* 122, 1081–1087. DOI: 10.1289/ehp.1307703

12 Occupational Heat Effects: A Global Health and Economic Threat

Colin D. Butler,[1] Jason K.W. Lee,[2,3] Subhashis Sahu[4] and Tord Kjellstrom[1,5]
[1]National Centre for Epidemiology and Population Health, Australian National University, Canberra, Australia; [2]Human Potential Translational Research Programme, Yong Loo Lin School of Medicine, National University of Singapore, Singapore; [3]Heat Resilience and Performance Centre, Yong Loo Lin School of Medicine, National University of Singapore, Singapore; [4]Ergonomics and Occupational Physiology Laboratory, Department of Physiology, University of Kalyani, West Bengal, India; [5]Health and Environment International Trust, Nelson, New Zealand

Abstract

This chapter reviews the past, present and future of occupational heat exposure, warning that the interaction of neoliberalism (market-driven economics), anthropogenic climate change and the urban heat island effect is a combination that already harms the health of millions of vulnerable people. We also review the potential and forecast economic aspects that climate change is likely to impose via its effects on working people, their families and their communities.

12.1 Climate Change is Increasing Workplace Heat Exposures for Many Millions of People

This chapter explores occupational health in the Anthropocene (the human-dominated era) (see Chapter 1), focusing especially on heat. We explore the history of the recognition of occupational heat as a problem and briefly outline cultural responses to limit adverse heat exposure at work. The basic physiology of what happens when a human body overheats was described in the previous chapter. Here, key points from it are repeated, with emphasis on problems specifically experienced by workers. We consider effects of the political economy, such as authoritarianism and neoliberalism (economics and politics excessively driven by market forces) on vulnerable groups, human rights and heat-related occupational risks. We also review the potential and forecast economic aspects that climate change is likely to impose via its impact on working people, their families and their communities. We also sketch some of the likely impacts on working people that will arise via non-heat effects, such as via mechanization, migration, conflict, rising food prices and infectious diseases. Finally, we describe interventions, both long-established and evolving, that seek to reduce the harm to occupational health and thus to economic and human well-being.

Anthropogenic factors, especially from greenhouse gas accumulation and from deforestation and urbanization, are causing hot seasons in hot places to become even hotter. The heat in many cities with a hot climate (temperatures regularly above 35°C) is intensified by the 'urban heat island effect' (Ebi *et al.*, 2021). This heat, which exacerbates and often exceeds the contribution from greenhouse gas forcing alone,

 Climate Change and Global Health: Primary, Secondary and Tertiary Effects, 2nd Edition (eds C.D. Butler and K. Higgs)
DOI: 10.1079/9781800620025.0012

is associated with the development of heat-trapping roads and structures (often dark coloured, even when they do not need to be, such as for roofs), waste heat from vehicles and buildings, and the clearance of green spaces, especially trees, in urban areas. Construction workers are at particularly high risk of heat stress as they often work outside in strong sunlight; cooling methods are difficult to apply (Al-Bouwarthan *et al.*, 2019).

In rural areas, a modelling study undertaken in the Berau regency of Kalimantan, Indonesia, documented a change in the mean maximum temperature of about 1°C, attributed to deforestation, between 2002 and 2018 (Wolff *et al.*, 2021).[1] An earlier study in the same region found that the maximum rise in ambient temperature was as high as 8.3°C (Masuda *et al.*, 2019). Wolff *et al.* (2021) attributed about 8% of all-cause mortality in this region, in 2018, to this additional heat. The same study concluded that, in deforested areas, unsafe work time due to heat had increased by about 20 min per day; with 2°C of additional warming (i.e. relative to 2018) this could increase to about 5 h of unsafe work per day.

By the start of this millennium harm to occupational health, as a result of climate change, was beginning to be appreciated as an important additional risk to the health and well-being of working people. It was also appreciated, though not sufficiently widely, that reducing work intensity, especially muscular, in order to minimize adverse health effects arising from excess heat and humidity involves a trade-off. Reduced effort reduces work output (Parsons, 2014) and thus lowers earnings (particularly if paid by output rather than by time), with consequent flow-on effects to vulnerable people, groups and their families (Kjellstrom, 2009). Importantly, also, the productivity of less muscular work, such as that performed in a non-air-conditioned office or factory, is reduced by excessive heat and humidity (Oppermann *et al.*, 2021).

In 2012 the first estimate of the global and national estimates of the economic consequences of reduced labour productivity resulting from increasing heat exposure due to climate change was released (DARA, 2012). It ranked the economic effects of climate change on labour productivity, by 2030, as of great importance. The literature on this topic is now slowly expanding (Dunne *et al.*, 2013, Kjellstrom *et al.*, 2019; Zhao *et al.*, 2021) and is discussed later in this chapter.[2] The chapter concludes with a brief discussion of modern interventions to improve occupational health in the Anthropocene.

12.2 A Short History of Occupational Heat Risks

Health hazards faced by workers from working in extreme heat (such as in agriculture, construction and mines) were described by Bernadino Ramazzini (1633–1714), the founder of occupational health (Felton, 1997; Kjellstrom *et al.*, 2019). In South Africa, starting in the 1880s, hundreds of thousands of black (Bantu) miners were forced to endure extremely hot and humid conditions in underground gold mines (Moeti, 1986). Recognition of heatstroke in these mines dates from at least 1924 (Wyndham, 1965). Interventions to reduce deaths were implemented as early as the 1930s, but may have been motivated more by a wish to maximize mining productivity and profitability, rather than to protect health.[3] A pioneering study in these mines, for the period 1956–1961, undertaken by a mining company medical officer, documented about 3.3 deaths/1000 miners/year due to fatal heat stroke when the wet-bulb globe temperature (WBGT) exceeded 34°C. At a WBGT between 31°C and 33°C, heat-related deaths were reduced but still averaged 0.7 deaths/1000 miners/year (Wyndham, 1969). Heat acclimatization before going underground (introduced as early as the 1930s) reduced the risk, but even after acclimatization two-thirds of new recruits were unable to carry out heavy labour in the hot work environment.

Occupational heat stress has also long been recognized among the military. Hollowell (2010) documented awareness of the risks of heat exposure among occupying British colonial forces both in 19th century India and in British troops in World War II, such as tank drivers engaged in desert warfare. High relative humidity impairs sweat evaporation, thus inhibiting heat loss, which is especially important for people working physically. Recognizing this, the US Army introduced the WBGT, seeking to protect marines and trainee soldiers from dangerous heatstroke

(Yaglou and Minard, 1957; Budd, 2008). This index combines air temperature, humidity, wind speed and heat radiation. Willett and Sherwood (2012) reported that the US military suspends all physical training and strenuous exercise when the WBGT exceeds 32°C. Also see Cheung *et al.* (2016) for more discussion of industrial and military guidelines for heat stress.

Elite athletes are also vulnerable to heat stress (Parsons, 2014). Cases of fatal exertional heatstroke have been reported in elite sports such as American football, rugby league, wrestling, and long-distance running (Ebi *et al.*, 2021). Heat-related concerns could be even greater for paralympic participants and aspirants, because of their vulnerability, both physiological and pharmaceutical. In some cases, sporting spectators may also be at risk, including vulnerable groups such as the elderly and pregnant women (Annear *et al.*, 2021).

12.3 Human Physiology and Performance Links to Ambient Heat Exposure

The normal core body temperature for humans is close to 37°C. However, humans, especially if young, fit and acclimatized, can safely operate without ill-effect at core temperatures of at least 39°C or even 40°C (Byrne *et al.*, 2006; Lee *et al.*, 2010). When core temperature is high, serious problems can arise, generally starting with heat exhaustion, characterized by nausea, dizziness, vomiting and ataxia. At higher bodily temperatures heatstroke can occur; this is potentially life-threatening and can be very rapid (Parsons, 2014).

If the external air temperature is greater than 37°C, physiological mechanisms can generally maintain the core body at a safe level, particularly by the evaporation of sweat, and by resting. However, as mentioned, sweat evaporation is less efficient at high relative humidity. Clothing also influences heat loss. Light clothing is best in most hot environments, but many jobs require special protective clothing, which adds to the heat stress (Bernard, 1999). The most recent examples of such protective dress emerged during the COVID-19 pandemic (Lee *et al.*, 2020).

Physical movement and activity (particularly muscular work) increases the body's core temperature and the extra heat is more difficult to dissipate in hot environments. If cooling methods in a workplace are insufficient and workers already wear light clothing, the only way for a working person to reduce heat stress is to take breaks or slow down their work to reduce their metabolic rate (Parsons, 2014). This reduces hourly work output and work productivity, with associated economic losses for the individual, the enterprise and the community (Kjellstrom, 2009; Dunne *et al.*, 2013). The quantitative reduction of work capacity and productivity in a specific labour-intensive work situation, rice farming, in India was documented by Sahu *et al.* (2013). They found that a change of WBGT from 26°C to 32°C reduced hourly productivity by 30% (Fig. 12.1). Even routine household activities can cause heat stress, necessitating a reduction of such activities on hot days. Commuting to and from work is also a daily source of heat exposure for many people. Children, the elderly and pregnant women are likely to be more sensitive to overheating than adults, and people carrying out heavy labour are particularly vulnerable (Parsons, 2014). In many countries female workers engaged in outdoor work are more vulnerable to heat stress, due to the lack of access to private sanitation facilities; consequently they are forced to drink less water and thus are more prone to dehydration, especially in hot conditions.

An important issue for work capacity, as well as clinical heat effects, is the profuse sweating that hot work involves and the associated risk of dehydration. Workers carrying out heavy labour can lose up to one litre of sweat per hour, yet in certain workplaces drinking water may not be readily available to replace this loss of fluid (Parsons, 2014). This is a common problem among, for example, sugarcane cutters in Central America (Hansson *et al.*, 2020), who, in many plantations, have to bring their own drinking water to work and do not carry sufficient quantities to maintain their fluid balance (10 l of water is heavy – 10 kg).

Human heat exposure and internal heat stress also lead to clinical effects, including injury, serious heatstroke and death. The symptoms and signs include heat exhaustion (a basis for the work capacity loss described above), heat rash, heat fainting (heat syncope), heatstroke and heat-induced death (see Chapter 11). As the

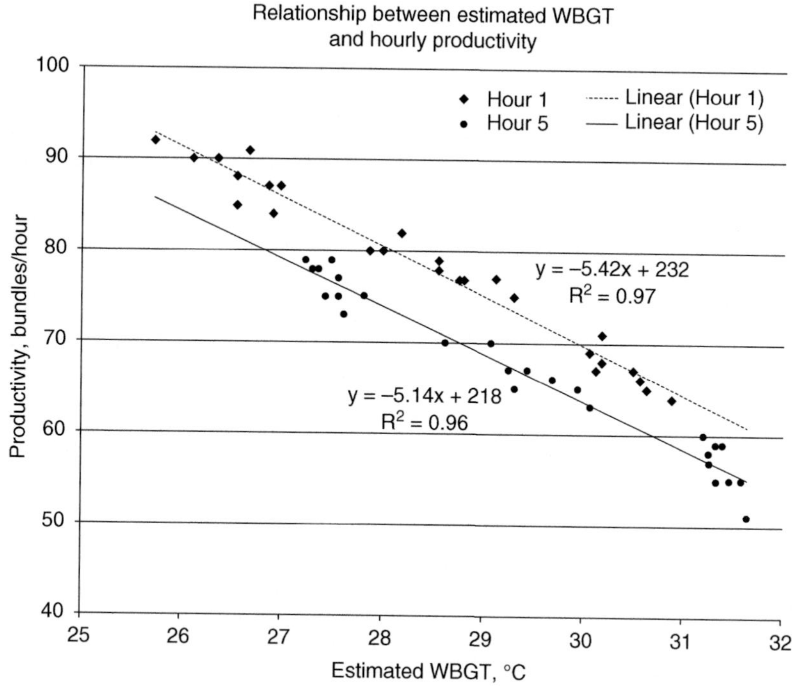

Fig. 12.1. Bundles of rice harvested per hour (productivity) at different environmental heat levels (WBGT) in West Bengal, India. Each point is a group average of 10–18 workers; hour 1 is the first work hour of the day and hour 5 is the 5th work hour (Sahu *et al.*, 2013).

global population ages (including workers), vulnerability to heat will increase, as diseases (with and without medication) compromise thermoregulation. Very serious clinical heat effects, including chronic kidney disease (Crowe *et al.*, 2013), heatstroke and death, are still reported among people carrying out work or heavy physical activity, for example construction workers, agricultural labourers, emergency workers, brickfield labourers and the military (Kjellstrom *et al.*, 2016).

12.4 The Political Economy of Occupational Health

The effects of excess heat, together with other adverse consequences to occupational health in the Anthropocene, operate in a variety of political environments. Example of this are slavery and the high rates of inequality that have long permeated human societies, including some that are Indigenous (Hayden, 2001). Although slavery is now illegal in most countries, and although the organization, agitation and (occasionally) revolution of workers and their advocates have improved pay, work hours and other conditions in many countries, such improvements are neither universal nor guaranteed. Bales (1999) estimated the global population of slaves at almost 30 million. A more recent calculation, from the same author, is over 40 million. Slaves (sometimes children) are found in diverse occupations, including fishing, shrimp farming, cocoa harvesting, brick manufacturing and mining, as well as domestic 'service' (Jackson *et al.*, 2019). This number includes an unknown number (probably millions) of bonded labourers, concentrated in parts of South Asia, some of whom toil indefinitely, to repay debt incurred by their parent or parents (James, 2019).[4]

There is a grey area between unequivocal slavery and the conditions of tens of millions (perhaps hundreds of millions) of grossly exploited workers. For example, many people, recruited from South Asia and elsewhere to work in the Middle East, pay in advance to secure their employment (Ter Haar, 2018). Others, innumerate or barely numerate, borrow money for their fares, incurring interest rates as high as 60% per annum (compounding) (Sinclair, 2019).

Upon arrival, conditions are rarely as good as advertised (Box 12.1).

Ter Haar (2018) quoted a media report claiming that, by 2015, 1200 deaths of workers had already been associated with constructing facilities for the 2022 FIFA (Fédération Internationale de Football Association) World Cup, held in Qatar. She compared this with a single worker's death involved in building the facilities for the 2012 Olympics, held in London. While many of these construction-related deaths in Qatar would have a heat-related cause (Pradhan *et al.*, 2019), almost all have a deeper layer of causation. This deeper layer is related to exploitation, the relative powerlessness of the workers involved, and of their advocates, and of an international system that limits meaningful foreign intervention in what is seen as the internal affairs of a sovereign nation.[5] The only force that seems to operate to counter this is international public opinion. In the case of the World Cup, a strong international campaign to embarrass FIFA and the Qatari government did lead to promised reform (Ter Haar, 2018). However, such gains may not be sustained; promises to improve workers' rights in the United Arab Emirates were unfulfilled (Box 12.1).

Outsourcing and offshoring practices, such as moving industries (manufacturing and service) from the global North to settings where wages (and workers' rights) are inferior, can reduce some forms of global inequality yet exacerbate regional and national inequality in high-income settings, such as in the US 'rust belts'. Conditions in some factories that supply goods primarily for consumption in the global North are notorious; these may be based both in China and in other parts of the global South. Examples include FoxConn, a company based in the Pearl River Delta regions of China, which supplies the computer company Apple (Lüthje and Butollo, 2017), and garment manufacturers such as in Dhaka, Bangladesh (Kates, 2019). Elements of the global supply chain for solar panels have also been traced to alleged slave labour performed by Uighurs, a persecuted Muslim minority in Xinjiang, China (Hendrix and Noland, 2021).[6]

Box 12.1

Sönmez *et al.* (2011), relying on a series of reports published by Human Rights Watch (HRW), gave a rare insight into the state-sponsored, systematic exploitation of vulnerable workers in the United Arab Emirates, particularly those involved in the construction of Dubai. They reported that negative press linked to the 2006 HRW report 'Building Towers, Cheating Workers' led Dubai's ruler, Sheikh Mohammed, to order the labour minister to institute immediate reforms structured around HRW's recommendations. However, follow-up in 2009 showed that virtually no reforms had been implemented. A law introduced in 2007 to improve workers' conditions had been abandoned. While helping to create Abu Dhabi's 'Island of Happiness' (Saadiyat Island), workers informed investigators that medical care for heatstroke, fever and workplace injuries consisted of male nurses handing out paracetamol, a generic painkiller and antipyretic. This was the treatment for a welder whose leg was injured by a grinding machine.

High-income settings, including in Singapore and Australia, also tolerate underpayment of vulnerable workers, such as students and those on temporary visas. Singapore has an estimated 300,000 low-wage migrant workers, in a total population of fewer than 5.5 million (Steinhauer, 2022). A recent report described the exploitation of migrant workers in Australia as rife (Coates *et al.*, 2023).

12.5 Long-Term Economic Impacts of Occupational Heat Related to Climate Change

Estimating the economic impacts of heat stress is increasingly recognized as significant for policy makers and employers. A systematic review of the economic impacts of heat on labour productivity identified four methods and concluded that global economic losses due to heat-related labour productivity are projected to range from 0.3% to 2.6% of global gross domestic product (GDP) in 2100, depending on the severity of warming (Zhao *et al.*, 2021). These studies identified the greatest losses as occurring in South and South-East Asia, sub-Saharan Africa and Central America.

Climate change, interacting with social factors, including competition among unskilled workers, and a willingness to exploit others (or at least to turn a blind eye to it) can be expected to increase the risk of morbidity and even of death for working people with limited ability, or economic disincentives, to take preventive action. These occupational heat effects may lead to reductions in work targets, with wider economic consequences, not only for the worker and their family, but also for employers and for the national and ultimately the global economy. An analysis by Kjellstrom *et al.* (2016) concluded that there is a risk that, later this century, many among the four billion people who live in hot areas worldwide will experience significantly reduced work capacity owing to climate change. They reported that, in some areas, 30–40% of annual daylight hours will become too hot for work to be carried out.

Although this chapter is located in the 'primary' or 'direct' health effect section of this book, we recognize that the impacts of heat on family and community economy will, if sufficiently severe, contribute to tertiary (systemic) effects on human health and well-being, perhaps of sufficient magnitude to contribute to localized or even widespread collapse and dislocation. By 2100, for example, unless greenhouse gas levels are quickly stabilized, thresholds of heat and humidity may be exceeded to such an extent that migration will become the only feasible adaptation strategy in some settings. Identified regions at such risk include parts of South Asia, coastal China and the Middle East. And, while workers will initially face disproportionate risks, entire populations will also be vulnerable (Pal and Eltahir, 2016; Im *et al.*, 2017; Kang and Eltahir, 2018). Such events will deliver devastating effects to the local economy, well before migration becomes inevitable.

Dunne *et al.* (2013) modelled past and future WBGTs (derived from historical climate data and Earth System Model projections combined with industrial and military guidelines) in order to assess an acclimated individual's occupational capacity to safely perform sustained labour under environmental heat stress. These authors estimated that environmental heat stress, over the past few decades, has reduced labour capacity by 10% in peak heat stress months. They forecast that, in most tropical and mid-latitude locations, under the highest heating scenario they considered (Representative Concentration Pathway 8.5), labour capacity would fall to 80% by 2050 and to less than 40% by 2200 in peak months. They noted that many uncertainties and caveats associated with these projections include climate sensitivity, climate warming patterns, carbon dioxide emissions, future population distributions and technological and societal change.

The growth and lower cost of internet-enabled sensing devices, including low-cost personalized monitoring, is enabling improved sensitivity and specificity of heat strain in workers (Nazarian *et al.*, 2021).

12.6 Non-Heat Effects of the Anthropocene on Occupational Health

Many adverse impacts on working people will arise in the Anthropocene via non-heat effects. The extent of these depends crucially on the severity of climate change and the avoidance of positive feedbacks in the Earth system, such as those that may lead to 'hothouse Earth' (Rockström *et al.*, 2023). Also crucial is the extent to which negative emission technologies can be successfully developed, as well as the viability and longevity of carbon offsets such as afforestation which appear vulnerable to climate change-aggravated wildfire.

The severity of the effects of climate change also depends on their interaction with human behaviour and other human traits, especially the capacity of our species to share resources if they grow scarce. Although there are many examples of impressive degrees of sharing within homogeneous groups, there is as yet little evidence (beyond aspirational, such as expressed by the Sustainable Development Goals) that the scale of sharing can be enlarged to the degree that will be needed if even moderately severe climate change occurs.

Many of these potential consequences are described elsewhere in this book. In the worst case they include greatly worsened migration (forced and voluntary), conflict, expanded undernutrition and even famine. Even if the worst case is avoided, these factors seem likely to depress, or at least fail to lift, the conditions for working people. Food prices are historically high (compared with recent decades), not only from extreme weather events, but also from other aspects of limits to growth, including the war in Ukraine and the high price of oil and other forms of energy. The COVID-19 pandemic, perhaps itself a manifestation of the Anthropocene, has also exacerbated the rise in hunger and a deterioration in working and living conditions. In some parts of the world the exposure to vector-borne diseases, to which climate change contributes, will increase.

Flooding, an extreme event likely to increase from climate change and other human-driven factors such as the paving of ground, may increase the risk of leptospirosis among agricultural workers (Saulnier *et al.*, 2017). The deaths of six Amazon employees in the US state of Kentucky, due to a tornado of record duration[7] in December 2021, illustrates that excessive risks to workers' health from extreme events are not confined to low-income countries (Hooker, 2021).

On the other hand, mechanization, automation, robotics and artificial intelligence are improving productivity in many settings around the world (Studley, 2021). These advances can also reduce repetitive, tedious, heavy and dangerous work. But there are many claims that these trends can rupture the lives of displaced workers, helping to generate increased inequality, leading to both un- and under-employment, with resultant harmful impacts on self-esteem, income and family life. A report by the Economic Policy Institute, a left-leaning think tank based in the USA, argued that such claims are greatly overstated, at least for the USA (Mishel and Bivens, 2017).

12.7 Actions to Prevent and Adapt to Climate Change in Order to Protect Occupational Health

In tropical areas at low altitude, where the majority of the global population live and work, heat exposure levels have long been recognized as capable of harming the health and well-being of working people engaged in many occupations. In response, many populations and cultures exposed to very hot weather have evolved cultural mechanisms, such as siestas,[8] earlier starting times, night work, and reduced expectation of excessive work during the hottest hours and the hottest months (Masuda *et al.*, 2019). Such adaptations may be vulnerable to reversal in the context not only of neoliberal globalization (LaDou, 2020), but also where workers are governed by authoritarian governments or industry-provided 'sponsorship' (kafala) (Ter Haar, 2018). Such risks are comparatively well-recognized in parts of the Middle East where the rights and organizing capacity of heat-exposed workers, especially migrant labourers, are already abysmal, as discussed above (Pradhan *et al.*, 2019; Sönmez *et al.*, 2011). Even in countries where slavery is banned, supply chains can easily involve quasi-slavery. Few populations are wholly innocent. The World Cup attracted billions of viewers. How many undertook any action to fairly compensate, or even acknowledge, the human exploitation involved? However, in recent decades, investment in sports science to protect and to enhance athletic performance has led to progress (Alhadad *et al.*, 2019). Insights from this sector can also be applied in occupational settings.

Despite efforts at adaptation, such as effective cooling and fluid-restoring methods, climate change is likely to force heat-exposed workers and householders to slow down their work and daily activities. Given a sufficiently just political and social environment, steps such as the provision of more shade, air conditioning and specially designed cooling clothes will help. Greener cities, with less heat-absorptive surfaces and more trees, will also assist. Solar energy-driven air-conditioning systems have been promoted as a way to reduce greenhouse gas emissions in hot places (Al-Yasiri *et al.*, 2022). Other innovative air-conditioning methods powered by photovoltaic (PV) cells have been proposed as being even more protective of the climate (Albatayneh *et al.*, 2021).

In many work situations, air-conditioning (AC) is not feasible (for instance, outdoors, or in large building sites), and, as mentioned, protective clothing may increase heat stress. In addition, there are important costs for AC, and, in most places, related electricity consumption still adds to greenhouse gas accumulation. Jay *et al.* (2021) also argued that excessive future reliance on AC further marginalizes the communities most vulnerable to the heat. However, the long-held recommendation (still endorsed by WHO) to avoid fans (which require less electricity than AC) when the air temperature exceeds 35°C (normal skin temperature) has been challenged (Morris *et al.*, 2021). On days of extreme heat and low humidity, it may be prudent to avoid fans until – and if – stronger evidence accrues.

In cities and other locations where extreme heat and humidity occur in waves, rather than persistently for months, well-communicated heat action plans with robust surveillance and monitoring will help to ameliorate adverse occupational heat effects (Jay *et al.*, 2021). In settings where high heat and humidity persist more or less continuously for many weeks, a different kind of plan is needed to protect workers' health. This is likely to involve the avoidance of physically demanding work for many hours each day. Perhaps people can adapt to work more at night, when it is cooler, though this will increase the risk of injuries, not only due to poor lighting but also because many people will not easily adjust to night work, especially if compensatory daytime rest and sleep are disrupted by intolerable heat and humidity, where AC quarters are unavailable or denied.

Olympic competitors may one day be protected by their relocation to cooler months (Smith *et al.*, 2016). But will the labourers and others who toil to construct the facilities used by athletes be similarly protected?

The greatest priority is to mitigate climate change. In a fairer world, without 'disposable' people (Bales, 1999), those with the means would pay more for both luxury (e.g. gold) and essential items (e.g. food) if their supply chains do not rely on the exploitation inevitable in the 'market forces'-dominated, neoliberal global economy. The rich would not make the implicit assumption that the health of the vulnerable is something that can reasonably be exchanged for a few years of wages – or for even less. In such a (fairer) world, the additional costs paid by consumers would not lodge in the accounts of 'middle men' but instead find their way to the pockets of workers. Nations would introduce and enforce minimal occupational standards, including safeguards against the hazards of excess heat and other physical risks, such as those likely to be magnified via climate change.

There is also a need to design and implement public and occupational health and safety programmes that protect individuals currently at risk; and to develop preventive policies and interventions (adaptation) through the design of urban areas and workplaces, as well as residential housing and commercial places, to reduce existing and future heat exposure in the locations where people live and work. Further evidence of local heat exposure and its effects is likely to be important in the development of policies and actions on heat prevention in an era of climate change.

Acknowledgements

This chapter is adapted from Kjellstrom, T., Lucas, R., Lemke, B. and Sahu, S. (2014) Occupational Heat Effects: A Global Health and Economic Threat Due to Climate Change. In: Butler, C.D. (ed.) *Climate Change and Global Health*. CABI, Wallingford, UK. pp. 38–44.

Notes

1 Yet extending up to 50 km into the undisturbed forest (Cohn *et al.*, 2019).

2 The website of ClimateCHIP.org is available at https://www.climatechip.org/ (accessed 7 June 2023). It is a rich source of data on heat stress and other health impacts of climate change.

3 'Heat stroke was now becoming a serious menace at City Deep mine, especially as still greater depths were contemplated. Further, during the period of acclimatization the average amount of work output per labourer per shift was equivalent to only one third of a normal shift. The financial position of the mine was such as to be profoundly affected by this loss in efficiency' (Wyndham, 1965).

4 Bonded labour was officially banned in India in 1976; nevertheless, it appears to persist (Samonova, 2022); it is also permitted in Pakistan, Nepal and Afghanistan.

5 Three trade unions, together with a migrant worker from Bangladesh, initiated unsuccessful legal action against FIFA in a Swiss court. They argued that FIFA should fulfil its responsibility to ensure better working circumstances for the migrant workers at World Cup construction sites (Ter Haar, 2018).

[6] Xinjiang accounts for nearly 50% of the global production of polysilicon, the material from which solar panels are built.
[7] The relationship of this tornado to anthropogenic climate change is uncertain. Pielke Jr. (2021) pointed out that the 2013 IPCC report concluded that 'observational trends in tornadoes, hail, and lightning associated with severe convective storms are not robustly detected' and also that the 'attribution of certain classes of extreme weather (e.g., tornadoes) is beyond current modelling and theoretical capabilities'. On the other hand, the effects of climate change may be accelerating; modelling will in some cases lag reality. The intensification of the 'waviness' of the jet stream brings cold Arctic air to unusually low latitudes, where it can interact with much warmer, adjacent air; this has been suggested as a possible amplifier of tornado activity (Moore and Fricker, 2020).
[8] A Spanish word meaning a rest in the hottest time of the day.

References

Albatayneh, A., Jaradat, M., Al-Omary, M. and Zaquot, M. (2021) Evaluation of coupling PV and air conditioning vs. solar cooling systems – case study from Jordan. *Applied Sciences* 11, 511. DOI: 10.3390/app11020511

Al-Bouwarthan, M., Quinn, M.M., Kriebel, D. and Wegman, D.H. (2019) Assessment of heat stress exposure among construction workers in the hot desert climate of Saudi Arabia. *Annals of Work Exposures and Health* 63, 505–520. DOI: 10.1093/annweh/wxz033

Alhadad, S.B., Tan, P.M.S. and Lee, J.K.W. (2019) Efficacy of heat mitigation strategies on core temperature and endurance exercise: A meta-analysis. *Frontiers in Physiology* 10, 71. DOI: 10.3389/fphys.2019.00071

Al-Yasiri, Q., Szabo, M. and Arici, M. (2022) A review of solar-powered cooling and air-conditioning systems for building application. *Energy Reports* 8, 2888–2907. DOI: 10.1016/j.egyr.2022.01.172

Annear, M., Kidokoro, T. and Shimizu, Y. (2021) Existential threats to the Summer Olympic and Paralympic Games? a review of emerging environmental health risks. *Reviews on Environmental Health* 36, 159–166. DOI: 10.1515/reveh-2020-0141

Bales, K. (1999) *Disposable People: New Slavery in the Global Economy*. University of California Press, Berkeley, Los Angeles, California; London.

Bernard, T.E. (1999) Heat stress and protective clothing: an emerging approach from the United States. *Annals of Occupational Hygiene* 43, 321–327. DOI: 10.1093/annhyg/43.5.321

Budd, G.M. (2008) Wet-bulb globe temperature (WBGT) – its history and its limitations. *Journal of Science and Medicine in Sport* 11, 20–32. DOI: 10.1016/j.jsams.2007.07.003

Byrne, C., Lee, J.K.W., Chew, S.A.N., Lim, C.L. and Tan, E.Y.M. (2006) Continuous thermoregulatory responses to mass-participation distance running in heat. *Medicine & Science in Sports & Exercise* 38, 803–810. DOI: 10.1249/01.mss.0000218134.74238.6a

Cheung, S.S., Lee, J.K.W. and Oksa, J. (2016) Thermal stress, human performance, and physical employment standards. *Applied Physiology, Nutrition, and Metabolism* 41, S148–S164. DOI: 10.1139/apnm-2015-0518 %M 27277564

Coates, B., Wiltshire, T. and Reysenbach, T. (2023) *Short-Changed: How to Stop the Exploitation of Migrant Workers in Australia.* Grattan Institute. Available at https://grattan.edu.au/report/short-changed-how-to-stop-the-exploitation-of-migrant-workers-in-australia/ (accessed 21 June 2023).

Cohn, A., Bhattarai, N., Campolo, J., Crompton, O., Dralle, D. *et al.* (2019) Forest loss in Brazil increases maximum temperatures within 50 km. *Environmental Research Letters* 14, 084047. DOI: 10.1088/1748-9326/ab31fb

Crowe, J., Roman, B., Wesseling, C., Robles, A., Kjellstrom, T. *et al.* (2013) Heat exposure in sugar cane harvesters in Costa Rica. *American Journal of Industrial Medicine* 56, 1157–1164. DOI: 10.1002/ajim.22204

DARA (2012) *Climate Vulnerability Monitor – A Guide to the Cold Calculus of a Hot Planet.* Development Assistance Research Associates, DARA International Foundation, Madrid, Spain. Available at: www.daraint.org (accessed 10 February 2022).

Dunne, J.P., Stouffer, R.J. and John, J.G. (2013) Reductions in labour capacity from heat stress under climate warming. *Nature Climate Change* 3, 563–566. DOI: 10.1038/nclimate1827

Ebi, K.L., Capon, A., Berry, P., Broderick, C., de Dear, R. *et al.* (2021) Hot weather and heat extremes: health risks. *The Lancet* 398, 698–708. DOI: 10.1016/S0140-6736(21)01208-3

Felton, J.S. (1997) The heritage of Bernardino Ramazzini. *Occupational Medicine* 47, 167–179. DOI: 10.1093/occmed/47.3.167

Hansson, E., Glaser, J., Jakobsson, K., Weiss, I., Wesseling, C. *et al.* (2020) Pathophysiological mechanisms by which heat stress potentially induces kidney inflammation and chronic kidney disease in sugarcane workers. *Nutrients* 12, 1639. DOI: 10.3390/nu12061639

Hayden, B. (2001) Richman, poorman, beggarman, chief: the dynamics of social inequality. In: Feinman, G.M. and Price, T.D. (eds) *Archaeology at the Millennium*. Springer, Boston, Massachusetts, pp. 231–272.

Hendrix, C. and Noland, M. (2021) *Economic diplomacy and genocide in Xinjiang*. Analysis from the East-West Center, No. 150, Honolulu, Hawaii. https://www.jstor.org/stable/resrep37725

Hollowell, D.R. (2010) Perceptions of, and reactions to, environmental heat: a brief note on issues of concern in relation to occupational health. *Global Health Action* 3, 3. DOI: 10.3402/gha.v3i0.5632.

Hooker, L. (2021) Amazon criticised over safety at tornado-hit warehouse. Available at: https://www.bbc.com/news/business-59641784 (accessed 14 December 2021).

Im, E.-S., Pal, J.S. and Eltahir, E.A.B. (2017) Deadly heat waves projected in the densely populated agricultural regions of South Asia. *Science Advances* 3, e1603322. DOI: 10.1126/sciadv.1603322

Jackson, B., Bales, K., Owen, S., Wardlaw, J. and Boyd, D.S. (2019) Analysing slavery through satellite technology: how remote sensing could revolutionise data collection to help end modern slavery. *Journal of Modern Slavery* 4, 169–199. Available at: https://www.nottingham.ac.uk/research/beacons-of-excellence/rights-lab/resources/academic-publications/2019/march/boyd-analysing-slavery.pdf (accessed 10 April 2022).

James, C. (2019) 'They don't play or run or shout … they're slaves': the first survey of children's literature on modern slavery. *Journal of Modern Slavery* 4, 143–168. Available at: https://slavefreetoday.org/journal_of_modern_slavery/v4i2a7_They_don%E2%80%99t_play_or_run_or_shout_They%E2%80%99re_slaves_The_First_Survey_of_Children%27s_Literature_on_Modern_Slavery_james.pdf (accessed 10 April 2022).

Jay, O., Capon, A., Berry, P., Broderick, C., de Dear, R. *et al.* (2021) Reducing the health effects of hot weather and heat extremes: from personal cooling strategies to green cities. *The Lancet* 398, 709–724. DOI: 10.1016/S0140-6736(21)01209-5

Kang, S. and Eltahir, E.A.B. (2018) North China Plain threatened by deadly heatwaves due to climate change and irrigation. *Nature Communications* 9, 2894. DOI: 10.1038/s41467-018-05252-y

Kates, M. (2019) Sweatshops, exploitation, and the case for a fair wage. *The Journal of Political Philosophy* 27, 26–47. DOI: 10.1111/jopp.12155

Kjellstrom, T. (2009) Climate change, direct heat exposure, health and well-being in low and middle income countries. *Global Health Action* 2, 1–4. DOI: 10.3402/gha.v2i0.1958

Kjellstrom, T., Briggs, D., Freyberg, C., Lemke, B., Otto, M. *et al.* (2016) Heat, human performance, and occupational health: A key issue for the assessment of global climate change impacts. *Annual Review of Public Health* 37, 97–112. DOI: 10.1146/annurev-publhealth-032315-021740

Kjellstrom, T., Lemke, B. and Lee, J. (2019) Workplace heat: an increasing threat to occupational health and productivity. *American Journal of Industrial Hygiene* 62, 1076–1078. DOI: 10.1002/ajim.23051

LaDou, J. (2020) A world of false promises: International Labour Organization, World Health Organization, and the plea of workers under neoliberalism. *International Journal of Health Services* 50, 314–323. DOI: 10.1177/0020731420917912

Lee, J.K.W., Nio, A.Q.X., Lim, C.L., Teo, E.Y.N. and Byrne, C. (2010) Thermoregulation, pacing and fluid balance during mass participation distance running in a warm and humid environment. *European Journal of Applied Physiology* 109, 887–898. DOI: 10.1007/s00421-010-1405-y

Lee, J., Venugopal, V., Latha, P.K., Alhadad, S.B., Leow, C.H.W. *et al.* (2020) Heat stress and thermal perception amongst healthcare workers during the COVID-19 pandemic in India and Singapore. *International Journal of Environmental Research and Public Health* 17, 8100. DOI: 10.3390/ijerph17218100

Lüthje, B. and Butollo, F. (2017) Why the Foxconn model does not die: production networks and labour relations in the IT industry in South China. *Globalizations* 14, 216–231. DOI: 10.1080/14747731.2016.1203132

Masuda, Y.J., Castro, B., Aggraeni, I., Wolff, N.H., Ebi, K. *et al.* (2019) How are healthy, working populations affected by increasing temperatures in the tropics? Implications for climate change adaptation policies. *Global Environmental Change* 56, 29–40. DOI: 10.1016/j.gloenvcha.2019.03.005

Mishel, L. and Bivens, J. (2017) *The zombie robot argument lurches on: there is no evidence that automation leads to joblessness or inequality*. Economic Policy Institute, Washington, DC. Available at: https://www.epi.org/publication/the-zombie-robot-argument-lurches-on-there-is-no-evidence-that-automation-leads-to-joblessness-or-inequality/ (accessed 21 June 2022).

Moeti, M. (1986) The origins of forced labor in the Witwatersrand. *Phylon* 47, 276–284. DOI: 10.2307/274623

Moore, T.W. and Fricker, T. (2020) Tornadoes in the USA are concentrating on fewer days, but their power dissipation is not. *Theoretical and Applied Climatology* 142, 1569–1579. DOI: 10.1007/s00704-020-03402-1

Morris, N.B., Chaseling, G.K., English, T., Gruss, F., Maideen, M.F.B. *et al.* (2021) Electric fan use for cooling during hot weather: a biophysical modelling study. *The Lancet Planetary Health* 5, e368–e377. DOI: 10.1016/S2542-5196(21)00136-4

Nazarian, N., Liu, S., Kohler, M., Lee, J.K.W., Miller, C. *et al.* (2021) Project Coolbit: can your watch predict heat stress and thermal comfort sensation? *Environmental Research Letters* 16, 034031. DOI: 10.1088/1748-9326/abd130

Oppermann, E., Kjellstrom, T., Lemke, B., Otto, M. and Lee, J.K.W. (2021) Establishing intensifying chronic exposure to extreme heat as a slow onset event with implications for health, wellbeing, productivity, society and economy. *Current Opinion in Environmental Sustainability* 50, 225–235. DOI: 10.1016/j.cosust.2021.04.00

Pal, J.S. and Eltahir, E.A.B. (2016) Future temperature in southwest Asia projected to exceed a threshold for human adaptability. *Nature Climate Change* 6, 197–200. DOI: 10.1038/nclimate2833

Parsons, K.C. (2014) *Human Thermal Environments: The Effects of Hot, Moderate and Cold Temperatures on Human Health, Comfort and Performance*, 3rd edn. CRC Press, Boca Raton, Florida. DOI: 10.1201/b16750

Pielke Jr., R. (2021) *The politics of our climate attribution obsession*. The Honest Broker. Available at: https://rogerpielkejr.substack.com/p/the-politics-of-our-climate-attribution (accessed 13 December 2021).

Pradhan, B., Kjellstrom, T., Atar, D., Sharma, P., Kayastha, B. *et al.* (2019) Heat stress impacts on cardiac mortality in Nepali migrant workers in Qatar. *Cardiology* 143, 37–48. DOI: 10.1159/000500853

Rockström, J., Gupta, J., Qin, D., Lade, S.J., Abrams, J.F. *et al.* (2023) Safe and just Earth system boundaries. *Nature* 619, 102–111. DOI: 10.1038/s41586-023-06083-8

Sahu, S., Sett, M. and Kjellstrom, T. (2013) Heat exposure, cardiovascular stress and work productivity in rice harvesters in India: implications for a climate change future. *Industrial Health* 51, 424–431. DOI: 10.2486/indhealth.2013-0006

Samonova, E. (2022) Human rights through the eyes of bonded labourers in India. *Journal of Modern Slavery* 7, 82–96. Available at: https://slavefreetoday.org/journal_of_modern_slavery/v7i2a5-Human-rights-through-the-eyes-of-bonded-labourers-in-india.pdf (accessed 4 May 2022).

Saulnier, D.D., Brolin Ribacke, K. and von Schreeb, J. (2017) No calm after the storm: a systematic review of human health following flood and storm disasters. *Prehospital and Disaster Medicine* 32, 568–579. DOI: 10.1017/S1049023X17006574

Sinclair, J. (2019) Strategic litigation as a tool to combat modern slavery. *Journal of Modern Slavery* 4, 46–65. Available at: https://slavefreetoday.org/journal_of_modern_slavery/v4i2a3_Strategic_Litigation_as_a_Tool_to_Combat_Modern_Slavery_sinclair.pdf (accessed 12 February 2022).

Smith, K., Woodward, A., Lemke, B., Otto, M., Chang, C.J. *et al.* (2016) The last summer Olympics? Climate change, health, and work outdoors. *The Lancet* 388, 642–644. DOI: 10.1016/S0140-6736(16)31335-6

Sönmez, S., Apostolopoulos, Y., Tran, D. and Rentrope, S. (2011) Human rights and health disparities for migrant workers in the UAE. *Health and Human Rights* 13, E17–E35. https://www.jstor.org/stable/10.2307/healhumarigh.13.2.17

Steinhauer, I. (2022) *Singapore's treatment of migrant workers makes 'racial harmony' impossible*. The Baines Report [student organization], Policy in Perspective from The LBJ School of Public Affairs, University of Texas, Austin. Available at: https://sites.utexas.edu/bainesreport/2022/01/10/singapores-treatment-of-migrant-workers-makes-racial-harmony-impossible/ (accessed 21 June 2023).

Studley, M. (2021) Onshoring through automation; perpetuating inequality? *Frontiers in Robotics and AI* 8, 634297. DOI: 10.3389/frobt.2021.634297

Ter Haar, B. (2018) FIFA, Qatar, Kafala: can the World Cup create a better world of work? *International Labor Rights Case Law* 4, 128–132. DOI: 10.1163/24056901-00401019

Willett, K.M. and Sherwood, S. (2012) Exceedance of heat index thresholds for 15 regions under a warming climate using the wet-bulb globe temperature. *International Journal of Climatology* 32, 161–177. DOI: 10.1002/joc.2257

Wolff, N.H., Zeppetello, L.R.V., Parsons, L.A., Aggraeni, I., Battisti, D.S. *et al.* (2021) The effect of deforestation and climate change on all-cause mortality and unsafe work conditions due to heat exposure in Berau, Indonesia: a modelling study. *The Lancet Planetary Health* 5(12), e882–e892 DOI: 10.1016/S2542-5196(21)00279-5

Wyndham, C.H. (1965) A survey of the causal factors in heat stroke and of their prevention in the gold mining industry. *Journal of the South African Institute of Mining and Metallurgy* 66, 125–155.

Wyndham, C.H. (1969) Adaptation to heat and cold. *Environmental Research* 2, 442–469. DOI: 10.1016/0013-9351(69)90015

Yaglou, C.P. and Minard, D. (1957) Control of heat casualties at military training centers. *AMA Archives of Industrial Health* 16, 302–316. Available at: https://www.cabdirect.org/cabdirect/abstract/19582900896 (accessed 10 April 2022).

Zhao, M., Lee, J.K.W., Kjellstrom, T. and Cai, W. (2021) Assessment of the economic impact of heat-related labor productivity loss: a systematic review. *Climatic Change* 167, 22. DOI: 10.1007/s10584-021-03160-7

13 A Great Disaster: The Floods of 2022 in Pakistan

Kerryn Higgs
School of Social Sciences, University of Tasmania, Hobart, Australia; Associate Member, Club of Rome, Winterthur, Switzerland

Abstract
The Pakistan floods of 2022 were an absolute catastrophe – for millions of people, across about a third of the country's area with houses and other buildings, crops and livestock swept away, submerged or ruined. Many factors contributed to the disaster, including poor infrastructure, neglected siltation, people living on floodplains and rich and powerful people who released water upstream to protect their own properties. Among the consequences was a public health crisis. Although there is some debate about the extent to which this unprecedented monsoon was affected by climate change, it is very likely that the intensity and longevity of the rain event played a significant part. As the climate continues to warm, the risk of such seasons will increase.

13.1 Introduction

Disasters are often natural, such as earthquakes and cyclones (also known as hurricanes or typhoons); most tectonic events remain so. However, human-caused climate change is a threat multiplier that has supercharged heatwaves, fires, storms and monsoon rains. The level of damage and loss is also affected by human-controlled factors, such as population density, engineering of natural systems such as rivers, and location of human settlements in vulnerable sites such as flood plains and steep hillsides.

Disasters often kill and injure people in the immediate term and leave challenging health problems in their wake. Devastation of the land and its ecosystems leads to huge losses where recovery takes several seasons or, as in the case of the catastrophic fires of 2019/20 in Australia, some areas may take decades before soil biota return, if at all; some rainforest areas which have not burned in recent millennia may not recover, but transition to a new ecological situation. When floods submerge agricultural land, direct losses to human welfare are usually calamitous.

13.2 Background Heating and Attribution

There is little dissent over the impact of global warming on the incidence and intensity of climate extremes. Even though some scientists still argue that it is not possible to attribute any specific heatwave or flood to climate change, it is widely accepted that both the intensity and the frequency of extreme weather events are rising as the background average temperature increases.

The Intergovernmental Panel on Climate Change (IPCC) (2021) noted that 'it is virtually certain that hot extremes (including heatwaves) have become more frequent and more intense

 Climate Change and Global Health: Primary, Secondary and Tertiary Effects, 2nd Edition (eds C.D. Butler and K. Higgs)
DOI: 10.1079/9781800620025.0013

across most land regions since the 1950s ... with high confidence that human-caused climate change is the main driver of these changes ... Some recent hot extremes observed over the past decade would have been *extremely unlikely* to occur without human influence on the climate system' [italics in original]. It also stated that 'projected changes in extremes are larger in frequency and intensity with every additional increment of global warming.' The mean global temperature (or what one might call 'background heat') is now approximately 1.2°C above that in pre-industrial times (usually taken to be 1850–1900) and heat extremes will continue to get hotter in step with rising average temperature. Research by Thompson *et al.* (2022) supported this conclusion.

Thompson *et al.* (2022), and Philip *et al.* (2022) at World Weather Attribution, calculated the percentage contribution of global warming to specific extreme events, with more confidence in results for heat extremes than for precipitation extremes. For example, Philip *et al.* (2022) argued that the 'heat dome' that settled over the Pacific north-west region of North America in 2021 would have been 'virtually impossible without human-caused climate change'. Friederike Otto, one of the authors, told the *Guardian*'s Damian Carrington (2022): 'It broke records by 5°C and really brought home to me that something that was impossible from a statistical point of view can now happen as a new extreme.'

As well as the impact of increasing background heat, it is estimated that the atmosphere holds 7% more water vapour for every extra degree Celsius (Allan and Soden, 2008); this will affect intensity of rainfall and may increase other aspects of patterns of rainfall, such as frequency. The IPCC (2021) rated an increase in heavy precipitation as a 'likely' outcome of climate change. One should remember that IPCC reports have to be signed off by all participating governments, so may be regarded as very conservative.

13.3 Prelude: Spring Heatwave

In the spring of 2022, north-west India and south-east Pakistan baked for more than two months with blistering heat that had arrived two to three months early; temperatures in March and April 2022 were very unusual for the time of year, although would be considered to be somewhat closer to normal if they had happened in later May or early June. India had the hottest March since records began 122 years ago. By May, even more extreme local temperatures were recorded, such as 49.5°C at Jacobobad (Pakistan) and 49.2°C in Delhi. The unusual features of this heatwave were that it went on for months and occurred at the 'wrong' time of year. According to Zachariah *et al.* (2022), climate change made the devastating early heat in south-east Pakistan and north-west India 30 times more likely than it would have been in pre-industrial times.

Although on a less cataclysmic scale than what was to come in Pakistan when the monsoon arrived, there were at least 90 deaths from this heatwave and an estimated 10–35% reduction in crop yields in Haryana, Uttar Pradesh, and Punjab. As a result, the Indian government had to reverse its earlier plan to supplement the global wheat supply, which had been reduced by the war in Ukraine (Zachariah *et al.*, 2022). Thompson and Kennedy-Asser (2022) noted: 'The impacts on human health are complex as both meteorological factors (how hot and humid it is) and socioeconomic factors (how people live and how they are able to adapt) come into play. We do know that heat stress can lead to long-term health issues such as cardiovascular diseases, kidney failure, respiratory distress and liver failure, though we will be unable to know exactly how many people will die in this heatwave due to the lack of necessary health data from India and Pakistan.'

According to Zachariah *et al.* (2022), the hot weather also melted glaciers in the high Hindu Kush and Himalayan (HKH) mountains to the north of Pakistan and in Nepal. They flow into the Indus and its numerous tributaries. Any impact the melting of glaciers might have had on the subsequent flooding is unknown.

13.4 The Mega Flood, June to August

Worse was to follow, especially for Pakistan, when the monsoon arrived in mid-June. Welcomed, as usual, for its cooling effects, the intensity and longevity of the 2022 monsoon turned out to be catastrophic. As the federal minister for climate change, Sherry Rehman, told a press conference in August: 'Pakistan is going through its 8th cycle of monsoon; normally the country has only three to four cycles.' She told reporters that two of the worst-hit provinces – Balochistan and Sindh – had received about 400% more than the 30-year average (Hussain, 2022). Inam Ul Haque (2022b), a retired major general who supervised flood rescue and recovery during the 2010 flood (the most extreme flood in Pakistan before 2022), compared the two events; he noted that, during the 2010 floods, 'Pakistan's Meteorological Department recorded 70 to 102% above normal rains nationwide. This time in just one city Khanpur (Punjab), the downpour was 1,483% above normal.' A labourer from a remote village in Sindh, whose son had been swept away, told *Al Jazeera*'s Abid Hussain (2022) that '[t]hings were not so bad even in the 2010 floods. We never had to leave our village but this time, everything is destroyed.'

Geographically, Pakistan is a relatively long and narrow country which stretches from the HKH mountains in the north down to the Arabian Sea. The Indus River flows more or less down the centre of the country, with many tributaries, a floodplain either side – very wide in the east – and further mountainous regions along the western border with Afghanistan. These mountains continue south and separate Balochistan from the Indus Valley. More than 7000 glaciers are found in Pakistan's high mountains.

The monsoon season often brings floods but, in 2022, the entire country was subject to repeated rain events, causing local flash flooding in towns and villages and along mountain streams while, simultaneously, the river and its tributaries kept rising as rain drenched both the mountainous regions upstream and the flood plains further south. With each pulse, the ground became more and more waterlogged with little capacity to absorb the rain. Ali Tauqeer Sheikh, a climate expert who has worked on Pakistani government advisory panels, told Hussain (2022): 'In 2010, the floods were riverine in nature, which means they mostly impacted areas around the Indus River and they were mostly predictable. This time, there are multiple types such as urban flood, flash flooding and floods caused by glacier burst.' Factors in the scope of the disaster, other than the actual monsoon, are discussed below.

Otto *et al.* (2023) explained that, in August, the Indus 'burst its banks across thousands of square kilometres ... [creating] a 100 km wide lake in the southern province of Sindh, which together with the neighbouring province of Balochistan constituted the worst affected region ... The record rainfall that fell between 1st and 31st August over Sindh and Balochistan was 726% and 590% of the usual totals for August.'

The enormity of the flood was described by Leo Sands (2022) from BBC News on 30 August. He reported that, according to Pakistan's climate minister, 'Literally, one-third of Pakistan is underwater right now, which has exceeded every boundary, every norm we've seen in the past ... It's all one big ocean, there's no dry land to pump the water out ... a crisis of unimaginable proportions.' Millions of houses were destroyed, many people died (estimated at more than 1500) and displaced people had to set up makeshift camps on sections of major highways that remained above water level or on other dry areas. Others found places in extremely crowded shelters provided by the government. Estimates of the number of displaced people varied, but many reports suggested that 33 million citizens were affected; Ul Haque (2022b) believed the true figure could be up to 50 million.

UNICEF's Gerida Binukila, speaking from Quetta in Balochistan (UN, 2022), advised that 'roads and bridges have been washed away; I've just come from the field and the water is not going anywhere ... There is no shelter ... people don't even have clothing ... One lady asked me, "Please, give me some clothing, I ran away two weeks ago." She is still wearing the same dress she wore two weeks ago because she cannot change. You just run with what you have on your back.' Binukila also reported speaking to Afghan refugees as they pointed out into the water and told her, 'That used to be my house, this used to be the school,' but all that she could see was just endless water. Officials warned that it could take 'up to six months for floodwaters to recede'.

UN (2022) staff also reported that 'the kind of healthcare that is required is not available' and that 'as had been feared, life-threatening illness and disease have now spread among displaced communities, including cerebral malaria, for which there is no available medicine'.

There were also grave problems in mounting rescues. Thousands of people who lived along streams in the mountains, such as in Swat and Khyber Pakhtunkhwa province, were ordered to evacuate but the authorities lacked enough helicopters. Trapped people were often endangered by landslides as well as rising stream water. Down river, Sindh province was so inundated that much of it was inaccessible; emergency workers struggled to reach people in danger of drowning as they were engulfed. 'There are no landing strips or approaches available ... our pilots find it difficult to land,' a Pakistani military official told the press (Sands, 2022). This situation was widespread in Balochistan as well. In addition, 'heavy rains and strong winds limit the use of helicopter and boats, compared to hovercraft, the work horse in 2010 floods'. But access to hovercraft was limited in 2022 (Ul Haque, 2022b).

Even after the rain had eased, people remained stranded for weeks or more; many received no food. As Dilawar *et al.* (2022) reported for *Bloomberg* on 8 September: 'For Pakistanis in still-flooded areas, the nightmare doesn't seem to end. On a recent trip to Sindh province, evacuation boats motored to villages where panicked people shouted from their rooftops. In some pockets, the water was 16 feet deep, according to officials. Many said government assistance was still piecemeal. A cluster of protesters blocked vehicles from crossing a crucial bridge. They screamed that their children were dying of hunger.'

13.5 Climate and the Monsoon

Otto *et al.* (2023) explained the difficulties of attempting to establish how much, if at all, global heating is contributing to extreme rainfall events such as the 2022 monsoon over Pakistan. They admitted that long-term variability and processes that are difficult to 'capture' leave room for uncertainty.

However, Otto *et al.* (2023) stated that '[t]he flooding occurred as a direct consequence of the extreme monsoon rainfall throughout the summer 2022 season, exacerbated by shorter spikes of very heavy rain particularly in August hitting the provinces Sindh and Balochistan'. They found that the 5-day maximum rainfall over the provinces Sindh and Balochistan is now about 75% more intense than it would have been had the climate not warmed by 1.2°C, whereas the 60-day rain across the basin is now about 50% more intense, meaning rainfall this heavy is now more likely to happen. They conceded that large uncertainties still exist regarding a causal link to climate change, due to variability of rainfall in this region and the short observational datasets they relied on.

Overall, they argued, many studies 'point towards an increase in the likelihood and intensity of both events ... We therefore conclude, for the 5 day rainfall extreme, where most models and observations we have analysed show that intense rainfall has become heavier as the world has warmed, that climate change indeed increased the rainfall intensity up to 50% as the best estimate in some models. For the 60 day event we also cannot exclude, that climate change is indeed responsible for all of the observed increase in heavy rain.'

Although the science of attribution is still in its infancy, residents and citizens of Pakistan during that summer, including the federal minister for climate change, were adamant that climate change played a significant part.

13.6 Complexity of Factors

While it is apparent that the level of the rainfall was unprecedented, there are many other factors that contributed to the disaster.

Human infrastructure such as homes, buildings, bridges and agricultural land are situated on or very close to the flood plains. In an era of climate change, what was once an occasional risk has become a clear and present danger and this applies to flood plains the world over. Catastrophic floods on the east coast of Australia in both 2021 and 2022 (during a La Niña event) created havoc on the many floodplains where Australians have built cities and suburbs.

As well as this inherent problem, the situation in Pakistan is compounded by inadequate infrastructure, lack of effective risk-reduction, and an outdated river management system. For example, the crucial Kotri (or Sukkar) barrage is 90 years old and the level of flooding in 2022 – and into the future – can be expected to exceed anything that was handled back in the 1940s. As anthropogenic climate change continues to heat the planet, the likelihood of further increases in extreme rainfall underlines the urgent need to reduce these vulnerabilities in Pakistan.

Given the high poverty rates of the country, these factors are as difficult to mitigate as the rain itself, which will only be mitigated if worldwide carbon emissions are drastically curtailed. As Ul Haque (2022b) suggested: 'With 2.6% of world's population, Pakistan contributes a minuscule 0.5% to global carbon emissions, compared to the US, with about 4% of the world's population contributing about 13%. So, the Global North has to face up to its moral, ethical and humanitarian responsibility' (see Chapter 2.ii). With the mighty Indus flowing down the centre of the whole country, it may prove impossible to fully mitigate floods of the scale seen in 2022. Nonetheless, there are many measures that could provide significant improvement.

Siltation and inadequate infrastructure

The 2010 flood deposited massive quantities of silt which had not been dredged away by 2022, leaving the river bed, the tributary streams and the manmade canals and drains with reduced capacity to hold water. Silt continues to be deposited as in any river basin with periods of high rainfall. According to Ul Haque (2022a), it is absolutely vital to de-silt and dredge the main channels, including the vast network of drainage channels designed to allow excess water to be diverted, stored where possible and any unusable excess ultimately vented to the ocean. The barrages and canals do serve this purpose, but many are in poor repair; they need re-construction, strengthening and capacity enhancement, as well as dredging. In Ul Haque's judgement, 'inefficient drainage' makes a significant contribution to flooding after rains such as those of 2022.

The problem is acute in Sindh. Ul Haque (2022a) pointed out that the traditional drains east of the river lack the capacity to mitigate a deluge of this magnitude. On the western side, he noted that the canal system was not fully completed and had been beset by 'corruption, capacity and alignment issues [which] mar the project after billions spent over years'. Influential landlords, who Ul Haque characterized as politically untouchable, blocked one of the main canals and meanwhile run-off from the Kirthar Range to the west, combined with the swollen Indus, had turned vast areas 'into an enormous lake, devastating villages and towns with an unseen ferocity'. The ferocity of the water also swept levees and embankments away, surprising some people who had no idea they would be flooded.

Ul Haque (2022a) also noted that, in upper Khyber Pakhtunkhwa, the River Swat was blocked by illegal constructions and that 'greedy hoteliers' encroached on the banks of other rivers, rivulets and streams. He also observed that unchecked and unregulated tourism was likely to create similar problems in other picturesque mountain areas such as Azad Kashmir and Gilgit Baltistan, as well as Balochistan, all of which are prone to flash floods and, in the case of Gilgit Baltistan, Glacial Lake Outburst Floods.

Ali Tauqeer Sheikh (Hussain, 2022) pointed to an overall lack of development planning exacerbating the impacts of climate change, which he described as a 'threat multiplier', since the frequency of heavy flooding is increasing. 'What we are seeing here in the country is development deficit. It is not only the excessive rain which is causing the issue, but rather the inadequate preparation and infrastructure.'

Population spilling onto flood plain and upstream releases

As observed by Ul Haque (2022b), many areas in Sindh province where the floodwaters of 2010 had overflowed were wasteland at that time. By 2022, as Pakistan's population increased, the 'wastelands' had been progressively occupied. And the overflows were not simply an inevitable consequence of the lie of the land. It was a policy practice 'in order to save barrages and other water works' to breach river and canal banks upstream to release excess water onto the erstwhile wastelands, without regard to the people now living there. Such actions were carried out by the Irrigation Department, or by 'politically untouchable influentials ... especially in lower Sindh ... Generally powerful landlords save their lands at the cost of have-nots' (Ul Haque, 2022b). *The Guardian*'s Shah Meer Baloch (2022) noted reports that 'protesters had blocked the Indus Highway, the only safe passage between Hyderabad and Karachi and northern Sindh, and the rest of Pakistan at Naseerabad ... claim[ing that] the local lawmakers have endangered the population by diverting floodwater'.

13.7 Damage and Loss: Infrastructure and Crops

The damage was cataclysmic and affected the entire country. Around 6700 km of road, 269 bridges and 1460 health facilities were destroyed, 18,590 schools were damaged, and millions of dwellings were washed away or ruined (Otto *et al.*, 2023).

Agriculture is central to Pakistan's economy, as well as being an indispensable element in feeding people locally. The sector employs almost 40% of Pakistanis and represents about a fifth of the economy (Otto *et al.*, 2023). The River Indus is both a blessing and, in times like these, a curse. Farmers have suffered especially harsh losses. One 70-year-old farmer from near Sukkur in Sindh province told Leo Sands of the BBC: 'Our crop spanned over 5000 acres on which the best quality rice was sown and is eaten by you and us ... All that is finished.'

According to Dilawar *et al.* (2022), around 800,000 farm animals had perished in the months since June, and Pakistan's food ministry had estimated crop losses at more than US$1.5 billion. Cotton, rice and onion fields became massive lakes. And, as noted above, given the paucity of dry land to pump to, these fields were likely to have remained underwater for weeks or months. In addition, vegetables, orchards and rice fields were damaged. Otto *et al.* (2023) estimated that 'some 18,000 square kilometres of cropland was ruined, including roughly 45% of the cotton crop – one of the nation's key exports'. They assessed the loss of food crops at US$2.3 billion and noted that the flood losses have compounded the food shortages already created by the war in Ukraine and the heatwave in the spring.

Ul Haque (2022b) estimated a production shortfall of around two million tons of rice and unquantified losses of vegetables such as onions, tomatoes and chilli, fruit (including date palms), wheat, pulses and milk. Sugar cane had been obliterated. He also noted that wheat storage facilities had been inundated and ruined, and that undrained floodwater would preclude many fields from winter cultivation.

13.8 Health Crisis

As already mentioned, lack of food and scarcity of dry places to shelter afflicted people across Sindh in particular; these realities underlay an immense health crisis. According to a survey conducted by the International Rescue Committee (IRC), very few people reported having access to basic services. 'On average, only 17% population have access to health services, 14% to food and livelihood, 11% to mobile networks while WASH[1] services, education, shelter, information services, legal assistance, hygiene supplies, women and child protection services' were all in even shorter supply (IRC, 2022). As noted above, people even demonstrated on highways to demand help for their hungry children.

The breakdown of sanitation and water supply across the country meant that hygiene and sanitation were inadequate in shelters where vast numbers of people were crowded together. The ubiquitous stagnant water provided a breeding ground for pathogens and vectors and, along with the crowding, led to an increased risk of disease incubating and spreading. Across Sindh and Balochistan, there were outbreaks of waterborne diseases such as diarrhoea and cholera, as well as skin and eye infections, malaria and fever (IRC, 2022). Many clinics had been swept away, medicines were scarce, and medical help was rarely available.

13.9 Conclusion

The Pakistan floods of 2022 were catastrophic for tens of millions of people. Nearly a third of Pakistan went underwater, and most of what had been there was destroyed (including dwellings, crops and farm animals). Factors such as neglected siltation and inadequate infrastructure played a role; people live and farm on floodplains and, due to the increasing population, even more people were living there than in the previous major flood of 2010; rich and powerful people released water upstream to protect their own properties. Pakistan is not a rich country and lacked most of the medicines, facilities and medical staff needed to address the resulting public health crisis.

Although there is some debate about the extent to which this unprecedented monsoon was affected by climate change, it is very likely that the intensity and longevity of the rain event played a significant part. As the climate continues to warm, the risk of such seasons will increase.

Note

[1] Water, sanitation and hygiene.

References

Allan, R.P. and Soden, B.J. (2008) Atmospheric warming and the amplification of precipitation extremes. *Science* 321, 1481–1484. DOI: 10.1126/science.1160787

Baloch, S.M. (2022) Pakistan floods death toll passes 1,000, say officials. *The Guardian*, 28 August 2022. Available at: https://www.theguardian.com/world/2022/aug/27/pakistan-floods-death-toll-nearing-1000-say-officials (accessed 18 July 2023).

Carrington, D. (2022) Revealed: how climate breakdown is supercharging toll of extreme weather. *The Guardian*, 5 August 2022. Available at: https://www.theguardian.com/environment/2022/aug/04/climate-breakdown-supercharging-extreme-weather (accessed 17 July 2023).

Dilawar, I., Chaudhary, A. and Schultz, K. (2022) How Pakistan's catastrophic floods spiraled into a nightmare. *Bloomberg*, 8 September 2022. Available at: https://www.bloomberg.com/news/features/2022-09-08/how-pakistan-s-catastrophic-floods-spiraled-into-a-nightmare (accessed 21 July 2023).

Hussain, A. (2022) Pakistan declares national emergency as flood toll nears 1,000. *Al Jazeera* 26 Aug 2022. Available at: https://www.aljazeera.com/news/2022/8/26/pakistan-floods (accessed 6 July 2023).

IPCC (2021) Summary for Policymakers. In: Climate Change 2021: The Physical Science Basis. Contribution of Working Group I to the Sixth Assessment Report of the Intergovernmental Panel on Climate Change. Cambridge University Press, Cambridge, UK, and New York. Available at: https://www.ipcc.ch/report/ar6/wg1/downloads/report/IPCC_AR6_WGI_SPM.pdf (accessed 29 January 2024). DOI: 10.1017/9781009157896.001

IRC (2022) *Early Needs Identification Flood Disaster Sindh, Pakistan*. International Rescue Committee. Available at: https://reliefweb.int/report/pakistan/early-needs-identification-flood-disaster-sindh-pakistan (accessed 21 July 2023).

Otto, F., Zacharia, M., Saeed, F., Siddiqi, A., Kamil, S. *et al.* (2023) Climate change increased extreme monsoon rainfall, flooding highly vulnerable communities in Pakistan. *Environmental Research Climate* 2, 025001. DOI: 10.1088/2752-5295/acbfd5

Philip, S., Kew, S., van Oldenborgh, G.J., Wenchang, Y., Vecchi, G. *et al.* (2022) Rapid attribution analysis of the extraordinary heat wave on the Pacific coast of the US and Canada in June 2021. *Earth System Dynamics* 13, 1689–1713. Available at: https://www.worldweatherattribution.org/wp-content/uploads/NW-US-extreme-heat-2021-scientific-report-WWA.pdf (accessed 17 July 2023). DOI: 10.5194/esd-13-1689-2022

Sands, L. (2022) Pakistan floods: One third of country is under water – minister. *BBC News*. Available at: https://www.bbc.com/news/world-europe-62712301 (accessed 13 July 2023).

Thompson, V. and Kennedy-Asser, A. (2022) India heatwave: why the region should prepare for even more extreme heat in the near future. *The Conversation*, 5 May 2022. Available at: https://theconversation.com/india-heatwave-why-the-region-should-prepare-for-even-more-extreme-heat-in-the-near-future-182452 (accessed 16 July 2023).

Thompson, V., Kennedy-Asser, A.T., Vosper, E., Lo, Y.T.E., Huntingford, C. *et al.* (2022) The 2021 western North America heat wave among the most extreme events ever recorded globally. *Science Advances* 8, 18. DOI: 10.1126/sciadv.abm6860

Ul Haque, I. (2022a) More on floods, aftermath and way forward. *The Express Tribune*, 8 September 2022. Available at: https://tribune.com.pk/story/2375396/more-on-floods-aftermath-and-way-forward (accessed 14 July 2023).

Ul Haque, I. (2022b) Mega flood-2022, effects and mitigation strategy. *The Express Tribune*, 15 September 2022. Available at: https://tribune.com.pk/story/2376678/mega-flood-2022-effects-and-mitigation-strategy (accessed 14 July 2023).

UN (2022) Pakistan floods: Six month wait for water to recede, warn relief agencies. *UN News*, 20 September 2022. Available at: https://news.un.org/en/story/2022/09/1127051 (accessed 13 July 2023).

Zachariah, M., Thanigachalam, A., AchutaRao, K., Saeed, F., Jha, R. *et al.* (2022) Climate change made devastating early heat in India and Pakistan 30 times more likely. *World Weather Attribution*. Available at: https://www.worldweatherattribution.org/wp-content/uploads/India_Pak-Heatwave-scientific-report.pdf (accessed 14 July 2023).

14 The Double-Whammy of Stoichiometric Imbalance: C–H–O and Minerals in Global Food Nutrition

Irakli Loladze

Bryan College of Health Sciences, Lincoln, Nebraska, USA; School of Mathematical and Statistical Sciences, Arizona State University, Tempe, Arizona, USA

Abstract

This chapter delves into the profound impact on global food nutrition caused by a growing elemental imbalance between carbon (C), hydrogen (H), oxygen (O) and essential minerals. The current yield-driven model of Big Agriculture and the escalating levels of atmospheric carbon dioxide both lead to a depletion of essential minerals in plants, constituting a double-whammy impact. This stoichiometric disparity is further accentuated by Big Food, which capitalizes on the overabundance of C–H–O compounds for profitability. The chapter also spotlights 'gap minerals' that are vital for humans but not necessarily for plants, and their inconsistent presence in crops contributes to widespread dietary deficiencies. To rectify the growing stoichiometric imbalance in our food supply, the chapter advocates adjusting monetary agricultural incentives to account for mineral density alongside yield.

14.1 Big Ag and Mineral Decline

Modern agriculture, fixated on crop yield, inadvertently fosters a nutritional deficit, of increasing but already large significance. The current economic incentives propel farmers to prioritize harvest quantity, emphasizing crops laden with carbon (C), hydrogen (H) and oxygen (O) – the elemental and sole constituents of carbohydrates and fats – but meagre in essential minerals and trace elements (collectively referred to as an ionome). The US Department of Agriculture's subsidies for grain production, particularly for corn, wheat and soybeans – cash crops that occupy over 80% of eligible crop areas in the US – further prop up the production of commodities rich in carbohydrates and fats, i.e. rich in C–H–O (CRS, 2019).

The issue is amplified in open-field farming, the main source of human caloric intake. Fertilizers based on nitrogen (N) are most used, while essential minerals such as calcium (Ca), magnesium (Mg), iron (Fe) and zinc (Zn) are usually ignored. These minerals, despite their importance for human health, are marginalized because they do not generally boost crop yields.

The Green Revolution further reinforced this trend and increased yields through innovations like dwarf grain varieties, but at the expense of mineral density (Fan *et al.*, 2008). Longitudinal studies show persistent and worrisome declines in the concentration of essential minerals in fruits, vegetables and grains over the past few decades (White and Broadley, 2005; Davis, 2009; Mayer *et al.*, 2021). The ensuing surge in the supply of 'empty' calories laden with C–H–O, but mineral poor, has deepened the elemental imbalance at the base of human nutrition – plants.

 Climate Change and Global Health: Primary, Secondary and Tertiary Effects, 2nd Edition (eds C.D. Butler and K. Higgs)
DOI: 10.1079/9781800620025.0014

14.2 Big Food Ag and Mineral Imbalance

Big Food has thrived on the back of Big Ag's oversupply of carbohydrates and fats. These purely C–H–O-based macronutrients are the cheapest raw calories available to Big Food. The industry systemically adds several forms of these compounds (e.g. sucrose, high-fructose corn syrup, partially and fully hydrogenated oils) to nearly all processed foods, with 'junk foods' being particularly rich in those. The longer shelf life, lower transportation costs, economies of scale in production and high-profit margins make them particularly attractive to Big Food. Thus, the market is inundated with calorie-dense but nutrient-poor processed foods that disproportionately and negatively affect low-income individuals because of their affordability and widespread availability (Drewnowski and Darmon, 2005; Atak *et al.*, 2023; Tan *et al.*, 2023)

While selective micronutrient fortification does occur, Big Food's undeniable contribution to human nutrition is an oversupply of 'empty' calories contributing to a dilution of essential mineral intake. Thus, the disparity between C-H-O and minerals in the food supply instigated by Big Ag is further widened by Big Food's profit-driven practices. Numerous studies have examined Big Food and its influence on human nutrition. This chapter, however, offers a stoichiometric lens, further detailed in the subsequent section. By emphasizing elemental balance, it provides a clear yet distilled perspective on this complex issue.

14.3 Carbon Dioxide Levels This Century and Stoichiometric Imbalance

By the turn of the century, the question of whether rising carbon dioxide (CO_2) levels could worsen mineral malnutrition in humans had remarkably not been raised in the literature. From the 1980s, experimental studies growing crops and wild plants under elevated CO_2 conditions identified alterations in the mineral composition of plant tissues, primarily in leaves (Porter and Grodzinski, 1984; Kuehny *et al.*, 1991; Fangmeier *et al.*, 1997). However, the small sample sizes rendered these findings inconclusive, leaving the impact of CO_2 on plant mineral content, especially in edible parts, uncertain (Gifford *et al.*, 2000). The absence of a clear pattern in the response of plant minerals to CO_2 made a conceptual interdisciplinary leap of tying such changes to human malnutrition challenging.

Two decades ago, Sterner and Elser (2002) introduced ecological stoichiometry[1], a framework rooted in the law of mass conservation, offering a fresh lens to ecological interactions. A hypothesis linking escalating CO_2 levels to the exacerbation of human malnutrition emerged within this framework (Loladze, 2002). This hypothesis accounted for the enhanced influx of C atoms into plants due to CO_2 stimulation, projected to create an elemental imbalance in plant tissues.

The imbalance was characterized by an increase in carbohydrate content and a decrease in mineral concentration compared with ambient CO_2 conditions. While plants, with their specialized structures of vacuoles and amyloplasts, are well-equipped to store surplus carbohydrates, this imbalance could still pose a significant threat to their consumers worldwide.

Incorporating the widespread issue of 'hidden hunger', or mineral undernutrition, particularly the deficiencies of Fe and Zn that already threatened over half the world's population, the hypothesis stressed its impact on human health. It asserted that the elemental imbalance induced by rising CO_2 levels would exacerbate mineral deficiencies, intensifying a then-growing global public health concern (Loladze, 2002).

While these conclusions were alarming, they were more broadly perceived as speculative due to unfamiliarity of plant physiologists with the then fledgling framework of ecological stoichiometry. The extremely limited extant data and the small sample sizes (typically three to eight replicates) employed in CO_2 experiments further seeded mistrust in the hypothesis. Consequently, the issue was neglected by funding agencies. A decade later, the situation pivoted when Nimesha Fernando generated a dataset of unprecedented size for her PhD research at the Australian Grains Free-Air Carbon dioxide Enrichment (AG-FACE) facility. With over 5000 new observations, on wheat minerals, Fernando achieved the highest statistical power for any FACE or enclosure CO_2 experiment, showing consistent and robust CO_2-induced mineral declines, including Fe and Zn (Fernando *et al.*, 2012a,b; Fernando *et al.*, 2014a,b). A high-profile study in *Nature* further brought the spotlight on the issue by generating smaller datasets on four other key crops and reusing Fernando's entire wheat dataset (Myers *et al.*, 2014). Regrettably, it failed to credit Fernando's groundbreaking work. Years later, an author correction was issued, partially rectifying the omission (Myers *et al.*, 2019).

Fernando's work was highlighted in a concurrent study that synthesized nearly all published open-field and enclosure experiments conducted across four continents. With over 15,000 observations, the study achieved a statistical power sufficient for uncovering a hidden, persistent, and systemic downshift in the plant ionome caused by rising CO_2 levels (Loladze, 2014). It showed that it was critically important to achieve a statistical power of at least 40%, and preferably over 80%, to detect meaningful changes in plant quality. It also showed that the type of experimental setup (e.g. very costly FACE or inexpensive enclosure) had only a marginal effect, with an average decline of minerals being 8% across crop and wild plant tissues, including edible parts. Coupled with a CO_2-induced increase in carbohydrate content, this decline mirrored the stoichiometric dichotomy observed in Big Ag and Big Food practices, exacerbating the elemental imbalance in human nutrition.

A gedanken[3] experiment in the study showed that declining crop mineral density can potentially lead to compensatory food consumption, contributing to human weight gain (Loladze, 2014). Later studies focusing on grains reinforced these broad mineral declines and provided some evidence for an increase in carbohydrates (Broberg *et al.*, 2017; Zhu *et al.*, 2018; Ben Mariem *et al.*, 2021).

Aside from minerals, other aspects of plant quality change under high CO_2. N levels, vital for agricultural yields and easily measurable, were often investigated in plants exposed to elevated CO_2. By the late 1990s, with ample data (in contrast to scarce data on minerals), it was clear that N and protein levels, which are closely correlated, decline under high CO_2 conditions in plant tissues, particularly in photosynthetic ones (Cotrufo *et al.*, 1998; Körner, 2000; Jablonski *et al.*, 2002). Taub *et al.* (2008) found that protein declines in edible parts of key crops, and Taub and Wang (2013) devised an innovative methodology to estimate their impact on human nutrition on a country-by-country basis, an approach later reused by Medek *et al.* (2016).

Vitamins, especially those composed of C, H, and O, display mixed responses, with some decreasing while others increase (Dong *et al.*, 2018; Zhu *et al.*, 2018). Carotenoids, including the vision- and memory-important xanthophylls, appear to decrease by ~15% under elevated CO_2 levels (Loladze *et al.*, 2019). Moreover, not all plant types react identically to increased CO_2. C_4[2] plants, for example, are less responsive, and leguminous plants might not experience substantial protein decline, offering some exceptions (Loladze, 2014; Myers *et al.*, 2014; Beach *et al.*, 2019).

Despite the complexity introduced by the thousands of chemical compounds in plants and the variability of responses among them, the stoichiometric framework provides a robust prediction: in this century, the cumulative concentration of C-H-O is very likely to increase while mineral and protein levels decrease in most field crops and wild plants. These global shifts in response to rising CO_2 levels underline the profound connection between our changing environment and its impact on world nutrition and health.

14.4 Gap Minerals

The array of essential chemical elements for plants does not fully encompass those needed by humans. While all elements necessary for plant health also benefit humans, the inverse is not universally true. Certain elements – namely sodium (Na), iodine (I), fluorine (F), lithium (Li), selenium (Se), chromium (Cr) and a few others – though non-essential for most plants, are vital for human health. Let us term these elements 'gap minerals' (Fig. 14.1).

Plants are inconsistent sources of minerals for humans. Although plant tissues partially reflect soil mineral availability (Sterner and Elser, 2002; Jones *et al.*, 2017) they offer no guaranteed minimum concentration for gap minerals such as I or F. This uncertainty has led to widespread dietary deficiencies and initiated extensive

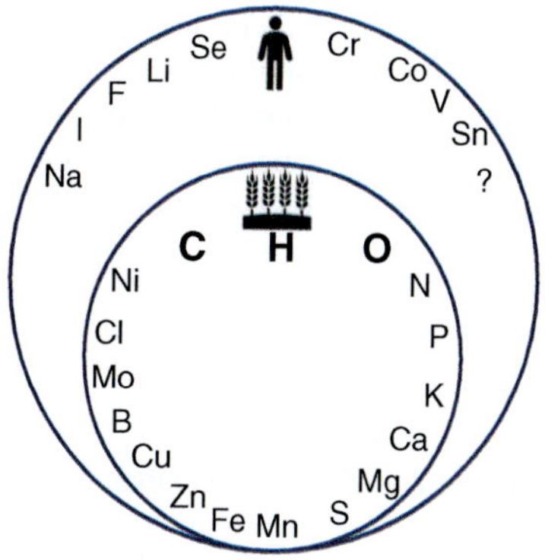

Fig. 14.1. Essential plant and human gap minerals. The inner circle represents 17 elements deemed essential for plant health, routinely included in commercial hydroponic solutions. The outer circle showcases the 'gap minerals' – elements vital for human health but not universally essential for all plants. Although these gap minerals may inadvertently be present in commercial fertilizers or hydroponic solutions as impurities or constituents of other compounds, they are not deliberately added as they do not enhance plant yields.

supplementation programmes for some but not all gap minerals, including iodized table salt, fluoridated water and toothpaste, and Li prescriptions for mental health.

These supplementation strategies have been largely effective. For example, iodized salt, introduced in the 1920s, almost eradicated iodine deficiency in countries with wide distribution. However, it persists among hundreds of millions where poverty or logistical factors have prevented its use, lowering the average IQ and economic potential of affected regions (Stein, 2009). Fluoridated water and toothpaste, launched in the 1950s, have significantly reduced tooth decay. Since the mid-20th century, Li prescriptions have revolutionized psychiatry, emerging as one of the most potent treatments for bipolar disorder.

These initiatives are confined to a few selected minerals, leaving human populations vulnerable and reliant on soil or water levels for these gap minerals. For instance, dietary deficiency of Se is common in regions with low soil Se, and areas with lower Li levels in tap water have higher suicide rates (Schrauzer, 2002; Jones *et al.*, 2017).

Moreover, even the minerals essential to plants may not be present in optimal concentrations for human nutrition. Examples such as Ca, Fe, Mg and Zn underscore this, with deficiencies in these elements posing a persistent global health concern (Stein, 2009).

14.5 Navigating the Double-Whammy Impact

The global food system is under the double-whammy impact of a carbohydrate surge, driven by both Big Ag and rising levels of CO_2. This twofold pressure exacerbates the elemental imbalance between plant and human needs, setting the stage for an even broader gap in this century between C–H–O and minerals. Consequently, the raw plant produce obtained by the Big Food industry starts off being less nutrient dense, a condition worsened by the industry's profit-driven overuse of C–H–O, which further reduces the mineral density of most processed foods.

Could we engineer a strategy to rectify this stoichiometric imbalance and enhance the mineral density of our crops? While grassroots movements, selective breeding and genetic engineering offer valuable solutions, an impactful change might well be orchestrated through a top-down approach (Ebi *et al.*, 2021). Altering financial incentives for farmers and producers, from large-scale operations to smallholders, could result in a healthier food supply. This might mean incentivizing higher mineral density in produce and imposing penalties for excessive heavy metal content.

The introduction of such financial rewards for stoichiometrically more balanced crops could spark a wave of innovation among farmers, fertilizer manufacturers and the soil management industry. Potential initiatives could encompass the use of more diverse fertilizers, incorporating 'gap' minerals into open fields, and nurturing healthier soil biota, such as mycorrhizae. Horticultural practices, including hydroponic technology, might also begin to see the advantages of introducing 'gap' minerals for the sake of human health, even if they do not directly boost yields.

This shift in focus from mere yield to nutritional quality and mineral density would usher in a more stoichiometrically balanced and healthier food supply. Such incentive-driven changes have the potential to send positive shockwaves through public health, possibly extending human lifespans and reducing the prevalence of obesity, diabetes and metabolic syndrome. While the fields of human nutrition, plant physiology, agriculture and changing environment are complex and separated by their disciplinary silos, the simple yet logical framework rooted in ecological stoichiometry allows for a holistic approach to the world's changing nutritional landscape.

Notes

[1] Ecological and biological stoichiometry considers how energy and the relative abundance of chemical elements influence organisms and their interactions among each other and the environment. It is rooted in the principles of mass conservation and mass balance at the atomic scale, acknowledging the inherent limitations of living systems to engage in nuclear reactions.

[2] Plants can be classified by variations in the biochemistry of their photosynthetic pathways; two main forms are called C_3 and C_4. Most edible crops are C_3 (the most ancient variety) but a few important C_4 crops also exist, including corn and sorghum.

[3] Gedanken or thought experiments are traditionally used in physics, where exactness of physical laws permits constructing logically tight arguments. While such experiments are not commonplace in life sciences, ecological stoichiometry provides a rigorous foundation to construct them.

References

Atak, S., Boye, A., Peciña, S. and Liu, Z.X. (2023) High-fat-sugar diet is associated with impaired hippocampus-dependent memory in humans. *Physiology & Behavior* 268, 114225. DOI: 10.1016/j.physbeh.2023.114225

Beach, R.H., Sulser, T.B., Crimmins, A., Cenacchi, N., Cole, J. *et al.* (2019) Combining the effects of increased atmospheric carbon dioxide on protein, iron, and zinc availability and projected climate change on global diets: a modelling study. *The Lancet Planetary Health* 3, e307–e317. DOI: 10.1016/S2542-5196(19)30094-4

Ben Mariem, S., Soba, D., Zhou, B., Loladze, I., Morales, F. *et al.* (2021) Climate change, crop yields, and grain quality of C3 cereals: A meta-analysis of [CO_2], temperature, and drought effects. *Plants* 10, 1052. DOI: 10.3390/plants10061052

Broberg, M., Högy, P. and Pleijel, H. (2017) CO_2-induced changes in wheat grain composition: meta-analysis and response functions. *Agronomy* 7, 32–32. DOI: 10.3390/agronomy7020032

Cotrufo, H., Ineson, P. and Scott, A. (1998) Elevated CO_2 reduces the nitrogen concentration of plant tissues. *Global Change Biology* 4, 43–54. DOI: 10.1046/j.1365-2486.1998.00101.x

CRS (Congressional Research Service) (2019) 2018 Farm Bill Primer: ARC and PLC Support Programs. IF11161. Available at: https://crsreports.congress.gov/product/pdf/IF/IF11161 (accessed 13 June 2023).

Davis, D.R. (2009) Declining fruit and vegetable nutrient composition: what is the evidence? *HortScience* 44, 15–19. DOI: 10.21273/HORTSCI.44.1.15

Dong, J., Gruda, N., Lam, S.K., Li, X. and Duan, Z. (2018) Effects of elevated CO_2 on nutritional quality of vegetables: a review. *Frontiers in Plant Science* 9, 924–924. DOI: 10.3389/fpls.2018.00924

Drewnowski, A. and Darmon, N. (2005) The economics of obesity: dietary energy density and energy cost. *The American Journal of Clinical Nutrition* 82, 265S–273S. DOI: 10.1093/ajcn/82.1.265S

Ebi, K.L., Anderson, C.L., Hess, J.J., Kim, S.H., Loladze, I. *et al.* (2021) Nutritional quality of crops in a high CO_2 world: an agenda for research and technology development. *Environmental Research Letters* 16, 064045. DOI: 10.1088/1748-9326/abfcfa

Fan, M.S., Zhao, F.J., Fairweather-Tait, S.J., Poulton, P.R., Dunham, S.J. *et al.* (2008) Evidence of decreasing mineral density in wheat grain over the last 160 years. *Journal of Trace Elements in Medicine and Biology* 22, 315–324. DOI: 10.1016/j.jtemb.2008.07.002

Fangmeier, A., Grüters, U., Högy, P., Vermehren, B. and Jäger, H.J. (1997) Effects of elevated CO_2, nitrogen supply and tropospheric ozone on spring wheat – II. Nutrients (N, P, K, S, Ca, Mg, Fe, Mn, Zn). *Environmental Pollution* 96, 43–59. DOI: 10.1016/S0269-7491(97)00013-4

Fernando, N., Panozzo, J., Tausz, M., Norton, R., Fitzgerald, G. *et al.* (2012a) Rising atmospheric CO_2 concentration affects mineral nutrient and protein concentration of wheat grain. *Food Chemistry* 133, 1307–1311. DOI: 10.1016/j.foodchem.2012.01.105

Fernando, N., Panozzo, J., Tausz, M., Norton, R.M., Fitzgerald, G.J. *et al.* (2012b) Wheat grain quality under increasing atmospheric CO_2 concentrations in a semi-arid cropping system. *Journal of Cereal Science* 56, 684–690. DOI: 10.1016/j.jcs.2012.07.010

Fernando, N., Panozzo, J., Tausz, M., Norton, R.M., Fitzgerald, G.J. *et al.* (2014a) Intra-specific variation of wheat grain quality in response to elevated [CO_2] at two sowing times under rain-fed and irrigation treatments. *Journal of Cereal Science* 59, 137–144. DOI: 10.1016/j.jcs.2013.12.002

Fernando, N., Panozzo, J., Tausz, M., Norton, R.M., Neumann, N. *et al.* (2014b) Elevated CO_2 alters grain quality of two bread wheat cultivars grown under different environmental conditions. *Agriculture, Ecosystems and Environment* 185, 24–33. DOI: 10.1016/j.agee.2013.11.023

Gifford, R.M., Barrett, D.J. and Lutze, J.L. (2000) The effects of elevated [CO_2] on the C:N and C:P mass ratios of plant tissues. *Plant and Soil* 224, 1–14. DOI: 10.1023/A:1004790612630

Jablonski, L.M., Wang, X. and Curtis, P.S. (2002) Plant reproduction under elevated CO_2 conditions: a meta-analysis of reports on 79 crop and wild species. *The New Phytologist* 156, 9–26. DOI: 10.1046/j.1469-8137.2002.00494.x

Jones, G.D., Droz, B., Greve, P., Gottschalk, P., Poffet, D. *et al.* (2017) Selenium deficiency risk predicted to increase under future climate change. *Proceedings of the National Academy of Science (USA)* 114, 2848–2853. DOI: 10.1073/pnas.1611576114

Körner, C. (2000) Biosphere responses to CO_2 enrichment. *Ecological Applications* 10, 1590–1619. DOI: 10.1890/1051-0761(2000)010[1590:BRTCE]2.0.CO;2

Kuehny, J.S., Peet, M.M., Nelson, P.V. and Willits, D.H. (1991) Nutrient dilution by starch in CO_2-enriched chrysanthemum. *Journal of Experimental Botany* 42, 711–716. DOI: 10.1093/jxb/42.6.711

Loladze, I. (2002) Rising atmospheric CO_2 and human nutrition: toward globally imbalanced plant stoichiometry? *Trends in Ecology and Evolution* 17, 457–461. DOI: 10.1016/S0169-5347(02)02587-9

Loladze, I. (2014) Hidden shift of the ionome of plants exposed to elevated CO_2 depletes minerals at the base of human nutrition. *eLife* 3, e02245. DOI: 10.7554/eLife.02245

Loladze, I., Nolan, J.M., Ziska, L.H. and Knobbe, A.R. (2019) Rising atmospheric CO_2 lowers concentrations of plant carotenoids essential to human health: a meta-analysis, *Molecular Nutrition & Food Research* 63, e1801047. DOI: 10.1002/mnfr.201801047

Mayer, A.-M.B., Trenchard, L. and Rayns, F. (2021) Historical changes in the mineral content of fruit and vegetables in the UK from 1940 to 2019: a concern for human nutrition and agriculture. *International Journal of Food Sciences and Nutrition* 73, 315–326. DOI: 10.1080/09637486.2021.1981831.

Medek, D.E., Schwartz, J. and Myers, S.S. (2016) Estimated effects of future atmospheric CO_2 concentrations on protein intake and the risk of protein deficiency by country and region. *Environmental Health Perspectives* 125, 1–8. DOI: 10.1289/EHP41

Myers, S.S., Zanobetti, A., Kloog, I., Huybers, P., Leakey, A.D.B. *et al.* (2014) Increasing CO_2 threatens human nutrition. *Nature* 510, 139–142. DOI: 10.1038/nature13179

Myers, S.S., Zanobetti, A., Kloog, I., Huybers, P., Leakey, A.D.B. *et al.* (2019) Author Correction: Increasing CO_2 threatens human nutrition. *Nature* 574, E14. DOI: 10.1038/s41586-019-1602-8

Porter, M.A. and Grodzinski, B. (1984) Acclimation to high CO_2 in bean: carbonic anhydrase and ribulose bisphosphate carboxylase. *Plant Physiology* 74, 413–416. DOI: 10.1104/pp.74.2.413

Schrauzer, G.N. (2002) Lithium: occurrence, dietary intakes, nutritional essentiality. *Journal of the American College of Nutrition* 21, 14–21. DOI: 10.1080/07315724.2002.10719188

Stein, A.J. (2009) Global impacts of human mineral malnutrition. *Plant and Soil* 335, 133–154. DOI: 10.1007/s11104-009-0228-2

Sterner, R.W. and Elser, J.J. (2002) *Ecological Stoichiometry: The Biology of Elements from Molecules to the Biosphere*. Princeton University Press, Princeton, New Jersey.

Tan, D., Drewnowski, A. and Lê, K.-A. (2023) New metrics of dietary carbohydrate quality. *Current Opinion in Clinical Nutrition and Metabolic Care* 26, 358–363. DOI: 10.1097/MCO.0000000000000933.

Taub, D.R., Miller, B. and Allen, H. (2008) Effects of elevated CO_2 on the protein concentration of food crops: a meta-analysis. *Global Change Biology* 14, 565–575. DOI: 10.1111/j.1365-2486.2007.01511.x

Taub, D.R. and Wang, X. (2013) Effects of carbon dioxide enrichment on plants. In: Pielke Sr., R. (ed.) *Climate Vulnerability*. Academic Press, San Diego, California. DOI: 10.1016/B978-0-12-384703-4.00404-4

White, P.J. and Broadley, M.R. (2005) Historical variation in the mineral composition of edible horticultural products. *The Journal of Horticultural Science and Biotechnology* 80, 660–667. DOI: 10.1080/14620316.2005.11511995

Zhu, C., Kobayashi, K., Loladze, I., Zhu, J., Jiang, Q. *et al.* (2018) Carbon dioxide (CO_2) levels this century will alter the protein, micronutrients, and vitamin content of rice grains with potential health consequences for the poorest rice-dependent countries. *Science Advances* 4, eaaq1012. DOI: 10.1126/sciadv.aaq1012

15 Temperature-related Rise in the Potential Malaria Burden in the Ethiopian Highlands. A Proposal for a Taxation Model to Address Climate Justice

Menno J. Bouma[1,2] and Colin D. Butler[3]
[1]*Independent scientist;* [2]*Instituto de Salud Global, Barcelona, Spain;*
[3]*National Centre for Epidemiology and Population Health, Australian National University, Canberra, Australia*

Abstract

Malaria is the world's most important vector-borne disease and is of particular importance in sub-Saharan Africa, with its large burden of morbidity and mortality. Advances in malaria control have re-awakened the old dream of its complete eradication. However, this possibility appears out of reach, leaving climate change of increasing importance, particularly in the densely populated African highlands. This chapter combines historical data for the burden of the disease of malaria in Ethiopia with a theory to develop a proposal to quantify the malaria burden attributable to climate change. The proposed malaria tax is based on the temperature-dependent changes in malaria risk and the sharing of annual costs related to malaria prevention and the residual malaria burden.

15.1 Malaria Elimination Efforts: an Expanding Disease Burden with Environmental Warming

The malaria parasite and its vectors are recognized to be temperature sensitive. Therefore, the disease should move uphill with global warming, similar to plants, animals and many other organisms (Parmesan and Yohe, 2003). This has indeed been substantiated, for both the parasite (Siraj *et al.*, 2014) and African vectors (Carlson *et al.*, 2023). However, renewed hope that the global scourge of malaria can be eliminated appeared to challenge the relevance of the long debate concerning the potential impact of climate change upon the world's most important vector-borne disease. Intensified international efforts to 'roll back malaria' in the late 1990s (Nabarro and Tayler, 1998) evolved into an international resolve to attempt global elimination (WHO, 2008). Eradication was earlier pursued between 1955 and 1969, but failed (Enayati and Hemingway, 2010). After World War II (WWII), the newly established World Health Organization (WHO) coordinated this endeavour with promising new tools for vector control (the insecticide DDT) and treatment (chloroquine). This attempt to include the African continent was abandoned after recognition that transmission there was too intense for the available means and resources (Nájera *et al.*, 2011).

Contrary to the time-limited design of the first eradication 'blitzkrieg', the renewed efforts accept a slower approach to 'shrink' the malaria

 Climate Change and Global Health: Primary, Secondary and Tertiary Effects, 2nd Edition (eds C.D. Butler and K. Higgs)
DOI: 10.1079/9781800620025.0015

map (Feachem *et al.*, 2010), with a strong focus on sub-Saharan Africa and its exceptionally high disease burden. After an initial striking decline in morbidity and mortality, progress has stalled since 2015 (Noor and Alonso, 2022); malaria has made a striking return in countries where control was temporarily interrupted (Oxborough, 2016). This setback illustrates the fragility of the modern eradication method and its achievements.

The current approach to protecting the African highlands relies heavily on insecticides, applied to bed-nets and sprayed intra-domestically to leave residual deposits on walls. Its success cannot be taken for granted in view of developing insecticide resistance, availability and cost of alternative vector-control methods and the need for uninterrupted financial resources to maintain coverage.

Remaining vigilant about the rising malaria risk at higher altitudes in Africa, with the highest population densities, remains essential, as we demonstrate in this chapter. An associated rise in costs to mitigate the disease burden and to prevent epidemic outbreaks appears inescapable. The requirement to continue preventive intervention is important, not only for the epidemic-prone regions at higher altitudes, but also particularly in areas with traditionally low endemicity. These areas are now more vulnerable, as populations haves lost protective immunity as a result of increased control and the pursuit of elimination over the past 15 years.

The fallacy of the zero-sum argument

Malaria transmission ends where temperatures are too low or too high for the parasite–vector combination to survive, thrive and live long enough to become infectious. Current malaria transmission models for the African tropics use cut-off values for altitude (and temperature) that assume an abrupt loss of transmission potential. This approach fails to do justice to one of the most striking of malaria's epidemiological features; its *gradual* decline with increasing altitude. This phenomenon is observed on all malarial continents (Bouma *et al.*, 1996; Siraj *et al.*, 2014). This relationship has remained robust over time; observed in antiquity (Gill, 1923), in the 19th century (Hirsch, 1883), in the African highland countries studied before the first eradication attempt (Cox *et al.*, 1999) and in recent surveys used to guide the second elimination drive (Bouma *et al.*, 2011).

In this chapter we use Ethiopia, with its densely populated highlands[1] and sparsely populated dry-lands, as an example to show how the known relation between altitude and temperature can be used to calculate the potential malaria burden when ambient temperatures rise, taking into account the altitudinal population distribution. This simplified malaria model for higher altitudes is valid, because many contextual determinants of malaria at lower altitudes lose their importance. Even rainfall, usually a critical component of transmission, does not require to be taken into explicit account, because rainfall increases with altitude in most highlands.

One of the arguments used to discount the role of rising temperatures for the global malaria disease burden is that, since very high temperatures inhibit malaria transmission (reducing suitability for vector and parasite), global heating will geographically simply shift the malaria belt. Thus, winners and losers are implied to balance (Lafferty, 2009; see Eikenberry and Gumel, 2018, for an overview of this debate). Aggregated costs and gains, in human and in financial terms, would be zero too, were this trade-off valid. In that case the net-zero result could exonerate international actors associated with anthropogenic warming from responsibility. However, when population densities are considered, the argument is false (Pascual and Bouma, 2009). The additional burden of disease from the expansion of malaria to highly populated highland areas will greatly exceed that of any reduction in transmission in the hot and drier zone of malaria's distribution, where human habitation is sparse. Furthermore, the hypothesis that malaria transmission is either extinguished or greatly curtailed by additional heat caused by global warming is not yet supported by compelling epidemiological evidence. This is in contrast to time-series studies – based on epidemiological data – that support the expansion at the cool end of the spectrum (Loevinsohn 1994; Tulu, 1996; Bouma, 2003; Zhou *et al.*, 2004; Pascual *et al.*, 2006) and in the spatial domain (Siraj *et al.*, 2014).

Malaria's presence in all human habitats, apart from the colder climes, is enabled by a multitude of vector species, each adjusted to a specific ecological niche. In very hot areas, if water is present, malaria parasites and vectors can thrive in seasonal and local niches, avoiding unfavourable high temperatures. Endemic oasis malaria was well-documented in the first part of the 20th century, with resurgences more recently (Deida *et al.*, 2019). Environmental warming may indeed shift the geographical range of an *Anopheles* species in arid regions if its adaptation fails (Fouet *et al.*, 2012). Yet, in that case, another vector can take its place. In Africa, the more heat-tolerant *An. arabiensis* (Lyons *et al.*, 2013) may in future replace *An. gambiae sensu stricto* (*s.s.*) where the latter now dominates. Indeed, the optimum temperature for the development of *An. arabiensis*, from egg to adult, is unchanged at 32°C. This contrasts with laboratory observations of *An. gambiae* which found an optimum at 25°C and a sharp decline in survival over 28°C (Mordecai *et al.*, 2013). *An. stephensi*, an arid-loving vector that evolved in South Asia, is also more heat tolerant than *An. gambiae* (Murdock *et al.*, 2016) and well-adapted to urban conditions. This species has recently colonized East Africa, including Ethiopia (Sinka *et al.*, 2020). It thus may add to the malaria burden in the nation's hottest lowlands. In arid regions of Africa, including in Ethiopia, transmission may shift from months that have become too hot to seasonally cooler months without lowering the annual malaria burden. In very hot and dry years, nomads and settled populations may shift to wetter and cooler places with improved food security, but with greater malaria risk.

The focus on the possible beneficial effects for malaria transmission as a result of harmful effects from heat on the malaria parasite, vector and thus transmission has persisted over the past decade. New laboratory findings (Shapiro *et al.*, 2017) suggest that the survival of infected mosquitoes (and their ability to transmit) is compromised by higher temperatures. The 26°C optimum temperature suggested for the parasite–vector combination indeed seems a more appropriate choice than the 30°C used by earlier modellers (Craig *et al.*, 1999). These findings regarding the optimum temperature refine an earlier estimate of 25°C (Mordecai *et al.*, 2013). Perhaps this lower optimum should not come as a surprise, as the parasite and vector share the same niche as the African equatorial human populations, with whom malaria has long co-evolved. The mean annual temperature of 45 meteorological stations in Central and Western equatorial Africa (between 6° N and 6° S latitude) is 24.8°C, which corresponds to 26.5°C at sea level.

The lowering of the estimated optimum temperature for malaria transmission shifted attention for the potentially beneficial aspects of warming away from the arid fringes of its distribution (where the argument of the asymmetric population distribution appears to have been appreciated) to Africa's endemic malaria regions. The suitability of transmission and the number of months suitable for transmission does indeed decline considerably with environmental warming, including if population densities are factored in by the use of the parameter 'person-months' of exposure. However, this decline in malaria suitability (Mordecai *et al.*, 2013; Ryan *et al.*, 2015) has no bearing on the disease burden. This is a crucial aspect, which is rather surprisingly misunderstood. Where transmission is intense, as in most parts of sub-Saharan Africa, a 90% reduction in the number of infectious bites a person receives during one year (say from 200 to 20) does not meaningfully reduce the prevalence (the intensity) of the disease (Beier *et al.*, 1999; Smith *et al.*, 2004). Although 'person-month' exposure accurately reflects the disease burden at the cold end of the spectrum (Craig *et al.*, 1999; Tanser *et al.*, 2003), this correspondence is absent in areas with higher levels of transmission suitability. Craig *et al.* (1999), using the Mapping Malaria Risk in Africa (MARA) model, found that maximum suitability for annual transmission corresponded with observed levels of suitability during five (consecutive) months in sub-Saharan Africa. If malaria transmission is possible for more than five months, the annual malaria burden does not actually increase.

Re-defining endemic malaria with 10–12 suitable transmission months, and seasonal malaria with 7–9 months (Ryan *et al.*, 2020) is incompatible with the observed epidemiology of malaria in Africa. Their projections that warming will lead to the rapid decline of malaria in the African lowlands over the 21st century is misleading.

In our analysis, we assess the potential 'unmitigated' burden using pre-1960s survey data in an era without significant intervention. This provides a baseline that can be used to assess the additional economic cost to control malaria in Ethiopia, and the attributable fraction of the price (morbidity and mortality) its people pay – currently and in future – for global warming. As malaria expansion appears predominantly restricted to the cool part of the temperature spectrum, often with higher population density, the global 'malaria price' from human-driven (or anthropogenic) climate change is unbalanced; it is not a zero-sum equation. Below we show the possibilities for a quantitative assessment, at a country level, of the global warming-related malaria burden. This can be used to estimate the costs and contributions Ethiopia could seek to recover from the global North to mitigate its malaria burden. If malaria is in fact eliminated, then such costs would disappear.

15.2 Ethiopia's Malaria 'Lapse' Rate and Disease Burden, Taking into Account the Population's Distribution by Altitude

The physical 'lapse' rate

The density of atmospheric molecules (and temperature) is inversely related to altitude. The rate at which temperatures decline (lapse) with altitude, in the absence of water vapour, is 9.8°C per kilometre; with Earth's water-laden atmosphere this is generally around 6.5°C per kilometre (Colman and Soden, 2021). In any region, this rate will vary, depending on the season and physical features of the terrain that may cause inversions of cold and warm air. In studies relating malaria to altitude, it is particularly important to consider changes in lapse rate in relation to rainfall and malaria transmission season(s). For Fig. 15.1 we use the average of the monthly lapse rates for months that are relevant to transmission. These rates are calculated from linear regression using monthly reported temperature data from meteorological stations with a known altitude. In the figure, the lapse rate has been estimated by the equation:

$$\textit{Temperature} = -0.0059\,m + 30.3\left(R^2 = 0.80\right)$$

(temperature in °C and m in metres above sea level). In this region, for every 182 m uphill, mean temperatures drop by 1°C; conversely an increase of 1°C (caused by human-driven climate change) would shift the malaria gradient (lapse rate; see below) by 182 m (Fig. 15.1).

The malaria lapse rate

Known climate sensitivities of the parasite and the vector are clearly visible on old malaria maps of sub-Saharan Africa. The Sahara desert is malaria-free; it is too dry, too hot and – probably of most importance – too sparsely populated. Mountainous East Africa and the temperate southern tip of the continent are currently too cold for transmission. On smaller spatial scales there is usually a multitude of local 'contextual determinants' (Casman and Dowlatabadi, 2002), relevant to understanding malaria transmission. Randolph (2009) nicely described:

> ... the mercurial epidemiology of each vector borne disease is the system specific product of complex, commonly non-linear, interactions between many disparate environmental factors. These include not only climate but also other abiotic conditions (e.g. land cover) and the physical structure of the environment (e.g. water sources), and furthermore biotic factors such as host abundance and diversity. ... In addition, socioeconomic factors drive the human living conditions and behaviour that determine the exposure to that risk, and the nutritional status and concomitant immunity that determines resistance to infection.

Gething *et al.* (2010) postulated that the decrease of global malaria in the past 50 years, despite rising temperatures, demonstrates that climate factors have lost most of their importance. However, despite the multitude of potential determinants, the epidemiology of highland malaria can be strikingly simplified; it continues to show strong altitude dependence (Bouma *et al.*, 2011). The gradual change of observed malaria with altitude, even before the transmission cycle of malaria was understood, was succinctly described by Hirsch (1883): 'As regards the relation of altitude and configuration of the ground to endemic and epidemic occurrence of malaria, the general law may be expressed by the formula, that the extent and severity of malarial diseases diminish *in proportion* as we ascend above the sea level' [emphasis added].

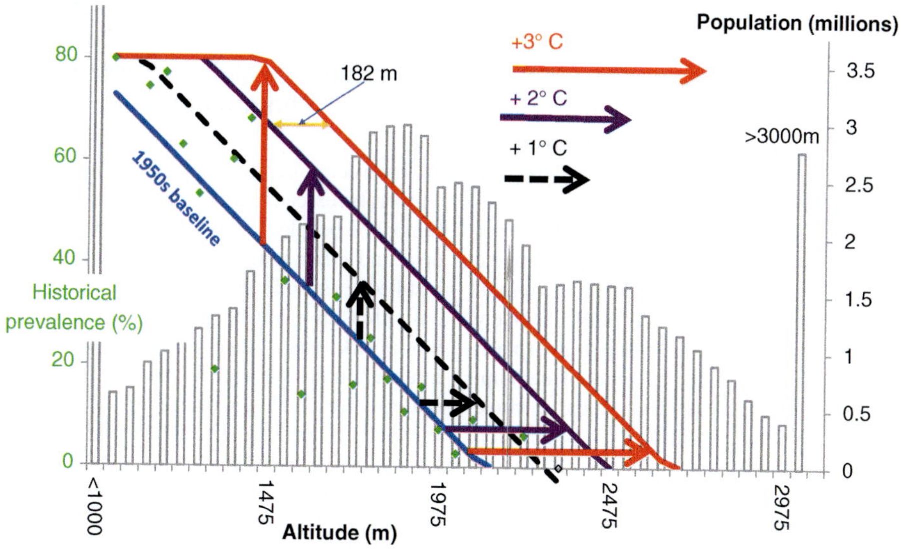

Fig. 15.1. The pre-control malaria lapse rate (prevalence with altitude) of the 1950s is shown by the blue band, with projected shifts with 1°C (black dashed line), 2°C (purple) and 3°C (red), based on a local temperature lapse rate of 182 m per °C. Columns represent the estimated rural population (2012 data) in 'bins', spaced at 50 m intervals, between 1000 and 3000 m. Arrows show increases in malarial territory (up to 546 m for 3°C of warming), and increases in the projected prevalence of malaria transmission with 1°C, 2°C and 3°C of warming. Green dots refer to spleen rates, a measure of prevalence using the average of historical observations for each 50 m bin. See text for more explanation of data sources and method.

Many studies have, in the past decades, shown the significance of (proportional) malaria decline with altitude in Africa (e.g. Drakeley *et al.*, 2005; Ernst *et al.*, 2006). Cox *et al.* (1999) recovered older observations on malaria prevalence (spleen and parasite rates) from the grey literature, including survey reports between WWII and the (original) eradication era from African highland countries. Survey locations were geo-referenced and compared with predicted values derived from the MARA model. This model estimates malaria suitability using combined monthly temperature and rainfall data (Craig *et al.*, 1999). Strikingly, the highlands in all countries studied displayed malaria lapse rates.[2] Variations in altitude at which malaria disappears may vary with latitude and vector preferences for indoor (warmer) vs outdoor (cooler) resting. The indoor resting preference of *An. funestus*, where it is prevalent in significant numbers close to permanent breeding sites near highland lakes, appears particularly to focally elevate the altitudinal limits of transmission. Fig. 15.1 uses the 51 data points recovered by Cox *et al.* (1999) for Ethiopia. These data points relating spleen rates to altitude were calculated, using linear regression, to determine the equation:

$$\text{prevalence (as spleen rate)} = -0.00675\,\text{m} + 141.9\,m$$

(in metres above sea level) ($R^2 = 0.73$). This formula was then used to plot the blue baseline in the figure. The population numbers for 21 altitude bins between 1025 m and 2325 m are also shown. The baseline maps the 'natural intensity' of malaria in an era and region without significant intervention, and before recognized anthropogenic climate change.[3]

15.3 Temperature-dependent Changes in the Potential Endemic Malaria Burden of Ethiopia

A country's potential malaria burden depends on the potential intensity of transmission that varies with altitude and the altitudinal population distribution. In Fig. 15.1 population numbers (in each bin representing a 50 m altitudinal increment) rise from 1000 to 1900 m, mirroring the population density. Because the malaria burden is considerably lower in urban areas, an ecosystem less conducive to the breeding of most vectors, we excluded the urban population from the assessment. For altitudes over 1900 m, population numbers and available land decrease, but population densities remain unchanged (not shown). We conservatively assume that the (point) prevalence of the spleen surveys (in children below 15 years of age) approximates the annual incidence for the whole population. If so, the product of the population size and the corresponding prevalence at each altitude (band) approximates the number of malaria cases that altitude band experiences each year. As the population living below 1000 m in Ethiopia is small, as a proportion (c. 7 million of c. 95 million in 2012), the sum of cases in the altitude bands where malaria occurs above 1000 m constitutes the bulk of the annual malaria burden for the nation. The baseline burden is calculated from the regression line between survey (spleen) data, the malaria lapse rate in the 1950s and the altitudinal population distribution. Using the physical lapse rate (182 m per °C) we can calculate the shift of the malaria lapse rate accordingly, for a 1, 2 or 3 degrees increase in ambient temperatures. These temperature increases correspond with an uphill shift of 182, 364 and 546 m (Fig. 15.1). Note not only that malaria occurs at increasing altitudes, but, more importantly, that the rise in burden is most striking in the bands that are most populated.

15.4 Conclusion: A Proposal to Quantify the Climate Change-Attributable Malaria Burden and Sharing of Costs

We have demonstrated that rising environmental temperatures lead to an increased malaria risk for Ethiopia's people. It therefore follows that Ethiopia and other malarial highland countries are justified to seek financial compensation for the costs that they will bear as a result of this additional risk. These funds could be provided from internationally allocated funds to address 'climate justice'. Table 15.1 compares an estimate of the malaria burden in the 1950s with the potential burden under different warming scenarios. Changes from the baseline are calculated as a percentage for 1, 2 and 3°C temperature increases, thus establishing a potential climate change-attributable burden (CCAB) in the absence of malaria control. The actual observed national malaria burden can be expressed as the sum of the expenditure to prevent malaria and the human cost resulting from malaria cases and deaths that were not prevented, i.e. the residual malaria costs.

Costs incurred by spraying houses with residual insecticides, by community use of (insecticide-treated) bed-nets, by methods to reduce vector breeding, and (possibly) the successful use of vaccines that prevent infection (now being tested) fall into the first category. The cost of treatment, treatment facilities (e.g. hospitals) and losses borne by the individual, the family and the society belong to the second category. The costs incurred for (attempted) control can simply be summed for a given year. Costs in the second category, particularly non-medical, are

Table 15.1. Malaria cases in Ethiopia per annum (m = million) with 1, 2 and 3°C increase, based on 2012 population estimates and 2002 data for the altitudinal distribution of the rural population.

Degrees of warming	Annual cases	Additional cases (95% CI) regression coeff. malaria lapse rate	Percentage change cases with base line CCAB	Proposed malaria tax as % of total malaria costs: preventive + 'residual'	Population (%) malaria free
Baseline	12.3 m				35.3 %
1°C		5.6 m (4.9–6.1 m)	46%	31.3%	22.6 %
2°C		11.2 m (9.7–12.2 m)	91%	47.7%	18.3 %
3°C		16.2 m (14.4–17.5 m)	132%	56.9%	10.7 %

CCAB, climate change-attributable burden; CI, confidence interval. Proposed malaria tax calculated as additional cases/(annual cases 1950s) + additional cases).

more difficult to establish, such as those associated with lost income, lost opportunities and lost lives (to a family and to society). Clearly, the cost incurred by effective prevention should reduce the cost in the second category.

We propose that the CCAB can be used to establish a fair fraction of the total malaria costs suffered by a country and can be used as a taxation model to establish a realistic share borne by a malarial country. With an increase of temperatures around 1.5°C already observed in the East African highlands, the current malaria risk has risen by around 70% from the baseline. This corresponds, when costs are shared between Ethiopia and countries (mainly in the global North) responsible for environmental warming, to an annual tax of approximately 40% of the costs (summing prevention and residual malaria costs incurred during one year). Note that the proposed malaria tax and the CCAB are ratios, only dependent on the temperatures (in relation to the 1950s baseline) and population distribution by altitude. However, the absolute sum of the costs will rise as the rural population expands, as has been forecast for the remainder of the 21st century. The sum of these costs will change over time and will depend on many parameters, such as biological developments (drug and insecticide resistance), technological developments (research breakthroughs) and consideration regarding the value of human life and the morbidity-related costs.

As a percentage of the actual costs, the proposed malaria tax (percentage) does not depend on scientific breakthroughs that improve cost-effectiveness (e.g. effective vaccines). However, the sharing of actual costs between the international community responsible and the malarial highland countries will provide incentives for the first party to provide sustainable funding for research and control. To minimize the sum of preventive and disease costs, both parties have incentives to cooperate closely to improve cost-effectiveness of residual malaria interventions. 'Ultimately, putting numbers on consequences of climate change will open the door to legal challenges' (Ananthaswamy, 2010); we hope that the quantification here provided may help galvanize the provision of funds to enable future protection of highland populations in malarial countries.

Acknowledgements

This chapter is adapted from Bouma, M.J. and Pascual, M. (2014) Global Warming and Malaria in Tropical Highlands – An Estimation of Ethiopia's 'Unmitigated' Annual Malaria Burden. In: Butler, C.D. (ed.) *Climate Change and Global Health*. CABI, Wallingford, UK, pp. 65–76.

We also wish to thank Dr Jonathan St H. Cox for generating the altitudinal distribution of the population in Ethiopia.

Notes

[1] Ethiopia is sometime called the 'roof of Africa'. Most of its major cities (including the capital, Addis Ababa) are located at elevations of at least 2000 m. However, the malaria burden falls largely on the rural population. See also Fig. 15.1.

[2] A measurable decline of malaria with increasing altitude.

[3] Although quinine has been used to treat malaria since the 19th century, its availability in Ethiopia until and even during the 1950s was limited. As the text mentions, both chloroquine and DDT were important tools of the first campaign to eradicate malaria (1955–1969). Ethiopia had one of Africa's few eradication programmes at that time; it started in the early 1960s, soon after the great 1958 epidemic.

References

Ananthaswamy, A. (2010) Time to blame climate change for extreme weather? *New Scientist* August 24: 14. Available at: https://www.newscientist.com/article/mg20727754-200-time-to-blame-climate-change-for-extreme-weather/ (accessed 22 August, 2023).

Beier, J.C., Killeen, G.F. and Githure, J.I. (1999) Short report: entomologic inoculation rates and *Plasmodium falciparum* malaria prevalence in Africa. *American Journal of Tropical Medicine and Hygiene* 61, 109–113. DOI: 10.4269/ajtmh.1999.61.109

Bouma, M.J. (2003) Methodological problems and amendments to demonstrate effects of temperature on the epidemiology of malaria. A new perspective on the highland epidemics in Madagascar, 1972–1989. *Transactions of The Royal Society of Tropical Medicine and Hygiene* 97, 133–139. DOI: 10.1016/s0035-9203(03)90099-x

Bouma, M., Dye, C. and van der Kaay, H. (1996) Falciparum malaria and climate change in the northwest frontier province of Pakistan. *American Journal of Tropical Medicine and Hygiene* 55, 131–137. DOI: 10.4269/ajtmh.1996.55.131

Bouma, M.J., Baeza, A., terVeen, A. and Pascual, M. (2011) Global malaria maps and climate change: a focus on East African highlands. *Trends in Parasitology* 27, 421–422. DOI: 10.1016/j.pt.2011.07.003

Carlson, C.J., Bannon, E., Mendenhall, E., Newfield, T. and Bansal, S. (2023) Rapid range shifts in African *Anopheles* mosquitoes over the last century. *Biology Letters* 19, 20220365. DOI: 10.1098/rsbl.2022.0365

Casman, E.A. and Dowlatabadi, H. (eds) (2002) *The Contextual Determinants of Malaria*. Routledge, New York.

Colman, R. and Soden, B.J. (2021) Water vapor and lapse rate feedbacks in the climate system. *Reviews of Modern Physics* 93, 045002. DOI: 10.1103/RevModPhys.93.045002

Cox, J., Craig, M., Le Sueur, D. and Sharp, B. (1999) *Mapping malaria risk in the highlands of Africa*. MARA/HIMAL Technical Report, London School of Hygiene & Tropical Medicine, London. Available at: https://idl-bnc-idrc.dspacedirect.org/bitstream/handle/10625/31899/117291.pdf

Craig, M., Snow, R. and le Sueur, D. (1999) A climate based distribution model of malaria transmission in sub-Saharan Africa. *Parasitology Today* 15, 105–111. DOI: 10.1016/S0169-4758(99)01396-4

Deida, J., Tahar, R., Khalef, Y.O., Lekweiry, K.M., Hmeyade, A. *et al.* (2019) Oasis malaria, Northern Mauritania. *Emerging Infectious Diseases* 25, 273–280. DOI: 10.3201/eid2502.180732

Drakeley, C.J., Carneiro, I., Reyburn, H., Malima, R., Lusingu, J.P.A. *et al.* (2005) Altitude-dependent and -independent variations in *Plasmodium falciparum* prevalence in northeastern Tanzania. *The Journal of Infectious Diseases* 191, 1589–1598. DOI: 10.1086/429669

Eikenberry, S.E. and Gumel, A.B. (2018) Mathematical modeling of climate change and malaria transmission dynamics: a historical review. *Journal of Mathematical Biology* 77, 857–933. DOI: 10.1007/s00285-018-1229-7

Enayati, A. and Hemingway, J. (2010) Malaria management: Past, present, and future. *Annual Review of Entomology* 55, 569–591. DOI: 10.1146/annurev-ento-112408-085423

Ernst, K.C., Adoka, S.O., Kowuor, D.O., Wilson, M.L. and John, C.C. (2006) Malaria hotspot areas in a highland Kenya site are consistent in epidemic and non-epidemic years and are associated with ecological factors. *Malaria Journal* 5, 78. DOI: 10.1186/1475-2875-5-78

Feachem, R.G.A., Phillips, A.A., Hwang,J., Cotter, C.,Wielgosz, B. *et al.* (2010) Shrinking the malaria map: progress and prospects. *The Lancet* 376, 1566–1578. DOI: 10.1016/S0140-6736(10)61270-6

Fouet, C., Gray, E., Besansky, N.J. and Costatinin, C. (2012) Adaptation to aridity in the malaria mosquito *Anopheles gambiae*, chromosomal inversion polymorphism and body size influence resistance to desiccation. *PLoS ONE* 7, e34841. DOI: 10.1371/journal.pone.0034841

Gething, P.W., Smith, D.L., Patil, A.P., Tatem, A.J., Snow, R.W. *et al.* (2010) Climate change and the global malaria recession. *Nature* 465, 342–346. DOI: 10.1038/nature09098

Gill, C.A. (1923) The relation of malaria to altitude. *Indian Journal of Medical Research* 11, 511–542.

Hirsch, A. (1883) *Geographical and Historical Pathology. Vol.1. Acute Infective Diseases*. The New Sydenham Society, London.

Lafferty, K.D. (2009) The ecology of climate change and infectious diseases. *Ecology* 90, 888–900. DOI: 10.1890/08-0079.1

Loevinsohn, M. (1994) Climatic warming and increased malaria incidence in Rwanda. *The Lancet* 343, 714–718. DOI: 10.1016/S0140-6736(94)91586-5

Lyons, C.L., Coetzee, M. and Chown, S.L. (2013) Stable and fluctuating temperature effects on the development rate and survival of two malaria vectors, *Anopheles arabiensis* and *Anopheles funestus*. *Parasites and Vectors* 6, 104. DOI: 10.1186/1756-3305-6-104

Mordecai, E.A., Paaijmans, K.P., Johnson, L.R., Balzer, C., Ben-Horin, T. *et al.* (2013) Optimal temperature for malaria transmission is dramatically lower than previously predicted. *Ecology Letters* 16, 22–30. DOI: 10.1111/ele.12015

Murdock, C.C., Sternberg, E.D. and Thomas, M.B. (2016) Malaria transmission potential could be reduced with current and future climate change. *Scientific Reports* 6, Article no. 27771. DOI: 10.1038/srep27771

Nabarro, D.N. and Tayler, E.M. (1998) The 'roll back malaria' campaign. *Science* 280, 2067–2068. DOI: 10.1126/science.280.5372.206

Nájera, J.A., González-Silva, M. and Alonso, P.L. (2011) Some lessons for the future from the global malaria eradication programme (1955–1969). *PLoS Medicine* 8, e1000412. DOI: 10.1371/journal.pmed.1000412

Noor, A.M. and Alonso, P.L. (2022) The message on malaria is clear: progress has stalled. *The Lancet* 399, 1777. DOI: 10.1016/S0140-6736(22)00732-2

Oxborough, R.M. (2016) Trends in US President's Malaria Initiative-funded indoor residual spray coverage and insecticide choice in sub-Saharan Africa (2008–2015): urgent need for affordable, long-lasting insecticides. *Malaria Journal* 15, 146. DOI: 10.1186/s12936-016-1201-1

Parmesan, C. and Yohe, G. (2003) A globally coherent fingerprint of climate change impacts across natural systems. *Nature* 421, 37–42. DOI: 10.1038/nature01286

Pascual, M. and Bouma, M.J. (2009) Do rising temperatures matter? *Ecology* 90, 906–912. DOI: 10.1890/08-0730.1

Pascual, M., Ahumada, J., Chaves, L., Rodo, X. and Bouma, M. (2006) Malaria resurgence in the East African highlands: Temperature trends revisited. *Proceedings of the National Academy of Science (USA)* 103, 5829–5834. DOI: 10.1073/pnas.0508929103

Randolph, S.E. (2009) Perspectives on climate change impacts on infectious diseases. *Ecology* 90, 927–931. DOI: 10.1890/08-0506.1

Ryan, S.J., McNally, A., Johnson, L.R., Mordecai, E.A., Ben-Horin, T. *et al.* (2015) Mapping physiological suitability limits for malaria in Africa under climate change. *Vector-Borne and Zoonotic Diseases* 15, 718–725. DOI: 10.1089/vbz.2015.1822

Ryan, S.J., Lippi, C.A. and Zermoglio, F. (2020) Shifting transmission risk for malaria in Africa with climate change: a framework for planning and intervention. *Malaria Journal* 19, 170. DOI: 10.1186/s12936-020-03224-6

Shapiro, L.L.M, Whitehead, S.A. and Thomas, M.B. (2017) Quantifying the effects of temperature on mosquito and parasite traits that determine the transmission potential of human malaria. *PLOS Biology* 15, e2003489. DOI: 10.1371/journal.pbio.2003489

Sinka, M.E., Massey, N.C., Longbottom, J., Hemingway, J., Moyes, C.L. *et al.* (2020) A new malaria vector in Africa: Predicting the expansion range of *Anopheles stephensi* and identifying the urban populations at risk. *Proceedings of the National Academy of Science (USA)* 117, 24900–24908. DOI: 10.1073/pnas.2003976117

Siraj, A.S., Santos-Vega, M., Bouma, M.J., Yadeta, D., Carrascal, D.R. *et al.* (2014) Altitudinal changes in malaria incidence in highlands of Ethiopia and Colombia. *Science* 343, 1154–1158. DOI: 10.1126/science.1244325

Smith, T., Killeen, G., Lengeler, C. and Tanner, M. (2004) Relationship between the outcome of *Plasmodium falciparum* infection and the intensity of transmission in Africa. *American Journal of Tropical Medicine and Hygiene* 71, 80–86. DOI: 10.4269/ajtmh.2004.71.80

Tanser, F.C., Sharp, B. and le Sueur, D. (2003) Potential effect of climate change on malaria transmission in Africa. *The Lancet* 362, 1792–1798. DOI: 10.1016/S0140-6736(03)14898-2

Tulu, A.N. (1996) Determinants of Malaria Transmission in the Highlands of Ethiopia: The Impact of Global Warming on Morbidity and Mortality Ascribed to Malaria. PhD thesis, London School of Hygiene and Tropical Medicine, London.

WHO (2008) *Global Malaria Control and Elimination: Report of a Technical Review.* World Health Organization, Geneva. Available at: https://www.who.int/publications/i/item/9789241596756 (accessed 8 August, 2023).

Zhou, G., Minakawa, N., Githeko, A.K. and Yan, G. (2004) Association between climate variability and malaria epidemics in the East African highlands. *Proceedings of the National Academy of Science (USA)* 101, 2375–2380. DOI: 10.1073/pnas.0308714100

16 Arboviruses, Vectors, Poverty and Climate Change

Colin D. Butler,[1] Cyril Caminade[2,3] and Andrew P. Morse[4]
[1]National Centre for Epidemiology and Population Health, Australian National University, Canberra, Australia; [2]Abdus Salam International Centre for Theoretical Physics, Trieste, Italy; [3]Department of Livestock and One Health, Institute of Infection, Veterinary and Ecological Sciences, University of Liverpool, UK; [4]School of Environmental Sciences, University of Liverpool, UK

Abstract
This chapter focuses on the history and public health impact of the world's five most significant arthropod-borne viruses (arboviruses) which cause dengue, chikungunya, yellow fever, Zika and Japanese encephalitis. These diseases occur in some high-income settings, but are only highly problematic in the global South. They thus have a bi-directional relationship to poverty (arboviruses can worsen poverty; poverty makes arboviruses harder to control). Also discussed are factors that link the distribution and severity of arboviral infections to environmental change, especially to the climate. Of these five arboviruses, two have proven vaccines, yet logistic factors associated with rapidly scaling up the supply and distribution for one (yellow fever), in times and places of need, means that it remains a major threat. Prospects towards and obstacles slowing the development of an efficient dengue fever vaccine are described. Also discussed are potentially important changes in arbovirus distribution, latitude, altitude and reservoir species.

16.1 Introduction and History

Arboviruses (**ar**thropod-**bo**rne viruses) cause diseases in humans and animals. Almost all are ribonucleic acid (RNA) based. Some arboviruses have had a significant influence on human affairs, and have caused disease and epidemics for centuries. Of all known arboviruses, the deadliest causes yellow fever, with a case fatality rate of 20–50% (Wilder-Smith and Monath, 2017). Brought (with its arthropod vector *Aedes aegypti*) from West Africa to the 'New World' via the slave trade in the 16th century, yellow fever soon established itself in South and Central America, as well as in ports of the USA and southern Europe (Brès, 1986; Clements and Harbach, 2017; Powell *et al.*, 2018). Pioneering work in Cuba established both its transmission by *Aedes* mosquitoes (c. 1880) and its ultra-microscopic (viral) nature (1898). Thus, yellow fever became the second infectious agent shown to include mosquitoes in its life cycle (by Finlay, following Manson's work on filariasis, as discussed by Nye and Gibson, 1997) and the first viral disease infecting humans (Oldstone, 2014).[1]

Yellow fever

The rigorous application of sanitary measures (including quarantine of infected patients in rooms screened from mosquitoes) and 'mosquito brigades' quickly enabled the eradication of yellow fever from Havana, Cuba, in 1901. At that time,

 Climate Change and Global Health: Primary, Secondary and Tertiary Effects, 2nd Edition (eds C.D. Butler and K. Higgs)
DOI: 10.1079/9781800620025.0016

this was an enormous public health success (Clements and Harbach, 2017) and soon facilitated the construction of the Panama Canal, a mammoth engineering feat thwarted until then by yellow fever (Brès, 1986).[2]

Building on this progress, the Rockefeller Foundation commenced a determined effort to completely eradicate yellow fever in South America in 1918, almost three decades before the widespread availability of the insecticide dichlorodiphenyltrichloroethane (DDT), a tool which only became easily available at the close of World War II (WWII) (Carter, 2008). Eradication was then thought feasible because the virus was believed to exist only in humans (i.e. not in other animals, including non-human primates such as monkeys), and to be transmitted only by *Aedes* mosquitoes. Sanitary measures, alone, led to its disappearance in many cities and much of Brazil. However, in May 1928, cases that appeared inexplicable occurred in Rio de Janeiro, Brazil (Soper, 1938). This epidemiological puzzle was solved by the finding that yellow fever has a forest (sylvatic) transmission, involving non-human primates, with vector transmission by the genus *Haemagogus* rather than *Aedes*.

Although this discovery ended the dream of its eradication, the development of an extremely efficient vaccine in the late 1930s (one dose probably confers lifelong immunity),[3] combined with ongoing vector-control campaigns, enabled impressive control of yellow fever for decades, in urban South America and elsewhere, wherever vaccination was enforced (Clements and Harbach, 2017). It was then long considered that such an effective vaccine should mean that yellow fever was no longer a priority arbovirus, even in the presence of worsening climate change. However, this optimism has proved to be premature, due to coalescing factors, particularly insufficient vaccine supply to the most vulnerable populations (Barrett, 2016). Underlying this is the persistent failure of many countries, including in sub-Saharan Africa, to 'develop' as once was promised by supporters of globalization and neoliberalism (de Rivero, 1999; George, 1999).

Control of the 2016 yellow fever outbreak in densely populated Luanda, the capital city of Angola, was complicated by insufficient provision of vaccines and difficulties administering them (Grobbelaar *et al.*, 2016; Wilder-Smith and Monath, 2017). A deep underlying factor for the re-emergence of yellow fever as a threat is the decline in the perceived value and prestige of public health measures in some countries. Complacency, funding fatigue and thus insufficient prestige of and funding for public health is another co-factor (Hemenway, 2010).

Curiously, yellow fever has never been shown to have been successfully transmitted in Asia (Kuno, 2020). Were yellow fever transmission in Asia to evolve, possibly as the inadvertent consequence of human actions,[4] then vaccine production would require enormous expansion, as well as other strategies to use existing supplies more effectively (Barrett, 2016).

Dengue fever

During WWII, in the India–Pacific conflict zone, control of another arboviral disease, dengue fever, became a research priority for both the Japanese and the Americans, each of which isolated a strain of the virus, in 1943 and 1945, respectively (Kuno, 2007; Messina *et al.*, 2014). Troop movements in this war facilitated its spread, including of different strains, which in turn led to the emergence of dengue haemorrhagic fever (Wilder-Smith and Gubler, 2008). Although dengue has a far lower mortality rate than yellow fever, the emergence (in the early 1950s) of its haemorrhagic form, with shock and occasional mortality, mainly in children, heightened its public health importance (Cohen and Halstead, 1966). This syndrome is now known to generally be triggered by reinfection with a second dengue viral (DENV) strain, i.e. any two of the four that are currently known (Brady and Hay, 2020).

In addition, multiple dengue strains have slowed development of an effective vaccine. Even though a prototype vaccine, Dengvaxia, has been developed and deployed (Thomas and Yoon, 2019), dengue still remains highly problematic.

Japanese encephalitis, chikungunya, Zika and others

Japanese encephalitis (JE), transmitted by some mosquitoes of the *Culex* genus, is another important arbovirus, today causing more disability-adjusted life years lost than any other arbovirus, owing to its frequent neurological sequelae (Turtle and Solomon, 2018). Despite the availability of many effective vaccines since the 1980s, an estimated 100,000 clinical cases occurred in 2015, in 24 countries of the Asia-Pacific region, mainly in South-East Asia (Quan *et al.*, 2020). Humans are 'dead end' hosts for JE; that is, they do not form a 'reservoir' from which an outbreak could spread. Instead, natural reservoirs for JE virus are water birds of the family Ardeidae (herons and egrets). This makes eradication of JE, in many settings, extremely unlikely.

More recently, globally significant outbreaks of chikungunya and Zika virus have also drawn attention to the ongoing challenges of arboviruses. Over 500 additional arboviruses have also been identified (Centers for Disease Control, 2021); one or more of these may also become a globally significant cause of morbidity (Table 16.1). In this chapter we focus on the most important known arboviruses, their vectors, and their relationship with poverty, governance and environmental change.

Table 16.1. Some key facts about some of the most important and interesting arboviruses.

Arbovirus	Genus (virus)	Vector (genus)	Notes
Bluetongue	Orbivirus	*Culicoides* Midges	Affects ruminants, not a zoonosis, but of economic and veterinary importance
Chikungunya[a]	Alphavirus	*Aedes*	*Ae. albopictus* is a very important vector
Crimean–Congo haemorrhagic fever	Orthonairo-virus	Ticks, humans	Wide distribution outside the Americas, South and East Asia; increased reports in Western Europe since 2010; sexual transmission reported
Dengue[a]	Flavivirus	*Aedes*	Wide tropical distribution
Japanese encephalitis[a]	Flavivirus	*Culex*, other genera	Humans are a 'dead end' host. Pigs and wild waterbirds serve as a reservoir for the virus
O'nyong'nyong	Alphavirus	*Anopheles*, *Mansonia*	Ugandan epidemic (1959–1962) affected about 2 million people in East Africa; no known transmission by *Aedes* probably limits expansion potential
Rift Valley fever	*Phlebovirus*	*Aedes*, *Culex*, direct contact	Important wild and livestock animal disease, also a zoonosis that can spread to humans in contact with viraemic animal material (butchers, slaughterhouse personnel and veterinarians at risk)
Ross River fever	Alphavirus	*Aedes*, *Culex*, up to 40 species	Proven expansion to Pacific islands, with endemic transmission despite absence of marsupial reservoirs (Flies *et al.*, 2018)
Sindbis virus	Alphavirus	*Culex*	Wide distribution (not in the Americas)
St Louis encephalitis	Flavivirus	*Culex*	Only in the Americas (to date)
Usutu virus	Flavivirus	*Culex*	Wide African distribution, recent introduction to Europe (including London). Humans dead-end host, avian reservoir
Venezuelan equine encephalitis	Alphavirus	Several genera	Only in the Americas (to date)
West Nile virus[a]	Flavivirus	*Culex*	Leading cause of viral encephalitis, worldwide, zoonosis
Yellow fever[a]	Flavivirus	*Aedes*, *Haemogogus*	Excellent 100% efficient vaccine, virus not found in Asia
Zika virus[a]	Flavivirus	*Aedes*, *Anopheles*, other	Vector-borne and sexual transmission occurs

[a]Arboviruses discussed in the text.

16.2 Arthropods and Their Human Infections

Arthropods are invertebrate animals, with jointed, external skeletons. They form the largest phylum in the animal kingdom. While some arthropods bite and sting (e.g. wasps) only a tiny percentage act as vectors, transmitting pathogens of concern to humans and other vertebrates. In addition to viruses, vector-borne pathogens include bacteria, sometimes called arbobacteria (Arbeitskreis Blut, 2009). Examples include *Yersinia pestis*, *Coxiella burneti*, protozoa (e.g. malaria, trypanosomiasis, babesiosis), helminths (e.g. *Dirofilaria immitis*) and filarial nematodes (e.g. *Wuchereria bancrofti*). In common, they all require a haematophagous (bloodsucking) arthropod (a mosquito, tick, flea,[5] fly or midge) for transmission between vertebrate hosts (Gubler, 1998).

Of these vectors, the mosquito genus *Aedes* is by far the most significant, particularly the species *Aedes aegypti*, the main vector capable of transmitting each of the four arboviruses of greatest contemporary concern: dengue, yellow fever, chikungunya and Zika. *Aedes* mosquitoes also transmit many other arboviruses (Table 16.1). However, to date, these remain restricted in range, or are under-surveyed and consequently under-detected (Weetman *et al.*, 2018).

Aedes, urbanization and evolution

Today, as for the last century, most arboviral transmission occurs in urban and peri-urban areas in tropical and subtropical countries, the preferred habitat of *Aedes aegypti* and *Aedes albopictus* species.[6] *Aedes* is a highly efficient vector and is well adapted for dispersal by humans. Its desiccation-resistant eggs are usually laid in containers above the water line (not in the water, like *Culex* or *Anopheles* mosquitoes). If rainfall is favourable, the eggs will be submerged by the rising water level in the container and the larvae (wrigglers) hatch to complete their development. If rainfall is unfavourable and the eggs stay dry, they can remain dormant for many months until submerged subsequently. *Aedes* mosquitoes demonstrate staggered hatching: a proportion of the eggs only hatch after a second (or even third) immersion. This guarantees the survival of at least some individuals in ephemeral conditions, such as those experienced in coconut shells on traditional Malay boats, where light rain might inundate the eggs without providing sufficient water for them to complete their development before it evaporates (Weinstein and Ravi, 2008).

The *Aedes* genus is highly susceptible to many arboviruses, feeds preferentially on humans, is active during the daytime at dawn and dusk, thrives in peri-domestic environments close to people, and often bites several people in a short period. It is (now) resistant to many insecticides (including DDT), and its daytime activity means that it evades exposure to insecticide-treated bed-nets designed for night-time biters such as *Anopheles* species. Alternative forms of insecticide exposure, such as through clothes and curtains, have so far proven unsuccessful (Brady and Hay, 2020). Overcrowding facilitates transmission via urban *Aedes* mosquitoes, further compounded by increased human-made larval habitats for mosquitoes, where piped water is unavailable (as is common in poor settings) and where people often use containers to store water, and where mosquito screens are rare.

In the increasingly fragmented Amazon forest, the zone of dengue transmission is extending westwards, with *Aedes*-driven outbreaks more likely, and of greater extent, during El Niño-associated droughts, where people rely more on stored water (Lowe *et al.*, 2020). Human-made water containers are also an important proximal cause for the spread of dengue and Zika in Brazilian favelas (Snyder *et al.*, 2017). As with yellow fever, the concerted efforts to control vector populations, once comparatively successful, have weakened, driven by institutional fatigue, insecticide resistance, less funding available for public health services, and urban poverty.

Urbanization in tropical, low-income settings, driven by human population growth and the pursuit of the economic opportunity that cities offer, is a deep underlying cause for this expansion (Braack *et al.*, 2018). Estimates of the population potentially exposed to arboviruses already exceed four billion, but they are based on geographical models that tend to overestimate the risk. Gwee *et al.* (2021) estimated 3.97 billion people at risk, just for dengue. The World Health Organization reported about 5.2 million cases in 2019 and estimated the annual dengue burden to range between 100 and 400 million cases (WHO, 2023). As populations across the developing world continue to urbanize, and as forests continue to be fragmented and 'developed', the health problem from arboviruses is likely to worsen. These issues are compounded by the risk that climate change will worsen poverty in many low-income settings, or at least impair governance capacity to combat arboviruses. At the same time, disabling diseases, including arboviruses, will potentially slow economic development in such settings (Short *et al.*, 2017).

Culex and *Anopheles*

Mosquitoes of the *Culex* genus are also important transmitters of arboviruses, including JE, which remains an important public health issue despite the existence of a vaccine. A number of other arboviruses can be transmitted by this genus, while others are borne by ticks (e.g. Crimean Congo haemorrhagic fever, tick-borne encephalitis) or midges (e.g. bluetongue, a disease mainly of ruminants). Bunyaviruses can be transmitted by mosquitoes and midges. Some arboviruses are important veterinary infections, and thus are of indirect relevance to human health (Maclachlan *et al.*, 2019). *Anopheles* mosquitoes, the only genus to transmit malaria, are comparatively poor vectors of arboviruses. However, o'nyong'nyong virus, an alphavirus close to chikungunya (Carissimo *et al.*, 2018) and JE, at least in India, are exceptions (Pearce *et al.*, 2018).

Climate, reservoirs, travel and migration

Several factors are required for successful arboviral transmission. Climatic conditions that are both appropriate for the biology and ecology of the vector and warm enough for virus replication (in the ectotherm[7] vector) are necessary – but insufficient. Within this eco-climatic 'envelope', biotic, socio-demographic, economic and immunological factors determine the possibility and intensity of arbovirus transmission. A viral source is required for transmission to occur. While most 'reservoirs' are in human or other vertebrate species (e.g. non-human primates, birds, or ruminants), vertical transmission (transmission of an arbovirus from an infected female mosquito to its offspring) is also a means by which arboviruses can persist in conditions that are otherwise unfavourable, such as during dry seasons in tropical regions, or to overwinter in temperate zones (Lequime and Lambrechts, 2014).

Some regions will have competent vectors but no transmission, because the area lacks a human or animal reservoir and/or vertical transmission. In such locations, viral importation via, for example, infected tourists or returning workers can initiate transmission, and even an outbreak. An example is the island of Madeira, an autonomous region of Portugal 1000 km west of Morocco, which experienced its first major dengue outbreak in 2012. Analysis showed that it was probably introduced by air travellers from Venezuela or Brazil (Wilder-Smith *et al.*, 2014). Human migration is not required if the virus is endemic.

Finally, there must be contact between infected vectors and a susceptible (non-immune) human population. These factors are driven and shaped by social, economic and environmental circumstances. Table 16.1 provides a summary of the most important arboviruses and their vectors, some of which are not otherwise discussed in this chapter.

16.3 Arboviruses, Poverty and Culture

Although disputed by some, we argue that poverty and localized inequality are particularly important non-climatic determinants of arbovirus transmission risk, enhancing population vulnerability in multiple ways. These include: limiting the capacity for public vector control programmes; the creation and promotion of urban microenvironments that are conducive to vector breeding, i.e. intended or unintended water collection containers, such as those created by litter or broken structures; overcrowding; and urban greening initiatives. Poverty and inequality can contribute to a high density of human hosts, inadequate health systems, unaffordable healthcare and poorer underlying population health.

Of course, arboviruses do not only infect poor people. Dengue is a recurrent problem in Cairns (northern Australia), Singapore (Sim *et al.*,

2020) and the southern fringes of the USA and Europe. However, the burden of disease from arboviruses in the global North has been greatly restricted, compared with that of Jakarta, Dhaka and Delhi, for example. Dengue is not endemic in Florida but it is in nearby Puerto Rico, an American dependency that lacks statehood and where public health has inexorably declined since the 1990s (Ramos *et al.*, 2022).

Nonetheless, the relationship between arboviral diseases and poverty is not straightforward. Bisanzio *et al.* (2021) found that West Nile virus was more common among socio-economically deprived parts of the US states bordering the Gulf of Mexico, while dengue, chikungunya and Zika were more commonly reported in nearby, better-off areas. They attributed this difference in part to the behaviour of the vector, pointing out that diseases transmitted by *Culex* species were more likely to thrive in contaminated water sources, sometimes from associated sewage overflow. They also suggested that the scarcity of health services in southern Alabama may have led to an under-reporting of these other viruses, while the lower mobility of the poorest fraction of the population may have reduced exposure to dengue, chikungunya and Zika diseases. These arboviral diseases have had little, if any, autochthonous (local) transmission reported in Alabama, although this has been reported repeatedly in the nearby US states of Florida and Texas.

A meta-analysis that included 36 studies, with over 106,000 study participants, examined the relationship between socio-economic status and infections caused by seven arboviruses (Grace *et al.*, 2022). These authors found a higher risk of arbovirus infection associated with markers of lower socio-economic position, including lower education, income poverty, low healthcare coverage, poor housing materials, interrupted water supply, non-white ethnicity and migration status. The authors were careful to point out that microscale study of the relationship between specific individual, household and community-level socio-economic position indicators and arbovirus infection risk is needed to help inform targeted public health interventions.

A systematic review of four arboviral diseases found a similar, mixed relationship between poverty and disease. The authors even challenged the view that arboviral illnesses are 'diseases of poverty' (Whiteman *et al.*, 2020). No doubt, in many parts of the global South, wealthy people are still vulnerable to arboviruses. However, to claim that arboviruses are not associated with poverty is unjustified. Two arboviral diseases that can be prevented by vaccines (yellow fever and JE) are extremely well controlled in high-income settings. Local transmission of dengue still occurs in Singapore, and parts of Australia, Europe and the USA (Gomes and Bahia, 2022), but these cases are rare and subject to intensive surveillance and control.

There is a risk that the poor will be ostracized and victimized, as sources of infection, including of arboviruses. Eduardo Paes, Rio de Janeiro's mayor from 2009 to 2016, blamed the collective tenement houses (*cortiços*) of central Rio for an outbreak of yellow fever, arguing that they were the 'source of the "poison" that caused the "black vomit", and as such, "dangerous classes" that must be eliminated' (Barbosa and Coates, 2021).

Culture and human behaviour can also play a role, such as the openness or resistance to biological controls for the vector, for example with predatory copepods[8] (Kay and Nam, 2005), or genetically engineered male mosquitoes (Qsim *et al.*, 2017). Household water storage practices, access to piped water, the number of households per property, condition of the house and patio, and knowledge and perceptions of dengue were found to be determinants of the dengue burden in urbanized parts of Ecuador (Ibarra *et al.*, 2013). Local belief systems may have helped to amplify the outbreak of chikungunya on La Réunion (a French 'overseas department' in the tropical Indian Ocean), as water in jars offering flowers in numerous shrines and cemeteries is likely to have facilitated local mosquito breeding. Such water vessels are numerous in some parts of Mauritius (Colwell-Chanthaphonh and de Salle-Essoo, 2014), an island in the tropical Indian ocean, 180 km to the north-east, and which is culturally similar.

Whole-of-community acceptance of the need to use personal sprays, as well as to reduce vectorial breeding sites, is also important to reduce arboviral infections (Weinstein and Ravi, 2008). An important element of Singapore's successful approach is the use of 'community engagement', and other techniques to encourage bottom-up participation of residents to check for mosquito breeding sites in and near homes (Sim *et al.*, 2020). Such approaches will be far less efficient in settings marked by inequality, resentment of elites and distrust of authorities.

16.4 Arboviruses and Climate Change

An influence of anthropogenic climate change in the transmission of vector-borne diseases, long hypothesized (Kovats *et al.*, 2001) and once widely questioned (Brisbois and Harris, 2010), is now generally accepted, not only because of temperature changes, but also via shifts in rainfall patterns and humidity (Robert *et al.*, 2020). Regardless of other factors, climate directly constrains vectorial capacity – the ability of a vector to transmit a virus. Within limits, a warmer, wetter climate increases this capacity, and thus the potential for arbovirus transmission, in several ways. These include by influencing the breeding cycle, survival and biting rate of the vector. The speed of arboviral development inside the warm-blooded animal (human or other vertebrate), the 'intrinsic incubation period', is not influenced by ambient temperature. However, crucially, warmer temperatures can accelerate the extrinsic incubation period (EIP). This is the time between the exposure of the cold-blooded mosquito (the vector) to the pathogen (e.g. a virus or protozoan such as malaria) and its replication and spread within the vector's body until it reaches the salivary glands, where it can be passed to another host during the next blood meal of the next cold-blooded vector (Tjaden *et al.*, 2013; Winokur *et al.*, 2020). A shorter EIP increases the opportunities for virus transmission during the lifetime of an infected mosquito, whose survival in the environment rarely exceeds a month. This is especially important where transmission is seasonal, because it can contribute to a much larger outbreak, by enabling additional infection cycles (Box 16.1).

If ambient temperatures fall outside the preferred vectorial 'climatic envelope', then mosquitoes may not survive long enough to become infectious and transmit disease. Excessive heat (Reinhold *et al.*, 2018), prolonged drought and (in some cases) flooding can reduce mosquito survival, though water storage used to adapt to drought can increase it in vulnerable settings. On the other hand, vectors may adapt to evolving environmental conditions over time (Tang *et al.*, 2018; Rose *et al.*, 2020). If so, this would call into question the dominant approach of applying present temperature-dependent life cycle rates (dynamical models) or climatic disease bounds (statistical models) to future climate conditions.

The findings of theoretical and empirical models of dengue and climate are broadly consistent, and in general representative of other arboviruses. For example, dengue transmission is highly sensitive to climate; relatively small increases in temperature (as little as 1°C) could lead to substantial increases in transmission potential and the potential length of the transmission season (Colón-González *et al.*, 2018).

Box 16.1

Jetten and Focks (1997) modelled dengue vectorial capacity (VC), defined as the average number of potentially infective contacts by a mosquito population per infectious person per year:

$$\mathrm{VC} = mbca^2P^n / -log^e(P)$$

where m is the number of female mosquitoes per person; b is the probability that an infectious mosquito transmits dengue while biting a susceptible human; c is the probability that a mosquito acquires a dengue infection while biting a viraemic human; a is the number of bites per person per day; P is the survival rate of the mosquito; and n is the EIP in days.

With a 2–4°C increase in temperature, both the range of transmission (latitude and altitude) and the length of the dengue transmission season increase, at least at the 'cool' border of the previous transmission zone. These changes are also likely to lead to initial human infections at a younger age, and this, in turn, increases the risk of haemorrhagic dengue, owing to a greater likelihood of multiple infections over a lifetime with different serotypes. Thus, an increased burden of the disease in individuals is likely, perhaps with an increased number of deaths from dengue (Jetten and Focks, 1997). This standard model has been shown to apply for other arboviruses.

Nonetheless, estimating a precise contribution of either past or future anthropogenic climate change to the rise in the incidence (or risk) of each arboviral disease is not our primary goal in this chapter. Such estimates depend on too many factors, some of which are problematic due to insufficient data, including historical records (Brady and Hay, 2020). Suffice to say that there is a growing literature, nearing consensus, showing that the contribution of climate change to vector-borne disease transmission, including of arboviruses, is significant (Short *et al.*, 2017; Caminade *et al.*, 2019). But climate change never acts alone; other important factors in the rise in the incidence of arbovirus infections include urbanization, population growth in the global South, poverty, trade, travel and insecticide resistance (Wilder-Smith and Monath, 2017; Brady and Hay, 2020).

Climate change is also likely to worsen the burden of some vector-borne diseases via other means, such as intensified natural disasters and their impacts on infrastructure (Boyce *et al.*, 2016). Earthquakes, another form of disaster, may also interact (Sorensen *et al.*, 2018).

Indirect (tertiary) effects of climate change include conflict, migration and expanded refugee camps. Each of these is likely to worsen the burden of arboviral diseases, along with any decrease in social and public health capacity, due in part to climate change or other reasons (Tidman *et al.*, 2021).

16.5 Dengue and Dengue-like Illness

Dengue is transmitted by the urban-dwelling mosquito *Aedes aegypti* and, less efficiently, by the colder-tolerant species *Aedes albopictus*. Dengue is the world's most important and frequent arboviral disease, currently present in more than 128[9] countries in the tropics and subtropics, yet poised to increase even further in terms of incidence and geography (Wilder-Smith, 2020). This expansion has long been predicted to be, in part, related to climate change (Kovats *et al.*, 2001).

Dengue has four genetically distinct serotypes, each in the genus *Flavivirus*. The ancestor for all four serotypes is thought to have emerged very recently, perhaps within the past millennium (Holmes and Twiddy, 2003). Each of the four serotypes of dengue may vary in their transmissibility, and the dominant strain may change over time within a given area (Oki and Yamamoto, 2012). Infection with any of these serotypes results in lifelong immunity to that specific form, but temporary (6 months to 2 years) cross-immunity to the others (Brady and Hay, 2020). Although many infections are asymptomatic, each of the four serotypes has been individually found to be responsible for dengue epidemics with more severe clinical manifestations (Murray *et al.*, 2013). These can range from a mild undifferentiated febrile syndrome to a severe form, characterized by plasma leakage that can degenerate into haemorrhagic fever, organ failure, shock and, occasionally, death. Crucially, patients with an infection of a second dengue serotype are at increased risk for severe symptoms.[10] Thus, areas with two or more dengue serotypes, such as some countries in South America and Asia, expose populations to greater risk than where there is only one serotype circulating. For example, a major dengue outbreak (DENV-1) occurred in 2005 in Singapore, which was followed by a DENV-2 outbreak in 2007 (Yung *et al.*, 2015). A detailed review of DENV serotypes circulating in Latin America is provided by Ramos-Castañeda *et al.* (2017).

Like yellow fever, dengue (and its vector) is thought to have crossed to the New World with enslavement (part of the 'Columbian exchange'); outbreaks were reported as far north as Philadelphia, USA, by 1780 (Gubler, 1998).[11] Unlike yellow fever, dengue is widely distributed in Asia and may even have originated there.[12] Like most arboviruses, each dengue serotype was originally a zoonosis. Each dengue serotype has a sylvatic (forest or jungle) cycle, involving non-human primates and mosquitoes. Sometimes, when arboviruses leave their original 'home' new sylvatic cycles are established (as with yellow fever in the Americas); however, dengue currently has no sylvatic cycle in the Americas, among other locations (Hanley *et al.*, 2013; Brady and Hay, 2020). In the case of dengue, a non-sylvatic cycle can easily be sustained in densely populated regions by the viral reservoir in humans.[13]

To date, effective dengue vector-control measures have proven neither scalable nor sustainable, at least when considered over a sufficient period. Such efforts are made immensely more difficult by poverty and other factors, such as conflict, that impair effective public health measures. Successful vector control over several years reduces population immunity through limiting viral exposure. If the vector subsequently returns, such populations are relatively serologically naïve and especially susceptible to an epidemic (Oki and Yamamoto, 2012). Obviously, though, it is better to adopt policies that endeavour to do this; if outbreaks recur, their extent can sometimes be curtailed. Furthermore, such gaps in population exposure are likely to reduce the risk of serious disease, arising from exposure to multiple strains.

Dengue has no effective antiviral treatment. Promising strategies such as the release of sterile male insects and use of *Wolbachia* (see Box 16.2, below) are still under development. The recent deployment of *Wolbachia*-infected *Ae. aegypti* in Yogyakarta, Indonesia, led to a protective efficacy of 77.1% and was similar for the four dengue serotypes (Utarini *et al.*, 2021). A safe and effective tetravalent dengue vaccine would appear to be the best intervention. Yet, to date, such vaccines remain problematic.[14] Their performance depends on serostatus. In those with a previous dengue infection, i.e. who are seropositive, efficacy is high and the vaccine is safe and long-lasting. However, in seronegative vaccinees, approximately three years after vaccination, the risk of developing severe dengue is increased (i.e. in this vaccinated individual) if they experience a natural dengue infection of any serotype. It thus appears to have an effect similar to that which arises in populations who experience an infection with a second serotype. The safety and acceptability of dengue vaccines will be greatly improved if a cheap and reliable test for serological status can be developed (Wilder-Smith, 2020).

It is estimated that the total number of dengue infections has increased approximately fivefold between 1990 and 2017, from about 23 million to 105 million people (95% uncertainty interval 63–159 million), mostly acquired in tropical and subtropical regions. Autochthonous transmission has, however, been reported in Italy (Lazzarini *et al.*, 2020) and France (Cochet *et al.* 2022). In 2017 dengue caused an estimated 40,000 deaths per annum (Zeng *et al.*, 2021).

16.6 Chikungunya

Chikungunya virus was first identified, in the 1950s (Lumsden, 1955), in eastern Africa, where it may also have originated, like many other arboviruses, in an enzootic sylvatic cycle involving wild primates and forest-dwelling *Aedes* mosquitoes (Braack *et al.*, 2018). Though rarely fatal, its common symptoms (fever, rash, arthralgia, myalgia and headache) complicate clinical diagnosis, including differentiation from other arboviruses.[15] Symptoms can last for many years following infection. Like dengue, yellow fever and Zika, its main vector is the *Aedes* genus.

While identified outbreaks of chikungunya occurred in Africa and South-East Asia in the second half of the 20th century, concern for the disease heightened following a large outbreak (approximately 266,000 people) in the volcanic Indian Ocean island of La Réunion in 2005–2006, affecting one-third of the population of this isolated outpost of the French state. Over 260 associated deaths were reported. Reaction by central French authorities was belated, though ultimately effective (Weinstein and Ravi, 2008). Entomological factors also contributed to amplify the epidemic. One was the gradual replacement, on the island, of *Aedes aegypti* by *Aedes albopictus*, an ecological phenomenon related to competition of larvae that may have been amplified by vigorous vector-control measures (to reduce the incidence of malaria) undertaken in the 1950s (Bagny *et al.*, 2009).[16] The second co-factor appears to have been a mutation enabling easier viral multiplication within *Aedes albopictus*, leading to a 100-fold increase in viraemia within the mosquito's salivary glands (Tsetsarkin *et al.*, 2007).[17]

Like dengue, chikungunya has multiple genotypes: East/Central/South African, West African and Asian. Following its spread to La Réunion it quickly reached the Asian continent (including India and Thailand) and then spread in 2013, via the Caribbean island of St Martin, to 43 countries or territories in the Americas. Chikungunya is now widespread globally.

16.7 Zika

In the past decade, Zika virus has emerged as the most recent global arbovirus. Like other arboviruses, it is often asymptomatic. When symptomatic, it mimics many other viral infections, including chikungunya and dengue. Symptoms can include fever (usually mild), rash, arthralgia, arthritis, myalgia, headache, conjunctivitis and oedema; severe cases requiring hospitalization are uncommon; fatalities are rare. It was first identified in 1947, from a sentinel[18] rhesus monkey in the Zika Forest, on the shores of Lake Victoria, in Uganda, and soon after from *Aedes africanus* mosquitoes collected in the same forest. Subsequent serological and other studies showed Zika to be widespread in eastern, western and southern Africa, at least to Mozambique.

The range of Zika expanded across the Pacific Ocean from 2007, reaching French Polynesia in 2013 (Musso *et al.*, 2019), Brazil (2015) and other parts of South and Central America. Its most serious complication was first noticed in South America: infections in pregnant women caused many to give birth to babies with a range of eye, cardiac and neurological anomalies, including microcephaly (Mlakar *et al.*, 2016). The incidence of microcephaly in Brazil was initially reported to have risen approximately 20-fold (Fauci and Morens, 2016), but subsequent analysis has led to lower estimates of this risk (Musso *et al.*, 2019). In French Polynesia, such complications were also retrospectively attributed to Zika exposure in 2013/14, especially in the first trimester of pregnancy (Cauchemez *et al.*, 2016). Zika infections have also been linked to Guillain-Barré syndrome in French Polynesia and South America.

In 2016 Fauci and Morens observed: 'Decades ago, African researchers noted that *Aedes*-transmitted Zika epizootics inexplicably tended to follow *Aedes*-transmitted chikungunya epizootics and epidemics'. They added: 'An analogous pattern began in 2013, when chikungunya spread pandemically from west to east, and Zika later followed.' Other investigations detected a distinct strain of Zika that had wide circulation across much of Asia since at least 1951, but it is plausible that this lineage was already present in Asia during World War II, misdiagnosed as the far more common dengue (Braack *et al.*, 2018). At this time, Zika was not known to be associated with microcephaly or other congenital abnormalities.

Zika has three distinct genotypes: West African, East African and Asian. To date, congenital birth defects have only been reported in cases of infection strains belonging to the Asian lineage, including those identified during a 2016 outbreak in Angola (Musso *et al.*, 2019). Similar to dengue, chikungunya and yellow fever, the expansion of Zika seems to have evolved from a sylvatic cycle (involving non-human primates and *Aedes* mosquitoes, with humans as incidental hosts) to one where transmission flourishes as a human–mosquito–human cycle in impoverished urban settings, involving *Aedes aegypti* and to a lesser extent *Aedes albopictus*. Although Zika virus has been isolated from other genera of mosquitoes, *Aedes* are the only proven vectors (Epelboin *et al.*, 2017).

16.8 Conclusion: Arboviruses, Development and the Future

Increased living standards, including improved infrastructure (piped water, better housing, disease surveillance, vector control programmes, and in some cases effective vaccines) have contributed greatly to the reduction in arbovirus burden, evident until (approximately) World War II. In the second half of the 20th century, at least during the heyday of optimism for malaria vector control, these gains seemed secure, perhaps with additional wins in the 21st century to follow. This progress was enabled by cheap available insecticides and the surge of public health optimism (and the profession's prestige) that accompanied and followed decolonization, and when 'overseas development assistance' had approval and cachet.

In recent decades this optimism has abated, for many reasons. One was the realization that malaria eradication, globally, was too difficult to attain. As spraying campaigns contracted (in part due to insecticide resistance and in part due to human resistance to these 'almost forever' chemicals) vector populations rebounded.[19] Trade, and an under-appreciation of its risks, probably worsened by a growing disregard for quarantine, led to the quasi-globalization of important vectors and many arboviruses, such as the introduction of *Aedes albopictus* to the Americas and Europe, and the spread of West Nile virus there. Inexorable urbanization in the hot cities of many 'developing' nations outpaced infrastructure provision, providing enormous new opportunities for *Aedes aegypti*, in particular. The attempt to follow the successful campaign to restrict yellow fever by doing the same for dengue has faltered, not least due to severe and ongoing problems with developing a safe and effective vaccine for this immunologically complicated arbovirus.

Then, the near-global tropical spread of both chikungunya and Zika not only caused additional grief, but also gave disturbing hints for the future. We do not predict the next arbovirus of global concern, but point out that evolution is a talented opponent, capable of ingenious surprises. The rapid chikungunya mutation (strain E1-226V) that enabled *Ae. albopictus* to transmit disease during the 2005/6 epidemic in La Réunion is a prime example. A research priority to enable the reduction of the disease burden of arboviruses is likely to include the examination of their geographical heterogeneity (Brady and Hay, 2020), and their immunological and ecological interactions, as well as further ways to strengthen surveillance and to reduce vectorial capacity.

Without a similarly strong realization of the role of poverty in arbovirus spread, success is in question. One way to assist this is by efforts to enhance 'behavioural immunity', i.e. ways to reduce arboviral spread, including information campaigns, citizen participation in reducing vector habitat, timely insecticide use, [20] and (perhaps) judicious quarantine.

Novel developments in arbovirus control may play a significant role in the future distribution and transmission intensity of the virus. These may take the form of an effective vaccine (perhaps using mRNA technology), or involve biological control measures that reduce mosquito population densities or interrupt virus transmission. One such emerging arena is infection of some mosquito vectors with *Wolbachia* bacteria (Box 16.2). However, while promising in the longer term, the effectiveness and sustainability of such technologies in poor-country settings are still uncertain.

The role of global climate change in the spread of arboviruses is difficult to quantify, but far from trivial. By facilitating vector survival in new temperate areas (even though reducing risk in some of the hottest regions), climate change is likely to deepen the arboviral disease burden by exposing non-immune populations, or perhaps by triggering a second or parallel infection among people with previously limited arboviral exposure. Even more profound (tertiary) effects of climate change, including widespread famine, conflict and population dislocation, have the potential to undermine effective public health, especially in poor settings that are vulnerable to

Box 16.2

Although the release of reproductively compromised or sterile insects has been successfully used to combat agricultural pests on a large scale, attempts to extend these methods to improve health have been problematic since the first attempts in the 1960s (Crawford *et al.*, 2020). Methods include irradiation (to sterilize male mosquitoes that are then released to mate), genetic engineering and the infection and release of males with the intracellular bacterium *Wolbachia pipientis*. Because female mosquitoes generally only mate once,[21] all these techniques promise a way to reduce vector populations. *Wolbachia*, once introduced into wild mosquito populations, is then transmitted, maternally, to their offspring. Eggs fertilized by the mating of *Wolbachia*-uninfected females with infected males die, whereas embryos from infected females survive, developing into *Wolbachia*-infected adults. This 'incompatible insect technique' is analogous to the sterile insect technique. *Wolbachia*-infected females have a reduced ability to transmit pathogens, either by reducing insect lifespan or lowering the ability of viruses and other pathogens to replicate within the insect (Hoffmann *et al.*, 2011). *Wolbachia* is a promising technique, with encouraging results in Vietnam, Yogyakarta, Indonesia (Utarini *et al.*, 2021) and Brazil. A 'mosquito factory' is planned for Brazil, to start operating in 2024 and to produce up to five billion infected mosquitoes per annum (Lenharo, 2023). Two years of careful observation at field sites near Kuala Lumpur, Malaysia, have shown that infected strains of *Aedes* still have reduced dengue transmission capacity there. Theoretically, *Wolbachia* should work to reduce other arboviral transmission too. Long-term disease suppression may be affected by evolutionary changes in the mosquito host, virus, or *Wolbachia* and by a public commitment to maintain surveillance and perform additional releases as required (Ross *et al.*, 2019).

arboviruses. Access to healthcare systems can easily be harmed by extreme weather events, while in some places conflict can cause health workers to flee, and even for clinics and hospitals to be attacked.

The burden of arboviruses is likely to exacerbate poverty, not only by known effects such as prolonged symptoms and sequelae but also (at least for Zika) by long-lasting congenital abnormalities. Indirect harm will also accrue via lack of capacity to work, or to attend school. Additional surprises, even shocks, are likely, as arboviruses continue to adapt and to evolve, in order to 'maximize' their own distribution and survival. Human societies and their leaders need to pay deeper attention to the social and ecological determinants of health, including arboviruses, underlining the importance of the One Health approach.

Acknowledgements

This chapter is adapted from Bambrick, H. and Hales, S. (2014) Dengue: Distribution and transmission dynamics with climate change. In: Butler, C.D. (ed.) *Climate Change and Global Health*. CABI, Wallingford, UK, pp. 77–84. We also wish to thank Professor Philip Weinstein.

Notes

[1] Following tobacco mosaic virus (a disease of plants) and foot-and-mouth disease, a devastating infection of cloven-hoofed animals, many of which are economically important, including cattle, sheep, pigs and camels.

[2] Transmission of malaria was also greatly reduced during the successful canal construction phase.

[3] Originally thought to convey immunity for 10 years.

[4] The apparent 'resistance' to YF in Asia may be an example of 'ecosystem immunity'. There are other possible examples: Kunjin virus has been hypothesized as preventing the establishment of West Nile virus in Australia; until 2022, JE was not known to have penetrated the Australian mainland, despite the presence of *Aedes* and amplifying hosts (pigs) (Weinstein and Stanhope, 2022); foxes have never been successfully established in the Australian island of Tasmania, despite multiple attempts. However, in 2016, a Chinese citizen, infected in Angola, died of yellow fever after return to China (Chen *et al.*, 2016).

[5] The main vector for plague.

[6] Some *Aedes* species prefer rural environments. *Ae. africanus* (the second leading yellow fever vector in Zambia) is mostly present in forests and wetlands. *Ae. vexans*, the vector of Rift Valley fever, is usually found in woodland pools and cultivated fields.

[7] Such organisms regulate their body temperature by exchanging heat with their surrounding environment.

[8] Copepods are crustacea that act as predators for mosquito larva. Their deliberate introduction in water vessels in Vietnam has been successfully used to reduce dengue transmission.

[9] Including 36 previously dengue-free countries in temperate regions (Gwee *et al.*, 2021).

[10] 'Infection with a heterotypic serotype is a major risk factor for developing severe dengue due to the mechanism of antibody-dependent enhancement. However, it has been shown that within-serotype antigenic variation is almost as broad as between-serotype variation, suggesting a more complex genetic and immunological organization of dengue. This remains one of several different hypotheses for why dengue exhibits such year-on-year variability in case numbers and disease severity' (Brady and Hay, 2020).

[11] Such historical outbreaks are based on clinical and epidemiological features; serological testing for dengue was impossible until the 1950s. Many infectious diseases may be confused with dengue, including chikungunya, Rift Valley fever, Ross River fever, West Nile fever, yellow fever, malaria, leptospirosis, pappataci fever (sandfly fever), scarlet fever, typhoid fever, measles and influenza.

[12] Cases clinically consistent with dengue were described in the Ch'in Dynasty (265–420 ce) (Murray *et al.*, 2013).

[13] '... four different types of DENV have established human transmission cycles that are ecologically and evolutionarily distinct from their sylvatic ancestors' (Hanley *et al.*, 2013).

[14] Dengue has four strains. Ideally a tetravalent immune response is desired, resulting from the administration of a live attenuated vaccine with a mixture of all four dengue serotypes. However, each component needs to 'independently result in four different monotypic immune responses that are solid to each serotype' (Wilder-Smith, 2020). This has, unfortunately, proven to be difficult to achieve.

[15] Halstead argued that the disease originally described as dengue was actually chikungunya (Weaver and Forrester, 2015).
[16] La Réunion is slightly north of the Tropic of Capricorn, a latitude sufficiently warm (at least at low altitudes) to normally favour *Ae. aegypti*.
[17] *Ae. aegypti* is still a highly effective vector for chikungunya.
[18] Intended to provide early warning for yellow fever.
[19] Vectorial resistance to insecticides, including pyrethroids, is occurring. Although DDT is often effective, there can be social resistance to this (Hecht, 2012), but its use may be justified in some cases.
[20] Especially at the beginning of the season for transmission, at least in cities with seasonal transmission.
[21] Depending on the species, between 0% and 20% of female mosquitoes will mate with multiple males.

References

Arbeitskreis Blut (2009) Arbobacteria – pathogens transmittable by arthropods. *Transfusion Medicine and Hemotherapy* 36, 62–78. DOI: 10.1159/000197341

Bagny, L., Delatte, H., Quilici, S. and Fontenille, D. (2009) Progressive decrease in *Aedes aegypti* distribution in Réunion island since the 1900s. *Journal of Medical Entomology* 46, 1541–1545. DOI. 10.1603/033.046.0644

Barbosa, L.M. and Coates, R. (2021) Resisting disaster chronopolitics: Favelas and forced displacement in Rio de Janeiro, Brazil. *International Journal of Disaster Risk Reduction* 63, 10244. DOI: 10.1016/j.ijdrr.2021.102447

Barrett, A.D.T. (2016) Yellow fever in Angola and beyond – The problem of vaccine supply and demand. *New England Journal of Medicine* 375, 301–303. DOI: 10.1056/NEJMp1606997

Bisanzio, D., Martello, E., Izenour, K., Stevens, K., Kaur, R. *et al.* (2021) Arboviral diseases and poverty in Alabama, 2007–2017. *PLoS Neglected Tropical Diseases* 15, e0009535. DOI: 10.1371/journal.pntd.0009535

Boyce, R., Reyes, R., Matte, M., Ntaro, M., Mulogo, E. *et al.* (2016) Severe flooding and malaria transmission in the western Ugandan highlands: Implications for disease control in an era of global climate change. *The Journal of Infectious Diseases* 214, 1403–1410. DOI: 10.1093/infdis/jiw363

Braack, L., Gouveia de Almeida, A.P., Cornel, A.J., Swanepoel, R. and de Jager, C. (2018) Mosquito-borne arboviruses of African origin: review of key viruses and vectors. *Parasites & Vectors* 11, 29. DOI: 10.1186/s13071-017-2559-9

Brady, O.J. and Hay, S.I. (2020) The global expansion of dengue: How *Aedes aegypti* mosquitoes enabled the first pandemic arbovirus. *Annual Review of Entomology* 65, 191–208. DOI: 10.1146/annurev-ento-011019-024918

Brès, P.L.J. (1986) A century of progress in combating yellow fever. *Bulletin of the World Health Organization* 64, 775–786. Available at: https://www.ncbi.nlm.nih.gov/pmc/articles/PMC2490971/ (accessed 23 March 2023).

Brisbois, B.W. and Harris, S. (2010) Climate change, vector-borne disease and interdisciplinarity research: social science perspectives on an environment and health controversy. *EcoHealth* 7, 425–438. DOI: 10.1007/s10393-010-0354-6

Caminade, C., McIntyre, K.M. and Jones, A.E. (2019) Impact of recent and future climate change on vector-borne diseases. *Annals of the New York Academy of Science* 1476, 157–173. DOI: 10.1111/nyas.13950

Carissimo, G., Pain, A., Belda, E. and Vernick, K.D. (2018) Highly focused transcriptional response of *Anopheles coluzzii* to O'nyong nyong arbovirus during the primary midgut infection. *BMC Genomics* 19, 526. DOI: 10.1186/s12864-018-4918-0

Carter, E.D. (2008) 'God bless General Perón': DDT and the endgame of malaria eradication in Argentina in the 1940s. *Journal of the History of Medicine and Allied Sciences* 64, 78–122. DOI: 10.1093/jhmas/jrn057

Cauchemez, S., Besnard, M., Bompard, P., Dub, T., Guillemette-Artur, P. *et al.* (2016) Association between Zika virus and microcephaly in French Polynesia, 2013–15: a retrospective study. *The Lancet* 387, 2125–2132. DOI: 10.1016/S0140-6736(16)00651-6

Centers for Disease Control (2021) *Arbovirus Catalog*. Available at: https://wwwn.cdc.gov/arbocat (accessed 19 October 2021).

Chen, Z., Liu, L., Lv, Y., Zhang, W., Li, J, *et al.* (2016) A fatal yellow fever virus infection in China: description and lessons. *Emerging Microbes & Infections* 5, 1–8. DOI: 10.1038/emi.2016.89

Clements, A.N. and Harbach, R.E. (2017) History of the discovery of the mode of transmission of yellow fever virus. *Journal of Vector Ecology* 42, 208–222. DOI: 10.1111/jvec.12261

Cochet, A., Calba, C., Jourdain, F., Grard, G., Durand, G.A. *et al.* (2022) Autochthonous dengue in mainland France, 2022: geographical extension and incidence increase. *Eurosurveillance* 27. DOI: 10.2807/1560-7917.Es.2022.27.44.2200818

Cohen, S.N. and Halstead, S.B. (1966) Shock associated with dengue infection: I. Clinical and physiologic manifestations of dengue hemorrhagic fever in Thailand, 1964. *The Journal of Pediatrics* 68, 448–456. DOI: 10.1016/S0022-3476(66)80249-4

Colón-González, F.J., Harris, I., Osborn, T.J., Steiner São Bernardo, C., Peres, C.A. *et al.* (2018) Limiting global-mean temperature increase to 1.5–2°C could reduce the incidence and spatial spread of dengue fever in Latin America. *Proceedings of the National Academy of Sciences (USA)* 115, 6243–6248. DOI: 10.1073/pnas.1718945115

Colwell-Chanthaphonh, C. and de Salle-Essoo, M (2014) Saints and evil and the wayside shrines of Mauritius. *Journal of Material Culture* 19, 253–277. DOI: 10.1177/13591835145400

Crawford, J.E., Clarke, D.W., Criswell, V., Desnoyer, M., Cornel, D. *et al.* (2020) Efficient production of male *Wolbachia*-infected *Aedes aegypti* mosquitoes enables large-scale suppression of wild populations. *Nature Biotechnology* 38, 482–492. DOI: 10.1038/s41587-020-0471-x

de Rivero, O. (1999) *The Myth of Development – The Non-Viable Economies of the 21st Century*. Zed Books, London.

Epelboin, Y., Talaga, S., Epelboin, L. and Dusfour, I. (2017) Zika virus: An updated review of competent or naturally infected mosquitoes. *PLoS Neglected Tropical Diseases* 11, e0005933. DOI: 10.1371/journal.pntd.0005933

Fauci, A.S. and Morens, D.M. (2016) Zika virus in the Americas – yet another arbovirus threat. *New England Journal of Medicine* 374, 601–604. DOI: 10.1056/NEJMp1600

Finlay, C.J. (1989) The mosquito hypothetically considered as the agent of transmission of yellow fever (first published 1881). In: Buck, C., Llopis, A., Nájera, E. and Terris, M. (eds) *The Challenge of Epidemiology. Issues and Selected Readings.* Pan American Health Organization, Washington, DC, pp. 60–66.

Flies, E.J., Lau, C.L., Carver, S. and Weinstein, P. (2018) Another emerging mosquito-borne disease? Endemic Ross River virus transmission in the absence of marsupial reservoirs. *BioScience* 68, 288–293. DOI: 10.1093/biosci/biy011

George, S. (1999) *The Lugano Report: On Preserving Capitalism in the Twenty-First Century*. Pluto Press, London.

Gomes, F.M. and Bahia, A.C. (2022) Implications of vector surveillance for arbovirus epidemiology in Miami-Dade County, Florida. *The Lancet Regional Health – Americas* 11, 100310. DOI: 10.1016/j.lana.2022.100310

Grace, M.P., Aisling, M.V., Luxi, Q., Nuria Sanchez, C., Julia, M.P. *et al.* (2022) Socioeconomic risk markers of arthropod-borne virus (arbovirus) infections: a systematic literature review and meta-analysis. *BMJ Global Health* 7, e007735. DOI: 10.1136/bmjgh-2021-007735

Grobbelaar, A.A., Weyer, J., Moolla, N., van Vuren, P.J., Moises, F. *et al.* (2016) Resurgence of yellow fever in Angola, 2015–2016. *Emerging Infectious Diseases* 22, 1854–1855. DOI: 10.3201/eid2210.160818

Gubler, D. (1998) Resurgent vector-borne diseases as a global health problem. *Emerging Infectious Diseases* 4, 442–450. DOI: 10.3201/eid0403.980326

Gwee, S.X.W., St John, A.L., Gray, G.C. and Pang, J. (2021) Animals as potential reservoirs for dengue transmission: A systematic review. *One Health* 12, 100216. DOI: 10.1016/j.onehlt.2021.100216

Hanley, K.A., Monath, T.P., Weaver, S.C., Rossi, S.L., Richman, R.L. *et al.* (2013) Fever versus fever: The role of host and vector susceptibility and interspecific competition in shaping the current and future distributions of the sylvatic cycles of dengue virus and yellow fever virus. *Infection, Genetics and Evolution* 19, 292–311. DOI: 10.1016/j.meegid.2013.03.008

Hecht, D.K. (2012) How to make a villain: Rachel Carson and the politics of anti-environmentalism. *Endeavour* 36, 149–155. DOI: 10.1016/j.endeavour.2012.10.004

Hemenway, D. (2010) Why we don't spend enough on public health. *New England Journal of Medicine* 362, 1657–1658. DOI: 10.1056/NEJMp1001784

Hoffmann, A.A., Montgomery, B.L., Popovici, J., Iturbe-Ormaetxe, I., Johnson, P.H. *et al.* (2011) Successful establishment of *Wolbachia* in *Aedes* populations to suppress dengue transmission. *Nature* 476, 454–459. DOI: 0.1038/nature10

Holmes, E.C. and Twiddy, S.S. (2003) The origin, emergence and evolutionary genetics of dengue virus. *Infection, Genetics and Evolution* 3, 19–28. DOI: 10.1016/S1567-1348(03)00004-2

Ibarra, A.M.S., Ryan, S.J., Beltrán, E., Mejía, R.I., Silva, M. *et al.* (2013) Dengue vector dynamics (*Aedes aegypti*) influenced by climate and social factors in Ecuador: Implications for targeted control. *PLoS ONE* 8, e78263. DOI: 10.1371/journal.pone.0078263

Jetten, T.H. and Focks, D.A. (1997) Potential changes in the distribution of dengue transmission under climate warming. *The American Journal of Tropical Medicine and Hygiene* 57, 285–297. DOI: 10.4269/ajtmh.1997.57.285

Kay, B. and Nam, V.S. (2005) New strategy against *Aedes aegypti* in Vietnam. *The Lancet* 365, 613–617. DOI: 10.1016/S0140-6736(05)17913-6

Kovats, R.S., Campbell-Lendrum, D.H., McMichael, A.J., Woodward, A. and Cox, J.S.H. (2001) Early effects of climate change: do they include changes in vector-borne disease? *Philosophical Transactions of the Royal Society of London. Series B: Biological Sciences* 356, 1057–1068. DOI: 10.1098/rstb2001.0894

Kuno, G. (2007) Research on dengue and dengue-like illness in East Asia and the Western Pacific during the first half of the 20th century. *Reviews in Medical Virology* 17, 327–341. DOI: 10.1002/rmv.545

Kuno, G. (2020) The absence of yellow fever in Asia: History, hypotheses, vector dispersal, possibility of YF in Asia, and other enigmas. *Viruses* 12, 1349. DOI: 10.3390/v1212134

Lazzarini, L., Barzon, L., Foglia, F., Manfrin, V., Pacenti, M. *et al.* (2020) First autochthonous dengue outbreak in Italy, August 2020. *Eurosurveillance* 25. DOI: 10.2807/1560-7917.Es.2020.25.36.2001606

Lenharo, M. (2023) Massive mosquito factory in Brazil aims to halt dengue. *Nature* 616, 637–638.

Lequime, S. and Lambrechts, L. (2014) Vertical transmission of arboviruses in mosquitoes: A historical perspective. *Infection, Genetics and Evolution* 28, 681–690. DOI: 10.1016/j.meegid.2014.07.025

Lowe, R., Lee, S., Martins Lana, R., Torres Codeço, C., Castro, M.C. *et al.* (2020) Emerging arboviruses in the urbanized Amazon rainforest. *BMJ* 371, m4385. DOI: 10.1136/bmj.m4385

Lumsden, W.H. (1955) An epidemic of virus disease in Southern Province, Tanganyika Territory, in 1952–53. II. General description and epidemiology. *Transactions of the Royal Society of Tropical Medicine and Hygiene* 49, 33–57. DOI: 10.1016/0035-9203(55)90081-x

Maclachlan, N.J., Zientara, S., Wilson, W.C., Richt, J.A. and Savini, G. (2019) Bluetongue and epizootic hemorrhagic disease viruses: recent developments with these globally re-emerging arboviral infections of ruminants. *Current Opinion in Virology* 34, 56–62. DOI: 10.1016/j.coviro.2018.12.005

Messina, J.P., Brady, O.J., Scott, T.W., Zou, C., Pigott, D.M. *et al.* (2014) Global spread of dengue virus types: mapping the 70 year history. *Trends in Microbiology* 22, 138–146. DOI: 10.1016/j.tim.2013.12.011

Mlakar, J., Korva, M., Tul, N., Popović, M., Poljšak-Prijatelj, M. *et al.* (2016) Zika virus associated with microcephaly. *New England Journal of Medicine* 374, 951–958. DOI: 10.1056/NEJMoa1600651

Murray, N.E.A., Quam, M.B. and Wilder-Smith, A. (2013) Epidemiology of dengue: past, present and future prospects. *Clinical Epidemiology* 5, 299–309. DOI: 10.2147/CLEP.S34440

Musso, D., Ko, A.I. and Baud, D. (2019) Zika virus infection – after the pandemic. *New England Journal of Medicine* 381, 1444–1457. DOI: 10.1056/NEJMra1808246

Nye, E.R. and Gibson, M.E. (1997) *Ronald Ross: Malariologist and Polymath*. MacMillan, London, p. 86.

Oki, M. and Yamamoto, T. (2012) Climate change, population immunity, and hyperendemicity in the transmission threshold of dengue. *PloS ONE* 7, e48258. DOI: 10.1371/journal.pone.0048258

Oldstone, M.B.A. (2014) History of virology. In*:* Schmidt, T. (ed.) *Encyclopedia of Microbiology.* Academic Press, Cambridge, Massachusetts, pp. 608–612.

Pearce, J.C., Learoyd, T.P., Langendorf, B.J. and Logan, J.G. (2018) Japanese encephalitis: the vectors, ecology and potential for expansion. *Journal of Travel Medicine* 25, S16–S26. DOI: 10.1093/jtm/tay009

Powell, J., Gloria-Soria, A. and Kotsakiozi, P. (2018) Recent history of *Aedes aegypti*: vector genomics and epidemiology records. *Bioscience* 68, 854–860. DOI: 10.1093/biosci/biy119

Qsim, M., Ashfaq, U.A., Yousaf, M.Z., Masoud, M.S., Rasul, I. *et al.* (2017) Genetically modified *Aedes aegypti* to control dengue: a review. *Critical Reviews in Eukaryotic Gene Expression* 27, 331–340. DOI: 10.1615/CritRevEukaryotGeneExpr.2017019937

Quan, T.M., Thao, T.T.N., Duy, N.M., Nhat, T.M. and Clapham, H. (2020) Estimates of the global burden of Japanese encephalitis and the impact of vaccination from 2000–2015. *eLIFE* 9, e51027. DOI: 10.7554/eLife.51027

Ramos, J.G.P., Garriga-López, A. and Rodríguez-Díaz, C.E. (2022) How is colonialism a sociostructural determinant of health in Puerto Rico? *AMA Journal of Ethics* 24, E305–312. DOI: 10.1001/amajethics.2022.305

Ramos-Castañeda, J., Barreto dos Santos, F., Martínez-Vega, R., Galvão de Araujo, J.M., Joint, G. *et al.* (2017) Dengue in Latin America: Systematic review of molecular epidemiological trends. *PLoS Neglected Tropical Diseases* 11, e0005224. DOI: 10.1371/journal.pntd.0005224

Reinhold, J.M., Lazzari, C.R. and Lahondère, C. (2018) Effects of the environmental temperature on *Aedes aegypti* and *Aedes albopictus* mosquitoes: A review. *Insects* 9, 158. DOI: 10.3390/insects9040158

Robert, M.A., Stewart-Ibarra, A.M. and Estallo, E.L. (2020) Climate change and viral emergence: evidence from Aedes-borne arboviruses. *Current Opinion in Virology* 40, 41–47. DOI: 10.1016/j.coviro.2020.05.001

Rose, N.H., Sylla, M., Badolo, A., Lutomiah, J., Ayala, D. *et al.* (2020) Climate and urbanization drive mosquito preference for humans. *Current Biology* 30, 3570–3579.e6. DOI: 10.1016/j.cub.2020.06.092

Ross, P.A., Turelli, M. and Hoffmann, A.A. (2019) Evolutionary ecology of *Wolbachia* releases for disease control. *The Annual Review of Genetics* 53, 93–116. DOI: 10.1146/annurev-genet-112618-043609

Short, E.E., Caminade, C. and Thomas, B.N. (2017) Climate change contribution to the emergence or re-emergence of parasitic diseases. *Infectious Diseases: Research and Treatment* 10, 1–7. DOI: 10.1177/1178633617732296

Sim, S., Ng, L.C., Lindsay, S.W. and Wilson, A.L. (2020) A greener vision for vector control: The example of the Singapore dengue control programme. *PLoS Neglected Tropical Diseases* 14, e0008428. DOI: 10.1371/journal.pntd.0008428

Snyder, R.E., Boone, C.E., Cardoso, C.A.A., Aguiar-Alves, F., Neves, F.P.G. *et al.* (2017) Zika: A scourge in urban slums. *PLoS Neglected Tropical Diseases* 11, e0005287. DOI: 10.1371/journal.pntd.0005287

Soper, F.L. (1938) Yellow fever: The present situation (October 1938) with special reference to South America. *Transactions of The Royal Society of Tropical Medicine and Hygiene* 32, 297–322. DOI: 10.1016/S0035-9203(38)90045-8

Sorensen, C.J., Borbor-Cordova, M.J., Calvello-Hynes, E., Diaz, A., Lemery, J. *et al.* (2018) Climate variability, vulnerability, and natural disasters: A case study of Zika virus in Manabi, Ecuador following the 2016 earthquake. *Geohealth* 1, 298–304. DOI: 10.1002/2017GH000104

Tang, C., Davis, K.E., Delmer, C., Yang, D. and Wills, M.A. (2018) Elevated atmospheric CO_2 promoted speciation in mosquitoes (Diptera, Culicidae). *Communications Biology* 1, 182. DOI: 10.1038/s42003-018-0191-7

Thomas, S.J. and Yoon, I.-K. (2019) A review of Dengvaxia®: development to deployment. *Human Vaccines & Immunotherapeutics* 15, 2295–2314. DOI: 10.1080/21645515.2019.1658503

Tidman, R., Abela-Ridder, B. and de Castañeda, R.R. (2021) The impact of climate change on neglected tropical diseases: a systematic review. *Transactions of The Royal Society of Tropical Medicine and Hygiene* 115, 147–168. DOI: 10.1093/trstmh/traa192

Tjaden, N.B., Thomas, S.M., Fischer, D. and Beierkuhnlein, C. (2013) Extrinsic incubation period of dengue: knowledge, backlog, and applications of temperature dependence. *PLoS Neglected Tropical Diseases* 7, e2207. DOI: 10.1371/journal.pntd.0002207

Tsetsarkin, K.A., Vanlandingham, D.L., McGee, C.E. and Higgs, S. (2007) A single mutation in chikungunya virus affects vector specificity and epidemic potential. *PLoS Pathogens* 3, e201. DOI: 10.1371/journal.ppat.0030201

Turtle, L. and Solomon, T. (2018) Japanese encephalitis – the prospects for new treatments. *Nature Reviews Neurology* 14, 298–313. DOI: 10.1038/nrneurol.2018.30

Utarini, A., Indriani, C., Ahmad, R.A., Tantowijoyo, W., Arguni, E. *et al.* (2021) Efficacy of *Wolbachia*-infected mosquito deployments for the control of dengue. *New England Journal of Medicine* 384, 2177–2186. DOI: 10.1056/NEJMoa2030243

Weaver, S.C. and Forrester, N.L. (2015) Chikungunya: Evolutionary history and recent epidemic spread. *Antiviral Research* 120, 32–39. DOI: 10.1016/j.antiviral.2015.04.0

Weetman, D., Kamgang, B., Badolo, A., Moyes, C.L., Shearer, F.M. *et al.* (2018) *Aedes* mosquitoes and aedes-borne arboviruses in Africa: Current and future threats. *International Journal of Environmental Research and Public Health* 15, 220. DOI: 10.3390/ijerph15020220

Weinstein, P. and Ravi, S. (2008) The failure of colonial 'distancing': Changing representations of the 2005–06 chikungunya epidemic in Réunion, France. *Singapore Journal of Tropical Geography* 29, 221–235. DOI: 10.1111/j.1467-9493.2008.00330.x

Weinstein, P. and Stanhope, J. (2022) Commentary: Japanese encephalitis virus as another emerging infectious disease: is a lack of epidemiological tools the pig in the room? *International Journal of Epidemiology* 51, 1418–1420. DOI: 10.1093/ije/dyac111

Whiteman, A., Loaiza, J.R., Yee, D.A., Poh, K.C., Watkins, A.S. *et al.* (2020) Do socioeconomic factors drive Aedes mosquito vectors and their arboviral diseases? A systematic review of dengue, chikungunya, yellow fever, and Zika Virus. *One Health* 11, 100188. DOI: 10.1016/j.onehlt.2020.100188

WHO (2023) *Dengue and severe dengue*. World Health Organization. Available at: https://www.who.int/news-room/fact-sheets/detail/dengue-and-severe-dengue (accessed 26 January 2024).

Wilder-Smith, A. (2020) Dengue vaccine development: status and future. *Bundesgesundheitsblatt – Gesundheitsforschung – Gesundheitsschutz* 63, 40-44. DOI: 10.1007/s00103-019-03060-3

Wilder-Smith, A. and Gubler, D.J. (2008) Geographic expansion of dengue: The impact of international travel. *Medical Clinics of North America* 92, 1377–1390. DOI: 10.1016/j.mcna.2008.07.002

Wilder-Smith, A. and Monath, T.P. (2017) Responding to the threat of urban yellow fever outbreaks. *The Lancet Infectious Diseases* 17, 248–250. DOI: 10.1016/S1473-3099(16)30588-6

Wilder-Smith, A., Quam, M., Sessions, O., Rocklov, J., Liu-Helmersson, J. *et al.* (2014) The 2012 dengue outbreak in Madeira: exploring the origins. *Eurosurveillance* 19, 20718. DOI: 10.2807/1560-7917.es2014.19.8.20718

Winokur, O.C., Main, B.J., Nicholson, J. and Barker, C.M. (2020) Impact of temperature on the extrinsic incubation period of Zika virus in *Aedes aegypti*. *PLoS Neglected Tropical Diseases* 14, e0008047. DOI: 10.1371/journal.pntd.0008047

Yung, C.-F., Lee, K.-S., Thein, T.-L., Tan, L.-K., Gan, V.C. *et al.* (2015) Dengue serotype-specific differences in clinical manifestation, laboratory parameters and risk of severe disease in adults, Singapore. *The American Journal of Tropical Medicine and Hygiene* 92, 999–1005. DOI: 10.4269/ajtmh.14-0628

Zeng, Z., Zhan, J., Chen, L., Chen, H. and Cheng, S. (2021) Global, regional, and national dengue burden from 1990 to 2017: A systematic analysis based on the global burden of disease study 2017. *EClinicalMedicine* 32, 100712. DOI: 10.1016/j.eclinm.2020.100712

17 Lyme Disease and Climate Change

Nick H. Ogden
Public Health Risk Sciences Division, National Microbiology Laboratory, Public Health Agency of Canada, Saint-Hyacinthe, Canada

Abstract
This chapter explores possible ways that climate and climate change may impact risks from Lyme disease by effects on: (i) the survival and life cycle of the tick vectors; (ii) populations of animal hosts of the ticks; (iii) transmission cycles of the bacterial cause of Lyme disease, *Borrelia burgdorferi sensu lato* (via effects on reservoir hosts and seasonal tick activity); (iv) dispersion of ticks and bacterium by animal hosts; and (v) human behaviour patterns that may bring them in contact with infected ticks. The current status of assessments of the impacts of climate change on Lyme disease risk and evidence for effects of climate change to date are described.

17.1 Introduction

Lyme disease is caused by spirochaetal bacteria of the *Borrelia burgdorferi sensu lato* (*s.l.*)[1] species complex, which are transmitted by hard-bodied ticks. Four species mostly cause human disease: *B. burgdorferi sensu stricto* (*s.s.*), *Borrelia afzelii*, *Borrelia garinii* and *Borrelia bavariensis*. The species complex is termed *B. burgdorferi* in the following.

Climate change may affect Lyme disease risk via four main outcomes: (i) changes to the geographical footprint of the occurrence of transmission cycles; (ii) changes in the abundance of infected ticks, and thus the level of risk, where Lyme disease transmission cycles already occur; (iii) changes to trajectories of bacterial evolution, with potential consequences for pathogenicity; and (iv) changes to human activities that alter the frequency with which humans are exposed to Lyme disease risk.

17.2 Ecology and Epidemiology of Lyme Disease

Lyme disease is the most commonly reported vector-borne disease in the northern temperate zone (Lindgren and Jaenson, 2006; Bacon *et al.*, 2008). Risk factors for infection are season, leisure activities, location of residence and occupations that bring people into contact with the woodland habitats that support tick vectors and animal hosts of *B. burgdorferi* (Ogden *et al.*, 2014a).

The bacterium is maintained in transmission cycles between wild-animal reservoir hosts (mostly rodents and birds) and immature stages of ticks of the *Ixodes ricinus*–*Ixodes persulcatus* complex (Fig. 17.1). This typical transmission cycle occurs globally for *B. burgdorferi* transmitted by *Ixodes pacificus* (western North America), *Ixodes scapularis* (central and eastern North America), *I. ricinus* (Western and Central Europe) and *I. persulcatus* (eastern Europe and Asia).

Climate Change and Global Health: Primary, Secondary and Tertiary Effects,
2nd Edition (eds C.D. Butler and K. Higgs)
DOI: 10.1079/9781800620025.0017

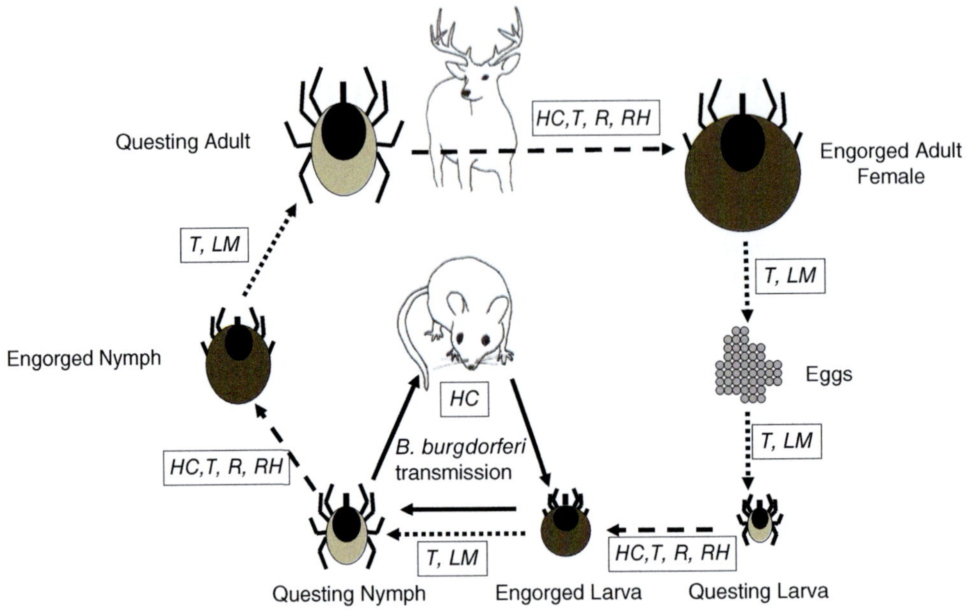

Fig. 17.1. The life cycle of the ixodid tick vectors of Lyme disease, and the cycle of *Borrelia burgdorferi* transmission from infected questing nymphs to vertebrate reservoir hosts, to larvae and then to nymphs through the moult. Dotted arrows indicate tick development processes; dashed lines indicate host-finding processes; while solid lines indicate the *B. burgdorferi* transmission cycle. Boxes indicate influences on these processes, including temperature (T), rainfall (R), air relative humidity (RH), soil/litter layer moisture (LM) and the abundance and species range of tick and reservoir hosts that comprise the host community (HC).

Adult ticks feed mostly on reservoir-incompetent deer and do not participate in transmission cycles (Gern, 2008).

The existence of transmission cycles depends on the presence of competent tick vectors and reservoir hosts and, therefore, on the environmental conditions that support them. Within these broad bounds, Lyme disease risk can vary due to a number of environmentally sensitive factors. These include the abundance and seasonal activity of ticks and the structure of the community of hosts for ticks and *B. burgdorferi* (Tsao, 2009).

Key species of the *B. burgdorferi* transmission cycle tend to be generalists. *Borrelia burgdorferi s.s.* has efficient reservoir hosts spanning several taxa (Margos *et al.*, 2011). The tick vectors are mostly host generalists; some hosts being important reservoirs, e.g. *Peromyscus* spp. mice in North America and *Apodemus* spp. in Europe. The geographical range of these tick vectors spans continents; they survive in a wide range of woodland habitats, as do deer, which are the only hosts of adult ticks. Consequently, *B. burgdorferi* transmission cycles are essentially generalist, can survive perturbation and have the capacity for successful invasion if climate change provides appropriate conditions (Kurtenbach *et al.*, 2006).

17.3 Possible Effects of Climate Change on Lyme Disease Risk

Direct effects of climate change on the tick vectors

Climate can impact the survival of tick vectors in four ways. First, extremes of temperature will kill ticks directly when they are not parasitizing hosts (when they are relatively protected by their host's body temperature regulation). In nature, the direct effects of temperature on survival are modulated by the habitat in which ticks are found. In the laboratory, *I. scapularis* ticks die rapidly at temperatures below −2°C, but when ticks are placed in containers in the field in locations where air temperatures reach −30°C, their mortality rates are similar to those in summer in woodlands where the ticks are endemic (Lindsay *et al.*, 1995; Ogden *et al.*, 2004). Presumably, the litter layer or surface soil layers insulate the ticks from extremes of temperature. Very high temperatures also reduce tick survival (Brownstein *et al.*, 2005). Secondly, low humidity is lethal for ticks if they cannot find refugia in which to rehydrate using their hygroscopic saliva (Bowman and Sauer, 2004), while ticks must return more frequently to the litter layer to rehydrate when humidity is low, depleting their finite energy reserves (Randolph, 2004). Thirdly, heavy rainfall inhibits host-seeking activity, reducing the likelihood that ticks successfully find a host (Randolph, 2004). Fourthly, the duration of development of ticks from one stage to the next, the duration of pre-oviposition in engorged adult females and the duration of development of eggs from laying to hatching all depend on temperature (temperature-independent, daylength-dependent diapause mechanisms also operate, but are idiosyncratic for different tick species and populations (Randolph, 2004; Ogden *et al.*, 2018)). Higher temperatures generally mean shorter life cycles, so, assuming constant per capita temperature-independent daily mortality rates, the higher ambient temperatures are, the higher is the likelihood that a larval tick survives to be an adult tick (Ogden *et al.*, 2014b). When not feeding, ticks reside in the litter layer, which buffers diurnal fluctuations in air temperature, and there is latency of several days in the ability of the tick's metabolism to respond to changing temperatures in terms of changes in development rates (Ogden *et al.*, 2004). This, combined with the capacity of ticks to return to litter layer refugia, means that extreme weather may have a lesser effect on ticks compared with mosquitoes. A further effect of climate change in many locations is an increased duration of the season of tick activity and thus increased risk in locations where Lyme disease risk already occurs (Gray *et al.*, 2009).

Effects of climate change on hosts and their communities

Climate change is likely to impact the geographical range and local abundance of hosts of *B. burgdorferi* and tick vectors by direct effects and by indirect effects on the resources that hosts need for their survival. However, as *B. burgdorferi* vectors and hosts are generalists, the core components of the transmission cycle will likely be resilient to long-term climatic changes and short-term climate variability. Individual species will likely be affected by climate change, producing local idiosyncratic variations in Lyme disease risk within a general pattern of northward expansion and southern contraction of geographical ranges (Kilpatrick *et al.*, 2017).

Effects on *B. burgdorferi* transmission cycles

Beyond the direct effects of climate and climate change on tick abundance, host range and abundance, and possibly survival of *B. burgdorferi* in ticks (high temperatures can reduce pathogen survival in ticks, although this has not been studied for *B. burgdorferi*), more subtle effects of climate change on transmission cycles have been suggested. In North America, different strains of *B. burgdorferi s.s.* are transmissible from infected rodent hosts for different periods of time (varying from a few weeks to the life of the rodent). For persistence of strains that are short-lived in their host, the seasonal activity of nymphs infecting the hosts must be coincident with the larvae to which infected hosts must transmit infection. Strains that are long-lived in their hosts are more resilient to asynchronous seasonal activity of nymphal and larval ticks (Kurtenbach *et al.*, 2006). Seasonality of ticks is partly dependent on temperature effects on development and host-seeking activity (Ogden *et al.*, 2018); and climate change is likely to affect seasonality, altering the geographical pattern of abundance of different strains. A further consequence of this is the potential for evolutionary trajectories to change from specialism (via adaptive radiation) to one of generalism, which may have unpredictable effects on emergence of pathogenic strains (Kurtenbach *et al.*, 2006). Additional complications are that different populations of ticks may behave differently in terms of questing activity. In North America, immature *I. scapularis* in populations in the north-east will quest high in the herbage, so they contact and feed on a wide range of reservoir hosts (as well as humans), but in populations in the south the immature ticks appear semi-nidicolous, feeding mostly on reservoir incompetent lizards and rarely biting people (Ginsberg *et al.*, 2021).

Effects of climate change on dispersion of ticks and *B. burgdorferi*

For actual changes in geographical range of ticks and *B. burgdorferi* to occur, these need to be dispersed by hosts. Terrestrial hosts of *B. burgdorferi* and ticks have limited capacity for dispersion. Maximal rodent home ranges are in the order of 1 km, and similar (< 5 km) values have been suggested for deer in North America (Leighton *et al.*, 2012). Migratory birds are very likely an important method of long-distance dispersion of ticks in Europe and North America, because northward migration routes and seasons are coincident with the activity of nymphal *I. ricinus* (in Europe) and *I. scapularis* (in North America). Importantly, this means that ticks are continually being dispersed far into northern areas, where they can seed reproducing tick populations as the climate warms and becomes more suitable for tick survival (Ogden *et al.*, 2008a).

Effects of climate change on human activities affecting Lyme disease incidence

How environmental hazard is realized as a real public health risk depends on the degree to which people come into contact with infected ticks by virtue of the ways in which people use the environment for their work, leisure, home building and other activities. In some parts of north-eastern USA, the design of residential areas means that there is peri-domestic risk from Lyme disease (Jackson *et al.*, 2006). In areas where Lyme disease risk already occurs, human activity and the likelihood of contacting ticks will most likely increase with climate change: seasons for work and leisure activities in woodland environments are likely to increase.

17.4 Assessments and Evidence for Effects of Climate Change on Lyme Disease Risk

Three assessments of the potential effects of projected climate change on Lyme disease risk in North America were roughly similar in their predictions for significant northward range expansion using statistical and simulation modelling approaches (respectively: Brownstein *et al.*, 2005; Ogden *et al.*, 2006, 2014b). One assessment suggested significant contraction of the southern range of risk (Brownstein *et al.*, 2005). In Europe, two assessments suggested northern expansion of the range of risk as well as expansion of the seasonal duration of risk (Lindgren and Jaenson, 2006; Gray *et al.*, 2009), one suggested the possibility of increasing altitude of risk (Gilbert, 2010) and one identified the complexity of predicting changes in Lyme disease risk when accounting for effects of climate change and landscape/land use change (Li *et al.*, 2019). Current evidence is clear: vector ticks and Lyme disease risk are expanding northwards in the northern hemisphere in a way anticipated in risk assessments. Examples include: the northward expansion of *I. ricinus* in Europe (Lindgren *et al.*, 2000), and *I. scapularis* in Canada (Ogden *et al.*, 2008b, 2010; Leighton *et al.*, 2012); altitudinal increases in tick abundance (Moràn Cadenas *et al.*, 2007); increasing Lyme disease incidence in Scotland and Canada (Ogden *et al.*, 2009; Slack *et al.*, 2011); evidence that new tick populations are setting up in areas where temperature conditions are more favourable (Ogden *et al.*, 2008b, 2010; Leighton *et al.*, 2012); and synchrony of the northward spread of *I. scapularis* in Canada with a period of climate warming (Ebi *et al.*, 2017). To date, however, few observed changes in tick and *B. burgdorferi* risk can be linked conclusively to climate change, even though this may be a very likely cause or co-factor.

17.5 Future Prospects

Further assessments of the impact of climate change on Lyme and other tick-borne zoonoses are needed as projections of future climate evolve. Surveillance to identify where Lyme disease risk is emerging or changing is essential to the detection and attribution of changes in risk to climate change (Ebi *et al.*, 2017).

Note

[1] Latin for 'in the broad sense'; *sensu stricto* – in the strict sense.

References

Bacon, R.M., Kugeler, K.J. and Mead, P.S. (2008) Surveillance for Lyme disease – United States, 1992–2006. *Morbidity and Mortality Weekly Report Surveillance Summaries* 57, 1–9. Available at: https://www.cdc.gov/mmwr/preview/mmwrhtml/ss5710a1.htm (accessed 30 May, 2023).

Bowman, A.S. and Sauer, J.R. (2004) Tick salivary glands: function, physiology and future. *Parasitology* 129, S67–S81. DOI: 10.1017/s0031182004006468

Brownstein, J.S., Holford, T.R. and Fish, D. (2005) Effect of climate change on Lyme disease risk in North America. *EcoHealth* 2, 38–46. DOI: 10.1007/s10393-004-0139-x

Ebi, K.L., Ogden, N.H., Semenza, J.C. and Woodward, A. (2017) Detecting and attributing health burdens to climate change. *Environmental Health Perspectives* 125, 085004. DOI: 10.1289/EHP1509

Gern, L. (2008) *Borrelia burgdorferi sensu lato*, the agent of lyme borreliosis: life in the wilds. *Parasite* 15, 244–247. DOI: 10.1051/parasite/2008153244.

Gilbert, L. (2010) Altitudinal patterns of tick and host abundance: a potential role for climate change in regulating tick-borne diseases? *Oecologia* 162, 217–225. DOI: 10.1007/s00442-009-1430-x

Ginsberg, H.S., Hickling, G.J., Burke, R., Ogden, N.H., Beati, L. *et al.* (2021) Why Lyme disease is common in the northern U.S., but rare in the south: the roles of host choice, host-seeking behavior, and tick density. *PloS Biology* 19, e3001066. DOI: 10.1371/journal.pbio.3001066

Gray, J.S., Dautel, H., Estrada-Peña, A., Kahl, O. and Lindgren, E. (2009) Effects of climate change on ticks and tick-borne diseases in Europe. *Interdisciplinary Perspectives on Infectious Diseases* 2009, 593232. DOI: 10.1155/2009/593232.

Jackson, L., Hilborn, E. and Thomas, J. (2006) Towards landscape design guidelines for reducing Lyme disease risk. *International Journal of Epidemiology* 35, 315–322. DOI: 10.1093/ije/dyi284

Kilpatrick, A.M., Dobson, A.D.M., Levi, T., Salkeld, D.J., Swei, A. *et al.* (2017) Lyme disease ecology in a changing world: consensus, uncertainty, and critical gaps for improving control. *Philosophical Transactions of the Royal Society B* 372, 20160117. DOI: 10.1098/rstb.2016.0117

Kurtenbach, K., Hanincová, K., Tsao, J., Margos, G., Fish, D. *et al.* (2006) Fundamental processes in the evolutionary ecology of Lyme borreliosis. *Nature Reviews Microbiology* 4, 660–669. DOI: 10.1038/nrmicro1475

Leighton, P.A., Koffi, J.K., Pelcat, Y., Lindsay, L.R. and Ogden, N.H. (2012) Predicting the speed of tick invasion: an empirical model of range expansion for the Lyme disease vector *Ixodes scapularis* in Canada. *Journal of Applied Ecology* 49, 457–464. DOI: 10.1111/j.1365-2664.2012.02112.x

Li, S., Gilbert, L., Vanwambeke, S.O., Yu, J., Purse, B.V. *et al.* (2019) Lyme disease risks in Europe under multiple uncertain drivers of change. *Environmental Health Perspectives* 127, 067010. DOI: 10.1289/EHP4615

Lindgren, E. and Jaenson, T.G. (2006) *Lyme borreliosis in Europe: Influences of Climate and Climate Change, Epidemiology, Ecology and Adaptation Measures*. WHO Regional Office for Europe, Copenhagen, Denmark.

Lindgren, E., Tälleklint, L. and Polfeldt, T. (2000) Impact of climatic change on the northern latitude limit and population density of the disease-transmitting European tick *Ixodes ricinus*. *Environmental Health Perspectives* 108, 119–123. DOI: 10.1289/ehp.00108119

Lindsay, L.R., Barker, I.K., Surgeoner, G.A., McEwen, S.A., Gillespie, T.J. and Robinson, J.T. (1995) Survival and development of *Ixodes scapularis* (Acari: Ixodidae) under various climatic conditions in Ontario, Canada. *Journal of Medical Entomology* 32, 143–152. DOI: 10.1093/jmedent/32.2.143

Margos, G., Vollmer, S.A., Ogden, N.H. and Fish, D. (2011) Population genetics, taxonomy, phylogeny and evolution of *Borrelia burgdorferi sensu lato*. *Infection, Genetics and Evolution* 11, 1545–1546. DOI: 10.1016/j.meegid.2011.07.022.

Moràn Cadenas, F., Rais, O., Jouda, F., Douet, V., Humair, P.F. *et al.* (2007) Phenology of *Ixodes ricinus* and infection with *Borrelia burgdorferi sensu lato* along a north- and south-facing altitudinal gradient on Chaumont Mountain, Switzerland. *Journal of Medical Entomology* 44, 683–693. DOI: 10.1603/0022-2585(2007)44[683:poirai]2.0.co;2

Ogden, N.H., Lindsay, L.R., Charron, D., Beauchamp, G., Maarouf, A. *et al.* (2004) Investigation of the relationships between temperature and development rates of the tick *Ixodes scapularis* (Acari: Ixodidae) in the laboratory and field. *Journal of Medical Entomology* 41, 622–633. DOI: 10.1603/0022-2585-41.4.622

Ogden, N.H., Barker, I.K., Beauchamp, G., Brazeau, S., Charron, D. *et al.* (2006) Investigation of ground level and remote-sensed data for habitat classification and prediction of survival of *Ixodes scapularis* ticks in habitats of southeastern Canada. *Journal of Medical Entomology* 43, 403–414. DOI: 10.1603/0022-2585(2006)043[0403:ioglar]2.0.co;2

Ogden, N.H., Lindsay, L.R., Hanincová, K., Barker, I.K., Bigras-Poulin, M. *et al.* (2008a) The role of migratory birds in introduction and range expansion of *Ixodes scapularis* ticks, and *Borrelia burgdorferi* and *Anaplasma phagocytophilum* in Canada. *Applied and Environmental Microbiology* 74, 1780–1790. DOI: 10.1128/AEM.01982-07

Ogden, N.H., St-Onge, L., Barker, I.K., Brazeau, S., Bigras-Poulin, M. *et al.* (2008b) Risk maps for range expansion of the Lyme disease vector, *Ixodes scapularis*, in Canada now and with climate change. *International Journal of Health Geographics* 7, 24. DOI: 10.1186/1476-072X-7-24

Ogden, N.H., Lindsay, L.R., Morshed, M., Sockett, P.N. and Artsob, H. (2009) The emergence of Lyme disease in Canada. *Canadian Medical Association Journal* 180, 1221–1224. DOI: 10.1503/cmaj.080148

Ogden, N.H., Bouchard, C., Kurtenbach, K., Margos, G., Lindsay, L.R., *et al.* (2010) Active and passive surveillance, and phylogenetic analysis of *Borrelia burgdorferi* elucidate the process of Lyme disease risk emergence in Canada. *Environmental Health Perspectives* 118, 909–914. DOI: 10.1289/ehp.0901766

Ogden, N.H., Artsob, H., Margos, G.M. and Tsao, J.I. (2014a) Non-rickettsial tick-borne bacteria and the diseases they cause. In: Sonenshine, D. and Roe, M. (eds) *Ticks and Tick-Borne Diseases Volume 2.* Oxford University Press, Oxford, UK, pp. 278–312.

Ogden, N.H., Radojevic, M., Wu, X., Duvvuri, V.R., Leighton, P.A. *et al.* (2014b) Estimated effects of projected climate change on the basic reproductive number of the tick vector of Lyme disease *Ixodes scapularis*. *Environmental Health Perspectives* 122, 631–638. DOI: 10.1289/ehp.1307799

Ogden, N.H., Pang, G., Ginsberg, H.S., Hickling, G.J., Burke, R.L. *et al.* (2018) Evidence for geographic variation in life-cycle processes affecting phenology of the Lyme disease vector *Ixodes scapularis* in the USA. *Journal of Medical Entomology* 55, 1386–1401. DOI: 10.1093/jme/tjy104

Randolph, S.E. (2004) Tick ecology: processes and patterns behind the epidemiological risk posed by ixodid ticks as vectors. *Parasitology* 129, S37–S65. DOI: 10.1017/s0031182004004925

Slack, G.S., Mavin, S., Yirrell, D. and Ho-Yen, D.O. (2011) Is Tayside becoming a Scottish hotspot for Lyme borreliosis? *Journal of the Royal College of Physicians of Edinburgh* 41, 5–8. DOI: 10.4997/JRCPE.2011.102

Tsao, J.I. (2009) Reviewing molecular adaptations of Lyme borreliosis spirochaetes in the context of reproductive fitness in natural transmission cycles. *Veterinary Research* 40, 36. DOI: 10.1051/vetres/2009019

18 Human Helminthiases and Climate Change: An Overview

Alexander J. Blum and Photini Sinnis
Bloomberg School of Public Health, School of Medicine, Johns Hopkins University, Baltimore, USA

Abstract

The effects of climate change on the incidence of human helminthiases are difficult to predict accurately. The burden of disease will depend on the continued exposure of susceptible hosts to viable populations of infectious helminths and their respective vectors (depending on mode of transmission). There are many contributing factors that will affect this degree of overlap, some of which are parasite-specific while others are unique to human–host populations and the multifactorial determinants of poverty. Broadly, warmer and more humid climates will favour the proliferation of geohelminth species up to certain temperature thresholds, above which extreme heat may diminish the viability of larvae. The incidence of vector-transmitted helminthiases will likely mirror the prevalence of their respective intermediate vectors. In sum, helminths are highly resilient parasites that have coevolved with humans for millennia and are likely to remain a significant contributor of morbidity and mortality among the world's most impoverished populations.

18.1 Introduction

Predicting the impacts of climate change on helminth populations – and their associated burdens of disease – quickly becomes a complex exercise in modelling and remains an evolving area of active research. Distinct from many other forms of infectious pathogens, helminths are complex eukaryotic organisms with intricate multi-stage life cycles that often involve reservoir populations, intermediate hosts, accidental hosts and zoonotic vectors. In the context of climate change, these nuances pose challenges to predicting changes in disease burden and distribution of helminthiases with any meaningful degree of accuracy or precision. The multitude of factors requisite for successful infective cycles all stand to be strongly influenced by climate change with variable and unpredictable positive and negative feedbacks on one another. As such, this overview chapter is largely focused on discussing predictive trends – rather than gross numerical estimates – based on the best currently available evidence and data.

It is also important to acknowledge the outsize role that socio-economic and geopolitical determinants of health will play in determining the global distribution of helminth infections in coming decades. By definition, the vast majority of human helminthiases are considered neglected tropical diseases (NTDs) by the World Health Organization (WHO) because they affect the world's most impoverished populations (Hotez, 2020). Existing disparities and inequities – both within and among societies – will be exacerbated by climate change, with the world's poorest and most vulnerable bearing the brunt of these hardships (Hotez, 2018). Widespread displacement

DOI: 10.1079/9781800620025.0018

and migration of human populations will likely amplify existing trends towards urbanization and concentration of populations (Hotez, 2017).

The consequences of these changes are difficult to predict. For example, an increased concentration of impoverished persons living in peri-urban slum environments could provide an unprecedented volume of susceptible human hosts vulnerable to infection and transmission. At the same time, large urban communities could receive readier access to clean water or mass drug administration – if there are coordinated systems-level efforts – by simplifying the logistical feasibility of delivery and implementation. Such progress would likely reduce the global burden of human helminthiases. In this setting, the breadth of possible outcomes is highly variable, especially when attempting to holistically consider environmental adaptability, compensatory mechanisms and public health interventions.

The remainder of this chapter is divided into two sections exploring the projected effects of climate change on (i) geohelminths and (ii) vector-transmitted helminthiases with a particular focus on filariasis, onchocerciasis and schistosomiasis.

18.2 Geohelminths: Strongyloides, Ascarias, Trichurias and Hookworm Infections

The life cycles of all primary geohelminth species include a soil-based larval stage, whose maturation outside the human host to a human-infective form is strongly influenced by precipitation, humidity and temperature (Weaver *et al.*, 2010; Parker *et al.*, 2015). For these and other variables, each species has a unique range of survivability, tolerance and ability to thrive, with key trends summarized in Table 18.1.

Generally, wetter, warmer climates with higher humidity are more favourable to the proliferation of *Strongyloides* spp., hookworms (*Ancylostoma* and *Necator* spp.), *Ascaris* spp. and *Trichuris* spp., though each soil-transmitted helminth has optimal ranges of temperature

Table 18.1. Primary climate factors affecting transmission of common human helminthiases

	Genera/species	Relevant characteristics	Key trends and related climate factors affecting viability
Geohelminths#	*Strongyloides* spp.	Threadworm	#Summarily favoured by warmer temperatures and higher humidity
	Ascaris spp. *Trichuris* spp. *Ancylostoma* spp. *Necator* spp.	Roundworm Whipworm Hookworm* Hookworm*	#Larval viability is generally diminished at temperatures above 38-40°C and below 8°C #*Hookworm larvae are less susceptible to land surface temperatures but more sensitive to desiccation. They may emerge as the dominant geohelminths
Vector-transmitted helminths	*Schistosoma* spp.	Vector: Freshwater snails	Trends likely to mirror population-level changes in freshwater snails Evaporation and depletion of existing watersheds could disrupt transmission cycles, though human resettlement patterns will likely reflect access to fresh water
	Wucheria spp.	Vector: Mosquitoes	Mosquitoes are likely to proliferate in regions with higher humidity and warmer temperatures Areas of excessive aridity or high water velocity (floodplains) may experience decreased transmission of mosquito-borne helminthiases
	Onchocerca spp.	Vector: Blackflies	Selected blackfly populations (i.e. *Similium* spp.) anticipated to thrive in warmer temperatures, which could augment transmission of *Onchocerca* spp.

and moisture (Weaver *et al.*, 2010; Beknazarova *et al.*, 2016). Broadly, one major implication of the geohelminths' shared preference for warm and wet climates is that regions experiencing increased rainfall or flooding secondary to the effects of climate change may experience a higher burden of disease, while environments that become exceedingly hot and arid could become inhospitable to helminthic larvae. For example, studies suggest that only a fraction of *Ascaris lumbricoides* and *Trichuris trichiura* larvae survive at land surface temperatures greater than 38°C (Brooker *et al.*, 2004; Brooker *et al.*, 2006) and hookworm larvae are diminished at 40°C (Brooker *et al.*, 2004). Similarly, *Strongyloides stercoralis* larvae are bound by the lower and upper limits of viability of 8°C and 40°C, respectively (Brooker and Bundy, 2013). Temperature increases exceeding these upper thresholds of 38–40°C could interrupt chains of transmission at a population level. Areas where these geohelminths are currently endemic may become unsuitable for their continued proliferation (Brooker *et al.*, 2004; Okulewiccz, 2017). In contrast, cooler areas that experience winters below 8°C do not currently support larval maturation but may become new areas of transmission as temperatures increase.

In the absence of massive civil engineering projects to re-design water supply systems, humans will likely depopulate regions with increasingly scarce access to water resources (UNESCO, 2020). Taken together, these trends suggest that there could be increased human–helminth cohabitation in overlapping water-rich geographical locales, which may perpetuate patterns of endemic transmission, especially in the absence of adequate water and sanitation infrastructure. This is especially true for *Strongyloides* spp., which thrive in overcrowded conditions, such as those that will likely characterize climate refugee settlements. If human and helminth migration patterns are likely to mirror projected water distribution trends, it is instructive to consider projected changes in global precipitation and humidity.

Relative to other geohelminths, hookworm larvae are more sensitive to desiccation but less susceptible to higher land surface temperatures (Brooker *et al.*, 2004; Weaver *et al.*, 2010). It is therefore reasonable to expect that hookworms may become the dominant geohelminth in regions experiencing higher temperatures with sustained or increased precipitation and humidity. In contrast, *A. lumbricoides* and *T. trichiura* may become more prevalent in regions with more moderate increases in temperature but declining levels of soil moisture (Brooker *et al.*, 2004). These geohelminth populations may also serve as important markers for understanding and predicting the distributions of other helminthiases as temperature, precipitation and floodwater distributions continue to change across the globe.

18.3 Vector-transmitted Helminths

The fundamental biology of vector-transmitted helminths directly links their incidence and prevalence to those of the intermediate vectors required for transmission. Lymphatic filariasis and onchocerciasis are both insect-borne helminthiases spread by select mosquito and blackfly species, respectively. Trematodes such as *Schistosoma* spp. rely on freshwater snails as intermediate hosts and therefore their transmission varies directly in tandem.

For lymphatic filariasis, the primary determinants favouring mosquito proliferation include tropical temperatures and precipitation patterns that facilitate small pools of standing water in which they breed (Martens *et al.*, 1995; Thompson *et al.*, 1996). Therefore, changes in lymphatic filariasis prevalence may mirror hookworm hotspots since both are favoured by warm and wet environments. Lengthened transmission seasons and in-migration of human populations to these geographical locales could further exacerbate the spread of mosquito-borne disease, including but not limited to lymphatic filariasis (Slater and Michael, 2012). Indeed, some estimates suggest that the global at-risk population for malaria and dengue – which are also mosquito-borne diseases – will increase 'by up to 4.7 billion people by 2070 relative to 1970–99' (Colón-González *et al.*, 2021). Distinct species of mosquitoes are responsible for most malaria and dengue transmission, but the trends mirror those predicted for mosquito populations that transmit lymphatic filariasis.

Selected blackfly (*Simulium*) populations that transmit *Onchocerca volvulus* appear well positioned to thrive in a warming world. One study focusing on West Africa found that a temperature increase of 3–7°C – consistent with projected changes by 2100 – would be ideal for blackfly vectors (Cheke *et al.*, 2015).

Warmer water temperatures and moderate–low water velocities favour freshwater snail populations, and thereby increased prevalence of schistosomiasis (Zhou *et al.*, 2008). As with hookworm and lymphatic filariasis, *Schistosoma* spp. will therefore be favoured by warmer locales that also have sustained rainfall or access to freshwater estuaries and waterways. At the same time, they may disappear from other regions, such as increasingly arid portions of North Africa that are projected to become increasingly susceptible to drought in coming decades (McCreesh *et al.*, 2015).

Human interventions such as dams and water irrigation systems are projected to become more prevalent as fresh water becomes an increasingly scarce resource for some regions due to climate change (UNESCO, 2020). These engineering projects can create waterway environments with warmer water temperatures and reduced water velocities, favouring expansion of snail populations (Short *et al.*, 2017).

Taken together, vector control efforts, preventive measures and improved access to prompt diagnosis and treatment will remain paramount in limiting morbidity and mortality attributable to vector-transmitted helminthiases.

18.4. Conclusion

All current indications suggest that climate change will yield net increases in the prevalence of geohelminths and vector-transmitted helminths, albeit with potential changes in geographical distribution based largely on temperature, precipitation and flooding patterns. However, the trends explored above are largely descriptive and based on current projected changes of environments, hosts and vectors. They do not include consideration of compensatory actions or human interventions, which could transform the currently projected narrative.

Each of the helminth infections explored in this brief chapter are preventable, curable and well understood from a clinical and pathophysiological perspective. Nonetheless, they continue to exert a significant toll on the world's poorest populations. By virtue of being neglected tropical diseases (CDC, 2020) that primarily affect impoverished communities, helminthiases rarely garner adequate funding or other resources to meaningfully disrupt transmission in endemic regions across the world. As climate change increases the proportion of people living in poverty due to forced migration and other factors, a larger percentage of the world's population could be impacted by these infections.

There are many interventions that could dramatically disrupt the predicted synergism between climate change and the burden of disease attributable to human helminthiases and other neglected tropical diseases. Specifically, increased investment in basic water and sanitation infrastructure, innovative vaccine platforms and novel strategies to control proliferation of insect vectors could all provide some measure of protection from the net effects of climate change. Successful strategies will need to be flexible and adaptable to a warming world, while also considering the changing circumstances facing those most affected by these and other neglected tropical diseases.

References

Beknazarova, M., Whiley, H. and Ross, K. (2016) Strongyloidiasis: a disease of socioeconomic disadvantage. *International Journal of Environmental Research and Public Health* 13, 517. DOI: 10.3390/ijerph13050517

Brooker, S.J. and Bundy, D.A.P. (2013) Soil-transmitted helminths (Geohelminths). In: Farrar, J., Hotez, P., Junghanss, T., Kang, G., Lalloo, D. *et al.* (eds) (2013) Soil-transmitted helminths (geohelminths). *Manson's Tropical Diseases*, 23rd edn, Elsevier Saunders, Philadelphia, Pennsylvania, pp. 766–794. DOI: 10.1016/C2010-0-66223-7

Brooker, S., Kabatereine, N.B., Tukahebwa, E.M. and Kazibwe, F. (2004) Spatial analysis of the distribution of intestinal nematode infections in Uganda. *Epidemiology and Infection* 132, 1065–1071. DOI: 10.1017/s0950268804003024

Brooker, S., Clements, A.C.A. and Bundy, D.A.P. (2006) Global epidemiology, ecology and control of soil-transmitted helminth infections. *Advances in Parasitology* 62, 221–261. DOI: 10.1016/S0065-308X(05)62007-6

CDC (updated 2020) *Neglected tropical diseases (NTDs).* Centers for Disease Control and Prevention. Available at: https://www.cdc.gov/globalhealth/ntd/diseases/index.html (accessed 10 July 2021).

Cheke, R.A., Basáñez, M.-G., Perry, M., White, M.T., Garms, R. *et al.* (2015) Potential effects of warmer worms and vectors on onchocerciasis transmission in West Africa. *Philosophical Transactions of the Royal Society B: Biological Sciences* 370, 20130559. DOI: 10.1098/rstb.2013.0559

Colón-González, F.J., Sewe, M.O., Tompkins, A.M., Sjödin, H., Casallas, A. *et al.* (2021) Projecting the risk of mosquito-borne diseases in a warmer and more populated world: a multi-model, multi-scenario intercomparison modelling study. *The Lancet Planetary Health* 5, e404–e414. DOI: 10.1016/S2542-5196(21)00132-7

Hotez, P.J. (2017) Global urbanization and the neglected tropical diseases. *PLoS Neglected Tropical Diseases* 11, e0005308. DOI: 10.1371/journal.pntd.0005308

Hotez, P.J. (2018) Chapter two – human parasitology and parasitic diseases: heading towards 2050. *Advances in Parasitology* 100, 29–38. DOI: 10.1016/bs.apar.2018.03.002

Hotez, P.J. (2020) What constitutes a neglected tropical disease? *PLoS Neglected Tropical Diseases* 14, e0008001. DOI: 10.1371/journal.pntd.0008001

Martens, W.J., Niessen, L.W., Rotmans, J., Jetten, T.H. and McMichael, A.J. (1995) Potential impact of global climate change on malaria risk. *Environmental Health Perspectives* 103, 458–464. DOI: 10.1289/ehp.95103458

McCreesh, N., Nikulin, G. and Booth, M. (2015) Predicting the effects of climate change on *Schistosoma mansoni* transmission in eastern Africa. *Parasites & Vectors* 8, 4. DOI: 10.1186/s13071-014-0617-0

Okulewicz, A. (2017) The impact of global climate change on the spread of parasitic nematodes. *Annals of Parasitology* 63, 15-20. DOI: 10.17420/ap6301.79

Parker, G.A., Ball, M.A. and Chubb, J.C. (2015) Evolution of complex life cycles in trophically transmitted helminths. I. Host incorporation and trophic ascent. *Journal of Evolutionary Biology* 28, 267–291. DOI: 10.1111/jeb.12575

Short, E.E., Caminade, C. and Thomas, B.N. (2017) Climate change contribution to the emergence or re-emergence of parasitic diseases. *Infectious Diseases: Research and Treatment* 10, 1178633617732296. DOI: 10.1177/1178633617732296

Slater, H. and Michael, E. (2012) Predicting the current and future potential distributions of lymphatic filariasis in Africa using maximum entropy ecological niche modelling. *PLoS ONE* 7, e32202. DOI: 10.1371/journal.pone.0032202

Thompson, D.F., Malone, J.B., Harb, M., Faris, R., Huh, O.K. *et al.* (1996) Bancroftian filariasis distribution and diurnal temperature differences in the southern Nile delta. *Emerging Infectious Diseases* 2, 234–235. DOI: 10.3201/eid0203.960313

UNESCO (2020) *The United Nations World Water Development Report 2020: Water and Climate Change*. United Nations Educational, Scientific and Cultural Organization, Perugia, Italy.

Weaver, H.J., Hawdon, J.M. and Hoberg, E.P. (2010) Soil transmitted helminthiases: implications of climate change and human behaviour *Trends in Parasitology* 26, 574–581. DOI: 10.1016/j.pt.2010.06.009.

Zhou, X.-N., Yang, G.-J., Yang, K., Wang, X.-H., Hong, Q.-B. *et al.* (2008) Potential impact of climate change on Schistosomiasis transmission in China. *American Journal of Tropical Medicine and Hygiene* 78, 188–194. DOI: 10.4269/ajtmh.2008.78.188

19 Water and Sanitation

Katrina J. Charles
School of Geography and the Environment, University of Oxford, UK

Abstract
Water and sanitation services that are safe, reliable and accessible are critical to ensure health and well-being. Climate change is impacting these services, with intensification of the water cycle with climate change and rising sea levels making it increasingly difficult to deliver and sustain a high level of service. This chapter explores the level of service needed to achieve health benefits and the ways in which climate change is impacting on health via disruptions to water and sanitation services. It then considers the different ways climate change is disrupting services and the implications for global health.

19.1 Introduction

Safe drinking water and sanitation are widely recognized as critical public health interventions and yet, in 2020, an estimated 2 billion people still lacked access to safely managed drinking water, and 3.6 billion people lacked access to safely managed sanitation (WHO and UNICEF, 2021). Anthropogenic climate change is eroding access and increasing the challenge of providing access to drinking water and sanitation services that can deliver and sustain health benefits. The areas with poorest access are also those most vulnerable to the impacts of climate change.

Access to water, sanitation and (hand-) hygiene (WASH) services are essential for maintaining health. Diarrhoeal diseases remained a leading cause of death in 2016, with lack of access to WASH contributing an estimated 60% of all diarrhoeal deaths (Prüss-Ustün *et al.*, 2019). Increasingly, the benefits of water and sanitation are recognized to extend beyond reduction of diarrhoeal diseases to a range of chronic diseases, both communicable and non-communicable.

Access to WASH services is often interrupted by extreme weather events. Significant droughts in recent years have highlighted the vulnerability of large piped-water systems such as in São Paulo, Brazil, and Cape Town, South Africa. Heavy rainfall is linked to disease outbreaks, including from damage to piped-water systems in developed countries. It is not just rainfall variability that affects water systems: a cold spell in 2021 in the US state of Texas led to pipe bursts; heatwaves increase demand for water from all sectors. Smaller water and sanitation systems, such as the non-piped water sources used by 35% of the global population and non-sewered sanitation used by 57% (WHO and UNICEF, 2021), are particularly vulnerable, due to a lack of capacity to manage climate risks. There is less evidence of the health outcomes from climate-related disruptions in WASH services for these smaller systems, which often serve particularly vulnerable populations. In such settings, the

 Climate Change and Global Health: Primary, Secondary and Tertiary Effects, 2nd Edition (eds C.D. Butler and K. Higgs)
DOI: 10.1079/9781800620025.0019

best evidence is often for the impacts on the water and sanitation systems themselves. For example, compelling data support the view that faecal contamination increases with heavy rainfall in boreholes and dug wells, but there is limited data on the consequent health impact. The indirect impacts of weather and climate change on water and sanitation access are much more complex. For example, weather changes people's behaviour in how they access, manage and store water. Competition over water resources, which will generally be intensified by droughts, exposes women and girls to increased gender-based violence.

This chapter will first explore the evidence for health benefits from water and sanitation services and the level of service needed to achieve those benefits. An increasing body of evidence suggests that, while infrastructure remains necessary to deliver water and sanitation services, it is not sufficient for good health outcomes. Also important are access, quality and reliability of these services. Secondly, the chapter will summarize the evidence for the climate change-related health impacts that can be attributed to disruptions to WASH services. Thirdly, due to limitations in available data linking health and climate, the chapter will present evidence for the mechanisms by which climate change already (and thus likely increasingly) impacts WASH services. The chapter will also illustrate the breadth of pathways connecting WASH and health impacts, and the difficulties in measuring them.

19.2 Water and Sanitation: Relationship to Health

Water-related diseases have been characterized by Bradley (in White *et al.*, 1972) and updated by Bartram and Hunter (2015) according to the mode of intervention, including diseases that are waterborne, water-access related, water-based, or involve water-related insect vectors or engineered water systems (Table 19.1). The burden of disease from inadequate access to water, sanitation and hygiene services was estimated to be 1.6 million deaths in 2016, 2.8% of all deaths, globally (Prüss-Ustün *et al.*, 2019). Much of the evidence for the links between water, sanitation and health has focused on waterborne diseases, particularly diarrhoea, which constitute approximately half of the deaths linked to inadequate WASH. The remainder were linked to a range of causes, including schistosomiasis, acute respiratory infections, undernutrition, malaria, soil-transmitted helminth infections and trachoma. Many more negative health outcomes were recognized but not quantified in the study, including from arsenicosis, fluorosis, dengue and legionellosis.

The understanding of the relationship between WASH and chronic health outcomes has broadened in recent years. A growing literature considers the relationship between impaired access to water and sanitation, environmental enteric (dys)function and stunting; however, these have not specifically been used in studies that also consider climatic factors. Environmental enteric dysfunction arises via gut inflammation, in turn caused by chronic exposure to enteric pathogens associated with poor drinking-water quality, poor hygiene and high faecal environmental contamination, not only from humans but also from animals. This inflammation leads to nutrient malabsorption and undernutrition, which increases susceptibility to further infections and the risk of stunting (Korpe and Petri, 2012). Lack of community access to sanitation is considered the second leading risk factor for stunting, after foetal growth restriction (Danaei *et al.*, 2016). A higher cumulative burden of diarrhoea in a child's life increases the risk of stunting (Checkley *et al.*, 2008). The impact of WASH interventions is influenced by other environmental exposure pathways, including animal faeces (Cumming *et al.*, 2019).

Other areas where the evidence of WASH-associated health impacts is expanding include musculoskeletal conditions, hypertension, psychosocial distress and cognitive impairment (Charles *et al.*, 2019). Hypertension is associated with salinity of drinking water and waterborne environmental toxins such as arsenic, cadmium and mercury. Psychosocial distress is associated with drinking-water contamination and water scarcity. Cognitive impairment is associated with intake of arsenic and lead. There is evidence that carrying water, for those who do not have access to water on premises (e.g. piped or from a well), is associated with pain, fatigue, perinatal health problems and violence against vulnerable people (Geere *et al.*, 2018);

Table 19.1. Water-related disease classification with indicative impact of climate on the mechanism relevant to each class and sub-class; many individual organisms will have specific relationships with temperature and rainfall that are not addressed here.

Class	Subclass	Examples	Main climate impacts associated with drinking water and sanitation
Waterborne – *diseases caused by the ingestion of pathogens or chemicals in water*	Infectious	*Cryptosporidium*, *Giardia*, typhoid, various viruses	Increases in drinking water associated with heavy rainfall and flooding, including flooding of sanitation systems
	Toxic chemicals	Arsenic, fluoride, lead, sodium	Sea-level rise and storm surges increasing salinity in coastal areas. Reduced dilution of pollutants in surface water. Groundwater chemistry changes mobilizing heavy metals
	Nutrient minerals	Fluoride	
Water access-related – *access to and use of water for personal, food and domestic hygiene*	Superficial	Trachoma, scabies	Increases when water is scarce, restricting hygiene and washing
	Intestinal	Shigella dysentery	Increases when water is scarce, restricting hygiene and washing
	Respiratory	Pneumonia	Reduced hygiene during water scarce periods associated with increases in respiratory diseases
	Hydration		Lack of water access due to disruption during floods or droughts
	Injury and violence		Water scarcity, such as during drought, is associated with longer water collection times which can lead to increased musculoskeletal disease, sexual violence, animal attacks and psychosocial distress. Heavy rainfall is associated with injuries from falls
Water-based – *agents of disease that pass part of their life cycle in water*	Contact	Leptospires, *Naegleria fowleri*	Increased temperature can promote growth of microorganisms such as *Naegleria fowleri*
	Ingested	Toxic algae and their toxins	Temperature and low rainfall increase algal blooms with increasingly large impacts on drinking water quality, and temporary closures of water treatment plants
Water-related insect vectors – *that breed in or bites near water*	Water biting	Gambian sleeping sickness	
	Water breeding	Onchocerciasis, malaria	Water storage in droughts associated with outbreaks of dengue
Engineered water system-associated – risks that arise in and are managed via engineered water systems	Inhaled	*Legionella*, radon	
	Ingested	Mycobacteria	
	Contact	Pseudomonas	

these impacts from water carriage are particularly the burden of women and girls. Lack of access to adequate sanitation also poses unique health risks for women and girls, including increased maternal mortality risk, urogenital tract infections, urinary incontinence, chronic constipation and psychosocial distress (Sahoo *et al.*, 2015).

In the remainder of this section, definitions of water and sanitation are explored with reference to the levels of service, as these are essential to understanding the health outcomes.

Defining the level of WASH service

Water and sanitation are widely recognized as beneficial for health; however, the type and magnitude of health benefits vary greatly, depending on the level of WASH service available and accessible. Historically, the relationship between WASH and health has been assessed using simple metrics as proxies for the service level, such as use of infrastructure. This approach is now recognized as insufficient; WASH service levels are increasingly used to account for differences in health outcomes. Understanding changing levels of WASH services is critical to evaluating the impact of climate change on health via WASH.

WASH service levels are commonly defined in accordance with the Joint Monitoring Programme's[1] ladders (Fig. 19.1). For drinking water, an improved water source (including the delivery system) is defined as one where the nature and construction of the infrastructure provide the potential to deliver safe water. Examples include piped household connections, public taps or standpipes, boreholes or tube wells, dug wells and natural springs (protected from contamination), rainwater, tanker trucks and bottled water. Three further aspects of service levels that are required for 'safely managed drinking water' are as follows.

- **Location of the water source** on premises and the time taken to collect water if off-premises. Location is primarily related to water access-related diseases due to the availability of water and the injury and violence risks associated with collecting water. On-premises water sources are associated with lower prevalence of helminth infections. Longer collection times, including distance and queueing, are associated with lower quantities of water being used, with implications for water access-related diseases, and increased faecal contamination during collection and transport (Overbo *et al.*, 2016). Longer collection

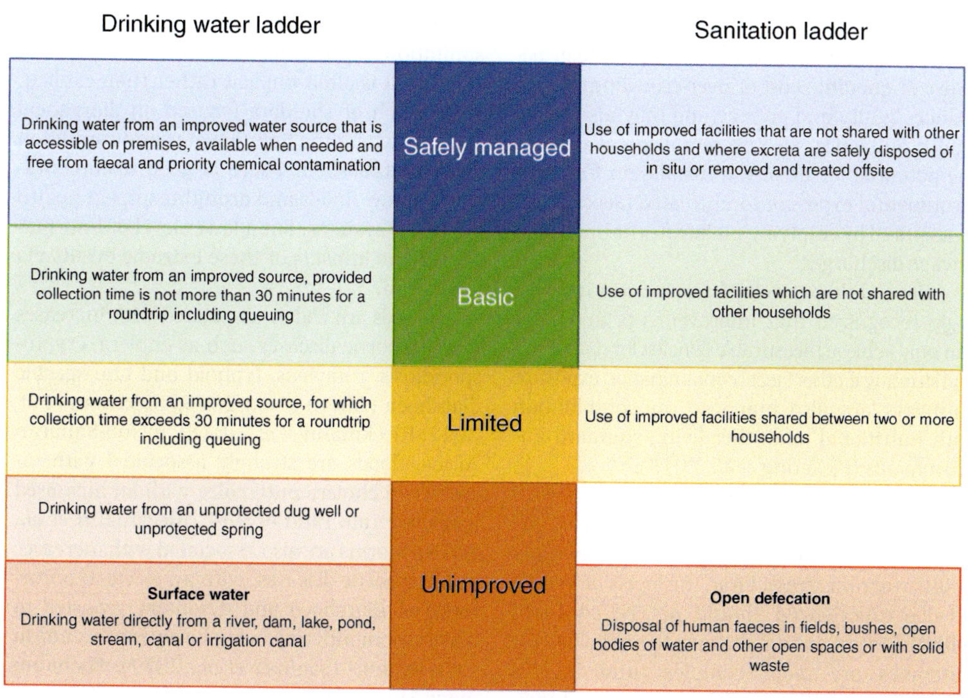

Fig. 19.1. Joint Monitoring Programme's drinking water and sanitation ladders, developed to track progress towards Sustainable Development Goal 6 (adapted from WHO and UNICEF, 2021).

times are likely to contribute to fatigue (in populations already vulnerable and often sicker) as well as to reduce the potential for learning and to earn money.

- **Availability when needed.** The health benefits from drinking safe water are lost very quickly if people have to revert to using unsafe water sources.
- **Water quality.** This is primarily related to waterborne diseases. Freedom from faecal contamination is the main indicator used to measure water quality, but toxic contaminants, such as with metals or synthetic chemicals, may be used where of local or national concern.

The service levels for sanitation start with improved facilities – those designed to hygienically separate excreta from human contact. Examples include flush/pour flush toilets connected to piped sewer systems, septic tanks or pit latrines, pit latrines with slabs (including ventilated pit latrines), and composting toilets (Fig. 19.2). Further service levels required for 'safely managed sanitation' refer to the sharing of toilets between households and the safe disposal of excreta through faecal sludge management. Shared sanitation facilities are associated with poorer hygiene conditions and partial access (e.g. because of queuing, cost or over-crowding) which reduce dignity and privacy and may also harm health outcomes. Safe excreta disposal reduces the potential for disease transmission from environmental exposure to untreated faeces, such as acquired by emptying pit latrines or untreated sewage discharges.

Beyond these levels of service, it is increasingly recognized that improvements in WASH can only achieve measurable benefits for diarrhoea and stunting if other faecal contamination exposure pathways are also managed, in combination with nutritional and other living standard improvements (Pickering *et al.*, 2019).

This categorization provides a useful framework to discuss the links between WASH services, health and climate. However, two key points warrant stress. First, the levels of WASH service experienced should not be confused with the intervention. Actions to achieve these outcomes are more complex, and involve changes in practices, policies, roles, responsibilities and behaviour, as well as technologies and infrastructure. Secondly, WASH service levels do not necessarily correlate with the resilience of a socio-ecological system to climatic events. Resilience is associated with infrastructure, but it is mediated by the governance of the system and other elements of social function. The data available on WASH systems is biased towards larger systems with better service, and often with increased resilience. Such sources (which generally provide greater data enabling higher statistical power) provide the strongest evidence for the impacts of climate and climate change. These conclusions may not necessarily be applicable to more disadvantaged populations.

19.3 Climate-related Impacts on Health

Water and sanitation services are vulnerable to changes in weather and climate and, therefore, to climate change. Evidence for health impacts is limited, due to the challenges of collecting data with sufficient spatial and temporal resolution, including during emergencies. Below, evidence for health impacts related to water and sanitation is presented. The role of water and sanitation is often implicit rather than explicit, with much of the data focused on diarrhoeal diseases where the link to water and sanitation is well-established. There is good documentation on how floods and droughts impact health via water systems, but a lower level of information of the impacts of these extreme events via sanitation.

Floods are widely associated with increases in waterborne diseases such as cholera, cryptosporidiosis, rotavirus, typhoid and non-specific diarrhoea (Ahern *et al.*, 2005; Alderman *et al.* 2012; Rieckmann *et al.*, 2018). In sub-Saharan Africa, floods are strongly associated with increases in cholera outbreaks, with an increased incidence rate ratio of 144 (Rieckmann *et al.*, 2018). Storms are also associated with increases in waterborne diseases, with an elevated occurrence of diarrhoea and dysentery reported to persist for months following heavy rain events, in some settings (Saulnier *et al.*, 2017). There are many studies that show that heavy rainfall increases diarrhoea in settings, including markedly

Fig. 19.2. **(A)** School in Bangladesh showing improved (pit) toilets and improved tube well (credit: Katrina Charles, REACH). **(B)** Shared toilet (pit latrine) in Kampala, Uganda (raised to avoid flooding) (credit: Katrina Charles, 3K-SAN). **(C)** Pit latrine slab, Kigali, Rwanda (credit: Aime Tsinda 3K-SAN). **(D)** Shared sanitation (improved pit latrines) in an informal settlement in Kampala, Uganda, also raised (credit: Kenan Okurut, 3K-SAN used with permission).

heavy rainfall after a dry period (Levy *et al.*, 2016). The health burdens from floods and rainstorms arise via multiple pathways. These often include contaminated drinking sources, overflows of untreated faecal matter from sanitation and sewerage systems, and disruptions in access to handwashing and sanitation facilities.

Evidence for adverse effects to WASH services, and thus to health, from droughts is more limited. The multiple definitions of drought and their nature as longer-term, slow-onset events complicate the attribution of their putative health impacts (Stanke *et al.*, 2013; Levy *et al.*, 2016). Although good evidence links adverse health impacts with water scarcity, the latter can also arise via social factors, for example unaffordability or infrastructure failure. This obscures the relationship between drought, health and climate change. More recently, analysis of cholera outbreaks in sub-Saharan Africa found an increased incidence rate during droughts (Rieckmann *et al.*, 2018; Charnley *et al.*, 2022). There is evidence that reduced water access during droughts is associated with increases in scabies (Stanke *et al.*, 2013; Asmall *et al.*, 2021). Dengue outbreaks have been associated with drought in Brazil and in East Africa (Stanke *et al.*, 2013; Asmall *et al.*, 2021) due to increased water storage near the household encouraging mosquito breeding (also see Chapter 16 for more on vectors).

The absence of data on the health impacts of major droughts in urban areas is a surprising gap in the literature. For the Cape Town (South Africa) drought in 2017/18 an increase in community-acquired *Pseudomonas aeruginosa* was identified, potentially related to biofilm growth associated with increased household water storage (John *et al.*, 2017). While this evidence is limited, scholars have highlighted many other plausible causal pathways linking drought and poorer health outcomes (Stanke *et al.*, 2013; Levy *et al.*, 2016).

Climate-related impacts have been noted to have had a potential impact on health outcomes in only two WASH randomized control trials (RCTs), in both cases linked to the impact of seasonal rains. In Pakistan, increased diarrhoea was reported during very heavy rainfall and flooding, with none of the handwashing or water treatment interventions effective during that period (Luby *et al.*, 2009). In Guatemala, a peak in diarrhoea incidence in a drinking-water treatment trial was observed to occur shortly after the onset of seasonal rains (Reller *et al.*, 2003). Few RCTs measure and report weather conditions. This may partly explain why longer RCTs are more likely to find a weaker health outcome, as weather variability is more likely to impact results (Waddington and Snilstveit, 2009).

19.4 The Impact of Weather and Climate Change on Water and Sanitation Services

The impact of climate change on water and sanitation systems is already evident: heavy rainfall and droughts are increasing in frequency, and global warming is projected to further intensify the global water cycle, including its variability, global monsoon precipitation and the severity of wet and dry events (IPCC, 2021). These changes in the water cycle will increase the difficulty in maintaining and improving the level of WASH services.

Large advances have been made in the previous decades in the management of water and sanitation. This, combined with evolving demands for water, complicates attribution of the impact of climate change on water and sanitation systems, and thus on health. Hence, much of the available evidence presented here relates to shorter-term changes in weather and extreme weather events. Additionally, in some areas, climate change will also impact on water and sanitation systems through sea level rise. This will reduce the availability of fresh water and, in some places, inundate existing WASH-related infrastructure. Long-term declines to groundwater recharge will also impair water availability (and quality); delayed rainy seasons also complicate effective water resources management.

The impact on the level of WASH services people will experience due to climate change is complex. As with many other impacts this depends on hazard, exposure and resilience. Here, impacts are categorized in three ways: (i) a complete loss of service, requiring an immediate shift to a different WASH system; (ii) a change in the level of WASH services that may be temporary or require longer-term adaptation; and (iii) changes in user behaviour that affect the level of WASH service experienced. The evidence of health effects, presented above, is related to floods and droughts in general. These data provide limited insight to the specific role of water and sanitation. Included below is the evidence for the health burden related to specific changes in the service level.

Loss of service

The most critical impacts of climate change on water and sanitation systems are those that result in a total loss of service, of sufficient duration to require an immediate alternative. These events have the potential to generate rapid health crises. When they occur at scale, at least in well-governed and well-resourced communities, they receive prompt and significant attention. Examples include destruction of infrastructure or loss of functionality such as caused by earthquakes, war or extreme weather events.

Floods and storms commonly lead to destruction of infrastructure, including for water and sanitation. Soil erosion due to flooding and heavy rainfall can undermine pipes and other equipment. High turbidity and flooded infrastructure can require shut downs of water supply (Charles *et al.*, 2010). During droughts, loss of system function can occur due to insufficient water. Decreases in dam storage and groundwater levels are widely reported in droughts, but direct data are lacking to assess the scale of impact on small water systems. As water sources dry up, people increasingly rely on fewer water sources, accelerating the frequency of breakdown at the remaining water sources. Longer-term shifts in water availability, such as from sea-level rise and declines in groundwater recharge, can also result in sudden changes to water availability requiring immediate alternatives.

When existing WASH services are lost, people need an alternative. In more resilient and wealthier communities, adverse health impacts can be minimized by rapid implementation of alternative WASH services, such as bottled water or temporary emergency water treatment as is used during flooding in Europe (Charles *et al.*, 2010). In many settings, however, people make do with a lower level of WASH services, relying on unimproved water supplies or sanitation. In droughts, this may necessitate travelling long distances for water, exposing women and girls particularly to musculoskeletal health risks, gender-based violence and the opportunity cost of lost time (Dometita, 2017). In some cases, drought will help stimulate migration to areas with more reliable water supplies, such as cities, placing additional pressure on infrastructure and health systems.

Change of WASH service level

Frequently, the impacts of climate change and weather are less obvious or less immediate than a total loss of WASH service. These changes can still have substantial health impacts, but they are less well understood, due to insufficient data and analysis. Temporary or long-term service interruptions may occur, and may become more frequent as extreme events increase. Two areas considered here are the reliability of the WASH service and water quality.

Decreasing reliability of WASH services has been associated with drought-associated water scarcity, but can also occur if there is any long-term trend to reduced water availability, such as diversion of water for irrigation. Lower water usage is not necessarily problematic for health; humans can safely adapt within limits. There are wide estimates for the amount needed per person per day. However, unreliable water availability requiring shifts in behaviour or temporary use of alternative sources *is* likely to harm health. Cape Town's Day Zero campaign (in response to its 2017/18 drought) illustrates impacts in a large system. In this case, water usage was restricted to 50 l per day (Ziervogel *et al.*, 2010). Decreasing water availability is also associated with increasing domestic water storage, which can easily provide breeding sites for water-related vectors. For example, an increased range of domestic water storage tanks in Australia could expand the range of the mosquito *Aedes aegypti*; in turn this has been suggested as driving postulated climate change-associated expansion of dengue in Australia (Beebe *et al.*, 2009).

Changes in water quality have direct impacts on health. Limited monitoring of water quality in disadvantaged areas is common, leading to increased risk of adverse health impacts. Three key mechanisms for changes with climate are: (i) increased faecal contamination associated with heavy rainfall events; (ii) increased salinity from sea level rise; and (iii) increased algal blooms. Increases in faecal contamination and diarrhoea with heavy rainfall are widely reported, including in piped-water systems in high-income countries (Hrudey and Hrudey, 2019). In low- and middle-income countries, faecal contamination is seasonal in many systems, indicating poor management of seasonal water quality risks, and it increases with extreme rainfall events (Kostyla *et al.*, 2015; Charles *et al.*, 2022). In Bolivia, poorer water quality due to flooding resulted in people switching water sources or avoiding drinking water, leading to increases in dehydration in children post-flood (Rosinger, 2018). Sea level rise is increasing salinity in coastal water supplies, through punctuated jumps associated with storm surges as well as more gradual changes (Murgatroyd *et al.*, 2021). Algal blooms in rivers and lakes are associated with drought, due to low flow, warm weather and increasing nutrients concentrations, but also with drought-breaking rainfall as nutrients are flushed off the land and into reservoirs (Khan *et al.*, 2017). These changes in water quality can be addressed through treatment; however, the additional management and financial costs, and the lack of monitoring data in smaller water systems, make treatment unlikely in many settings.

Change in user behaviour

People respond to changes in weather, and to the perceived risks to water associated with weather, in ways that also affect the service levels they experience. These behaviours are generally highly contextual, depending on the existing level of service and the climate, and will change with changing service levels. It is important to be aware of these in considering how to manage the health impacts of climate change via water and sanitation. Seasonal changes in household treatment of water, hygiene practices and community water management practices have been observed in Bangladesh, Nepal and Tanzania, impacting the level of faecal contamination in drinking water (Charles *et al.*, 2022). Seasonal variability in choices of water sources is reported, not just from necessity, but from preference for rainwater when available, with implications for financial sustainability of managed water supplies (Hoque and Hope, 2020).

Limitations of the evidence

The health benefits and impacts of water and sanitation are mediated by many contextual factors. WASH interventions may interrupt some causal pathways for waterborne diseases, but other factors also influence health outcomes, such as nutrition, parasitic and other chronic infections, and environmental exposures to animal and human faeces (including exposure via bare feet – see Chapter 18). Most evidence evaluates improvements in just one aspect of the level of service, such as water treatment or water quantity. RCTs have been argued to be a poor tool to evaluate WASH interventions, due to these complexities and the biases they create from only doing research in settings where it is feasible to achieve cluster randomization (Schmidt, 2014). As mentioned above, they rarely track the influence of climate.

The impact of climate change and even of weather on WASH services is understudied. The evidence presented here is derived from the exploration of research questions related to other issues, rather than to systematic approaches to investigate the effect of climate change. This creates a large gap in the knowledge base needed to understand the scale of vulnerability to climate changes. Due to insufficient attention at the design stage, existing studies pose questions over how data should be interpreted. For example, the nationally representative surveys of drinking-water quality that track progress toward the Sustainable Development Goals are usually undertaken in the dry season; repeat surveys may not be comparable if weather conditions differ (Bain *et al.*, 2021; Charles and Greggio, 2021). Furthermore, data on changes in service levels are limited, complicating estimates for the frequency or magnitude of impacts.

This chapter has focused on WASH in households, but WASH access is critical in all parts of people's lives, including work, school and healthcare facilities. Evidence is even more limited in these settings.

19.5 Conclusion: Building Climate Resilience for Water and Sanitation Services

Building resilience in water and sanitation services is essential to manage the threats to health from climate change. Despite the limitations in the evidence presented above, a high level of water and sanitation services is undoubtedly critical to health (Cumming *et al.*, 2019). However, advancing and maintaining universal access to safely managed drinking water and sanitation will become increasingly challenging with the variability in the water cycle and the potential for rapid migration of people in response to climate events. Many of the other effects of climate change (unless quickly addressed) are also likely to impair social capacity to deliver and maintain adequate WASH services.

An improved resilience of WASH services also requires better access to more granular climate information at the appropriate scale, more evidence on the impact of climate on WASH systems, shifts within organizations to build capacity to utilize this information to inform water management decisions, and the financing to support these changes (Murgatroyd *et al.*, 2021). A range of tools have been developed to support WASH managers to respond to these changing risks, such as the World Health Organization's climate-resilient water safety planning approach (WHO, 2017). Resilience can be improved through strategic infrastructure design, but it will not be achieved without addressing the management of that infrastructure to ensure that a high level of service is maintained, and without addressing the socio-political issues that restrict continuous use of those services. Critically, changes are needed to ensure that operational finance is available, not just capital finance to install infrastructure.

Finally, considerable progress has been made in the past two decades to provide access to WASH. The challenge now is to adapt and maintain those systems to increase resilience of the systems to extreme events, sea level rise and longer-term changes in climate, and ensure resilience of new infrastructure. Even without invoking 'civilization breakdown' scenarios, poorly controlled climate change, interacting with other aspects of limits to growth, threatens to generate immense difficulties.

Note

[1] More information on the Joint Monitoring Programme, an initiative of WHO and UNICEF, is available at https://washdata.org/

References

Ahern, M.J., Kovats, R.S., Wilkinson, P., Few, R. and Matthies, F. (2005) Global health impacts of floods: epidemiological evidence. *Epidemiological Reviews* 27, 36–45. DOI: 10.1093/epirev/mxi004

Alderman, K., Turner, L.R. and Tong, S. (2012) Floods and human health: A systematic review. *Environment International* 47, 37–47. DOI: 10.1016/j.envint.2012.06.003

Asmall, T., Abrams, A., Röösli, M., Cissé, G., Carden, K. *et al.* (2021) The adverse health effects associated with drought in Africa. *Science of The Total Environment* 793, 148500. DOI: 10.1016/j.scitotenv.2021.148500

Bain, R., Johnston, R., Khan, S., Hancioglu, A. and Slaymaker, T. (2021) Monitoring drinking water quality in nationally representative household surveys in low- and middle-income countries: cross-sectional analysis of 27 multiple indicator cluster surveys 2014–2020. *Environmental Health Perspectives* 129, 97010. DOI: 10.1289/ehp8459

Bartram, J. and Hunter, P. (2015) Bradley classification of disease transmission routes for water-related hazards. In: Bartram, J. (ed.) *Routledge Handbook of Water and Health*. Routledge, London, pp. 20–37.

Beebe, N.W., Cooper, R.D., Mottram, P. and Sweeney, A.W. (2009) Australia's dengue risk driven by human adaptation to climate change. *PLoS Neglected Tropical Diseases* 3, e429. DOI: 10.1371/journal.pntd.0000429

Charles, K.J. and Greggio, E. (2021) Invited perspective: beyond national water quality surveys: improving water quality surveillance to achieve safe drinking water for all (Sustainable Development Goal 6.1). *Environmental Health Perspectives* 129, 091301. DOI: 10.1289/EHP10064

Charles, K., Pond, K., Pedley, S., Hossain, R. and Jacot-Guillarmod, F. (2010) *Vision 2030: the resilience of water supply and sanitation in the fact of climate change.Technology Projection Study*. Robens Institute, University of Surrey, Guildford, UK. Available at: https://www.ircwash.org/resources/vision-2030-resilience-water-supply-and-sanitation-face-climate-change-technology (accessed 31 January 2024)

Charles, K.J., Nowicki, S., Thomson, P. and Bradley, D. (2019) Water and health. In: Dadson, S.J., Garrick, D.E., Penning-Rowsell, E.C., Hall, J.W., Hope, R. *et al.* (eds.) *Water Science, Policy, and Management.* Wiley-Blackwell, Hoboken, New Jersey, pp. 97–116.

Charles, K., Howard, G., Prats, E.V., Gruber, J., Alam, S. *et al.* (2022) Infrastructure alone cannot ensure resilience to weather events in drinking water supplies. *Science of the Total Environment* 813, 151876. DOI: 10.1016/j.scitotenv.2021.151876

Charnley, G.E.C., Kelman, I. and Murray, K.A. (2022) Drought-related cholera outbreaks in Africa and the implications for climate change: a narrative review. *Pathogens and Global Health* 116(1), 3–12. DOI: 10.1080/20477724.2021.1981716

Checkley, W., Buckley, G., Gilman, R.H., Assis, A.M., Guerrant, R.L. *et al.* (2008) Multi-country analysis of the effects of diarrhoea on childhood stunting. *International Journal of Epidemiology* 37, 816–830. DOI: 10.1093/ije/dyn099

Cumming, O., Arnold, B.F., Ban, R., Clasen, T., Esteves Mills, J. *et al.* (2019) The implications of three major new trials for the effect of water, sanitation and hygiene on childhood diarrhea and stunting: a consensus statement. *BMC Medicine* 17, 173. DOI: 10.1186/s12916-019-1410-x

Danaei, G., Andrews, K.G., Sudfeld, C.R., Fink, G., McCoy, D.C. *et al.* (2016) Risk factors for childhood stunting in 137 developing countries: a comparative risk assessment analysis at global, regional, and country levels. *PLoS Medicine* 13, e1002164. DOI: 10.1371/journal.pmed.1002164

Dometita, M.L.M. (2017) *Beneath the Dryland: Kenya Drought Gender Analysis*. Available at: www.oxfam.org (accessed 27 September, 2021).

Geere, J.-A.L., Cortobius, M., Geere, J.H., Hammer, C.C. and Hunter, P.R. (2018) Is water carriage associated with the water carrier's health? A systematic review of quantitative and qualitative evidence. *BMJ Global Health* 3, e000764. DOI: 10.1136/bmjgh-2018-000764

Hoque, S.F. and Hope, R. (2020) Examining the economics of affordability through water diaries in coastal Bangladesh. *Water Economics and Policy* 06, 1950011. DOI: 10.1142/s2382624x19500115

Hrudey, S.E. and Hrudey, E.J. (2019) Common themes contributing to recent drinking water disease outbreaks in affluent nations. *Water Supply* 19, 1767–1777. DOI: 10.2166/ws.2019.051

IPCC (2021) Summary for Policymakers. In: Masson-Delmotte, V., Zhai, P., Pirani, A., Connors, S.L., Péan, C. *et al.* (eds) *Climate Change 2021: The Physical Science Basis*. Contribution of Working Group I to the Sixth Assessment Report of the Intergovernmental Panel on Climate Change. Cambridge University Press, Cambridge, UK, pp. 1–41.

John, T.-J., Lalla, U., Taljaard, J.J., John, K.G., Slabbert, J. *et al.* (2017) An outbreak of community-acquired Pseudomonas aeruginosa pneumonia in a setting of high water stress. *QJM* 110, 855–856. DOI: 10.1093/qjmed/hcx148

Khan, S.J., Deere, D., Leusch, F.D.L., Humpage, A., Jenkins, M. *et al.* (2017) Lessons and guidance for the management of safe drinking water during extreme weather events. *Environmental Science: Water Research & Technology* 3, 262–277. DOI: 10.1039/c6ew00165c

Korpe, P.S. and Petri, W.A. (2012) Environmental enteropathy: critical implications of a poorly understood condition. *Trends in Molecular Medicine* 18, 328–336. DOI: 10.1016/j.molmed.2012.04.007

Kostyla, C., Bain, R., Cronk, R. and Bartram, J. (2015) Seasonal variation of fecal contamination in drinking water sources in developing countries: A systematic review. *Science of The Total Environment* 514, 333–343. DOI: 10.1016/j.scitotenv.2015.01.018

Levy, K., Woster, A.P., Goldstein, R.S. and Carlton, E.J. (2016) Untangling the impacts of climate change on waterborne diseases: a systematic review of relationships between diarrheal diseases and temperature, rainfall, flooding, and drought. *Environmental Science & Technology* 50, 4905–4922. DOI: 10.1021/acs.est.5b06186

Luby, S.P., Agboatwalla, M., Bowen, A., Kenah, E., Sharker, Y. *et al.* (2009) Difficulties in maintaining improved handwashing behavior, Karachi, Pakistan. *American Journal of Tropical Medicine and Hygiene* 81, 140–145. DOI: 10.4269/ajtmh.2009.81.140

Murgatroyd, A., Charles, K.J., Chautard, A., Dyer, E., Grasham, C. *et al.* (2021) *Water Security for Climate Resilience Report: A Synthesis of Research from the Oxford University REACH programme*. REACH, University of Oxford, UK.

Overbo, A., Williams, A.R., Evans, B., Hunter, P.R. and Bartram, J. (2016) On-plot drinking water supplies and health: a systematic review. *International Journal of Hygiene and Environmental Health* 219, 317–330. DOI: 10.1016/j.ijheh.2016.04.008

Pickering, A.J., Null, C., Winch, P.J., Mangwadu, G., Arnold, B.F. *et al.* (2019) The WASH Benefits and SHINE trials: interpretation of WASH intervention effects on linear growth and diarrhoea. *The Lancet Global Health* 7, e1139–e1146. DOI: 10.1016/S2214-109X(19)30268-2

Prüss-Ustün, A., Wolf, J., Bartram, J., Clasen, T., Cumming, O. *et al.* (2019) Burden of disease from inadequate water, sanitation and hygiene for selected adverse health outcomes: An updated analysis with a focus on low- and middle-income countries. *International Journal of Hygiene and Environmental Health* 222, 765–777. DOI: 10.1016/j.ijheh.2019.05.004

Reller, M.E., Mendoza, C.E., Lopez, M.B., Alvarez, M., Hoekstra, R.M. *et al.* (2003) A randomized controlled trial of household-based flocculant-disinfectant drinking water treatment for diarrhea prevention in rural Guatemala. *American Journal of Tropical Medicine and Hygiene* 69, 411–419. DOI: 10.4269/ajtmh.2003.69.411

Rieckmann, A., Tamason, C.C., Gurley, E.S., Rod, N.H. and Jensen, P.K.M. (2018) Exploring droughts and floods and their association with cholera outbreaks in Sub-Saharan Africa: a register-based ecological study from 1990 to 2010. *American Journal of Tropical Medicine and Hygiene* 98, 1269–1274. DOI: 10.4269/ajtmh.17-0778

Rosinger, A.Y. (2018) Household water insecurity after a historic flood: Diarrhea and dehydration in the Bolivian Amazon. *Social Science & Medicine* 197, 192–202. DOI: 10.1016/j.socscimed.2017.12.016

Sahoo K.C., Hulland K.R.S., Caruso B.A., Swain R., Freeman M.C. *et al.* (2015) Sanitation-related psychosocial stress: A grounded theory study of women across the life-course in Odisha, India. *Social Science & Medicine* 139, 80–89. DOI: 10.1016/j.socscimed.2015.06.031

Saulnier, D.D., Brolin Ribacke, K. and von Schreeb, J. (2017) No calm after the storm: a systematic review of human health following flood and storm disasters. *Prehospital and Disaster Medicine* 32, 568–579. DOI: 10.1017/S1049023X17006574

Schmidt, W.-P. (2014) The elusive effect of water and sanitation on the global burden of disease. *Tropical Medicine and International Health* 19, 522–527. DOI: 10.1111/tmi.12286

Stanke, C., Kerac, M., Prudhomme, C., Medlock, J. and Murray, V. (2013) Health effects of drought: a systematic review of the evidence. *PloS Currents*, 5. DOI: 10.1371/currents.dis.7a2cee9e980f91ad7697b570bcc4b004

Waddington, H. and Snilstveit, B. (2009) Effectiveness and sustainability of water, sanitation, and hygiene interventions in combating diarrhoea. *Journal of Development Effectiveness* 1, 295–335. DOI: 10.1080/19439340903141175

White, G.F., Bradley, D.J. and White, A.U. (1972) *Drawers of Water: Domestic Water Use in East Africa*. The University of Chicago Press, Chicago, Illinois.

WHO (2017) Climate-resilient water safety plans: Managing health risks associated with climate variability and change. World Health Organization, Geneva. Available at: http://apps.who.int/iris/bitstream/handle/10665/258722/9789241512794-eng.pdf (accessed 28 September 2021).

WHO and UNICEF (2021) *Progress on Household Drinking Water, Sanitation and Hygiene 2000–2020*. Available at: https://data.unicef.org/resources/progress-on-household-drinking-water-sanitation-and-hygiene-2000-2020/ (accessed 26 January 2024).

Ziervogel, G., Shale, M. and Du, M. (2010) Climate change adaptation in a developing country context: The case of urban water supply in Cape Town. *Climate and Development* 2, 94–110. DOI: 10.3763/cdev.2010.0036

20 Air Pollution, Fires, Climate Change and Health

Colin D. Butler[1] **and Ivan C. Hanigan**[2]
[1]*National Centre for Epidemiology and Population Health, The Australian National University, Canberra, Australia;* [2]*Faculty of Health Science and School of Population Health, Curtin University, Perth, Australia*

Abstract
Air pollution from particulate matter and noxious gases, generated mainly by burning fossil fuels and biomass, has enormous short- and long-term adverse health effects, not only to the respiratory system (as long recognized) but also to other bodily systems. The impacts include cardiovascular, neurological and immune; maternal and infant health can also be harmed. Climate change, chiefly caused by the high and still rising concentrations of three invisible gases, can itself be considered a consequence of air pollution. Rising greenhouse gases, together with human-induced landscape change, interacts with particulate air pollution by increasing the frequency and intensity of wildfires – a reinforcing feedback. Yet, particulate air pollution acts to partially mask global heating, which shows how complex the issue is. Air pollution can be categorized as ambient, indoor and household, associated with different driving forces and with different impacts. Reducing all forms of air pollution is an urgent and 'wicked' problem.

20.1 Introduction

There is now widespread acknowledgement that the burden of disease attributable to air pollution rivals other major global health risks, such as unhealthy diets and tobacco smoking (WHO, 2021). This recognition is surprisingly recent; it has been slowed by the time needed to identify, measure and gain popular acceptance for this concept, and also by political opposition. Commensurate action to reduce major air pollutants is an ongoing struggle, now mostly playing out in the global South. This chapter sketches some of the main issues involved, before turning its focus to fires – which contribute to air pollution – and climate change.

Climate change, chiefly caused by the high and still rising concentrations of the three invisible greenhouse gases carbon dioxide (CO_2), methane (CH_4) and nitrous oxide (N_2O), can itself be considered a consequence of a form of air pollution. These three transparent gases do not harm health at their normal concentrations but, if allowed to increase, they can lead to global warming and a wide range of subsequent health impacts. In fact, small quantities of these 'greenhouse' gases are essential for life on Earth, to maintain a habitable temperature. While first theorized in the 19th century, official recognition that climate change could become extremely problematic dates to 1965, when a US government speculated on the need to 'deliberately' cause 'countervailing climatic changes' (Keith, 2000).

Climate change interacts with air pollution in many ways, including via wildfires. This is discussed later in this chapter.

 Climate Change and Global Health: Primary, Secondary and Tertiary Effects, 2nd Edition (eds C.D. Butler and K. Higgs)
DOI: 10.1079/9781800620025.0020

20.2 Air Pollution Components

Components of air pollution are often classified as 'primary', such as nitrogen and sulfur oxides (NOx and SOx), carbon monoxide (CO), volatile organic compounds (VOCs) and methane (CH_4), and 'secondary', examples of which include ozone (O_3) and organic aerosols. Other contributors to air pollution include particulate matter (PM) – which can be primary or secondary – and synthetic halocarbons, such as chlorofluorocarbons. Halocarbons also help trap heat, while the effects of PM on climate change are complex but generally lead to cooling, due to interaction between aerosols and cloud droplets.

Ozone

A gas of public health concern is O_3. Although its characteristic odour was recognized by the ancient Greeks,[1] O_3 was identified in the 19th century as a new molecular form of oxygen. In its tropospheric[2] setting it is a major component of smog, especially photochemical[3] (Sillman, 1993). Tropospheric O_3 concentrations rise due to photochemical reactions involving precursors like VOCs and NOx. These reactions occur more rapidly during periods of elevated solar insolation and air temperature. Inhaled O_3 is highly reactive and causes damage to biological molecules through oxidation. O_3 has long been recognized as an irritant that reduces lung capacity, and some studies show it may – by itself – cause premature mortality (Zanobetti and Schwartz, 2011; Huangfu and Atkinson, 2020).

As human exposure to O_3 is rarely solitary, its harmful health effect is difficult to disentangle (Chen *et al.*, 2007). There are also interactions with other air pollution sources; for example, tropospheric O_3 production is enhanced by bushfire smoke (Jaffe and Wigder, 2012) and dust. As climate change will, in many places, enhance smoke, dust, heat and VOCs, it follows that tropospheric O_3 will also increase. Furthermore, tropospheric O_3 pollution has been shown to harm crop growth, for example in China (Feng *et al.*, 2019) and some cultivars of African food staples (Hayes *et al.*, 2020).

O_3 also occurs high in the atmosphere (stratosphere). In this setting, it is not a pollutant, but an essential and beneficial element of the Earth system. Produced through natural atmospheric chemistry processes, it protects the biosphere from excessive ultraviolet radiation. However, this high altitude O_3 'layer' can also be thinned by interaction with halocarbons such as chlorofluorocarbons. This insight, confirmed in the 1980s, led to effective action, industrial substitution for most O_3-depleting substances, and stabilization of this problem, unlike that of climate change (Crutzen, 2002).

Particulate matter

The importance of PM to health has long been recognized but the term is comparatively recent. Previous names for PM include 'smoke', 'soot', 'fumes', 'haze' and 'dust'. The term smog was introduced to indicate a mixture of PM and other atmospheric variables. Formally, PM is the sum of all solid and liquid particles suspended in air. It is PM that obscures visibility. Much PM is carbonaceous, i.e. carbon-containing. PM spans at least four orders of magnitude (1–10,000 nm) and has a large range of chemical compositions. Their size ranges from ultrafine (< 1 μm aerodynamic diameter) (PM_1), fine (1–2.5 μm) ($PM_{2.5}$), to coarse (2.5–10μm) (PM_{10}).

PM comes from many forms of combustion (including all forms of fossil fuels) as well as dust from industrial and natural processes, such as from deserts and windblown sea salt. This can bind with other material as it ages, which is of concern in coastal urban areas. PM can also arise by atmospheric chemistry, i.e. from reactions involving inorganic (sulfur dioxide (SO_2), NO_x and ammonia (NH_3)) and organic (VOC) gases. Smoke from wildfires typically contains smaller particles than from the combustion of fossil fuels, with a higher proportion of $PM_{2.5}$ and PM_1 than PM_{10} (Xu *et al.*, 2020). Wildfire-sourced PM may also include acids (both organic and inorganic compounds).

PM is generally considered the main harmful component of air pollution (EPA, 2022) but many PM-related studies struggle with the issue of mixed co-pollutants. It may be that studies that rely on PM as the key exposure variable are actually capturing information about a mix of pollutants, leading to subtle variations in the aetiological role attributed to it.

Visible particles (i.e. some far larger than PM_{10}) may be trapped on the mucosa in the outer layer of the human airway, with their deeper penetration of the lungs reduced by the clearing action of cilia – fine hairs that line the upper airway and move particles upwards by their motion. Particles with a diameter less than 10 μm are too fine to be trapped and thus reach the alveoli – the air sacs that adjoin the membrane across which oxygen (O_2) and CO_2 diffuse into the bloodstream. Even finer particles ($PM_{2.5}$) can cross the lung–bloodstream barrier, i.e. they can enter into blood vessels (behaving like a gas) and thus reach most cells in the body, where they can trigger inflammatory responses. PM may even cross the blood–brain barrier, with adverse neurological impacts (Kim *et al.*, 2020).

PM also contributes to changes in the Earth's energy balance both by absorption of radiation (e.g. black carbon) and by the scattering of light and heat. When PM mixes with clouds it can increase cloud reflectivity and duration with a cooling effect, partially offsetting greenhouse gas-induced warming (Villanueva *et al.*, 2022). Many of the ecosystem effects of air pollutants arise from the deposition of PM on plant surfaces, in both dry and wet weather. Efforts to control PM will probably dominate air pollution legislation in the coming decades (Fowler *et al.*, 2020).

20.3 Air Pollution and Health: History and Regulation

Air pollution is reported to have impacted health in ancient Rome, due to the inhalation of smoke from the burning of animal and vegetable oils for light, and the burning of wood, other vegetal materials and animal dung for cooking and heating. Anthracosis (a black discoloration of the bronchial mucosa) has also been reported in Egyptian and Peruvian mummies (Capasso, 2000). In China, smoke and noxious gases from the burning of coal were recognized as problematic during the Song Dynasty (ruling from 960 to 1279) (Fowler *et al.*, 2020).

Brimblecombe (1977) reported that two enquiries into air pollution in London were held in the 1280s, and that despite this the problem steadily worsened, not only due to the burning of coal (at first for the production of lime, rather than for use within houses, which were initially heated by wood) but also from the stench of sewage. Smoke from tobacco is an indoor form of air pollution which was introduced to Europe from the Americas in the 16th century, as part of the Columbian exchange. Due to the immensity of this topic, and also that WHO does not classify tobacco (cigarette) smoking as an environmental health issue, this topic is not otherwise discussed in this book, including in this chapter. We note that the health effects of tobacco exposure interact with individual susceptibility, and often magnify the health effects of air pollution from other sources due to co-morbidities.

By the mid-17th century, the statistician and pioneering epidemiologist John Graunt attributed the higher mortality in London, compared with rural areas, to air pollution (Brimblecombe, 1977). Brimblecombe also noted that a reduction in light, due to air pollution in London, was mentioned by the diarist John Evelyn in 1661. Fowler *et al.* (2020) reported that, as early as the 17th century, observers were speculating that different components of air pollution may have different effects, including from acid, sulfur, antimony and lead.

The first attempts to measure and record air quality systematically were undertaken in England, in the second half of the 19th century. From that time, good records of coal consumption also began to be kept in England and Wales. By the mid-19th century domestic coal consumption in England and Wales exceeded 60 million tonnes, as the Industrial Revolution rapidly progressed (Fig. 20.1).

In 1878, the UK government conducted a Royal Commission on noxious vapours. Its recommendations were ineffectual and, in any case, not enforced, as the government prioritized industry and economic growth (Fowler *et al.*, 2020). In 1905, at the London Public Health Congress, the word 'smog' was introduced. The term was conceived to combine the words 'smoke' and 'fog'. Today, smog is recognized to include both PM and (transparent) gaseous components (Fowler *et al.*, 2020).

Attempts to regulate air pollution in the 20th century

In the early 20th century new laws to try to reduce air pollution were introduced, not only in Europe and the USA, but also in parts of the British

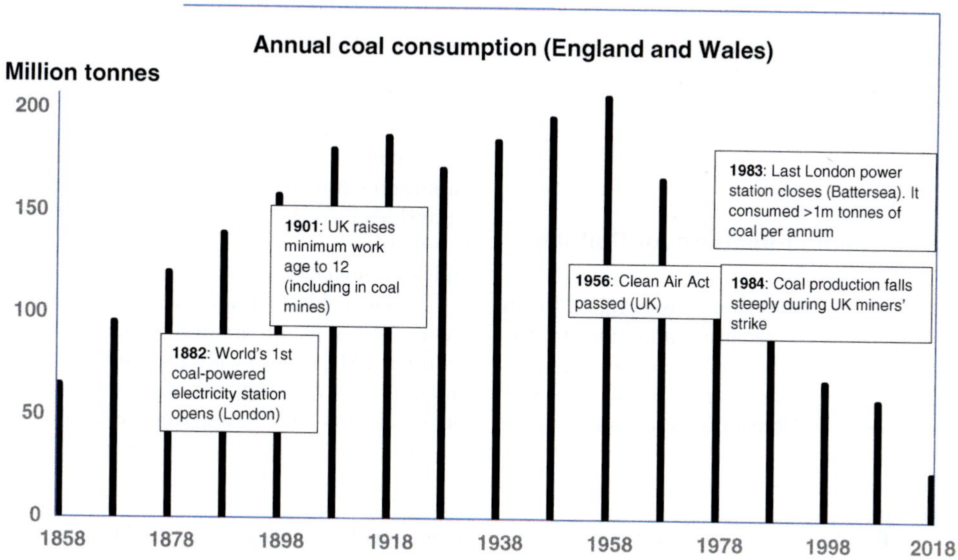

Fig. 20.1. Annual coal consumption in England and Wales, averaged by decade from 1853–1862 to the period 2013–2021. Data available at: https://www.gov.uk/government/statistical-data-sets/historical-coal-data-coal-production-availability-and-consumption (accessed 10 October 2022).

empire. The motivation to do this was strengthened by air pollution disasters in three industrialized countries, in which the health harm from smog was undeniable. The first was in 1930 in the Meuse Valley, Belgium, a narrow valley and thus susceptible to temperature inversion layers (Nemery *et al.*, 2001). Eighteen years later, in 1948, a choking smog in the steel town of Donora, Pennsylvania, USA (also in a valley) took dozens of lives. Both occurrences left thousands literally gasping for breath, while the Donora event led the United States Public Health Service to enter the arena of air pollution policy (Snyder, 1994). In 1952, in London, the 'great smog' resulted in the premature death of an estimated 12,000 people (Bell and Davis, 2001). It stimulated passage of the Clean Air Act in 1956 but for some time after this Act was passed the UK government resisted further efforts to control air pollution. The environmental historian Berridge (2006), for example, noted that 'although government was wary about the smoking and lung cancer case as a policy issue, it was infinitely preferable to air pollution. That was the issue which government did not want reopened.' Acid rain and forest decline, each linked to air pollution, also generated legislation that improved air quality, including in western Europe, North America and Japan (Reis *et al.*, 2012). In Ireland, coal bans in several cities resulted in immediate and sustained decreases in particulate concentrations, especially in winter. This was associated with reduced respiratory mortality (Dockery *et al.*, 2013).

Attempts to regulate air pollution in the 21st century

As emissions of SO_2, NO_x and VOCs declined in the global North (in some cases by more than 50% from their peaks), those in many parts of Asia increased, as they followed the precedent set by global North and as their economies and populations increased (Fowler *et al.*, 2020). Some of this pollution was to produce goods to be used in the global North, i.e. some air pollution was effectively 'outsourced' to China and the global South (Wiedmann *et al.*, 2020). Globally, emissions of SO_2 have declined modestly since their peak in the late 1990s, but this is not yet the case for NOx and VOCs (Hoesly *et al.*, 2018). From 2000 to 2014 the most heavily $PM_{2.5}$ polluted areas have been found to be in north-western Africa, eastern China and northern India, followed by parts of the Middle East (Li *et al.*, 2018). Some of this is over the Sahara Desert, apparently from locally sourced dust (Fayiga *et al.*, 2018).

20.4 Health Effects of Air Pollution

In 1993 the 'six cities studies' research in North America brought renewed attention to the human health effects of 'inhalable' and fine particulate matter (Dockery *et al.*, 1993). This was considered groundbreaking because it was the first large air pollution study to control carefully for smoking and other known risk factors. Today, air pollution is recognized as the largest environmental health risk by the World Health Organization (WHO) which, in 2016, estimated that 7 million deaths annually could be attributed to air pollution, second only to tobacco smoking as a cause of chronic disease (Campbell-Lendrum and Prüss-Ustün, 2019). This burden is believed to be primarily caused by exposure to $PM_{2.5}$.

Lelieveld *et al.* (2020) provided an even larger estimate of excess mortality, exceeding that of tobacco smoking. They reported almost 9 million premature deaths per year, with each person losing, on average, 2.9 years of life expectancy (95% confidence interval 2.3–3.5). They reported that mortality from air pollution was dominated by populations in East and South Asia (67%), followed by Africa (11%) and Europe (9%).

Air pollution affects many bodily systems, but the most important are respiratory and cardiovascular. PM exposure contributes to acute lower respiratory infections in children younger than five, an important cause of death and morbidity in the global South (Odo *et al.*, 2022). Tobacco smoke exposure, of course, adds to their likelihood and burden. Other respiratory diseases attributed in part to air pollution include lung cancer, asthma and chronic obstructive airways disease. Cardiovascular conditions that are associated with air pollution include ischaemic heart disease and stroke (Campbell-Lendrum and Prüss-Ustün, 2019). Prolonged exposure to low concentrations of air pollutants can also be harmful, as can short-term exposure to high concentrations, including by causing respiratory inflammation (Kampa and Castanas, 2008).

Recognition of the adverse impact of air pollution on cardiovascular health is comparatively recent; even more recently, studies have shown that air pollution harms cognitive functioning, school performance and neurological function and increases the risk of dementia and Parkinson's disease (Chen *et al.*, 2017). These relationships hold even after adjustment for confounders such as inadequate nutrition, poverty and household illiteracy.

Renal health, immune function and maternal and infant health are also vulnerable to air pollution. $PM_{2.5}$ exposure has been linked with as much as 20% of the global burden of chronic kidney disease (CKD), mostly in low- and middle-income countries (Al-Aly and Bowe, 2020). Silica and other components of air pollution released from burning sugar cane stubble may interact with heat, dehydration and pesticide exposure to explain part of the burden of chronic kidney disease of unknown origin (Johnson *et al.*, 2019). This is a major cause of premature disease in such settings, with devastating harm to sufferers and their families. A systematic review of eight studies undertaken in four countries (two of them in the global South), involving over 1.7 million births, concluded that maternal exposure to wildfire smoke during late pregnancy is associated with lower birth weight and a heightened risk of prematurity (Amjad *et al.*, 2021).

20.5 Household, Indoor and Ambient Air Pollution

For many years, air pollution was primarily classified as either 'indoor' (such as from cooking fires using non-gaseous or solid fuels – coal, vegetation including wood and dung – in poorly ventilated settings) or 'outdoor' ('ambient') from dust, motor vehicles, the burning of coal for heat or for electricity and from cooking and other fires in the open. For example, an estimated 500 million tonnes of crop residues on the Indo-Gangetic plain, in India, are burned each winter and autumn. This coincides with and contributes to the peak of Delhi's annual 'great smog' (Basu, 2019).

In many settings in the global South, e.g. South Asia and parts of Africa and South America, the distinction between indoor and outdoor pollution has been criticized because of the fallacy that indoor air pollution in the global South can be greatly alleviated simply by better ventilation (Pillarisetti *et al.*, 2022). If the quality of outside air is very poor, then a ventilated home will allow this pollution to combine with that generated indoors. Supporters of this competing view have proposed the term 'household associated pollution' (HAP) as an alternative to indoor air pollution. Supporters argue that wider use of the term HAP may reduce this misunderstanding. They propose, instead, that the solution lies in cleaner indoor fuel, together with altered farming practices and other methods to reduce ambient air pollution, for example by the incorporation of scrubbing devices in the chimneys of coal-fired power stations.

The virtue of the term HAP (instead of indoor air pollution) may be difficult for some in the global North to appreciate. In richer countries, chimneys and other forms of ventilation mean that indoor air quality is generally acceptable, even if solid fuel such as wood is burned inside. But in the global South, in some settings, even an exemplary chimney cannot guarantee reasonable indoor air quality. This is primarily because of the persistently foul quality of outdoor air in some places. For example, in November 2017, the locally derived air quality index (AQI)[4] in Delhi, India (a conurbation of over 30 million), exceeded 999 µg/m^3, the upper limit of the measuring device. These levels are far above the WHO's recommended daily air quality guideline levels, which for $PM_{2.5}$ is 15 µg/m^3 and for PM_{10} is 45 µg/m^3 (WHO, 2021).

In contrast, mean daily $PM_{2.5}$ in Sydney, Australia, between 2010 and 2020 ranged from ~1 to 122 µg/m^3 (Jegasothy *et al.*, 2023). This study period included the 2019/20 fire season, which saw forest fires of unprecedented size and intensity sweep across south-eastern Australia, generating the highest levels of fire smoke-related $PM_{2.5}$ yet recorded in Sydney. In another Australian example, following an effective intervention in Launceston, Tasmania, the mean wintertime concentration of PM_{10} fell from 44 µg/m^3 (1994–2000) to 27 µg/m^3 (2001–2007) (Johnston *et al.*, 2013). In comparison with the levels routinely observed in Delhi, these concentrations in Australia are very mild.

In the global North, during episodes of severe wildfire smoke, the closing of windows is sometimes recommended to reduce indoor air pollution. This can be effective, though less so if smoke persists for days. However, in many places in the tropics, heat and the lack of air conditioning makes exposure to cooler outside air essential, especially at night. Furthermore, in settings where people do much manual work, most time is passed outside. Chimneys in low-income countries, if maintained, can effectively direct air polluted by a cooking fire to the outside. Unfortunately, despite persistent and long-held beliefs and hopes, trials with ventilation-improved stoves in the global South have not been shown to benefit health (Smith and Pillarisetti, 2018). Cooking with gas reduces HAP, but still drives GHG emissions.

A strategy that has been proposed instead is the electrification of cooking, especially if power is generated by renewable energy technology such as wind, solar or hydro, none of which cause significant air pollution. Unfortunately, this approach remains far beyond reach for billions of people, although the 'leapfrog' technology of induction stoves – which are 50% more efficient and 50% faster (as well as safer and longer-lived) than old-style electric stoves – has been suggested as promising, in parts of Asia and in countries with abundant hydro power, such as Ecuador. To date, such hopes remain largely unfulfilled. For example, Gould (2021) reported that induction stoves remain rarely used in Ecuador, due to broken parts and anxiety over running costs.

In 2018, Smith and Pillarisetti pointed out that population growth in low-income settings is outpacing the growth of cleaner fuels such as gas, and that as many as 3 billion people still rely on solid, dirty fuel, a similar number to that of 30 years earlier. Five years later, little seems to have changed but, although HAP is approximately stable in South Asia, it has been reported as declining in East Asia and much of sub-Saharan Africa (Fisher *et al.*, 2021). On the other hand, these authors find that ambient air pollution is projected to continue to increase in Africa.

In the global North, there is a strong case for conserving the term 'indoor air pollution'. Better sealed or 'tighter' housing (to reduce energy costs and in part to adapt to climate change) is associated with improved insulation, especially in settings with cold winters. However, these also increase exposure to indoor pollutant risks, including from gas stoves and heaters and from radon (Walls, 2016), $PM_{2.5}$, CO, SO_2, NOx, NH_3, polycyclic aromatic hydrocarbons including formaldehyde, VOCs (Spiru and Simona, 2017) and CO_2 (which at levels of as low as 1000 parts per million is harmful to health) (Jacobson *et al.*, 2019). Exposures vary with factors such as building materials, the size of the dwelling and its age.

Climate change interacts with air pollution (as more conventionally defined) in many ways. One mechanism is via wildfires; this is discussed in the next section.

20.6 Wildfires

Earth has been called a 'fire planet' with a chronicle of combustion since the onset of terrestrial life (Pyne, 2020). James Lovelock, one of the parents of Earth system science, long ago argued that an oxygen concentration much higher than its current – and long-standing – level (21%) would lead to far more fires; even to ones ignited by wet lightning, i.e. thunderstorms accompanied by substantial rain (Watson *et al.*, 1978). Once started, wildfires are determined by fuel load, weather, topography and fire-fighting effort. High temperatures, low humidity, strong winds and dry soil and vegetation can lead to superfires, some of which create their own weather, and even terrifying phenomena such as fire tornados (Lareau *et al.*, 2022).

Fire has shaped the evolution of humans and their hominin forerunners and contemporaries. It has been hypothesized that active lava flows in the African Rift Valley may have provided consistent but isolated sources of fire, providing specific adaptive pressures and opportunities to small, isolated groups of pre-human fire creatures (Medler, 2011). Its use by humans and their related species, including *Homo erectus* and Neanderthals, may date to 400,000 years ago (Rosell and Blasco, 2019). But the use of fire is not exclusively mammalian; at least three bird species in northern Australia also use fire, opportunistically, to aid their hunting (Bonta *et al.*, 2017). Ancient uses of fire by humans extended beyond light, warmth, protection, cooking and hunting to lithic industry – the fashioning of stone to make tools (Scott *et al.*, 2016).

Many landscapes have evolved with fire; some plants only germinate with fire or smoke. Fire recycles nutrients and can provide light to new generations of plants. However, although humans and fire have always interacted, increasing encroachment by human populations into wildlands creates additional challenges and tensions between competing interests (Scott *et al.*, 2016). Fires, especially when large and hot, can harm not only infrastructure, but also livestock, soil and other ecosystem components. Human lives can be lost and survivors traumatized (Adu *et al.*, 2022). Smoke from large fires not only affects human health (to be discussed) but can also taint crops such as wine grapes, and affect both tropospheric and stratospheric ozone (Cicerone, 1994).

Types of fire

Fires have different names in different settings; descriptors include 'wild', 'forest', 'bush', 'brush', 'scrub' and 'grass' fires. Wildfires can be defined as uncontrolled combustion in a vegetated environment whether a forest, grassland, or in peat; they are generally classified into three forms, depending on their location of combustion, in relation to the ground. Crown fires are terrifying and spectacular; they move quickly in and among the tops of trees, with adjacent tree tops (separated by air) suddenly igniting. Surface fires typically crawl along forest floors; they do the least harm and are the easiest to manage. The third form, ground fires (also known as smouldering and underground fires), can be very hard to extinguish, cause extensive air pollution, emit very large amounts of carbon and are difficult to detect and suppress. All three kinds of fires can occur simultaneously and each may accelerate climate change, particularly by reducing the carbon store in vegetation, which instead moves to the atmosphere and oceans (Bowman *et al.*, 2020; Rein and Huang, 2021).

Wildfire smoke components

Wildfire smoke is a complex mix of gases and PM, the composition of which varies with the kind of fuel. Gaseous components may include

CO_2, CO, complex hydrocarbons and NOx. Trace minerals, heavy metals and other toxic and carcinogenic compounds can also be airborne in smoke, especially if there is combustion of buildings and other manufactured substances, including plastic (Kizer, 2020).

Dry fuels emit more CO_2 than ones that smoulder, which produce more CO, as combustion in such fires is less complete (Rein and Huang, 2021). Wildfire smoke also contains aerosols of black and brown carbon. The exact health effects from different fire components are complex and poorly characterized.

20.7 Wildfires, Smoke and Health

Smoke from wildfires can travel thousands of kilometres and has the potential to generate widespread adverse health impacts. Smoke from the 2021 Siberian fires drifted across the Arctic and North Pole, eventually covering more than 5 million square kilometres (Dixon, 2021). Yet, carbon emissions from Arctic fires in 2021 (16 million tonnes) were only 27% of those of 2020 (58 million tonnes, the record year to date (Copernicus, nd). Smoke from fires in South-East Asia, especially from the clearing of forests for palm oil, regularly affects Kuala Lumpur (Malaysia) and Singapore, even in some non-El Niño years. These fires often involve smouldering peat. In Singapore, daily information on air quality has been available since 2005, but data for Kuala Lumpur have been far more guarded, due to political sensitivities (Forsyth, 2014).

The adverse health effects of smoke from wildfires may even rival the morbidity and mortality from being burned, although direct exposure to the trauma of wildfire (including in cases of forced, often panicky evacuation) is likely to have an additional mental health burden which makes comparisons difficult. For example, in 2016 an unseasonably early and intense wildfire, which eventually took more than two months to extinguish, destroyed 2400 homes in northern Alberta, Canada. It caused the emergency evacuation, only two days after the fire's ignition, of 88,000 people, almost the entire population of the 'tarsands'[5] city of Fort McMurray (Adu *et al.*, 2022).

Five years later, the town of Lytton (population circa 1000), in the adjacent Canadian province of British Columbia, was destroyed by fire. This followed a week of freakishly hot weather, reported to have been made 150 times more likely by the existing levels of greenhouse gases (GHGs) than without them (Schiermeier, 2021). In this week more than 500 excess deaths occurred (mostly from heat) and 180 wildfires in the region were reported. The peak temperature (49.6°C) recorded at Lytton was the highest ever reported in Canada at that time. Lytton's inhabitants were evacuated just before their houses were burned (Schiermeier, 2021).

Such events cause significant psychological trauma to survivors, including post-traumatic stress disorder (PTSD). One year after the Fort McMurray fire, more than a third of evacuees had clinically significant psychological symptoms, including insomnia, depression, anxiety and substance use, in addition to PTSD (Belleville *et al.*, 2021). A later study of evacuees concluded that increasing age and the presence of PTSD symptoms were independently associated with 'low resilience' in 2021 (Adu *et al.*, 2022). These researchers defined resilience as personal capacity to handle inimical life experiences based on their mindset, resourcefulness and social support from family, friends and the community at large. Fort McMurray students, aged from 11 to 19, also showed significant psychological distress (Gao *et al.*, 2023). PTSD symptoms, anxiety, insomnia and depression have also been reported after wildfires in many other locations, including in Australia, the USA and Greece (Psarros *et al.*, 2017; Gao *et al.*, 2023).

Fires can also cause enormous harm to wildlife; in the eastern Australian fires of 2019/20 up to 3 billion animals are estimated to have died (van Eeden and Dickman, 2023) (Fig. 20.2).

Fire and climate change

There is a widespread perception that anthropogenic climate change is worsening fires. This seems plausible, not only because there are increasing reports of megafires, but also made likely by altered precipitation patterns and rising temperatures. Increased fire 'weather' has been reported as globally pervasive between 1979 and 2019 due to climate change; however, the reported burned area in this time has actually decreased, mainly due to less burning in African savannahs (Jones *et al.*, 2022). Bowman *et al.*

Fig. 20.2. A critically endangered regent honeyeater (*Anthochaera phrygia*) against the smoke-shrouded sky, during Australia's 'Black Summer'. Fires raged for weeks, encroaching on the national capital, Canberra, in early 2020. As well as directly harming human health such as via impaired air quality and damage to housing and other infrastructure, the fires had a devastating impact on plant and animal life and therefore, indirectly, on the well-being of humans who care about and depend on these other species. ©Raw Shorty (used with permission).

(2020) argued that the scarcity of adequate historical data means it is too soon to conclude that events such as the Australian 2019/20 bushfire crisis reflect increasing global fire activity related to climate change. They also pointed out that plantations of highly flammable species such as *Pinus* and *Eucalyptus* contributed to ferocious fires in Chile (2017) and Portugal (2018). In many areas, on different continents, the regular burning of country by Indigenous people has also been interrupted, or even lost altogether, which makes very large fires more likely (Bowman *et al.*, 2020). It may be more accurate to say that the Anthropocene is altering fire behaviour, rather than climate change on its own (Fig. 20.3). On the other hand, as with many aspects of climate change, the search for sufficient data to ensure a statistically 'certain' link between climate change and fires may lead to a 'type 2 error', i.e. a failure to detect a true effect (Butler *et al.*, 2016).

Nevertheless, some aspects of climate change appear to be having adverse effects upon wildfires. Fire seasons are starting earlier and ending later in both the northern and southern hemispheres (Bowman *et al.*, 2020) and thus are increasingly overlapping (Richardson *et al.*, 2022). This complicates fire suppression, where considered necessary, when firefighters and their equipment, sometimes including large aircraft capable of dumping fire retardants on large wildfires, are not available for transfer to the opposite hemisphere.

Wildfires are encroaching on and remaining longer at high latitudes, including parts of the conifer forests in Russia and Canada. So called 'zombie' fires can smoulder for months in peat, even when blanketed by snow in the Arctic Circle during winter, then reignite in early spring when temperatures rise (Irannezhad *et al.*, 2020). Fires in the Mediterranean regions are predicted to increase due to climate change (Turco *et al.*, 2018); this will have adverse flow-on effects for tourism.

Although often started accidentally or deliberately by human activity, wildfires appear to

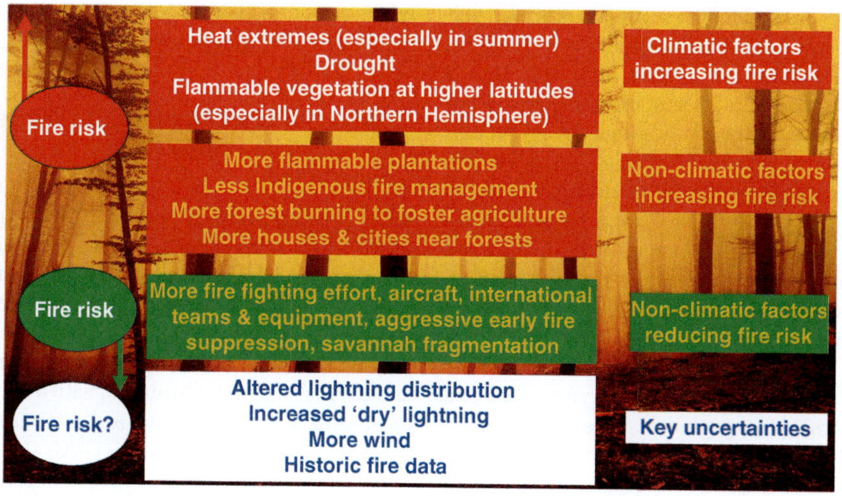

Fig. 20.3. Many factors influence the risk of fire, both 'wild' and deliberately started by humans. Some of these factors are related to climate change; we cannot identify any climate change-related factor that definitely reduces wildfire risk, but not all factors that increase fire risk are climate change-related.

be increasingly caused by 'dry' lightning.[6] There is evidence for this in both Australia (Canadell *et al.*, 2021) and the Arctic (Kharuk *et al.*, 2022). There is less certainty as to whether dry lightning is itself increasing, although Watson *et al.* (1978) considered it to be a rare event, and it no longer seems to be. The routine observation of lightning activity by satellites and networks of ground sensors is beginning to enable a more detailed understanding of the role of lightning activity as a driver of fire (Jones *et al.*, 2022). The pattern of 'wet' lightning may also be changing, with a projected decrease in the tropics (Xu *et al.*, 2020). Global wind speeds – another fire driver – are reported to have increased since 2010, after three decades of stilling (Xu *et al.*, 2020).

Fires and the carbon cycle, including methane

As climate change intensifies, methods seeking to reduce atmospheric and oceanic carbon grow in importance. The three main ways to do this are to: (i) reduce GHG emissions (preferably to zero, as fast as possible); (ii) preserve or enhance natural carbon sinks, such as forests, mangroves and peatlands; and (iii) to remove carbon from the atmosphere and to then safely sequester it for centuries, if not millennia (Fuhrman *et al.*, 2023). Here, we discuss the second pathway in more detail. CO_2 removal is discussed in Section 20.8.

Bowman *et al.* (2020) found that between 1997 and 2016, global mean CO_2 emissions from wildfires approximated 22% of those from burning fossil fuels. Deep-burning peat fires in Indonesia in 1997/98 (during a strong El Niño) released approximately 0.95 Gt of carbon, equivalent to ~15% of global fossil fuel emissions at that time (Turetsky *et al.*, 2015). Ongoing fires in this region, especially during La Niña periods, release less, but are still of global significance. South-East Asia continues to be clouded by severe haze every few years, particularly from smouldering peat fires (Rein and Huang, 2021). Exposure to smoke, particularly in the tropics, was estimated by Johnston *et al.* (2012) as contributing to over 300,000 premature deaths per year. The fires in the eucalyptus forests of south-eastern Australia in 2019/20, which covered an area approximately 2.5 times the size of Belgium, have been estimated as having released between 0.52 and 0.87 gigatonnes of carbon (GtC) (van der Velde *et al.*, 2021). On the plus side, although fires are already burning increasing areas of Arctic peat and tundra, this harmful depletion of terrestrial global carbon stores may be partially compensated by northerly forest expansion (Kharuk *et al.*, 2022).

The great, biodiverse and culturally rich rainforests of the Amazon and Congo river basins are also at risk, in part from fire. After a pause, Amazonian deforestation (mostly illegal) has recently increased, although the re-election of President Lula in 2022 may partly reverse this trend (Fearnside, 2023). Unlike eucalyptus forest, Amazonian species have little or no evolutionary fire adaptation (Lapola *et al.*, 2023). The Amazon generates between a third and a half of its own rainfall, by recycling the moisture that falls on it by 5 to 6 times (Lovejoy and Nobre, 2018). Thus, deforestation is likely to reduce rainfall, leading to increased dryness, fire and other forms of degradation. Lovejoy and Nobre (2018) argued that eastern, southern and central Amazonia are at risk of shifting to a non-forest ecosystem at a threshold of 20–25% deforestation, far lower than earlier estimates of 40%. Already, almost 40% of the surviving forest is degraded in some way, including by severe drought; Brando *et al.* (2020) forecast that intensifying Amazon fire regimes, combined with climate change, will double the area burned by wildfires, affecting up to 16% of the region's forests by 2050. Fires also contribute to degradation of forests in the Congo Basin, the world's second largest rainforest (Jiang *et al.*, 2023). Already, the Congo basin rainforest is a net carbon source during the dry season.[7]

Another reinforcing feedback involving climate change and fires has recently been recognized, concerning CH_4, the second most important GHG. The atmospheric concentration of CH_4 has increased steeply since about 2007, following a poorly understood pause which lasted for several years. Global warming is thought to be driving increased CH_4 emissions, through processes such as accelerating microbial activity in wetlands (Feng *et al.*, 2022) and by liberating carbon stored in tundra (Plaza *et al.*, 2019). In addition, CO, an emission of wildfires (especially from peat), is reducing atmospheric hydroxyl (OH) levels, a molecule sometimes called the atmospheric 'detergent'. Its reduction has been postulated as helping to elevate CH_4 (Zhao *et al.*, 2020).

Finally, Saide *et al.* (2016) postulated a causal link between wildfires in Central America and tornadoes in North America. These workers have suggested that particles produced by wildfires increase the vertical sheer in atmospheric wind speeds, which in turn makes tornadoes more likely and more severe.

20.7 Aerosol Masking: A 'Wicked' Problem

In 1974 scientists announced the detection of a 12% loss of solar radiation over the preceding four decades, based on observations taken in the Sinai Peninsula (Cohen *et al.*, 2004). Around that time the new discipline of climate science was trying to understand why the then newly created global temperature data, with a duration of several decades, showed a decline, evident since the 1940s, despite rising GHGs (Peterson *et al.*, 2008; Wild, 2009). This phenomenon – which also spawned a subsequently distorted and exaggerated myth that a new Ice Age might be looming (Peterson *et al.*, 2008) – is now understood to be caused by a reduction in the heating effect of GHGs, due to particulate air pollution. This phenomenon has several names, including 'global dimming' and 'aerosol masking' (Shindell and Smith, 2019). Its chemistry is complex, as two major components of PM – dust and black carbon – have a warming effect, especially by their deposition on glaciers and snow, where their dark colour hastens melting (Di Mauro, 2020) (Table 20.1).

There has long been recognition that as efforts to reduce particulate air pollution advance, additional warming will therefore result (Xu *et al.*, 2018). The net effect depends not only on the effect of each component, but whether they are reduced at the same rate. Regional variations in the rate of decline add to this complexity.

Aerosols from human activities are currently estimated to be offsetting (masking) as much as one-third of the effect of GHGs. The

Table 20.1. The predicted contribution of reducing forms of particulate air pollution to warming in the 21st century, at the global level. Adapted from Xu and Ramanathan (2017).

Mitigation measure	Change by 2050 (°C)	Change by 2100 (°C)
Black carbon	–0.2	–0.3
Aerosol unmasking	+0.3	+0.6
Cumulative effect	+0.1	+0.3

extent of this effect is uncertain, and there may be regional variation (Li *et al.*, 2022). Despite this uncertainty, it is accepted that a trade-off exists between the imperative to improve human health by lowering air pollution, and the risk of additional warming from so doing (Polonik *et al.*, 2021).

In recent years some progress has been made in stabilizing and even (in some parts of the world) reducing particulate air pollution. Fears that this will accelerate climate change are being used by advocates for greater geoengineering research, as an emergency response (Doherty *et al.*, 2023). In an enlightened world, with a longer-term outlook by policy makers – and with more regard for those who are more economically and socially vulnerable – the existence of this 'wicked'[8] problem would be used to accelerate the transition to clean energy, particularly in the global South. We do not live in such an enlightened world. Rather than generate and implement policies that will lower GHG emissions, generations of policy makers, from almost all countries, have continued to subsidize fossil fuels, and to act as though climate change and many other aspects of planetary overload are non-urgent. A key strategy for procrastination has been the 'carbon offset' industry and the lure of technologies to remove CO_2. If these fail, a last resort is seen, called geoengineering. The next section provides a brief introduction to these aspects.

20.8 Carbon Offsets, Carbon Dioxide Removal and Geoengineering

Carbon offsets involve the exchange of money between polluters (who continue polluting) and individuals and corporations who promise to do something to improve or protect carbon storage. A common mechanism by such parties is to avoid deforestation but, even if they do, this will lead to comparatively little carbon sequestration – though it will slow GHG accumulation. At best, offsetting results in no change in total emissions, but as theory predicts and experience shows, this best result is rarely attainable. There is increasing criticism of this industry (McAffee, 2022), which has also been described as 'climate colonialism' (Bachram, 2004).

A second key strategy is carbon dioxide removal (CDR). One proposed method is by afforestation, but the potential for this is limited not only because of the imperative to grow food, but because such forests themselves risk burning. Another CDR strategy is called bioenergy with carbon capture and storage. While a key strategy proposed at the 2015 Paris conference, this too remains expensive, theoretical and extensively criticized (Anderson, 2015). Other methods are direct air capture with carbon storage, enhanced weathering, biochar and direct ocean capture with carbon storage (Fuhrman *et al.*, 2023). CDR faces enormous challenges. In 2023, Jonathan Foley, chief executive officer of Project Drawdown, claimed on Twitter that 'all of the engineered carbon removal on Earth has so far removed about 60–80 thousand tonnes of CO_2, mainly as biochar. That's about 40 seconds of current emissions'. In addition, the prospect of CDR and the illusion of effective offsets worsens moral hazard.[9]

Geoengineering by altering planetary albedo

David Keith, an advocate for geoengineering, has defined this as intentional, large-scale environmental manipulation in order to reduce undesired anthropogenic climate change. Several methods have been proposed, including an attempt to increase Earth's albedo (reflectiveness), by scattering particulate matter into the stratosphere (Keith, 2000; Doherty *et al.*, 2023). Such proposals have long been intensely controversial, not least as effects on the hydrological cycle and on human health and other aspects of well-being are so poorly understood (Wild, 2009; Ricke, *et al.*, 2023). Although aerosols wash out of the atmosphere within a few days, their use – if successful at lowering temperature – could quickly become addictive, since the cessation of the frequent replacement of aerosols could generate an almost instant heating response. The prospect of geoengineering also creates additional 'moral hazard', as has long been recognized. A more straightforward and much safer form of geoengineering is to increase albedo by the use of lighter colours on roofs and other urban surfaces (Lontorfos *et al.*, 2018).

20.9 Conclusion

The damage to health from air pollution is of public health concern globally, and especially in highly polluted hotspots. Clean air is an essential life support system for humans, who inhale air many times each minute. Although ancient, the scale and complexity of air pollution has greatly increased in the past century, due to factors such as population increase (especially in the global South), industrialization, and the near-global spread of the combustion of fossil fuels, which are intensely polluting. Progress has been made in estimating the numerous forms of adverse impact to human health that arises from exposure to pollutants, including $PM_{2.5}$.

Vegetation fires, both deliberate and 'natural', add to the problem. Such fires may be worsened by climate change, although some uncertainty still surrounds this hypothesis. Air pollution is especially severe in the global South, a setting in which billions of people rarely see any stars (due to constant haze), and where limited visibility is widely accepted as normal. Although gradual improvements in the determinants of air pollution are emerging, the pace and scale of their implementation, especially in the global South, remains far too low. There is also a non-trivial risk that insufficient action to reduce GHG emissions will trigger reinforcing feedbacks, such as increased fires in the great surviving carbon sinks, which make escape from our dilemma even more difficult.[10]

Notes

[1] A peculiar smell following lightning strikes has been recorded for over 2000 years. In 1840 the German-Swiss chemist Schönbein coined the word ozone (from the Greek *ozein*, 'to smell') in recognition that the passage of electricity through oxygen creates a new gas (Stolarski, 2001).

[2] The troposphere is the atmospheric layer near ground level, including that which we breathe.

[3] Chemical reactions in the atmosphere associated with light.

[4] The AQI translates air quality data into numbers and (in some cases) colours to help people understand when to take action to protect their health against air pollution (EPA, 2022). Unfortunately, no globally standardized definition of this is agreed upon.

[5] 'Tarsands' are a mixture of sand, clay, water and bitumen. It requires enormous energy to convert their fossil fuel content to a useable form. Tarsands are generally mined in open pits, like coal.

[6] Canadell *et al.* (2021) defined this as lightning associated with 2.5 mm or less of rain.

[7] In the wet season the Congo forests still absorb more carbon than they emit. However, satellite observations show that the atmospheric CO_2 over the Congo basin is ~2 parts per million higher than the regional background during June–August, primarily due to biomass burning and significantly reduced photosynthetic activities during the dry season.

[8] The term 'wicked' problem dates, at least, to Rittel and Webber (1973). They argued that some problems lack solutions in the sense of definitive and objective answers. Since then, the term has been increasingly used to describe intractable problems, including climate change and other issues pertinent to sustainability (Lönngren and van Poeck, 2021).

[9] Moral hazard was, originally, an economic term to describe the possibility that an insured person might take an imprudent risk. A wider meaning is that individuals, societies and even entire peoples or governments may accept higher risks than is prudent, because of a belief that they are protected. This is, arguably, a moral issue because it may lead to actions that prove extremely harmful (including to future generations) and which could have been avoided.

[10] This chapter was completed before the extraordinary extent of the 2023 Canadian fire season was appreciated. Carbon emissions from these fires were 450%–709% above the 2003–2022 average (Friedlingstein *et al.*, 2023). In our view this is a disturbing event that underscores the risk of a harmful reinforcing feedback in the Earth system.

References

Adu, M.K., Eboreime, E., Shalaby, R., Sapara, A., Agyapong, B. *et al.* (2022) Five years after the Fort McMurray wildfire: Prevalence and correlates of low resilience. *Behavioral Sciences* 12, 96. DOI: 10.3390/bs12040096

Al-Aly, Z. and Bowe, B. (2020) Air pollution and kidney disease. *Clinical Journal of the American Society of Nephrology* 15, 301–303. DOI: 10.2215/CJN.16031219

Amjad, S., Chojecki, D., Osornio-Vargas, A. and Ospina, M.B. (2021) Wildfire exposure during pregnancy and the risk of adverse birth outcomes: A systematic review. *Environment International* 156, 106644. DOI: 10.1016/j.envint.2021.106644

Anderson, K. (2015) Talks in the city of light generate more heat. *Nature* 528, 437. DOI: 10.1038/528437a

Bachram, H. (2004) Climate fraud and carbon colonialism: the new trade in greenhouse gases. *Capitalism Nature Socialism* 15, 5–20. DOI: 10.1080/1045575042000287299

Basu, M. (2019) The great smog of Delhi. *Lung India* 36, 239–240. DOI: 10.4103/lungindia.lungindia_363_18

Bell, M.L. and Davis, D.L. (2001) Reassessment of the lethal London fog of 1952: novel indicators of acute and chronic consequences of acute exposure to air pollution. *Environmental Health Perspectives* 109, 389–394. DOI: 10.1289/ehp.01109s3389

Belleville, G., Ouellet, M.C., Lebel, J., Ghosh, S., Morin, C.M. *et al.* (2021) Psychological symptoms among evacuees from the 2016 Fort McMurray wildfires: a population-based survey one year later. *Frontiers in Public Health* 9, 655357. DOI: 10.3389/fpubh.2021.655357

Berridge, V. (2006) The policy response to the smoking and lung cancer connection in the 1950s and 1960s. *The Historical Journal* 49, 1185–1209. DOI: 10.1017/S0018246X06005784

Bonta, M., Gosford, R., Eussen, D., Ferguson, N., Loveless, E. *et al.* (2017) Intentional fire-spreading by 'fire-hawk' raptors in Northern Australia. *Journal of Ethnobiology* 37, 700–718. DOI: 10.2993/0278-0771-37.4.700

Bowman, D.M.J.S., Kolden, C.A., Abatzoglou, J.T., Johnston, F.H., van der Werf, G.R. *et al.* (2020) Vegetation fires in the Anthropocene. *Nature Reviews Earth & Environment* 1, 500–515. DOI: 10.1038/s43017-020-0085-3

Brando, P.M., Soares-Filho, B., Rodrigues, L., Assuncao, A., Morton, D. *et al.* (2020) The gathering firestorm in southern Amazonia. *Science Advances* 6, eaay1632. DOI: 10.1126/sciadv.aay1632

Brimblecombe, P. (1977) London air pollution, 1500–1900. *Atmospheric Environment (1967)* 11, 1157–1162. DOI: 10.1016/0004-6981(77)90091-9

Butler, C.D., Hales, S., Beggs, P.J. and Ebi, K.L. (2016) From Paris towards 1.5 degrees C. In: Butler, C.D. (ed.) *Climate Change and Global Health.* CABI, Wallingford, UK, pp. 293–305.

Campbell-Lendrum, D. and Prüss-Ustün, A. (2019) Climate change, air pollution and noncommunicable diseases. *Bulletin of the World Health Organization* 97, 160–161. DOI: 10.2471/BLT.18.224295

Canadell, J.G., Meyer, C.P., Cook, G.D., Dowdy, A., Briggs, P.R. *et al.* (2021) Multi-decadal increase of forest burned area in Australia is linked to climate change. *Nature Communications* 12, 6921. DOI: 10.1038/s41467-021-27225-4

Capasso, L. (2000) Indoor pollution and respiratory diseases in Ancient Rome. *The Lancet* 356, 1774. DOI: 10.1016/S0140-6736(05)71971-1

Chen, T.-M., Kuschner, W.G., Gokhale, J. and Shofer, S. (2007) Outdoor air pollution: Ozone health effects. *The American Journal of the Medical Sciences* 333, 244–248. DOI: 10.1097/MAJ.0b013e31803b8e8c

Chen, H., Kwong, J.C., Copes, R., Tu, K., Villeneuve, P.J. *et al.* (2017) Living near major roads and the incidence of dementia, Parkinson's disease, and multiple sclerosis: a population-based cohort study. *The Lancet* 389, 718–726. DOI: 10.1016/S0140-6736(16)32399-6

Cicerone, R.J. (1994) Fires, atmospheric chemistry, and the ozone layer. *Science* 263, 1243–1244. DOI: 10.1126/science.263.5151.1243

Cohen, S., Liepert, B. and Stanhill, G. (2004) Global dimming comes of age. *Eos* 85, 362–362. DOI: 10.1029/2004EO380004

Copernicus (nd) *European State of the Climate 2021*. Available at: https://climate.copernicus.eu/esotc/2021/arctic-wildfires (accessed 17 June 2023).

Crutzen, P.J. (2002) Geology of mankind. *Nature* 415, 23. DOI: 10.1038/415023a

Di Mauro, B. (2020) A darker cryosphere in a warming world. *Nature Climate Change* 10, 979–980. DOI: 10.1038/s41558-020-00911-9

Dixon, R. (2021) Siberia's wildfires are bigger than all the world's other blazes combined. *The Washington Post.* Available at: https://www.washingtonpost.com/world/2021/08/11/siberia-fires-russia-climate/ (accessed 17 June 2023)

Dockery, D.W., Pope, C.A., Xu, X., Spengler, J.D., Ware, J.H. *et al.* (1993) An association between air pollution and mortality in six US cities. *New England Journal of Medicine* 329, 1753–1759. DOI: 10.1056/NEJM199312093292401

Dockery, D., Rich, D., Goodman, P., Clancy, L., Ohman-Strickland, P. *et al.* (2013) Effect of air pollution control on mortality and hospital admissions in Ireland. *Research Report (Health Effects Institute)* 176, 3–109. PMID: 24024358

Doherty, S.J., Rasch, P.J., Wood, R., Haywood, J., Forster, P.M. *et al.* (2023) An open letter regarding research on reflecting sunlight to reduce the risks of climate change. Available at: https://climate-intervention-research-letter.org/ (accessed 16 March 2023).

EPA (2022) *Particulate Matter (PM) Standards – Policy Assessments from Current Review*. Environmental Protection Agency, Research Triangle Park, North Carolina. Available at: https://www.epa.gov/naaqs/particulate-matter-pm-standards-policy-assessments-current-review-0 (accessed 27 April 2023).

Fayiga, A.O., Ipinmoroti, M.O. and Chirenje, T. (2018) Environmental pollution in Africa. *Environment, Development and Sustainability* 20, 41–73. DOI: 10.1007/s10668-016-9894-4

Fearnside, P.M. (2023) The outlook for Brazil's new presidential administration. *Trends in Ecology & Evolution* 38, 387–388. DOI: 10.1016/j.tree.2023.01.002

Feng, L., Palmer, P.I., Zhu, S., Parker, R.J. and Liu, Y. (2022) Tropical methane emissions explain large fraction of recent changes in global atmospheric methane growth rate. *Nature Communications* 13, 1378. DOI: 10.1038/s41467-022-28989-z

Feng, Z., De Marco, A., Anav, A., Gualtieri, M., Sicard, P. *et al.* (2019) Economic losses due to ozone impacts on human health, forest productivity and crop yield across China. *Environment International* 131, 104966. DOI: 10.1016/j.envint.2019.104966

Fisher, S., Bellinger, D.C., Cropper, M.L., Kumar, P., Binagwaho, A. *et al.* (2021) Air pollution and development in Africa: impacts on health, the economy, and human capital. *The Lancet Planetary Health* 5, e681–e688. DOI: 10.1016/S2542-5196(21)00201-1

Forsyth, T. (2014) Public concerns about transboundary haze: A comparison of Indonesia, Singapore, and Malaysia. *Global Environmental Change* 25, 76–86. DOI: 10.1016/j.gloenvcha.2014.01.013

Fowler, D., Brimblecombe, P., Burrows, J., Heal, M.R., Grennfelt, P. *et al.* (2020) A chronology of global air quality. *Philosophical Transactions of the Royal Society A. Mathematical, Physical and Engineering Sciences* 378, 20190314. DOI: 10.1098/rsta.2019.0314

Friedlingstein, P., O'Sullivan, M., Jones, M.W., Andrew, R.M., Bakker, D.C.E. *et al.* (2023) Global carbon budget 2023. *Earth System Science Data* 15, 5301–5369. DOI: 10.5194/essd-15-5301-2023

Fuhrman, J., Bergero, C., Weber, M., Monteith, S., Wang, F.M. *et al.* (2023) Diverse carbon dioxide removal approaches could reduce impacts on the energy–water–land system. *Nature Climate Change* 13, 341–350. DOI: 10.1038/s41558-023-01604-9

Gao, Y., Huang, W., Yu, P., Xu, R., Yang, Z. *et al.* (2023) Long-term impacts of non-occupational wildfire exposure on human health: A systematic review. *Environmental Pollution* 320, 121041. DOI: 10.1016/j.envpol.2023.121041

Gould, C.F. (2021) As the Smoke Clears: Assessing the Air Pollution and Health Benefits of a Nationwide Transition to Clean Cooking Fuels in Ecuador. PhD thesis, Columbia University, New York. DOI: 10.7916/d8-m7dg-yx85

Hayes, F., Sharps, K., Harmens, H., Roberts, I. and Mills, G. (2020) Tropospheric ozone pollution reduces the yield of African crops. *Journal of Agronomy and Crop Science* 206, 214–228. DOI: 10.1111/jac.12376

Hoesly, R.M., Smith, S.J., Feng, L., Klimont, Z., Janssens-Maenhout, G. *et al.* (2018) Historical (1750–2014) anthropogenic emissions of reactive gases and aerosols from the Community Emissions Data System (CEDS). *Geoscientific Model Development* 11, 369–408. DOI: 10.5194/gmd-11-369-2018

Huangfu, P. and Atkinson, R. (2020) Long-term exposure to NO_2 and O_3 and all-cause and respiratory mortality: A systematic review and meta-analysis. *Environmental International* 144, 105998. DOI: 10.1016/j.envint.2020.105998

Irannezhad, M., Liu, J., Ahmadi, B. and Chen, D. (2020) The dangers of Arctic zombie wildfires. *Science* 369, 1171. DOI: 10.1126/science.abe1739

Jacobson, T.A., Kler, J.S., Hernke, M.T., Braun, R.K., Meyer, K.C. *et al.* (2019) Direct human health risks of increased atmospheric carbon dioxide. *Nature Sustainability* 2, 691–701. DOI: 10.1038/s41893-019-0323-1

Jaffe, D.A. and Wigder, N.L. (2012) Ozone production from wildfires: A critical review. *Atmospheric Environment* 51, 1–10. DOI: 10.1016/j.atmosenv.2011.11.063.

Jegasothy, E., Hanigan, I.C., Van Buskirk, J., Morgan, G.G., Jalaludin, B. *et al.* (2023) Acute health effects of bushfire smoke on mortality in Sydney, Australia. *Environment International* 171, 107684. DOI: 10.1016/j.envint.2022.107684

Jiang, X., Albright, R., Creecy, E., Li, K.-F., Liang, M.-C. *et al.* (2023) Congo basin rainforest is a net carbon source during the dry season. *Earth and Space Science* 10, e2022EA002644. DOI: 10.1029/2022EA002644

Johnson, R.J., Wesseling, C. and Newman, L.S. (2019) Chronic kidney disease of unknown cause in agricultural communities. *The New England Journal of Medicine* 380, 1843–1852. DOI: 10.1056/NEJMra1813869

Johnston, F.H., Henderson, S.B., Chen, Y., Randerson, J.T., Marlier, M. *et al.* (2012) Estimated global mortality attributable to smoke from landscape fires. *Environmental Health Perspectives* 120, 695–701. DOI: 10.1289/ehp.1104422

Johnston, F.H., Hanigan, I.C., Henderson, S.B. and Morgan, G.G. (2013) Evaluation of interventions to reduce air pollution from biomass smoke on mortality in Launceston, Australia: retrospective analysis of daily mortality, 1994–2007. *BMJ* 346, e8446. DOI: 10.1136/bmj.e8446

Jones, M.W., Abatzoglou, J.T., Veraverbeke, S., Andela, N., Lasslop, G. *et al.* (2022) Global and regional trends and drivers of fire under climate change. *Reviews of Geophysics* 60, e2020RG000726. DOI: 10.1029/2020RG000726

Kampa, M. and Castanas, E. (2008) Human health effects of air pollution. *Environmental Pollution* 151, 362–367. DOI: 10.1016/j.envpol.2007.06.012

Keith, D.W. (2000) Geoengineering the climate: history and prospect. *Annual Review of Energy and Environment* 25, 245–284. DOI: 10.1146/annurev.energy.25.1.245

Kharuk, V.I., Dvinskaya, M.L., Im, S.T., Golyukov, A.S. and Smith, K.T. (2022) Wildfires in the Siberian Arctic. *Fire* 5, 106. DOI: 10.3390/fire5040106

Kim, H., Kim, W.-H., Kim, Y.-Y. and Park, H.-Y. (2020) Air pollution and central nervous system disease: A review of the impact of fine particulate matter on neurological disorders. *Frontiers in Public Health* 8, 575330. DOI: 10.3389/fpubh.2020.575330

Kizer, K.W. (2020) Extreme wildfires – a growing population health and planetary problem. *JAMA* 324, 1605-1606. DOI: 10.1001/jama.2020.19334

Lapola, D.M., Pinho, P., Barlow, J., Aragão, L.E.O.C., Berenguer, E. *et al.* (2023) The drivers and impacts of Amazon forest degradation. *Science* 379, eabp8622. DOI: 10.1126/science.abp8622.

Lareau, N.P., Nauslar, N.J., Bentley, E., Roberts, M., Emmerson, S. *et al.* (2022) Fire-generated tornadic vortices. *Bulletin of the American Meteorological Society* 103, E1296–E1320. DOI: 10.1175/BAMS-D-21-0199.1

Lelieveld, J., Pozzer, A., Pöschl, U., Fnais, M., Haines, A. *et al.* (2020) Loss of life expectancy from air pollution compared to other risk factors: a worldwide perspective. *Cardiovascular Research* 116, 1910–1917. DOI: 10.1093/cvr/cvaa025

Li, J., Wang, N., Wang, J. and Li, H. (2018) Spatiotemporal evolution of the remotely sensed global continental $PM_{2.5}$ concentration from 2000-2014 based on Bayesian statistics. *Environmental Pollution* 238, 471–481. DOI: 10.1016/j.envpol.2018.03.050

Li, J., Carlson, B.E., Yung, Y.L., Lv, D., Hansen, J. *et al.* (2022) Scattering and absorbing aerosols in the climate system. *Nature Reviews Earth & Environment* 3, 363–379. DOI: 10.1038/s43017-022-00296-7

Lönngren, J. and van Poeck, K. (2021) Wicked problems: a mapping review of the literature. *International Journal of Sustainable Development & World Ecology* 28, 481–502. DOI: 10.1080/13504509.2020.1859415

Lontorfos, V., Efthymiou, C. and Santamouris, M. (2018) On the time varying mitigation performance of reflective geoengineering technologies in cities. *Renewable Energy* 115, 926–930. DOI: 10.1016/j.renene.2017.09.033

Lovejoy, T.L. and Nobre, C. (2018) Amazon tipping point. *Science Advances* 4, eaat2340. DOI: 10.1126/sciadv.aat2340

McAfee, K. (2022) Shall the American Association of Geographers endorse carbon offsets? Absolutely not! *The Professional Geographer* 74, 171–177. DOI: 10.1080/00330124.2021.1934879

Medler, M.J. (2011) Speculations about the effects of fire and lava flows on human evolution. *Fire Ecology* 7, 13–23. DOI: 10.4996/fireecology.0701013.

Nemery, B., Hoet, P.H.M. and Nemmar, A. (2001) The Meuse Valley fog of 1930: an air pollution disaster. *The Lancet* 357, 704–708. DOI: 10.1016/S0140-6736(00)04135-0

Odo, D.B., Yang, I.A., Dey, S., Hammer, M., van Donkelaar, A. *et al.* (2022) Ambient air pollution and acute respiratory infection in children aged under 5 years living in 35 developing countries. *Environment International* 159, 107019. DOI: 10.1016/j.envint.2021.107019

Peterson, T.C., Connolley, W.M. and Fleck, J. (2008) The myth of the 1970s global cooling scientific consensus. *Bulletin of the American Meteorological Society* 89, 1325–1327. DOI: 10.1175/2008BAMS2370.1

Pillarisetti, A., Ye, W. and Chowdhury, S. (2022) Indoor air pollution and health: Bridging perspectives from developing and developed countries. *Annual Review of Environment and Resources* 47, 197–229. DOI: 10.1146/annurev-environ-012220-010602

Plaza, C., Pegoraro, E., Bracho, R., Celis, G., Crummer, K.G. *et al.* (2019) Direct observation of permafrost degradation and rapid soil carbon loss in tundra. *Nature Geoscience* 12, 627–631. DOI: 10.1038/s41561-019-0387-6

Polonik, P., Ricke, K. and Burney, J. (2021) Paris agreement's ambiguity about aerosols drives uncertain health and climate outcomes. *Earth's Future* 9, e2020EF001787. DOI: 10.1029/2020EF001787

Psarros, C., Theleritis, C., Economou, M., Tzavara, C., Kioulos, K.T. *et al.* (2017) Insomnia and PTSD one month after wildfires: evidence for an independent role of the 'fear of imminent death'. *International Journal of Psychiatry in Clinical Practice* 21, 137–141. DOI: 10.1080/13651501.2016.1276192

Pyne, S.J. (2020) From Pleistocene to Pyrocene: Fire replaces ice. *Earth's Future* 8, e2020EF001722. DOI: 10.1029/2020EF001722

Rein, G. and Huang, X. (2021) Smouldering wildfires in peatlands, forests and the arctic: Challenges and perspectives. *Current Opinion in Environmental Science & Health* 24, 100296. DOI: 10.1016/j.coesh.2021.100296

Reis, S., Grennfelt, P., Klimont, Z., Amann, M., ApSimon, H. *et al.* (2012) From acid rain to climate change. *Science* 338, 1153–1154. DOI: 10.1126/science.122651

Richardson, D., Black, A.S., Irving, D., Matear, R.J., Monselesan, D.P. *et al.* (2022) Global increase in wildfire potential from compound fire weather and drought. *npj Climate and Atmospheric Science* 5, 23. DOI: 10.1038/s41612-022-00248-4

Ricke, K., Wan, J.S., Saenger, M. and Lutsko, N.J. (2023) Hydrological consequences of solar geoengineering. *Annual Review of Earth and Planetary Sciences* 51, 447–470. DOI: 10.1146/annurev-earth-031920-083456

Rittel, H. and Webber, M. (1973) Dilemmas in a general theory of planning. *Policy Sciences* 4, 155–169. DOI: 10.1007/BF01405730

Rosell, J. and Blasco, R. (2019) The early use of fire among Neanderthals from a zooarchaeological perspective. *Quaternary Science Reviews* 217, 268–283. DOI: 10.1016/j.quascirev.2019.03.002

Saide, P.E., Thompson, G., Eidhammer, T., da Silva, A.M., Pierce, R.B. *et al.* (2016) Assessment of biomass burning smoke influence on environmental conditions for multiyear tornado outbreaks by combining aerosol-aware microphysics and fire emission constraints. *Journal of Geophysical Research: Atmospheres* 121, 17. DOI: 10.1002/2016JD02505

Schiermeier, Q. (2021) Climate change made North America's deadly heatwave 150 times more likely. *Nature*. DOI: 10.1038/d41586-021-01869-0

Scott, A.C., Chaloner, W.G., Belcher, C.M. and Roos, C.I. (2016) The interaction of fire and mankind: Introduction. *Philosophical Transactions of the Royal Society B: Biological Sciences* 371, 20150162. DOI: 10.1098/rstb.2015.0162

Shindell, D. and Smith, C.J. (2019) Climate and air-quality benefits of a realistic phase-out of fossil fuels. *Nature* 573, 408–411. DOI: 10.1038/s41586-019-1554-z

Sillman, S. (1993) Tropospheric ozone: The debate over control strategies. *Annual Review of Energy and the Environment* 18, 31–56. DOI: 10.1146/annurev.eg.18.110193.000335

Smith, K. and Pillarisetti, A. (2018) Household air pollution from solid cookfuels and its effects on health. In: Mock, C.N. *et al.* (eds) *Injury Prevention and Environmental Health*, 3rd edn. International Bank for Reconstruction and Development / The World Bank, Washington, DC. Available at: https://www.ncbi.nlm.nih.gov/books/NBK525225/ (accessed 14 March 2023).

Snyder, L.P. (1994) 'The death-dealing smog over Donora, Pennsylvania': industrial air pollution, public health policy, and the politics of expertise, 1948–1949. *Environmental History Review* 18, 117–139. DOI: 10.2307/3984747

Spiru, P. and Simona, P.L. (2017) A review on interactions between energy performance of the buildings, outdoor air pollution and the indoor air quality. *Energy Procedia* 128, 179–186. DOI: 10.1016/j.egypro.2017.09.039

Stolarski, R.S. (2001) History of the study of atmospheric ozone. *Ozone: Science & Engineering* 23, 421–428. DOI: 10.1080/01919510108962025

Turco, M., Rosa-Cánovas, J.J., Bedia, J., Jerez, S., Montávez, J.P. *et al.* (2018) Exacerbated fires in Mediterranean Europe due to anthropogenic warming projected with non-stationary climate-fire models. *Nature Communications* 9, 3821. DOI: 10.1038/s41467-018-06358-z

Turetsky, M.R., Benscoter, B., Page, S., Rein, G., van der Werf, G.R. *et al.* (2015) Global vulnerability of peatlands to fire and carbon loss. *Nature Geoscience* 8, 11–14. DOI: 10.1038/ngeo2325

van der Velde, I.R., van der Werf, G.R., Houweling, S., Maasakkers, J.D., Borsdorff, T. *et al.* (2021) Vast CO_2 release from Australian fires in 2019–2020 constrained by satellite. *Nature* 597, 366–369. DOI: 10.1038/s41586-021-03712-y van Eeden, L.M. and Dickman, C.R. (2023) Estimating the number of wild animals affected by Australia's 2019–20 wildfires. In: Van Leeuwen, S., Wintle, B.A., Woinarski, J.C.Z., Rumpff, L. and Legge, S.M. (eds) *Australia's Megafires: Biodiversity Impacts and Lessons from 2019–2020*. CSIRO Publishing, Melbourne, Australia, pp. 154–166.

Villanueva, D., Possner, A., Neubauer, D., Gasparini, B., Lohmann, U. *et al.* (2022) Mixed-phase regime cloud thinning could help restore sea ice. *Environmental Research Letters* 17, 114057. DOI: 10.1088/1748-9326/aca16d

Walls, K. (2016) Ensuring adaptation to climate change does not increase health risks from the built environment. *Journal of Environmental Engineering & Ecological Science* 5, 1. DOI: 0.7243/2050-1323-5-1

Watson, A., Lovelock, J.E. and Margulis, L. (1978) Methanogenesis, fires and the regulation of atmospheric oxygen. *Biosystems* 10, 293–298. DOI: 10.1016/0303-2647(78)90012-6

WHO (2021) *Global Air Quality Guidelines. Particulate matter (PM2.5 and PM10), Ozone, Nitrogen Dioxide, Sulfur Dioxide and Carbon Monoxide*. World Health Organization, Geneva. Available at: https://apps.who.int/iris/handle/10665/345329 (accessed 27 April 2023).

Wiedmann, T., Lenzen, M., Keyßer, L.T. and Steinberger, J.K. (2020) Scientists' warning on affluence. *Nature Communications* 11, 3107. DOI: 10.1038/s41467-020-16941-y

Wild, M. (2009) Global dimming and brightening: A review. *Journal of Geophysical Research: Atmospheres* 114, D00D16. DOI: 10.1029/2008JD011470

Xu, Y. and Ramanathan, V. (2017) Well below 2°C: Mitigation strategies for avoiding dangerous to catastrophic climate changes. *Proceedings of the National Academy of Sciences (USA)* 114, 10315–10323. DOI: 10.1073/pnas.161848111

Xu, Y., Ramanathan, V. and Victor, D.G. (2018) Global warming will happen faster than we think. *Nature* 564, 30–32. DOI: 10.1038/d41586-018-07586-5

Xu, R., Yu, P., Abramson, M.J., Johnston, F.H., Samet, J.M. *et al.* (2020) Wildfires, global climate change, and human health. *New England Journal of Medicine* 383, 2173–2181. DOI: 10.1056/NEJMsr2028985

Zanobetti, A. and Schwartz, J. (2011) Ozone and survival in four cohorts with potentially predisposing diseases. *American Journal of Respiratory and Critical Care Medicine* 184, 836–841. DOI: 10.1164/rccm.201102-0227OC

Zhao, Y., Saunois, M., Bousquet, P., Lin, X., Berchet, A. *et al.* (2020) On the role of trend and variability in the hydroxyl radical (OH) in the global methane budget. *Atmospheric Chemistry and Physics* 20, 13011–13022. DOI: 10.5194/acp-20-13011-2020

21 Climate Change and its 'Tertiary' Effects: Thinking Systemically in a World of Limits

Colin D. Butler
National Centre for Epidemiology and Population Health, The Australian National University, Canberra, Australia

Abstract
This chapter reviews several classifications of the health effects associated with climate change, arguing that one of the oldest forms ('primary', 'secondary' and 'tertiary' effects) is not only the simplest, but the most powerful, as it recognizes a hierarchy of scale, in terms of both the resultant burden of disease and complexity of causal pathways. The chapter insists that such effects interact with a range of environmental and social ('eco-social') determinants, withstanding simplistic criticisms of 'environmental determinism'. Potentially, climate change is catastrophic to humanity.

21.1 Introduction

This chapter serves as an introduction to Section 5 of this book; the 'tertiary' effects of climate change upon health. The other chapters in this section concern famine, population displacement, conflict and collapse. It is argued that these are potentially catastrophic. Although intimations of these effects are already evident, their full scale is (hopefully) decades in the future. If civilization acts very quickly to slow greenhouse gas accumulation, and/or if it is very lucky, then these consequences may be avoided.

The author recognizes that, to many people, consideration of such dire consequences may seem ridiculous. I appeal to the reader to consider the unthinkable. However, it is also recognized that climate change never acts alone; it is a risk multiplier. There are always co-factors, both social and environmental ('eco-social'). Ecological factors are here conceptualized as a subset of the wider environment, which includes non-living components, including the depletion of non-renewable resources such as of oil, phosphorus and rare earths. In contrast, ecological factors refer here to living elements of the Earth system, including biodiversity and many resources that are theoretically renewable, such as fish stocks and rainforests.

21.2 Primary, Secondary and Tertiary Health Effects of Climate change

When the publisher of this book approached me to edit the first edition, I agreed, provided that the book could be based on a then recently revived three-tiered classification of the effects of climate change on health. These ranks were called primary, secondary and tertiary. Of these three, I have long regarded the third (tertiary) category as the most important. Of course, all conceptual models simplify reality; none are perfect. Yet, this three-way model is preferred in this book. This chapter seeks to explain why.

DOI: 10.1079/9781800620025.0021

History of the primary, secondary and tertiary framework

This classification was first presented at a meeting in 1992 at the Royal Society of Medicine, in London, and published the following year (Haines and Parry, 1993) (Box 21.1). A second paper using these terms was published in the same month. One of the authors of the latter (Haines *et al.*, 1993) was Tony McMichael, while the third and final author was Paul Epstein. Thus, three of the four authors involved with the conceptualization of this framework were closely associated with the first edition of this book. However, although that first edition used a three-way framework, it did not contain a detailed outline of the concepts, nor a history of the terms' usage.

A three-way classification of the health effects of climate change (without the terms primary, secondary and tertiary) appeared in the first edited book on the topic (McMichael *et al.*, 1996) and in each of the chapters in the 4th and 5th Intergovernmental Panel on Climate Change (IPCC) reports (Confalonieri *et al.*, 2007; Smith *et al.*, 2014). These focused on 'direct' and 'indirect' effects, as a substitute for 'primary' and 'secondary', with very limited discussion of a third category. The 5th IPCC report (i.e. Smith *et al.*, 2014) did discuss a third effect class, in which was included undernutrition, population displacement and violent conflict, as well as their mental health effects (Box 21.2).

Box 21.1

'The *primary* effects of climate change on health will be caused by the *direct* effects of raised temperature. The *secondary* effects will be due to the impact of changes in temperature and precipitation on food and water supply and on determinants of disease such as *vectors ... Tertiary* effects will be experienced as a result of the effects of changes in the availability of food and water, shelter and changing patterns of disease on population movements, including *refugees*, on the probability of *conflict* and on political and economic systems ... it is likely that the secondary and tertiary impacts will outweigh the importance of the primary effects' (Haines and Parry, 1993) [italics added].

A four-part classification of the health effects of climate change has also been published (Butler *et al.*, 2005). It also used different terms, but added an additional effect category, which it called 'multi-state failure' (Fig. 21.1)

The terms primary, secondary and tertiary (i.e. in relation to the health effects of climate change) were revived in 2010 (Butler and Harley, 2010). In the meantime, more complex conceptualizations of the health effects of climate change evolved and dominated. For example, Patz *et al.* (2000) identified five main effects.[1] Portier *et al.* (2010), in an 80-page report, described 11 effect routes.[2] More recently, Watts *et al.* (2015) listed ten main kinds of health consequences, evolving via numerous interacting paths.[3]

No classification is perfect, but it is here argued that the less complex forms (especially the earliest description in Box 21.1) have several advantages. This categorization has consequently been adopted in each edition of this book, with the suggestion that mental health effects be considered as 'cross-cutting'. In addition, this categorization suggests a hierarchy, not only of the burden of disease, but also of the complexity of causation. I argue, for example, that population displacement will be both more important and more causally complex than poisoning, listed by

Box 21.2

'Climate change affects health in three ways; 1) Directly, such as the mortality and morbidity (including "heat exhaustion") due to extreme heat events, floods, and other extreme weather events in which climate change may play a role; 2) Indirect impacts from environmental and ecosystem changes, such as shifts in patterns of disease carrying mosquitoes and ticks, or increases in waterborne diseases due to warmer conditions and increased precipitation and runoff; and 3) indirect impacts mediated through societal systems, such as undernutrition and mental illness from altered agricultural production and food insecurity, stress and undernutrition and violent conflict caused by population displacement, economic losses due to widespread "heat exhaustion" impacts on the workforce, or other environmental stressors, and damage to health care systems by extreme weather events' (Smith *et al.*, 2014).

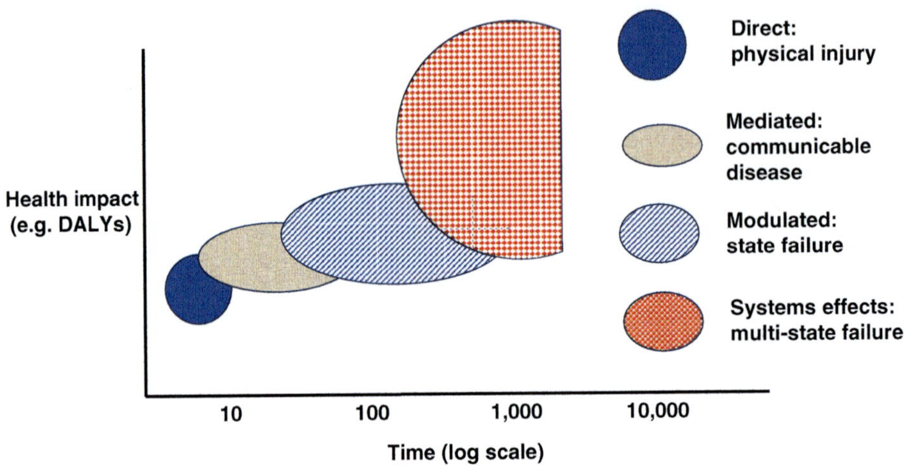

Fig. 21.1. Characteristics of four kinds of adverse health effects due to eco-social changes. DALYs (disability-adjusted life years lost) are a commonly used measure of health impact that reflects chronic illness as well as premature death. This figure avoids the names 'primary', 'secondary' and 'tertiary' but substitutes three names for very similar concepts ('direct', 'mediated', 'modulated'). Source: adapted from Butler *et al.* (2005) (Used with permission).

Watts *et al.* (2015) as one of the eight main forms of health effects.

The restriction of the third category to impacts 'mediated through societal systems' in the 5th IPCC report (Smith *et al.*, 2014) is arguable. This is because social factors also influence (either to aggravate or to ameliorate) the burden of disease in direct and indirect effects (i.e. primary and secondary), such as by community-initiated early-warning systems (for heat or disasters), or of interventions to reduce the burden of vector-borne diseases (see Chapters 15–18).

The 6th IPCC health report (Cissé *et al.*, 2022) did include a variant of the primary, secondary and tertiary framework, but only in context of Indigenous health (Bowles, 2015). Its discussion of tertiary effects focused on mental health effects, malnutrition and conflict. Population displacement and famine were not discussed at all in this part of that report, although 'migration and displacement in a changing climate' was allocated its own section, elsewhere.

In summary, despite variation, most forms of the three-way categorization include the health effects of population displacement, undernutrition and violence in this third effect class. In this book it is argued that these effects may, ultimately, be the most important (Figs 21.1 and 21.2). None of these consequences are listed as health impacts by the more complex causal frameworks, although Watts *et al.* (2015) did list two of them as 'social mediating factors'. Of course, population displacement, famine and conflict are drivers of health consequences, rather than named diseases. Nonetheless, it is argued that their health consequences are of such importance that they warrant grouping in their own category, here called 'tertiary'.

21.3 Thinking the Unthinkable

Oreskes and Conway (2013), looking back from an imagined future in which civilization as we now know it has collapsed, reflected on the bias most scientists had and still have towards a 'type II' error (i.e. waiting too long to make an unorthodox conclusion). If by (say) 2200 several dozen cities have become substantially uninhabitable, or if World War III (or IV) has transpired, in part from climate change, then there may be wide agreement that unaddressed climate change was a risk not worth taking.

McMichael (2013) called the tendency to not think hard enough about cause as 'epidemiologizing', or looking under the lamp post, because that is where the light is, even if the keys are not likely to be. Similarly, I appeal to readers to think hard about consequences of unabated climate change; not as an expression of prurience

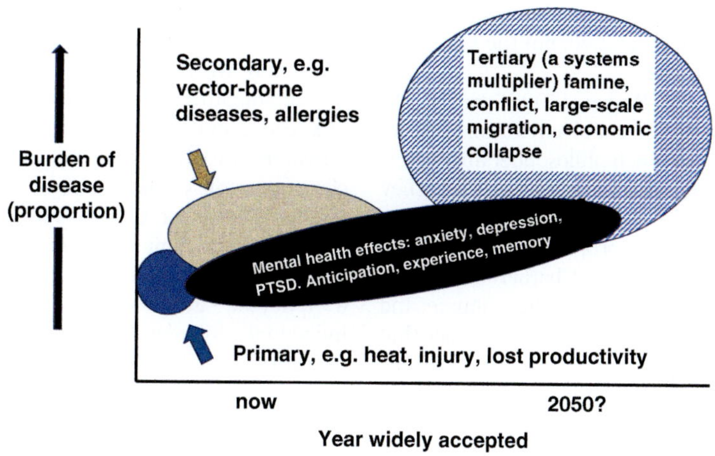

Fig. 21.2. This conceptual diagram compares the likely burden of disease from the primary, secondary and tertiary effects of eco-socially mediated environmental change (especially to the climate) with the time at which the effects are likely to be widely accepted as causally related by the general and even the scientific community. Mental health effects can arise from all forms of effects. PTSD = post traumatic stress disorder. Source: adapted from Butler *et al.* (2014).

or what Michael Mann (2021) criticized as 'doom porn', but to become motivated to try to stop the unmanageable.

Today, it is still possible to hope that the worst case can be avoided. However, I believe it is increasingly realistic to fear that reality will be worse for billions than most optimists claim. Even if nuclear war is avoided, and even if the next El Niño does not lead to record global high temperatures, there are now too many 'wild cards' in play to be optimistic. We can still hope for a 'muddle through' world but we cannot take this scenario at face value. Figure 21.2 postulates that although a wide acceptance of the impact of 'tertiary' effects still lies in the future, it is already less speculative than a decade ago.

In addition, the health effects of primary and secondary events are no longer speculative. Two major European (France and Russia) heatwaves (primary effects) since 2000 have killed over 100,000 people. Both extreme events are likely to have been contributed to by climate change. Secondary effects, such as changes to vector-borne diseases, probably have a lower burden of disease, but such effects are also no longer controversial, as they were as recently as 2010 (Brisbois and Harris, 2010).

Tertiary effects such as the contribution of anthropogenic climate change to the conflicts in Sudan and Syria, though still regarded as speculative by some, are increasingly accepted as valid, including by many scientists. Yet, a recent proposal to update the estimated burden of disease of climate change (Carlson *et al.*, 2023) appeared to entirely overlook the prospect of civilization collapse, and it also ignored the three main forms of tertiary effects here discussed. This is consistent with most of the general health literature, which remains too optimistic (Butler, 2018). Tertiary events have the potential to cause a burden of disease at least an order of magnitude higher than that from heatwaves or from allergies, vector-borne disease, or even new viral reassortments fuelled by temperature change (Carlson *et al.*, 2022).

21.4 'Eco-social' Causes: Refuting Accusations of Two Forms of Environmental Determinism

While there is a strong view in the literature that droughts and other forms of environmental change are a contributing factor to intrastate conflict, including the civil war in Syria (Gleick, 2017; Kelley *et al.*, 2017), there are occasional claims that scholarship on eco-climatic change is a form of 'environmental determinism'.

Environmental determinism has two main forms: one is its historical association with racism and colonialism; the other is its allegation of causal oversimplification. Associations with either form are here rejected, but the topic warrants brief discussion.

Coombes and Barber (2005) traced the concept of this first form of environmental determinism to the French philosopher and historian Baron de Montesquieu (1689–1755). They stated that in one of his works, published in 1748, de Montesquieu proposed that characteristics of European, Asian and 'barbarian' cultures were 'intimately connected' to their climate; and that 'human nature was thus a construct that could be related to the environment by scientific laws'. Peet (1985) described this kind of environmental determinism as the 'first version of modern geography' and proposed that the field was used to 'legitimate imperial conflict and conquest'. Coombes and Barber (2005) continued this theme, for example by stating that the 'imperial powers of the West sought to justify their colonial activities with a glut of pseudo-Darwinian proofs of their superiority'. Thus, the concept of environmental determinism is, today, justifiably tainted.

This book has 'global health' in the title. Some of its contents, and many of its contributors, explicitly acknowledge, discuss and seek to remedy past and ongoing harm to human (and non-human) well-being due to colonialism and other forms of exploitation. Any suggestion that work which considers the relationships between environmental (or ecological, including climatological) change and society is necessarily a disguised form of racism or colonialism is explicitly rejected. The negative effects of climate change will tend to accrue disproportionately to countries and people near the equator and those with fewer resources to adapt, in part due to a history of colonial oppression. Failure to confront this reality will perpetuate and worsen inequalities.

Populations in high-income settings are also being harmed by climate change. An example is the 'urban firestorm', believed to be triggered by an electrical fault, that destroyed about 1000 homes in Denver, Colorado (USA) on New Year's Eve, 2021. This was brought about by the combination of an unusually dry and warm autumn and urban planning that underestimated the danger of fast and strong (well over 100 km per hour) down-slope winds in suburban areas (Wallace-Wells, 2022).

The second form of environmental determinism is the implication, sometimes explicit, that many writers, perhaps even the majority, who attempt to discuss the causal pathways that link eco-climatic change to complex social outcomes are oversimplifying. Geographer and climate scientist Mike Hulme, founder of the Tyndall Centre for Climate Change Research, used the term 'climate determinism', which he defined as work that 'elevates' climate to 'become a – if not *the* – universal predictor (and cause) of individual physiology and psychology and of collective social organisation and behaviour' (Hulme, 2011) [italics original]. Hulme did, however, accept that there is an opposite extreme, which he called 'climate indeterminism'. He argued that there is a need to find a balance between these extremes. The author agrees with this need for balance, but disagrees that the scholarly literature, at least in recent decades, has many examples of the former, even if occasional quotes from mainstream media can be cherrypicked to suggest otherwise.

The main evidence Hulme provided in support of his proposition of oversimplification was media headlines, such as 'climate change is likely to cause wars'. In turn, he traced this headline to the book by the German Advisory Council on Global Change (2008) called *Climate Change as a Security Risk*. However, this report championed (and perhaps coined) the concept of climate change as a 'risk amplifier'. The key idea this term intends to convey is the antithesis of 'determinism', because the word 'amplify' clearly acknowledges that there are many concurrent causal factors for conflict and insecurity.

An analogy involves fire. A spark may be necessary for a large fire, but it is insufficient; 'tinder' is required. Many conflicts have occurred (and no doubt will continue to do so) without the spark of climate change. This does not mean that climate change is irrelevant, either to the conflicts in Darfur (Sudan), Syria, or the future. To minimize the risk of tertiary effects it is vital to reduce the tinder and the risk of sparks, only one of which need be climatic.[4]

Causal oversimplification is also explicitly claimed in an editorial in *Nature*, which asserted that 'climate change is never the sole cause of war, violence, unrest or migration' (Anonymous, 2018). I certainly agree with this; the disagreement by Butler and Kefford (2018) and others is with the implicit claim in this editorial that such oversimplification exists, or is even common. Gleick *et al.* (2018) wrote: 'Most researchers take pains to describe climate and weather merely as "contributing factors" to some conflicts' (see Chapter 24, this volume, for more examples).

The meteorologist Hubert H. Lamb (1913–1997) is widely acknowledged as a parent of research into the nexus of climate, ecology and society. He commented that climatology is a discipline that needs to 'pull in every kind of evidence, from every field of learning' (Martin-Nielsen, 2015). I agree. In the view of the author, all recent scholarly authors whose work is of relevance (including Jared Diamond and Norman Myers) acknowledge the complexity of the causal relationships between climate change (past or future), ecology and society, even if, at times, occasional words have been used out of context in an attempt to make the counter argument.

No doubt mainstream media is at times guilty of oversimplification. Perhaps the primary cause of ambiguity is miscommunication about the level of causation. Only a few consumers of the mass media are academics; exaggeration and simplification sell copy, and can be promoted by nationalist politicians and others whose motivation may be to provoke partisan allegiance. The scientific literature should try to avoid exaggeration and oversimplification. But there is no evidence, in the scientific literature, of a significant return of environmental determinism.

It would be misleading to argue that climate change caused any particular conflict without reference to other factors which also contributed to that conflict. However, if climate change increases the risk of many different conflicts across many countries over a prolonged period, an increase in conflict globally over decades compared with a counterfactual world without climate change is a statistical near-certainty. At this level of analysis, the contention that climate change causes conflict is reasonable.

21.5 IPCC Timidity

Another general point is worth discussing. There is a perennial debate regarding climatic reconstructions during the Holocene, including terms such as the Little Ice Age, the Medieval Warm Period and the Roman Climatic Optimum (Büntgen and Hellmann, 2014; Haldon *et al.*, 2018). One reason for this debate is that reconstructions of past temperatures necessarily rely on a tiny spatial subset of proxies for climatic data. For example, how do we 'know' if a temperature record relying on Greenland ice-core records accurately reflects temperature changes in New Zealand?

In an attempt to reduce such uncertainty, Marcott *et al.* (2013) used 73 'globally distributed' temperature records to reconstruct average global temperature over the past 11,300 years. Even so, no assemblage of these data can ever be complete. In 2021, prior to the release of the 6th IPCC report, it became clear that climate models widely used in the second half of the second decade of this century were too pessimistic (Voosen, 2021).[5] A paper from 2014 seemed to presage this problem, noting that the apparent decline in average global temperature from the middle Holocene (about 6000 years ago) until the onset of anthropogenic warming is puzzling, pointing to 'potentially significant biases in both the seasonality of the proxy reconstruction and the climate sensitivity of current climate models' (Liu *et al.*, 2014).

On the other hand, IPCC models have long been criticized, by some, as too optimistic, not only for understating amplifying physical feedbacks, such as sea level rise (Brysse *et al.*, 2013; Siegert *et al.*, 2020) but also because of their conservative (optimistic) assumptions concerning social consequences (Butler, 2005; Oreskes and Conway, 2013). Oppenheimer *et al.* (2019) pointed out that the demand for consensus may lead to scientists avoiding controversial issues, and the exclusion of experts whose opinions are known to be 'controversial' (but may nevertheless have pertinent expertise). Consensus reports such as the IPCC may consciously or unconsciously avoid reporting on extreme outcomes (Bendell, 2018) (Box 21.3). Oppenheimer *et al.* (2019) called this 'erring on the side of least drama'. In short, the push for agreement and caution may undermine other important goals, including inclusivity, accuracy and comprehension.

Box 21.3

Professor (now Sir) Robert Watson co-chaired the Millennium Ecosystem Assessment (MEA) from 2002 to 2005 (MEA, 2003). Watson was also influential in some IPCC reports, as well as other multi-authored scientific assessments. Early in the Millennium Assessment (MA) process, Watson advised MA authors (including this author) that he did not want the MA reports to mimic, in tone, those from the WorldWatch Institute, led by Lester Brown. Brown was, particularly in the last quarter of the 20th century, a globally recognized pioneer for his work linking ecological and demographic change to future human well-being. Rather than publishing in scholarly journals, Brown relied mainly on his own institute, producing, for example, annual 'State of the World' reports published by Norton, in New York. Brown's tone was, like that of Norman Myers, frequently criticized as too alarmist.

Yet, in 2021, Watson co-wrote an article in *The Conversation*, arguing that the concept of 'net zero' is a dangerous trap (Dyke *et al.*, 2021). The essay opened: 'Sometimes realisation comes in a blinding flash. Blurred outlines snap into shape and suddenly it all makes sense. Underneath such revelations is typically a much slower-dawning process. Doubts at the back of the mind grow. The sense of confusion that things cannot be made to fit together increases until something clicks. Or perhaps snaps.' Dyke *et al.* continued: 'Collectively we three authors of this article must have spent more than 80 years thinking about climate change. Why has it taken us so long to speak out about the obvious dangers of the concept of net zero? In our defence, the premise of net zero is deceptively simple – and we admit that it deceived us.'

Implicitly, Watson and his colleagues are now warning that the conservatism of the scientific assessments, including the IPCC and the MEA, may have contributed to the insufficient progress that has been made in mitigating the risk to human well-being from adverse climate and ecological change.

Despite uncertainties and complexities, the relationship between climate change and social changes, including famine, migration, conflict and even the decline and fall of whole states, cannot and should not ignored. To wait for complete certainty risks a type II error, as described above. Such an error places all of us at risk.

Finally, it is acknowledged that political links operate between ecological factors and social outcomes, in complex 'systemic' feedback loops. Rudolf Virchow, a father of social medicine, famously said that 'politics is nothing but medicine at a larger scale' (Mackenbach, 2009). Virchow did not wish to completely disentangle politics from public health. Nor does this author; furthermore, environmental factors and political consequences cannot be fully disentangled[6] (Butler and Kefford, 2018). To ignore climatic and ecological factors, including on the ground of 'securitization", could even be criticized as 'social determinism'.

21.6 Conclusion

Numerous classifications have attempted to describe the health effects of climate change. There is general agreement about the importance of direct and indirect health consequences (analogous to primary and secondary health effects) such as from heatwaves, disasters, vector-borne diseases and allergies. There appears near consensus that mental health effects of climate change are important, but there is less agreement as to whether such effects are direct, indirect, or belong in another category. Although several variants of a threefold health effect classification have been used, this chapter has argued that a form closely modelled on the earliest three-way classification (i.e. primary, secondary and tertiary effects) has many strengths, although it too is not perfect. The chapter has also discussed uncertainty in climatic and other forms of data, and the issue of alleged environmental determinism.

Notes

1 Temperature-related morbidity and mortality; health effects of extreme weather events (storms, tornadoes, hurricanes and precipitation extremes); air-pollution-related health effects; water- and foodborne diseases; and vector- and rodent-borne diseases.

[2] Asthma, respiratory allergies and airway diseases, cancer, cardiovascular disease and stroke, foodborne diseases and nutrition, heat-related morbidity and mortality, human developmental effects, mental health and stress-related disorders, neurological diseases and disorders, vector-borne and zoonotic diseases, waterborne diseases and 'weather-related morbidity and mortality'.
[3] Mental illness, undernutrition, allergies, cardiovascular disease, infectious diseases, injuries, respiratory diseases, poisoning (Watts *et al*., 2015, listed in Fig. 2 on p. 1867). In addition, in Fig. 1 (p. 1863), they listed harmful algal blooms, and vector-borne diseases as broad effect categories. The same figure described loss of habitat, poverty, mass migration, violent conflict and 'other social determinants of health' as 'social mediating factors'.
[4] Nevertheless, most conflicts involve competition over limited environmental resources, such as the Rwandan genocide (1994), which clearly did not involve climatic changes.
[5] Most recent models project a 'climate sensitivity' (warming expected from a doubling of carbon dioxide concentration to 560 parts per million) of 2°C to 4.5°C. In 2020, an influential paper that largely eschewed models calculated a likely climate sensitivity of between 2.6°C and 3.9°C. However, many of the new models from leading centres show warming of > 5°C. It is interesting, perhaps prescient, that Lamb consistently warned against over-reliance on numerical climate models.
[6] For example, political disputes, potentially including war, triggered by scarcity such as of food or water, due to drought or flood.

References

Anonymous (2018) Don't jump to conclusions about climate change and civil conflict. *Nature* 554, 275–276. DOI: 10.1080/00263206.2013.850076

Bendell, J. (2018) *Deep Adaptation: A Map for Navigating Climate Tragedy*. IFLAS Occasional Papers Vol. 2. Institute for Leadership and Sustainability, University of Cumbria, Ambleside, UK. Available at: http://www.lifeworth.com/deepadaptation.pdf (accessed 22 December, 2021).

Bowles, D.C. (2015) Climate change and health adaptation: Consequences for indigenous physical and mental health. *Annals of Global Health* 81, 427–431. DOI: 10.1016/j.aogh.2015.06.004

Brisbois, B.W. and Harris, S. (2010) Climate change, vector-borne disease and interdisciplinarity research: social science perspectives on an environment and health controversy. *EcoHealth* 7, 425–438. DOI: 10.1007/s10393-010-0354-6

Brysse, K., Oreskes, N., O'Reilly, J. and Oppenheimer, M. (2013) Climate change prediction: Erring on the side of least drama? *Global Environmental Change* 23, 327–337. DOI: 10.1016/j.gloenvcha.2012.10.008

Büntgen, U. and Hellmann, L. (2014) The Little Ice Age in scientific perspective: cold spells and caveats. *Journal of Interdisciplinary History* xliv, 353–368. DOI: 10.1162/JINH_a_00575

Butler, C.D. (2005) Peering into the fog: ecologic change, human affairs and the future. *EcoHealth* 2, 17–21. DOI: 10.1007/s10393-004-0090-x

Butler, C.D. (2018) Climate change, health and existential risks to civilization: A comprehensive review (1989–2013). *International Journal of Environmental Research and Public Health* 15, 2266. DOI: 10.3390/ijerph15102266

Butler, C.D. and Harley, D. (2010) Primary, secondary and tertiary effects of the eco-climate crisis: the medical response. *Postgraduate Medical Journal* 86, 230–234. DOI: 10.1136/pgmj.2009.082727

Butler, C.D. and Kefford, B.J. (2018) Climate and conflict: magnifying risks. *Nature* 555, 587. DOI: 10.1038/d41586-018-03795-0

Butler, C.D., Corvalán C.F. and Koren. H.S. (2005) Human health, well-being and global ecological scenarios. *Ecosystems* 8(2), 153–162. DOI: 10.1007/s10021-004-0076-0

Butler, C.D., Bowles, D.C., McIver, L. and Page, L. (2014) Mental health, cognition and the challenge of climate change. In: Butler, C.D. (ed.) *Climate Change and Global Health*. CABI, Wallingford, UK, pp. 251–259.

Carlson, C.J., Albery, G.F., Merow, C., Trisos, C.H., Zipfel, C.M. *et al.* (2022) Climate change increases cross-species viral transmission risk. *Nature* 607, 555–562. DOI: 10.1038/s41586-022-04788-we

Carlson, C.J., Alam, M.S., North, M.A., Onyango, E. and Stewart-Ibarral, A.M. (2023) The health burden of climate change: A call for global scientific action. *PLoS Climate* 2, e0000126. DOI: 10.1371/journal.pclm.0000126

Cissé, G., McLeman, R., Adams, H., Aldunce, P., Bowen, K. *et al.* (2022) Health, wellbeing, and the changing structure of communities. In: Pörtner, H.-O., Roberts, D.C., Tignor, M., Poloczanska, E.S., Mintenbeck, K.

et al. (eds) *Climate Change 2022: Impacts, Adaptation, and Vulnerability. Contribution of Working Group II to the Sixth Assessment Report of the Intergovernmental Panel on Climate Change.* Cambridge University Press, Cambridge, UK, pp. 1041–1170.

Confalonieri, U., Menne, B., Akhtar, R., Ebi, K.L., Hauengue, M. *et al.* (2007) Human health. In*:* Parry, M.L., Canziani, O.F., Palutikof, J.P., Linden, P.J.V.D. and Hanson, C.E. (eds) *Climate Change 2007: Impacts, Adaptation and Vulnerability. Contribution of Working Group II to the Fourth Assessment Report of the Intergovernmental Panel on Climate Change.* Cambridge University Press, Cambridge, UK, pp. 391–431.

Coombes, P. and Barber, K. (2005) Environmental determinism in Holocene research: causality or coincidence? *Area* 37.3, 303–311. DOI: 10.1111/j.1475-4762.2005.00634.x

Dyke, J., Watson, R. and Knorr, W. (2021) Climate scientists: concept of net zero is a dangerous trap. *The Conversation*. Available at: https://theconversation.com/climate-scientists-concept-of-net-zero-is-a-dangerous-trap-157368 (accessed 26 April, 2021).

German Advisory Council on Global Change (2008) *Climate Change as a Security Risk*. Earthscan, London.

Gleick, P.H. (2017) Climate, water, and conflict: Commentary on Selby *et al.* (2017). *Political Geography* 60, 248–250. DOI: 10.1016/j.polgeo.2017.06.009

Gleick, P.H., Lewandowsky, S. and Kelley, C. (2018) Critique of conflict and climate analysis is oversimplified. *Nature* 555, 587. DOI: 10.1038/d41586-018-03794-1

Haines, A. and Parry, M.L. (1993) Climate change and human health. *Journal of the Royal Society of Medicine* 86, 707–711. DOI: 10.1177/014107689308601212

Haines, A., Epstein, P.R. and McMichael, A.J. (1993) Global health watch: monitoring impacts of environmental change. *The Lancet* 342, 1464–1469. DOI: 10.1016/0140-6736(93)92937-O

Haldon, J., Elton, H., Huebner, S.R., Izdebski, A., Mordechai, L. *et al.* (2018) Plagues, climate change, and the end of an empire: A response to Kyle Harper's The Fate of Rome (1): Climate. *History Compass* 16, e12508. DOI: 0.1111/hic3.12508

Hulme, M. (2011) Reducing the future to climate: A story of climate determinism and reductionism. *Osiris* 26, 245–266. DOI: 10.1086/661274

Kelley, C., Mohtadi, S., Cane, M., Seager, R. and Kushnir, Y. (2017) Commentary on the Syria case: Climate as a contributing factor. *Political Geography* 60, 245–247. DOI: 10.1016/j.polgeo.2017.06.013

Liu, Z., Zhu, J., Rosenthal, Y., Zhang, X., Otto-Bliesner, B.L. *et al.* (2014) The Holocene temperature conundrum. *Proceedings of the National Academy of Sciences (USA)* 111, E3501–E3505. DOI: 10.1073/pnas.1407229111

Mackenbach, J.P. (2009) Politics is nothing but medicine at a larger scale: reflections on public health's biggest idea. *Journal of Epidemiology and Community Health* 63, 181–184. DOI: 10.1136/jech.2008.077032

McMichael, A.J. (2013) Impediments to comprehensive research on climate change and health. *International Journal of Environmental Research and Public Health* 10, 6096–6105. DOI: 10.3390/ijerph10116096

McMichael, A.J., Haines, A., Slooff, R. and Kovats, S. (eds) (1996) *Climate Change and Human Health.* World Health Organization, World Meteorological Organization and United Nations Environment Programme, Geneva.

Mann, M.E. (2021) *The New Climate War: The Fight to Take Back Our Planet*. Scribe, London.

Marcott, S.A., Shakun, J.D., Clark, P.U. and Mix, A.C. (2013) A reconstruction of regional and global temperature for the past 11,300 years. *Science* 339, 1198–1201. DOI: 10.1126/science.1228026

Martin-Nielsen, J. (2015) Ways of knowing climate: Hubert H. Lamb and climate research in the UK. *WIREs Climate Change* 6, 465–477. DOI: 10.1002/wcc.349

MEA (2003) *Ecosystems and Human Well-being. A Framework for Assessment*. Millennium Ecosystem Assessment. Island Press, Washington, DC.

Oppenheimer, M., Oreskes, N., Jamieson, D., Brysse, K., O'Reilly, J. *et al.* (2019) *Discerning Experts. The Practices of Scientific Assessment for Environmental Policy*. University of Chicago Press, Chicago, Illinois.

Oreskes, N. and Conway, E.M. (2013) The collapse of Western civilization: a view from the future. *Daedalus* 142, 40–58. DOI: 10.1162/DAED_a_00184

Patz, J.A., McGeehin, M.A., Bernard, S.M., Ebi, K.L., Epstein, P.R. *et al.* (2000) The potential health impacts of climate variability and change for the United States: executive summary of the report of the health sector of the U.S. National Assessment. *Environmental Health Perspectives* 108, 367–376. DOI: 10.1289/ehp.00108367

Peet, R. (1985) The social origins of environmental determinism. *Annals of the Association of American Geographers* 75, 309–333. DOI: 10.1111/j.1467-8306.1985.tb00069.x

Portier, C.J., Thigpen Tart, K., Carter, S.R., Dilworth, C.H., Grambsch, A.E. *et al.* (2010) *A Human Health Perspective on Climate Change: A Report Outlining the Research Needs on the Human Health Effects of Climate Change*. Environmental Health Perspectives and the National Institute of Environmental Health Sciences, Research Triangle Park, North Carolina. DOI: 10.1289/ehp.1002272

Siegert, M., Alley, R.B., Rignot, E., Englander, J. and Corell, R. (2020) Twenty-first century sea-level rise could exceed IPCC projections for strong-warming futures. *One Earth* 3, 691–703. DOI: 10.1016/j.oneear.2020.11.002

Smith, K., Woodward, A., Campbell-Lendrum, D., Chadee, D., Honda, Y. *et al.* (2014) Human health: impacts, adaptation, and co-benefits. In: Field, C.B., Barros, V. and Dokken, D.J. (eds) *Climate Change 2014: Impacts, Adaptation and Vulnerability, Contribution of Working Group II to the Fifth Assessment Report of the Intergovernmental Panel on Climate Change.* Cambridge University Press, Cambridge, UK, and New York, pp. 709–754.

Voosen, P. (2021) UN climate panel confronts implausibly hot forecasts of future warming. *Science* 373, 474–475. DOI: 10.1126/science.373.6554.474

Wallace-Wells, D. (2022) The return of the urban firestorm. *New York Magazine*. Available at: https://nymag.com/intelligencer/2022/01/colorado-saw-the-return-of-the-urban-firestorm.html (accessed 2 January 2022).

Watts, N., Adger, W.N., Agnolucci, P., Blackstock, J., Byass, P. *et al.* (2015) Health and climate change: policy responses to protect public health. *The Lancet* 386, 1861–1914. DOI: 10.1016/S0140-6736(15)60854-6

22 Famine, Hunger, Food Prices and Climate Change

Colin D. Butler
National Centre for Epidemiology and Population Health, The Australian National University, Canberra, Australia

Abstract
For over 30 years researchers have warned that climate change will affect global food security. This chapter, after reviewing evidence, concludes that these concerns are valid, via means such as increased heat, drought, flooding and other alterations in rainfall patterns and intensity. However, many other factors are, to date (and probably for the near future), far more important than climate change in the causation of current food insecurity and high global food prices; these include the rising cost of energy and an entrenched system of global inequality, sustained by the neoliberalism inherent in the Sustainable Development Goals. In some especially vulnerable areas (e.g. the Horn of Africa and Madagascar) climate change is probably, already, an important co-factor for existing famine. As climate change intensifies, its impact on food security and famine is likely to increase, especially if the price of energy remains high, or rises further, which also seems likely.

> And he gave it for his opinion that he that could make two ears of corn or two blades of grass to grow upon a spot of ground where only one grew before would deserve better of mankind and do more essential service to his country than the whole race of politicians put together.
> (Swift, 1729)

22.1 Introduction: Famine, Population Growth and the Cold War

In 1966, successive monsoon failures were threatening food security in the chronically impoverished, densely populated (57 million in 1967) caste-afflicted northern Indian state of Bihar, which at that time included Jharkhand, now an adjoining state. A contributing cause was the limited availability of the then emerging technology of tube wells to extract groundwater (Brass, 1986). Grain production in Bihar in 1966/67 was approximately 50% lower than in the exceptionally good season of 1964/65 (Mellor and Gavian, 1987; Dyson and Maharatna 1992). Social factors contributing to Bihar's food scarcity included the state's extreme inequality and its high proportion of landless labourers, many of whom were 'scheduled' castes, Indigenous tribals and Muslims. Ramalingaswami *et al.* (1971) reported that nearly 30% of those who worked the land in Bihar were themselves landless. They also described desperation sufficient 'that people gathered around railway sidings and grain chutes in order to collect individual pieces of grain from the dust'.

India was an important Asian democracy, then as now. Non-aligned in the Cold War, it was a prize ally for either of that undeclared conflict's

 Climate Change and Global Health: Primary, Secondary and Tertiary Effects, 2nd Edition (eds C.D. Butler and K. Higgs)
DOI: 10.1079/9781800620025.0022

main protagonists. Winning the hearts and minds of the people of the global South (then usually called the 'Third World') was US policy, including through a programme established in the 1950s and dubbed 'food for peace' in 1960 by John F. Kennedy, as he campaigned for the US presidency. On hearing of the Bihar situation,[1] senior US state officials recommended emergency grants of American grain. However, Kennedy's successor, Lyndon Johnson, prevaricated, instead authorizing incremental shipments in exchange for evidence that India's family planning programme was being strengthened (Connelly, 2006). In total, as much as 20% of the US annual wheat crop was given to India (Dyson and Maharatna, 1992). Thanks to these shipments, and to an effective Indian government response, the number of deaths from starvation in Bihar was greatly restrained; far less than in either the Bengal famine during World War II or the subsequent famine in Maharashtra (India) in the 1970s (Dyson and Maharatna, 1992).

Famine in China – with no Western intervention

At the time that the USA was providing grain relief to India, intended mainly for Bihar, its highest officials may not then have known of the comparatively recent famine in newly Communist China, where, between 1958 and 1961, the 'great leap forward' was accompanied by death by starvation of at least 30 million people (Bernstein, 2006). Lloyd Boyd Orr, founding director of the United Nations' Food and Agricultural Organization (FAO), had been one of the rare Westerners invited to visit China during that famine. Orr (then aged about 80), along with other dignitaries, including the President of the Royal Society, reportedly accepted at face value Chinese assurances of adequate food (Becker, 1996).

Food security and peace

Orr had been awarded the Nobel Peace Prize in 1949 for his pioneering work in nutrition. He was deeply aware of the links between food insecurity and possible conflict. As FAO director, Orr had tried to promote world peace and food security via the establishment of the World Food Board and expanded agricultural productivity (Staples, 2003).

Two decades later, the Norwegian Nobel committee again affirmed their recognition of the link between food security and the avoidance of conflict by awarding the highest peace honour to the agricultural scientist, Norman Borlaug. Borlaug is the main parent of the 'Green Revolution', a term adopted because it was thought that greater food supply could lower the attraction of Communism and thus avert the 'Red' (communist) Revolution that many policy makers feared (Gaud, 1968).

22.2 The Green Revolution and Its Partial Conquest of Famine

The Green Revolution, the origins of which can be traced to at least the 1930s, was achieved by the long-established techniques of selective plant breeding and strong international and interdisciplinary cooperation (Boyer *et al.*, 2013). Improved seeds enabled far more food to be grown in a given area, provided that there was adequate fertilizer and water (Pingali, 2012). In theory, more intense cultivation of land might contribute to increased protection of nature (by reducing the amount of land needed to grow food).

Although the early decades of this agricultural revolution did not eliminate hunger, it was spectacularly successful in positioning the 'global plough' (food supplies) before the 'global stork' (human numbers) (Fig. 22.1). World hunger declined, both as a proportion and, from 1970 to 1996, absolutely. While a few famines following the near-miss in Bihar (most notably in newly independent, war-ravaged Bangladesh in 1974 and in Ethiopia in 1984), their scale was historically modest.

Without the Green Revolution it is difficult to argue with the proposition that the widespread famines, perceived by many as otherwise plausible as the 20th century lengthened, could have been averted (Tauger, 2003). Today's human population – over eight billion – could not exist. However, the dependency of the Green Revolution on monocultures and fossil fuels, especially as the basic source of pesticides and nitrogen fertilizer, has long been criticized (Shiva,

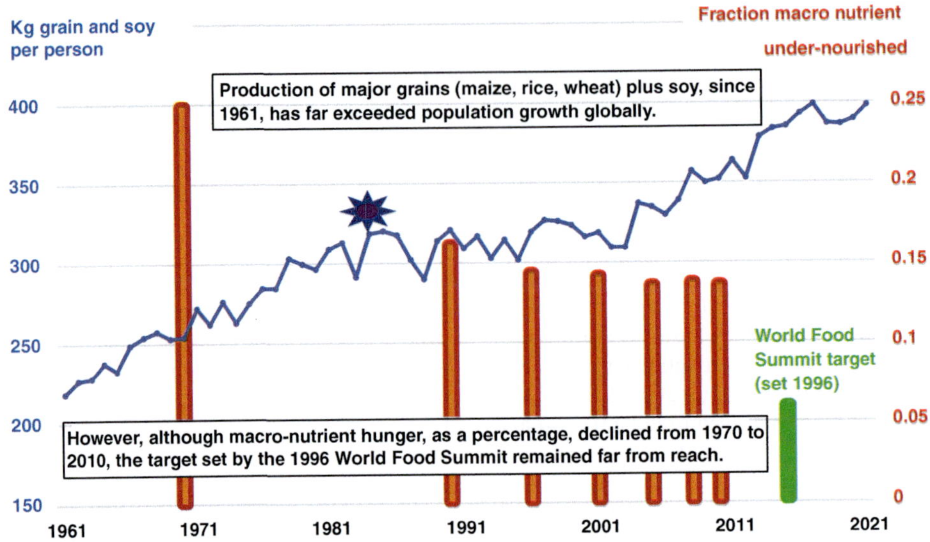

Fig. 22.1. Between 1961 and 2021, grain (and soy) production increased far faster than population, due principally to the Green Revolution, irrigation and the use of additional land on which to grow crops. The blue line shows the three main grains per person (globally): wheat, maize and rice, plus soybeans. Soybeans are included because they share important characteristics with grains: durability, transportability. They are also produced in vast quantities. Macronutrient hunger (i.e. calories), as a percentage (red bars), initially fell, but the optimism of the World Food Summit (1996) target, which aimed to lower the number of hungry people by 2015 to 420 million, proved far out of reach. FAO *et al.* (2022) reported that approximately 10% of the global population was calorie undernourished. The calculation method was changed compared with earlier measures; nevertheless the number who are chronically hungry and food insecure remains far larger than was once hoped. The star signifies 1984, the year of US President Reagan's comment re population. Data: https://www.fao.org/faostat/ and several 'State of World Food Insecurity' reports.

1991). Subramanian (2015) argued that the benefit of its improved seeds is also overstated; he instead attributes yield growth, at least in India in the 1960s, more to investment in tubewells, thus providing adequate irrigation, including during drought. Other critics point out that the industrialized agriculture of the Green Revolution favoured large over small farmers, and helped to intensify inequality (Aga, 2021). Perkins (1990) argued that this consequence was not only foreseeable, but also that, by 1956 at the latest, it had been brought to the attention of the Rockefeller Foundation – a major supporter of the Green Revolution – which ignored it. On the other hand, there was clearly a widespread expectation that more total food would 'trickle down' to the poor, reducing the risk of social upheaval, including support for economic practices that challenged capitalism.

While the Green Revolution may be necessary to eliminate regional and global hunger, it was never sufficient; many other mechanisms are required to reach that elusive goal. This absence of success is likely to persist, not least as those with substantial economic and political power (e.g. Bill Gates) deny the need for fundamental reform of capitalism (Hickel, 2019). Instead, supporters and beneficiaries of immense global inequality have resorted to a series of high-profile, unenforceable promises to reduce global hunger, the most recent of which is the Sustainable Development Goal (SDG) to 'end hunger'[2] by 2030 (UN, nd). The Alliance for a Green Revolution in Africa (AGRA), launched in 2006 and largely funded by the Bill and Melinda Gates Foundation, promised 'to double yields and incomes for 30 million farming households by 2020'. It failed dismally; in nations in which AGRA operates, the number of people suffering hunger is reported to have increased by 30% (Mkindi *et al.*, 2020).

The undeclared strategy seems to be: 'when one target fails, set a new one – between one and two decades away'. Complicating this issue (but also consistent with the theory, here advanced, that what matters is the pretence of concern rather than the reality) is the fact that the FAO has repeatedly changed the way that it measures hunger. In addition, published estimates of hunger, made in earlier decades, have been expunged from FAO's more recent publications on this topic. For example, the data used to estimate hunger in Fig. 22.1 were obtained from a series of FAO books, published between 1999 and 2012, called *The State of Food Insecurity in the World*. In 2012, the method used to measure hunger was substantially revised (Lappé *et al.*, 2013; Butler, 2015). Since then, yet another revision in FAO methodology and reporting has made the comparison of energy undernutrition over decades even more difficult. However, all reports, including the most recent, find that well in excess of 700 million people are chronically hungry (FAO *et al.*, 2022). Also, as many as two billion additional people lack adequate micronutrients, especially iron and zinc (Camaschella, 2019). Their absence is of profound importance for many capacities that determine health, income, status and respect, including stamina, cognitive potential, learning ability, immunity and height (Ekholuenetale *et al.*, 2022).

22.3 Borlaug and Population

In his Nobel Peace Prize acceptance speech, Borlaug (1970) warned:

> The Green Revolution has won a temporary success in man's war against hunger and deprivation; it has given man a breathing space. If fully implemented, the revolution can provide sufficient food for sustenance during the next three decades. But the frightening power of human reproduction must also be curbed; otherwise the successes of The Green Revolution will be ephemeral only.

This warning was ignored. In 1984, US President Ronald Reagan became the first US president to declare explicitly that the importance of human population size had been overestimated (Reagan and Mondale, 1984). In so doing, he reversed the position of his immediate four predecessors (Donaldson, 1990). Although Reagan's comment was at a time when hunger was declining (Fig 22.1), this is no longer the case. Although total food production per person has continued to increase in recent decades and is now at a record high, the crude, unadjusted figures are deceptive. For example, more than one-third of the US maize crop is diverted to biofuels. While a small fraction of its food value is recovered and fed to cattle as 'distiller's grain', this quantity of maize could, instead, feed millions of people if supplied directly to them. If grain, soy, sugar and other foods lost from the human diet because of diversion to biofuel were subtracted, the increase in per capita food production would be less impressive. Compounding this is the problem of the increasing quantity of edible calories in food, particularly in soy, that is diverted to supply livestock, including farmed fish. At the same time, the global population continues to rise by as many as 80 million per annum, an annual increment probably exceeding that in 1968, when global population increase peaked as a percentage.

The incidence of famine is again increasing, especially in sub-Saharan Africa (SSA), including parts of the Democratic Republic of Congo, Kenya, Madagascar, Niger, Somalia and Sudan. Sometimes claims are made that SSA has the potential to be transformed to a 'grain breadbasket' because of large gaps between current low yields and yield potential with good management, and vast land resources with adequate rainfall. However, rigorous evaluation of the extent to which soils can support high, stable yields has been limited by lack of data on rootable soil depth of sufficient quality and spatial resolution. Investigators have concluded that while SSA could produce a modest maize surplus, this requires rootable soil depths comparable to that of other major breadbaskets, such as the US Corn Belt and South American Pampas, and that this is doubtful. Therefore, producing surplus grain for export will require the expansion of crop area in regions where soil depth and rainfall are sufficient, and where negative impacts on biodiversity can be minimized (Guilpart *et al.*, 2017).

Other problems confront food security in SSA. In Somalia, the Islamist group Al-Shabaab has worsened the human toll of food insecurity by tactics such as intimidating, kidnapping and killing famine relief workers (Maxwell *et al.*, 2012). Yet, in nearby Somaliland, a quasi-independent breakaway nation, little, if any, famine occurred at all, although it too has experienced severe drought.

Increased food per capita means, at least in theory, that surpluses elsewhere can be used to avert famine, including in nations (e.g. Somalia) that are consistent recipients of food aid. However, such assistance, unless accompanied by other means that promote self-reliance (such as the nurturing of export industries), is unlikely to generate prosperity. In the case of Somalia, prospects of food and other forms of self-sufficiency have also been harmed by the plundering of its fisheries by many foreign nations, enabled in part by an ineffectual national government (Sumaila and Bawumia, 2014).

Famine in parts of Asia, even beyond North Korea, is again possible. Food security in Afghanistan is parlous (Oskorouchi and Sousa-Poza, 2021), while a shortfall of funding in 2023 forced a cut in food aid for Rohingya refugees in Bangladesh, from vouchers worth US$12 per month to US$10 per month (WFP, 2023a).

22.4 Inequality, Vegetarianism and Reducing Food Waste

A complete global sharing of food resources, with the total abandonment of using grain and other crops such as soy to feed livestock, would eliminate not only famine but also hunger and micronutrient deficiency, making the Green Revolution unnecessary, as Lappé and Collins (1971) advocated in their book, *Diet for a Small Planet*. However, as with alternatives to the Green Revolution, I am unaware of scholars, then or now, who view complete vegetarianism as a viable or total solution to world hunger. Totally vegetarian diets can be healthy for many (provided that vitamin B12 supplements are available and used), but are neither culturally preferred nor ecologically possible in many places. Nor, probably, are they physiologically adequate for many populations. Genetic variations such as haemochromatosis are advantageous for people with limited iron intake (Naugler, 2008). Those lacking this or similar genes may be at a disadvantage if forced to be vegetarian, but less consumption of animal-sourced food, by at least a billion people, will deliver health as well as climatic benefits (McMichael *et al.*, 2007).

Arguments for a completely equal distribution of global food resources are as utopian as are calls for the equal sharing of wealth – and, for that matter, the SDGs. While excessive unfairness cannot be tolerated indefinitely in either human society or many animal groups, most (if not all) human and most animal groups are, none the less, characterized by many forms of inequality, such as of power, experience, strength and ability. A large reduction of human inequality, on a scale sufficient to generate adequate food for all, may be partly achievable, not least as being overfed is harmful to health and sustainability.

Furthermore, much food is lost from human ingestion due to pests and poor storage. The custom of eating offal has vanished among most affluent populations, leading to a colossal waste of nutrients. In rich countries, enormous quantities of edible food are thrown out of supermarkets, restaurants and homes (Chen *et al.*, 2020). As countries become more affluent this trend is likely to worsen. In India and elsewhere, surplus grain is spoiled due to its poor storage.

22.5 Famines Have Ecological *and* Social Causes

West Bengal, an adjoining state to Bihar, was the birthplace of Amartya Sen, famous for his work on famine (Sen, 1981). As a sensitive child in Calcutta (today Kolkata) he witnessed the catastrophe of the Bengal famine during World War II, which claimed 2–3 million lives (Sur, 2020). Sen showed that harvests in the year of the greatest death toll (1942) exceeded that of 1941. In 1998 Sen was awarded the Nobel Prize in Economics, partly for his economically sophisticated discussion of what he called food 'entitlement'; that is, that the deeper cause for a lack of food is insufficient money and other means to acquire it (e.g. land to grow food, or capacity to borrow money).

Although some writers claim or imply that the dominant cause for all or most recent famines is social (Patel, 2012), ecological factors also contribute, such as crop diseases and extreme weather events. Conflict can also be an important co-factor in some settings, including the Bengali famine, when the import of rice from Burma, then recently invaded by the Japanese, was made impossible. Further, Tauger (2003) not only reviewed evidence that a very severe fungal disease reduced the Bengali rice harvest in 1942, but also stated that the harvest data Sen relied on were excessively optimistic, due to the crude assumptions that estimators used. Davis (2001), whose book *Late Victorian Holocausts* recounted how more entitled populations, whether in India,[3] China or parts of Africa, thrived while others starved, implicitly recognized an eco-climatic contribution in his subtitle *El Niño Famines and the Making of the Third World.*

Many others also support the view that most famines have mixed causation. Mallory (1926) wrote: 'There are some famines that are due almost certainly to "natural" causes (i.e. drought, locusts, plant disease), and there are *scarcely any to which natural phenomena do not contribute*' [italics added]. Brass (1986), discussing the Bihari food shortage of 1966/67, described proximate and remote causes, both ecological and social.

The prolonged famine in the Democratic People's Republic of Korea in the 1990s (Moon, 2009) also had ecological co-factors. While the hermit kingdom's cruel and repressive regime had numerous policies which repelled foreign aid and assistance, a series of poor harvests, some associated with torrential rains and others with unusually high temperatures, also played a part. Typhoons as well as the indifference of the invading Japanese also contributed to the lesser-known famine in Vietnam (1944/45), which killed up to 2 million people (Huff, 2019). Earlier in the 20th century, the Ukrainian famine, while largely a socially caused event resulting from the policies of Stalinist Russia, was also exacerbated by a drought (Krawchenko, 1989) (Fig. 22.2). Natural disasters played a role in the great Chinese famine, although by far the main cause was social, including a reduction in the area of grain sown and the 'wind of exaggerated output' (i.e. overstated harvest estimates) (Bernstein, 2006).

At a finer scale, the social and environmental factors that can generate abundant locally available food do not guarantee either local ingestion or good utilization of ingested nutrients. In the agriculturally rich, populous Indian state of Uttar Pradesh, up to 60% of children were reported as chronically undernourished by Black *et al.* (2008). Over a decade later Swaminathan *et al.* (2019) reported little improvement. This dismal loss of human potential arises not only because of inadequate food but also, in too many cases, through poor hygiene, dirty water and

Fig. 22.2. The plaque reads '7,000,000. Dedicated to victims of the 1932–1933 enforced famine – genocide in Ukraine', Canberra, Australia (photo C.D. Butler).

limited sanitation. This leads to excess disease, such as diarrhoea, fevers and malabsorption, including from parasites (Fig. 22.3). It is not only the ingestion of food but also access to toilets and clean water that rests on 'entitlement'.

Furthermore, even if a population overall has sufficient entitlement to obtain and ingest food of adequate quality, quantity and variety, adequate nutrition for less favoured members of that group is not automatic. Many groups discriminate against females and infants (including, perhaps, by excessively prolonged exclusive breast feeding) (Subramanian *et al.*, 2023) or preferentially give food to earners rather than dependents, even if dependents are ill.

Crucially, discrimination against less powerful individuals, and sometimes entire populations, is particularly likely if food supplies are stressed by environmental factors such as drought, flooding or, as in parts of the Sudan, fear of random attack by bombers operated by the national government or its opposition. In parts of Africa, physical insecurity has favoured the planting of the less nutritious root crop, cassava, which remains edible below ground for up to 36 months

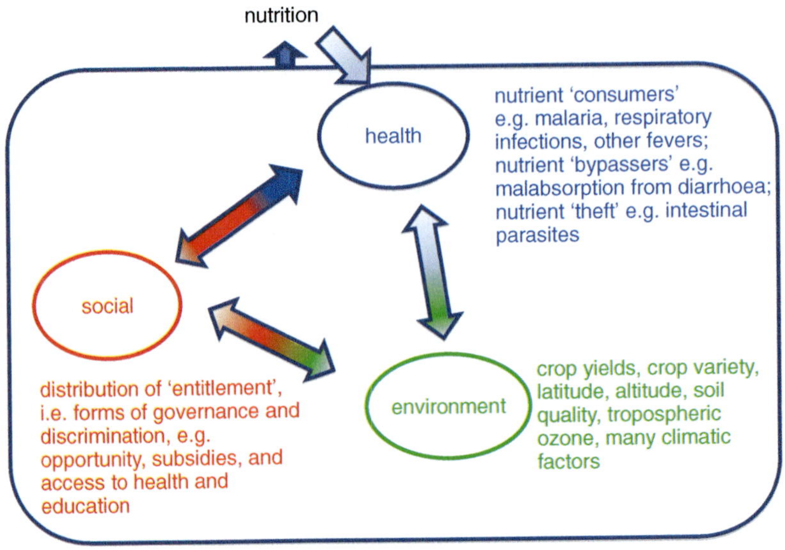

Fig. 22.3. This complex system (here conceptualized as within the blue border) influences nutrition - but the components of this system are also influenced by nutrition. Many relationships are bidirectional. For example, poor population health status can impair social function (e.g. by ineffective education).

and is easier to hide from marauding enemies than, for example, maize (Rosenthal and Ort, 2011).

In summary, attributing famine solely to social causes is simplistic. Even if humans were as egalitarian and cooperative as worker bees or ants, equal ingestion of food and sufficient nutrients to avoid famine is not always possible during times of extreme generalized food scarcity. However, there is evidence that Indigenous Australians maintained low population numbers in part so that they could easily avoid famine or other forms of undue hardship during seasons of extreme drought (Butler, 2012).

This section has argued that, more often than not, famines are caused by an interaction of eco-climatic causes (for example, drought, flood and plant diseases such as potato blight) and social causes (such as poor governance, occupation, conflict and discrimination against minorities and vulnerable individuals). This interplay between eco-climatic and political phenomena is a characteristic of tertiary events, as outlined elsewhere in this book. Famines and large-scale food insecurity, even without frank starvation, also have the potential to harm health on a far larger scale than any of the primary or secondary effects discussed in previous sections.

The next part of this chapter outlines the evidence of effects of climate change on crop growth. It is plausible that, as the other tertiary effects of climate change worsen (i.e. large-scale population dislocation and conflict), governance could increasingly fail, also contributing to famine.

22.6 Climate Change, Crops and Inequality

There is a rich scientific literature on climate change and food production, dating for almost three decades. All plants have an optimal band of environmental conditions, including temperature, moisture and nutrients (Agnolucci *et al.*, 2020). In cold regions, higher temperatures will improve yields provided that there is adequate water, soil, nutrients, etc. Carbon dioxide (CO_2) can also be conceptualized as a nutrient. Beyond this 'envelope' of conditions, which of course varies for different crops, yields will fall (for example, millet is more heat tolerant than wheat; rice requires considerable water). On balance, therefore, agricultural production in tropical (hot) countries – which are disproportionately poor – is likely to be harmed by anthropogenic climate change, while countries at higher latitudes may benefit. This point was repeatedly stressed in the early scientific literature on climate change and nutrition (Butler, 2009a). The first major paper in this field (Rosenzweig and Parry, 1994) stated in its abstract that 'developing countries are likely to bear the brunt of the problem'.

Intensifying inequality, as a result of climate change, was also central to a subsequent letter in the same journal (Pittock *et al.*, 1994) which stated that 'Rosenzweig and Parry's data suggest that climate change will significantly increase the present unequal distribution of food supplies, thereby exacerbating the problem'. Yet, an accompanying (commissioned) commentary, described by Pittock *et al.* (1994) as 'optimistic', overlooked this issue entirely (Reilly, 1994). It concluded that 'although the location of food production may shift,[4] global food prospects do not worsen significantly. There seems to be no "disaster threshold" for global food production.'

Butler (2009b) proposed that this more optimistic view inappropriately influenced global policy, delaying rapid and effective response to the threat of climate change for food security. For example, there is little evidence that FAO considered climate change as an important problem until 2003, when this issue was featured at the 29th session of the FAO Committee on World Food Security. If the burden of risk had been reversed – that is, that crop growth in wealthier nations was likely to disproportionately suffer – would such delay have occurred?

Pittock *et al.* (1994) also questioned the strength of the carbon fertilization effect (CFE), then widely considered as likely to significantly offset harm to crop growth from other elements of climate change. It is now well accepted that the CFE for 'C4' crops[5] (e.g. corn, sorghum, sugarcane and millet) is minimal, other than in drought conditions (Ainsworth and Long, 2021). Reassuringly, a recent review of 186 independent studies of 18 C3 crops found that elevating CO_2 to 580 parts per million (ppm), compared with about 425 ppm in 2023, leads to an approximate rise in yield of about 18%, including for wheat, rice and soy, in 'non-stress' conditions (Ainsworth and Long, 2021). On the other hand, Wang *et al.* (2020) found that the strength of the CFE for vegetation photosynthesis has declined at the global scale. They suggested that the cause may be insufficient plant nutrients such as nitrogen and phosphorus. It is now also accepted that the CFE lowers the concentrations of many important micronutrients in many crops, including iron, zinc and nitrogen (Ebi and Loladze, 2019). Finally, Pittock *et al.* (1994) observed that climate change is likely to harm crop growth through several mechanisms excluded from the models, such as changed storm intensity, heavier rainfall and altered climatic variability.

22.7 Rising Food and Energy Prices and Climate Change

For centuries, the long-term trend of food prices has declined (Naylor and Falcon, 2010). Since 1961, the FAO has calculated an index of food prices (Fig. 22.4) which show a fall in price from its inception until the late 1980s, apart from the first oil crisis, in 1974. This was followed by a prolonged trough, lasting almost two decades. Butler (2002) termed this era the 'cornucopian enchantment', a period when policy makers shared increasing complacency, lulled by low prices for oil and food. However, by 2007, the era of cheap food was closing, surprising observers who had ignored, forgotten, or never heard of the warnings delivered not only by Borlaug but also by the World Scientists' Warning to Humanity (Union of Concerned Scientists, 1992).

In its 2014 report, the Intergovernmental Panel on Climate Change, speculated (as did this chapter in the first edition of this volume) that global food prices are now more sensitive to weather-related impacts than previously (Porter *et al.*, 2014). The latter authors also noted that the rise in food prices is increasingly associated with the price of crude oil, disguising identification of any additional climate change signal. High energy prices, especially oil, are a well-recognized determinant of food prices, as they increase the price of fertilizers, pesticides, food shipping and many other agricultural inputs. The diversion of potentially edible food to biofuel is also elevating food prices and is thus also contributing to global food insecurity (Subramaniam *et al.*, 2019).

Nevertheless, the most recent IPCC report found, with 'high confidence', that the impacts of climate change are already hindering efforts to meet human needs, and that factors associated

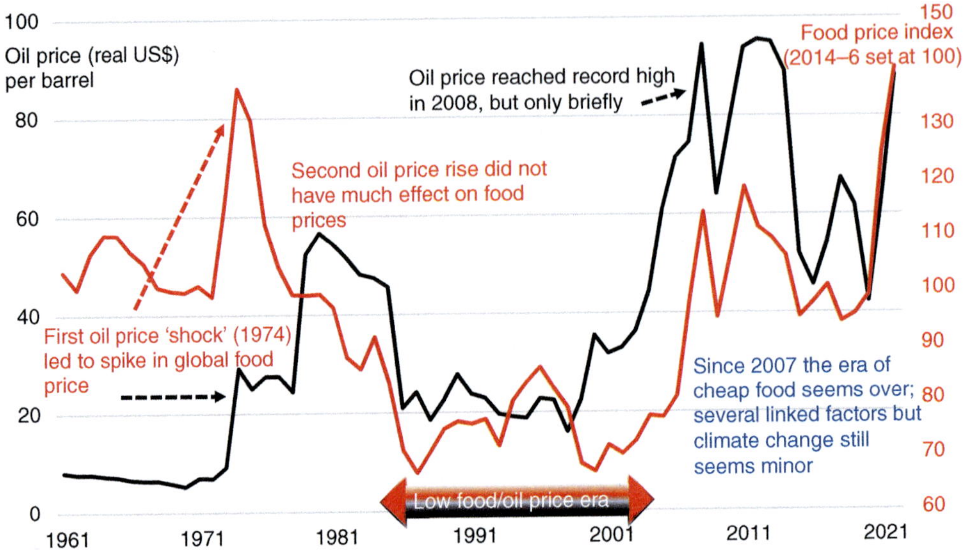

Fig. 22.4. There is a complex relationship between the prices of food and oil. Although food per person has continued to increase (see Fig. 22.1), its price has risen since 2007, in parallel with the price of oil. The war in Ukraine has elevated fuel and food prices, but even before then prices of each were historically high; after that war ends prices are unlikely to return to the level of 20 years ago. Future wars are also possible. While climate change is also likely to have contributed to some of the rise in food prices (especially 2010–2012) and will contribute to future food prices, other aspects of limits to growth (especially the high cost of energy, but also diminishing yield gaps[6]) appear a more important determinant in the recent past and near future. Data: FAO, World Bank (https://thedocs.worldbank.org).

with warming (including rainfall variability) have slowed the growth of agricultural productivity over the past half century, especially in mid and low latitudes (Kerr *et al.*, 2022).

Bearing in mind that the production of abundant food, globally, has never come close to ending macronutrient hunger, the world food system may, in the past, have had enough spare capacity to absorb temporary supply shocks caused by extreme weather events in ways that did not noticeably increase either hunger or the world food price. For example, the extraordinary European heatwave of 2003 lowered crop yields there, but at a time when food prices were historically low; that heatwave had no discernible impact on the global food price. Less than a decade later, however, the Russian and Ukrainian heatwave of 2010 led to a 66% increase in the global price of wheat within 2 months and to a smaller rise in other grain prices (Naylor and Falcon, 2010).

In 2011/12 world food prices reached what was then a record high for the period since 1961, apart from (briefly) during the first oil shock. In that time the USA (the world's biggest producer of food) also experienced a severe drought and heatwave. Although similar levels of dryness in the USA have occurred at other times in the past millennium (Boyer *et al.*, 2013), its conjunction with such heat may be unprecedented, and thus attributable in part to elevated greenhouse gas levels (Butler, 2014; Porter *et al.*, 2014). While there is debate as to whether climate change worsened the 2010 Russian heatwave, there is increasing support for the proposition that the burden of proof concerning the contribution of climate changes to such events should be reversed. That is, some climate scientists now argue that climate change should be accepted as a contributor to these extreme events until proved otherwise (Trenberth, 2011).

Famine in Somalia and Madagascar

The Somali famine appears to be causally related to climate change. Since 1980, the Indian Ocean has warmed especially fast, in association with increased precipitation over the tropical Indian

Ocean. Williams and Funk (2011) found that, since 1980, this has suppressed convection over tropical eastern Africa, decreasing precipitation during the 'long rains' season of March–June. Funk (2020) again warned of worsening drought in the Horn of Africa, in Kenya, Ethiopia and Somalia, associated with La Niña. In 2023 the World Food Programme (WFP) reported that, following five consecutive poor rainy seasons, the humanitarian crisis in the Horn of Africa has indeed deepened, with 22 million people being food insecure and 5.1 million acutely malnourished children, in drought-affected areas of the three countries.

The WFP's Drought Response Plan for the Horn of Africa (23 pages) does not mention terms such as fertility, contraception, or population pressure (WFP, 2023b). If the WFP had done so they would have inevitably provoked outrage, including accusations of victim blaming. Nonetheless, although the total fertility rate in Somalia was reported as almost six in 2022, the absence of discussion of population pressure in this region creates a cause for ongoing immiseration in this region. Maurice King, to whom this book is co-dedicated, often spoke of his preference for 'benign' rather than 'malignant' uproar (Abbasi, 1999). Raising family planning in the context of climate change and food security is certain to provoke benign uproar. Ignoring it, however, is likely to lead to more famine, refugees and conflict.

A famine in southern Madagascar has been suggested by the WFP, following four years of drought, as 'possibly the world's first climate-induced famine' (Schlein, 2021). However, Rice (2022) pointed out that one of the famine's roots lies in what he asserted was a deliberate attempt by French occupiers, using biological warfare (the cochineal insect) to destroy the population's main defence against famine: prickly pear, useful as a barricade and a drought-resistant food, for both cattle and humans. While Middleton (1999) fell shy of alleging this insect's deliberate introduction, she vividly described the cactus's value (introduced to the island centuries before) and the tragedy of its loss, which only benefited the French. As with many 'tertiary' events, causation of this famine is mixed. But climate change is a likely co-factor, as is high fertility – though, at 3.9, this total fertility rate is far lower than in Somalia.

Modelling climate change and famine

The modelling of past and future agroclimatic effects is a formidable challenge, and the consideration of health impacts harder still. Existing agro-climate models are excessively simple and biased toward the optimistic end of the spectrum (Butler, 2010; Gornall *et al.*, 2010). Although there is still debate as to whether climate change has reduced food production to date (Rizzo *et al.*, 2022), this remains an increasingly likely reality for the future, even without runaway climate change (Richards *et al.*, 2023).

Current models also poorly incorporate increased extremes, including rainfall intensity, sea level rise, saline intrusion, glacial melting and the possibilities of monsoon weakening and intensification of the El Niño Southern Oscillation and other ocean currents and atmospheric oscillations. They also omit the effect of climate change on mycotoxins, and on crop and animal diseases (Butler, 2010).

The capacity of ingested food to provide adequate nutrition is influenced by other factors, such as the level of physical activity (not just paid work, but unpaid labour such as hauling water, sometimes over hilly terrain) and by states of health and disease that lower appetite, nutrient absorption and metabolic rates. This further complicates estimates not only of health, but also of famine, as opposed to total nutrient production, which is already very difficult.

Finally, many other factors, such as tropospheric ozone and visible air pollution, influence crop growth, as does availability of sufficient pollinators, and negative episodes such as locust swarms[7] and fungal diseases. In some settings, insufficient human labour can reduce harvests. Unseasonable frosts may be climate related, triggered by less cloudy skies at crucial times. Wildfires can burn crops and their smoke can taint grapes.

22.8 Fisheries

Climate change is also predicted to have complex, generally adverse effects on fisheries, including by changing the pattern of ocean currents, thus redistributing marine productivity, especially to higher latitudes (Mendenhall *et al.*, 2020).

The oceans are warming to a depth of at least 2 km, driving some fish deeper (where there is less oxygen) and others polewards (Cheng *et al.*, 2022). In eight out of ten ocean regions, a declining trend in the global phytoplankton concentration since 1899, estimated by satellite and *in situ* observations, has been linked with warming sea surface temperatures (Boyce *et al.*, 2010). More recent work has questioned satellite interpretation of oceanic chlorophyll; phytoplankton may be more resilient than feared (Behrenfield *et al.*, 2016). However, high future temperatures have been forecast as leading to a mismatch between phytoplankton blooms and fish spawning phenology (Asch *et al.*, 2019).

Increasing ocean acidity and climate change and nitrogen-associated deoxygenated zones will also harm future marine productivity. Furthermore, ocean acidification associated with increased CO_2 concentrations interferes with the development of a wide range of aquatic species. It is already harming coral reef systems (Hill and Hoogenboom, 2022) and further stressing fish stocks already in decline. Finally, increased conflict among humans – by far the largest oceanic predator – is also a likely consequence (Mendenhall *et al.*, 2020).

22.9 Solutions

Increased famines, consequent in part to climate change, are inevitable, unless a profound revolution occurs not only in global policy and practices, but also in how scholars, activists and the media think and report about the causes of famine. A recurrent theme in this book, and of central relevance to this chapter, is a call to loosen the taboo on the discussion of human numbers, and their rate of increase in low-income settings. Increasing the availability of contraception in many parts of the global South should not be viewed as some form of draconian or neocolonial intervention; it would instead provide a powerful mechanism to improve human capability, including by giving a better chance for the poor to escape the poverty trap. Any idea, perhaps unconscious, that high fertility in places like Madagascar or Bihar will provide a reserve army of people who will then struggle more effectively for a fairer share of resources and opportunity is tragically flawed. Mothers who exclusively breastfeed for six months benefit their infant's health, and also increase the spacing of their children (Kramer and Kakuma, 2012). However, exclusive feeding beyond this age may keep children alive, but in some cases risk maternal health, perhaps setting the stage for repeat behaviour in similarly impoverished future generations.

Many of the SDGs are admirable, but their insistence on increased economic growth, as conventionally defined, reveals a fatal, neoliberal flaw (Hickel, 2019). In addition, given the global North's repeated history of breaking or fudging promises, it is most unlikely that the SDGs will be meaningfully achieved, including its ludicrous pledge to *eliminate* hunger by 2030 [italics added]. If under-addressed, climate change will lead to increasingly severe effects on food security, and hence on global health and nutrition. Rising food prices stimulate social unrest and can easily threaten governments, often leading to more repression, as occurred after the Arab Spring, which was itself partly brought on by higher food prices associated with extreme weather events. Conflict can also be triggered by competition for scarce materials, including food and fertile soil. While climate change requires adaptation, mitigation is a far more important priority (Anderson *et al.*, 2020).

Much can be done to reduce the enormous waste of food, whether pre-harvest (such as losses on the field), post-harvest (e.g. eaten by rodents) or post-processing (e.g. thrown out from supermarkets or wasted after purchase). Food fed to people suffering from chronic diarrhoea can be utilized better if their illnesses can be reduced by better healthcare, adequate sanitation and clean water. Overfed people, especially if consuming animal products raised on grain and soy, can and should reduce their consumption of these products, with benefits to their health as well as to the climate (McMichael *et al.*, 2007).

There are calls for a new agricultural revolution, to be fostered by further investment. While this might help at the margins, too much hope and money has already been wasted on this strategy, including in Africa (Mkindi *et al.*, 2020). It is vital that the determinants of fertility in low-income populations are altered in ways that will increase child survival and the effective demand for education and lower population growth. The final chapter of this book will discuss other general principles of solution. Adapting crops to cope with drought, flood and inundation from the sea is feasible to a point, but only to thresholds that may soon be exceeded. A holistic, global approach is required, which integrates society, education, health care, equity and technology.

Notes

[1] Some analysts (e.g. Pielke Jr. and Linnér, 2019) claim that the scale of this famine was exaggerated. However, they acknowledge that donated grain helped to reduce its severity.

[2] Target 2.1: 'By 2030, end hunger and ensure access by all people, in particular the poor and people in vulnerable situations, including infants, to safe, nutritious and sufficient food all year round.' Target 2.2: 'By 2030, end all forms of malnutrition, including achieving, by 2025, the internationally agreed targets on stunting and wasting in children under 5 years of age, and address the nutritional needs of adolescent girls, pregnant and lactating women and older persons.'

[3] India's colonial ruler regularly appropriated Indian-grown food for consumption in Britain, including during severe Indian famines. On the other hand, the rapid spread of the Indian rail network – a then recent British invention – helped relieve Indian famine.

[4] Anderson *et al.* (2020) reviewed evidence to show that growing seasons, yields and new areas for crops were improving at higher latitudes, including in Russia.

[5] Plants can be classified by variations in the biochemistry of their photosynthetic pathways; two main forms are known, called C_3 and C_4. Most edible crops are C_3 (the most ancient form) but a few important C_4 crops also exist.

[6] The yield gap is the difference between the maximum theoretical yield and the actual yield.

[7] Cyclone-associated flooding in the Arabian Peninsula was a factor in the large locust swarm in north-east Africa at the end of 2019 (Salih *et al.*, 2020).

References

Abbasi, K. (1999) King in a maverick style. *BMJ* 319, 942. DOI: 10.1136/bmj.319.7215.942

Aga, A. (2021) Farm protests in India are writing the Green Revolution's obituary. *Scientific American*. Available at: https://www.scientificamerican.com/article/farm-protests-in-india-are-writing-the-green-revolutions-obituary/ (accessed 13 May 2023).

Agnolucci, P., Rapti, C., Alexander, P., De Lipsis, V., Holland, R.A. *et al.* (2020) Impacts of rising temperatures and farm management practices on global yields of 18 crops. *Nature Food* 1, 562–571. DOI: 10.1038/s43016-020-00148-x

Ainsworth, E.A. and Long, S.P. (2021) 30 years of free-air carbon dioxide enrichment (FACE): What have we learned about future crop productivity and its potential for adaptation? *Global Change Biology* 27, 27–49. DOI: 10.1111/gcb.15375

Anderson, R., Bayer, P.E. and Edwards, D. (2020) Climate change and the need for agricultural adaptation. *Current Opinion in Plant Biology* 56, 197–202. DOI: 10.1016/j.pbi.2019.12.006

Asch, R.G., Stock, C.A. and Sarmiento, J.L. (2019) Climate change impacts on mismatches between phytoplankton blooms and fish spawning phenology. *Global Change Biology* 25, 2544–2559. DOI: 10.1111/gcb.14650.

Becker, J. (1996) *Hungry Ghosts: China's Secret Famine*. Henry Holt, New York.

Behrenfeld, M.J., O'Malley, R.T., Boss, E.S., Westberry, T.K., Graff, J.R. *et al.* (2016) Revaluating ocean warming impacts on global phytoplankton. *Nature Climate Change* 6, 323–330. DOI: 10.1038/nclimate2838

Bernstein, T.P. (2006) Mao Zedong and the famine of 1959–1960: A study in wilfulness. *The China Quarterly* 186, 421–445. DOI: 10.1017/S0305741006000221

Black, R.E., Allen, L.H., Bhutta, Z.A., Caulfield, L.E., de Onis, M. *et al.* (2008) Maternal and child undernutrition: global and regional exposures and health consequences. *The Lancet* 371, 243–260. DOI: 10.1016/S0140-6736(07)61690-0

Borlaug, N. (1970) Nobel Prize acceptance speech. Available at: https://www.nobelprize.org/prizes/peace/1970/borlaug/acceptance-speech/ (accessed 13 May 2023).

Boyce, D.G., Lewis, M.R. and Worm, B. (2010) Global phytoplankton decline over the past century. *Nature* 466, 591–496. DOI: 10.1038/nature09268

Boyer, J.S., Byrne, P., Cassman, K.G., Cooper, M., Delmer, D. *et al.* (2013) The U.S. drought of 2012 in perspective: a call to action. *Global Food Security* 2, 139–143. DOI: 10.1016/j.gfs.2013.08.002

Brass, P.R. (1986) The political uses of crisis: the Bihar famine of 1966–1967. *The Journal of Asian Studies* 45, 245–267. DOI: 10.2307/2055843

Butler, C.D. (2002) Inequality and Sustainability. PhD thesis, The Australian National University, Canberra, Australia. Available at: https://openresearch-repository.anu.edu.au/handle/1885/46233 (accessed 3 September 2022).

Butler, C.D. (2009a) Food security in the Asia-Pacific: climate change, phosphorus, ozone and other environmental challenges. *Asia Pacific Journal of Clinical Nutrition* 18, 590–597. https://pubmed.ncbi.nlm.nih.gov/19965353/

Butler, C.D. (2009b) Food security in the Asia-Pacific: Malthus, limits and environmental challenges. *Asia Pacific Journal of Clinical Nutrition* 18, 577–584. https://pubmed.ncbi.nlm.nih.gov/19965351/

Butler, C.D. (2010) Climate change, crop yields, and the future. *SCN News* 38, 18–25.

Butler, C.D. (2012) Population trends and the environment. In: Friis, R.H. (ed.) *Praeger Handbook of Environmental Health*. Praegar, Westport, Connecticut, pp. 215–231.

Butler, C.D. (2014) Famine, hunger, society and climate change. In*:* Butler, C.D. (ed.) *Climate Change and Global Health.* CABI, Wallingford, UK, pp. 124–134.

Butler, C.D. (2015) Revised hunger estimates accelerate apparent progress towards the MDG hunger target. *Global Food Security* 5, 19–24. DOI: 10.1016/j.gfs.2015.04.002

Camaschella, C. (2019) Iron deficiency. *Blood* 133, 30–39. DOI: 10.1182/blood-2018-05-815944

Chen, C., Chaudhary, A. and Mathys, A. (2020) Nutritional and environmental losses embedded in global food waste. *Resources, Conservation and Recycling* 160, 104912. DOI: 10.1016/j.resconrec.2020.104912

Cheng, L., von Schuckmann, K., Abraham, J.P., Trenberth, K.E., Mann, M.E. *et al.* (2022) Past and future ocean warming. *Nature Reviews Earth & Environment* 3, 776–794. DOI: 10.1038/s43017-022-00345-1

Connelly, M. (2006) Population control in India: prologue to the emergency period. *Population and Development Review* 32, 629–667. DOI: 10.1111/j.1728-4457.2006.00145.x

Davis, M. (2001) *Late Victorian Holocausts: El Niño Famines and the Making of the Third World*. Verso, London.

Donaldson, P.J. (1990) On the origins of the United States government's international population policy. *Population Studies* 44, 385–399. DOI: 10.1080/0032472031000144816

Dyson, T. and Maharatna, A. (1992) Bihar famine, 1966–67 and Maharashtra drought, 1970–73. The demographic consequences. *Economic and Political Weekly* 27, 1325–1332. https://www.jstor.org/stable/4398553

Ebi, K.L. and Loladze, I. (2019) Elevated atmospheric CO_2 concentrations and climate change will affect our food's quality and quantity. *The Lancet Planetary Health* 3, e283–e284. DOI: 10.1016/S2542-5196(19)30108-1

Ekholuenetale, M., Okonji, O.C., Nzoputam, C.I. and Barrow, A. (2022) Inequalities in the prevalence of stunting, anemia and exclusive breastfeeding among African children. *BMC Pediatrics* 22, 333. DOI: 10.1186/s12887-022-03395-y

FAO, IFAD, UNICEF, WFP and WHO (2022) *The State of Food Security and Nutrition in the World 2022*. Food and Agriculture Organization, Rome.

Funk, C. (2020) Ethiopia, Somalia and Kenya face devastating drought. *Nature* 586, 645. DOI: 10.1038/d41586-020-02698-3

Gaud, W.S. (1968) *The Green Revolution: Accomplishments and Apprehensions*. Address before the Society for International Development, Washington, DC, 8 March. Available at: http://www.agbioworld.org/biotech-info/topics/borlaug/borlaug-green.html (accessed 12 May 2023).

Gornall, J., Betts, R., Burke, E., Clark, R., Camp, J., Willett, K. *et al.* (2010) Implications of climate change for agricultural productivity in the early twenty-first century. *Philosophical Transactions of the Royal Society B* 365, 2973–2989. DOI: 10.1098/rstb.2010.0158

Guilpart, N., Grassini, P., van Wart, J., Yang, H., van Ittersum, M.K. *et al.* (2017) Rooting for food security in Sub-Saharan Africa. *Environmental Research Letters* 12, 114036. DOI: 10.1088/1748-9326/aa9003

Hickel, J. (2019) The contradiction of the sustainable development goals: Growth versus ecology on a finite planet. *Sustainable Development* 27, 873–884. DOI: 10.1002/sd.1947

Hill, T.S. and Hoogenboom, M.O. (2022) The indirect effects of ocean acidification on corals and coral communities. *Coral Reefs* 41, 1557–1583. DOI: 10.1007/s00338-022-02286-z

Huff, G. (2019) Causes and consequences of the great Vietnam famine, 1944–5. *The Economic History Review* 72, 286–316. DOI: 10.1111/ehr.12741

Kerr, R.B., Hasegawa, T., Lasco, R., Bhatt, I., Deryng, D. *et al.* (2022) Food, fibre and other ecosystem products. In: Pörtner, H.-O., Roberts, D.C., Tignor, M., Poloczanska, E.S., Mintenbeck, K. *et al.* (eds) *Climate Change 2022: Impacts, Adaptation, and Vulnerability. Contribution of Working Group II to the Sixth Assessment Report of the Intergovernmental Panel on Climate Change.* Cambridge University Press, Cambridge, UK, pp. 713–906.

Kramer, M.S. and Kakuma, R. (2012) Optimal duration of exclusive breastfeeding. *Cochrane Database of Systematic Reviews*, CD003517. DOI: 10.1002/14651858.CD003517

Krawchenko, B. (1989) The great famine of 1932–3 in Soviet Ukraine: Causes and consequences. *Critique* 17, 137–146. DOI: 10.1080/03017608908413328

Lappé, F.M. and Collins, J. (1971) *Diet for a Small Planet*. Institute for Food and Development Policy, San Francisco, California.

Lappé, F.M., Clapp, J., Anderson, M., Broad, R., Messer, E. *et al.* (2013) How we count hunger matters. *Ethics & International Affairs* 27, 251–259. DOI: 10.1017/S0892679413000191

Mallory, W.H. (1926) *China: Land of Famine*. American Geographical Society, New York.

Maxwell, D., Haan, N., Gelsdorf, K. and Dawe, D. (2012) The 2011–12 famine in Somalia: introduction to the special edition. *Global Food Security* 1, 1–4. DOI: 10.1016/j.gfs.2012.07.007

McMichael, A.J., Powles, J., Butler, C.D. and Uauy, R. (2007) Food, livestock production, energy, climate change and health. *The Lancet* 370, 1253–1263. DOI: 10.1016/S0140-6736(07)61256-2

Mellor, J.W. and Gavian, S. (1987) Famine: causes, prevention, and relief. *Science* 235, 539–545. DOI: 10.1126/science.235.4788.53

Mendenhall, E., Hendrix, C., Nyman, E., Roberts, P.M., Hoopes, J.R. *et al.* (2020) Climate change increases the risk of fisheries conflict. *Marine Policy* 117, 103954. DOI: 10.1016/j.marpol.2020.103954

Middleton, K. (1999) Who killed 'Malagasy cactus'? Science, environment and colonialism in southern Madagascar (1924–1930). *Journal of Southern African Studies* 25, 215–248. DOI: 10.1080/030570799108678

Mkindi, A., Maina, A., Urhahn, J., Koch, J., Bassermann, L. *et al.* (2020) *False Promises. The Alliance for a Green Revolution in Africa (AGRA)*. Rosa Luxemburg Stiftung, Berlin. Available at: https://www.rosalux.de/en/publication/id/42635/false-promises-the-alliance-for-a-green-revolution-in-africa-agra (accessed 14 May 2023).

Moon, W.J. (2009) The origins of the great North Korean famine: Its dynamics and normative implications. *North Korean Review* 5, 105–122. https://www.jstor.org/stable/43910265

Naugler, C. (2008) Hemochromatosis: a Neolithic adaptation to cereal grain diets. *Medical Hypotheses* 70, 691–692. DOI: 10.1016/j.mehy.2007.06.020

Naylor, R.L. and Falcon, W.P. (2010) Food security in an era of economic volatility. *Population and Development Review* 36, 693–723. DOI: 10.1111/j.1728-4457.2010.00354.x

Oskorouchi, H.R. and Sousa-Poza, A. (2021) Floods, food security, and coping strategies: Evidence from Afghanistan. *Agricultural Economics* 52, 123–140. DOI: 10.1111/agec.12610

Patel, R.C. (2012) Food sovereignty: power, gender, and the right to food. *PLoS Medicine* 9, e1001223. DOI: 10.1371/journal.pmed.1001223

Perkins, J.H. (1990) The Rockefeller Foundation and the Green Revolution, 1941–1956. *Agriculture and Human Values* 7, 6–18. DOI: 10.1007/BF01557305

Pielke Jr., R. and Linnér, B.-O. (2019) From Green Revolution to green evolution: A critique of the political myth of averted famine. *Minerva* 57, 265–291. DOI: 10.1007/s11024-019-09372-7

Pingali, P.L. (2012) Green Revolution: impacts, limits, and the path ahead. *Proceedings of the National Academy of Sciences (USA)* 109, 12302–12308. DOI: 10.1073/pnas.0912953109

Pittock, A.B., Whett, P. and Wang, Y. (1994) Climate and food supply. *Nature* 371, 25–25. DOI: 10.1038/371025a0

Porter, J.R., Xie, L., Challinor, A.J., Cochrane, K., Howden, S.M. *et al.* (2014) Food security and food production systems. In: Field, C.B., Barros, V.R., Dokken, D.J., Mach, K.J., Mastrandrea, M.D. *et al.* (eds) *Climate Change 2014: Impacts, Adaptation, and Vulnerability. Part A: Global and Sectoral Aspects. Contribution of Working Group II to the Fifth Assessment Report of the Intergovernmental Panel on Climate Change.* Cambridge University Press, Cambridge, UK and New York, pp. 485–533.

Ramalingaswami, V., Deo, M., Guleria, J., Malhotra, K., Sood, S. *et al.* (1971) Studies of the Bihar famine of 1966–67. In: Blix, G., Hofvander, Y. and Vahlquist, B. (eds) *Famine: Nutrition and Relief Operations.* Swedish Nutrition Foundation, Uppsala, Sweden, pp. 94–110.

Reagan, R. and Mondale, W. (1984) Debate on Foreign Policy. *The New York Times*, October 24, B4-B6.

Reilly, J. (1994) Crops and climate change. *Nature* 367, 118–119. DOI: 10.1038/367118a0

Rice, S. (2022) Madagascar's famine is more than climate change. *Georgetown Journal of International Affairs*, Georgetown University, Washington, DC. Available at: https://gjia.georgetown.edu/2022/01/24/madagascars-famine-is-more-than-climate-change/#:~:text=Climate%20change%20has%20brought%20historic,since%20colonialism%2C%20off%20the%20hook (accessed 15 May, 2023).

Richards, C., Gauch, H. and Allwood, J. (2023) International risk of food insecurity and mass mortality in a runaway global warming scenario. *Futures* 150, 103173. DOI: 10.1016/j.futures.2023.103173

Rizzo, G., Monzon, J.P., Tenorio, F.A., Howard, R., Cassman, K.G. *et al.* (2022) Climate and agronomy, not genetics, underpin recent maize yield gains in favorable environments. *Proceedings of the National Academy of Science (USA)* 119, e2113629119. DOI: 10.1073/pnas.211362911

Rosenthal, D.M. and Ort, D.R. (2011) Examining cassava's potential to enhance food security under climate change. *Tropical Plant Biology* 5, 30–38. DOI: 10.1007/s12042-011-9086-1

Rosenzweig, C. and Parry, M.L. (1994) Potential impact of climate change on world food supply. *Nature* 367, 133–138. DOI: 10.1038/367133a0

Salih, A.A.M., Baraibar, M., Mwangi, K.K. and Artan, G. (2020) Climate change and locust outbreak in East Africa. *Nature Climate Change* 10, 584–585. DOI: 10.1038/s41558-020-0835-8

Schlein, L. (2021) WFP: Madagascar may be suffering world's first climate-induced famine. *Voice of America (VOA) News.* Available at: https://www.voanews.com/a/wfp-madagascar-may-be-suffering-world-s-first-climate-induced-famine/6296676.html (accessed 21 August 2022).

Sen, A.K. (1981) *Poverty and Famines: An Essay on Entitlement and Deprivation*. Clarendon Press, Oxford, UK, and New Delhi, India.

Shiva, V. (1991) *The Violence of the Green Revolution. Third World Agriculture, Ecology and Politics*. Third World Network and with ZED Books, Penang Malaysia and London.

Staples, A.L.S. (2003) To win the peace: the Food and Agriculture Organization, Sir John Boyd Orr, and the World Food Board proposals. *Peace and Change* 28, 495–523. DOI: 10.1111/1468-0130.00273

Subramanian, K. (2015) Revisiting the Green Revolution. Irrigation and Food Production in Twentieth-Century India. PhD thesis, King's College, London. Available at: https://kclpure.kcl.ac.uk/portal/en/theses/revisiting-the-green-revolution(c7fedff0-862a-4fdb-aef8-0526f78772b5).html (accessed 16 May 2023).

Subramaniam, Y., Masron, T.A. and Azman, N.H.N. (2019) The impact of biofuels on food security. *International Economics* 160, 72–83. DOI: 10.1016/j.inteco.2019.10.003

Subramanian, S.V., Ambade, M., Sharma, S., Kumar, A. and Kim, R. (2023) Prevalence of zero-food among infants and young children in India: patterns of change across the States and Union Territories of India, 1993–2021. *eClinicalMedicine* 58. DOI: 10.1016/j.eclinm.2023.101890

Sumaila, U.R. and Bawumia, M. (2014) Fisheries, ecosystem justice and piracy: A case study of Somalia. *Fisheries Research* 157, 154–163. DOI: 10.1016/j.fishres.2014.04.009

Sur, B. (2020) Amartya Sen and famine: Why Bengalis are 'uncomfortable' with the economist. *Counterview*. Available at: https://www.counterview.net/2020/11/amartya-sen-and-famine-why-bengalis-are.html (accessed 14 May 2023).

Swaminathan, S., Hemalatha, R., Pandey, A., Kassebaum, N.J., Laxmaiah, A. *et al.* (2019) The burden of child and maternal malnutrition and trends in its indicators in the states of India: the Global Burden of Disease Study 1990–2017. *The Lancet Child & Adolescent Health* 3, 855–870. DOI: 10.1016/S2352-4642(19)30273-1

Swift, J. (1729) *A Modest Proposal. For Preventing the Children of Poor People in Ireland from Being a Burden to their Parents or Country, and for Making Them Beneficial to the Public*. Available at: http://art-bin.com/art/omodest.html#hit (accessed 23 February 2023).

Tauger, M.B. (2003) Entitlement, shortage and the 1943 Bengal famine: Another look. *The Journal of Peasant Studies* 31, 45–72. DOI: 10.1080/0306615031000169125

Trenberth, K.E. (2011) Attribution of climate variations and trends to human influences and natural variability. *WIREs Climate Change* 2, 925–930. DOI: 10.1002/wcc.142

UN (nd) *Sustainable Development Goals*. United Nations, New York. Available at: https://www.un.org/sustainabledevelopment/hunger/ (accessed 16 May 2023).

Union of Concerned Scientists (1992) *World Scientists' Warning to Humanity*. Cambridge, Massachusetts. Available at: http://www.ucsusa.org/about/1992-world-scientists.html (accessed 14 May 2023).

Wang, S., Zhang, Y., Ju, W., Chen, J.M., Ciais, P. *et al.* (2020) Recent global decline of CO_2 fertilization effects on vegetation photosynthesis. *Science* 370, 1295–1300. DOI:10.1126/science.abb7772

WFP (2023a) *Lack of funds forces WFP to cut rations for Rohingya in Bangladesh*. World Food Programme, Rome. Available at: https://www.wfp.org/news/lack-funds-forces-wfp-cut-rations-rohingya-bangladesh (accessed 27 February 2023).

WFP (2023b) Regional Drought Response Plan for the Horn of Africa. World Food Programme, Rome. Available at: https://www.wfp.org/publications/regional-drought-response-plan-horn-africa-2023 (accessed 14 May 2023).

Williams, A.P. and Funk, C.A. (2011) Westward extension of the warm pool leads to a westward extension of the Walker circulation, drying eastern Africa. *Climate Dynamics* 37(11–12), 2417–2435. DOI: 10.1007/s00382-010-0984-y

23 Climate Change, Migration and Health

Colin D. Butler[1] and Devin C. Bowles[2]
[1]*National Centre for Epidemiology and Population Health, The Australian National University, Canberra, Australia;* [2]*Faculty of Health, University of Canberra, Canberra, Australia*

Abstract
This chapter highlights some of the key issues that link climate change, migration and health. It is strongly recognized that climate change does not operate independently of history or politics. Nonetheless, climate change is increasingly likely to act as 'risk multiplier', contributing to an increase in population displacement, irrespective of the name that is applied to such people (such as climate migrants). After describing general principles that drive migration, the chapter focuses on two case studies from Central and South America: migrant caravans fleeing poverty and instability in Guatemala, Honduras and El Salvador and the egress of millions of Venezuelans. Even if climate change cannot be sufficiently curtailed, some of the co-factors that help drive migration can be improved. However, this prospect currently seems faint, given the dominant ideology of neoliberalism and its denial of limits to growth.

> When global warming takes hold, there could be as many as 200 million people overtaken by sea-level rise and coastal flooding, by disruptions of monsoon systems and other rainfall regimes, and by droughts of unprecedented severity and duration.
>
> (Norman Myers, 2002)

23.1 Introduction: Migration, Evolution and the Human Condition

Humans, like all other animals, inhabit ecological niches that provide resources essential for life and reproduction. Numerous species modify environments in ways that improve their suitability, but the human capacity to do this greatly exceeds the scale of any other species. Moreover, some humans and their hominin ancestors seem never to have been content to stay in one place, even if verdant. For hundreds of millennia, *Homo sapiens* has explored the next valley and, technology permitting, crossed the next river and even the next ocean (Flannery, 1994). Via means that remain unknown, but which involved an extensive[1] sea journey, humans crossed the Wallace line, leading to the occupation of Sahul (now mainly Australia and New Guinea) at least 50 millennia ago (Bird *et al.*, 2019).

The struggle to gain, control and consume resources, in order either to improve comfort and security or to escape danger, is central to the human story. The decision, whether of an individual or a group, to stay or to leave can be conceptualized as the product of four interacting forces, which we here call 'glue' (attachment), 'push', 'pull' and 'fend' (deterrence) (Woodruff *et al.*, 2006) (Fig. 23.1). Once humans have a comfortable niche, a territory they call home, they will seek to defend and enjoy it. Groups may, at times,

DOI: 10.1079/9781800620025.0023

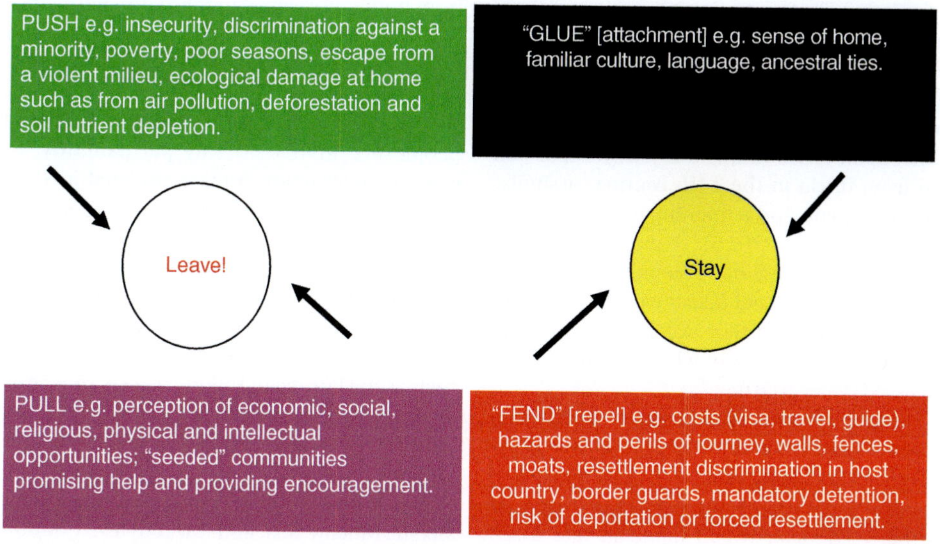

Fig. 23.1. In addition to 'push' and 'pull' (each of which helps to stimulate the wish to leave, or to migrate), powerful factors (here called 'glue' and 'fend') deter such a decision.

willingly share their territory, but only if they do not expect any substantial or prolonged decrease in access to its resources. Many times in human history, and almost universally in recent millennia, such resources have been defended, not always successfully, rather than freely shared. Some countries still welcome selective 'skilled' in-migration, such as Australia and the USA. In 2015 and 2016 Germany accepted approximately 1 million Syrian asylum seekers (Keita and Dempster, 2020). Israel welcomes Jewish immigrants but is hostile to non-Jewish asylum seekers (Raijman, 2020). However, most refugees languish in camps; they may be skilled, but not with attributes that are in sufficient demand. The mix of skills, identities and other attributes is not left to chance; in-migration is perceived as a zero sum game.[2]

At times, attachment to place is strengthened not only by personal memories but also by knowledge of and reverence for ancestors who are tied to that location. The 'trail of tears' was thus called to reflect the trauma the survivors of the Cherokee Nation felt on being forced from their ancestral homeland in south-east USA (Perdue, 1989). Departure from resource-rich and memory-laden habitats entails effort and risk; attachment reduces this prospect. In Ukraine, some people, many of them older, have chosen to risk death rather than to flee from the Russians (de Freytas-Tamura, 2022). Some populations, with less means to migrate (for example because of frailty, extreme poverty or lack of skills and knowledge), have been characterized as 'trapped' (Foresight, 2011; Adams, 2016; Ayeb-Karlsson *et al.*, 2018). Even for such people, thresholds exist beyond which migration becomes essential, for example if whole urban neighbourhoods are destroyed by bombing, as occurred in several Syrian cities, including Homs (Azzouz, 2020).

Physical and social environments continually change, including through co-evolutionary processes.[3] Increasing human numbers and exploitative practices can partially or substantially degrade formerly lush and attractive environments, contributing to push factors that propel migration, or at least the wish to emigrate. Sometimes, environmental factors, such as the 130 m rise in sea level, distributed over 6000 years from the start of the Holocene (Lambeck *et al.*, 2014), or dust-vegetation feedbacks during and at the end of the green Sahara[4] (Tierney *et al.*, 2017), must have overwhelmed any socially mediated wish or capacity to remain, generating a threshold that made migration essential for survival. Social factors can operate and evolve faster than physical conditions, creating circumstances where dominant groups challenge, discriminate against and, in some cases, actively endanger and expel vulnerable groups, who must choose between migration and survival, or, less often, choose assimilation. The migration of a comparatively small number of Jews from Nazi Germany, immediately before World War II, is one such example of expulsion, but very few (if any) Jews could choose assimilation once the Nazis took power. In an earlier example of Jewish expulsion, that from Iberia in the 14th century, assimilation, conversion and remaining in place was far from rare (Gorsky, 2015). A more recent case was the willingness of some members of the Islamic State to accept Yazidi women into their group, provided that they accepted the extreme interpretation of Islam which their captors professed (Vale, 2020).

Another recent illustration of how social factors operate faster than purely 'natural' physical factors (including climate change) and which may drive migration is the conflict in Tigray, in northern Ethiopia, which flared in late 2020 (Deckers *et al.*, 2020). Although resources there are tight, on a per capita basis, and were worsened by the locust swarm in late 2019/20 (Salih *et al.*, 2020), peace prevailed until November 2020. The conflict, refugee flight and hunger were then reported as exacerbated by the blocking of food aid, the destruction of crops, the killing of livestock and the prevention of crop planting (Anonymous, 2021; Anonymous, 2022).

Today, most fertile parts of the world are either lightly inhabited forests (Amazon, Congo and in Indonesia), cleared and densely populated and farmed using human labour (such as most of rural India, China, Rwanda), or highly mechanized but sparsely populated (such as in the US mid-west). Although humans also live in deserts, at sea, in Antarctica, on the icefields, among high-latitude conifer forests and even on a single permanent (crowded) space station, all these settings have low human population densities; the permanent survival of large numbers of people in each of these locations is impossible and in any case depends on the supply of food and other resources from elsewhere. There are also limits to technology and to human fortitude. All locations (even outside Earth's atmosphere) have some desirable resources, such as sunshine or a great view, but the obvious and daunting physical harshness of many locations acts as a natural deterrent.

Not all inhabited locations are hostile to immigrants and visitors. Numerous pull factors lure prospective migrants and temporary workers from homes of relative (or absolute) hardship to more desirable locales, such as from the Malabar Coast of India to the Middle East as 'guest' workers, or to the USA from the highlands of Guatemala (Blitzer, 2019). In some cases, these additional people are invited to stay permanently, for example if bringing traits considered as desirable, such as a wanted skill, cheap labour, or a large bank account.

However, in our highly unequal and increasingly crowded world, socially created fend factors are increasing and are now commonplace. Almost all nations (or coalitions of nations, such as the European Union) are separated from other states by controlled borders. The UK is currently trying to send asylum seekers to Rwanda (d'Orsi, 2022). Australia is surrounded by a 'moat', a function that is also performed by the Mediterranean Sea and the English Channel. Some countries (including China and India, the two most populous) maintain internal travel controls, restricting mobility and sometimes rights to those more favoured, such as Hindus in Assam and other parts of its north-east (Paulose, 2019). India was reported, in 2021, to have completed construction of a fortified fence along 76% of its perimeter with Bangladesh, forming what has been called the 'largest human cage in the world'. In 2011 a 15-year-old girl, Felani Khatun, was shot and killed along this border, with her body left to hang for several hours, contributing to deterrence (Sur, 2014). Some attempted migrants are snared in conditions of frank or quasi slavery en route to the global North, such as in the failed state of Libya (Amnesty International, 2017; Mafu, 2019). It seems unlikely that many who become trapped fully appreciate the risk that they take (i.e. when starting their journey), but the push conditions (such as for Rohingya escaping from Rakhine in north-west Myanmar) are so strong that such desperate journeys may still be commenced (Samarasekera, 2021).

The USA–Mexico border is increasingly fenced and patrolled, with frequent detention and deportation, including for unaccompanied children (Vives, 2020). In response to the arrival of 1.2 million asylum seekers to the European Union in 2015, member nations quickly followed their own paths; Germany was generous, but Austria and Hungary were openly hostile. Hungary erected a razor-wire fence to impede entrance (Betts, 2019). Australia and some southern European countries have engaged in the 'tow back' of boats carrying asylum seekers, to international waters and to north Africa (Amnesty International, 2017; Tazreiter, 2017, 2019). Such strategies are not exclusive to European-dominated countries. In 2015 a crudely termed game of 'human ping pong', involving Thai, Malaysian and Indonesian authorities, left about 8000 people (mainly Muslim Rohingya) temporarily stranded at sea, of whom at least 350 perished in the ocean. This 'game' involved repairing the vessels of would-be immigrants, providing them with food and water, and escorting them into 'international' waters (Ghráinne, 2017).

23.2 Causality, Migration and Environmental Change

Piguet (2013) identified a conceit, arising in the second half of the 20th century among many social scientists, that modern society 'increasingly develops outside nature' (Beck, 2010) and that 'the history of mankind is the history of the conquest of nature'. Piguet also noted that 'natural phenomena' were only marginally considered in the 'behaviourist' approach to migration which developed in the 1960s and which dominated for the next few decades. Ecological push factors for migration were acknowledged (by some) but relegated (we argue naïvely) to 'primitive' status (Petersen, 1958). This denial of environmental co-factors for migration can be seen as an overreaction to the 'determinist' geographical literature that arose in the 19th century and which was used to justify European colonialism and, later, the Nazi concept of 'lebensraum', which motivated the German invasion of Poland and Russia in World War II (Smith, 1980; Herwig, 1999). However, without in any way supporting the Nazi interpretation, the rejection of environmental determinants for migration should be recognized as hopelessly outdated in the Anthropocene. Humanity (at first the poor, largely in the global South, but later – potentially – everyone) will pay an increasing price for this recent complacency (Butler, 2024).

As evidence of climate change and other aspects of limits to growth accumulated (see Chapter 8), pioneering estimates in the late 20th century by the United Nations Environment Programme (UNEP) (El-Hinnawi, 1985) and Norman Myers struggled to revive environmental consciousness among policy makers and migration scholars. In the health literature, pioneers who claimed global environmental change would adversely impact health were scarce; exceptions included the ecologists Sargent (1972) and Boyden (1972), and the ecologically inspired epidemiologist McMichael, whose 1993 book *Planetary Overload* forecast an increasingly full world, in which migration (or at least internment in camps) would increase.

Piercing this complacency remains difficult. Even as recently as 2013, Piguet characterized Myers as 'alarmist'. Myers (1993) had projected 150 million displaced by 2050, due, in part, to environmental change. In 2022 the United Nations High Commission for Refugees (UNHCR) reported that this number (albeit with an increasing complexity of categories) exceeded 100 million people for the first time (UNHCR, 2022a) (Fig. 23.2).

Similar to all tertiary consequences (see Chapter 21), we stress that the causation of migration is multi-factorial. We are not claiming that all of these 100 million (plus) people are driven to leave solely because of environmental change. However, environmental factors are implicit or explicit in virtually all such movements, including, in the case of Venezuela, due to its low remaining supply of easily refinable oil, an environmental resource (see below). In making this claim, we reject any characterization as environmental (or climate) 'determinists', nor are we climate 'reductionists' (Hulme, 2011), climate 'indeterminists' or 'social determinists' (Butler and Kefford, 2018). Instead, we argue, the causal factors that influence the decision to migrate, either voluntarily or through force, are a co-mingling of environmental and social factors, including those that are political. We will illustrate this in the two case studies.

Norman Myers (1993) clearly understood this causal complexity when he wrote:

> It is often difficult to differentiate between refugees driven by environmental factors and those driven by economic problems. In certain instances, cross-border refugees, notably those with moderate though tolerable economic circumstances, are pulled by opportunity for a

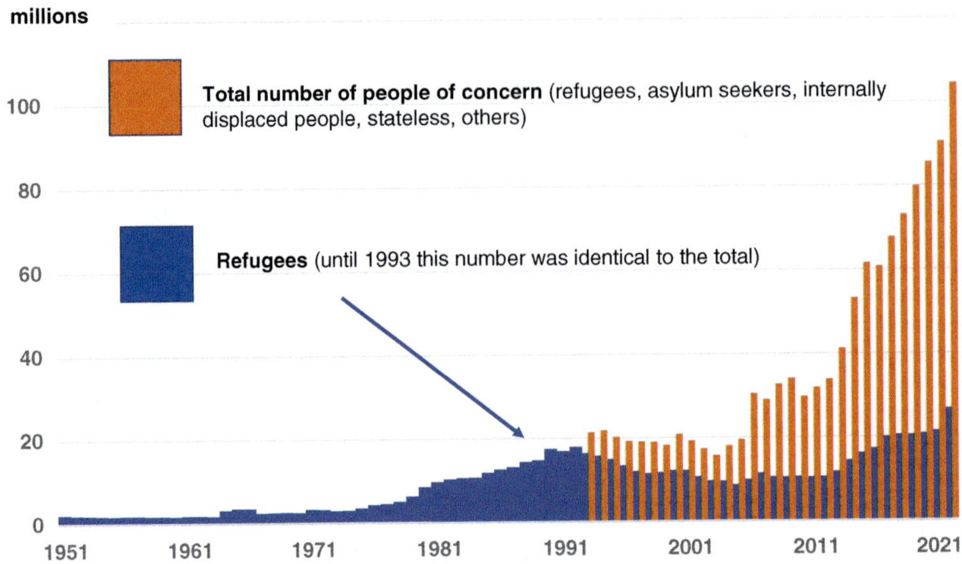

Fig. 23.2 Although slightly declining following its foundation in 1951, the number of people 'of concern' to the UNHCR has grown enormously. As a percentage of the world population this number has risen from its nadir of 0.05% (around 1960) to 1.3% in 2022. Data sources: https://www.unhcr.org/refugee-statistics, https://population.un.org/wpp/.

> better economic life elsewhere rather than pushed by environmental destitution. This description ostensibly applies to many Hispanics heading for the United States. But those people who migrate because they suffer poverty are frequently driven by root factors of environmental degradation; indeed, it is their environmental plight as much as any other factor that makes them economically impoverished.

Other recent authors have also recognized mingled causation. Black *et al.* (2011a) proposed five drivers of migration: economic, political, social, demographic and environmental. The same authors also noted that 'migration influenced by environmental change cannot be separated from migration associated with a range of other drivers' (Black *et al.*, 2011b).

As McMichael (1999) observed, the perception of causation is influenced by the 'focal depth' or pre-existing bias of the observer. Some will see 'politics' as the key driver, others may focus on economics or even place climate change in that role. We do not give pre-eminence to any single driver.[5] However, we stress that a habitable climate is key to remaining in place. While humans can and do adapt to environmental change, sometimes enabling them to remain at home in the face of considerable stress, there are clear limits to this in many cases.

23.3 Migration and Past Climate Change

Resource scarcity and migration have numerous causes, often interlinked, one of which is an unfavourable climate (Ljungqvist *et al.*, 2021); another is sea level rise. The early Holocene and the green Sahara have been mentioned; cooling associated with volcanic eruptions is thought to have driven large-scale social events including migration and the collapse of the western Roman empire (Harper, 2017). The onset of the Little Ice Age was an important cause in the decline of harvests in the 14th century, which may then have contributed to the calamitous Black Death, via large-scale undernutrition (DeWitte and Slavin, 2013). Following it, migration to villages that had lost significant numbers of people was sometimes welcomed. Some villages were completely abandoned.[6]

Between 1500 and 1800, numerous poor harvests occurred in Eurasia, contributing to famine, warfare and localized population decline in China and Europe (Zhang *et al.*, 2011). These also contributed, indirectly, to the invasion and colonization of the 'New World', as want, greed and new technology assisted Europeans to intensify what Engerman and Sokoloff (2012) called 'a grand long-term campaign to extract material and other advantages from underpopulated or underdefended territories'. The enslavement and indentured labour of millions of Africans, Indigenous Americans and Asians also facilitated trans-continental European migration and the import of food and other resources to Europe. These imports and profits facilitated the support of large populations remaining in Europe, as well as the sophistication and lives of gross excess for a tiny fraction of people, politically or socially close to the ruling families of European empires.

23.4 Migration and Future Climate Change

Climate change is increasingly redistributing resources, including food, water and shipping lanes. The scale of near-future (e.g. to 2100) migration due to anthropogenic climate and other aspects of human-driven environmental change has long been contested, as sketched above. This debate continues, including the issue of 'securitization' (also see Chapters 21 and 24). Participants in this debate such as Boas *et al.* (2018) argue that the extent of 'climate-induced migration' is exaggerated, used as 'a common rationale' to 'strengthen and protect national and regional borders in the Global North'. They criticize what they call 'climate migration myths'. They also criticize the UN Security Council for its warning of 'mass climate migration and the subsequent risk of aggravating conflicts'.

The 31 contributors to Boas *et al.* (2018) (with 22 academic affiliations, of which only two are in the global South) acknowledged that 'the potential for climate change to disrupt livelihoods and threaten lives is real'; but they did not mention that, even in 2018 when their provocative paper was published, the number of displaced people was nearing 70 million (UNHCR, 2022b). Nor did they discuss the existence of strong fend signals in parts of the global South, including in Malaysia and South Asia.[7] Although they mentioned the strong deterrent policies introduced by Australia, they failed to note that these measures were introduced well before climate cause was claimed to be a significant contributor to existing migration, even by commentators such as Myers.

They also ignored the 'risk amplification' concept (see Chapter 24), even though a comment by one of the contributors (Richard Black) on co-mingled causation is quoted above. A wish to protect one's assets, including home, is almost universal. This may be called 'securitization', and it may be part of neoliberal thinking. But this does not mean that climate change is not a risk multiplier for tertiary events, even if these authors correctly identified an over-simplification of language in some of the policy dialogue.

Later in his career, Myers (2002) provided an even more pessimistic forecast, of 200 million displaced when global warming 'takes hold'. Even this may prove conservative, although that still depends on the extent to which greenhouse gas accumulation can be curtailed, and on the proximity of reinforcing feedbacks that may trigger runaway warming.

For example, with average global temperature rise still under 1.5°C above pre-industrial levels, parts of South Asia are already experiencing wet-bulb temperatures (Tw) (which integrate humidity with temperature) at the upper limits of labour productivity (Tw 32°C) and even human survivability (Tw 35°C) (Saeed *et al.*, 2021). If, as currently seems inevitable, global temperatures continue to rise, parts of South Asia may thus become increasingly unable to support its enormous population. Lethal heat and humidity might be alleviated by large-scale adaptation measures such as the provision of reliable air-conditioned shelters, extended siestas, and perhaps the availability of air-conditioned farming equipment. But even if such adaptations can be provided on sufficient scale, any downturn in crop production (due to extreme weather) is also likely to add to the push factor, driving migration. There are also very concerning forecasts, for the same reasons, concerning the habitability of parts of south-west Asia (Pal and Eltahir, 2016) and the north China Plain (Kang and Eltahir, 2018).

Another study found that 190 million people (90% confidence interval 150–250 million) currently occupy land below the projected high-tide lines for 2100, under a low-carbon-emission scenario (Kulp and Strauss, 2019). The same study estimated that 110 million people already live below the high-tide line, mostly protected by levee banks or in elevated dwellings, such as on stilts. While the future migration of many of these people is thus not inevitable, an uncertain proportion of them will be forced to move, including because of soil salinization from sea level rise, a process already under way in (for example) Bangladesh and the Gambia. The study found that the majority of people living on implicated land were in developing countries across Asia, and that, by 2100, chronic coastal flooding or permanent inundation threatens areas occupied by more than 10% of the current populations of Bangladesh, Vietnam and many small island developing states.

There is considerable and growing risk that a 'low carbon emissions' future will not be attained (Anderson, 2015). Fears are strengthening for the extent of ice melt from the 'Third Pole' of the Himalayan glaciers (Rounce *et al.*, 2023). The Greenland and Antarctic ice shelves continue to lose mass. However, although DeConto *et al.* (2021) described ice cliff calving (in Greenland and Antarctica) as a 'wildcard', they forecast that, even without rapid carbon dioxide stabilization, the contribution of Antarctic ice retreat to sea level rise by 2100 is likely to be restricted to a modest 0.2 m compared with 2000. These authors identified a strong negative feedback, in which expanded sea ice suppresses surface atmospheric warming, slowing ice sheet melt. If these optimistic assumptions are correct, substantial sea level rise will be delayed until after 2100, although it may then become unstoppable.

More pessimistically, van de Wal *et al.* (2022) provide a 'high end' forecast of 1.6 m sea level rise by 2100, under a 'business as usual' scenario (IPCC's RCP8.5/SSP5-8.5 assumptions). Due to inertia, this could translate to over 10 m of sea level rise by 2300, a date relevant to our descendants. Hansen *et al.* (2016, 2023) also warned that the dominant literature remains far too optimistic, with insufficient incorporation of feedbacks. In their interpretation, even a rise of 2°C could generate several metres of sea level rise within 50 years. One mechanism which they postulate for this is an alteration in the Southern Ocean currents, reducing the sea ice that currently helps protect land-borne Antarctic ice. Warnings from Hansen, perhaps the most insightful climatologist alive, should surely be heeded.

However, the broader literature is more sanguine, finding that, on the whole, sea level rise by 2100 should not, by itself, generate an unmanageable degree of population displacement. Yet, sea level rise is far from the only driver of human displacement related to anthropogenic climate change; it will drive other push factors, including heat, humidity, diminished local food production and conflict. Additionally (but excluded from most discussion), most areas vulnerable to climate change-enhanced out-migration, including parts of sub-Saharan Africa, have comparatively high rates of population growth (Bryant *et al.*, 2009).

Gemenne (2011) discussed one forecast that is much higher than that of Myers. This is a report by Christian Aid (2007) which forecast up to 1 billion 'forcibly displaced' people by 2050, of whom 300 million would be directly displaced 'as a result of environmental disruptions'. A range of other numbers were mentioned, with the comment that they were 'received with great scepticism in the scholarly community, and great interest in the media and policy-making community'. Gemenne discussed several categories of climate refugees, focusing on sea level rise, tropical cyclones, water stress and disasters. One paper he discussed forecast 'forcible displacement' of up 187 million people by 2100 if temperatures are 4°C higher than pre-industrial levels. Gemenne criticized this as 'excessively deterministic', in part because it does not consider adaptive strategies. However, Gemenne's review contained no discussion of the possibility that extreme heat might influence migration, nor that climate change might influence conflict, and thus migration.

Other impacts of climate change are likely to induce more variable migration, both spatially and temporally. To date, rapid-onset disasters have led mostly to temporary, internal migration, with victims often returning to their original site, even if precarious, in part as it is familiar and in part because they lack the resources to move further away. People are also vulnerable to a sequence of 'compound' disasters, such as repeated typhoons. Even faced with multiple disasters, many people cannot move permanently.

Repeated catastrophes (like multiple illnesses) can deepen poverty, creating socio-economic traps from which escape, including by migration, is almost impossible (Benveniste *et al.*, 2022). Those unable to migrate may be most at risk.

The UNHCR (2022b) listed over 6.6 million Syrians who were refugees in 2019. Additionally, approximately 1 million Syrians have been granted permanent asylum in Germany (Keita and Dempster, 2020). The degree to which Syrian refugees are also climate refugees is disputed (see Chapter 24) but only the most conservative critic could argue that there is no relationship.

Lower-income countries are especially vulnerable to climate change. Pathways include decreased economic opportunities, which, depending on cultural and historical factors, may create harsher political realities, including ethnic tension and diminishing prospects for security and health. In turn, as climate change makes life harder for such populations, it will increase the push factors described above, driving both internal and external migration, both to nearby cities and to developed countries.

We next discuss two case studies from Central and South America; migrant caravans fleeing poverty and instability, and the egress of millions of Venezuelans.

23.5 The Dry Corridor

Three impoverished and adjoining Central American nations – Guatemala, Honduras and El Salvador – are sometimes called the 'northern triangle' or the 'dry corridor'. Examination of migration from this region requires analysis of internal and international politics, economics, cultural factors and environmental factors, each of which shapes and is shaped by the other factors. Honduras has been called the quintessential 'banana republic' (Bucheli, 2008). Movements towards democracy and greater fairness in all three countries have long been suppressed, particularly by a sequence of military dictators supported by US governments. For example, in 1954 the US Central Intelligence Agency played a prominent role in the overthrow of Guatemalan President Jacobo Guzmán (Holland, 2005).

All of these countries, in recent years, have been sources of attempted and sometimes successful 'irregular' emigration to the USA. This movement of people, including 'migrant caravans', has been linked to climate variability and to climate change, most prominently in a report co-authored by four agencies, including the World Food Programme (WFP *et al.*, 2017). This report implied that a severe drought, starting in 2014, associated with El Niño, was worsened by elevated greenhouse gas concentrations and thus helped to trigger a significant increase in attempted migration. Although most intercepted migrants were young men (aged in their 20s) over 50,000 unaccompanied children were recorded by US authorities in 2014. Several journalists also identified climate change as an aggravating factor (Blitzer, 2019; Markham, 2019).

Analysis of factors in this dry corridor reveals clear push, pull and fend factors; all of these influence attempted out-migration. The fact that millions from this region do not attempt to emigrate points also to glue determinants, at least for those who are older, disabled and frail.

Pull factors

The existence of and attraction towards the (comparatively nearby) USA is well understood by many individuals and communities in the dry corridor, for example via remittances from successful emigrees (approximately 10% of Guatemala's gross domestic product) and the construction of empty mansions to which emigrants hope to return eventually (Blitzer, 2019). Family reunification (i.e. with previously successful emigrees) has also been identified as a pull factor (WFP *et al.*, 2017).

Push factors

Recognized push factors include persistent poverty, undernutrition, long-standing inequality, poor governance, high population growth and decades of imperialism (Bucheli, 2008). Poor harvests, whether from drought, frost, soil nutrient exhaustion or from plant diseases, are an important factor, in largely mountainous terrain reliant on rainfall, where the best land is occupied by the United Fruit company. A hungry season already exists in places, even in good years. Climate change, at least in some parts of the Guatemalan highlands, is already thought to be worsening crop yields, including through intensified droughts, floods and in some cases more heat and humidity (Milan and Ruano, 2014; Imbach *et al.*, 2017).

Going deeper, push factors include gang violence, colonization and centuries of discrimination towards the Indigenous Mayans (Bucheli, 2008). Several prominent female environmental activists have been murdered in these nations, including at least two who were Indigenous (Tran *et al.*, 2020). It is also claimed that gangs in this region are engaged in a form of human trafficking, not for international purposes (such as for sex workers, forced labour or to provide human organs) but instead to service the needs of gangs (Boerman and Golob, 2021).

Fend factors

For many years the USA has implemented increasingly harsh measures to reduce migration from Mexico and countries to its south. These measures have evolved from border patrols to an attempt to create a wall along the entire 3145 km border. People who are captured crossing the border (sometimes in tunnels) are returned. Children who cross (and who are captured) have not only been forcibly separated from their parents but have even lost contact with them, apparently due to administrative incompetence (Katkov, 2020).

23.6 The Ongoing Venezuelan Crisis

In the late 20th century Venezuela was widely considered, by many orthodox analysts, as a nation with a promising future, due to its high oil reserves. In the 1950s and 1960s, it was one of the 20 wealthiest countries in the world measured in per capita income (Venezuelan and Ausman, 2019). It was, once, the world's largest exporter of oil, and is still listed by the Organization of the Petroleum Exporting Countries (OPEC) as the nation with the highest oil reserves. However, most of this remaining energy resource is in the form of 'extra heavy' oil, the extraction and refining of which is extremely energy intensive, yielding 'net energy' no higher than 6 and perhaps as little as 1.5 (Youngquist, 1997). This means that its extraction is only financially profitable when the price of oil is extremely high (Figs 23.3 and 23.4). In parallel to

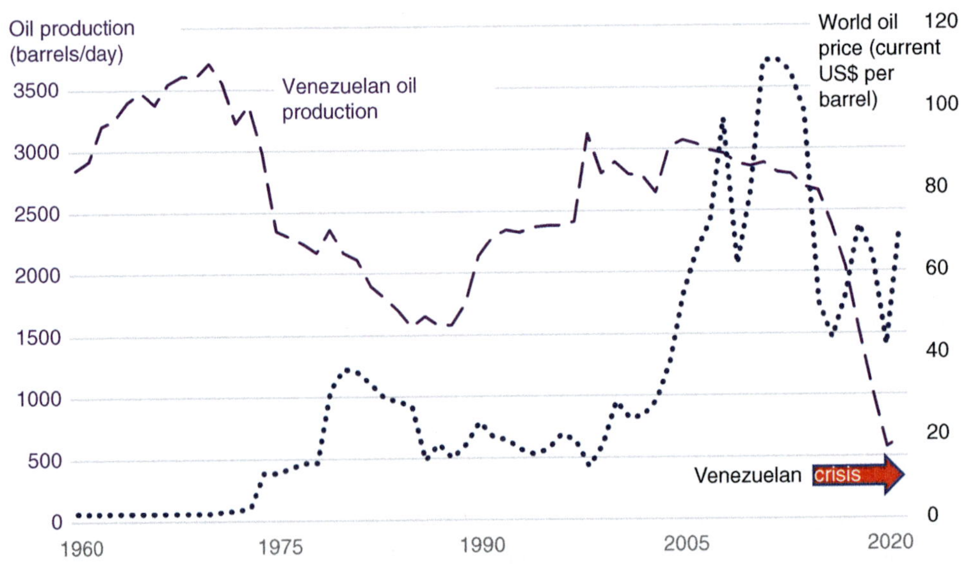

Fig. 23.3. Helping to explain the crisis in Venezuela. World oil price 1961–2022 (current US$ per barrel) compared with total Venezuelan oil production. When prices fell in 2015 most of the remaining (heavy) oil could not be profitably extracted and exported; revenue fell sharply, intensifying the general crisis, including of emigration. Data: https://asb.opec.org/data/ASB_Data.php, https://ourworldindata.org/grapher/crude-oil-prices.

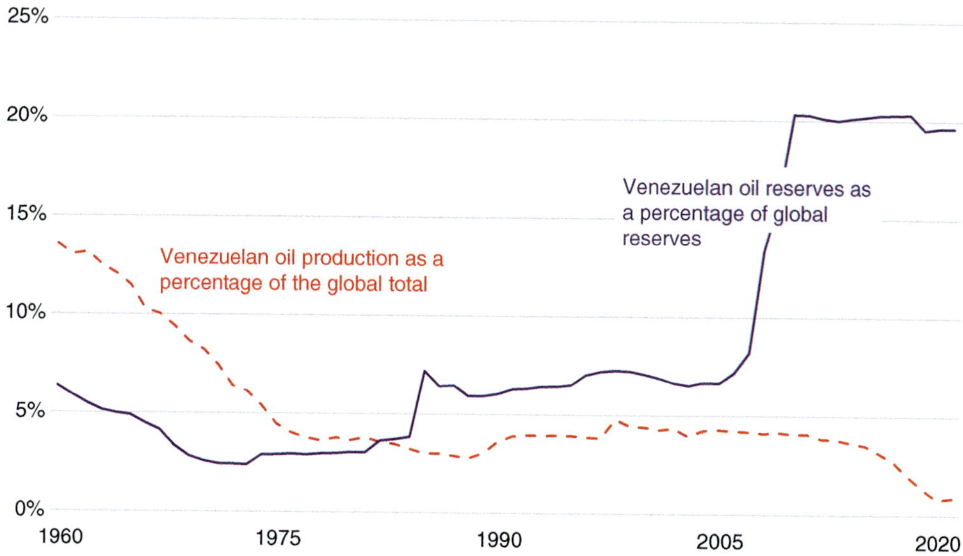

Fig. 23.4. OPEC attributes an enormous fraction of remaining world oil reserves to Venezuela. However, this part of the chart is extremely misleading, as it takes no account of the 'net energy' in these reserves.

the decline in oil production, agricultural, manufacturing and industrial production fell, as did health parameters. Violence increased, and people fled. Debate about the legitimacy of a new president contributed to its paralysis (Ahlborn and Falleti, 2019).

Venezuela now has the third highest ratio of displaced people, as a percentage of its total population (13%), following Syria (27%) and South Sudan (17%) (UNHCR, 2022b). As of August 2023, over 7 million Venezuelans were displaced, living in other countries as refugees, migrants and asylum-seekers (R4V, 2024). Although there is intense (and justifiable) criticism of the failure of successive elected Venezuelan leaders and governments to convert the country's earlier oil bonanza into quasi-sustainable prosperity (Venezuelan and Ausman, 2019), the recent impossibility of extracting heavy oil profitably seems almost completely overlooked in the literature.

While climate change is not a significant contributing factor for this crisis, it is still relevant, especially since the production of heavy oil is carbon-intensive. As Venezuelan natural capital declined, so too did its human carrying capacity (see Chapter 9, this volume). Although some Venezuelan exiles have reached the USA, most are now living in adjoining parts of South America and the Caribbean islands of Aruba and Curacao (UNHCR, 2022b).

23.7 Development, Climate Refugees and Law

Mathew (2021) argued that the strategy of the global North, as reflected by the United Nations in its Declaration for Refugees and Migrants (2016), the Global Compact on Refugees (2018) and the Global Compact for Safe, Orderly and Regular Migration, rested on the opinion that 'development' in sending countries will lessen migratory push factors. She argued that, although the international community claims (e.g. in the Sustainable Development Goals) that it will 'leave no one behind', the costs of care for displaced people will continue to be borne unevenly, by host countries in the Global South.

> The mantra of leaving no one behind may mask a determination within the Global North to provide just, or not even 'enough' for refugees on the premise that these 'others' must be demobilized (in the case of refugees and asylum seekers) or contracted to do the humanitarian heavy lifting (in the case of host states) under the guise of development.
>
> (Mathew, 2021).

The UNHCR (2022b) observed that of the 89 million people displaced at the end of 2021, 83% were hosted in 'low and middle income' countries. This high percentage appears to support Mathew's argument. However, most of the increase in displacement in 2022 are Ukrainians, sheltered in Russia and the European Union, thus lowering that percentage slightly. Asylum seekers comprised less than 6% of this total, although it is likely that there would be more if host countries, especially in the global North, were less hostile to them.

Global food and energy prices are rising, and so are many other costs of living, as global limits to growth become increasingly obvious and close (Butler, 2017). In many impoverished nations (largely tropical), birth rates remain high, complicating their escape from poverty (Lutz *et al.*, 2019). Rural poverty intensifies the push for rural to urban migration, expanding the size of 'unplanned' settlements and deepening urban poverty. Per capita economic growth, whether defined conventionally, or adjusted for changes in natural and other forms of capital, is slowing in many parts of the world. In many places, pre-existing disparities and inequalities are deepening. The factors that determine migration are continually changing; overall, however, increased demand and more attempts to migrate appear inevitable.

Though recognizing that climate change can be a driver of displacement, McAdam (2011) argued that advocacy for a new treaty to address climate change-related movement is misplaced, though he admitted that this could have political benefits. Fornalé and Doebbler (2017) noted that, irrespective of a specific category for climate migrants, their protection already falls under the jurisdiction of the UNHCR, especially if they are displaced internationally. In theory, climate migrants can be defined as those who emigrate in a world with climate change compared with the 'counterfactual' – the same world without climate change. In practice, this is almost impossible to calculate.

The UNHCR identified several 'root causes' of displacement, including poverty and conflict. It called for the international community to address such causes. However, recent UNHCR documents (perhaps most, or even all) lack discussion of the contribution of high fertility, or of population pressure, to the exceedance of regional carrying capacity. Butler (2017) argued that the scale of forced population displacement is a manifestation of 'regional overload'. Recent UNHCR reports include extensive discussion of climate change as a driver of displacement, including via the impact of climate change on conflict. Nonetheless, the UNHCR does not support the term 'climate refugee', which, as it points out, is not recognized by international law. Such a term is likely to remain unpalatable, as its acceptance would help remind overconsuming populations of their role, and might also hint at moral responsibility and legal liability.

More positively, in 2020 the UN Human Rights Committee (UNHRC) issued a landmark ruling in which it recognized, for the first time, that forcibly returning a person to a place where their life would be at risk due (in part) to the adverse effects of climate change may violate their right to life under article 6 of the International Covenant on Civil and Political Rights (Sinclair-Blakemore, 2020). However, the threshold for legal recognition remains very high.

23.8 Could Looming 'Underpopulation' in Some Areas Solve the Problem of Population Displacement?

There has long been concern, expressed by some writers and politicians, over declining birth rates in parts of the global North, including Italy and Japan (Morgan, 2003). In 2023 the population of China was reported as declining for the first time, while Japan's Prime Minister warned that declining fertility risked the continued functioning of Japan (Anonymous, 2023). Russian President Putin has also been reported as expressing concern about the 'dire' demographic conditions of high mortality and low fertility in 'his' vast nation (Frejka and Zakharov, 2013). These fears, in part, reflect a 'mercantilist' view of population size, in which the total size of a population is thought to correlate with political power, prosperity and national prestige.

If numbers were all that mattered then any looming or actual population 'shortfall' in such countries could easily be compensated for by adjusting the fend and pull settings, permitting the desired in-flow of migrants. For low-paid jobs that require manual labour and a work ethic, but which are increasingly shunned by citizens of richer nations, such 'valves' have long been managed. For example, Pacific Islanders are welcomed as guest workers in Australia. The incomes they earn when employers meet their legal obligations are generally sufficient to send funds (remittances) to their home country.

Many high-income countries also admit highly skilled migrants from low-income settings, such as to work in healthcare or, if they are sufficiently wealthy, to establish businesses. Although the integration of such migrants may take generations, many do become accepted; some will be highly valued. However, even if they return significant funds to their original country, this form of highly skilled migration constitutes a 'brain drain', particularly if the costs of training such people are borne by the sending country. In addition, at least in poor nations with high fertility (e.g. Guatemala and Nepal), there is little evidence that remittances allow such populations to escape poverty, even if some individuals or families are greatly helped. Furthermore, inequality may be deepened (Salike *et al.*, 2022).

There has long been speculation that additional farmers may be welcomed in a warming Asian Russia (east of the Ural Mountains) where the soil may be sufficiently good to establish large-scale agriculture later this century (Parfenova *et al.*, 2019). In theory, assuming Russian fertility remains low, such immigrants could originate from (and even be welcomed from) China or South Asia, each of which (under high carbon emission scenarios) could generate millions of displaced people with farming skills within a few decades. The potential for large-scale migration to other countries of high latitude (e.g. Canada, Scandinavia) is not as promising from the perspective of soil quality, but could also occur.

Even if large numbers of in-migrants are welcomed and successfully integrate in these high-latitude settings, the potential scale of such resettlement is still likely to be small compared with the need. Countries will, in coming decades, continue to seek to control and manage immigration. Perceived national interest will trump the well-being of those who are displaced, although, at times, such interests may align for a small number of people. Nations such as Japan and Italy are particularly xenophobic; rebounding fertility in such countries seems more likely than admission of tens of millions of culturally distinct people. Circumstances for large numbers of displaced people, should a 'breakdown world' evolve, are likely to be far worse than the present, especially if that leads to a marked reduction in the capacity of the UNHRC and its companion organizations to care for the people in their current remit.

23.9 Conclusion

The prospect for future climate change continues to darken. The number of displaced people (however defined) continues to rise, while the evidence of climate change is ever stronger. Although the concept of climate and other forms of environmental migration has, in recent decades, been contested, there is today growing acceptance of its validity. Decisions to migrate have complex causation, and classifications of migration as purely 'economic', 'social' or 'climatic' lack intellectual rigour. Here, we prefer the term eco-social to describe these complex and interacting determinants. We stress that the absence of a category of purely 'climatic' migrants does not negate the validity of climate change as a major driver of migration. This is likely to be increasingly recognized, but probably not by courts.

The destiny of displaced people, the number of which already exceeds 100 million, and which is likely to increase greatly in size, is a test for civilization and 21st century ethics. These people are not only disproportionately vulnerable, but are also disproportionately low emitters of greenhouse gases. The health of donor and recipient populations could also be worsened by excess migration, as discussed in the following section. Put simply, will the 'talent drain' be fully compensated by remittances?

Indications to date suggest that the global North will fail this test, for migrants and themselves, though some nations with low birth rates or new agricultural areas opened by climate change may be more receptive to culturally dissimilar settlers. Most displaced people will continue to be housed in camps in low-income settings, as they are now. Their wider integration, even in the global South, is also likely to be resisted, such as is already apparent in Bangladesh (for Rohingya from Myanmar) and in north-east India (for Muslim Bengalis), among other places such as South Africa (Mangu, 2019). Instead, a trickle of displaced people is likely to be tolerated in the global North, either arriving as legal immigrants, or participating in the informal economy after entering illegally, i.e. by ignoring and evading the fend signal. Some of these people (or their descendants) may live fulfilling lives, but others will struggle to integrate and endure harsh discrimination, often with the fear of expulsion. In the North, reporting of the lives of the millions who remain in camps in the South will continue to be minimal; also minimal will be analysis of the deep, underlying causes of their fate. An illustrated article by Gustin (2022) is a partial exception but it focuses almost exclusively on climatic drivers of migration.

Recognition of the dimensions of this evolving failure could galvanize action both to mitigate climate change and to put in place the more enlightened policies and practices that are keenly required. Limited aid, from governments and by individuals (mostly in the global North), has been and will continue to be provided for food and other basic necessities, at least in the near future. Such funds ameliorate catastrophes such as famine and genocide among displaced people. However, postponing calamity is likely to become increasingly difficult, even if the bleak forecasts of Hansen *et al.* (2023) remain unrealized in this century. A world with over 100 million displaced already exists; very few in the North perceive this as the catastrophic failure of global policy that it represents. Helping to perpetuate, and indeed to worsen, this dismal position is the fact that the organizations who distribute this aid and, with few exceptions, the UN as a whole, are inhibited from freely analysing the deeper determinants of this dilemma. As the military historian Sir Antony Beevor recently lamented in an interview by Sally Warhaft (Beevor, 2022: '[O]ne of the things that nobody really has the courage or even the ability to discuss properly and to discuss openly ... what is going to be the future?'

Acknowledgements

This chapter is adapted from Bowles, D.C., Reuveny, R. and Butler, C.D. (2014) Moving to a better life? Climate, migration and population health. In: Butler, C.D. (ed.) *Climate Change and Global Health*. Wallingford, UK, pp. 135–143.

Notes

1 At least one component of which approached 100 km.

2 In this context this means most governments (supported by the majority of their population) believe that additional immigrants, unless possessing skills or other wanted attributes (e.g. financial), will cause net harm to the 'host' population.

3 Social processes can affect the physical landscape, e.g. the transformation of hilly landscapes to watered rice terraces requires coordinated, cooperative human effort; physical events can affect society, for example drought of sufficient severity can cause the abandonment of settlements. The recent civil war in Syria had physical as well as social causes, with social and physical consequences. And so on.

4 During the green Sahara (11,000 to 5000 years before present), high amounts of rainfall supported diverse vegetation, permanent lakes and human populations. A prolonged 'pause' in these conditions 8000 years ago coincided with the temporary abandonment of sites occupied by Neolithic humans.

5 Zhang *et al.* (2011) disagreed, concluding that 'climate change was the ultimate cause, and climate-driven economic downturn was the direct cause, of large-scale human crises in preindustrial Europe and the Northern Hemisphere'. This claim may be valid for the period of their analysis (1500–1800). Today, however, the degree of anthropogenic climate change that has occurred should not be given an equivalent explanatory power. Nonetheless, it is an increasing factor, and its contribution to human crises is likely to increase. Runaway climate change, were it to occur, would overwhelm civilization.

6 Thousands of villages were abandoned in the later Middle Ages (1300–1500). The Black Death was far from the sole cause (Dyer, 1982). The extent of depopulation has been challenged (Pennington, 2023).

7 Bangladesh has accepted hundreds of thousands of Rohingya, but these people are denied citizenship, are housed in appalling conditions and are funded by the international community (Milton *et al.*, 2017).

References

Adams, H. (2016) Why populations persist: mobility, place attachment and climate change. *Population and Environment* 37, 429–448. DOI: 10.1007/s11111-015-0246-3

Ahlborn, S. and Falleti, T. (2019) *The Venezuelan crisis, explained: A Q&A*. University of Pennsylvania, Philadelphia. Available at: https://penntoday.upenn.edu/news/venezuelan-crisis-explained (accessed 8 January 2023).

Amnesty International (2017) *Libya's dark web of collusion. Abuses against Europe-bound refugees and migrants*. Available at: https://www.amnesty.org/download/Documents/MDE1975612017ENGLISH.PDF (accessed 27 September 2022).

Anderson, K. (2015) Talks in the city of light generate more heat. *Nature* 528, 437. DOI: 10.1038/528437a

Anonymous (2021) Wielding hunger as a weapon. *The Economist* 438, 10.

Anonymous (2022) Northern Ethiopia: conflict and food insecurity. *Crop Monitor*. Available at: https://cropmonitor.org/index.php/cmreports/food-security-conflict-reports/ (accessed 6 June, 2023).

Anonymous (2023) Japan's ageing population poses urgent risk to society, says PM. *The Guardian*. Available at: https://www.theguardian.com/world/2023/jan/23/japans-ageing-population-poses-urgent-risk-to-society-says-pm (accessed January 25 2023).

Ayeb-Karlsson, S., Smith, C.D. and Kniveton, D. (2018) A discursive review of the textual use of 'trapped' in environmental migration studies: The conceptual birth and troubled teenage years of trapped populations. *Ambio* 47, 557–573. DOI: 10.1007/s13280-017-1007-6

Azzouz, A. (2020) 2011 Reflections on a ruined homeland. *City* 24, 178–194. DOI: 10.1080/13604813.2020.1739414

Beck, U. (2010) Remapping social inequalities in an age of climate change: For a cosmopolitan renewal of sociology. *Global Networks* 10, 165–181. DOI: 10.1111/j.1471-0374.2010.00281.x

Beevor, A. (2022) *Russia, Ukraine and the lessons of history*. Interviewed by Sally Warhaft. ABC (Australian Broadcasting Corporation). Available at: https://www.abc.net.au/radionational/programs/bigideas/russia-ukraine-and-the-lessons-of-history/14111548 (accessed 4 January 2023).

Benveniste, H., Oppenheimer, M. and Fleurbaey, M. (2022) Climate change increases resource-constrained international immobility. *Nature Climate Change* 12, 634–641. DOI: 10.1038/s41558-022-01401-w

Betts, A. (2019) Nowhere to go. *Foreign Affairs* 98, 122–133.

Bird, M.I., Condie, S.A., O'Connor, S., O'Grady, D., Reepmeyer, C. *et al.* (2019) Early human settlement of Sahul was not an accident. *Scientific Reports* 9, 8220. DOI: 10.1038/s41598-019-42946-9

Black, R., Adger, W.N., Arnell, N.W., Dercon, S., Geddes, A. and Thomas, D. (2011a) The effect of environmental change on human migration. *Global Environmental Change* 21, Supplement 1, S3–S11. DOI: 10.1016/j.gloenvcha.2011.10.001

Black, R., Adger, W.N., Arnell, N.W., Dercon, S., Geddes, A. and Thomas, D. (2011b) Migration and global environmental change (editorial). *Global Environmental Change* 21, Supplement 1, S1–S2. DOI: 10.1016/j.gloenvcha.2011.10.005

Blitzer, J. (2019) How climate change is fuelling the U.S. border crisis. *The New Yorker*. Available at: https://www.newyorker.com/news/dispatch/how-climate-change-is-fuelling-the-us-border-crisis (accessed 27 September 2022).

Boas, I., Farbotko, C., Adams, H., Sterly, H., Bush, S. *et al.* (2018) Climate migration myths. *Nature Climate Change* 9, 901–903. DOI: 10.1038/s41558-019-0633-3

Boerman, T. and Golob, A. (2021) Gangs and modern-day slavery in El Salvador, Honduras and Guatemala: A non-traditional model of human trafficking. *Journal of Human Trafficking* 7, 241–257. DOI: 10.1080/23322705.2020.1719343

Boyden, S. (1972) The environment and human health. *Medical Journal of Australia* 116, 1229–1234.

Bryant, L., Carver, L., Butler, C.D. and Anage, A. (2009) Climate change and family planning: least developed countries define the agenda. *Bulletin of the World Health Organization* 87, 852–857. DOI: 10.2471/BLT.08.0625

Bucheli, M. (2008) Multinational corporations, totalitarian regimes and economic nationalism: United Fruit Company in Central America, 1899–1975. *Business History* 50, 433–454. DOI: 10.1080/00076790802106315

Butler, C.D. (2017) Limits to growth, planetary boundaries, and planetary health. *Current Opinion in Environmental Sustainability* 25, 59–65. DOI: 10.1016/j.cosust.2017.08.002

Butler, C.D. (2024) Climate change and human health: Primary, secondary, and tertiary effects. In: Letcher, T.M. (ed.) *Living with Climate Change*. Elsevier, Amsterdam, The Netherlands, pp. 213–240. DOI: 10.1016/B978-0-443-18515-1.00007-1

Butler, C.D. and Kefford, B.J. (2018) Climate and conflict: magnifying risks. *Nature* 555, 587. DOI: 10.1038/d41586-018-03795-0

Christian Aid (2007) *Human Tide: The Real Migration Crisis*. Christian Aid, London. Available at: https://reliefweb.int/report/colombia/human-tide-real-migration-crisis-christian-aid-report (accessed 22nd June 2021).

d'Orsi, C. (2022) Outsourcing asylum seekers: the case of Rwanda and the UK. *The Conversation*. Available at: https://theconversation.com/outsourcing-asylum-seekers-the-case-of-rwanda-and-the-uk-180973 (accessed 21 June 2022).

Deckers, S., Nyssen, J. and Lanckriet, S. (2020) Ethiopia's Tigray region has seen famine before: why it could happen again. *The Conversation*. Available at: https://theconversation.com/ethiopias-tigray-region-has-seen-famine-before-why-it-could-happen-again-150181 (accessed 20 August 2022).

DeConto, R.M., Pollard, D., Alley, R.B., Velicogna, I., Gasson, E. *et al.* (2021) The Paris Climate Agreement and future sea-level rise from Antarctica. *Nature* 593, 83–89. DOI: 10.1038/s41586-021-03427-0

de Freytas-Tamura, K. (2022) What it's like when your loved ones won't leave Ukraine. *The New York Times*. Available at: https://www.nytimes.com/2022/04/06/nyregion/ukraine-older-people.html (accessed 13 June 2023).

DeWitte, S. and Slavin, P. (2013) Between famine and death: England on the eve of the Black Death – Evidence from paleoepidemiology and manorial accounts. *The Journal of Interdisciplinary History* 44, 37–60. DOI: 10.1162/JINH_a_00500.

Dyer, C. (1982) Deserted medieval villages in the west Midlands. *The Economic History Review* 35, 19–34. DOI: 10.2307/2595101

El-Hinnawi, E. (1985) *Environmental Refugees*. United Nations Environment Programme (UNEP), Nairobi, Kenya.

Engerman, S.L. and Sokoloff, K.L. (2012) Colonisation and development. *Economic History of Developing Regions* 27, S28–S40. DOI: 10.1080/20780389.2012.658660

Flannery, T.F. (1994) *The Future Eaters*. Reed Books, Melbourne, Australia.

Foresight (2011) *Migration and Global Environmental Change: Future Challenges and Opportunities*. The Government Office of Science, London.

Fornalé, E. and Doebbler, C.F.J. (2017) UNHCR and protection and assistance for the victims of climate change. *The Geographical Journal* 183, 329–335. DOI: 10.1111/geoj.12193

Frejka, T. and Zakharov, S. (2013) The apparent failure of Russia's pronatalist family policies. *Population and Development Review* 39, 635–647. DOI: 10.1111/j.1728-4457.2013.00631.x

Gemenne, F. (2011) Why the numbers don't add up: a review of estimates and predictions of people displaced by environmental changes. *Global Environmental Change* 21, Supplement 1, S41–S49. DOI: 10.1016/j.gloenvcha.2011.09.005

Ghráinne, B.N. (2017) The 2015 Andaman Sea boat 'crisis': Human rights and refugee law considerations. In: Salomon, S., Heschl, L., Oberleitner, G. and Benedek, W. (eds) *Blurring Boundaries: Human Security and Forced Migration.* Brill, Leiden, The Netherlands, pp. 123–134.

Gorsky, J. (2015) New Christians. *Exiles in Sepharad: The Jewish Millennium in Spain.* University of Nebraska Press, Lincoln, Nebraska, pp. 57–66.

Gustin G. (2022) Climate change is driving millions to the precipice of a 'raging food catastrophe', *Inside Climate News*. Available at: https://hiiraan.com/news4/2022/Dec/189067/climate_change_is_driving_millions_to_the_precipice_of_a_raging_food_catastrophe.aspx (accessed 24 December 2022).

Hansen, J., Sato, M., Hearty, P., Ruedy, R., Kelley, M. *et al.* (2016) Ice melt, sea level rise and superstorms: evidence from paleoclimate data, climate modeling, and modern observations that 2°C global warming could be dangerous *Atmospheric Chemistry and Physics* 16, 3761–3812. DOI: 10.5194/acp-16-3761-2016

Hansen, J.E., Sato, M., Simons, L. Mazarenko, L.S., Sangha, I. *et al.* (2023) Global warming in the pipeline. *Oxford Open Climate Change* 3. DOI: 10.1093/oxfclm/kgad008

Harper, K. (2017) *The Fate of Rome*. Princeton University Press, Princeton, New Jersey.

Herwig, H.H. (1999) Geopolitik: Haushofer, Hitler and lebensraum. *Journal of Strategic Studies* 22, 218–241. DOI: 10.1080/01402399908437762

Holland, M. (2005) Private sources of U.S. Foreign Policy: William Pawley and the 1954 coup d'Etat in Guatemala. *Journal of Cold War Studies* 7, 36–73. DOI: 10.1162/1520397055012442

Hulme, M. (2011) Reducing the future to climate: A story of climate determinism and reductionism. *Osiris* 26, 245–266. DOI: 10.1086/661274

Imbach, P., Beardsley, M., Bouroncle, C., Medellin, C., Läderach, P. *et al.* (2017) Climate change, ecosystems and smallholder agriculture in Central America: an introduction to the special issue. *Climatic Change* 141, 1–12. DOI: 10.1007/s10584-017-1920-5

Kang, S. and Eltahir, E.A.B. (2018) North China Plain threatened by deadly heatwaves due to climate change and irrigation. *Nature Communications* 9, 2894. DOI: 10.1029/2019GL083686

Katkov, M. (2020) Parents of 545 children separated at U.S.-Mexico border still can't be found. *NPR* (National Public Radio), Washington, DC. Available at: https://www.npr.org/2020/10/21/926031426/parents-of-545-children-separated-at-u-s-mexico-border-still-cant-be-found (accessed January 8 2023).

Keita, S. and Dempster, H. (2020) *Five years later, one million refugees are thriving in Germany*. Center for Global Development (CGDEV), Washington, DC. Available at: https://www.cgdev.org/blog/five-years-later-one-million-refugees-are-thriving-germany (accessed 8 January 2023).

Kulp, S.A. and Strauss, B.H. (2019) New elevation data triple estimates of global vulnerability to sea-level rise and coastal flooding. *Nature Communications* 10, 4844. DOI: 10.1038/s41467-019-12808-z

Lambeck, K., Rouby, H., Purcell, A., Sun, Y. and Sambridge, M. (2014) Sea level and global ice volumes from the Last Glacial Maximum to the Holocene. *Proceedings of the National Academy of Sciences (USA)* 111, 15296–15303. DOI: 10.1073/pnas.1411762111

Ljungqvist, F.C., Seim, A. and Huhtamaa, H. (2021) Climate and society in European history. *WIREs Climate Change* 12, e691. DOI: 10.1002/wcc.691

Lutz, W., Crespo Cuaresma, J., Kebede, E., Prskawetz, A., Sanderson, W.C. *et al.* (2019) Education rather than age structure brings demographic dividend. *Proceedings of the National Academy of Sciences (USA)* 116, 12798–12803. DOI: 10.1073/pnas.1820362116

Mafu, L. (2019) The Libyan/Trans-Mediterranean slave trade, the African Union, and the failure of human morality. *SAGE Open* 9(1). DOI: 10.1177/2158244019828849

Mangu, A.M. (2019) Xenophobia and migration in post-apartheid South Africa: myths and realities. *African Journal of Democracy and Governance* 6, 44–72. DOI: 10.10520/EJC-17c9cdf107

Markham, L. (2019) How climate change is pushing Central American migrants to the US. *The Guardian*. Available at: https://www.theguardian.com/commentisfree/2019/apr/06/us-mexico-immigration-climate-change-migration (accessed 22 April, 2023).

Mathew, P. (2021) Killing asylum softly or leaving no one behind? The New York declaration and global compacts in a divided world. *Globalizations*. DOI: 10.1080/14747731.2021.1974207

McAdam, J. (2011) Swimming against the tide: Why a climate change displacement treaty is not the answer. *International Journal of Refugee Law* 23, 2–7. DOI: 10.1093/ijrl/eeq045

McMichael, A.J. (1993) *Planetary Overload. Global Environmental Change and the Health of the Human Species*. Cambridge University Press, Cambridge, UK.

McMichael, A.J. (1999) Prisoners of the proximate: loosening the constraints on epidemiology in an age of change. *American Journal of Epidemiology* 149, 887–897. DOI: 10.1093/oxfordjournals.aje.a009732

Milan, A. and Ruano, S. (2014) Rainfall variability, food insecurity and migration in Cabricán, Guatemala. *Climate and Development* 6, 61–68. DOI: 10.1080/17565529.2013.857589

Milton, A.H., Rahman, M., Hussain, S., Jindal, C., Choudhury, S. *et al.* (2017) Trapped in statelessness: Rohingya refugees in Bangladesh. *International Journal of Environmental Research and Public Health* 14, 942. DOI: 10.3390/ijerph14080942

Morgan, S.P. (2003) Is low fertility a twenty-first-century demographic crisis? *Demography* 40, 589–603. DOI: 10.1353/dem.2003.0037

Myers, N. (1993) Environmental refugees in a globally warmed world. *BioScience* 43, 752–761. DOI: 10.2307/1312319

Myers, N. (2002) Environmental refugees: a growing phenomenon of the 21st century. *Philosophical Transactions of the Royal Society B* 357, 609–613. DOI: 10.1098/rstb.2001.0953

Pal, J.S. and Eltahir, E.A.B. (2016) Future temperature in southwest Asia projected to exceed a threshold for human adaptability. *Nature Climate Change* 6, 197–200. DOI: 10.1038/nclimate2833

Parfenova, E., Tchebakova, N. and Soja, A. (2019) Assessing landscape potential for human sustainability and 'attractiveness' across Asian Russia in a warmer 21st century. *Environmental Research Letters* 14, 065004. DOI: 10.1088/1748-9326/ab10a8

Paulose, R.M. (2019) A new dawn? Statelessness and Assam. *Groningen Journal of International Law & Literature* 7, 99–111. DOI: 10.4103/0019-5278.1569

Pennington, H. (2023) Maybe, maybe not. Pathogens and pandemics have played a huge part in shaping human history right up to COVID-19 — but their exact effects remain highly debatable. *Nature* 616, 654–655. DOI: 10.1038/d41586-023-01285-6

Perdue, T. (1989) Cherokee women and the trail of tears. *Journal of Women's History* 1, 14–30. DOI: 10.1353/jowh.2010.003

Petersen, W. (1958) A general typology of migration. *American Sociological Review* 23, 256–266. DOI: 10.2307/2089239

Piguet, E. (2013) From 'primitive migration' to 'climate refugees': The curious fate of the natural environment in migration studies. *Annals of the Association of American Geographers* 103, 148–162. DOI: 10.1080/00045608.2012.696233

R4V (2024) *Refugees and Migrants from Venezuela*. UNHCR and IOM Regional Inter-Agency Coordination Platform for Refugees and Migrants from Venezuela. Available at: https://www.r4v.info/en/home (accessed 17 April 2024).

Raijman, R. (2020) *A warm welcome for some: Israel embraces immigration of Jewish diaspora, sharply restricts labor migrants and asylum seekers*. Migration Policy Institute, Washington, DC. Available at: https://www.migrationpolicy.org/article/israel-law-of-return-asylum-labor-migration (accessed 6 June, 2023).

Rounce, D.R., Hock, R., Maussion, F., Hugonnet, R., Kochtitzky, W. *et al.* (2023) Global glacier change in the 21st century: Every increase in temperature matters. *Science* 379, 78–83. DOI: 10.1126/science.abo1324

Saeed, F., Schleussner, C.-F. and Ashfaq, M. (2021) Deadly heat stress to become commonplace across South Asia already at 1.5°C of global warming. *Geophysical Research Letters* 48, e2020GL091191. DOI: 10.1029/2020GL091191

Salih, A.A.M., Baraibar, M., Mwangi, K.K. and Artan, G. (2020) Climate change and locust outbreak in East Africa. *Nature Climate Change* 10, 584–585. DOI: 10.1038/s41558-020-0835-8

Salike, N., Wang, J. and Regis, P. (2022) Remittance and its effect on poverty and inequality: A case of Nepal. *NRB Economic Review* 34, 1–29. DOI: 10.3126/nrber.v34i2.49430

Samarasekera, U. (2021) Relocated Rohingya face uncertain health-care provision. *The Lancet* 397, 460. DOI: 10.1016/S0140-6736(21)00296-8

Sargent, F. (1972) Man–environment – Problems for public health. *American Journal of Public Health* 62, 628–633. DOI: 10.2105/AJPH.62.5.628

Sinclair-Blakemore, A. (2020) *Teitiota v New Zealand: A Step Forward in the Protection of Climate Refugees under International Human Rights Law?* Oxford Human Rights Hub, Oxford, UK. Available at: https://ohrh.law.ox.ac.uk/teitiota-v-new-zealand-a-step-forward-in-the-protection-of-climate-refugees-under-international-human-rights-law/ (accessed 7 June, 2023).

Smith, W.D. (1980) Friedrich Ratzel and the origins of Lebensraum. *German Studies Review* 3, 51–68. DOI: 10.2307/1429483

Sur, M. (2014) Divided bodies: Crossing the India-Bangladesh border. *Economic and Political Weekly* 49, 31–35. Available at: https://www.jstor.org/stable/24479355

Tazreiter, C. (2017) The unlucky in the 'lucky country': asylum seekers, irregular migrants and refugees and Australia's politics of disappearance. *Australian Journal of Human Rights* 23, 242–260. DOI: 10.1080/1323238X.2017.1372039

Tazreiter, C. (2019) Narratives of crisis migration and the power of visual culture. In: Menjívar, C. (ed.) *The Oxford Handbook of Migration Crises*. Oxford University Press, Oxford, UK, pp. 619–634.

Tierney, J.E., Pausata, F.S.R. and deMenocal, P.B. (2017) Rainfall regimes of the green Sahara. *Science Advances* 3, e1601503. DOI: 10.1126/sciadv.1601503

Tran, D., Martínez-Alier, J., Navas, G. and Mingorría, S. (2020) Gendered geographies of violence: a multiple case study analysis of murdered women environmental defenders. *Journal of Political Ecology* 27, 1189–1212. DOI: 10.2458/v27i1.23760

UNHCR (2022a) *More than 100 million people are forcibly displaced 2022*. UN High Commissioner for Refugees, Geneva. Available at: https://www.unhcr.org/refugee-statistics/insights/explainers/100-million-forcibly-displaced.html (accessed 7 September 2022).

UNHCR (2022b) *Global Trends. Forced Displacement in 2021*. UN High Commissioner for Refugees, Geneva. Available at: https://www.unhcr.org/62a9d1494/global-trends-report-2021 (accessed 7 September 2022).

Vale, G. (2020) Liberated, not free: Yazidi women after Islamic State captivity. *Small Wars & Insurgencies* 31, 511–539. DOI: 10.1080/09592318.2020.1726572

van de Wal, R.S.W., Nicholls, R.J., Behar, D., McInnes, K., Stammer, D. *et al.* (2022) A high-end estimate of sea level rise for practitioners. *Earth's Future* 10, e2022EF002751. DOI: 10.1029/2022EF002751

Venezuelan, A. and Ausman, J. (2019) The devastating Venezuelan crisis. *Surgical Neurology International* 10, 1–6. DOI: 10.25259/SNI_342_2019

Vives, L. (2020) Child migration in the US and Spain: Towards a global border regime? *International Migration* 58, 29–44. DOI: 10.1111/imig.12704

WFP *et al.* (2017) *Food Security and Emigration. Why people flee and the impact on family members left behind in El Salvador, Guatemala and Honduras. Research report.* World Food Programme (WFP) Regional Bureau for Latin America and the Caribbean, Panama City, in collaboration with Inter-American Development Bank (IDB), International Fund for Agricultural Development (IFAD), International Organization for Migration (IOM) and Organization of American States (OAS). Available at: https://docs.wfp.org/api/documents/WFP-0000019629/download (accessed 5 January 2023).

Woodruff, R.E., McMichael, A.J., Butler C.D. and Hales, S. (2006) Action on climate change: the health risks of procrastinating. *Australian and New Zealand Journal of Public Health* 30, 567–571. DOI: 10.1111/j.1467-842X.2006.tb00788.x

Youngquist, W. (1997) *GeoDestinies: The Inevitable Control of Earth Resources Over Nations and Individuals*. National Book Co., Portland, Oregon. Cited by: Ehrenfeld, D. (2005) Recalculating future oil reserves. *Science* 309, 54–56. DOI: 10.1126/science.309.5731.5

Zhang, D.D., Lee, H.F., Wang, C., Li, B., Pei, Q., Zhang, J. *et al.* (2011) The causality analysis of climate change and large-scale human crisis. *Proceedings of the National Academy of Sciences (USA)* 108, 17296–17301. DOI: 10.1073/pnas.1104268108

24 Climate Change, Conflict, Complexity and Health

Colin D. Butler,[1] Mark Braidwood[2] and Devin C. Bowles[3]
[1]National Centre for Epidemiology and Population Health, The Australian National University, Canberra, Australia; [2]Department of Family Medicine, Queens University, Kingston, Ontario, Canada; [3]Faculty of Health, University of Canberra, Canberra, Australia

Abstract
This chapter provides a selective literature review of the theoretical reasons that link resource scarcity, social capacity, conflict and climate change. It finds that many outcomes associated with climate change are likely to amplify the risk of conflict, particularly via shortages of food, water and arable land. It is strongly recognized that climate change does not operate independently of history or politics, contrary to the claims sometimes advanced by critics. The chapter examines two case studies involving civil war: Darfur, Sudan (1980s–2005), and Syria (2012–present). It finds that scarcity, partly induced by environmental factors, was an important driver for each. Climate change, intersecting with the co-factors that drive conflict, has the potential to trigger devastating harm to human health.

> We depend upon the Nile 100 percent in our life. So, if anyone, at any moment, thinks to deprive us of our life, we shall never hesitate [to go to war] because it is a matter of life or death (Egyptian President Anwar Sadat, 1978).
>
> (Gleick, 1991)

24.1 Introduction

As we ascend through primary, secondary and tertiary effects of climate change, causal complexity increases. The more the effects of climate change interact with confounding factors, the more these factors influence eventual outcomes in pathways that are ever more entangled. This makes the prediction of tertiary (systemic) effects of climate change, including conflict, difficult at timescales such as years or even decades. It also means that critics of the causal links have numerous opportunities for dissent. Yet tertiary effects have the greatest potential impact on human health.

Further complications arise from multiple interactions between climate change and its co-factors that impede adaptation. The potential for conflict, associated with and fed partially by climatic factors, illustrates these points.

Causes of conflict

The dominant view at the heart of the thinking of both the military and science is that conflict (armed and otherwise), competition and evolution are linked to resource scarcity (Sullivan *et al.*, 2007; Zhang *et al.*, 2011). Competition between individuals and between species (as well as co-operation among allies) is central to evolutionary theory, as reflected in the title 'Arms races between and within species' (Dawkins *et al.*, 1979).

 Climate Change and Global Health: Primary, Secondary and Tertiary Effects, 2nd Edition (eds C.D. Butler and K. Higgs)
DOI: 10.1079/9781800620025.0024

In the first decade of this millennium, as the likely severity of anthropogenic climate change became clearer, a range of military and political analysts warned that climate change is likely to act to increase the risk of armed conflict, between and within states (McDonald, 2013).[1] For simplicity, this group of authors are here called 'eco-social'. However, many critics (here called 'social') instead disputed any positive causal connection between climate change and conflict, arguing that proponents (of such a causal link) overlook political perspectives, including historical grievances, which are claimed as of more importance. At times, eco-social theorists have been denigrated as 'neo-Malthusian', or as being 'overly deterministic, since it removes violent conflict from its local social and political contexts' (Koubi, 2019).

An example of such a 'strawman' assertion was provided by Brown (2010) who wrote that 'data therefore contradict assertions that eco-scarcity or climate change *alone* provoked the outbreak of conflict' [emphasis added]. Brown did not provide any reference to support his assertion that eco-social authors are mono-causal. For Reeves (quoted in Butler, 2007) the 'true culprit' in Darfur (Sudan) was the National Islamic Front, which was 'bent on expanding its political base through ethnic cleansing, using terror as a tactic'. Reeves also claimed that 'this is not about competition over resources'. Koubi (2019) provided another incorrect claim of a mono-causal explanation, stating that 'many academics and policy makers attribute the civil conflict in Syria to the drought that occurred in late 2000'. Although this author cited several eco-social authors, she did not actually justify this assertion with any specific reference.

In contrast to the suggestion that eco-social authors think mono-causally, *all* the authors discussed here outline pathways that are far more complex (Homer-Dixon and Blitt, 1998; Kahl, 2006; United Nations Environment Programme (UNEP), 2007; Zhang *et al.*, 2007, 2011; Mazo, 2009; Butler and Kefford, 2018; Gleick *et al.*, 2018). For example, the UNEP report (358 pages) explicitly acknowledges 'many factors that contribute to conflict in Sudan ... have *little or no link to the environment or natural resources*' [emphasis added]. 'These include political, religious, ethnic, tribal and clan divisions, economic factors, land tenure deficiencies and historical feuds. In addition, where environment and natural resource management issues are important, they are generally contributing factors only, not the sole cause for tension.'

While eco-social authors do argue that climate change will contribute to resource scarcity, they also identify, at least to a degree, how this will in turn exacerbate and interact with social, economic, historical and political factors. Contrary to the implication alleged as prevalent by the sceptically entitled paper 'One effect to rule them all' (Buhaug *et al.*, 2014), these scholars are careful to note that characterization of any single factor as the conflict 'trigger' is problematic.

Instead, eco-social researchers argue that a mosaic of consequences of climate change increases conflict risk. They point out that, in many parts of the global South, climate change will interact with comparatively high rates of population growth, further aggravating the scarcity of renewable and non-renewable resources, especially food, water and arable land. Foreseeable consequences include economic stress, trade imbalances, deepening inequality and legions of under- and unemployed youth. Such factors are likely to reduce tax revenue, weaken governance and help generate terrorism and extremism; affecting, for example, states such as Yemen, Sudan, South Sudan and Ethiopia, each of which is currently experiencing civil war. All these factors can reinforce each other, further increasing the risk of conflict. Autocrats may arise who are adept at creating and exploiting any perceived grievances amplified not only by climate change but, in some cases, also by actions intended to slow its pace.

Much military thinking is similarly nuanced, conceiving of anthropogenic climate change as a 'risk amplifier' (Schubert *et al.*, 2007) or a 'threat multiplier'[2] (Sullivan *et al.*, 2007), interacting with existing challenges to peace, making them harder to solve. Although this concept has been largely ignored by social analysts, an exception is Abrahams and Carr (2017), who criticized risk amplification as 'unidirectional' and were silent about the temporal and spatial scales by which climate change may interact with conflict. These authors also explained that climate change may provoke opportunities for peacebuilding, and that adaptation strategies, if implemented in ways that ignore equity, may worsen conflict. This is valid, though not exclusive to the group we have characterized as stressing 'social' causes for conflict. Indeed, the observation that responses to climate change, in an effort to adapt, can worsen conflict is entirely consistent with the conception of climate change as a threat multiplier.

Irrespective of theoretical debates, a significant fraction of the world's militaries are engaged with planning for the contingency of growing resource scarcity aggravated by climate change and other factors (Schwartz and Randall, 2003; Sullivan *et al.*, 2007; Paskal, 2010). Countering these trends may require peacekeeping missions by the United Nations and by relatively stable countries, while other countries may see more value in bolstering their own local defences and capacity. Either way, the task load of militaries is expected to increase (Sullivan *et al.*, 2007).

In another explicit challenge to the purpose of this chapter, some members of the social explanatory group writers have warned of 'securitization', arguing that even to discuss the possibility of a link between climate change and conflict could generate a 'self-fulfilling prophecy' (Gleditsch, 2012). However, ignoring the potential for conflict or refusing to consider it as a possible outcome is naïve; if academics and the general public fail to engage in this debate, then the vacuum of ideas will be filled solely by the military.

Consideration and thus apprehension of the tertiary dimensions of climate change, including its influence on conflict, may motivate community support for climate change mitigation, perhaps even playing a role akin to that of the peace movement during the Cold War. Lastly, understanding such mechanisms could inform the design of treaties and institutions that can play a role in mitigating the potential for conflict.

24.2 Mechanisms and Conflict Thresholds

The 'threat multiplier' conceptualization of climate change can be likened to that of an epileptic seizure threshold. The brain is a highly complex system; precise prediction of seizures is difficult but many factors (e.g. alcohol and lack of sleep) lower the threshold at which a seizure may occur. Conversely, reversal of these factors (and some anti-epileptic drugs) can increase the threshold and make a seizure less likely.

Similarly, the complex, interdependent nature of conflict makes precise temporal predictions difficult. However, we can say that many of the effects of climate change will lower conflict thresholds, making violence more likely, but in ways influenced greatly by regional factors.

Mechanisms by which climate change is likely to harm human security include interruptions in access to water, food, land and infrastructure, such as from disasters including flooding. Greater human insecurity does not automatically mean increased conflict; security or its lack is embedded within a socio-political, historical and economic context that also influences the likelihood of conflict. In some cases, people choose to cooperate rather than fight over limited resources. Others may lack the means to engage in conflict, and simply accept reduced access to limited resources. But where climate change weakens the integrity of social systems, conflict-promoting synergies may occur.

States and institutions have a major influence on the conflict threshold, due to their role as a buffer against reduction in human security and providing non-violent forms of redress. A history of cooperation is important, because the political knowledge and capital of previous episodes can be applied to new challenges.

24.3 Climate Change and Conflict in History

Although the scale, speed, direction and duration of modern global climate change have no precise historical analogue, previous periods of climate change are instructive (McMichael, 2012). Large-scale conflict in the archaeological record is often associated with natural changes in climate and resultant resource scarcity, particularly of food and water.

For instance, in south-west USA, no large-scale conflict is evident in the archaeological record for millennia after humans first inhabited the area. Until about 1150 CE, this region experienced climatic conditions favourable to agriculture, allowing population expansion. But then, a series of droughts lowered agricultural productivity and led to hardship, as demonstrated by evidence in the skeletal remains of adults and by increased child and infant mortality. The first evidence of intergroup conflict emerges at this time, with people building defensible dwellings, including the famous cliff houses, and settling in larger groups for protection. Houses and villages were burned and destroyed. There was an increase in the number of skeletons with signs of violent deaths, including embedded arrows, cut marks and missing skulls. Substantial depopulation followed, lowering resource pressure and reducing violence (Haas, 2009). Environmental scarcity has contributed to conflict in many other areas, including the 'Wild Nile', during the early Holocene (Close, 1996).

These examples are corroborated by some statistical studies. Zhang *et al.* (2007, 2011) reviewed evidence that natural climatic fluctuations decreased agricultural productivity in several continents, thus influencing economic downturn, war, famine and population reduction, including the European 17th century 'general crisis'. Their analysis also supported the proposition that state-level organization and technological advancement cannot always eliminate the risk of conflict posed by climate change, demonstrating a perennial human dependence upon a stable climate.

24.4 Modern Climate Change and Conflict

Arctic

The far north is a 'hotspot' of anthropogenic climate change. Rapidly warming temperatures and melting ice are increasing access to important resources, including fossil fuels and minerals. Some of these resources are under the sea, in areas contested by major powers such as Russia, the USA, and even China (Mazo, 2009; Young, 2021). Until recently, open conflict in the Arctic appeared less likely than nearer the equator; this calculus has altered given the Russian invasion of Ukraine in 2022.

Food-related conflict

While climate change will enhance access to non-renewable resources in some high-latitude countries and enhance their agricultural productivity in the medium term, per capita food resources of the planet are likely to decrease further. Anthropogenic climate change may already have contributed to higher food prices, and – as it intensifies – this is likely to happen again. Over the longer term, unless rapidly checked by emergency decarbonization, sea level rise (sometimes coupled with land subsidence) will likely displace millions of people and harm, or render useless, some of the most fertile agricultural land, such as in Bangladesh and Vietnam (Sullivan *et al.*, 2007; Boretti, 2020).

Countries that employ a high proportion of their population directly in agriculture have fewer resources with which to adapt and fewer alternative industrial and post-industrial sources of income than developed countries. Countries rich in natural resources (the so-called resource curse) are sometimes poorly governed and often home to significant ethnic, religious and linguistic diversity, which can interact easily with these other factors, intensifying social segmentation, opposition and tension (Kaplan, 1994).

Case Study 1: Darfur, Sudan

The recent history of Sudan and South Sudan is characterized by extensive conflict. Small-scale tribal violence, mostly provoked by competition for resources such as land or cattle, has been documented for decades in Darfur, western Sudan. However, early this century, this smouldering conflict erupted as a ferocious, 'ambiguously genocidal' series of atrocities, with an estimated 200,000–500,000 deaths in a population of 6 million (de Waal, 2005; Lippman, 2007; UNEP, 2007). In this example, elements discussed above are at play; environmental stressors interact with social, political and economic factors to lower the conflict threshold and to trigger the horror of war.

The UNEP report attributed the resultant brutality, in part, to an estimated 50–200 km southerly shift of the Sahara Desert, occurring gradually since the 1930s. Although the concept and science of desertification is complex[3] (Sterk and Stoorvogel, 2020) even critics of Sahelian drought 'narratives' agree that a marked reduction in rainfall occurred in the Sahel in the 1970s and 1980s (Gangneron *et al.*, 2022).

The decline in rain in the eastern Sahel (including Darfur)[4] has been suggested as in part due to anthropogenic factors for which the global North is mainly responsible, including particulate air pollution over Europe (Lelieveld *et al.*, 2002) and rising ocean temperatures (Giannini and Kaplan, 2019). Heat stress (also related to greenhouse gas accumulation) has been linked to an impaired recovery of Sahelian ecosystems, following drought (Liu and Xue, 2020).

De Waal (2005, 2007), UNEP (2007) and Mazo (2009) (among numerous others) identified many non-climatic factors in the Darfur conflict (Fig. 24.1), including continuing high population growth, worsened by an influx of tens of thousands of refugees. Darfur constituted a complex 'ethnic mosaic', rather than a 'binary polarized "Arab"–"African" identity divide' (de Waal, 2005). Other factors also included neglect of Darfur by the central government. De Waal (2007) describes how contact of 'failed nomads' with Chadian militia, 'for whom pillage had become a way of life since the mid-1970s' helped create the Janjawiid[5] phenomenon. Also contributing were an influx of automatic weapons and corruption. Lippman (2007) pointed out that the conflict occurred in the context of a small portion of the Sudanese population benefiting from oil income. Militias destroyed villages and forests. Both the drought and conflict eroded the natural resource base, creating a positive (reinforcing) feedback loop. In an effort to retain disproportionate power in an ethnically divided country, the Sudanese government used oil revenues to pay militias and provided them with modern weaponry and communications technology.

Ban Ki-Moon (2007), when UN Secretary General, identified climate change as an underlying factor for the Darfur conflict. De Waal also conceded that climate change, too, was a plausible contributor to the Darfur conflict, though clearly not the whole explanation (Box 24.1).

We find this anecdote compelling, especially as De Waal is generally known to stress political (social) determinants of conflict, rather than ecological.[6] Our argument can be expressed very simply. A mosaic of pre-existing tensions in Darfur was inflamed by a reduction in food and other natural resources following a reduction in rain that was itself in part due to environmental factors arising from the global North. These physical environmental determinants acted as a risk amplifier leading to the ambiguous genocide. The tragedy of Darfur could have occurred with no known anthropogenic climatic element

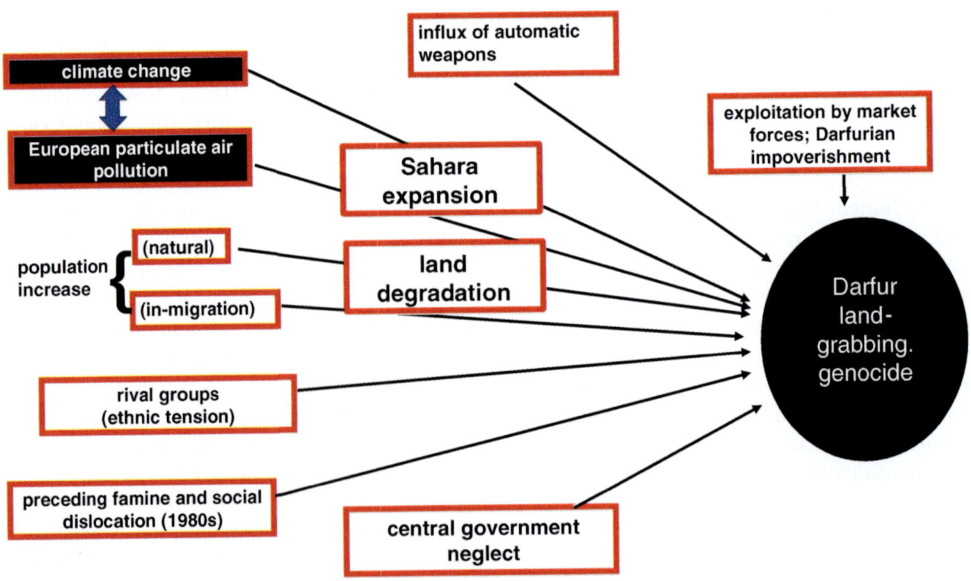

Fig. 24.1. There are many causal co-factors for the genocidal conflict in Darfur in the 2000s; most are shown in this figure. Environmental factors, including climate change, is a highly plausible co-factor.

Box 24.1.

Faris (2007) described how de Waal told him a story that, he said, 'keeps coming back to him'.

While travelling in Darfur in the mid-1980s de Waal met a bedridden, almost blind Arab sheikh named Hilal Abdalla, 'who said he was noticing things he had never seen before. Sand blew into fertile land, and the rare rain washed away alluvial soil. Farmers who had once hosted his tribe and his camels were now blocking their migration; the land could no longer support both herder and farmer. Many tribesmen had lost their stock and scratched at millet farming on marginal plots. The God-given order was broken.'

The sheikh said that he feared for the future. De Waal recalled: 'The way the world was set up since time immemorial was being disturbed, and it was bewildering, depressing. And the consequences were terrible.' In 2003 a 1.93 m Arab with a commanding presence was listed by the US State Department as a suspected war criminal. De Waal recognized him as Musa Hilal, son of the almost blind sheikh he had met decades before.

(as the case with the Rwandan genocide); this does not invalidate the risk amplification argument,[7] nor does the time lag.[8]

The Darfur war was brought to an uneasy end in 2007, with a peace-keeping mission brokered by the UN Security Council, involving UN and African Union forces. The harm it caused to health was immense. Over 2 million displaced civilians were forced into camps for safety and access to basic resources. The Sudanese government slowed the flow of food and supplies, and Sudan became an extremely dangerous location for humanitarian workers (Lippman, 2007). This conflict also crossed borders; at times the fighting extended into Chad and the Central African Republic (UNEP, 2007).

In 2022, a team concluded that the combined impact of warming and oceanic oscillations was likely to worsen future droughts in Sudan and the whole Nile River Basin (Mahmoud *et al.*, 2022). We cannot be complacent about the long-term prospects for peace in this region. Darfur is, again, a place of violence, as is the whole country.

Case Study 2: Syria

In 2010, the substantial rise in the price of wheat, itself triggered by a severe heatwave in Russia and Ukraine, contributed to the 'Arab Spring' (Johnstone and Mazo, 2011). In turn, this combination of unrest and hope helped initiate the Syrian civil war, a catastrophic event with global repercussions that persist. As with Darfur, multiple social factors contributed, including economic liberalization (neoliberalism), inequality and youth unemployment (Ide, 2018) (Fig. 24.2). But so, too, did a very severe drought (Kelley *et al.*, 2015).

A vigorous debate in the literature subsequently occurred concerning the extent to which climate change was a co-factor for the Syrian conflict. Authors in group B claimed that the drought could have occurred naturally, but neglected the potential link between the Syrian conflict, the Arab spring and the 2010 heatwave (e.g. Ide, 2018). These authors also either overlooked or dismissed the conflict threshold conceptual framework used in this chapter, instead accusing less conservative analysts of oversimplification. For a counter-argument, see Gleick (2017) and Butler and Kefford (2018).

Other locations

Other locations suggested as at risk from increasing climate change-associated conflict include the Levant, Somalia and Afghanistan (Schleussner *et al.*, 2016) and the districts around Lake Chad (north-eastern Nigeria and three adjoining nations) (Ehiane and Moyo, 2022) and Lake Turkana (northern Kenya). In each of these settings ethnic tensions coexist with conflict. Investigators in Kenya also identified an influx of high-powered weapons as a factor. Lake Turkana was once known as a free garden (Waila *et al.*, 2018). Population growth is a co-factor for conflict in all of these regions, including the two case studies.

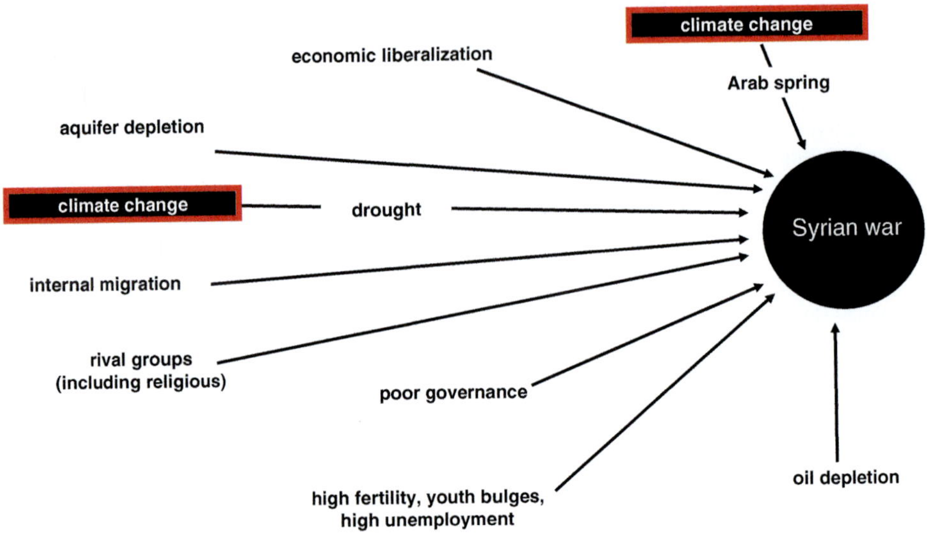

Fig. 24.2. There are many causal co-factors for the civil war in Syria; major ones are shown in this figure. Neoliberalism contributed to aquifer depletion, as did the drought and population increase. Climate change is a likely contributor via two pathways.

24.5 Emergent properties of conflict and health induced by climate change

We identify three categories of effects from conflict induced (in part) by climate change (Table 24.1). The direct effects of conflict on health, such as from trauma, are obvious. The indirect health effects of individual conflicts (e.g. resurgent infectious diseases) are also well described (Connolly *et al.*, 2004). However, in addition, a third category of health effect is evident if conflict becomes sufficiently widespread. This is largely undescribed.

A world with a higher level of conflict than today is likely to see increased military spending, with the risk of new arms races, even in countries not directly involved in conflict. Since governmental and societal resource allocation is only partially discretionary, increased defence expenditure necessitates an opportunity cost, including lower spending on health care and other health determinants such as education. A parallel shift from humanitarian to military foreign aid is also likely (Klare, 2008).

Increased xenophobia and other forms of intolerance are likely in situations of increased scarcity (Keeley, 1996; Gelfand *et al.*, 2011). This could intensify protectionism, decrease trade and deepen global economic inefficiencies and inequality. Equatorial countries, poor in human and financial resources, appear the most vulnerable.

Conflict induced, in part, by climate change is itself likely to breed additional violence. Modern conflicts in Africa readily cross state boundaries. Sometimes, apparently minor wars can entangle external powers, as occurred during World War I. Conflict is more likely to occur when countries feel insecure or rivals increase military spending. Conflict can also give rise to militaristic governments more likely to use force than diplomacy when confronted with foreign and domestic problems. Conflict erodes the norms and institutions that promote peace. Climate change can also worsen the impacts of existing conflict in unpredictable ways, such as by excessive heat or a flood further harming conditions in a refugee camp.

Minimizing negative health impacts

If we understand climate change as a conflict threat amplifier, which can lower conflict thresholds, this may help decision makers design policies and programmes to minimize the potential impact on human health. Historical studies demonstrate that interruptions to food supply correlate strongly with conflict, so enhancing

Table 24.1. Effects on health of conflict induced by climate change (first two sections adapted from Connolly *et al.*, 2004).

Direct effects	Battle deaths and injuries of combatants
	Civilian deaths and injuries from weapons of war, potentially including weapons of mass destruction
	Civilian deaths, injuries and undernutrition due to targeting of civilian food supplies and infrastructure
Indirect effects of single or repeated conflict	Reduced monitoring for new and re-emerging disease
	Reduced vector control programmes (e.g. resurgence of African trypanosomiasis)
	Increases in crowded, unsanitary conditions in refugee camps, leading to increased possibility for disease transmission
	Reduced agricultural productivity and economic output, leading to impaired nutritional status
	Increased rates of sexual violence, leading to psychological trauma and spread of sexually transmitted illnesses
	Flight or killing of medical personnel
	Shift to deliberate planting of less nutritious root crops, for example cassava, to protect from theft or destruction
Indirect effects of generalized increase in conflict due to climate change	Shift in domestic spending from public health measures to defence
	Decreased humanitarian aid, with potential for concomitant increase in military aid
	Increased xenophobia and protectionism – especially of food and other critical resources
	Reduced economic stability in conflict areas, with global effects
	Increasing regional and global instability, leading to cascading conflict

resilience in the agricultural sector is important. Strong states and institutions are also vital for promoting peaceful outcomes. Finally, if we can improve understanding and prediction of conflict, we can be better prepared, for example with emergency relief or pre-emptive diplomacy.

The challenges are grave. One concern is that the rate and scale of climate change will overwhelm our institutions, undermining many of our assumptions and previous studies regarding the relationship between environmental scarcity and conflict. The Intergovernmental Panel on Climate Change (IPCC) is increasingly perceived as conservative concerning temperature, rainfall change, polar ice melt and sea level rise. The IPCC also tends to understate the risk of ecological tipping points and, even more, of social tipping points, such as the decision or need to migrate.

24.6 Conclusion

Conflict driven by climate change is a major health risk, yet still underappreciated, including in the health literature. This is despite the appeal by some of the world's militaries for civil society to engage in ways to promote peace, in the context of increasing risk (Bowles *et al.*, 2015). However, we have stressed that climate change does not, in and of itself, cause conflict.

Adaptation is likely to be only partially successful for every situation, including preventing conflict. Many actors may react to increased conflict by increasing military spending, further withdrawing resources from public health. The already locked-in climate change appears sufficient to increase the likelihood of conflict. Without urgent mitigation on a massive scale (which currently appears unlikely) the average global temperature may rise far beyond 2°C above the pre-industrial level, accompanied by intensified heatwaves, droughts, floods and other factors that lower agricultural productivity. As per capita resources decline beyond critical thresholds, conflict in some regions will be virtually inevitable.

Improved mechanisms for maintaining peace and resolving conflict are vital. Adaptation is essential, but even more so is strong and urgent action to mitigate climate change. Diplomacy, reduced international inequity and genuine global solidarity in solving the climate crisis also need to be nurtured to lower conflict thresholds.

Notes

[1] There is also a large literature on the association between heat – made worse by climate change – and interpersonal violence; e.g. see Levy *et al.* (2017).
[2] To us these terms are synonymous. Of interest, they each became prominent in 2007, the same year as the UN Security Council first focused explicitly on climate change as a thematic topic (Anonymous, 2021). The language used at the UN seems to prefer threat or risk 'multiplier'.
[3] For example, satellite-derived greening trends may occur as grassland is converted to scrubland, something interpreted as impoverishment by local people.
[4] Rainfall records in Darfur are especially scarce for such a vast region, and may also be precarious. For example, one of only three long-term rain records is in Al Geneina (West Darfur's administrative capital), a city normally of half a million people reported as '70% emptied' in 2023, as civil war again terrifies many Sudanese (Elbagir, 2023).
[5] Arab militia, often on horseback. Although the etymology is unclear, one suggestion is that the name is derived from the Arabic *jinnī* (spirit) and *jawad* (horse) (https://www.britannica.com/topic/Janjaweed).
[6] De Waal (2018) has even used the term 'Malthusian zombie' to deprecate explanations for famine that are partly ecological. We note, however, that Darwin and Wallace (1858) likened evolutionary principles to 'the doctrine of Malthus'.
[7] Trenberth (2011) and Hansen *et al.* (2012) argued that *all* extreme weather events are influenced by climate change (including the 2010 heatwave in Russia and Ukraine). Their proposal can be extended to conflict.
[8] At the time the Darfur conflict fully emerged, the drought was probably improving; it is possible (though unclear) that this translated into better conditions for pastoralists. However, complex tertiary phenomena are unlikely to emerge simultaneously or instantly; for example, people and communities may harbour grievances for many years, even across multiple generations. The expression of such grievances may be delayed until it becomes 'rational', i.e. when the holder of the grudge calculates that the chance of 'successful payback' is maximized.

References

Abrahams, D. and Carr, E.R. (2017) Understanding the connections between climate change and conflict: Contributions from geography and political ecology. *Current Climate Change Reports* 3, 233–242. DOI: 10.1007/s40641-017-0080-z

Anonymous (2021) *The UN Security Council and Climate Change*. Security Council Report. Available at: https://www.securitycouncilreport.org/atf/cf/%7B65BFCF9B-6D27-4E9C-8CD3-CF6E4F-F96FF9%7D/climate_security_2021.pdf (accessed 28 August, 2023).

Boretti, A. (2020) Implications on food production of the changing water cycle in the Vietnamese Mekong Delta. *Global Ecology and Conservation* 22, e00989. DOI: 10.1016/j.gecco.2020.e00989

Bowles, D.C., Butler, C.D. and Morisetti, N. (2015) Climate change, conflict, and health. *Journal of the Royal Society of Medicine* 108, 390–395. DOI: 10.1177/0141076815603234

Brown, I.A. (2010) Assessing eco-scarcity as a cause of the outbreak of conflict in Darfur: A remote sensing approach. *International Journal of Remote Sensing* 31, 2513–2520. DOI: 10.1080/01431161003674592

Buhaug, H., Nordkvelle, J., Bernauer, T., Böhmelt, T., Brzoska, M. *et al.* (2014) One effect to rule them all? A comment on climate and conflict. *Climatic Change* 127, 391–397. DOI: 10.1007/s10584-014-1266-1

Butler, C.D. and Kefford, B.J. (2018) Climate and conflict: magnifying risks. *Nature* 555, 587. DOI: 10.1038/d41586-018-03795-0

Butler, D. (2007) Darfur's climate roots challenged. *Nature* 447, 1038. DOI: 10.1038/4471038b

Close, A.E. (1996) Plus ça change. The Pleistocene–Holocene transition in Northeast Africa. In: Strauss, L.G., Eriksen, B.V., Erlandson, J.M. and Yesner, D.R. (eds) *Humans at the End of the Ice Age*. Plenum Press, New York and London, pp. 43–60.

Connolly, M.A., Gayer, M., Ryan, M.J., Salama, P., Spiegel, P. *et al.* (2004) Communicable diseases in complex emergencies: impact and challenges. *The Lancet* 364, 1974–1983. DOI: 0.1016/S0140-6736(04)17481-3

Darwin, C. and Wallace, A. (1858) On the tendency of species to form varieties; and on the perpetuation of varieties and species by natural means of selection. *Journal of the Proceedings of the Linnean Society of London. Zoology* 3, 45–62. DOI: 10.1111/j.1096-3642.1858.tb02500.x

Dawkins, R., Krebs, J.R., Maynard Smith, J. and Holliday, R. (1979) Arms races between and within species. *Proceedings of the Royal Society of London. Series B. Biological Sciences* 205, 489–511. DOI: 10.1098/rspb.1979.0081

de Waal, A. (2005) Deep down in Darfur: Nothing is as we are told in Sudan's killing fields. (Book review of "Darfur: The Ambiguous Genocide" by Gerard Prunier). *Review of African Political Economy* 106, 653–658. DOI: 10.1080/03056240500467146

de Waal, A. (2007) Is climate change the culprit for Darfur? *African Arguments*, Royal African Society, London. Available at: https://africanarguments.org/2007/06/is-climate-change-the-culprit-for-darfur/ (accessed 12 August 2022).

de Waal, A. (2018) The end of famine? Prospects for the elimination of mass starvation by political action. *Political Geography* 62, 184–195. DOI: 10.1016/j.polgeo.2017.09.004

Ehiane, S. and Moyo, P. (2022) Climate change, human insecurity and conflict dynamics in the Lake Chad region. *Journal of Asian and African Studies* 57, 1677–1689. DOI: 10.1177/00219096211063817

Elbagir, Y. (2023) Sudan: Mass graves in West Darfur's capital haunt remaining residents. *Sky News*. Available at: https://news.sky.com/story/i-buried-my-sister-54-days-after-she-was-killed-mass-graves-in-west-darfurs-capital-haunt-remaining-residents-12941508 (accessed 19 August 2023).

Faris, S. (2007) The real roots of Darfur. *The Atlantic Monthly*. Available at: http://www.theatlantic.com/doc/200704/darfur-climate (accessed 17 September 2022).

Gangneron, F., Pierre, C., Robert, E., Kergoat, L., Grippa, M. *et al.* (2022) Persistence and success of the Sahel desertification narrative. *Regional Environmental Change* 22, 118. DOI: 10.1007/s10113-022-01969-1

Gelfand, M.J., Raver, J.L., Nishii, L., Leslie, L.M., Lun, J. *et al.* (2011) Differences between tight and loose cultures: a 33-nation study. *Science* 332, 1100–1104. DOI: 0.1126/science.1197754

Giannini, A. and Kaplan, A. (2019) The role of aerosols and greenhouse gases in Sahel drought and recovery. *Climatic Change* 152, 449–466. DOI: 10.1007/s10584-018-2341-9

Gleditsch, N.P. (2012) Whither the weather? Climate change and conflict. *Journal of Peace Studies* 49, 3–9. DOI: 10.1177/0022343311431288

Gleick, P.H. (1991) Environment and security: the clear connections. *Bulletin of the Atomic Scientists* 47, 16–22. DOI: 10.1080/00963402.1991.11459956

Gleick, P.H. (2017) Climate, water, and conflict: Commentary on Selby *et al.* 2017. *Political Geography* 60, 248–250. DOI: 10.1016/j.polgeo.2017.06.009

Gleick, P.H., Lewandowsky, S. and Kelley, C. (2018) Critique of conflict and climate analysis is oversimplified. *Nature* 555, 587. DOI: 10.1038/d41586-018-03794-1

Haas, J. (2009) The origins of war and ethnic violence. In: Carman, J. and Harding, A. (eds) *Ancient Warfare: Archaeological Perspectives*. The History Press, Cheltenham, UK, pp. 11–24.

Hansen, J., Sato, M. and Ruedy, R. (2012) Perception of climate change. *Proceedings of the National Academy of Sciences (USA)* 109, 14726–14727. DOI: 10.1073/pnas.1205276109

Homer-Dixon, T. and Blitt, J. (1998) Introduction: a theoretical overview. In: Homer-Dixon, T. and Blitt, J. (eds) *Ecoviolence: Links Among Environment, Population, and Security*. Rowman and Littlefield Publishers, Inc., Lanham, Maryland, pp. 1–18.

Ide, T. (2018) Climate war in the Middle East? Drought, the Syrian civil war and the state of climate-conflict research. *Current Climate Change Reports* 4, 347–354. DOI: 10.1007/s40641-018-0115-0

Johnstone, S. and Mazo, J. (2011) Global warming and the Arab Spring. *Survival* 53, 11–17. DOI: 10.1080/00396338.2011.571006

Kahl, C.H. (2006) *States, Scarcity and Civil Strife in the Developing World*. Princeton University Press, Princeton, New Jersey.

Kaplan, R.D. (1994) The coming anarchy. *Atlantic Monthly* 273(2), 44–76. Available at: https://www.theatlantic.com/magazine/archive/1994/02/the-coming-anarchy/304670/ (accessed 6 July 2022).

Keeley, L.H. (1996) *War Before Civilization*. Oxford University Press, New York.

Kelley, C.P., Mohtadi, S., Cane, M.A., Seager, R. and Kushnir, Y. (2015) Climate change in the Fertile Crescent and implications of the recent Syrian drought. *Proceedings of the National Academy of Sciences (USA)* 112, 3241–3246. DOI: 10.1073/pnas.1421533112

Ki-Moon, B. (2007) A climate culprit In Darfur. United Nations Secretary-General (via *The Washington Post*). Available at: https://www.un.org/sg/en/content/sg/articles/2007-06-16/climate-culprit-darfur (accessed 20 July 2023).

Klare, M.T. (2008) *Rising Powers, Shrinking Planet: The New Geopolitics of Energy*. Henry Holt and Co., New York.

Koubi, V. (2019) Climate change and conflict. *Annual Review of Political Science* 22, 343–360. DOI: 10.1146/annurev-polisci-050317-070830

Lelieveld, J., Berresheim, H., Borrmann, S., Crutzen, P.J., Dentener, F.J. *et al.* (2002) Global air pollution crossroads over the Mediterranean. *Science* 298, 789–791. DOI: 10.1126/science.1075457

Levy, B.S., Sidel, V.W. and Patz, J.A. (2017) Climate change and collective violence. *Annual Review of Public Health* 38, 241–257. DOI: 10.1146/annurev-publhealth-031816-044232

Lippman, M. (2007) Darfur: the politics of genocide denial syndrome. *Journal of Genocide Research* 9, 193–213. DOI: 10.1080/14623520701368594

Liu, Y. and Xue, Y. (2020) Expansion of the Sahara Desert and shrinking of frozen land of the Arctic. *Scientific Reports* 10, 4109. DOI: 10.1038/s41598-020-61085-0

Mahmoud, S.H., Gan, T.Y., Allan, R.P., Li, J. and Funk, C. (2022) Worsening drought of Nile basin under shift in atmospheric circulation, stronger ENSO and Indian Ocean dipole. *Scientific Reports* 22, 8049. DOI: 10.1038/s41598-022-12008-8

Mazo, J. (2009) Darfur: The first modern climate-change conflict. *The Adelphi Papers* 49, 73–86. DOI: 10.1080/19445571003755538

McDonald, M. (2013) Discourses of climate security. *Political Geography* 33, 42–51. DOI: 10.1016/j.polgeo.2013.01.002

McMichael, A.J. (2012) Insights from past millennia into climatic impacts on human health and survival. *Proceedings of the National Academy of Sciences (USA)* 109, 4730–4737. DOI: 10.1073/pnas.1120177109

Paskal, C. (2010) *Global Warring: How Environmental, Economic and Political Crises Will Redraw the World Map*. Palgrave Macmillan, New York.

Schleussner, C.-F., Donges, J.F., Donner, R.V. and Schellnhuber, H.J. (2016) Armed-conflict risks enhanced by climate-related disasters in ethnically fractionalized countries. *Proceedings of the National Academy of Sciences (USA)* 113, 9216–9221. DOI: 10.1073/pnas.1601611113

Schubert, R., Schellnhuber, H.J., Buchmann, N., Epiney, A., Grießhammer, R. *et al.* (2007) *World in Transition: Climate Change as a Security Risk. Summary for Policy-Makers*. Earthscan, London.

Schwartz, P. and Randall, D. (2003) *An Abrupt Climate Change Scenario and Its Implications for United States National Policy*. Jet Propulsion Laboratory, Pasadena, California.

Sterk, G. and Stoorvogel, J.J. (2020) Desertification–scientific versus political realities. *Land* 9, 156. DOI: 10.3390/land9050156

Sullivan, G.R., Bowman, F., Ferrell, L.P.J., Gaffney, P.G.I., Kern, P.J. *et al.* (2007) *National Security and the Threat of Climate Change*. The Centre for Naval Analyses (CAN) Corporation, Alexandria, Virginia. Available at: https://www.cna.org/archive/CNA_Files/pdf/national%20security%20and%20the%20threat%20of%20climate%20change.pdf (accessed 15 February 2024).

Trenberth, K.E. (2011) Attribution of climate variations and trends to human influences and natural variability. *WIREs Climate Change* 2, 925–930. DOI: 10.1002/wcc.142

UNEP (2007) *Sudan: Post-conflict Environmental Assessment*. United Nations Environment Programme, Nairobi, Kenya. Available at: https://www.unep.org/explore-topics/disasters-conflicts/where-we-work/sudan/sudan-post-conflict-environmental-assessment (accessed 6 July 2022).

Waila, J.M., Mahero, M.W., Namusisi, S., Hoffman, S.J. and Robertson, C. (2018) Outcomes of climate change in a marginalized population: An ethnography on the Turkana pastoralists in Kenya. *American Journal of Public Health* 108, S70–S71. DOI: 10.2105/AJPH.2017.304063

Young, O.R. (2021) Arctic futures – future Arctics? *Sustainability* 13, 16. DOI: 10.3390/su13169420

Zhang, D.D., Brecke, P., Lee, H.F., He, Y.-Q. and Zhang, J. (2007) Global climate change, war, and population decline in recent human history. *Proceedings of the National Academy of Sciences (USA)* 104, 19214–19219. DOI: 10.1073/pnas.0703073104

Zhang, D.D., Lee, H.F., Wang, C., Li, B., Pei, Q., Zhang, J. *et al.* (2011) The causality analysis of climate change and large-scale human crisis. *Proceedings of the National Academy of Sciences (USA)* 108, 17296–17301. DOI: 10.1073/pnas.1104268108

25 Collapse: The Climate Endgame

Peter Stoett[1] and Rob White[2]
[1]*Faculty of Social Science and Humanities, Ontario Tech University, Oshawa, Ontario, Canada;* [2]*School of Social Sciences, University of Tasmania, Hobart, Tasmania, Australia*

Abstract

This chapter discusses the increasingly evident toll of the climate crisis in terms of the prospects of civilizational collapse. It examines the rapidity of change, collapse scenarios currently permeating the literature and popular culture, and pathways to avoid them. A new interdisciplinary discipline investigating the origins, process and prospects of climate change-induced human systems collapse is emerging. We cannot understand the links explored between climate and human health in the present volume without a conceptual understanding of the threat of socio-ecological system collapse which is constantly lurking in the background of climate science, advocacy, diplomacy, art and other forms of human action and expression. Collapse is not precise; it is more helpful to think in terms of degrees of collapse based on human perceptions, in addition to the very real ecosystem collapse that is emerging. The chapter also discusses some political currents impacted by, and impacting, collapse discourses.

25.1 Introduction

The present climate emergency from other natural and human historical phenomena is distinguished by the obsolescence of traditional categorizations of scale (temporal and geographical), effect (direct and indirect) and consequence (a biodiversity meltdown and potential human-extinction event). The threat of collapse emerges from popular literature, film and theatre; but it is also a permanent feature of climate science (and, more broadly, the science of global environmental change) and is an increasingly visible fixture of climate diplomacy. It lurks in the background of all things climate-related today.

This trend – stemming from the rapidity of biophysical change – also challenges typical thinking about stages and development. The sequence of change is becoming increasingly difficult to timeline: prediction and realization are becoming simultaneous; the transition from fear and an initial experience of social dislocation and conflict due to climate change (e.g. climate-induced migration) at one end runs straight towards the demise of civilization (e.g. radical diminishment of stable social formation, or even its possibility) at the other. These profound transformations are occurring at different rates in different places, but the central conceptual difficulty is that they are occurring now. We need only witness the methane being released with rapidly thawing permafrost in the northern regions; the recent firestorms in Siberia, Canada, the Greek Islands, Australia, California and elsewhere; severe and extended droughts across Africa and mid-western North America; the diminishing area available to Micronesian islanders, where culture is undergoing an existential

 Climate Change and Global Health: Primary, Secondary and Tertiary Effects, 2nd Edition (eds C.D. Butler and K. Higgs)
DOI: 10.1079/9781800620025.0025

threat; and many other physical manifestations of climate change with profound human impacts (Hofmann, 2014; WMO, 2020a, 2020b; IPCC, 2021).

We are thus rapidly moving from the conditions of 'slow crisis' (in which observed change is exponential, though broadly predictable) to 'fast crisis' (in which the speed of change is accelerating beyond our limited capacity to both predict and respond). This leads inexorably to visions of ecological, societal and even civilizational and species collapse that could follow from such a profound disruption to the natural equilibrium that has characterized the Holocene. (There have been civilizational collapses earlier in the Holocene, of course, but none at a global scale.) This chapter discusses the rapidity of the change we are confronting, several scenarios of collapse that are currently permeating the literature and popular culture, and pathways to avoid them. We are not, by any means, attaching the stark label of inevitability to any vision of collapse; however, social and cultural change is certainly inevitable if we are to avoid any of them. We note also that, sadly enough, a new academic discipline has been emerging in recent decades: an interdisciplinary discipline investigating the origins, process and prospects of climate change-induced human systems collapse (see Butler, 2018; Richards *et al.*, 2021), noted in a recent publication as a 'dangerously unexplored topic' (Kemp *et al.*, 2022). This chapter is another initial contribution to this literature but, more directly, should be seen as an expression of deep concern regarding the most pernicious of links between climate and human health explored in the present volume.

25.2 The Rapidity of Change

The alarming fact about climate change is not that climatic conditions change over time but the rate at which the changes occur. As demonstrated in scientific literature and Intergovernmental Panel on Climate Change (IPCC) documentation (Mann, 2021; IPCC, 2023), the increased pace of global warming began in earnest with the advent of the capitalist industrial revolution in Europe and has then accelerated more quickly in the past 50 years. The result has been climate disruption across the global ecology, as indicated in substantive changes in temperature, greenhouse gases, ocean heat content, sea levels, the cryosphere, sea ice and ice sheets.

The effects of global heating have already been devastating for many and are the subject of numerous health reports as well as regular updates by *The Lancet* (Romanello *et al.*, 2022). The modelling of the health consequences of climate change has become ever more sophisticated and complicated, again due to the shifts in and overlaps between direct and indirect effects, and the difficulties in rigidly assigning labels such as primary, secondary and tertiary to what are unparalleled trends and situations. Hierarchy is continually challenged in practice by simultaneity and the exacerbating confluence of diverse effects. The numerous pathways to socio-ecological collapse are less modelled (Butler, 2018). Cumulative and compounding harms affect the scope and scale of unfolding harms and they mercilessly feed into each other. The consequences of climate changes and disruptions are putting health at increasing risk in numerous ways – from disaster-related events such as fires and floods, and the proliferation of vector-borne diseases (Machalaba *et al.*, 2017), through to food insecurity and population displacement due to climate variability and extreme weather. Most visibly, novel circumstances and climate-related trends exacerbate existing health and environmental problems. For example, there is a notable link between pollution, climate change and health (WMO, 2020a):

> An estimated 4.2 million premature deaths every year globally are linked to ambient air pollution, mainly from heart disease, stroke, chronic obstructive pulmonary disease, lung cancer, and acute respiratory infections in children. New models estimate that fossil-fuel-related greenhouse gas emissions account for about 65% of the excess mortality rate attributable to air pollution. Furthermore, climate and weather strongly affect where and when air pollution occurs. Weather systems influence the movement and dispersion of air pollutants in the atmosphere through the action of winds, vertical mixing, and precipitation, all of which are likely to alter in a changing climate. Air pollution, stratospheric ozone depletion, persistent pollutants and climate change are interlinked problems.

Importantly, part of the problem in responding to such a phenomenon is the rapidity of its onset and the sheer scale of its social impact. The COVID-19 pandemic provided a perfect case in point with respect to this; but so too do the recent extensive fires in Canada, Australia, Siberia, California, Turkey and Greece, all of which have likewise contributed to air pollution and its health-related effects.

Indeed, the question of time, the rapid rate of change and where this is occurring are especially troubling from health prevention and social cohesion perspectives. Climate-related changes, such as the melting of the northern tundra and transformations in ocean currents, are far outstripping many IPCC estimates, including the intensity of heating in specific locales. How significant transnational biophysical changes are translating into regional contexts (for example, changing temperature of ocean currents along the east coast of Tasmania, Australia, affecting fish species) reinforces the global/local nature of the problem. Universal processes have both general and specific effects and consequences. These have ramifications, as well, for how and if collapse occurs.

25.3 Visions of Collapse

There are many historical narratives of societal collapse related to environmental degradation, both anthropogenically caused and those that are not caused by humans, and not all of them are mutually exclusive. Butler examines several of these in an earlier chapter in this volume (see also Fagan, 2008; Arbesman, 2011; Acemoglu and Robinson, 2012). But 'collapse' should not be taken as a definite term. For present purposes, this section outlines a possible sequence of degrees of collapse, which might be considered stages, associated with contemporary climate change.

Fear of collapse

Many individuals and collectives worldwide are at this stage. This can be described as 'climate collapse anxiety syndrome' at the community, social and individual levels. This is a social psychological phenomenon that derives from several sources. For example, arguably, most religious texts have some form of end-of-days vision; we certainly experienced genuine concern that all human life would cease during the height of the Cold War; and it is not unusual to fear death beyond one's own as we care deeply for our families, communities and even humanity as a whole. Mixed with what has been popularly labelled 'Apocalypse fatigue', fatalism can set in and people may become complacent about the inevitability of future collapse, discovering a new reason to embrace inaction or falling into serious depression. This can also affect children in deleterious ways; as A.J. McMichael (2017, p. 67) wrote: 'Children, too, are vulnerable to anxieties and stress disorders. As public awareness of climate change and its consequences grows, many children will face special stresses from long-term insecurities and anxiety about climate change ... Chronic anxiety in childhood causes baseline neuro-hormonal changes with increased blood levels of cortisol ... [possibly increasing rates of] type II diabetes' (see also Strazdins *et al.*, 2010).

In some circumstances this is a preliminary stage of collapse itself (but that partly depends upon the speed of the climate-related changes taking place). Fear of collapse can be useful if it generates spiritual or political energy to fight against the tide, but it can also be debilitating, and no doubt is open to political manipulation (Mann, 2021). One wonders whether former civilizations that collapsed due to imbalances of various types had similar communal forms of anxiety. At present a new subfield of pop psychology, including self-help books, is springing forth as we try to cope with climate collapse anxiety (Marshall, 2015; Stoknes, 2015). Some of this literature echoes earlier work on 'eco-psychology' by Theodore Roszak (1992) and others. One thing is certain: there is a mental health challenge to climate change that extends beyond the response to material stress. Moreover, Indigenous people and other victims of genocidal policies may be even more familiar with the communal psychological challenges associated with existential threats to one's physical and cultural life, and the threat of collapse associated with climate change is adding to this historic burden.

Socio-economic collapse

Socio-economic collapse is characterized by political instability, social distress, prolonged conflict (with periodic spikes) and mass migrations. Indeed, forced, large-scale migration has often been at the forefront of climate-related fears, many of them unwarranted, but many of them presently under way. Climate change (entangled with other factors, such as land use change, civil conflicts, severe disruptions in food supply) is seen to lead to mass migrations, mostly from Africa to Europe, Latin America to the USA, Pacific Islands to Australia, or low-lying Asia (such as Bangladesh) to other parts of Asia, though there is no predetermined route (Abel *et al.*, 2019). The upsurge in asylum seeking as well as illegal migration in the 2020s is more related to conflict and other immediate threats to human security (such as economic drivers), but no doubt climate change has operated as a threat multiplier.

Compounding this perceived threat to security, there is already an impact from economic collapse due to supply shortages of food and potable water and other natural resources (including rare minerals used in climate change adaptation/technological innovation). Impacts also include health system overloads in areas permeated with forest fire smoke or scarred by dehydration due to drought, local as well as global epidemics, and hurricane-struck seaside communities and those experiencing other sorts of disaster. All these could, in regional fashion, push governance capacity over the limits of effective social control and result in what would probably be best described as quarantined chaos (see, for example, Lebanon). Privatized security is replacing public security as the scramble for survival becomes a marketable commodity-driven enterprise. Whether or not the state persists in protecting private wealth to the remarkable degree it has done over the past 300 years or so is another question.

One ubiquitous chord here is that we lack the appropriate physical or socio-legal infrastructure – technological, engineered, societal, legislative – to deal with rapidly changing conditions. This vision of collapse may or may not involve conflict within and between states or smaller communities, but it certainly entails the lack of fairness and social justice in terms of access to resources and human security that rapid change reveals.

Civilizational collapse

Civilizational collapse can refer to the end of empires, or even the entire nation-state system, accompanied by radical cultural shifts. At a yet deeper level, there are concerns that society as we know it – civilization, to use a broader term – will itself collapse. This goes beyond the ability of infrastructure to handle changing conditions and would include the entire breakdown of food supply chains, potential large-scale warfare, unalterable vulnerability to extreme weather, and other threats to human security so grave that social organization becomes unimaginable; in short, a post-apocalyptic society such as that envisioned in the late Cormac McCarthy's book, *The Road* (McCarthy, 2006). Moral and ethical renewal under such circumstances might look rather Hobbesian in character (Hobbes, 1651) – life that is 'solitary, poor, nasty, brutish and short'. Disorder and anarchy best describe what is left after such a collapse. Many international relations theorists have characterized the international system in this manner, arguing that it explains the violence inherent in war-prone systems when a rules-based order breaks down. But this vision extends the premise to the local community level as well.

There is the possibility that the entire nation-state system itself could collapse as national governments splinter, regional factions engage in resource conflicts, high-tech military weaponry becomes less effective under conditions of unpredictable weather, and private wealth becomes exasperated by the limited effectiveness of international law. The result is heavily fortified enclaves of privilege in the midst of immense dislocation, suffering and death (White, 2014). Ideologies such as liberalism that are based on some form of universalizing (teleological) narrative of progress would make little sense in such a prolonged human condition.

Another scenario is that humans will revisit some form of the hunter-gatherer society after the ravages of climate change wipe out most of the population and shift us back to Pleistocenic conditions and agriculture – at least as we conceive it now (Gowdy, 2020). This scenario obviously places little faith in the trope that technology will enable societal survival, but at least suggests that *Homo sapiens* as a species could survive a dramatic climatic shift, or even evolve into a new species that has adapted to climate change.

Human extinctions: species collapse

The extinction scenario is ostensibly good for Earth but bad for us humans and whatever civilizational progress that we have made or could make (Finneron-Burns, 2017). At its most extreme, the ultimate collapse would entail the final extinction of *Homo sapiens* as a species. The survival of the Earth itself is not at stake; Earth will abide. While the Intergovernmental Science-Policy Platform on Biodiversity and Ecosystem Services (IPBES) reported that countless other species – one million or more – may also be rendered extinct as climate change proceeds (IPBES, 2019; see also IPBES–IPCC, 2021), it is the extinction of humans that, for most, characterizes the final, irreversible collapse of all that has gone on before (at least from an anthropocentric perspective). When a bestselling book suggests we are heading toward an 'uninhabitable earth', the author is not referring to the survival chances of cockroaches or catfish, though it is self-evident that the collapse of ecosystems will follow the destruction of biodiversity, thus paving the way to possible human extinction (Wallace-Walls, 2019). This has led to imaginative work hypothesizing the nature of a natural world without humans (Weisman, 2007). Of course, there are other, often linked, existential threats to the human race, from pathogens through to nuclear warfare, but climate change is probably as strong a candidate at this stage. Existential risk prevention has grown into a field of risk analysis in its own right (Matheny, 2007; Bostrom and Ćirković, 2008; Bostrom, 2013; Beard *et al.*, 2020).

These scenarios of collapse will probably frame the discussions surrounding climate change for years to come, although the intensity and rapidity of change is affecting the tempo and content of collapse itself. Yet, even today, political leaders of the hegemonic nation-states appear to believe that time is not of the essence when it comes to climate responses, though the timelines for preventive, mitigating action have now shrunk to less than one decade (IPCC, 2021; see also White, 2018; Kramer, 2020). Global warming is preventable. Human-driven climate change is not immutable. But powerful interests are making it inevitable. Even the visions of collapse outlined above have been critiqued as driven by ontologies of power that limit our range of alternative futures; Mitchell and Chaudhury (2020), for example, counter 'white apocalyptic visions' with 'BIPOC futurisms', i.e. those that feature Black, Indigenous and People of Colour. The discussion of collapse runs the risk of making collapse seem either inevitable or simply frames potential pathways of escape bereft of critical analysis of the agency and structural power of those who are doing the most to drive the climate crisis in the first place.

Business as usual politics

Contrary to what is needed, therefore, nation-states are largely ignoring corporate-generated harm and colluding directly in profit-enhancing but planet-destroying activities. This occurs wherever the corporate form is dominant, and capitalism is entrenched. In other words, it happens everywhere.

For example, the International Energy Agency (IEA) argued recently that the energy sector holds the key to averting the worst effects of climate change by following a net zero pathway built upon renewables (IEA, 2021). According to the IEA, in order to reach a vital milestone on this pathway, there should be no new oil and gas fields approved and no new coal mines or mine extensions. Yet, '[g]lobal fossil fuel consumption subsidies increased by 50% over the past 3 years, reaching a peak of almost US$430 billion in 2018' (Watts *et al.*, 2019, p. 1836). In Australia, a 2021 report found that 'Every year Australian governments provide subsidies worth billions to fossil fuel producers and major users .... these subsidies cost state, territory and federal governments in Australia $10.3 billion in 2020–21' (Campbell *et al.*, 2021, p. 10).

Meanwhile, also in Australia, on the day of the release of the IEA report, the federal government announced that it would spend up to AUS$600 million building a new gas-fired power station in New South Wales, with the owners of the Tomago aluminium smelter, including Rio

Tinto, the main beneficiary (Tingle, 2021; Gooley, 2021). All of this is not surprising, given that the Australian coronavirus economic recovery committee was stacked with business leaders from the fossil fuel industries, and from gas in particular (Morgan and Long, 2020). As of 2023, a change in government has not stopped approval of new mining licences or government subsidies to the most polluting industries in Australia. The fossil fuel industry also retains inordinate political power in Canada, the USA, Russia and elsewhere; to many politicians, fears of the collapse of these industries are as animating as broader fears of the collapse of civilization.

All of this leads back to the centrality of politics in discussions about 'what is to be done' about climate change. Whyte (2020, p. 125) pointed out that:

> ... the rate of the political economy of speed matters. It matters because governments can and do generally intervene in ways that regulate the speed of production, distribution and consumption. Indeed, this is the importance of the *political* sphere. The rate at which we are destroying the planet is regulated by a process that is still largely determined by politics.

Solutions to environmental degradation that rely on corporations (a form of 'green capitalism') quickly take us down the cul-de-sac of continued growth (van der Velden and White, 2021). Time and space for 'business as usual' must be radically and urgently diminished if social and ecological justice is to be achieved. Fundamentally, this is about democracy, social need and the public interest. Our personal and social health and well-being depend upon these.

25.4 Fighting for the Future

The risks and threats to ecosystems, humans, flora and fauna posed by climate change are considerable and growing. The damage caused by global heating is already being felt in the form of extreme weather events, increased competition for dwindling natural resources, outbreaks of viral and other diseases, further extinctions of species, and intensification of social conflict (Crank and Jacoby, 2015; White, 2018; Heydon, 2020; Watts *et al.*, 2020). Social inequality and environmental injustice will undoubtedly be the drivers of continuous conflicts for many years to come, as the most dispossessed and marginalized of the world's population suffer the brunt of food shortages, undrinkable water, climate-induced migration and general hardship in their day-to-day lives. Even sincere efforts to adapt to or curb climate change can themselves threaten ecosystems or encourage crimes against humanity (Stoett and Omrow, 2021).

The contemporary politics of the climate endgame continues to exhibit the fortress mentality among the dominant world powers (White, 2014), in ways that parallel the roll-out of the COVID-19 vaccine worldwide. They also promulgate continued reliance upon private sector 'solutions' to problems systematically created and sustained by the global capitalist system. These are characterized by monopolistic and authoritarian decision making by billionaire owners and entrepreneurs, and appeal to technological fixes that are both undemocratic and unrealistic. Privately financed geoengineering is just around the corner; it will be presented as the potential remedy to the environmental sickness of potential catastrophic collapse (Morton, 2016). Regardless of their physical accuracy, scenarios of collapse are becoming visualized as imminent rather than hypothetical. This could lead to fatalism as well as an unrestricted licence to replicate the blunt application of political power in tandem with growth-seeking investment decisions.

It appears that the world is heading down the path of ultra-right political authoritarianism and contrarianism, and the demolition and disintegration of democracy in its many forms and guises (Cooper, 2021; White, 2023). Variations of green capitalism may be intended to mitigate the worst aspects of climate change, but climate reform has generally been subservient to existing power relations – that is, those precisely based on the continued exploitation of humans and environments. In the midst of these political and economic realities, things are progressing so quickly and at such magnitude that the harms caused are collapsing strategic categories of understanding and of intervention. The net result is that global warming and its consequences are continuing apace, with the dire 'consequences of the consequences' affecting health and well-being everywhere on the planet.

The shortfalls of authoritarian capitalism are exemplified by the limitations in the pandemic response and the systemic failures to address climate change. They are also evident in the democratic deficit – the feeling that people are not being listened to and that their interests are being ignored by governments. The profound inequalities of the economic system are translating into the populist politics of the demagogue, whose splinter politics promise much but ultimately reflect the sectoral interests of a few, not the majority (Cooper, 2021; van der Velden and White, 2021). Yet under conditions of rapidly approaching collapse, it is difficult to see how non-violent political order can effectively resist the turn to authoritarianism. The Chinese state may serve as a model for many as it bluntly manages citizen expectations alongside environmental policy development with unprecedented technological social control.

Popular struggle and climate action

The majority worldwide are bound to suffer from climate-related health problems and existential insecurities now and into the near future, with the minority elite prolonging their health privilege for a period thereafter. To change this situation, there has to be transformative change in the relations that structure and drive the global political economy. Fundamentally, decisions about the crucial pillars of social life – air, energy, food, shelter, security, water – are too important to be left in private hands (the corporations and the billionaires who own them) or a secretive state apparatus (information-controlling bureaucrats and self-serving politicians): governance must be public, open, cooperative and democratic.

Central to this is taking command of the central levers of the economy – taking the private capital that owns and controls these levers and putting them into public hands. This entails a form of nationalization that incorporates 'nation building' (such as infrastructure projects) and 'socialization' of essential industries and services (such as finance, housing and manufacturing). As environmental activists might put it, the four elements – earth, air, water and sun (energy) – constitute 'the commons' that needs to be put into democratic hands (van der Velden and White, 2021).

People want change. We have the innovative skills and knowledge to foster this (i.e. ecologically benign technologies and techniques). We have the desire for a better 'home' – we like Planet A. But we cannot achieve transformation without fundamentally changing the ownership structure of the present political economy. Our future is inextricably tied to that of the ruling classes of Earth – one way or the other. In the meantime, climate change-induced collapse will continue to animate both despair and action, acquiescence and revolutionary thought.

References

Abel, G.J., Brottrager, M., Cuaresma, J.C. and Muttarak, R. (2019) Climate, conflict and forced migration. *Global Environmental Change* 54, 239–249. DOI: 10.1016/j.gloenvcha.2018.12.003

Acemoglu, D. and Robinson, J.A. (2012) *Why Nations Fail: The Origins of Power, Prosperity, and Poverty*. Crown Publishers, New York.

Arbesman, S. (2011) The life-spans of empires. *Historical Methods: A Journal of Quantitative and Interdisciplinary History* 44, 127–129. DOI: 10.1080/01615440.2011.577733

Beard, S., Rowe, T. and Fox, J. (2020) An analysis and evaluation of methods currently used to quantify the likelihood of existential hazards. *Futures* 115, 102469. DOI: 10.1016/j.futures.2019.102469

Bostrom, N. (2013) Existential risk prevention as global priority. *Global Policy* 4, 15–31. DOI: 10.1111/1758-5899.12002

Bostrom, N. and Ćirković, M.M. (2008) *Global Catastrophic Risks*. Oxford University Press, Oxford, UK.

Butler, C.D. (2018) Climate change, health and existential risks to civilization: A comprehensive review (1989–2013). *International Journal of Environmental Research and Public Health* 15, 2266. DOI: 10.3390/ijerph15102266

Campbell, R., Littleton, E. and Armistead, A. (2021) *Fossil Fuel Subsidies in Australia: Federal and State Government Assistance to Fossil Fuel Producers and Major Users 2020–21.* The Australia Institute, Canberra, Australia.

Cooper, L. (2021) *Authoritarian Contagion: The Global Threat to Democracy*. Bristol University Press, Bristol, UK.

Crank, J. and Jacoby, L. (2015) *Crime, Violence, and Global Warming*. Routledge, London.

Fagan, B.M. (2008) *The Great Warming: Climate Change and the Rise and Fall of Civilizations*. Bloomsbury Press, New York.

Finneron-Burns, E. (2017) What's wrong with human extinction? *Canadian Journal of Philosophy* 47, 327–343. DOI: 10.1080/00455091.2016.1278150

Gooley, C. (2021) Federal government will spend $600 million on new Kurri Kurri gas plant in the NSW Hunter Valley. *ABC News*, 18 May 2021. Available at: https://www.abc.net.au/news/2021-05-18/federal-government-commits-600m-for-kurri-kurri-gas-plant/100147956 (accessed August 24, 2021).

Gowdy, J. (2020) Our hunter-gatherer future: climate change, agriculture and uncivilization. *Futures* 115, 102488. DOI: 10.1016/j.futures.2019.102488

Heydon, J. (2020) *Sustainable Development as Environmental Harm: Rights, Regulation, and Injustice in the Canadian Oil Sands*. Routledge, London.

Hobbes, T. (1651) *Leviathan: Or The Matter, Forme, and Power of A Common-Wealth Ecclesiasticall and Civill*. Republished 2010 (Shapiro, I., ed.), Yale University Press, New Haven, Connecticut.

Hofmann, R. (2014) Culturecide in changing Micronesian climates? About the unintentionality of climate change. *The International Journal of Human Rights* 18(3), 350–368. DOI: 10.1080/13642987.2014.914707

IEA (2021) *Net Zero by 2050: A Roadmap for the Global Energy Sector*. International Energy Agency, Paris.

IPBES (2019) *Global Assessment Report on Biodiversity and Ecosystem Services*. Intergovernmental Science-Policy Platform on Biodiversity and Ecosystem Services, Bonn, Germany. DOI: 10.5281/zenodo.3831673 [full assessment report] and 10.5281/zenodo.3553579 [summary for policymakers]

IPBES–IPCC (2021) *IPBES–IPCC Co-sponsored Workshop Report on Biodiversity and Climate Change*. Intergovernmental Science-Policy Platform on Biodiversity and Ecosystem Services, Bonn, and Intergovernmental Panel on Climate Change, Geneva. DOI: 10.5281/zenodo.4782538.

IPCC (2021) *Climate Change 2021: The Physical Basis*. Intergovernmental Panel on Climate Change, Geneva. Cambridge University Press, Cambridge, UK.

IPCC (2023) *Synthesis Report of the IPCC Sixth Assessment Report (AR6), Summary for Policymakers*. Intergovernmental Panel on Climate Change, Geneva.

Kemp, L., Xu, C., Depledge, J., Ebi, K.L., Gibbins, G. *et al.* (2022) Climate endgame: exploring catastrophic climate change scenarios. *Proceedings of the National Academy of Sciences (USA)* 119, e2108146119. DOI: 10.1073/pnas.2108146119

Kramer, R. (2020) *Carbon Criminals, Climate Crimes*. Rutgers University Press, New Brunswick, New Jersey.

Machalaba, C., Romanelli, C. and Stoett, P. (2017). Global environmental change and emerging infectious diseases: Macrolevel drivers and policy responses. In: Bouzid, M. (ed.) *Examining the Role of Environmental Change on Emerging Infectious Diseases and Pandemics*. IGI Global, Hershy, Pennsylvania, pp. 24–67.

Mann, M. (2021) *The New Climate War: The Fight to Take Back Our Planet*. Scribe Publications, Sydney, Australia.

Marshall, G. (2015) *Don't Even Think About It: Why Our Brains are Wired to Ignore Climate Change*. Bloomsbury, New York.

Matheny, J. (2007) Reducing the risk of human extinction. *Risk Analysis* 27, 1335–1344. DOI: 10.1111/j.1539-6924.2007.00960.x

McCarthy, C. (2006) *The Road*. Alfred Knopf, New York.

McMichael, A.J. (2017) *Climate Change and the Health of Nations*. Oxford University Press, Oxford, UK.

Mitchell, A. and Chaudhury, A. (2020) Worlding beyond 'the' 'end' of 'the world': white apocalyptic visions and BIPOC futurisms. *International Relations* 34, 309–332. DOI: 10.1177/0047117820948936

Morgan, E. and Long, S. (2020) Coronavirus economic recovery committee looks set to push Australia towards a gas-fired future. *ABC News Analysis*. Available at: https://www.abc.net.au/news/2020-05-13/coronavirus-recovery-to-push-australia-towards-gas-future/12239978 (accessed 24 August 2021).

Morton, O. (2016) *The Planet Remade: How Geoengineering Could Change the World*. Princeton University Press, Princeton, New Jersey.

Richards, C.E., Lupton, R.C. and Allwood, J.M. (2021) Re-framing the threat of global warming: an empirical causal loop diagram of climate change, food insecurity and societal collapse. *Climate Change* 164, 49. DOI: 10.1007/s10584-021-02957-w

Romanello, M., Di Napoli, C., Drummond, P., Green, C., Kennard, H. *et al.* (2022) The 2022 report of the Lancet Countdown on health and climate change: health at the mercy of fossil fuels. *The Lancet* 400, 1619–1654. DOI: 10.1016/S0140-6736(22)01540-9

Roszak, (1992) *The Voice of the Earth*. Simon and Schuster, New York.

Stoett, P., and Omrow, D.A. (2021) *Spheres of Transnational Ecoviolence: Environmental Crime, Human Security, and Justice*. Palgrave Macmillan, London.

Stoknes, P.E. (2015) *What We Think About When We Try Not to Think About Global Warming: Toward A New Psychology of Climate Change*. Chelsea Green, White River Junction, Vermont.

Strazdins, L., Friel, S., McMichael, A.J, Woldenberg Butler, S. and Hanna, E. (2010) Climate change and health in Australia: likely futures, new inequalities? *International Public Health Journal* 2, 493–500. Available at: https://openresearch-repository.anu.edu.au/handle/1885/38778 (accessed 30 May, 2023).

Tingle, L. (2021) The Government's backing of a new power plant is a gas-fired distraction. *ABC News Analysis*. Available at: https://www.abc.net.au/news/2020-09-19/government-backing-power-plant-gas-fired-distraction/12677222 (accessed 26 August 2021).

van der Velden, J. and White, R. (2021) *The Extinction Curve: Growth and Globalisation in the Climate Endgame*. Emerald Publishing, London.

Wallace-Wells, D. (2019) *The Uninhabitable Earth: Life After Warming*, Penguin Random House, New York.

Watts, N., Amann, M., Arnell, N., Ayeb-Karlsson, S., Belesova, K. *et al.* (2019) The 2019 report of The Lancet Countdown on health and climate change: ensuring that the health of a child born today is not defined by a changing climate. *The Lancet* 394, 1836–1878. DOI: 10.1016/S0140-6736(19)32596-6

Watts, N., Amann, M., Arnell, N., Ayeb-Karlsson, S., Beagley, J. *et al.* (2020) The 2020 report of the Lancet Countdown on health and climate change: responding to converging crises. *The Lancet* 397, 129–170. DOI: 10.1016/S0140-6736(20)32290-X

Weisman, A. (2007) *The World Without Us*. Thomas Dunne Books, St Martin's Press, New York.

White, R. (2014) Environmental insecurity and fortress mentality. *International Affairs* 90, 835–851. DOI: 10.1111/1468-2346.12143

White, R. (2018) *Climate Change Criminology*. Bristol University Press, Bristol, UK.

White, R. (2023) Conceptions of ecocide and challenges for social transformation. *Current Issues in Criminal Justice* 35(3), 312–323. DOI: 10.1080/10345329.2023.2203272

Whyte, D. (2020) *Ecocide: Kill the Corporation Before it Kills Us*. Manchester University Press, Manchester, UK.

WMO (2020a) *WMO Provisional Statement on the State of the Global Climate in 2019*. World Meteorological Organization, Geneva.

WMO (2020b) *State of the Global Climate 2020*. World Meteorological Organization, Geneva.

26 Climate Change and Global Mental Health

Fiona Charlson,[1,2] Rebecca Patrick[3] and Cybele Dey[4]
[1]*Queensland Centre for Mental Health Research, Wacol, Queensland, Australia;*
[2]*School of Public Health, The University of Queensland, Brisbane, Australia;*
[3]*Melbourne School of Population and Global Health, University of Melbourne, Melbourne, Australia;* [4]*School of Clinical Medicine, Faculty of Medicine and Health, University of New South Wales, Sydney, Australia*

Abstract
This chapter explores the mental health issues arising from the primary, secondary and tertiary effects of climate change (a conceptual framework generally used in this book). The chapter examines the complex causal pathways to a range of climate-related mental health outcomes, insisting on a socio-ecological approach to understanding the drivers of and actions for protecting mental health. It identifies and discusses populations that are most at risk of climate-related mental health impacts with an acknowledgement that climate change is a driver and amplifier of existing inequities and, consequently, any populations experiencing inequities in mental health outcomes, or their determinants, should be considered as at risk, including marginalized groups within high-income countries. The chapter highlights co-benefit approaches to promoting mental health, including nature-based healthcare, inclusion of Indigenous ways of knowing and being in mental healthcare and climate-resilient development.

26.1 Introduction

Anthropogenic climate change (henceforth 'climate change' or 'climate' unless otherwise noted) is already having a significant impact on global mental health. This chapter demonstrates that mental health is a cross-cutting issue arising from many of the primary, secondary and tertiary effects of climate change. It provides a typology of mental health outcomes highlighting priority populations and places including young people and low-income countries. In conclusion, it draws attention to promising strategies for protecting population mental health and promoting community resilience in the context of climate change.

26.2 Mental Health Outcomes

Several adverse mental health-related outcomes have been associated with climate change in the literature. These range from psychological distress that may simply be part of the normal human experience to severe and clinically significant mental health outcomes that require specialist mental health care (Fig. 26.1).

Emotional responses felt in relation to the direct or indirect experiences of climate change are common and well-documented, and they are increased when responses from leaders or authorities are perceived as inadequate (Hickman *et al.*, 2021). Whilst it is important to note that emotional responses are considered part of the

DOI: 10.1079/9781800620025.0026

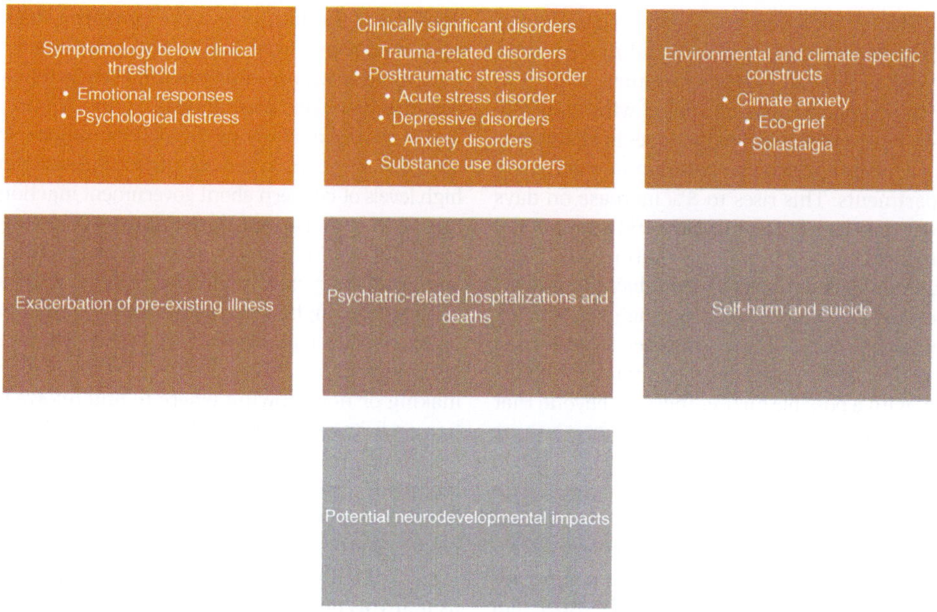

Fig. 26.1. Spectrum of mental health-related outcomes associated with climate change impacts.

normal human experience, they have been linked to more serious impacts on well-being and mental health, and are seen in both high-income countries in the global North and in low- and middle-income countries and communities in the global South (Lawrance *et al.*, 2022). Emerging climate change-specific terms, such as 'climate anxiety', are receiving increasing attention and research is under way to understand overlap and distinctions in symptomology between climate anxiety and other anxiety disorders (Crandon *et al.*, 2022). Another is solastalgia, which, while not exclusive to climate change, can stem from its impacts. Solastagia describes a profound sense of missing home while being home, as one's home (particularly the natural environment) is gradually, but irretrievably, destroyed and especially affects those with deep connection to place, including Indigenous people, and is exacerbated by a sense of powerlessness (Albrecht *et al.*, 2007; Galway *et al.*, 2019).

Perhaps the most researched mental health outcomes are trauma-related disorders that develop in response to an acute and extreme event, such as bushfire or flood. These include acute stress disorder, which typically resolves spontaneously in the months after exposure to the trauma, and post-traumatic stress disorder (PTSD), which is a chronic and disabling mental health outcome (Bryant *et al.*, 2014). Increases in repeated exposure to trauma, as with more frequent and severe extreme weather events, heightens risk of PTSD, and worsens outcomes (French *et al.*, 2019). Substance-use disorders also increase in the months and years following traumatic events and, in turn, contribute to poorer outcomes for people who also have other mental health disorders (Bryant *et al.*, 2014). There is emerging evidence of neurodevelopmental and psychiatric consequences for pre-school-aged children born following exposure of their mother to extreme weather events such as floods or bushfires during pregnancy (Nomura *et al.*, 2022).

Mood disorders (e.g. depression) and anxiety disorders are relatively common mental disorders that have well-established links to a range of climate change-related events at all stages of life; including extreme weather events, air pollution, unemployment, forced migration, food and water insecurity and disruptions to education (Charlson *et al.*, 2021). People are increasingly presenting with somatic symptom disorders following climate-related extreme events. Climate change can thus amplify already existing inequities in important determinants of global mental health, particularly in many low-income settings.

Hotter weather, even modest single-day increases in usually warmer and more humid regions, such as ambient temperatures of 31.4°C (75th percentile), are associated with clinically and statistically significant increases in all categories of mental health presentation to emergency departments. This rises to 8% increase on days of extreme heat (at or above the 95th percentile) compared with optimal temperature days, noting that this was observed at ambient temperatures just below 40°C (Nori-Sarma *et al.*, 2022). Heatwaves, where hotter weather is prolonged for 2 or more days, have substantial health burden, with a possible further increase beyond that of a heatwave's constituent days, for younger people and females (Guo *et al.*, 2017; Florido Ngu *et al.*, 2021; Liu *et al.*, 2021). Increases are seen in severe mental distress, suicidal ideation and suicide attempts, and agitation or distress from exacerbation of psychosis or delirium superimposed on dementia (Lawrance *et al.*, 2021; Liu *et al.*, 2021). Associated increased risk of death has also been seen in patients with psychosis, dementia, those on prescribed medications for mental illness and with substance misuse (Dodgen *et al.*, 2016; Liu *et al.*, 2021).

26.3 Priority Populations

This section outlines populations we identify as being most at risk in terms of the mental health impacts of climate change, including children and young people, Indigenous peoples, people from low-and middle-income countries, women and farmers (Charlson *et al.*, 2021; Crandon *et al.*, 2022). As discussed, climate change is a driver and amplifier of existing inequities; consequently, populations experiencing inequities in mental health or its determinants are at particular risk, even in high-income countries.

Children and young people

A recent editorial in the journal *The Lancet, Child and Adolescent Health* showed that the psychological burdens of climate change are affecting the lives and mental health of young people across the globe, including in low- and middle-income countries (Sharpe and Davison, 2022). Fear about the future and the perceived inadequate response from adults and governments contribute to a sense of distress and betrayal that negatively impacts mental health (Anonymous, 2021). A global study of 10,000 young people aged 16–25 years in ten countries underscored high levels of concern about government inaction and inter-generational dissonance (Hickman *et al.*, 2021). Similarly, a study of about 1800 young people from 37 countries, from all regions of the globe, highlighted how many are keen to be involved in climate policy processes yet face barriers such as feeling excluded from decision making or not knowing where to find information on how to participate (Rost *et al.*, 2021).

Young people express negative eco-emotions including stress, anxiety, hopelessness, powerlessness, and worry about the impact of climate change on their future (Burke *et al.*, 2018; Hayes *et al.*, 2018; Gunasiri *et al.*, 2022). In Australia, as many as one in five Australian young people aged 18–34 years are experiencing anxiety related to climate change, with impacts on cognitive and functional impairment affecting their work, family and social life (Patrick *et al.*, 2022).[1]

Crandon and colleagues (2022) proposed a model that depicts factors that may influence climate anxiety for a child or adolescent across a range of social–ecological systems. It is theorized that climate anxiety results from an interplay of individual factors: the micro- (family, peers), meso- (school, community, local physical environment), exo- (government, media, global physical environment) and macro- (culture) systems surrounding children and adolescents (Fig. 26.2). This structure of systems is also subject to climatic changes over time, which evolve depending on the responses of individuals, families, schools, communities, media, governments and cultures. These include the potential for transformational responses where communities mobilize, connect and engage with healthy behaviours to become more resilient both individually and at organizational and government levels to mitigate against climate change and prepare for unavoidable increases in hazards. Conversely, where individuals and communities become more isolated and disconnected, existing capacities become insufficient to meet increasing disruptions and demands (Doppelt, 2017; Lawrance *et al.*, 2022).

Children and adolescents are at particular risk because of their rapidly developing brains, vulnerability to disease, and limited capacity to avoid or adapt to threats and impacts (Fig. 26.3). This age group is particularly impacted by hotter weather and humidity, with increases in self-harm and suicidality for children and youth (Charlson *et al.*, 2021), and emerging evidence points to increases in PTSD, anxiety disorders and depression following exposure to disasters increased by

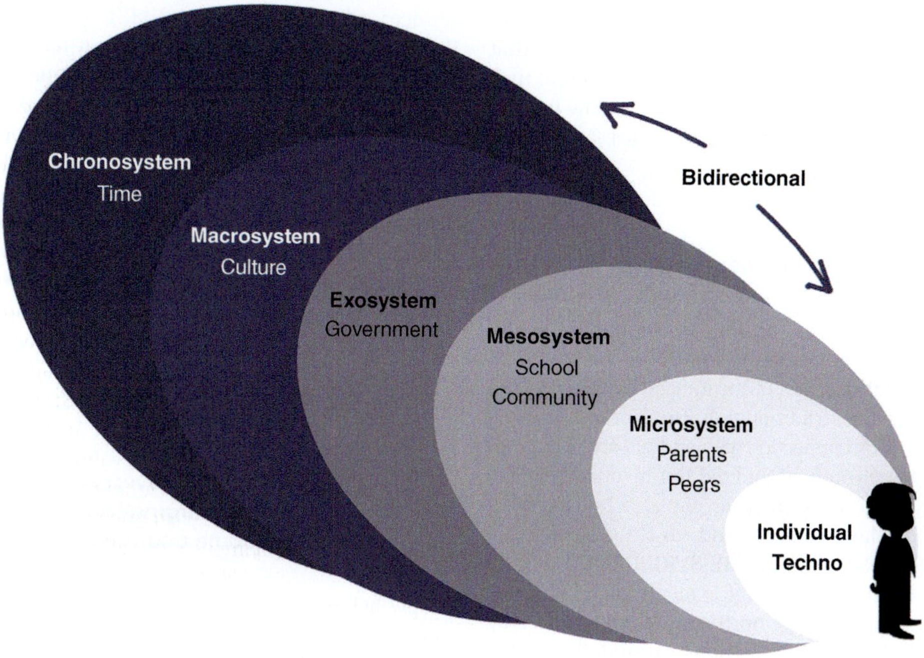

Fig. 26.2. A social–ecological framework of climate anxiety for children and adolescents. Adapted from Crandon *et al.* (2022).

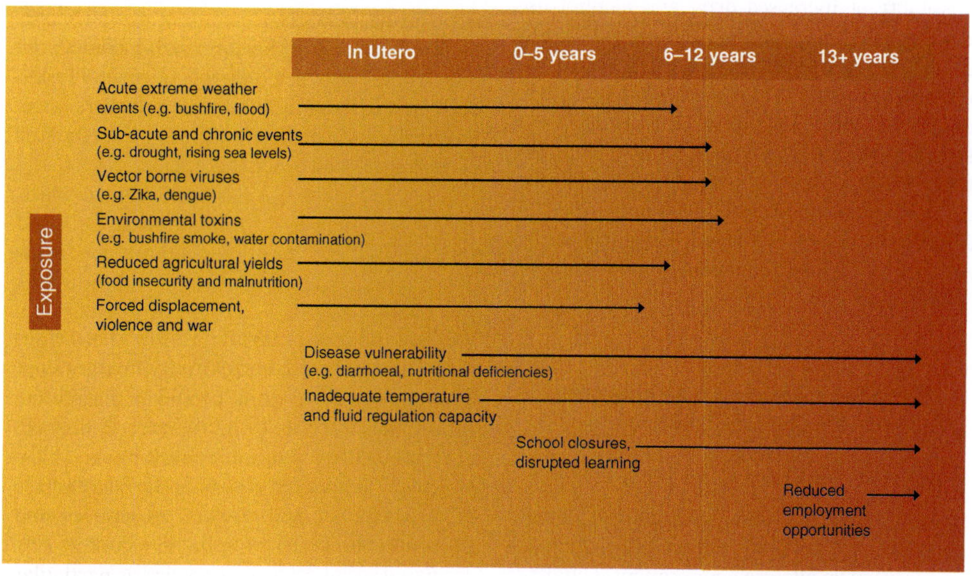

Fig. 26.3. Climate change related threats to neurodevelopment. Adapted from Vergunst and Berry (2021).

climate change (Dodgen *et al.*, 2016). They are also more likely to worry about climate change than any other age group. Drawing on a developmental life-course perspective, we show that climate change-related threats can additively, interactively and cumulatively increase psychopathology risk from conception onwards; that these effects are already occurring; and that they constitute an important threat to healthy human development worldwide (Vergunst and Berry, 2021).

Indigenous people

Leading scholars on Indigenous mental health identify climate change impacts as continuation and intensification of the impacts of colonization, noting that traditional cultural practices often focused on connection with and/or care of the natural environment and did not involve the main drivers of anthropogenic climate change: widespread land clearing and fossil fuel extraction (HEAL Network and CRE-STRIDE, 2021). There is an emerging body of evidence of the impacts of climate change on the emotional well-being of Indigenous populations around the world. Studies of North American Indigenous communities report that climate and environmental changes increase family stress, enhance the possibility of increased drug and alcohol use, amplify previous traumas and mental health stressors, and may increase the risk of suicidal ideation (Cunsolo Willox *et al.*, 2013). Studies conducted with Aboriginal and Torres Strait Islander peoples from Australia also highlight the environmental impacts of climate change on emotional well-being, including increased community distress from compromising the connection to country (McNamara and Westoby, 2011; Rigby *et al.*, 2011).

Low- and middle-income countries and small island developing states

Many low- and middle-income countries, predominantly in the global South, rely on the environment for subsistence; adverse environmental changes consequently expose these populations to significant emotional and psychological distress (Cooper *et al.*, 2019). Drought and subsequent loss of crops and livestock in Kilifi County, Kenya, has led to impacts on population mental health, with insufficient support being provided by health services (Sheriff and Mash, 2022). The 2017 flooding events in Bangladesh negatively affected food security and the mental health of women in rural areas (Gepp *et al.*, 2022). Low-lying coastal areas, particularly those of small island developing states and river deltas, are at risk, vulnerable to the impacts of climate change and (often) groundwater extraction. For example, coastal villages in the Solomon Islands and Tuvalu report that climate change-induced sea level rise provokes fear, worry and distress at a personal and community level (Asugeni *et al.*, 2015; Gibson *et al.*, 2019; Gibson *et al.*, 2020).

People living in poverty or otherwise marginalized in high-income countries

People living in poverty or who are otherwise marginalized in high-income countries, such as culturally and linguistically diverse populations are at increased risk of mental health harms from climate change. Lack of secure employment, housing, education and healthcare and discrimination are all social determinants of mental health, for people affected by climate in high-income countries. These social determinants of mental health and climate change act as threat multipliers where the combination of factors is greater than the sum of their parts (Lawrance *et al.*, 2022).

Gender

Women and post-pubertal girls are at increased risk of mental health harms from climate change (Stone *et al.*, 2022), and people of non-binary gender may experience increased risk through marginalization and discrimination. While biological factors for post-pubertal increases in impacts of heat and disaster on females may contribute, there is increasing evidence of gender-based social determinants of climate change mental health harms. These include increased exposure of post-pubertal females to physical and sexual violence with climate change-related increases in hotter weather, floods, bushfires, storms and with forced migration, carer responsibilities, reduced access to education and greater housing and income insecurity (Dodgen *et al.*, 2016; van Daalen *et al.*, 2020). Interestingly, there is also the potential for improving solutions to climate change-related problems through greater involvement of women (Figueiredo and Perkins, 2013; Amhair, 2021).

Farmers

Increasing climate variability and unpredictability have been demonstrated to impact across the whole agricultural community, in research from low-, middle- and high-income countries (Sorgho *et al.*, 2020; Yazd *et al.*, 2020; Berman *et al.*, 2021). Drought is of particular concern, leading to significant levels of psychological distress, anxiety, depression and suicide in farmers; furthermore, it may erode the social and economic bases on which entire farming communities depend (Acharibasam and Anuga, 2018; Austin *et al.*, 2018).

Pre-existing mental illness

People living with existing or sub-threshold mental illness may face increased risk to their mental health from climate change. Hotter weather poses a particular risk to people with schizophrenia, dementia and substance-use disorders, with increased risk of mental health presentation and hospitalization. People with pre-existing post-traumatic stress disorder or substance-use disorder are at increased risk of mental health disorder following disasters such as floods or bushfires. Those taking prescribed or non-prescribed psychotropic medications (including hypnotics, antidepressants and antipsychotics) are at increased risk of delirium, heatstroke and death during hotter weather, thought to be linked both to medication-related changes in thermoregulation and to risk of lowered serum sodium (Bouchama *et al.*, 2007; Hoffmann *et al.*, 2016, Page *et al.*, 2012).

26.4 Pathways to the Mental Health Impacts of Climate Change

Climate change is expected to impact mental health via a range of direct ('primary'), indirect ('secondary') and systemic ('tertiary') pathways (Fig. 26.4). Direct pathways include trauma imposed by bushfires and other extreme events. Indirect pathways operate through a range of social, political, economic and environmental determinants of mental health such as poverty, unemployment and housing, amplifying existing inequities in the global mental health. Crucially, extreme events are no longer isolated events and a compounding and cumulative effect on mental health is being observed in many parts of the world. Hence, the concept of 'recovery' may not be a feasible approach to addressing mental health.

Primary impacts

The literature links mental health outcomes to several direct climate-related exposures – heat, humidity, rainfall, drought, wildfires and floods. Extreme weather events are linked to the direct experience of trauma and ongoing disruption to families and communities through loss and lack of access to safe and secure homes, livelihoods, schools, and recreational and cultural facilities (Dodgen *et al.*, 2016; Lawrance *et al.*, 2021). Climate change-driven increases in frequency and/or intensity of such events mean that communities may have insufficient time and resources to recover before facing another disaster, thus compounding the mental health impacts of food insecurity, underemployment and unemployment (Dodgen *et al.*, 2016). The rates of psychological trauma consequent to such events have been reported to exceed those of physical trauma by 40 to 1 in some settings (Bryant *et al.*, 2014; Hayes *et al.*, 2018).

Heat

Systematic reviews have identified strong evidence of associations between heat and increased risk for a range of adverse mental health outcomes, including mood disorders, schizophrenia, suicide and substance misuse (Jennings *et al.*, 2020;

Primary

- Heat
- Rainfall and humidity
- Fires
- Floods
- Storms

Secondary

- Drought
- Drought
- Sea level rise
- Land degradation
- Air pollution
- Physical health
- Water and food security
- Sea level rise
- Land degradation
- Air pollution
- Physical health
- Water and food security

Tertiary

- Displacement and migration, loss of connection to land
- Health system pressures and reduced access to mental health care
- Reduced economic productivity
- Reduced financial security
- Strain on community cohesion and social capital
- Loss of nature spaces and connection to nature
- Conflict and political instability
- Family violence/gender-based violence

Fig. 26.4. Climate change impacts which are associated with mental health, grouped into three forms of effect.

Liu *et al.*, 2021). There is evidence that even modest levels of elevated daily mean temperature are associated with increased mental ill health, as seen in emergency hospital presentations and weekly deaths from suicide (Nori-Sarma *et al.*, 2022). Conversely, mild, cool weather appears protective (Liu *et al.*, 2021). Higher humidity, measured as the average daily vapour pressure over the previous four weeks, appears to be an additive risk to mental illness (Ding *et al.*, 2016). Lack of access to overnight cooling, which can result in insomnia, also negatively impacts mental health (Wong *et al.*, 2016; Obradovich *et al.*, 2017; Liu *et al.*, 2021).

Floods

Several studies have examined the mental health impacts of flooding. The evidence from Australia and the UK point to persistent, long-term psychological morbidity, particularly for respondents still displaced and for socio-economically marginalized groups (Matthews *et al.*, 2019; Mulchandani *et al.*, 2020).

Wildfires

A strong body of literature supports the impact of wildfires on mental health. Findings of a systematic review have shown an increased rate of PTSD, depression and generalized anxiety with both acute and long-term negative impacts on mental and emotional well-being. Several associated risk factors have been identified, the most significant being characteristics of the wildfire trauma itself (To *et al.*, 2021). The effects from smoke are discussed in the next section.

Secondary impacts

Drought

Worsening droughts are a well-established consequence of climate change and cause system-wide stressors, including agricultural and economic losses. Mental illness in those experiencing drought is more likely when accompanied by social isolation and the absence of support to affected communities (McMichael *et al.*, 2012; Dodgen *et al.*, 2016; Clayton, 2020).

Land degradation

Increases in frequency and intensity of droughts and repeated extreme weather events have exacerbated the effects of urbanization, deforestation and land overuse on reduced soil quality and land desertification. Land degradation negatively impacts subsistence living, farmers' livelihoods and surrounding communities. Subsequent increases in migration have been observed, which can further deepen negative impacts on mental health (Dodgen *et al.*, 2016).

Smoke and other forms of air pollution

Potential mental health impacts of fire smoke are unclear. However, prolonged smoke events have been linked to negative impacts on mental health, due to extended time indoors, and disruptions to livelihood and land-based activities (Dodd *et al.*, 2018). Furthermore, particulate air pollution is associated with increased rates of autism, attention deficit hyperactivity disorder, disruptive behaviour, dementia and depression and anxiety. Particulates are likely to adversely affect cognitive functioning across the lifespan (Haase 2017).

Sea level rise

Sea level rises have already resulted and will increasingly result in forced migration away from coastal communities – and subsequent loss of land, culture and identity – with significant mental health and well-being effects (Kelman *et al.*, 2021). Climate refugees are already experiencing the loss of homes in coastal communities impacted by sea level rise, including in Alaska, Northern Canada and islands in the Pacific (Wetzel *et al.*, 2012; Northern Alaska University, 2022).

Displacement and migration

Forced migration due to sea level rise, drought, land degradation and the need to flee repeated extreme weather events is already occurring and projected to increase substantially over the coming decades and even centuries, unless greenhouse gas (GHG) emissions can be rapidly stabilized (McMichael *et al.*, 2012; Dodgen *et al.*, 2016). In 2020, the UN High Commissioner for Refugees officially recognized the contribution of climate change and disasters to refugee status (Harper *et al.*, 2022a). The mental health impacts of displacement and forced migration are further worsened when safe refuges are unavailable, children lack parents or carers, and recognized human rights including access to education, the right to work and cultural practices are restricted or absent (McMichael *et al.*, 2012; Dodgen *et al.*, 2016; Clemens *et al.*, 2020).

Tertiary impacts

System-wide impacts

The drivers and pathways to climate-related mental well-being can be found in social structures and systems of power (Charlson *et al.*, 2021; Patrick *et al.*, 2022). A systems thinking approach is required to understand the complex and often unpredictable pathways to climate-related mental health outcomes (Fig. 26.5). Locating both the drivers (risk and protective factors) and solutions to climate-related mental well-being as 'external to' interpersonal characteristics enables action to be taken on many levels. Factors including ecological (e.g. access to green spaces), political (e.g. government inaction), social (e.g. intersectional inequalities), digital (e.g. misinformation and social media) and commercial (e.g. capitalism and consumerism) determine mental health outcomes (Patrick *et al.*, 2022). Fig. 26.5 is part of a systems map which highlights underlying structural issues impacting young people's climate-related mental well-being. The factors identified demonstrate that a rupture of 'social contract' leads to anger and mistrust in government. Mistrust in government feeds fear of the future, creating a flow-on effect where powerlessness increases anxiety. Capitalism and consumerism reinforce one another and generate the conditions for government inaction and increasing emissions.

Water and food security

The inability to access basic necessities, such as clean water and adequate food, harms mental and physical health. Lack of clean water particularly affects the mental health of women, potentially linked to the need for water for menstrual hygiene (Dodgen *et al.*, 2016; Lawrance *et al.*, 2021).

Economic impacts

Whilst climate change will restrict the availability of necessities such as food and clean water, a more insidious challenge is reduced economic productivity occurring through multiple consequences, including extreme heat, land degradation, disrupted opportunities to work following extreme weather events and the vicious cycle of untreated or inadequately treated mental illness

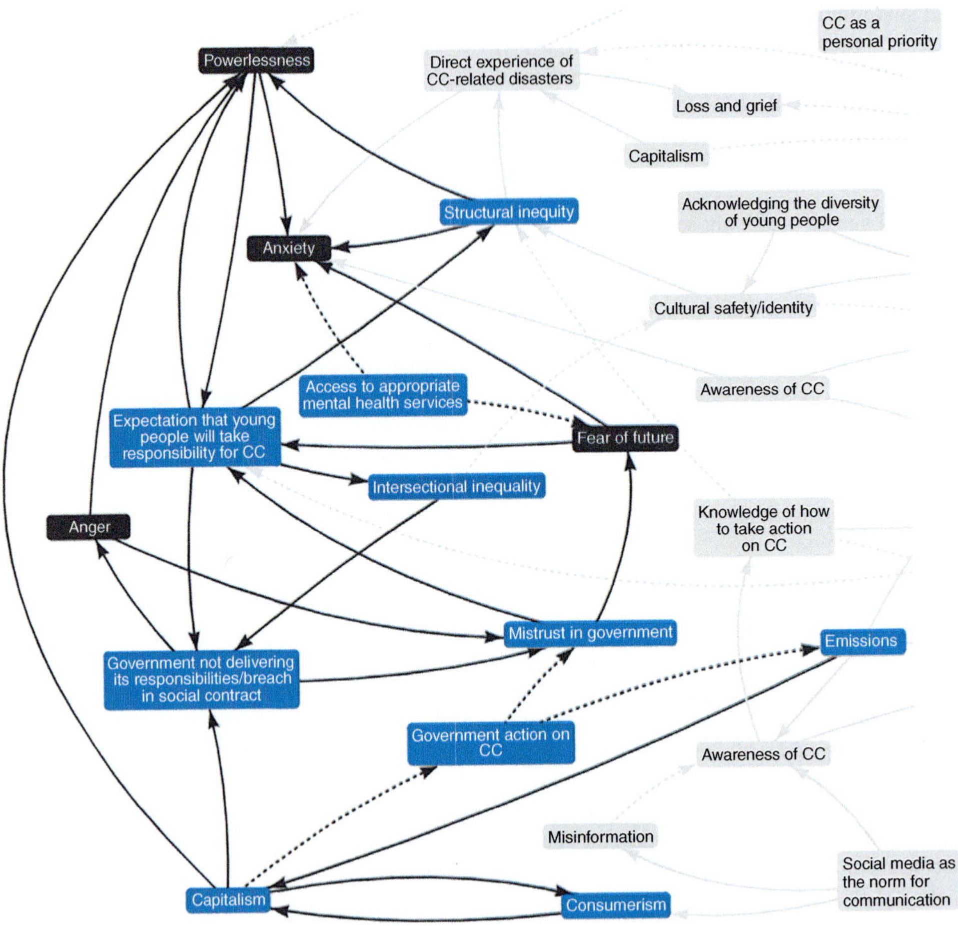

Fig. 26.5. Systems map of climate-related mental well-being among young people in Australia (simplified), sourced from Noy *et al.* (2022).

itself. Even for those not directly impacted, the flow-on effects of reduced economic productivity and urbanization can reduce financial security at a population level. This may potentially increase rates of suicide, which are strongly correlated with economic difficulties (Cianconi *et al.*, 2020).

Loss of connection to land, sea and nature

Experiencing the loss of important places, including one's home, natural environment and community, can cause distress and mental suffering, also known as 'solastalgia' (Albrecht *et al.*, 2007; Richardson, 2022). Grief and distress following the loss of one's home may be particularly severe for people whose culture and spirituality have been deeply connected to ancestral places for millennia (Dodgen *et al.*, 2016). Evidence suggests that time in nature and access to green and blue spaces are generally associated with better mental well-being and general health (White *et al.*, 2021). Conversely, lack of contact with nature spaces in childhood is associated with an increased risk of mental disorders, including mood and anxiety disorders, schizophrenia and substance abuse in adolescence and adulthood (Engemann *et al.*, 2019).

Conflict and political instability

Increasing evidence indicates that climate change is causally associated with conflict and political instability, generally in combination with other causal factors (see Chapter 21). Increased temperatures and extremes of precipitation with their associated consequences, including resultant scarcity of cropland and other key environmental resources, are major pathways by which climate change leads to conflict and instability (Levy *et al.*, 2017). Increased interpersonal violence has also been linked to heat (Mahendran *et al.*, 2021), extreme weather events, and threats to water and food security which aggravate risks to mental health through exposure to traumatic experiences, lack of basic services and forced migration (Abel *et al.*, 2019).

Strain on community cohesion and social capital

Loss of social connectedness is a frequent and pervasive consequence of climate change's direct and indirect impacts. Significant community resources, including grassroots organizations and groups, are often disrupted, overwhelmed, or unable to function, following climate-driven extreme weather events. A vicious cycle may arise whereby the disruption of social capital is a risk factor for mental disorders, which further reduces social capital (Di Giorgi *et al.*, 2020).

26.5 Co-benefits and Solutions

This chapter concludes by highlighting available evidence of promising strategies to protect and promote mental well-being and community resilience in the context of climate change. These approaches may be valid on both scales from the individual through to the population, including nature-based healthcare, health co-benefits of taking climate action and climate-resilient development.

Nature and mental well-being

Mental health is a state of mental well-being extending beyond the absence of mental disorder, which is necessary for people to learn, work, have meaningful relationships and engage with their community and cope with life stressors (WHO, 2022). The idea that contact with green and natural spaces is intrinsic to good human health is posited by the biophilia concept, i.e. that nature contact is essential for physiological and psychological well-being (Gullone, 2000). Human health is inextricably linked to the life-giving resources and ecological processes of a healthy natural environment – harming nature is to risk ourselves (Townsend and Ebden, 2010). Natural settings provide psychosocial, cultural, spiritual and recreational benefits arising from our sense of connection to place and country (Corvalan *et al.*, 2005). Many Indigenous communities have lived in harmony with their natural environments for millennia, appreciating Earth as a living entity, taking collective responsibility for its stewardship and valuing connection to country as essential for well-being (IUHPE, 2019).

The mental health benefits of nature connection are well-documented (Capaldi *et al.*, 2014; Shanahan *et al.*, 2016). Engaging with nature has been linked to lower incidence of mental illnesses such as depression and anxiety, improved cognitive function including concentration and attention restoration, and better sleep and immune function (Townsend and Weerasuriya, 2010; Shanahan *et al.*, 2016; Twohig-Bennett and Jones, 2018). Nature-based healthcare interventions, such as Green or Social Prescriptions, capitalize on these benefits and use structured programmes designed to engage participants with nature (Marx and More, 2022). Indeed, nature-based healthcare may have the potential to bring about significant co-benefits in the context of climate-driven mental health issues through intentionally promoting human–environment interactions (Robinson and Breed, 2019). Green prescriptions can contribute both reactively (outdoor healthcare intervention to address trauma, distress and anxiety) and proactively (prevention initiatives designed to build community resilience) through public health initiatives that simultaneously enhance the natural environment (Robinson and Breed, 2019). Such prescriptions, be they in the form of greenspace exercise programmes or more contemporary approaches such as environmental volunteering, have the potential to synergistically promote pro-environmental attitudes, social inclusion, mental health, emotional well-being, connection to place and environmental protection outcomes (Robinson and Breed, 2019; Marx and More, 2022; Patrick *et al.*, 2022). These synergistic outcomes are known as health co-benefits – strategies that reduce GHG emissions or offer other climate change mitigation and adaptation benefits whilst simultaneously promoting human health outcomes (Smith *et al.*, 2014).

Mental well-being benefits of climate action

There is an emerging body of evidence, predominantly from high-income countries, that acting on climate change can reduce personal stress, increase optimism and in turn reduce adverse climate-related mental health impacts (Jennings *et al.*, 2020; Patrick *et al.*, 2021). There is also evidence that Indigenous peoples' ways of knowing and caring for their ancestral homes bring additional mental well-being, as well as practical solutions, but is often missing from research on this topic (Dudgeon *et al.*, 2023). A direct experience of a climate-related event has been linked to a greater knowledge of and stronger awareness of climate change as a phenomenon. Whilst such experience elevates concern and distress, it can also generate a stronger sense of responsibility to act on climate change and greater psychological adaptation and behavioural involvement to address it (Reser *et al.*, 2012). Directly experiencing a climate event may, in some settings, change understanding, promote self-efficacy and motivate engagement in, and action on, environmental issues (Reser *et al.*, 2012; Fung and Adams, 2017). Furthermore, collective action on climate change can help to enable and sustain feelings of hope (Gunasiri *et al.*, 2022). In an Australian study of eco-emotions, eco-anger was found to predict better mental health outcomes than eco-anxiety and eco-depression, with the latter two eco-emotions found to be less adaptive and linked to lower well-being (Stanley *et al.*, 2021). Importantly, however, there are cross-cultural differences between eco-emotions and pro-environmental behaviours (Agoston *et al.*, 2022). In a European study it was established that for German participants empathy for victims was more likely to motivate action, whereas for French participants guilt and sadness motivated action (Caillaud *et al.*, 2019). The authors linked this to climate change being valued as a moral issue in Germany and a political issue in France (Caillaud *et al.*, 2019). Research from the Arctic region has found that Indigenous-led activities to care for ancestral lands can improve adaptation to climate change and benefit mental well-being for participants. Conversely, inability to access ancestral lands due to climate change has been found to magnify the traumas [of residential schools and assimilation] through colonization (Lebel *et al.*, 2022).

Moderators of climate-related mental well-being

Various factors and settings can moderate the relationship between climate change impacts and mental health outcomes. These factors can either promote positive mental health effects (e.g. sense of place, strong relationships within communities and families) when they are present, or, when absent, act as barriers to good mental health outcomes (e.g. inter-generational dissonance, government inaction on climate change) and psychosocial adaptation (Hayes *et al.*, 2019). Often these factors are one side of the same coin. For example, the framing of messages and information about climate change on social media can influence perceptions, response and emotions (O'Brien *et al.*, 2018). Systematically communicating stories of climate change as catastrophic or as 'Armageddon' can contribute to feelings of helplessness and create a sense of despair, which can in turn lead to apathy and disillusion (Ojala and Bengtsson, 2018). However, constructive framings of climate change can facilitate active rather than naïve hope, with realistic engagement, and more productive strategies for coping (O'Brien *et al.*, 2018; Ojala and Bengtsson, 2018). Social media is a key form of communication for many young people globally but social engagement and access to knowledge concerning the climate narrative is a knife with two blades. It can influence their sense of agency, generate optimism and be an accessible, powerful platform for their voice and to organize collective action. It may also generate a sense of responsibility for the future that can feel overwhelming. Positively, social media can enable engagement with diverse audiences, knowledge mobilization and transfer whilst linking geographically and culturally diverse communities (Pearce *et al.*, 2019; Gunasiri *et al.*, 2022).

A study conducted in Nunatsiavut region of Labrador, Canada, found that protective factors for climate-related mental well-being among young people included strong communities, sense of place or being on land, and relationships with family and friends (MacDonald *et al.*, 2015).

Promising co-benefit approaches

Table 26.1 provides a framework for individual through to population-based strategies for protecting and promoting mental health and well-being in a climate-impacted world. For example, at the population level, there is growing international focus on climate-resilient development,

Table 26.1. Strategies and approaches to protecting and promoting climate related mental health well-being. Adapted from WHO (1986), Cissé *et al.* (2022) and Ma *et al.* (2022).

Domain	Reorient health services	Enable education and skill development	Support community/sector	Create supportive settings and environments
Health/medical sector	Universal access to mental healthcare; clinical psychology/ counselling and referral pathways, e.g. trauma/grief individual and group therapy; climate sensitive screening and risk assessment protocols, outreach services to priority communities	Training and workforce capacity building – mental health workers, youth workers, educators etc.	Local, regional, national and international network, e.g. Climate Psychiatry Alliance, Psychologists for a Safe Climate	Population mental health surveillance systems and research; climate-related mental health public policy including gender inclusive climate policy; decolonizing mental healthcare
Nature-based interventions	Place-specific mental health infrastructure and nature therapy; nature-based healthcare, 'green' prescriptions	Nature-based health-promotion programmes; Connection to country initiatives	Climate action – e.g. environmental volunteering	National preventive health framework – nature/ecological health determinants
Disaster planning and recovery	Disaster planning and response health services – social work, family violence workers	Culturally appropriate and climate-specific mental health resources: e.g. children, environmental workers, climate refugees	Community-led disaster recovery initiatives	Multi-sectoral initiatives that enhance social and economic well-being; disaster action and scenario planning – mental health
Health co-benefit approach	Sustainable healthcare service delivery	Online content/ campaigns, e.g. evidence-based social media for mental health and environmental co-benefits	Co-led and co-designed campaigns and networks, e.g. Indigenous led, youth led, such as school strikes	Urban and regional planning – access to nature, green space, heat and flood protection; climate resilient and sensitive development

which provides a framework by which policies and practice can link with overarching sustainable development goals for a future that protects both humans and the planet and an approach that recognizes human agency and capacity to influence the future. Climate change mitigation and adaptation are key elements of this approach and inform integrated solutions and decision making that can generate mutual benefits, as well as introduce co-benefits within development policies (Denton *et al.*, 2014; Patrick *et al.*, 2021). For example, and as outlined previously, gender inequalities are reflected in higher vulnerability and exposure to impacts of climate change; addressing them is key to protecting mental health and well-being among women and increasing adaptive capacities of communities. A climate-resilient development approach underscores the importance of incorporating gender in scenarios assessing future climate impacts and highlighting the relevance of addressing gender inequalities in climate policies (Andrijevic *et al.*, 2020). Other examples where climate-resilient development and planning have demonstrable mental health outcomes are outlined by Cissé *et al.* (2022) in the 2022 IPCC report, and by Harper *et al.* (2022b). For example, advanced warning for flooding in the UK was linked to reduced mental health impacts; planning for temporary shelters for people displaced by flooding in China was linked to lower rates of anxiety, depression and PTSD; and, in the Arctic, rainwater harvesting initiatives that promoted water security had demonstrable co-benefits for mental health.

26.6 Conclusion

Climate change is already impacting global mental health via diverse and complex pathways. Children and young people, Indigenous communities, women, people living in poverty or with mental illness and low-income countries, especially in the global South, are at particular risk and face significant challenges now and into the future. Yet, many of those facing the greatest impacts are stepping forward with innovative solutions, including nature-based healthcare, health co-benefits of taking climate action and climate-resilient development. A key limiting factor is the need for high-income countries, predominantly from the global North, to work constructively alongside them. There are opportunities both to protect global mental health through improved climate mitigation, in proportion to the enormous benefits to mental health of doing so, and to collaborate on responses drawing on Indigenous and scientific knowledge while research continues to improve responses to the impacts that can no longer be prevented. The drivers of and pathways to mental health outcomes are complex and require systems-level responses spanning preventive programmes through medical care service provision.

Note

[1] A nation-wide Australian survey (n = 5483) of adults, approximately representative across sex, age, location, state and area disadvantage

References

Abel, G.J., Brottrager, M., Cuaresma, J.C. and Muttarak, R. (2019) Climate, conflict and forced migration. *Global Environmental Change* 54, 239–249. DOI: 10.1016/j.gloenvcha.2018.12.003

Acharibasam, J.W. and Anuga, S.W. (2018) Psychological distance of climate change and mental health risks assessment of smallholder farmers in Northern Ghana: Is habituation a threat to climate change? *Climate Risk Management* 21, 16–25. DOI: 10.1016/j.crm.2018.04.002

Ágoston, C., Csaba, B., Nagy, B., Kőváry, Z., Dúll, A. *et al.* (2022) Identifying types of eco-anxiety, eco-guilt, eco-grief, and eco-coping in a climate-sensitive population: A qualitative study. *International Journal of Environmental Research and Public Health* 19, 2461. DOI: 10.3390/ijerph19042461

Albrecht, G., Sartore, G.M., Connor, L., Higginbotham, N., Freeman, S. *et al.* (2007) Solastalgia: the distress caused by environmental change. *Australasian Psychiatry* 15, S95–98. DOI: 10.1080/10398560701701288

Amhair, L. (2021) A gender approach to climate change resilience. *Arribat-International Journal of Human Rights* 1, 204–211. Published by CNDH (National Human Rights Council of Morocco).

Andrijevic, M., Crespo Cuaresma, J., Lissner, T., Thomas, A. and Schleussner, C.-F. (2020) Overcoming gender inequality for climate resilient development. *Nature Communications* 11, 1–8. DOI: 10.1038/s41467-020-19856-w

Anonymous (2021) A climate of anxiety. *The Lancet Child and Adolescent Health* 5, 91. DOI: 10.1016/S2352-4642(21)00001-8

Asugeni, J., MacLaren, D., Massey, P.D. and Speare, R. (2015) Mental health issues from rising sea level in a remote coastal region of the Solomon Islands: current and future. *Australasian Psychiatry* 23, 22–25. DOI: 10.1177/1039856215609767

Austin, E.K., Handley, T., Kiem, A.S., Rich, J.L., Lewin, T.J. *et al.* (2018) Drought-related stress among farmers: findings from the Australian rural mental health study. *Medical Journal of Australia* 209, 159–165. DOI: 10.5694/mja17.01200

Berman, J.D., Ramirez, M.R., Bell, J.E., Bilotta, R., Gerr, F. *et al.* (2021) The association between drought conditions and increased occupational psychosocial stress among U.S. farmers: An occupational cohort study. *Science of the Total Environment* 798, 149245. DOI: 10.1016/j.scitotenv.2021.149245

Bouchama, A., Dehbi, M., Mohamed, G., Matthies, F., Shoukri, M. *et al.* (2007) Prognostic factors in heat wave-related deaths: a meta-analysis. *Archives of Internal Medicine* 167, 2170–2176. DOI: 10.1001/archinte.167.20.ira70009

Bryant, R.A., Waters, E., Gibbs, L., Gallagher, H.C., Pattison, P. *et al.* (2014) Psychological outcomes following the Victorian Black Saturday bushfires. *Australian & New Zealand Journal of Psychiatry* 48, 634–643. DOI: 10.1177/0004867414534476

Burke, S.E.L., Sanson, A.V. and Van Hoorn, J. (2018) The psychological effects of climate change on children. *Current Psychiatry Reports* 20, 35. DOI: 10.1007/s11920-018-0896-9

Caillaud, S., Krauth-Gruber, S. and Bonnot, V. (2019) Facing climate change in France and Germany: different emotions predicting the same behavioral intentions? *Ecopsychology* 11, 49–58. DOI: 10.1089/eco.2018.0060

Capaldi, C., Dopko, R. and Zelenski, J. (2014) The relationship between nature connectedness and happiness: a meta-analysis. *Frontiers in Psychology* 5, 976. DOI: 10.3389/fpsyg.2014.00976

Charlson, F., Ali, S., Benmarhnia, T., Pearl, M., Massazza, A. *et al.* (2021) Climate change and mental health: a scoping review. *International Journal of Environmental Research and Public Health* 18, 1–38. DOI: 10.3390/ijerph18094486

Cianconi, P., Betrò, S. and Janiri, L. (2020) The impact of climate change on mental health: a systematic descriptive review. *Frontiers in Psychiatry* 11, 74. DOI: 10.3389/fpsyt.2020.00074

Cissé, G., McLeman, R., Adams, H., Aldunce, P., Bowen, K. *et al.* (2022) Health, wellbeing, and the changing structure of communities. In: Pörtner, H.-O., Roberts, D.C., Tignor, M., Poloczanska, E.S., Mintenbeck, K. *et al.* (eds) *Climate Change 2022: Impacts, Adaptation, and Vulnerability. Contribution of Working Group II to the Sixth Assessment Report of the Intergovernmental Panel on Climate Change*. Cambridge University Press, Cambridge, UK, pp. 1041–1170.

Clayton, S. (2020) Climate anxiety: Psychological responses to climate change. *Journal of Anxiety Disorders* 74, 102263. DOI: 10.1016/j.janxdis.2020.102263

Clemens, V., von Hirschhausen, E. and Fegert, J.M. (2020) Report of the Intergovernmental Panel on Climate Change: implications for the mental health policy of children and adolescents in Europe – a scoping review. *European Child & Adolescent Psychiatry* 31, 701–713. DOI: 10.1007/s00787-020-01615-3

Cooper, S., Hutchings, P., Butterworth, J., Joseph, S., Kebede, A. *et al.* (2019) Environmental associated emotional distress and the dangers of climate change for pastoralist mental health. *Global Environmental Change* 59, 101994. DOI: 10.1016/j.gloenvcha.2019.101994

Corvalan, C., Hales, S. and McMichael, A.J. (eds) (2005) *Ecosystems and Human-wellbeing: Health Synthesis: A Report of the Millennium Ecosystem Assessment*. World Health Organization, Geneva. Available at: https://apps.who.int/iris/handle/10665/43354 (accessed 27 July 2022).

Crandon, T.J., Scott, J.G., Charlson, F.J. and Thomas, H.J. (2022) A social–ecological perspective on climate anxiety in children and adolescents. *Nature Climate Change* 12, 123–131. DOI: 10.1038/s41558-021-01251-y

Cunsolo Willox, A., Harper, S.L., Ford, J.D., Edge, V.L., Landman, K. *et al.* (2013) Climate change and mental health: an exploratory case study from Rigolet, Nunatsiavut, Canada. *Climatic Change* 121, 255–270. DOI: 10.1007/s10584-013-0875-4

Denton, F., Wilbanks, T.J., Abeysinghe, A.C., Burton, I., Gao, Q. *et al.* (2014) Climate-resilient pathways: adaptation, mitigation, and sustainable development. In: Field, C.B., Barros, V.R., Dokken, D.J., Mach, K.J.,

Mastrandrea, M.D. *et al.* (eds) *Climate Change 2014: Impacts, Adaptation, and Vulnerability. Part A: Global and Sectoral Aspects*. Contribution of Working Group II to the Fifth Assessment Report of the Intergovernmental Panel on Climate Change. Cambridge University Press, Cambridge, UK, pp. 1101–1131.

Di Giorgi, E., Michielin, P. and Michielin, D. (2020) Perception of climate change, loss of social capital and mental health in two groups of migrants from African countries. *Annali Dell'Istituto Superiore Di Sanità* 56, 150–156. DOI: 10.4415/ANN_20_02_04

Ding, N., Berry, H.L. and Bennett, C.M. (2016) The Importance of humidity in the relationship between heat and population mental health: Evidence from Australia. *PLoS ONE* 11, e0164190. DOI: 10.1371/journal.pone.0164190

Dodd, W., Scott, P., Howard, C., Scott, C., Rose, C. *et al.* (2018) Lived experience of a record wildfire season in the Northwest Territories, Canada. *Canadian Journal of Public Health* 109, 327–337. DOI: 10.17269/s41997-018-0070-5

Dodgen, D., Donato, D., Kelly, N., La Greca, A., Morganstein, J. *et al.* (2016) Mental health and well-being. In: *The Impacts of Climate Change on Human Health in the United States: A Scientific Assessment*. US Global Change Research Program, Washington, DC, pp. 218–246. DOI: 10.7930/J0R49NQX

Doppelt, B. (2017) *Transformational Resilience: How Building Human Resilience to Climate Disruption Can Safeguard Society and Increase Wellbeing*. Routledge, London. DOI: 10.4324/9781351283885

Dudgeon, P., Bray, A., Walker, R., Wright, M. and Sutherland, S. (2023) Caring for country: Indigenous, well-being, law, and environmental justice. In: Moore, R. (ed.) *Climate Change and Mental Health Equality*. Springer Publishing, New York. Available at: https://timhwb.org.au/wp-content/uploads/2023/03/16.3.23-FINAL-CARING-FOR-COUNTRY2-1.pdf (accessed 31 May, 2023).

Engemann, K., Pedersen, C.B., Arge, L., Tsirogiannis, C., Mortensen, P.B. *et al.* (2019) Residential green space in childhood is associated with lower risk of psychiatric disorders from adolescence into adulthood. *Proceedings of the National Academy of Sciences (USA)* 116, 5188–5193. DOI: 10.1073/pnas.180750411

Figueiredo, P. and Perkins, P.E. (2013) Women and water management in times of climate change: participatory and inclusive processes. *Journal of Cleaner Production* 60, 188–194. DOI: 10.1016/j.jclepro.2012.02.025

Florido Ngu, F., Kelman, I., Chambers, J. and Ayeb-Karlsson, S. (2021) Correlating heatwaves and relative humidity with suicide (fatal intentional self-harm). *Scientific Reports* 11, 22175. DOI: 10.1038/s41598-021-01448-3

French, C.E., Waite, T.D., Armstrong, B., Rubin, G.J., Beck, C.R. *et al.* (2019) Impact of repeat flooding on mental health and health-related quality of life: a cross-sectional analysis of the English national study of flooding and health. *BMJ Open* 9, e031562. DOI: 10.1136/bmjopen-2019-031562

Fung, C.Y. and Adams, E.A. (2017) What motivates student environmental activists on college campuses? An in-depth qualitative study. *Social Sciences* 6, 134. DOI: 10.3390/socsci6040134

Galway, L.P., Beery, T., Jones-Casey, K. and Tasala, K. (2019) Mapping the solastalgia literature: A scoping review study. *International Journal of Environmental Research and Public Health* 16, 2662. DOI: 10.3390/ijerph16152662

Gepp, S., Waid, J.L., Brombierstäudl, D., Kader, A., Müller-Hauser, A.A. *et al.* (2022) Impact of unseasonable flooding on women's food security and mental health in rural Sylhet, Bangladesh: a longitudinal observational study. *The Lancet Planetary Health* 6, Special Issue S14. DOI: 10.1016/s2542-5196(22)00276-5

Gibson, K., Haslam, N. and Kaplan, I. (2019) Distressing encounters in the context of climate change: Idioms of distress, determinants, and responses to distress in Tuvalu. *Transcultural Psychiatry* 56, 667–696. DOI: 10.1177/1363461519847057

Gibson, K.E., Barnett, J., Haslam, N. and Kaplan, I. (2020) The mental health impacts of climate change: Findings from a Pacific Island atoll nation. *Journal of Anxiety Disorders* 73, 102237. DOI: 10.1016/j.janxdis.2020.102237

Gullone, E. (2000) The biophilia hypothesis and life in the 21st century: increasing mental health or increasing pathology? *Journal of Happiness Studies* 1, 293–322. DOI: 10.1016/j.janxdis.2020.102237

Gunasiri, H., Wang, Y., Watkins, E.-M., Capetola, T., Henderson-Wilson, C. *et al.* (2022) Hope, coping and eco-anxiety: Young people's mental health in a climate-Impacted Australia. *International Journal of Environmental Research and Public Health* 19, 5528. DOI: 10.3390/ijerph19095528

Guo, Y., Gasparrini, A., Armstrong, B.G., Tawatsupa, B., Tobias, A. *et al.* (2017) Heat wave and mortality: a multicountry, multicommunity study. *Environmental Health Perspectives* 125, 087006. DOI: 10.1289/EHP1026

Haase, E. (2017) *Air Pollution*. Climate Psychiatry Alliance. Available at: https://www.climatepsychiatry.org/air-pollution (accessed 24 June, 2022).

Harper, A., Hernandez, C., Svobodova, V. and Schachter, B. (2022a) *Climate Change, Displacement and Human Rights*. United Nations High Commission for Refugees, Geneva. Available at: https://www.

unhcr.org/en-us/publications/brochures/6242ea7c4/climate-change-displacement-human-rights.html (accessed 24 June, 2022).

Harper, S.L., Cunsolo, A. and Clayton, S. (2022b) Including mental health as part of climate change impacts and adaptation assessment: A critical advance in IPCC AR6. *PLoS Climate* 1, e0000033. DOI: 10.1371/journal.pclm.0000033

Hayes, K., Blashki, G., Wiseman, J., Burke, S. and Reifels, L. (2018) Climate change and mental health: risks, impacts and priority actions. *International Journal of Mental Health Systems* 12, 28. DOI: 10.1186/s13033-018-0210-6

Hayes, K., Berry, P. and Ebi, K.L. (2019) Factors influencing the mental health consequences of climate change in Canada. *International Journal of Environmental Research and Public Health* 16, 1583. DOI: 10.3390/ijerph16091583

HEAL Network and CRE-STRIDE (2021) *Climate change and Aboriginal and Torres Strait Islander health*. Discussion paper, prepared for Lowitja Institute, Melbourne, Australia, by Healthy Environments and Lives (HEAL) Network and Centre for Research Excellence in Strengthening Systems for Indigenous Health Care Equity (CRE-STRIDE). Available at: https://www.lowitja.org.au/content/Image/Lowitja_ClimateChangeHealth_1021_D10.pdf (accessed January 31, 2023). DOI: 10.48455/bthg-aj15

Hickman, C., Marks, E., Pihkala, P., Clayton, S., Lewandowski, R.E. *et al.* (2021) Climate anxiety in children and young people and their beliefs about government responses to climate change: a global survey. *The Lancet Planetary Health* 5, e863–e873. DOI: 10.1016/S2542-5196(21)00278-3

Hoffmann, M.S., Oliveira, L.M., Lobato, M.I. and Belmonte-de-Abreu, P. (2016) Heat stroke during long-term clozapine treatment: should we be concerned about hot weather? *Trends in Psychiatry and Psychotherapy* 38, 56–59. DOI: 10.1590/2237-6089-2015-0066

IUHPE (2019) *Rotorua Statement: Waiora – Indigenous peoples' statement for planetary health and sustainable development*. 23rd IUHPE World Conference on Health Promotion – Waiora: Promoting Planetary Health and Sustainable Development for All Proceedings. International Union for Health Promotion Education, Saint-Maurice, France and Montreal, Canada. Available at: https://www.iuhpe2019.com/PicsHotel/iuhpe/Brochure/Indigenous%20Statement%20for%20Plenary%20Revised.pdf (accessed 22 August, 2022).

Jennings, N., Fecht, D. and De Matteis, S. (2020) Mapping the co-benefits of climate change action to issues of public concern in the UK: A narrative review. *Lancet Planetary Health* 4, 424–433. DOI: 10.1016/S2542-5196(20)30167-4

Kelman, I., Ayeb-Karlsson, S., Rose-Clarke, K., Prost, A., Ronneberg, E. *et al.* (2021) A review of mental health and wellbeing under climate change in small island developing states (SIDS). *Environmental Research Letters* 16, 033007. DOI: 10.1088/1748-9326/abe57d

Lawrance, E., Thompson, R., Fontana, G. and Jennings, N. (2021) *The impact of climate change on mental health and emotional wellbeing: current evidence and implications for policy and practice*. Grantham Institute, Briefing paper No. 36. Imperial College, London. Available at: https://www.imperial.ac.uk/grantham/publications/all-publications/the-impact-of-climate-change-on-mental-health-and-emotional-wellbeing-current-evidence-and-implications-for-policy-and-practice.php (accessed 22 August, 2022).

Lawrance, E.L., Thompson, R., Newberry Le Vay, J., Page, L. and Jennings, N. (2022) The impact of climate change on mental health and emotional wellbeing: A narrative review of current evidence, and its implications. *International Review of Psychiatry* 34, 443–498. DOI: 10.1080/09540261.2022.2128725

Lebel, L., Paquin, V., Kenny, T., Fletcher, C., Nadeau, L. *et al.* (2022) Climate change and Indigenous mental health in the Circumpolar North: a systematic review to inform clinical practice. *Transcultural Psychiatry* 59, 312–336. DOI: 10.1177/13634615211066698

Levy, B.S., Sidel, V.W. and Patz, J.A. (2017) Climate change and collective violence. *Annual Review of Public Health* 38, 241–257. DOI: 10.1146/annurev-publhealth-031816-044232

Liu, J., Varghese, B.M., Hansen, A., Xiang, J., Zhang, Y. *et al.* (2021) Is there an association between hot weather and poor mental health outcomes? A systematic review and meta-analysis. *Environment International* 153, 106533. DOI: 10.1016/j.envint.2021.106533

Ma, T., Moore, J. and Cleary, A. (2022) Climate change impacts on the mental health and wellbeing of young people: A scoping review of risk and protective factors. *Social Science & Medicine* 301, 114888. DOI: 10.1016/j.socscimed.2022.114888

MacDonald, J.P., Cunsolo Willox, A., Ford, J.D., Shiwak, I. and Wood, M. (2015) Protective factors for mental health and well-being in a changing climate: Perspectives from Inuit youth in Nunatsiavut, Labrador. *Social Science & Medicine* 141, 133–141. DOI: 10.1016/j.socscimed.2015.07.017

Mahendran, R., Xu, R., Li, S. and Guo, Y. (2021) Interpersonal violence associated with hot weather. *The Lancet Planetary Health* 5, e571–e572. DOI: 10.1016/S2542-5196(21)00210-2

Marx, V. and More, K.R. (2022) Developing Scotland's first green health prescription pathway: A one-stop shop for nature-based intervention referrals. *Frontiers in Psychology* 13, 1–14. DOI: 10.3389/fpsyg.2022.817803

Matthews, V., Longman, J., Berry, H.L., Passey, M., Bennett-Levy, J. *et al.* (2019) Differential mental health impact six months after extensive river flooding in rural Australia: A cross-sectional analysis through an equity lens. *Frontiers in Public Health* 7, 367. DOI: 10.3389/fpubh.2019.00367

McMichael, C., Barnett, J. and McMichael, A.J. (2012) An ill wind? Climate change, migration and health. *Environmental Health Perspectives* 120, 646–654. DOI: 10.1289/ehp.1104375

McNamara, K.E. and Westoby, R. (2011) Solastalgia and the gendered nature of climate change: an example from Erub Island, Torres Strait. *Ecohealth* 8, 233–236. DOI: 10.1007/s10393-011-0698-6

Mulchandani, R., Armstrong, B., Beck, C.R., Waite, T.D., Amlôt, R. *et al.* (2020) The English national cohort study of flooding & health: psychological morbidity at three years of follow up: psychological morbidity at three years of follow up. *BMC Public Health* 20, 321. DOI: 10.1186/s12889-020-8424-3

Nomura, Y., Newcorn, J.H., Ginalis, C., Heitz, C., Zaki, J. *et al.* (2022) Prenatal exposure to a natural disaster and early development of psychiatric disorders during the preschool years: stress in pregnancy study. *Journal of Child Psychology and Psychiatry* 64, 1080–1091. DOI: 10.1111/jcpp.13698

Nori-Sarma, A., Sun, S., Sun, Y., Spangler, K.R., Oblath, R. *et al.* (2022) Association between ambient heat and risk of emergency department visits for mental health among us adults, 2010 to 2019. *JAMA Psychiatry* 79, 341–349. DOI: 10.1001/jamapsychiatry.2021.4369

Northern Alaska University (2022) *Climate Change: Realities of Relocation for Alaska Native Villages*. Northern Arizona University, Flagstaff, Arizona. Available at: https://www7.nau.edu/itep/main/tcc/Basic/Ak_inupiaqnw_rel (accessed 22 August, 2022).

Noy, S., Capetola, T., Henderson-Wilson, C., Hayward, J., Allender, S. *et al.* (2022) *Keeping the Future in Mind: Climate-Related Mental Wellbeing Among Young People in Australia*. School of Health and Social Development/Institute for Health Transformation, Deakin University, Melbourne, Australia. Available at: https://www.healthpromotion.org.au/vic-documents/2605-keeping-the-future-in-mind-climate-related-mental-health-and-young-people (accessed 9 October, 2022).

O'Brien, K., Selboe, E. and Hayward, B. (2018) Exploring youth activism on climate change: dutiful, disruptive, and dangerous dissent. *Ecology and Society* 23, 42. DOI: 10.5751/ES-10287-230342

Obradovich, N., Migliorini, R., Mednick, S.C. and Fowler, J.H. (2017) Nighttime temperature and human sleep loss in a changing climate. *Science Advances* 3, e1601555. DOI: 10.1126/sciadv.1601555

Ojala, M. and Bengtsson, H. (2018) Young people's coping strategies concerning climate change: Relations to perceived communication with parents and friends and proenvironmental behaviour. *Environment and Behaviour* 51, 907–935. DOI: 10.1177/0013916518763894

Page, L.A., Hajat, S., Kovats, R.S. and Howard, L.M. (2012) Temperature-related deaths in people with psychosis, dementia and substance misuse. *British Journal of Psychiatry* 200, 485–490. DOI: 10.1192/bjp.bp.111.100404

Patrick, R., Armstrong, F., Capon, A., Bowen, K., Lo, S.N. *et al.* (2021) Health promotion in the Anthropocene: the ecological determinants of health. *Medical Journal of Australia* 214, S22–S26. DOI: 10.5694/mja2.51020

Patrick, R., Snell, T., Gunasiri, H., Garad, R., Meadows, G. *et al.* (2022) Prevalence and determinants of mental health related to climate change in Australia. *Australian & New Zealand Journal of Psychiatry* 57, 710–724. DOI: 10.1177/00048674221107872

Patrick, R., Noy, S., Henderson-Wilson, C., Hayward, J., Allender, S. *et al.* (2023) Minding our futures: Understanding climate-related mental wellbeing using systems science. *Health and Place* 84, 8. DOI: 10.1016/j.healthplace.2023.103134

Pearce, W., Niederer, S., Özkula, S.M. and Sánchez Querubín, N. (2019) The social media life of climate change: Platforms, publics, and future imaginaries. *Wiley Interdisciplinary Reviews: Climate Change* 10, e569. DOI: 10.1002/wcc.569

Reser, J.P., Bradley, G.L., Glendon, A.I., Ellul, M.C. and Callaghan, R. (2012) Public risk perceptions, understandings and responses to climate change and natural disasters in Australia, 2010 and 2011. National Climate Change Adaptation Research Facility Gold Coast, Griffith University, Brisbane, Australia. Available at: https://research-repository.griffith.edu.au/bitstream/handle/10072/49216/84369_1.pdf (accessed 22 August, 2022).

Richardson, A. (2022) *Missing home while being home. Solastalgia*. Climate Psychiatry Alliance. Available at: https://www.climatepsychiatry.org/missing-home (accessed 24 June, 2011).

Rigby, C.W., Rosen, A., Berry, H.L. and Hart, C.R. (2011) If the land's sick, we're sick: the impact of prolonged drought on the social and emotional well-being of Aboriginal communities in rural New South Wales. *Australian Journal of Rural Health* 19, 249–254. DOI: 10.1111/j.1440-1584.2011.01223.x

Robinson, J.M. and Breed, M.F. (2019) Green prescriptions and their co-benefits: Integrative strategies for public and environmental health. *Challenges* 10, 9. DOI: 10.3390/challe10010009

Rost, L., Cooke, J. and Fergus, I. (2021) *Reimagining Climate Education and Youth Leadership: Survey Report*. Plan International Australia. Available at: https://www.plan.org.au/wp-content/uploads/2021/07/Reimagining-Climate-Education-and-Youth-Leadership-ENG.pdf (accessed 31 May, 2023).

Shanahan, D., Bush, R., Gaston, K., Lin, B., Dean, J. *et al.* (2016) Health benefits from nature experiences depend on dose. *Medicine Scientific Reports* 6, 28551. DOI: 10.1038/srep28551

Sharpe, I. and Davison, C.M. (2022) A scoping review of climate change, climate-related disasters, and mental disorders among children in low- and middle-income countries. *International Journal of Environmental Research and Public Health* 19, 2896. DOI: 10.3390/ijerph19052896

Sheriff, M. and Mash, R. (2022) Climate change and primary health care in Chakama, Kilifi County, Kenya. *African Journal of Primary Health Care & Family Medicine* 14, e1–e3. DOI: 10.4102/phcfm.v14i1.3670

Smith, K.R., Woodward, A., Campbell-Lendrum, D., Chadee, D., Honda, Y. *et al.* (2014) Human health: impacts, adaptation, and co-benefits. In: Field, C.B., Barros, V. and Dokken, D.J. (eds) *Climate Change 2014: Impacts, Adaptation, and Vulnerability*. Cambridge University Press, Cambridge, UK, and New York, pp. 709–754.

Sorgho, R., Mank, I., Kagone, M., Souares, A., Danquah, I. *et al.* (2020) 'We will always ask ourselves the question of how to feed the family': Subsistence farmers' perceptions on adaptation to climate change in Burkina Faso. *International Journal of Environmental Research and Public Health* 17, 7200. DOI: 10.3390/ijerph17197200

Stanley, S.K., Hogg, T.L., Leviston, Z. and Walker, I. (2021) From anger to action: Differential impacts of eco-anxiety, eco-depression, and eco-anger on climate action and wellbeing. *The Journal of Climate Change and Health* 1, 100003. DOI: 10.1016/j.joclim.2021.100003

Stone, K., Blinn, N. and Spencer, R. (2022) Mental health impacts of climate change on women: a scoping review. *Current Environmental Health Reports* 9, 228–243. DOI: 10.1007/s40572-022-00346-8

To, P., Eboreime, E. and Agyapong, V.I.O. (2021) The impact of wildfires on mental health: A scoping review. Behavioural Science, 11. DOI: 10.3390/bs11090126

Townsend, M. and Ebden, M. (2010) Natural environments. In: Keleher, H. and Macdougall, C. (eds) *Understanding Health: A Social Determinants Approach*. 2nd edn. Oxford University Press, Melbourne, Australia, pp. 41–58.

Townsend, M. and Weerasuriya, R. (2010) *Beyond blue to green: The benefits of contact with nature for mental health and well-being*. Beyond Blue Ltd, Melbourne, Australia. Available at: https://www.deakin.edu.au/__data/assets/pdf_file/0004/310747/Beyond-Blue-To-Green-Literature-Review.pdf (accessed 22 August, 2022).

Twohig-Bennett, C. and Jones, A. (2018) The health benefits of the great outdoors: A systematic review and meta-analysis of greenspace exposure and health outcomes. *Environmental Research* 166, 628–637. DOI: 10.1016/j.envres.2018.06.030

van Daalen, K., Jung, L., Dhatt, R. and Phelan, A.L. (2020) Climate change and gender-based health disparities. *The Lancet Planetary Health* 4, e44–e45. DOI: 10.1016/S2542-5196(20)30001-2

Vergunst, F. and Berry, H.L. (2021) Climate change and children's mental health: A developmental perspective. *Clinical Psychological Science* 10, 767–785. DOI: 10.1177/21677026211040787

Wetzel, F.T., Kissling, W.D., Beissmann, H. and Penn, D.J. (2012) Future climate change driven sea-level rise: secondary consequences from human displacement for island biodiversity. *Global Change Biology* 18, 2707–2719. DOI: 10.1111/j.1365-2486.2012.02736.x

White, M.P., Elliott, L.R., Grellier, J., Economou, T., Bell, S. *et al.* (2021) Associations between green/blue spaces and mental health across 18 countries. *Scientific Reports* 11, 1–12. DOI: 10.1038/s41598-021-87675-0

WHO (1986) *The 1st International Conference on Health Promotion, Ottawa, 1986*. World Health Organization, Geneva. Available at: https://www.who.int/teams/health-promotion/enhanced-wellbeing/first-global-conference (accessed 8 October, 2022).

WHO (2022) *Mental health: strengthening our response*. World Health Organization, Geneva. Available at: https://www.who.int/news-room/fact-sheets/detail/mental-health-strengthening-our-response (accessed February 1, 2023).

Wong, M.M., Brower, K.J. and Craun, E.A. (2016) Insomnia symptoms and suicidality in the national comorbidity survey–adolescent supplement. *Journal of Psychiatric Research* 81, 1–8. DOI: 10.1016/j.jpsychires.2016.06.004

Yazd, S.D., Wheeler, S.A. and Zuo, A. (2020) Understanding the impacts of water scarcity and socio-economic demographics on farmer mental health in the Murray-Darling Basin. *Ecological Economics* 169, 106564. DOI: 10.1016/j.ecolecon.2019.106564

27 Nutrition, Soil Organic Carbon and Sustainability: Multiple Benefits of Agriculture Regeneration

Robyn Alders,[1,2,3,4] Kate Wingett,[5] Rosemary A. McFarlane,[6] Stewart Sutherland,[7] Justin Borevitz[2,8] and Namukolo Covic[9]

[1]Development Policy Centre, Australian National University, Canberra, Australia; [2]Institute for Climate, Energy and Disaster Solutions, Australian National University, Canberra, Australia; [3]Global Health Programme, Chatham House, London, UK; [4]Kyeema Foundation, Brisbane, Australia; [5]Faculty of Medicine and Health, University of Sydney, Sydney, Australia; [6]Health Research Institute, University of Canberra, Canberra, Australia; [7]College of Health & Medicine, Australian National University, Canberra, Australia; [8]Research School of Biology, Australian National University, Canberra, Australia; [9]International Livestock Research Institute, Addis Ababa, Ethiopia

Abstract

Regenerative agricultural practices can contribute to addressing the triple burden of: (i) multiple forms of malnutrition and diet-related diseases, (ii) a changing environment and climate that (iii) makes current food systems unsustainable; but significant investment in resources and efforts to accelerate progress ahead of the climate clock is essential. Commitment by and collaboration between public and private sectors and civil society is required to deliver enabling environments that facilitate: (i) the production of food using the principles of regenerative, climate-resilient agriculture; (ii) valuing food according to its natural nutrient density in addition to weight and/or volume; (iii) promotion of the efficient utilization of food by tailoring nutrition by gender, age, reproductive and health status; (iv) valuing nutrients in waste products and enabling their recycling; and (v) restructuring healthcare services to place a higher value on the contributions of agriculture and farmers to preventive medicine.

27.1 Introduction

Humanity is at a crossroads as we seek to deliver optimal and sustainable diets for the 9 billion people and billions of domestic animals predicted by 2050. The Sustainable Development Goals (SDGs), launched in 2015, encourage us to rethink the linkages between agriculture, nutrition security and health and to review associated policy frameworks. The United Nations (UN) Food Systems Summit process in 2020 and 2021 has further added much-needed attention to the requirement that diets serve not just nutrition and human health needs but also the planet. Rethinking our current food and nutrition challenges in the context of sustainable development requires revisiting the historical drivers that contributed to our current circumstances. This process assists with identifying feasible and efficient options that can deliver healthy and sustainable diets for all as well as healthy landscapes that are resilient to climate change.

 Climate Change and Global Health: Primary, Secondary and Tertiary Effects, 2nd Edition (eds C.D. Butler and K. Higgs)
DOI: 10.1079/9781800620025.0027

Globally, we are challenged by a triple burden (Box 27.1). Regenerative agricultural practices[1] offer us a partial pathway to address each of these three challenges, provided that there is significant investment of resources and effort to accelerate progress ahead of the climate clock. Some low- and middle-income countries (LMICs) are already making efforts in this direction with support from the Food and Agriculture Organization of the United Nations (FAO). Others must be supported to move in this direction rather than retrace the route of richer parts of the world over the past 100 years where intensive agricultural production has contributed to the loss of soil organic carbon (Herzfeld *et al.*, 2021) and biodiversity. The global North must also change course, installing 'nature-based solutions' that can reduce further agricultural greenhouse gas (GHG) emissions and sequester carbon dioxide in vegetation and soil.

27.2 The Way Agriculture Has Provided More Food: A Historical Review

The way humans acquire food has changed over time, becoming more extractive, intensive and polluting, especially over the past century. Dietary patterns have also changed, from more to less diverse, accompanied by a crisis-level loss of biodiversity, driven primarily by unsustainable production practices. But history also offers lessons on the value of regenerative production practices; these can be leveraged to help alleviate the triple burden.

From early history, wild animals and plants played a critical role in the emergence of Earth's most dominant mammalian species: *Homo sapiens*. Utilization of wild animal-source foods is reported to be one contributing factor to our species' large and complex brain and was critical to nutritional health, development and early expansion (Milton, 2003). Archaeological finds place the beginning of agriculture well before 7000 BCE. Since then, a growing human population has been sustained through the domestication of a small number of plant and animal species for food and, more recently, the industrialization of agricultural systems. This has occurred with little account taken of human nutritional requirements and natural capital (Wang *et al.*, 2009).

Many traditional and Indigenous wild foods are reported to have higher micronutrient concentrations than their domesticated counterparts (Cordain *et al.*, 2005). The manner in which foods were obtained from the wild inherently provided a diverse diet. Increased agricultural production and associated economic productivity metrics came with a significant reduction in the diversity of foods eaten. For example, before colonization in 1788, the diet of Indigenous Australians consisted of 150–750 different species, depending on regional fauna and flora. By the 21st century, dietary diversity of Aboriginal and Torres Strait Islander people has shrunk to around 45 different foods (Pascoe, 2014; Newton, 2016). The FAO has stated that, of 250,000 identified edible plant species, humanity had managed to reduce the 7000 that had been used for food to primarily three major staple cereals: maize, wheat and rice. The complexity of the past diet reflects variety and suggests better nutrition; this simplification reflects biodiversity loss and ecosystem degradation and risks the future security of food and nutrition.

In parallel with narrowing of the plant food basket, intensification of livestock production systems has steadily increased since the mid-1880s. In such systems, animals are no longer integrated into mixed crop–livestock farms[2] but are grazed or housed at high density, the latter dependent on dedicated feed production (Kirchhelle, 2018). Intensive, often 'landless' farms now dominate our global livestock food systems in terms of the quantity produced, while hundreds of millions of small-scale producers continue to raise livestock in smaller numbers in support of household livelihoods and local community health and well-being (FAO, 2020). These developments have disrupted production practices that formerly offered regenerative options for managing ecosystem degradation (Lal, 2020). Intensively

Box 27.1. The global triple burden.

1. There are multiple forms of malnutrition (wasting, stunting, micronutrient deficiencies, overweight/obesity) and diet-related diseases like diabetes, high blood pressure and some forms of cancer (FAO *et al.*, 2021).
2. There is a crisis of climate change and other forms of environmental degradation
3. Our current food systems are unsustainable.

raised livestock, aquaculture and crop production are reportedly associated with declining nutritional quality of the end product (Cordain *et al.*, 2005; Wang *et al.*, 2009). Agricultural intensification has also contributed to the emergence, spread and maintenance of new and established disease agents, as well as to antimicrobial and pesticide resistance (Kirchhelle, 2018) and negative impacts on animal welfare (Moore *et al.*, 2021).

There are significant environmental costs associated with intensification. Horticulture and grain intensification have been dependent on high input use, such as phosphorus and nitrogen fertilizers, herbicides and pesticides, antimicrobial agents, energy costs of mechanization and compaction and exposure of soils, resulting in loss and degradation (Stoate *et al.*, 2001; Cribb, 2017). The pesticides used affect soil ecosystems, impacting soil biological diversity and thus soil health. Reduced ecological connectivity, in addition to inorganic pesticides, may negatively impact natural pollinators[3] and pest control as well as broader biodiversity, including many non-pollinating insects that are vital in the food chain. There are some potential benefits of intensification arising from feeding livestock high-quality feed, with improved feed conversion efficiency and reduced methane emissions compared with high-roughage feed grazed under non-regenerative management practices. The total GHG emissions have increased as a result of concentrating large numbers of livestock in small areas. However, GHG emissions are not the only consideration.

Intensification has been accompanied by increasing globalization of the food system, longer value chains and decreased farmer and producer negotiating power (FAO, 2017). Increasing yields, i.e. quantity of food, has been essential to support financial security for farmers who sell into value chains where food is traded based on weight or volume; the natural nutrient density of the products is generally ignored (Alders *et al.*, 2018). This has gradually resulted in a high dependency on a few staple foods. These changes to the food system have increased competition between humans and animals for the few dominant human-edible crops. It is estimated that livestock consume one-third of global cereal production[4], requiring approximately 40% of arable land (Mottet *et al.*, 2017). Berners-Lee *et al.* (2018) argued that there is no nutritional case for feeding human-edible crops to animals and that a failure to heed this message will require a 119% increase in edible crops grown by 2050. By employing sustainable and circular bio-economy principles, Van Zanten *et al.* (2019) proposed that livestock production systems that convert grass and food system by-products into edible food and manure can contribute significantly to human nutrition, while simultaneously reducing negative environmental impacts.

A combination of increased productivity, waste reduction, environmentally protective and regenerative food production is required to complement a change in dietary patterns. Ultimately, sustainably sourced, high-quality, nutrient-dense food is required to deliver healthy diets while safeguarding future food production, including building increased resilience in the face of climate change. This is the ideal, gold standard that is explored in the remainder of this chapter.

27.3 Supporting the Water Cycle

Water, soil and environmental health have to be managed better for humanity to sustain nutrition and ecosystem health, to contribute to climate mitigation and in turn long-term food security and nutrition. Agriculture is particularly sensitive to the intensifying water cycle. Climate change is increasing both evaporation and transpiration, leading to acceleration of drying, drought and subsequent fire intensity, while, at the same time, increased moisture in the air is leading to increased storm intensity and flooding (IPCC, 2021). This puts crop and pasture lands at further risk as the cleared trees and woodlands no longer provide the shade and rain-infiltration buffering capacity that heterogeneous ecosystems do. However, rehydration projects worldwide – for example the Loess Plateau in China (Liu *et al.*, 2021) and in Australia (Mulloon Institute, 2022) – are showing how slowing and spreading rainfall can recharge aquifers and the root zone, thus increasing 'effective' precipitation and reducing or even reversing erosion. This provides near-term resilience to floods and droughts. Furthermore, planting tree rows and shelter belts along ridge, creek and contour lines as agro-forestry can protect crop and pastureland (silvo-pasture). This enables mixed farms to support multiple commodities and natural areas to withstand the increasing stress of climate change.

27.4 Soil Health Favours Long-term Food Security and Climate Mitigation Through Carbon Sequestration

The biosphere holds a tremendous carbon reserve, or stock, in vegetation (> 500 gigatonnes of carbon (GtC)) and soils (> 2000 GtC) which even exceeds proven fossil fuel reserves (Hepburn *et al.*, 2019). Soil carbon levels are higher in colder regions, where life moves slowly, while relatively more carbon is above ground in the tropics. Over 100 GtC has been lost from soils since sedentary agriculture commenced; avoiding further loss is very important, as soil carbon takes a long time to recover (Sanderman *et al.*, 2017). At the same time, the terrestrial biosphere cycles a tremendous carbon flux of ~123 GtC/year captured and used by plants in photosynthesis but then released again by plants (60 GtC/year) and soils (60 GtC/year). By contrast, it takes humanity a decade to cumulatively emit this amount of carbon from the combustion of fossil fuels. So, adjusting Earth systems via shifting global carbon (and water) cycles can have a large planetary effect, particularly as agriculture is practised on over half of the habitable land area.

Under current practices, agriculture is a large direct source of GHG emissions (including potent methane and nitrous oxide) but is also responsible for indirect emissions via clearing of natural vegetation whose carbon-capturing capacity is eliminated. Enhancing Earth systems by using regenerative agriculture may potentially significantly slow, or even reverse, global warming, by re-sequestering much or even all of > 100 GtC that has been lost.[5] This potential has been demonstrated at smaller scale via Indigenous methods of food sourcing. For example, integrative analyses in Ecuador identified four important ancestral farming technologies that are highly applicable for wider adoption in support of climate change adaptation: terraces, raised beds and two types of reservoirs that use sloping land to catch water and channel it to sites where water accumulates (Carrasco-Torrontegui *et al.*, 2021). Such agricultural practices must be implemented alongside transformative changes in our energy system, so that renewables electrify homes and other buildings, and are used for transport and industrial processes. Using agriculture for soil and ecosystem regeneration comes with many additional benefits to food and nutritional security, by providing habitats for diverse species while capturing, storing and filtering water. The crop rotational practices of regenerative agriculture include more diverse crop species; this also benefits dietary diversity and, in turn, potential nutrition and health outcomes. A study by Montgomery *et al.* (2022) on ten farms in the USA found that plant- and animal-source food grown under regenerative practices contained, on average, more magnesium, calcium, potassium and zinc and more vitamins, including B1, B12, C, E and K.

The challenge of achieving nutrition-sensitive agriculture and value chains in the face of climate change, weather variability, diminishing agricultural lands, declining soil health and increasing human and domesticated animal populations presents an unparalleled opportunity to rethink how we produce food to maintain healthy ecosystems and keep people well-nourished and healthy (Alders *et al.*, 2018). Soil is the base of the food chain, where rainwater is absorbed and nutrients are recycled to support plant growth that in turn supports animal life, including humans. Plants capture carbon dioxide from the air via photosynthesis and contribute to building soil via root and microbial growth as well as by adding organic matter through biomass litter decay. Healthy soils host an intricate food web of micro- and macro-organisms and buffer natural cycles of drought, fire and flood. Living soils store and cycle more carbon than degraded landscapes, while allowing greater water infiltration. Nutrients, mainly nitrogen and phosphorus, are made available via decomposition of plant matter, liberation from subsoil rocks or external amendments, including fertilizers. The latter have lifted yields as part of the Green Revolution, but come with environmental costs, including water eutrophication and GHG emissions, when over-applied. In general, healthy soils under regenerative management (i.e. that exclude the use of inorganic fertilizer) lead to better productivity, nutrient density and yield stability, enabling a return on the near-term investment required to restore them to a highly active state through careful management (Montgomery et al., 2022).

27.5 Not All Agricultural Practices Are Created Equal for Ecosystem Management and Nutrition Security

Nutritious, safe food is essential to good health; yet, many current agricultural practices are significant contributors to GHG emissions, biodiversity loss and land and water pollution (FAO, 2017; Mbow et al., 2019). The focus of agriculture in recent decades has been boosting production by adding more land into production and lifting productivity by increasing the weight/volume produced per hectare. Livestock and plant cultivars have been selected for their high yielding ability; often land has been fertilized and irrigated to extend the high yield and to compensate for soils that would otherwise be increasingly depleted of nutrients needed by plants. Globally, this has led to decreased diversity of food crop species in the past 50 years and increased homogeneity in food supplies. At the national level, there is more energy, protein, fat and food by weight available per capita (Khoury et al., 2014) but distribution of this food has been inequitable. At the product level, this has led to the 'dilution effect', where rapid growth has resulted in reduced micronutrient concentrations in both plant- and animal-source foods (Marles, 2017). Numerous studies have found a downward trend in mineral concentrations in fruit, vegetables and grains over the past few decades (Cordain et al., 2005; Thomas, 2007). It is important to note the uncertainties when interpreting the results of these retrospective, comparative studies, including temporal differences in sampling, analytical methods, the environment where plants were grown and the sources of data (Thomas, 2007). Further research into the impacts of agricultural practices on soil and food nutrient content, such as that being undertaken at the Centre for Regenerative Agriculture and Resilient Systems (Karas, 2022), is vital to better understand this complex issue.

Concurrently, malnutrition in all its forms is affecting every country on Earth. Unhealthy diet has been recently reported to contribute six of the top ten risk factors for the global burden of disease (Murray et al., 2020). Yet for 3 billion people, healthy diets are simply unaffordable (FAO et al., 2021). Unhealthy diets are one of four major risk factors driving the rise of non-communicable diseases, including diabetes mellitus, cancer and cardiovascular disease (FAO et al., 2021). Deficiencies of iron, vitamin A and iodine are a global public health problem, particularly for vulnerable nations and groups such as children and women of reproductive age. Agriculture is contributing to solutions. Currently available micronutrient-enriched crops by biofortification include iron beans, zinc and iron pearl millet, and vitamin A sweet potatoes, maize and cassava; however, micronutrient deficiencies rarely occur alone and the long-term benefits of this approach remain to be confirmed in a number of cases. Additional consequences of the emergence of sedentary agriculture thousands of years ago and its recent intensification have been significant for women. The females of our species have received less attention in relation to their nutritional requirements. The consequent impacts on women and girls of a shift to consumption of food with diminished nutritional profiles include: (i) the emergence of gender-based division of labour (Kohn, 2017); (ii) declining power of women; (iii) decreased height of women relative to men (i.e. enhanced sexual dimorphism) (Dyble et al., 2015; Gray and Wolfe, 1980); and (iv) increasing prevalence of anaemia in women (Metz et al., 1971).

In addition to these global changes, high-input agriculture has had an effect at the regional level on micronutrient availability. For example, repeated application of rock phosphate-based fertilizers to agricultural land leads to accumulation of cadmium in soils. Due to cadmium contamination in New Zealand and Australia, offal from many older cattle and sheep cannot enter the human food chain (New Zealand Government, 2020; Australian Government, 2020). Australia and New Zealand rank second and third, respectively, in global sheep meat production and Australia ranks fourth in global cattle meat production. Liver is the most nutrient-dense edible product of livestock and is a rich dietary source of iron and vitamin A; kidneys are lean, with relatively higher iron content than muscle meat. Cadmium contamination of soils has lessened, but continues to appreciably reduce the availability of key micronutrients from livestock value chains. With the increasing promotion of organic fertilizers, cadmium levels in organic fertilizers such as biosolids also need to be managed. In LMICs, monitoring of heavy metals and other potentially harmful substances is not routine; as fertilizer and other chemical use increase with agricultural intensification, the potential harmful impacts of widespread offal consumption remain largely unknown and call for investigation.

27.6 Agriculture for Regeneration Offers Greater Sustainability for Soil Health, Nutrition and the Planet

The term regenerative agriculture has been used to describe food and natural fibre production systems that not only conserve soil and water, biodiversity and ecosystem services on agricultural land but also remediate damage (Newton *et al.*, 2020). A key feature is the promotion of biologically active soils to drive these functions and to sequester a globally significant quantity of carbon, which would otherwise be in the atmosphere or ocean.

This term has been increasingly applied to a number of agricultural practices, some pre-dating the recent use of this term. The following examples reflect the diversity of land under agriculture and the complexity of challenges to be met (Giller *et al.*, 2021). Agro-ecology, agro-forestry and permaculture promote farming systems that mimic self-sustaining and regulating ecosystems. These typically integrate livestock to eat pests, convert waste and provide fertilizer as well as ultimately becoming a food product or producer. Organic and biodynamic methods are frequently practised in parallel. Natural sequence farming and keyline farming, developed in Australia, a dry continent with fragile soils, focus on slowing the flow of water through the landscape and the reversal of dryland salinity (Iles, 2021). The rotation of livestock so that pulses of intense grazing pressure keep grass at peak nutrition, reduce weeds, input manure and sequester carbon is a central feature of holistic management and grazing optimization. In grain production, direct drilling or no-till agriculture maintains groundcover on the soil at all times and prevents topsoil loss and degradation. This has also been referred to as conservation agriculture. Additives such as 'biochar', a stable charcoal produced by anaerobic pyrolysis of biomass, also improve soil health and carbon sequestration capacity.[6] Indigenous and traditional agricultural practices embody many of these features, including biodiversity conservation (Kuhnlein, 2015; Saxena *et al.*, 2016).

In LMICs, FAO has promoted agro-ecology to increase socially equitable food systems within which people can exercise food sovereignty alongside environmental sustainability of food production. FAO has provided support for smallholders, who often have strong links to traditional agricultural practices and low external inputs, aiming to adapt production methods and guide cross-sectoral and multiscale governance and public policies towards sustainable agriculture and food systems (FAO, 2018). Agro-ecology is now centrally concerned with the empowerment and equity of smallholder producers within the food system. Five hundred million smallholders are estimated to feed around 2 billion people, yet are overrepresented amongst the poor and undernourished (Parmentier, 2014). These smallholder producers are critical in meeting the increased global demand for food but require support to access markets and infrastructure and build climate resilience. Population growth will be greatest in sub-Saharan Africa and Asia, where smallholders contribute the majority of food produced and human undernutrition levels remain higher than the global average. Drawing on food sovereignty, regenerative agriculture and nature-based solutions (Iseman and Miralles-Wilhelm, 2021), agro-ecology provides an example of global food system transformation. However, time and appropriate support are required to reduce reliance on imported food products (Noort *et al.*, 2022), to assist small-scale farmers who adopted more intensive agricultural practices (as recommended by most international development programmes during the second half of the 20th century), and to transition/return to more nature-based management practices (John and Babu, 2021).

In many high-income countries, there has been no specific government support for regenerative or sustainable agriculture. Niche markets and eco-labelling have provided some opportunities for regenerative farmers and producers to realize costs. For many farmers and producers, growing interest in regenerative agricultural practices is a response to frustration at rising input costs, including oil and chemicals, declining farm production, and concern for the future:

farm succession and resilience against climate change. Adoption of these more sustainable methods is associated with increased well-being (Brown *et al.*, 2021). However, while there are many anecdotal and cross-fence farm comparisons of improved profitability and productivity, data are difficult to find (Giller *et al.*, 2021). In temperate zones, conversion from highly intensive agriculture to regenerative systems typically results in a reduction of yields (de Ponti *et al.*, 2012; Knapp and van der Heijden, 2018). The yield gap may be closer to parity in different regions or with different crops, particularly when measured over long timeframes allowing for soil regeneration and increased drought resilience (Hepperly *et al.*, 2006). Plateauing and declining crop yields under conventional, high-input agriculture provide a shifting comparison.

Slowing the increase in global mouths to feed, broad adoption of healthy and sustainable diets (Willett *et al.*, 2019) and equitable distribution of food are required to achieve a social as well as environmentally sustainable global food system. These are essential transformations irrespective of whether all current global food production requirements can be met through regenerative agriculture or other forms of land sharing (expansion) or land sparing (intensification) (Balmford *et al.*, 2019).

27.7 Call for a Food Alliance for Regeneration

Broad on-farm adoption and consumer support for regenerative agriculture requires further evidence, education and supportive policies. SDG target 2.4 ('to ensure sustainable food production systems that progressively improve land and soil quality') provides direction. In the European Union, social equity and environmental protection and regeneration are at the heart of the European Union's Farm-to-Fork strategy, part of the European Green Deal (Montanarella and Panagos, 2021).

Evidence suggests that what is good for environmental health is also good for human nutrition and health (Saxena *et al.*, 2016). There are physical and mental health benefits from eating healthy, nutritious food. Agricultural and health professionals working together have the potential to achieve optimal outcomes. Indigenous knowledge custodians and creators, engineers and food scientists are also required to achieve nutrient recycling and innovative processing of food with healthy nutritional profiles, to ensure stable supply across the seasons and years. Realigning policy drivers in the agricultural, health and economic sectors will be a complex but worthy endeavour which will require active collaboration with social scientists, including economists and political scientists, working across sectors and disciplines.

The importance of establishing resilient, functional linkages between regenerative food production systems, public health nutrition and global health has been highlighted during the COVID-19 pandemic where rates of malnutrition and food insecurity have risen dramatically (Food Secure Canada, 2020; World Bank, nd).

27.8 Conclusion

Commitment by and collaboration between public and private sectors and civil society is vital to ensure the delivery of enabling environments that facilitate: (i) the production of food using the principles of regenerative, climate-resilient agriculture; (ii) valuing food according to its natural nutrient density in addition to weight and/or volume; (iii) promoting the efficient utilization of food by tailoring nutrition by gender, age, reproductive and health status; (iv) valuing nutrients in waste products and enabling their recycling; and (v) restructuring healthcare services to place a higher value on the contributions of agriculture and farmers to preventive medicine.

Possibly most important will be a return to more 'mindful' practices of agriculture, land stewardship and the balancing of health and nutrition for all. One option for achieving this would be adopting the Indigenous Australians' 'pre-colonization' situation of actively teaching 'traditional food knowledge', linking intimately with responsibility to the land, taking only as much as needed and allowing nature to regenerate itself (Divarkarla *et al.*, 2020).

Acknowledgements

Thanks go to the researchers, producers and farmers in many parts of the world who have given freely of their time and expertise over the years.

Notes

[1] 'Regenerative agriculture describes holistic farming systems that, among other benefits, improve water and air quality, enhance ecosystem biodiversity, produce nutrient-dense food, and store carbon to help mitigate the effects of climate change. ... These farm systems are designed to work in harmony with nature, while also maintaining and improving economic viability' (Newton *et al.*, 2020).
[2] Mixed livestock–cropping systems usually occur in areas favourable for crop production (Alders *et al.*, 2021).
[3] Natural pollinators are insects such as bees, butterflies, beetles and flies, as well as some birds such as hummingbirds and mammals such as bats.
[4] More than three-quarters (77%) of global soy is fed to livestock for meat and dairy production. Most of the rest is used for biofuels, industry or vegetable oils (Ritchie *et al.*, 2021).
[5] Sequestration potential of soil C.
[6] Biochar sequestration potential 0.3 to 2 Gt CO_2/year (Hepburn *et al.*, 2019).

References

Alders, R.G., Ratanawongprasat, N., Schönfeldt, H. and Stellmach, D. (2018) A planetary health approach to secure, safe, sustainable food systems: workshop report. *Food Security* 10, 489–493. DOI: 10.1007/s12571-018-0780-9

Alders, R.G., Campbell, A., Costa, R., Guèye, E.F., Hoque, M.A. *et al.* (2021) Livestock across the world: diverse animal species with complex roles in human societies and ecosystem services. *Animal Frontiers* 11(5), 20–29. DOI: 10.1093/af/vfab047

Australian Government (2020) *Meat Notice 2020/3: Establishment sourcing of stock to comply with importing country requirements for cadmium levels in offal*. Department of Agriculture, Canberra. Available at: https://www.agriculture.gov.au/export/controlled-goods/meat/elmer-3/notices/2020/mn20-03 (accessed 18 October 2021).

Balmford, B., Green, R.E., Onial, M., Phalan, B. and Balmford, A. (2019) How imperfect can land sparing be before land sharing is more favourable for wild species? *Journal of Applied Ecology* 56, 73–84. DOI:10.1111/1365-2664.13282

Berners-Lee, M., Kennelly, C., Watson, R. and Hewit, C.N. (2018) Current global food production is sufficient to meet human nutritional needs in 2050 provided there is radical societal adaptation. *Elementa: Science of the Anthropocene* 6, 52. DOI: 10.1525/elementa.310

Brown, K., Schirmer, J. and Upton, P. (2021) Regenerative farming and human wellbeing: Are subjective wellbeing measures useful indicators for sustainable farming systems? *Environmental and Sustainability Indicators* 11, 100132. DOI: 10.1016/j.indic.2021.100132

Carrasco-Torrontegui, A., Gallegos-Riofrío, C.A., Delgado-Espinoza, F. and Swanson, M. (2021) Climate change, food sovereignty, and ancestral farming technologies in the Andes. *Current Developments in Nutrition* 5, 54–60. DOI: 10.1093/cdn/nzaa073

Cordain, L., Eaton, S.B., Sebastian, A., Mann, N., Lindeberg, S. *et al.* (2005) Origins and evolution of the Western diet: health implications for the 21st century. *The American Journal of Clinical Nutrition* 81, 341–354. DOI: 10.1093/ajcn.81.2.341

Cribb, J. (2017) *Surviving the 21st Century: Humanity's Ten Great Challenges and How We Can Overcome Them*. SpringerLink, Cham, Switzerland.

de Ponti, T., Rijk, B. and Van Ittersum, M.K. (2012) The crop yield gap between organic and conventional agriculture. *Agricultural Systems* 108, 1–9. DOI: 10.1016/j.agsy.2011.12.004

Divakarla, S.P., Aroni, R. and Sutherland, S. (2020) Transmission of traditional food knowledge: experiences and perspectives of young adults. *Mahika Kai Journal* 1, 96–129. DOI: 10.34900/mk.v1i1.1155

Dyble, M., Salali, G.D., Chaudhary, N., Page, A., Smith, D. *et al.* (2015) Sex equality can explain the unique social structure of hunter-gatherer bands. *Science* 348, 796–798. DOI: 10.1126/science.aaa5139

FAO (2017) *The Future of Food and Agriculture – Trends and Challenges*. Food and Agriculture Organization, Rome. Available at: https://www.fao.org/3/i6583e/i6583e.pdf (accessed 4 October 2021).

FAO (2018) *The 10 Elements of Agroecology: Guiding the Transition to Sustainable Food and Agricultural Systems*. Food and Agriculture Organization, Rome. Available at: https://www.fao.org/3/I9037EN/i9037en.pdf (accessed 4 October 2021).

FAO (2020) *Sustainable Small-scale Livestock Production*. Food and Agriculture Organization, Rome. Available at: https://www.fao.org/3/ca7660en/CA7660EN.pdf (accessed 4 October 2021).

FAO, IFAD, UNICEF, WFP and WHO (2021) *The State of Food Security and Nutrition in The World 2021. Transforming Food Systems for Food Security, Improved Nutrition and Affordable Healthy Diets for All*. Food and Agriculture Organization, Rome. DOI: 10.4060/cb4474en

Food Secure Canada (2020) COVID-19 highlights food system inequities: Time for systemic change. Food Secure Canada science blog, 18 March. Food Secure Canada, Montreal. Available at: https://foodsecurecanada.org/resources-news/news-media/covid-19-highlights-food-system-inequities-time-systemic-change (accessed 4 October 2021).

Giller, K. E., Hijbeek, R., Andersson, J. A. and Sumberg, J. (2021) Regenerative agriculture: An agronomic perspective. *Outlook on Agriculture* 50, 13–25. DOI: 10.1177/0030727021998063

Gray, J.P. and Wolfe, L.D. (1980) Height and sexual dimorphism of stature among human societies. *American Journal of Physical Anthropology* 53, 441–456. DOI: 10.1002/ajpa.1330530314

Hepburn, C., Adlen, E., Beddington, J., Carter E.A., Fuss S. *et al.* (2019) The technological and economic prospects for CO_2 utilization and removal. *Nature* 575, 87–97. DOI: 10.1038/s41586-019-1681-6

Hepperly, P.R., Douds, D. and Seidel, R. (2006) The Rodale Institute farming systems trial 1981 to 2005: Long-term analysis of organic and conventional maize and soybean cropping systems. In: Raupp, J., Pekrun, C., Oltmanns, M. and Kopke, U. (eds) *Long-term Field Experiments in Organic Farming*. International Society of Organic Agriculture Research, Berlin, pp. 15–31.

Herzfeld, T., Heinke, J., Rolinski, S. and Müller, C. (2021) Soil organic carbon dynamics from agricultural management practices under climate change. *Earth Systems Dynamics* 12, 1037–1055. DOI: 10.5194/esd-12-1037-2023

Iles, A. (2021) Can Australia transition to an agroecological future? *Agroecology and Sustainable Food Systems* 45, 3–41. DOI: 10.1080/21683565.2020.1780537

IPCC (2021) *Technical Summary. Climate Change 2022: Impacts, Adaptation and Vulnerability*. Working Group II Contribution to the IPCC Sixth Assessment Report. The Intergovernmental Panel on Climate Change, Geneva. Available at: https://www.ipcc.ch/report/ar6/wg2/ (accessed 25 July 2022).

Iseman, T. and Miralles-Wilhelm, F. (2021) *Nature-based Solutions in Agriculture: The case and pathway for adoption*. Food and Agriculture Organization, Rome, and The Nature Conservancy, Arlington, Virginia. Available at: https://www.fao.org/documents/card/en/c/cb3141en/ (accessed 6 April 2022). DOI: 10.4060/cb3141en

John, D.A. and Babu, G.R. (2021) Lessons from the aftermaths of Green Revolution on food system and health. *Frontiers of Sustainable Food Systems* 22, 644559. DOI: 10.3389/fsufs.2021.644559

Karas, S. (2022) Could regenerative agriculture increase the nutritional quality of our food? Center for Regenerative Agriculture and Resilient Systems, California State University, Chico. Available at: https://www.csuchico.edu/regenerativeagriculture/blog/nutrient-density.shtml (accessed 28 March 2022).

Khoury, C.K., Bjorkman, A.D., Dempewolf, H. and Struik, P.C. (2014) Increasing homogeneity in global food supplies and the implications for food security. *Proceedings of the National Academy of Sciences (USA)* 11, 4001–4006. DOI: 10.1073/pnas.1313490111

Kirchhelle, C. (2018). Pharming animals: a global history of antibiotics in food production (1935–2017). *Palgrave Communications* 4, Article 96. DOI:10.1057/s41599-018-0152-2

Knapp, S. and van der Heijden, M.G.A. (2018) A global meta-analysis of yield stability in organic and conservation agriculture. *Nature Communications* 9, 3632. DOI: 10.1038/s41467-018-05956-1

Kohn, L. (2017) *Pristine Affluence: Daoist Roots in the Stone Age*, 1st edn. Three Pines Press, St Petersburg, Florida.

Kuhnlein, H.V. (2015) Food system sustainability for health and well-being of Indigenous Peoples. *Public Health Nutrition* 18, 2415–2424. DOI: 10.1017/S1368980014002961

Lal, R. (2020) Integrating animal husbandry with crops and trees. *Frontiers in Sustainable Food Systems* 4, 113. DOI: 10.3389/fsufs.2020.00113

Liu, F., Kang, P., Zhu, H., Han, J., and Huang, Y. (2021) Analysis of spatiotemporal groundwater-storage variations in China from GRACE. *Water* 13, 2378. DOI: 10.3390/w13172378

Marles, R.J. (2017) Mineral nutrient composition of vegetables, fruits and grains: The context of reports of apparent historical declines. *Journal of Food Composition and Analysis* 56, 93–103. DOI: 10.1016/j.jfca.2016.11.012

Mbow, C., Rosenzweig, C., Barioni, L.G., Benton, T.G., Herrero, M. *et al.* (2019) Food security. In: Shukla, P.R., Skea, J., Calvo Buendia, E., Masson-Delmotte, V., Pörtner, H.-O. *et al.* (eds) *Climate Change and Land: An IPCC Special Report on Climate Change, Desertification, Land Degradation, Sustainable Land Management, Food Security, and Greenhouse Gas Fluxes in Terrestrial Ecosystems*. Intergovernmental Panel on Climate Change, Geneva, pp. 437–550. Available at: https://www.ipcc.ch/site/assets/uploads/sites/4/2021/02/08_Chapter-5_3.pdf (accessed 4 October 2021).

Metz, J., Hart, D. and Harpending, C. (1971) Iron, folate, and vitamin B12 nutrition in a hunter-gatherer people: a study of the !Kung Bushmen. *The American Journal of Clinical Nutrition* 24, 229–242. DOI:10.1093/ajcn/24.2.229

Milton, L. (2003) The critical role played by animal source foods in human (Homo) evolution. *The Journal of Nutrition* 133, 3886S–3892S. DOI: 10.1093/jn/133.11.3886S

Montanarella, L. and Panagos, P. (2021) The relevance of sustainable soil management within the European Green Deal. *Land Use Policy* 100, 104950. DOI: 10.1016/j.landusepol.2020.104950

Montgomery, D.R., Biklé, A., Archuleta, R., Brown, P. and Jordan, J. (2022) Soil health and nutrient density: preliminary comparison of regenerative and conventional farming. *PeerJ* 10, e12848. DOI: 10.7717/peerj.12848

Moore, T.C., Fong, J., Hernández, A.M.R. and Pogreba-Brown, K. (2021) CAFOs, novel influenza, and the need for One Health approaches. *One Health* 13, 100246 DOI: 10.1016/j.onehlt.2021.100246

Mottet, A., de Haan, C., Falcucci, A., Tempio, G., Opio, C. and Gerber, P. (2017) Livestock: On our plates or eating at our table? A new analysis of the feed/food debate. *Global Food Security* 14, 1–8. DOI: 10.1016/j.gfs.2017.01.001

Mulloon Institute (2022) *Landscape rehydration plan*. Mullon Institute, Bungendore, NSW, Australia. April 2022. Available at: http://www.mulloonconsulting.com.au/landscape-rehydration-plans/ (accessed 23 July 2022).

Murray, C.J.L., Aravkin, A.Y., Zheng, P., Abbafati, C., Abbas, K.M. *et al.* (2020) Global burden of 87 risk factors in 204 countries and territories, 1990–2019: a systematic analysis for the Global Burden of Disease Study 2019. *The Lancet* 396, 1223–1249. DOI: 10.1016/S0140-6736(20)30752-2

Newton, J. (2016) *The Oldest Foods on Earth: A History of Australian Native Foods with Recipes.* New South Publishing, Sydney, Australia.

Newton, P., Civita, N., Frankel-Goldwater, L., Bartel, K. and Johns, C. (2020) What is regenerative agriculture? A review of scholar and practitioner definitions based on processes and outcomes. *Frontiers in Sustainable Food Systems* 4, 194. DOI: 10.3389/fsufs.2020.577723

New Zealand Government (2020) *Monitoring cadmium in New Zealand soils*. Available at: https://www.mpi.govt.nz/funding-rural-support/environment-and-natural-resources/land-and-soil-health/monitoring-cadmium-in-nz-soils/ (accessed 18 October 2021).

Noort, M.W.J., Renzetti, S., Linderhof, V., du Rand, G.E., Marx-Pienaar, N.J.M.M. *et al.* (2022) Towards sustainable shifts to healthy diets and food security in Sub-Saharan Africa with climate-resilient crops in bread-type products: A food system analysis. *Foods* 11, 135 DOI: 10.3390/foods11020135

Parmentier, S (2014) *Scaling-up Agroecological Approaches: What, Why and How?* Oxfam- Solidarité, Brussels. Available at: https://www.gaiafoundation.org/app/uploads/2017/09/Agroecology_Scaling-up_agroecology_what_why_and_how_-OxfamSol-FINAL.pdf (accessed 25 July 2022).

Pascoe, B. (2014) *Dark Emu*, 1st edn. Magabla Books, Broome, Australia.

Ritchie, H., Spooner, F. and Roser, M. (2021) *Forests and Deforestation.* Published online at OurWorldInData.org. Available at: https://ourworldindata.org/forests-and-deforestation (accessed 11 February, 2023).

Sanderman, J., Hengl, T. and Fiske, G.J. (2017) Soil carbon debt of 12,000 years of human land use. *Proceedings of the National Academy of Science (USA)* 114, 9575–9580. DOI: 10.1073/pnas.1706103114

Saxena, K.A., Fuentes, C.X., Herbas, R.G. and Humphries, D.L. (2016) Indigenous food systems and climate change: Impacts of climatic shifts on the production and processing of native and traditional crops in the Bolivian Andes. *Frontiers in Public Health* 4, 20. DOI: 10.3389/fpubh.2016.00020

Stoate, C., Boatman, N.D., Borralho, R.J., Rio Carvalho, C., de Snoo, G.R. *et al.* (2001) Ecological impacts of arable intensification in Europe. *Journal of Environmental Management* 63, 337–365. DOI: 10.1006/jema.2001.0473

Thomas, D. (2007) The mineral depletion of foods available to us as a nation (1940–2002): A Review of the 6th Edition of McCance and Widdowson. *Nutrition and Health* 19, 21–55. DOI: 10.1177/026010600701900205

Van Zanten, H.H.E., Van Ittersum, M.K. and de Boer, I.J.M. (2019) The role of farm animals in a circular food system. *Global Food Security* 21, 18–22. DOI: 10.1016/j.gfs.2019.06.003

Wang, Y., Lehane, C., Ghebremeskel, K. and Crawford, M.A. (2009) Modern organic and broiler chickens sold for human consumption provide more energy from fat than protein. *Public Health Nutrition* 13, 400–408. DOI: 10.1017/S1368980009991157

Willett, W., Rockström, J., Loken, B., Springmann, M., Lang, T. *et al.* (2019) Food in the Anthropocene: the EAT–Lancet Commission on healthy diets from sustainable food systems. *The Lancet* 393, 447–492. DOI: 10.1016/S0140-6736(18)31788-4

World Bank (nd) *Food Security and COVID-19*. World Bank, Washington, DC. Available at: https://www.worldbank.org/en/topic/agriculture/brief/food-security-and-covid-19 (accessed 11 October 2021).

28 Climate Change, Breastfeeding and Health

Julie P. Smith
National Centre for Epidemiology and Population Health, The Australian National University, Canberra, Australia

Abstract
This chapter addresses the environmental impacts and implications of commercial milk formula feeding of infants in the context of the increasingly evident human health risks from climate change. These risks disproportionately threaten women, infants and children, yet the feeding of infants and young children, including by the substitution of breastfeeding with formula, is largely absent from climate change discussions. The chapter describes current global trends in diets for infants and young children, and the climate and health implications if breastfeeding were to become vestigial. Transformational pathways to a more healthy and sustainable food system for infants and young children are also discussed. These include a greater appreciation of ecological and feminist economic thought.

> Good health is the living manifestation of a species' biological compatibility with its ecosystem,
>
> (Tony McMichael, 1991)

28.1 Introduction

Global commercial milk formula (CMF) sales are booming, particularly in China (Baker *et al.*, 2016), rising from US$1.5 billion to US$55.6 billion annually between 1978 and 2019 (Baker *et al.*, 2021b). There is a global transition towards children starting life on a diet of commercial milk formulas (Baker *et al.*, 2016). A futuristic scenario for the USA refers to signs that breastfeeding is already vestigial (Whittington, 2015). For reasons elaborated below, this trend contributes to a dire future for planetary health, as well as for human health and well-being.

Infancy and early childhood are the most vulnerable periods of human life for illness and death. Furthermore, early nutrition establishes the trajectory for lifetime health and development. Despite this, the climate change implications of the food system for infant and young children food systems are under-recognized. Climate change-related risks for children include heat stress, climate-sensitive infections and impacts on nutrition and food security (Nakstad *et al.*, 2022).

Humans are mammals. Mammals evolved as a class over 70 million years and are mostly defined by suckling their young (Dettwyler, 1995; Fildes, 1995; Sellen, 2007). Breastfeeding was the main nutrition for infants and young children to at least 12 months of age and typically continued for 3 or more years (Eaton *et al.*, 1994; Loponte and Mazza, 2021). Complex and dynamic components of mother's milk (Kunz *et al.*, 1999) reduce infection risk, strengthen the immune system (Hanson, 2004; Volk and Atkinson, 2013) and promote normal development of body organs, including the nervous system and the brain (Smith and Harvey, 2011).

 Climate Change and Global Health: Primary, Secondary and Tertiary Effects, 2nd Edition (eds C.D. Butler and K. Higgs)
DOI: 10.1079/9781800620025.0028

Breastfeeding and lactation result in endocrinal changes that not only assist with mother–child bonding, but also delay the first post-partum ovulation and thus assist with child spacing and maternal reproductive health. While traditional breastfeeding populations have around 160 menstrual cycles during their lifetime to aged 60, modern women have around 450 (Eaton *et al.*, 1994). In populations without modern contraception, breastfeeding prevents up to half of pregnancies, an important role in reducing the contribution of population growth to climate change (Victora *et al.*, 2016), improving lifelong reproductive health and, for some, avoiding additional hardship.

Most attention in climate change discussions focuses on fossil fuels, yet it is increasingly recognized that food systems are an important driver of contemporary interrelated health and ecological crises. Their transformation is thus a key opportunity to improve planetary as well as human health. An important element of transformative change is moving away from animal-based diets, while 'first-food' systems[1] represent an underexplored aspect of global food system transformations (Dadhich *et al.*, 2021).

CMF is an ultra-processed dairy food. Many are unaware that commercial milk formula is powdered cow's milk, removed at the expense of calves. Mammalian milk is species-specific, so this 'cross-species nursing' has profound effects on child health, survival and development. As the evolved first-food system for the human infant (Smith, 2015; Baker *et al.*, 2021a), a decline in breastfeeding has far-reaching implications for both mother and child. For example, children deprived of breastmilk have substantially greater rates and severity of malnourishment, infectious illnesses such as diarrhoea and pneumonia, and higher mortality, including from sudden infant death syndrome (SIDS). Children who are not breastfed also have higher risk of chronic disease and reduced cognitive development (Horta *et al.*, 2015; Victora *et al.*, 2016; Meek and Noble, 2022). Mothers who end breastfeeding prematurely have higher rates of diabetes and breast cancer (Victora *et al.*, 2016; Ciampo and Ciampo, 2018). The risk of reproductive cancers in Western women may be up to 100-fold greater compared with prehistoric populations (Eaton *et al.*, 1994).

Breastmilk substitutes (BMSs) emerged from the mid-19th century as a niche product for the small proportion of children unable to breastfeed for socio-economic or biological reasons (Fildes, 1995). In the 1930s, infant formulas received medical endorsement as US physicians, notably paediatricians, aligned with the infant formula industry to oppose unregulated forms of commercial infant foods, argued as unsafe (Greer and Apple, 1991). From the 1950s, BMSs were widely promoted, including through health systems, disrupting breastfeeding worldwide (Fig. 28.1 for long-term trends in Norway and Australia). By 1973, World Bank nutritionist Alan Berg (1973) was warning that:

> An unusual depletion in the crude oil reserves of an oil producing country of Asia or Latin America would be termed a crisis. Its economic and social implications would be so apparent that actions to reverse the trend would be awarded high priority. Yet a comparable crisis, involving a valuable natural resource and losses in the hundreds of millions of dollars, is going virtually unnoticed in many of the poor countries of the world. The resource is human breast milk, and the loss is caused by the dramatic and steady decline of maternal nursing in recent decades.

By 2019 fewer than half of infants globally were exclusively breastfed for 6 months. Since 2000, the prevalence of breastfeeding beyond 12 months is reportedly declining in low- and lower-middle-income countries and CMF has substantially replaced animal milk feeding for children aged 6–23 months in upper-middle-income countries (Neves *et al.*, 2021). The World Health Organization (WHO) and the United Nations Children's Emergency Fund (UNICEF) recommend 6 months of exclusive breastfeeding with continued breastfeeding to 2 years 'and beyond' (M&C Saatchi World Services, 2022). Although infant formula is sometimes necessary, the contemporary level of use of this ultra-processed food is now far beyond what is medically necessary (WHO, 2009).

While other complex factors affect infant and young child feeding practices, key drivers of CMF sales include maternal employment trends, healthcare policies and practices, and industry marketing (Baker *et al.*, 2021a), including through health channels. The industry exercises its immense power through influencing policies and healthcare practices and practitioners in

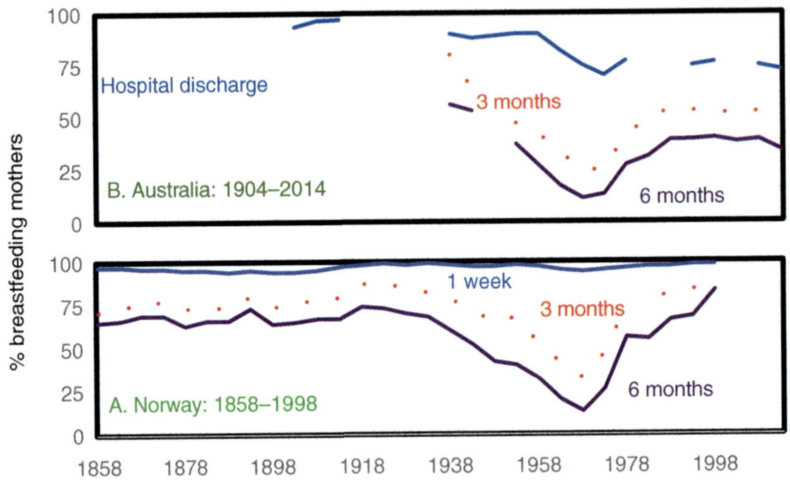

Fig. 28.1. Historical breastfeeding rates in (**A**) Norway and (**B**) Australia. Note that the time axes are not identical. Sources: Rosenberg (1989), Helsing and Haggkvist (2012) and Smith (1999).

directions that support sales and weaken policies that protect breastfeeding (Baker *et al.*, 2021a; Baker *et al.*, 2023). The perverse role of health systems and health professionals in promoting commercial milk formula products is increasingly scrutinized, and reveals the pervasive influence of milk formula companies (Nguyen *et al.*, 2021; Qassin *et al.*, 2021; Russ *et al.*, 2021; M&C Saatchi World Services, 2022; Baker *et al.*, 2023).

Below it is argued that this shift towards less breastfeeding – involving infant and child diets based on ultra-processed foods such as commercial milk formula products – is both maladaptive and archetypal of how capitalist processes drive climate change. It is also argued that well-established, cost-effective and globally scalable interventions exist. If implemented, these could feasibly, rapidly and equitably reverse the trend towards increased CMF use.

28.2 Climate Change and Commercial Milk Formula

The 20th century displacement of breastfeeding by CMF in early nutrition is relevant to all three pillars of climate change: mitigation, adaptation and resilience (Smith, 2019). It is also relevant to the cause of climate change. Evidence on the environmental implications of CMF has been building for decades. Social activists have long considered scientific evidence as essential to increase environmental awareness and investments to address the impact of formula feeding.

Environmental impacts of milk formula products fall into many categories (Linnecar *et al.*, 2014; Radford, 1992). They include global warming potential, water use, water and air pollution and contamination, zoonotic pathogens, antimicrobial resistance, land use change and biodiversity loss, and soil degradation. The energy and landfill aspects of processing, packaging and transporting formula are also relevant (Pope *et al.*, 2021). As noted, the effect of infant feeding practices on child spacing is also pertinent, as human population growth is a driver of climate change.

Mitigation of climate change

Countering the global expansion of CMF use directly lowers greenhouse gas (GHG) emissions and demand on water resources. Several studies have estimated the carbon footprint (CFP) of milk formula (Karlsson *et al.*, 2019; Dadhich *et al.*, 2021; Andresen *et al.*, 2022). Some cover the full range of CMF products for children aged 0–36 months – 'special' or 'standard infant' formula targeting infants aged 0–6 months, and 'follow-up' or 'toddler' formula marketed for children aged 6–36 months (Cadwell *et al.*, 2020;

Dadhich *et al.*, 2021); others cover the whole product life cycle for standard infant formula, and compare it with exclusive breastfeeding as recommended (Karlsson *et al.*, 2019; Andresen *et al.*, 2022). The CFP of pumped breastmilk (Amonkar *et al.*, 2019) and donor human milk banking (Shenker *et al.*, 2022) have also been estimated.

CMF has twice the CFP of breastfeeding (Pope *et al.*, 2021). Exclusive infant formula feeding for 6 months generates over 200 kg CO_2 equivalent (CO_2e)[2] or more of GHG emissions (Karlsson *et al.*, 2019; Andresen *et al.*, 2022), and requires safe and reliable access to water that is otherwise provided by breastfeeding (Aguayo *et al.*, 2001). Over a full product life cycle, a kilogram of CMF generates around 11–14 kg CO_2e of GHG emissions, a blue water footprint of 699 litres, and an overall water footprint including green and black water impacts which exceeds 5000 litres (Karlsson *et al.*, 2019; Andresen *et al.*, 2022). Including the vitamins and minerals in CMF would add up to 1.3 kg CO_2e/kg of BMS powder. Cow's milk was the main contributor to these environmental footprints.

Increased maternal food intake for breastfeeding can generate GHG emissions, but the CFP is half that for CMF under most dietary composition scenarios (Pope *et al.*, 2021). Based on an average Norwegian diet and theoretical energy requirements, breastfeeding was estimated to generate 1.58 kg CO_2e per 1 kg of breastmilk, compared with 12.65 CO_2e/kg of GHG generated by a kilogram of infant formula powder, or 2.02 kg CO_2e/kg as ready-to-feed infant formula. The impact of breastfeeding is decreased by eating plant-based foods rather than increasing meat intake to meet any additional nutrition needs during lactation. In reality, a lactating mother might need more or less than this to sustain breastfeeding, as energy needs during lactation are also met from changes in metabolism and fat accumulated during pregnancy.

Clearly, barriers to breastfeeding damage the environment. Lack of policy support for paid maternity leave may result in separation of mothers from their infants, requiring some women to express their milk or use a breast-pump instead of being able to breastfeed their child. A recent study estimated that fully feeding a child with breastmilk obtained using an electric pump generated around 1.858 kg CO_2e/day or 339 kg CO_2e for a 6-month period of exclusively breastmilk pumping (Amonkar *et al.*, 2019). This was higher than the CFP impact for formula milk feeding, largely due to consumer-stage GHGs such as packaging and waste. Likewise, maternity care services may generate high CFP unless breastfeeding is well supported (WHO, 2018). In Ireland fewer than half of its 62,053 infants were exclusively breastfed in hospital in 2017, and an audit showed that 60% of the 787,057–946,220 bottles of ready-to-use formula purchased by maternity services ended up as waste (Nurses Climate Challenge Europe, 2022). An increasing number of neonatal units use donor human milk (DHM) rather than CMF products, but an analysis of the CFP of donor milk banking in the UK recently calculated that the total CFP of DHM was 9.36 kg CO_2e per litre (Shenker *et al.*, 2022) because of emissions from transporting the milk from donors. These examples reinforce the fact that public policies and healthcare practices can be key drivers of food CFP for this age group.

CMF sales are expanding rapidly especially in low- and middle-income countries, making the large scale of these environmental impacts at country level an important consideration (Baker *et al.*, 2016). The CFP from CMF sales in just six Asia Pacific countries (Australia, South Korea, China, Malaysia, India, Philippines) contributed 2.9 million tonnes of CO_2e in 2012, equivalent to 6 billion miles of car travel (Dadhich *et al.*, 2015; Dadhich *et al.*, 2021). The projected 2017 level of GHG emissions for CMF retailed in China alone was over 4 million tonnes of CO_2e. Much of this impact arose from toddler formula sales, despite WHO advice that these products are nutritionally unnecessary and potentially harmful (WHO, 2013). Meanwhile, average GHG emissions per child aged 0–3 years in North America are at least 59 kg CO_2e, and sales of CMF in the USA, Canada and Mexico generated at least 1 million tonnes of CO_2e emissions (Cadwell *et al.*, 2020). In the UK the carbon emission savings from supporting mothers to breastfeed is akin to taking 50,000–77,500 cars off the road each year (Joffe *et al.*, 2019) based on savings of 95–153 kg CO_2e per baby (Karlsson *et al.*, 2019).

Adaptation

Breastfeeding helps maintain safe and healthy nutrition of infants and young children in the face of growing climate change challenges to food security and health (Dadhich, 2019; Joffe *et al.*, 2019; Smith, 2019). This includes the immunological importance of maternal antibodies in human milk as a means of adaptation to climate-sensitive infections. As recently emphasized by WHO and UNICEF, breastfeeding acts as a baby's first immunization (van den Elsen *et al.*, 2022), protecting them from common childhood illnesses. The deaths of more than 820,000 children a year from pneumonia, diarrhoeal illness and undernutrition could be prevented by achieving WHO recommendations for breastfeeding (Victora *et al.*, 2016). Antibodies in maternal milk provide some protection against many diseases, including malaria (Brazeau *et al.*, 2016), an important climate-sensitive disease, and measles (Binns and Lee, 2019). Of particular significance in the context of climate risks is the contribution of breastfeeding to reducing the harms of heat stress in infants (Nakstad *et al.*, 2022).

During the COVID-19 pandemic more infants were estimated to die from lack of breastfeeding due to changed hospital protocols (separating newborns from their mothers) than from COVID-19 (Rollins *et al.*, 2021). Especially in resource-poor settings, the benefits of breastfeeding far exceed the risks and harms of specific infections such as HIV, HLTV1 and Zika virus, and advances in knowledge and treatment may favour the continuation of breastfeeding (Binns and Lee, 2019).

Resilience

Breastfeeding also plays a role in climate resilience while supporting maternal and child health equity (Roberts *et al.*, 2013). A global market for CMF offers little guarantee of domestic supply, and disruptions even in relatively affluent country settings such as the USA are becoming more common. Infants relying on CMF are food insecure, yet breastfeeding has the shortest supply chain of all infant feeding methods. Food insecurity can result not only from supply chain disruptions, but also from changed family income and reduced purchasing power during economic crises. Globally, and within countries, the least well-off are particularly exposed to climate change risks, yet often lack access to supportive healthcare and employment environments for breastfeeding (Smith, 2019). Children who are not breastfeeding are particularly vulnerable in disasters and emergencies, when food and water may be scarce, or contaminated, and utilities such as electricity unavailable. Despite this, the planning and regulatory environment for infant and young child feeding in emergencies is underdeveloped (Smith and Iellamo, 2020); companies quickly exploit the situation to introduce milk formula to breastfeeding populations, and policy or health provider responses have been belated, confused and misdirected in many cases (Gribble *et al.*, 2020; Vu Hoang *et al.*, 2020; Latorre *et al.*, 2021; Rollins *et al.*, 2021). In 2022, UNICEF and WHO called for increased resources for breastfeeding policies and programmes, especially for the most vulnerable families living in emergency settings (UNICEF/WHO, 2022).

28.3 Connecting Women and Breastfeeding with Capitalism and Climate Change

Ecofeminists connect capitalism's failure to value women's unpaid labour with climate change. Some include breastfeeding in this critique. Breastfeeding raises important discussions within feminism and some warn that women, rather than the CMF industry, will be held responsible for mitigating against climate risks (Williams *et al.*, 2019). Nevertheless, expanding CMF markets signals a lost opportunity to address flaws in market-oriented institutions and corporate accountabilities that harm women, children and the environment, and urgently need to be reversed in a gender-just way (Zadkovic *et al.*, 2021).

Leading ecofeminist thinker Greta Gaard (2013) highlighted the connections between the oppression of women (and animals) and environmental destruction, noting (as did Marilyn Waring and many others) that the invisibility of the productivity of all three is enforced in economic statistics such as gross domestic product (GDP):

> On the international market the United Nations System of National Accounts (UNSNA) has no method of accounting for nature's own production or destruction until the products of nature enter the cash economy, nor does it account for the majority of the work done by women ... both women and nature do not 'count' in the international market economy.

There has been growing recognition since the 1960s that this internationally accepted economic accounting framework is not fit for purpose (Stiglitz *et al.*, 2018). Boundaries marked by the UN System of National Accounts (UNSNA) mean that GDP does not capture how ecosystems and non-market household production are supporting the economic well-being of humans. GDP significantly overstates economic growth performance, because of shifts over time from household to market economic activity and because it counts market-sector activities that deplete the environment as economic growth and ignores the environmental harms of market activity, such as pollution (van de Ven *et al.*, 2018). Clean-up costs are counted only when they are monetized and then are counted as boosting GDP.

Likewise, GDP critics have argued that policies expanding national income without accounting for the productivity of all producers, including women, result in inequitable distribution and exploitation of women (Messac, 2018). Feminist economists have provided strong critiques of the imposed invisibility of women's productivity (Waring, 1988; Folbre, 1991; Benería *et al.*, 2016), and argue that women's social reproduction work subsidizes the costs of the market economy (Federici, 1975). Marilyn Waring, a principal founder of feminist economics, asked in 1988 why the productivity of women's breasts in promoting cars or selling pornography is counted in conventional economic statistics, yet their primary function of feeding human infants is not. Waring's scathing critique of this system equally emphasized the environmental harm caused by excluding environmental degradation and depletion from its scope. Omitting non-market production from economic measures results in gender-biased measurement of economic development and growth, and misleads policymakers on priorities (Himmelweit, 2002). Waring (1988) characterized the UNSNA system as 'applied patriarchy'.

Ecofeminists expand these arguments to encompass the invisibility of the environment and lack of recognition of 'eco-system services' and the enforced productivity of animals in key measures of economic progress and well-being. In *Making Milk*, Gaard and others draw attention to the brutal exploitation and separation of cows and calves in the industrial dairy industry to facilitate profitable and environmentally destructive commodification and commercialization of their milk (Cohen and Otomo, 2017).

Both the Earth's climate and breastfeeding have been disrupted by the rise of industrial capitalism (Zadkovic *et al.*, 2021). As Tony McMichael often observed, for the past half a century such disruptions were kept out of view as economic policies remained transfixed on increasing the size of the monetized economy.

Breastfeeding and human milk are excluded from GDP. When breastfeeding is replaced by commercial milk formula, GDP records an economic gain (Smith and Ingham, 2005). Yet commercial baby food is counted and its environmental footprint ignored (Smith, 2004). Governments allow CMF marketing to expand production and sales and undermine breastfeeding, but fail to prioritize paid maternity leave policies that protect breastfeeding, women's health and economic security. Similarly, policy inertia resulting in inadequate maternity care and insufficient paid maternity leave has created opportunities for firms to profit from exporting human milk from Asia to the USA across thousands of 'food miles' (Smith, 2015).

A 2009 review of the SNA commissioned by the French President acknowledged that there were major biases in the GDP measure of material well-being (Stiglitz *et al.*, 2009; van de Ven, 2019). The Commission cited breastmilk as exemplifying a distortion from excluding the unpaid household economy from GDP, and noted that such omissions had major policy consequences. A decade on, SNA accounting still fails to account meaningfully for ecosystem or non-market household production (European Communities (EC) *et al.*, 2009).

GDP has, for decades, significantly overstated economic performance (Hawrylyshyn, 1976; Smith, 1982). In 2014, Norwegian national accounts researchers (Aslaksen and Koren, 2014) reported that much of the country's apparent economic progress between 1972 and 1990 (a period when household production transferred to the market sectors) was, in effect, a mirage. GDP overstated real economic progress by more than 20% by excluding unpaid household work. Excluding the value of unpaid household production also understated the rise in economic inequality, such as among single parents (Aslaksen and Koren, 1996). Investigations by the Organisation for Economic Co-operation and Development (OECD) showed that, after adjusting for marketization of childcare, GDP growth in countries including the USA, UK and Canada would be 0.1–1.8

percentage points less than measured in official statistics since the 1980s (van de Ven *et al.*, 2018).

Several studies also estimated the amount and monetary value of the milk which breastfeeding mothers produce for their babies each year (Smith, 2017). A new 'Mothers' Milk Tool' (Smith *et al.*, 2023a) quantifies the volume of breastmilk and value of breastfeeding at national and global levels. It also calculates how much is lost if breastfeeding prevalence is low. Such data – currently provided in national food statistics only in Norway – provides better scientific information for public policy and budgeting decisions and reduces the invisibility of women's productivity. For example, the Tool calculates that, globally, mothers make at least 35.6 billion litres of milk each year but more than a third of potential production, around 22 billion litres, is 'lost', due to lack of adequate breastfeeding support.

As each non-breastfed child uses a total of around 21 kg of milk formula for the first 6 months, lost milk estimates point to the scale of the environmental consequences of declining breastfeeding (Smith, 2013). Akin to the Mothers' Milk Tool is another innovation, the Green Feeding Tool, which is being developed to generate climate action in the area of infant and young child feeding. This instrument will help calculate the carbon and water footprints if breastfeeding declines and CMF expands in diets for infants (FHI Solutions, 2022). In Nepal, for example, a high prevalence of breastfeeding means that the carbon footprints from CMF use are currently very low, but advocacy is needed to ensure that the necessary policy investments are funded to prevent declines in breastfeeding during economic 'development' (Smith *et al.*, 2023b).

28.4 Conclusion

Formula feeding is a maladaptive practice, exemplifying and contributing to contemporary global environmental and population health challenges. Breastfeeding protection, support and promotion is a beneficial and just response to concerns about disease burdens and climate change that minimizes environmental harms and is an important constraint on population growth. As well as helping to mitigate climate change, and providing some protection against climate-related health risks, breastfeeding populations are more resilient in disasters and emergencies. Effective and cost-effective policies and interventions exist to improve opportunities for breastfeeding and reduce unnecessary use of formula. Recognizing and responding to the health and environmental costs of not breastfeeding and the value of breastfeeding requires fundamental changes in the measurement framework and systems of economic metrics, and a redirection of financing, economic resources and investments towards building a sustainable, healthy and affordable first-food system (Smith *et al.*, 2024).

Implementing such measures presents a rare opportunity both to reduce the greenhouse gas problem and to improve human nutrition, health and health equity, but requires greater awareness, political commitment and collaboration across key social movements, which has been lacking to date. The vast power and influence of the commercial milk formula industry, deeply embedded in contemporary capitalism, remains the key barrier.

Notes

[1] First-food systems have been defined as those that provision foods for infants and young children aged 0–36 months, including breastmilk, commercial BMS, other non-human milks, and home-prepared or commercial complementary foods (Baker, 2020).

[2] Carbon dioxide equivalent (CO_2e) is used to describe different GHGs in a common unit and signifies equivalent global warming impact for different quantities or types of greenhouse gases.

References

Aguayo, V.M., Ross, J., Saunero, R., Torrez, A. and Johnston, R. (2001) Monetary value of breast milk in Bolivia. *Revista Panamericana de Salud Publica* 10, 249–256. DOI: 10.1590/s1020-49892001001000005

Amonkar, Y., Chowdhury, N., Song, Y., Wu, J.S., Vaidya, P. *et al.* (2019) Life cycle GHG emission comparison of infant nursing using breast milk versus formula. *Journal of Environmental Accounting and Management* 7, 61–75, DOI: 10.5890/jeam.2019.03.005

Andresen, E.C., Hjelkrem, A.-G.R., Bakken, A.K. and Andersen, L.F. (2022) Environmental impact of feeding with infant formula in comparison with breastfeeding. *International Journal of Environmental Research and Public Health* 19, 6397. DOI: 10.3390/ijerph19116397

Aslaksen, I. and Koren, C. (1996) Unpaid household work and the distribution of extended income: The Norwegian experience. *Feminist Economics* 2, 65–80. DOI: 10.1080/135457096100017077766

Aslaksen, I. and Koren, C. (2014) Reflections on unpaid household work, economic growth, and consumption possibilities. In: Bjornholt, M. and Mckay, A. (eds) *Counting on Marilyn Waring*. Demeter Press, Bradford, Canada, pp. 57–71.

Baker, P. (2020) First-food systems transformations and the ultra-processing of infant and young child diets: The determinants, dynamics and consequences of the global rise in commercial milk formula consumption. *Maternal & Child Nutrition* 17, e13097. DOI: 10.1111/mcn.13097

Baker, P., Smith, J., Salmon, L., Friel, S., Kent, G. *et al.* (2016) Global trends and patterns of commercial milk-based formula sales: is an unprecedented infant and young child feeding transition underway? *Public Health Nutrition* 19, 2540–2550. DOI: 10.1017/S136898001600117

Baker, P., Santos, T., Neves, P.A., Machado, P., Smith, J. *et al.* (2021a) First-food systems transformations and the ultra-processing of infant and young child diets: The determinants, dynamics and consequences of the global rise in commercial milk formula consumption. *Maternal & Child Nutrition* 17, e13097. DOI: 10.1111/mcn.13097

Baker, P., Russ, K., Kang, M., Santos, T.M., Neves, P.A.R. *et al.* (2021b) Globalization, first-foods systems transformations and corporate power: a synthesis of literature and data on the market and political practices of the transnational baby food industry. *Global Health* 17, 58. DOI: 10.1186/s12992-021-00708-1

Baker, P., Smith, J.P., Garde, A., Grummer-Strawn, L.M., Wood, B. *et al.* (2023) The political economy of infant and young child feeding: confronting corporate power, overcoming structural barriers, and accelerating progress. *The Lancet* 401, 503–524. 10.1016/S0140-6736(22)01933-X

Benería, L., Berik, G. and Floro, M. (2016) *Gender, Development and Globalization: Economics As If All People Mattered*. Routledge, Abingdon, UK

Berg, A. (1973) *The Nutrition Factor. Its Role in National Development*. The Brookings Institution, Washington, DC.

Binns, C. and Lee, M.K. (2019) Public health impact of breastfeeding. In: Binns, C. and Lee, M.K. (eds) *Oxford Research Encyclopedia of Global Public Health*. Oxford University Press, Oxford, UK. DOI: 10.1093/acrefore/9780190632366.013.66

Brazeau, N.F., Tabala, M., Kiketa, L., Kayembe, D., Chalachala, J.L. *et al.* (2016). Exclusive breastfeeding and clinical malaria risk in 6-month-old infants: A cross-sectional study from Kinshasa, Democratic Republic of the Congo. *American Journal of Tropical Medicine and Hygiene* 95, 827–830, DOI: 10.4269/ajtmh.16-0011

Cadwell, K., Blair, A., Turner-Maffei, C., Gabel, M. and Brimdyr, K. (2020) Powdered baby formula sold in North America: Assessing the environmental impact. *Breastfeeding Medicine* 15, 671–679. DOI: 10.1089/bfm.2020.0090

Ciampo, L.A.D. and Ciampo, I.R.L.D. (2018) Breastfeeding and the benefits of lactation for women's health. *Revista Brasileira de Ginecologia e Obstetrícia* 40, 354–359. DOI: 10.1055/s-0038-1657766

Cohen, M. and Otomo, Y. (eds) (2017) *Making Milk*. Bloomsbury, London.

Dadhich, J.P. (2019) *GreenFeeding to Achieve Global Nutrition Targets 2025*. International Baby Food Action Network & Breastfeeding Promotion Network of India, Delhi. Available at: https://www. bpni. org/ wp- content/ uploads/ 2018/ 11/ Green-Feeding-RC- CarbonFootprint-10-Asian-Countries. pdf (accessed 7 March, 2023).

Dadhich, J.P., Smith, J.P., Iellemo, A. and Suleiman, A. (2015) *Report on Carbon Footprints Due to Milk Formula: A study from select countries of Asia-Pacific region*. BPNI/IBFAN Asia, New Delhi, India.

Dadhich, J.P., Smith, J.P., Iellamo, A. and Suleiman, A. (2021) Climate change and infant nutrition: Estimates of greenhouse gas emissions from milk formula sold in selected Asia Pacific countries. *Journal of Human Lactation* 37, 314–322. DOI: 10.1177/0890334421994769

Dettwyler, K.A. (1995) A time to wean: the hominid blueprint for the natural age of weaning in modern human populations. In: Stuart-Macadam, P. and Dettwyler, K.A. (eds) *Breastfeeding: Biocultural Perspectives.* Aldine De Gruyter, New York, pp. 39–74.

Eaton, S.B., Pike, M.C., Short, R.V., Lee, N.C. *et al.* (1994) Women's reproductive cancers in evolutionary context. *The Quarterly Review of Biology* 69, 353–367. DOI: 10.1086/418650

EC, IMF, OECD, UN and WB (2009) *System of National Accounts 2008*. European Communities, International Monetary Fund, Organisation for Economic Co-operation and Development, United Nations and World Bank. Available at: https://unstats.un.org/unsd/nationalaccount/sna2008.asp (accessed 10 July 2023).

Federici, S. (1975) *Wages Against Housework*. Falling Wall Press, Bristol, UK.

FHI Solutions (2022) *How Breastfeeding Helps Mitigate Climate Change*. Family Health International Solutions, Durham, North Carolina. Available at: https://www.fhisolutions.org/how-breastfeeding-helps-mitigate-climate-change/ (accessed 5 June 2023).

Fildes, V. (1995) The culture and biology of breastfeeding: an historical review of Western Europe. In: Stuart-Macadam, P. and Dettwyler, K.A. (eds) *Breastfeeding: Biocultural Perspectives*. Aldine De Gruyter, New York, pp. 102–126.

Folbre, N. (1991) The unproductive housewife: Her evolution in nineteenth-century economic thought. *Signs* 16, 463–484. DOI: 10.1086/494679

Gaard, G. (2013) Toward a feminist postcolonial milk studies. *American Quarterly* 65, 595–618. DOI: 10.1353/aq.2013.0040

Greer, F.R. and Apple, R.D. (1991) Physicians, formula companies, and advertising. A historical perspective. *American Journal of Diseases of Children* 145, 282–286. DOI: 10.1001/archpedi.1991.02160030050019

Gribble, K., Mathisen, R., Ververs, M.T. and Coutsoudis, A. (2020) Mistakes from the HIV pandemic should inform the COVID-19 response for maternal and newborn care. *International Breastfeeding Journal* 15, 67. DOI: 10.1186/s13006-020-00306-8

Hanson, L.A. (2004) *Immunobiology of Human Milk*. Pharmasoft Publishing, Amarillo, Texas.

Hawrylyshyn, O. (1976) The value of household services: a survey of empirical estimates. *Review of Income and Wealth* 22, 101–131. DOI: 10.1111j.1475-4991.1976.tb01146.x

Helsing, E. and Haggkvist, A.-P. (2012) *Understanding Breastfeeding and How to Succeed*. Hale Publishers, Amarillo, Texas.

Himmelweit, S. (2002) Making visible the hidden economy: The case for gender-impact analysis of economic policy. *Feminist Economics* 8, 49–70. DOI: 10.1080/13545700110104864

Horta, B.L., Loret de Mola, C. and Victora, C.G. (2015) Breastfeeding and intelligence: A systematic review and meta-analysis. *Acta Paediatrica* 104, 14–19. DOI: 10.1111/apa.13139

Joffe, N., Webster, F. and Shenker, N. (2019) Support for breastfeeding is an environmental imperative. *BMJ* 367, l5646. DOI: 10.1136/bmj.l5646

Karlsson, J.O., Garnett, T., Rollins, N.C. and Roos, E. (2019) The carbon footprint of breastmilk substitutes in comparison with breastfeeding. *Journal of Cleaner Production* 222, 436–445. DOI: 10.1016/j.jclepro.2019.03.043

Kunz, C., Rodriguez-Palmero, M., Koletzko, B. and Jensen, R. (1999) Nutritional and biochemical properties of human milk, Part I: General aspects, proteins, and carbohydrates. *Clinics in Perinatology* 26, 307–333. DOI: 10.1016/S0095-5108(18)30055-1

Latorre, G., Martinelli, D., Guida, P., Masi, E., De Benedictis, R. *et al.* (2021) Impact of COVID-19 pandemic lockdown on exclusive breastfeeding in non-infected mothers. *International Breastfeeding Journal* 16, 36. DOI: 10.1186/s13006-021-00382-4

Linnecar, A., Gupta, A., Dadhich, J. and Bidla, N. (2014) *Formula for Disaster*. IBFAN (International Baby Foods Action Network) Asia and BPNI (Breastfeeding Promotion Network of India). Available at: http://ibfan.org/docs/FormulaForDisaster.pdf (accessed 7 March 2023).

Loponte, D. and Mazza, B. (2021) Breastfeeding and weaning in Late Holocene hunter-gatherers of the lower Paraná wetland, South America. *American Journal of Physical Anthropology* 176, 504–520. DOI: 10.1002/ajpa.24381

M&C Saatchi World Services (2022) *The Impact of Marketing of Breast-Milk Substitutes on Infant Feeding Decisions and Practices: Report of a Multi-Country Research Study Commissioned by The World Health Organization and The United Nations Children's Fund*. World Health Organization (WHO), Geneva, and UNICEF (United Nations Children's Fund), New York. Available at: https://apps.who.int/iris/bitstream/handle/10665/354094/WHO-UHL-MCA-22.01-eng.pdf (accessed 5 March 2023).

McMichael, A.J. (1991) Global warming, ecological disruption and human health: the penny drops. *Medical Journal of Australia* 154, 499–501. DOI: 10.5694/j.1326-5377.1991.tb119437.x [See also: Butler, C.D., Dixon, J. and Capon, A.G. (eds) (2015) *Health of People, Places and Planet: Reflections based on*

Tony McMichael's four decades of contribution to epidemiological understanding. ANU Press, Canberra, Australia, pp. 335–337. DOI: 10.5694/j.1326-5377.1991.tb142316.x]
Meek, J.Y. and Noble, L. (2022) Section on Breastfeeding. Policy statement: Breastfeeding and the use of human milk. *Pediatrics* 150, e2022057988. DOI: 10.1542/peds.2022-057988
Messac, L. (2018) Outside the economy: Women's work and feminist economics in the construction and critique of national income accounting. *The Journal of Imperial and Commonwealth History* 46, 552–578. DOI: 10.1080/03086534.2018.1431436
Nakstad, B., Filippi, V., Lusambili, A., Roos, N., Scorgie, F. *et al.* (2022) How climate change may threaten progress in neonatal health in the African region. *Neonatology* 119, 1–8. DOI: 10.1159/000525573
Neves, P.A.R., Vaz, J.S., Maia, F.S., Baker, P., Gatica-Domínguez, G. *et al.* (2021) Rates and time trends in the consumption of breastmilk, formula, and animal milk by children younger than 2 years from 2000 to 2019: analysis of 113 countries. *The Lancet Child & Adolescent Health* 5, 619–630. DOI: 10.1016/s2352-4642(21)00163-2
Nguyen, T.T., Tran, H.T.T., Cashin, J., Nguyen, V.D.C., Weissman, A. *et al.* (2021) Implementation of the code of marketing of breast-milk substitutes in Vietnam: Marketing practices by the industry and perceptions of caregivers and health workers. *Nutrients* 13, 2884. DOI: 10.3390/nu13082884
Nurses Climate Challenge Europe (2022) *The interconnection of environment, climate change, and infant nutrition*. Queen's Nursing Institute Scotland, Edinburgh. Available at: https://www.qnis.org.uk/wp-content/uploads/2022/07/2022-05-03-NCCEurope-climate-smart-infantfeeding-part1.pdf (accessed 5 March 2023).
Pope, D.H., Karlsson, J.O., Baker, P. and McCoy, D. (2021) Examining the environmental impacts of the dairy and baby food industries: Are first-food systems a crucial missing part of the healthy and sustainable food systems agenda now underway? *International Journal of Environmental Research and Public Health* 18, 12678. DOI: 10.3390/ijerph182312678
Qassin, S., Homer, C.S.E. and Wilson, A.N. (2021) Funding received from breastmilk substitute manufacturers and policy positions of national maternity care provider associations: an online cross-sectional review. *BMJ Open* 11, e050179. DOI: 10.1136/bmjopen-2021-050179
Radford, A. (1992) The ecological impact of bottle feeding. *Breastfeeding Review* 11, 204–208. Available at: https://abm.me.uk/wp-content/uploads/Ecological-Impact-of-Bottlefeeding.pdf (accessed 4 March 2023).
Roberts, T.J., Carnahan, E. and Gakidou, E. (2013) Can breastfeeding promote child health equity? A comprehensive analysis of breastfeeding patterns across the developing world and what we can learn from them. *BMC Medicine* 11, 254. DOI: 10.1186/1741-7015-11-254
Rollins, N., Minckas, N., Jehan, F., Lodha, R., Raiten, D. *et al.* (2021) A public health approach for deciding policy on infant feeding and mother-infant contact in the context of COVID-19. *The Lancet Global Health* 9, e552–e557. DOI: 10.1016/S2214-109X(20)30538-6
Rosenberg, M. (1989) Breast-feeding and infant mortality in Norway 1860–1930. *Journal of Biosocial Science* 21, 335–348. DOI: 10.1017/S0021932000018034
Russ, K., Baker, P., Byrd, M., Kang, M., Siregar, R.N. *et al.* (2021) What you don't know about the codex can hurt you: How trade policy trumps global health governance in infant and young child nutrition. *International Journal of Health Policy Management* 10, 983–997. DOI: 10.34172/ijhpm.2021.109
Sellen, D.W. (2007) Evolution of infant and young child feeding: implications for contemporary public health. *Annual Review of Nutrition* 27, 123–148. DOI: 10.1146/annurev.nutr.25.050304.092557
Shenker, N., Tennant, M., Rickard, L., Huppertz, P., Cheung, T. *et al.* (2022) Understanding and mitigating the carbon footprint of human milk banking. In: *Proceedings of the Climate Change Working Group*. BMJ Publishing Group and Royal College of Paediatrics and Child Health, London, p. A313. BMJ Archives of Disease in Childhood 107. DOI: 10.1136/archdischild-2022-rcpch.505
Smith, J.P. (1982) *The Value of Household Work in Australian National Product*. Treasury Seminar Series, June 1987. Department of Economic History, Australian National University, Canberra, Australia
Smith, J.P. (1999) Human milk supply in Australia. *Food Policy* 24, 71–91. DOI: 10.1016/S0306-9192(99)00033-0
Smith, J. (2004) Mothers' milk and markets. *Australian Feminist Studies* 19, 369–379. DOI: 10.1080/0816464042000278003
Smith, J.P. (2013) 'Lost milk?': Counting the economic value of breast milk in gross domestic product. *Journal of Human Lactation* 29, 537–546. DOI: 10.1177/0890334413494827
Smith, J.P. (2015) Markets, breastfeeding and trade in mothers' milk. *International Breastfeeding Journal* 10, 9. DOI: 10.1186/s13006-015-0034-9

Smith, J.P. (2017) Markets in mothers' milk: Virtue or vice, promise or problem? In: Cohen, M. and Otomo, Y. (eds) *Making Milk*. Bloomsbury Publishing, London, pp. 117–137.

Smith, J.P. (2019) A commentary on the carbon footprint of milk formula: harms to planetary health and policy implications. *International Breastfeeding Journal* 14, 49. DOI: 10.1186/s13006-019-0243-8

Smith, J.P. and Harvey, P.J. (2011) Chronic disease and infant nutrition: is it significant to public health? *Public Health Nutrition* 14, 279–289. DOI: 10.1017/S1368980010001953

Smith, J.P. and Iellamo, A. (2020) Wet nursing and donor human milk sharing in emergencies and disasters: A review. *Breastfeeding Review* 28, 7–23. Available at: https://search.informit.org/doi/abs/10.3316/informit.480674778174237 (accessed 5 March 2023).

Smith, J.P. and Ingham, L.H. (2005) Mothers' milk and measures of economic output. *Feminist Economics* 11, 41–62. DOI: 10.1080/1354570042000332605

Smith, J.P., Iellamo, A., Nguyen, T.T. and Mathisen, R. (2023a) The volume and monetary value of human milk produced by the world's breastfeeding mothers: Results from a new tool. *Frontiers in Public Health* 11, 1152659. DOI: 10.3389/fpubh.2023.1152659

Smith, J.P., Borg, B., Iellamo, A., Nguyen, T.T. and Mathisen, R. (2023b) Innovative financing for a gender-equitable first-food system to mitigate greenhouse gas impacts of commercial milk formula: investing in breastfeeding as a carbon offset. *Frontiers in Sustainable Food Systems* 7, 1155279. DOI: 10.3389/fsufs.2023.1155279

Smith, J.P., Baker, P., Long, A., Mathisen, R., Rollins, N. and Waring, M. (2024) A proposal to recognize investment in breastfeeding as a carbon offset. *Bulletin of the World Health Organization* 102, DOI: 10.2471/BLT.23.290210

Stiglitz, J.E., Sen, A. and Fitoussi, J.P. (2009) *The Measurement of Economic Performance and Social Progress Revisited; Reflections and Overview*. French Observatory of Economic Conditions, Economics Research Center, Paris.

Stiglitz, J.E., Fitoussi, J.-P. and Durand, M. (2018) *Beyond GDP*. OECD (Organisation for Economic Co-operation and Development) Publishing, Paris.

UNICEF/WHO (2022) *Joint Statement by UNICEF Executive Director Catherine Russell and WHO Director-General Dr Tedros Adhanom Ghebreyesus on the occasion of World Breastfeeding Week*. Available at: https://www.unicef.org/serbia/en/press-releases/joint-statement-unicef-executive-director-catherine-russell-and-who-director-general#:~:text=%E2%80%9CProtecting%2C%20promoting%2C%20and%20supporting,development%20of%20millions%20of%20infants.

van de Ven, P. (2019) *It's About Time for a Change: Broadening the SNA Framework to Account for Sustainability and Well-being*. Comisión Económica para América Latina y et Caribe (CEPAL: Economic Commission for Latin America and the Caribbean). Available at: https://www.cepal.org/sites/default/files/presentations/broadening-sna-framework-account-sustainability-well-being.pdf (accessed 7 March 2023).

van de Ven, P., Zwijnenburg, J. and Queljoe, M.D. (2018) *Including Unpaid Household Activities: An Estimate of its Impact on Macro-Economic Indicators in The G7 Economies and The Way Forward*. Organisation for Economic Co-operation and Development, Paris. Available at: https://www.oecd-ilibrary.org/economics/including-unpaid-household-activities_bc9d30dc-en (accessed 9 November 2020).

van den Elsen, L.W.J., Kollmann, T.R. and Verhasselt, V. (2022) Microbial antigen in human milk: a natural vaccine? *Mucosal Immunology* 15, 1058–1059. DOI: 10.1038/s41385-022-00561-4

Victora, C.G., Bahl, R., Barros, A.J., Franca, G.V., Horton, S. *et al.* (2016) Breastfeeding in the 21st century: epidemiology, mechanisms, and lifelong effect. *The Lancet* 387, 475–490. DOI: 10.1016/s0140-6736(15)01024-7

Volk, A.A. and Atkinson, J.A. (2013) Infant and child death in the human environment of evolutionary adaptation. *Evolution and Human Behavior* 34, 182–192. DOI: 10.1016/j.evolhumbehav.2012.11.007

Vu Hoang, D., Cashin, J., Gribble, K., Marinelli, K. and Mathisen, R. (2020) Misalignment of global COVID-19 breastfeeding and newborn care guidelines with World Health Organization recommendations. *BMJ Nutrition, Prevention and Health* 3, 339–350. DOI: 10.1136/bmjnph-2020-000184

Waring, M. (1988) *Counting for Nothing*. Allen & Unwin, Wellington, New Zealand.

Whittington, A. (2015) Exploring the future of infant feeding: Scenarios about the future of parenting, infant nutrition and consumer habits in the United States of America. *Journal of Future Studies* 19, 13–28. Available at: https://jfsdigital.org/2015-2/vol-19-no-2-mar-2015/articles/exploring-the-future-of-infant-feeding-scenarios-about-the-future-of-parenting-infant-nutrition-and-consumer-habits-in-the-united-states-of-america/ (accessed 19 June, 2023).

WHO (2009) *Acceptable medical reasons for use of breast-milk substitutes*. World Health Organization, Geneva. Available at: http://apps.who.int/iris/bitstream/handle/10665/69938/WHO_FCH_CAH_09.01_eng.pdf (accessed 7 March 2023).

WHO (2013) *Information concerning the use and marketing of follow-up formula*. World Health Organization, Geneva.

WHO (2018) *Implementation Guidance. Protecting, promoting and supporting breastfeeding in facilities providing maternity and newborn services: the revised baby-friendly hospital initiative*. World Health Organization, Geneva.

Williams, E.J., Maia, S., McNee, R., Wishart, K., Murphy, C. and Cannon, E. (2019) Reducing the environmental impact of formula must not jeopardise women's reproductive rights. *BMJ* 367, l6328. DOI: 10.1136/bmj.l6328

Zadkovic, S., Lombardo, N. and Cole, D.C. (2021) Breastfeeding and climate change: Overlapping vulnerabilities and integrating responses. *Journal of Human Lactation* 37, 323–330. DOI: 10.1177/0890334420920223

29 Disasters, Education, Public Health and Climate Change

Ruth Irwin
Independent scholar, Australia and New Zealand

Abstract
Health practitioners have important roles to reduce the impact of climate change, and also to advocate for a more 'sustainable' and equitable civilization. However, to be effective, education aimed at medical and public health students concerning these issues needs to improve its engagement with fundamental political and economic questions. This chapter describes some of the devastating consequences of the 2023 flooding event, in New Zealand's North Island, to illustrate some of the key principles that are needed to deliver effective global health education, including resilience and social cohesion, each of which is placed at risk by the now dominant paradigm of neoliberalism, which fosters denialism, doomerism and 'prepping'. Instead, doctors and other health workers need encouragement, direction and encounters with role models who can challenge power and work for the public good.

29.1 Introduction

Climate change is here. For every degree Celsius the average planetary temperature rises, the atmosphere carries 7% more moisture. This increase in evaporation and higher levels of rainfall stimulate more dynamic wind and ocean currents. We have left the Holocene – a period of relatively placid climate stability in which humanity thrived, and civilizations blossomed for over 12,000 years (Irwin, 2010). We have spawned the unstable Anthropocene, a new climatic epoch, more reminiscent of the extinction period of dinosaurs.

These changes are decimating animal and insect populations. Not only severe weather events but also rapid changes in ecological capacity create event cascades. Hotter weather is good for some creatures but bad for others. Some insects like the new conditions, but overall there has been notable decrease in insect volumes in recent years, all over the world, largely driven by habitat loss and the exuberant use of new pesticides – few of which are adequately tested – as well as climate change (Shiva, 1989; Goulson, 2019). This lack of food source low in the food chain will impact birds, lizards, small mammals and also marsupials (IPBES, 2019). Their lower numbers will have an effect on larger carnivores. Marine food chains are also being undermined (Testa *et al.*, 2022).

Humanity is not immune to these ecological changes. The industrialization of agriculture makes it easy for some to ignore local ecological conditions (Heidegger, 1954). Global trade means that local communities in fortunate settings can import fruit, vegetables and other consumables, making proximate environmental degradation less relevant in the short term. But this alienation has a heavy cost. The lack of awareness of the impacts of climate change and other forms of pollution of the ecological commons

 Climate Change and Global Health: Primary, Secondary and Tertiary Effects, 2nd Edition (eds C.D. Butler and K. Higgs)
DOI: 10.1079/9781800620025.0029

over the past 200 years is escalating to unprecedented proportions. Climate events that were 'once in a hundred' or 'once in a thousand' years are coming with increased ferocity and frequency; as a species, humanity needs to mitigate and adapt in order to build resilience.

The warming planet is producing two kinds of impact, each with adverse consequences for human health and its service providers. Long-term food security is becoming more precarious, for subsistence communities as well as highly 'modern' ones. Persistent, low-level stressors make sudden climate events even harder to accommodate. Severe storms and enormous forest fires produce injury and illness. In widespread flooding, the mixture of flood waters with effluent overflows makes all flood-affected areas more likely to be contaminated and diseased.

In the New Zealand summer floods in 2023, five months of rain fell in three days (NIWA, 2023). Floodwaters polluted by sewerage swept over crops, writing off 20–30% of local produce. Major landslides took out dozens of houses in several suburbs, and two firemen, trying to rescue inhabitants, were crushed to death. Multiple roads were closed for days, and in some cases will be for years. The northern train line had over 50 landslides, one so large that it will take over three months to dig out the mud. The wine-producing Esk Valley suffered an enormous deluge, which produced 2–3 metres of silt that swept over cars and houses and contaminated the land. It may take 10 years for this large fertile valley to come back into production. This community has lost its housing, infrastructure, roads and livelihoods. Such events have disastrous effects on those most affected – lost houses, land, driveways, and billions of dollars of damage to public infrastructure, including roads, bridges, water and electricity supply, Wi-fi, phone lines, buildings, furnishings, electronics, and so on. Some will opt to stay and rebuild from scratch; others will join the mass movement of climate-affected internal and external migrants.

For the health system, climate events will become (un)expected black swan phenomena, a little like the occasional industrial catastrophe or multi-vehicle pile-up. Hospitals are essential for civil defence communication and a crucial element of the emergency response to injury, illness and longer-term mental health reactions. It is also increasingly likely that hospitals will be the hub for communications and civil defence. The level of these catastrophes is vaster than government services, even in parts of the global North, like New Zealand, a well-ordered, still-affluent nation. Communication is at risk, threatened by power outages and even the collapse of telephone and Wi-fi. It is the community itself that generates the heroic figures and tireless unseen provision that helps recovery and resilience in the face of disaster.

Climate change discourse is built around three intersecting categories: mitigation, adaptation and resilience. All three need to be considered in the education of public health professionals concerning climate change, as are the emotional responses that climate change provokes.

Mitigation, adaptation and resilience

Mitigation is the reduction of emissions, to reduce further exacerbation of the ferocity and turbulence of global warming. Without mitigation, it is increasingly likely that the Thwaites glacier will continue to melt, releasing the plug that stops land-based Antarctic ice melt sufficient to raise the ocean by 2.3 m (Davis *et al.*, 2023). Likewise, global warming continues to melt ice in the Arctic, including land-based glaciers in Greenland, which could raise sea levels by 782 +/– 135 mm per year (Box *et al.*, 2022). Adaptation encompasses the steps needed to respond to existing environmental changes, including changes in ecosystems, sea level rise, increased erosion, flooding and risk of larger-scale and hotter fires. Resilience is the ability of the community to activate and respond to disaster without collapsing. An aspect of this is to recognize the emotional consequences of climate change, which can be challenging to populations shaped by modern neoliberal discourses of the rugged and rational individual, with few community interdependencies. All three of these lenses are necessary to teach climate change and global public health effectively.

29.2 Climate Education is Necessarily Political

There is increasing recognition that health practitioners, including medical doctors, have important roles to reduce the impact of climate change. For example, a conference hosted by the World Health Organization in 2016 called for 'mainstreaming climate change and health topics into medical and public health training'. In 2017, UNESCO (UN Educational, Scientific and Cultural Organization) encouraged educators to prepare professions to be informed about climate change and to be able to play their part to engage with the current challenges and build a more 'environmentally sustainable' future (Madden *et al.*, 2018). Some also recognize a role to advocate for a more 'sustainable' and equitable civilization.

However, the growing catastrophe of climate events demonstrates that public health authorities must also take mitigation measures seriously, aiming to reduce emissions within the health sector (Pencheon and Wight, 2020), and thus lessen the risk of passing tipping points and further climatic-linked disaster. Public understanding of mitigation is also a vital element of the public learning to cope with climate change, as it brings focus, direction and effective agency to people engaging in resilience measures.

Warmer temperatures will increase the evaporation factors in speed and extent, drying out soils and inhibiting soil complexity. Sea levels are rising, partly because warmer water expands, and partly from glacier- and land-based polar ice melt. The strength of the Gulf Stream, the North Atlantic segment of the major thermohaline current, and one of Earth's major oceanic systems for redistributing heat, may be in decline (van Western *et al.*, 2024). If so, this could have profound, though still poorly understood, effects on climate, which could include paradoxical cooling in parts of Europe (Rahmstorf, 2020). Similarly, the overturning current in the Australia–Antarctic basin is also showing signs of weakening (Gunn *et al.*, 2023).

It is important to understand the science of climate change. The continued accumulation of carbon dioxide and other greenhouse gases in the atmosphere pushes the planet closer to irreversible tipping points (see Chapter 1). It is not simply a matter of learning to endure increased storm volatility. Adaptation is insufficient. Mitigation and structural societal change are also vital to eliminate continuing greenhouse gas accumulation.

Adaptation and resilience

Enormous vulnerability is associated with both persistent and sudden climate change events. Isolation, poverty and homelessness escalate risk factors. The homeless have almost nowhere to go. During the COVID-19 pandemic, homelessness, at least in some parts of the global North, was better addressed than usual, but in a sudden climate event, the homeless are especially vulnerable. Public amenities that are purposefully left open and available for emergency accommodation are important for the chronic and suddenly dispossessed. Climate education needs to engage with social justice in new ways, as the vulnerability of the marginalized is increasingly precarious. This has international and local implications.

Climate change is the malignant child of unrestrained free market capitalism, often called neoliberalism, and is also the product of modernity and the rise of industrial economies. Education about climate change, including for the health community, needs to engage with the root philosophy of economic theory, as this informs normative policy and practice, which is no longer fit for purpose (if it ever was). Both traditional 'left-wing' and 'right-wing' politics have long been committed to growing industrial production, ignoring emissions and other harm to social and environmental commons. Economic growth floats on global population growth, especially of the expanding consumer class (amongst other factors). Education about global health should also grapple with claims that concerns about population have been undermined by neoliberalism; this applies not only to total numbers, but also to the harmful effects of rapid growth in population size in low-income settings, such as Niger and other parts of the Sahel (Butler and Higgs, 2018). Both standpoints are uncomfortable territory for many educators, including, we suggest, those who teach medical students in institutions in the global North, many of whom are comparatively privileged, and who aspire to carbon-intensive lifestyles (Skinner, 2021; Singh *et al.*, 2022). It is much more palatable to advocate for less carbon-intensive hospitals than for less carbon-intensive doctors.

In countries dominated by neoliberalism (i.e. most of the global North), double parent working, consumer aspirations and other aspects of 'atomization' have effectively squeezed out public amenities such as community halls. This is one symptom of the lack of cohesion and capacity for fun, and of voluntary activities in the community, from life-long learning at higher education and schools, to limited social contact via community dances or other social events. As a result, large sectors of the community in the North have very limited or almost entirely no contact with other local people except through shopping. The elderly and health-compromised make up another sector of the community that can be very vulnerable during a large-scale extreme event influenced by climate change. Public health needs to consider these sectors of the community in terms of emergency shelter, and also in more long-term ways, to build community cohesion and resilience so that locals are aware of the need to look out for each other.

A lesson from serious storms is that civil defence cannot always come to the rescue. A couple of men near North Island's Esk Valley in New Zealand got up during the first night of heavy rain in February 2023, tracked down some tools and a small 'tinny' boat with an outboard motor, and were on the water throughout the night, rescuing families and animals from the roofs of houses. In one case, occupants were trapped in the roof space, but rescued in time, by pluck, ingenuity and the community spirit that neoliberalism endangers. Something similar happened in Lismore in Australia in 2022.

Resilience is also about long-term community building, awareness of vulnerable people and planning for safer communal accommodation in the face of disaster. This might be in existing large public spaces, such as libraries, community centres, shopping malls, or the *wharenui* (traditional large Māori community houses). Preferably people will better understand the ecological vulnerabilities of where they live; river deltas and flood plains that have been built on and not flooded for centuries are now at risk. Forest fires occur in rainforests. Landslides and erosion on steep land are much more likely. Significant sea level rise is expected. A careful analysis of what is likely to happen in a locality according to the types of geology and hydrology (and fire risk) is increasingly important.

Building community assets on high stable ground with independent power supply and communication systems that can accommodate anyone who might be affected in a climate emergency event is an important element of resilience and adaptation. Resilient communities will develop their own communication networks – ones that are not exclusively electronic, since Wi-fi and phone lines are vulnerable. Considering safe clean water supply, and food, are also important elements of family and community preparation. This level of preparation is part of adequate adaptation to permanently changed circumstances. The more engaged members of the community are in developing their own adaptation assets and infrastructure, the more effective and empowered they are in the face of calamity. The more politically activated they are, the more effective and resilient they feel (Bright and Eames, 2021). But such solutions are much harder in localities of pervasive poverty, and also where local inequality, fuelled by extractive economics, is rampant.

29.3 Denial, Doomers and Preppers

If public health requires better community cohesion and more awareness of local effects of climate change, then events and networks to promote climate education are also necessary. The receptivity of people in the community to climate messaging varies considerably, depending on their existing knowledge and emotional response to the climate catastrophe. The topic is depressing, and responses vary from enthusiastic motivation for change at a variety of levels from political activism to personal preparation, to depression, inertia or denial (Bright and Eames, 2021; Chapman *et al.*, 2017).

Enthusiasm needs direction to channel, and various local non-government associations can facilitate motivated volunteers. But at a deeper level, understanding the ways in which our current modern economic system contributes to continuing emissions growth is vital for focusing effective response, and building genuine mitigation, adaptation and resilience. Without direction, the community has dissolved into despair, which shows up as denialists, hedonists, preppers and doomers.

Climate denial is finally waning. Increasingly, media outlets, community, government and business organizations are ignoring denialists and getting on with mitigation and adaptation. Denial is, in our opinion, an early stage of grief. The ferocity of some people's denial may be related to their fear. Furthermore, denialists tend to be people deeply committed to benefiting from the status quo: often older, white, male and comparatively privileged. Their psychology is straightforward but effectively shifting them from their anxiety is challenging. Empathetic interpretation of their denialism as fear may be the best strategy; denialists are deeply committed to their views.

Denialists are under increasing pressure to engage with mitigation, as the organizations that they are entrenched within pivot to accommodate the Paris Agreement on climate change (UNFCCC, 2016) and the new legislation that is emerging in response. For those agents of mitigation and adaptation, working with denialist colleagues is exhausting and disheartening. It is important for educationalists to recognize that denialists are changing as fast as they are able, even if they continue to obstruct, weaken, or make token gestures towards mitigation. Their resistance, however, is on the ebb.

Related to denial is the more recent phenomena of doomers. People in this section of the community are well educated on the causes and impacts of climate change, are certain that it is 'too late' to change the course of events and feel that civil society is headed towards complete collapse. The psychology of collapse has two most common responses. The first is the hedonism of carrying on 'partying, while Rome burns', so to speak, which involves knowingly ignoring the ecological consequences of continued tourism, flying and excessive consumerism. The second is 'prepping' by buying rural land and building the equivalent of a nuclear bunker. Multiple uber-rich billionaires are taking the second route, as described by Rushkoff (2018), who was interviewed by five billionaires who wanted advice about how to keep their security staff under control, once money was useless, 'after the event'. Doomers lack the imagination to change the normative infrastructure of neoliberal globalization, despite understanding that it is creating devastating and toxic pollution. They have abandoned technological fixes such as 'carbon capture' which, after almost 50 years, has not succeeded at scale. But they cannot imagine a sophisticated global economy without emissions. This dilemma is producing a psychological collapse, as deep as the environmental one. It is becoming clearer that even the wealthiest individuals on the planet cannot survive without a deep level society change. Educationally, this doomer stage might be necessary to create the widespread courage necessary to consider what facets of globalized consumerism and the machinery of industrialism are producing emissions, and which are far older and less harmful.

Mitigation, resilience and adaptation are all needed to resolve the dilemma of climate change, and they need to go hand in hand. This means understanding the fundamental drivers to emissions growth, in order to effectively do something about it. Effective action requires a good understanding of what needs to change. It is the most promising mode of avoiding the mental health pitfalls of anxiety, fear and collapse that plague the denialists, doomers and preppers.

29.4 Effective Action

Practical adaptation within the parameters available is fairly simple – many of us can prepare by storing clean water and long-life food, installing emergency power generators and trying not to open the doors of refrigerators unless vital. But such measures assume that the house is still standing, not washed away or burned to the ground. Public health education on mitigation, adaptation and resilience could usefully contribute at this practical level. But to leverage truly effective activity in the public domain, deeper understanding of existing political theory and emergent discourses is required.

The global economic system remains enmeshed in the rhetoric of economic growth. John Maynard Keynes argued that Marx's analysis of class warfare, where the proletariat would rise up and dethrone the wealthy capitalist owners, could be averted, if the working classes could also benefit from the riches of capitalist industrial production. The promise of low levels of exponential economic growth meant that the working classes could hope for a better tomorrow, if not within their own lifetimes, at least for their children (Keynes, 1936; Glahe, 1991). Supporters of Keynes worked to strengthen the state provision of healthcare, education and welfare, aiming to distribute some of the rapidly increasing wealth generated by mass production. Unfortunately, resource exhaustion (Higgs, 2014) and growth in emissions are closely correlated with that curve, and by now the small annual increments have exponentially grown and become massive ones (Bank of International Settlements, 2022).

The commitment to zero emissions has focused understandably on the source of energy. Fossil fuels have underpinned capitalist industrialism (Altvater, 2007) and their versatile, mobile and incredibly dense energy has effectively replaced manual labour. As Hagans (2020) explained:

> One barrel of crude oil can perform about 1700 kW h of work. A human laborer can perform about 0.6 kW h in one workday (IIER, 2011). Simple arithmetic reveals it takes over 11 years of human labor to do the same work potential in a barrel of oil. Even if humans are 2.5× more efficient at converting energy to work, the energy in one barrel of oil substitutes approximately 4.5 years of physical human labor.

One of the most intensive industrial powerhouses, Germany, led the way to converting to renewable energy in the 1990s and 2000s. Germany encouraged private individuals to put up solar panels and get paid to feed into the national grid. In other countries this transition has been slower, as lobbying from coal-fired energy suppliers and existing centralized power distribution meant that governments actively discouraged solar input from distributed housing through unfavourable tariffs. However, as a result of continuously growing economic demand for energy, even in Germany, today, the economy remains strongly reliant on coal and gas.

Nevertheless, the sway of the fossil fuel industry, like tobacco a decade or so earlier, is being perceived as progressively more dubious and immoral. In sport, art and theatre, oil sponsorship is gradually being rejected. Government subsidies and kickbacks to fossil fuel generation and mining are under increased public scrutiny, in some settings.

The renewable generation of electricity holds much promise. The price of wind turbines and solar panels has reduced to a point claimed to be cheaper than coal generation in many settings. There are still substantial problems with archaic centralized distribution systems, and the political power of centralized generators that discourage the development of decentralized generation. But these issues are gradually getting resolved.

29.5 Conclusion

Education to deal with climate change needs a clear-eyed look at the economic premises of 'growth' (Altvater, 2002). A fundamental narrative of modern classical and neoliberal economics needs to be challenged. This is the assumption that the profit motive is indispensable, and that the purpose of all innovation and efficiency is to produce profit, rather than to reduce the generation of energy (and goods). Tied to this are assumptions about human behaviour that erroneously and simplistically attribute all decisions to hedonistic self-serving utility maximizing. Crucially, without these assumptions, the function of the Invisible Hand that sets the price of the perfect market, where the costs are reflected in the lowest price point and profit is minimal, cannot take place. If people make market decisions based on factors other than self-interested hedonism, the Invisible Hand will be 'warped'. In these ways hedonism and exponential growth are hard baked into the narrative assumptions of the current system.

A primary requirement for education about climate change and public health is to identify the structural problems of modernity. This can help prepare and motivate people to enact change. Resilience requires a direction for action, which helps people to focus and find the drive to challenge existential problems with deep change solutions. Public health is implicated at every level of these changes.

References

Altvater, E. (2002) The growth obsession. In: Panitch, L. and Leys, C. (eds) *Socialist Register; A World of Contradictions*. Merlin Press, Frenwood Publishing and Monthly Review Press, London, pp. 73–92.

Altvater, E. (2007) The social and natural environment of fossil capitalism. *Socialist Register* 43, 37–59. Available at: https://socialistregister.com/index.php/srv/article/view/5857 (accessed 12 June 2023).

Bank of International Settlements (2022) *OTC derivatives statistics at end-June 2022*. Available at: https://www.bis.org/publ/otc_hy2211.pdf (accessed 21 April 2023).

Box, J., Hubbard, A., Bahr, D., Colgan, W., Fettweis, X. *et al.* (2022) Greenland ice sheet climate disequilibrium and committed sea-level rise. *Nature Climate Change* 12, 808–813. DOI: 10.1038/s41558-022-01441-2

Bright, M.L. and Eames, C. (2021) From apathy through anxiety to action: emotions as motivators for youth climate strike leaders. *Australian Journal of Environmental Education* 38, 13–25. DOI: 10.1017/aee.2021.22

Butler, C.D. and Higgs, K. (2018) Health, population, limits and the decline of nature. In: Marsden, T. (ed.) *The Sage Handbook of Nature.* Sage, London. pp. 1122–1149.

Chapman, D.A., Lickel, B. and Markowitz, E. (2017) Reassessing emotion in climate change communication. *Nature Climate Change* 7, 850–852. DOI: 10.1038/s41558-017-0021-9

Davis, P.E.D., Nicholls, K.W., Holland, D.M., Schmidt, B.E., Washam, P. *et al.* (2023) Suppressed basal melting in the eastern Thwaites Glacier grounding zone. *Nature* 614, 479–485. DOI: 10.1038/S41586-022-05586-0

Glahe, F.R. (1991) *Keynes's 'The General Theory of Employment, Interest, and Money': A Concordance*. Rowman & Littlefield, London.

Goulson, D. (2019) The insect apocalypse, and why it matters. *Current Biology* 29, R967–R971. DOI: 10.1016/j.cub.2019.06.069

Gunn, K., Rintoul, S., England, M. and Bowen, M. (2023) Recent reduced abyssal overturning and ventilation in the Australian Antarctic Basin. *Nature Climate Change* 13, 537–544. DOI: 10.1038/s41558-023-01667-8

Hagans, N.J. (2020) Economics for the future – Beyond the superorganism. *Ecological Economics* 169, 1–16. DOI: 10.1016/j.ecolecon.2019.10652

Heidegger, M. (1954) *Die Frage nach der Technik. In: Vorträge und Aufsatze* [Lectures and Essays]. [Translated into English by W. Lovitt (1977) in: *The Question Concerning Technology and Other Essays*. Harper and Row, New York, pp. 3–35.]

Higgs, K. (2014) *Collision Course: Endless Growth on a Finite Planet*. MIT Press, Cambridge, Massachusetts.

IIER (2011) *Green Growth – an Oxymoron?* Institute for Integrated Economic Growth, Horgen, Switzerland. Available at: https://www.iier.us/content/green-growth-oxymoron (accessed 4 February 2024).

IPBES (2019) *Global Assessment Report on Biodiversity and Ecosystem Services*. Intergovernmental Platform on Biodiversity and Ecosystem Services, Bonn, Germany. Available at: https://www.ipbes.net/global-assessment (accessed 21 April, 2023).

Irwin, R. (2010) Reflections on modern climate change and finitude. In: Irwin, R. (ed.) *Climate Change and Philosophy; Transformational Possibilities*. Continuum International Publishing Group, London and New York, pp. 48–72.

Keynes, J.M. (1936) *The General Theory of Employment, Interest, and Money*. Palgrave Macmillan, London.

Madden, D.L., McLean, M. and Horton, G.L. (2018) Preparing medical graduates for the health effects of climate change: an Australasian collaboration. *Medical Journal of Australia* 208, 291-292e1. DOI: 10.5694/mja17.01172

NIWA (2023) *Auckland suffers wettest month in history*. National Institute of Water and Atmospheric Research, Auckland, New Zealand. Available at: https://niwa.co.nz/news/auckland-suffers-wettest-month-in-history (accessed 21 April 2023).

Pencheon, D. and Wight, J. (2020) Making healthcare and health systems net zero. *BMJ* 368, m970. DOI: 10.1136/bmj.m970

Rahmstorf, S. (2020) *New studies confirm weakening of the Gulf Stream circulation (AMOC)*. RealClimate blog, Washington, DC. Available at: https://www.realclimate.org/index.php/archives/2020/09/new-studies-confirm-weakening-of-the-gulf-stream-circulation-amoc/ (accessed 23 April 2022).

Rushkoff, D. (2018) *Survival of the Richest; the wealthy are plotting to leave us behind*. OneZero. Available at: https://onezero.medium.com/survival-of-the-richest-9ef6cddd0cc1 (accessed 21 April 2023).

Shiva, V. (1989) *The Violence of the Green Revolution, Third World Agriculture, Ecology, and Politics.* Research Foundation for Science and Ecology, Dehradun, India. Part available at: https://www.proquest.com/openview/e17430b1fedf8a0f1eb975e152df0e4e/1?cbl=1056335&pq-origsite=gscholar&parentSessionId=X7CO8qVH49Zj9G%2BE9bQJ0Qz9Pip47WZC9R1E%2F0rAgL4%3D

Singh, A., Hadfield, J., Gale, J. and Shaw, C. (2022) Doctors' travel in the Anthropocene. *New Zealand Medical Journal* 135, 88–93.

Skinner, J.R. (2021) Doctors and climate change: First do no harm. *Journal of Paediatrics and Child Health* 57, 1754–1758. DOI: 10.1111/jpc.15658

Testa, G., Neira, S., Giesecke, R. and Piñones, A. (2022) Projecting environmental and krill fishery impacts on the Antarctic Peninsula food web in 2100. *Progress in Oceanography* 206, 102862. DOI: 10.1016/j.pocean.2022.102862

UNFCCC (2016) *UN Climate Change Conference (COP 21), Paris*. United Nations Framework Convention on Climate Change, Bonn (Germany), Rio de Janeiro (Brazil) and New York. Available at: https://unfccc.int/process-and-meetings/the-paris-agreement (accessed 21 April 2023).

van Westen, R.M., Kliphuis, M. and Dijkstra, H.A. (2024) Physics-based early warning signal shows that AMOC is on tipping course. Science *Advances* 10, eadk1189. DOI: 10.1126/sciadv.adk1189

30 Communication and Climate Change

Rebecca M. Colvin
Crawford School of Public Policy, The Australian National University, Canberra, Australia

Abstract

This chapter outlines a relational approach to communicating about climate change. It highlights the social aspects of communication which sit alongside informational aspects and shape how communications are received. The chapter outlines the social identity approach as a theoretical framework to helpfully guide effective communication on climate change, by highlighting the 'us and them' dynamics at play in communication. Strategies for adopting a relational approach to climate change communication are proposed, including a focus on in-group messaging and navigation of inter-group divisions.

30.1 Introduction

Effective communication about climate change is not so much a matter of knowing the right words to say as it is a matter of understanding the social context within which communication occurs. A message about climate change delivered by someone who is trusted will be received very differently than exactly the same message delivered by someone who is not trusted. This is effectively a relational approach to communication. This chapter provides a brief introduction to this relational way of approaching effective climate communication, including introduction to some (useful) theory and strategies.

30.2 The Relational Approach to Climate Communication

Most people engage with climate change as everyday people, neither as climate advocates nor as climate adversaries. They are not required by work to communicate about climate change, nor have they taken it upon themselves to do the same. These people share the privilege of only having to view climate change from their own perspective. They are not seeking to improve climate science literacy, explore policy options, or inform others about how they can adapt to a climate-changed world. And, because of this, there is no obligation on them to understand how other people might think about climate change, and therefore what it takes to communicate effectively with them.

The people who do communicate about climate change, however, carry a heavy responsibility. This includes those who communicate for work and those who in their personal lives seek to improve general awareness and galvanize effective responses to climate change. Unlike the everyday people, these communicators do not have the privilege of viewing climate change only from their own perspective. Instead, they have a responsibility to understand the multiple

 Climate Change and Global Health: Primary, Secondary and Tertiary Effects, 2nd Edition (eds C.D. Butler and K. Higgs)
DOI: 10.1079/9781800620025.0030

perspectives from which climate change can be viewed if they are to be effective in their communication efforts.

Communicators are therefore in a very different position to everyday people when it comes to thinking about climate change. These different positions can be understood with the aid of an analogy from the art world. Everyday people get to view the issue of climate change as if they are viewing a three-dimensional sculpture. They can plant their feet in the ground, view the artwork from their perspective, and not worry that other people, who are viewing the same work from a different perspective, see it differently. Communicators, on the other hand, do not get to do this. Unlike viewing a three-dimensional sculpture from a single position, communicators have to be able to see all sides of the issue at once; the multiple and different perspectives held by multiple and different people. To communicators, engaging with the issue of climate change is less like viewing a sculpture and more like viewing a cubist artwork (Fig. 30.1). In the cubist approach, the subject of the artwork is pulled apart and broken up so that all perspectives are on the canvas and visible. The cubist approach trades off the comfort and familiarity of viewing a sculpture from one's own position and replaces that with the discomfort and discordance of viewing the subject in all its complexity.

It is this cubist-like view of the issue of climate change that underpins a relational approach to climate change communication. Communicators need to understand all different perspectives and how they are proffered by all different experiences, interests, identities, values and ideologies. And they need to understand how who they are interacts with all these dimensions in the act of communicating. This is because communication is not just about transmitting information from one person to another. It is also constitutive, in that the act of communicating creates meaning about the subject of communication (Cox, 2013). In this way, in addition to the factual accuracy and usefulness of the information being shared, the effectiveness (or otherwise) of communication is strongly shaped by social aspects.

The social aspects of communication are nicely illustrated by conflict over a very large-scale

Fig. 30.1. Effective climate communication demands a cubist approach. (Left): Bust of Julius Caesar. Photograph by Thad Zajdowicz, 2018. (Right): Figures by Henri Le Fauconnier, 1913. Both images: Public Domain (CC0 1.0: https://creativecommons.org/publicdomain/zero/1.0/).

wind energy development proposal in Australia. Opponents of the wind energy proposal objected to the project being described as a 'wind farm' as was the case in official communications, media coverage, and day-to-day conversation. To the opponents, 'wind farm' was not a neutral term that simply described a renewable energy technology. They felt the term falsely implied that wind turbines belonged in an agrarian landscape and that, instead, the project should be described as a 'wind factory', because to them the proposal would have meant industrialization of the agrarian landscape (Colvin, 2016). When the term 'wind farm' was used, it did more than transmit information about turbines and electricity. It also served to indicate whether the communicator felt the turbines belonged in the agrarian landscape, and by extension whether the grounds for opposition were understood and acknowledged. So, the terms 'wind farm' and 'wind factory' came to signal how different people related to the issue at hand and, importantly, how people related to each other; whether they were one of 'us' or one of 'them'.

30.3 Useful Theory: The Social Identity Approach

Theoretically, the relational approach to communication is well served by the insights provided by the social identity approach, as it can be broadly considered the science of 'us and them'. The social identity approach is a meta-theoretical perspective on the importance of groups in social life, inclusive of two social psychological theories: self-categorization theory and social identity theory (Brown, 2019). Self-categorization theory explains how people come to identify with social groups such that the group identity becomes part of their concept of their self. Social identity theory focuses on the relations between groups, once identities are present (Haslam, 2000). Important to the social identity approach is the difference between in-groups and out-groups. In-groups are the social identity groups with which a person identifies, making the in-group social identity psychologically relevant to their sense of self (Abrams and Hogg, 2004). Out-groups are the identity groups with which a person does not identify.

The social identity approach emphasizes the dynamic and context-dependent nature of social identity and intergroup relations. The legacy of past interactions between groups provides the context for future intergroup relations (Colvin *et al.*, 2015). Context makes some social identities salient and others dormant (Cruwys *et al.*, 2016), meaning that people can see others as more or less like an in-group member, depending on the context (Mason *et al.*, 2014). Social norms – the expected and accepted beliefs and behaviours for a particular identity group in a given social setting – are primed by context and in turn shape context (Nyborg, 2018). The approach acknowledges a number of dynamics that tend to play out in intergroup settings, where in-groups are aware of their differences with out-groups (Colvin, 2020), summarized in Fig. 30.2:

- **In-group favouritism** promotes trust and charity within the in-group.
- **Out-group hostility** promotes distrust and suspicion of the out-group.
- **Stereotyping** of the out-group exacerbates hostility, leads to dehumanization, and can result in the justification of direct or indirect harm to out-groupers.
- **Leadership polarization** emerges when in-group leaders adopt extreme positions that favour differentiation from the out-group over moderate representation of the in-group.
- **Consensus-seeking** behaviour (often described as 'group think') occurs when in-group members favour agreement with fellow in-groupers over critical appraisal of their claims.
- **Shared norms** (beliefs and behaviours) develop and spread readily within in-groups, often becoming markers of group belonging that may be performed for symbolic rather than material reasons.

30.4 Strategies for a Relational Approach to Climate Communication

Social identity and intergroup dynamics emphasize that in-group messengers will be more effective than out-group messengers (Fielding *et al.*, 2020). For climate communication, this can mean focusing on building relationships

Fig. 30.2. Social identity and intergroup dynamics relevant to a relational approach to climate communication.

with key members of a target in-group so that they can communicate with other in-groupers, rather than attempting to communicate as an out-grouper (Badullovich, 2022). In fact, when communication comes solely from a single identity group, it can create challenges that go beyond being unable to reach out-groupers. When an issue such as climate change is perceived to be the domain of some identity groups but not others, it can lead to the issue being bundled with the identity such that hostility to the identity is transferred to the issue, too (Colvin *et al.*, 2020). In the context of climate change, research has shown that because 'environmentalists' were the early messengers about the need for climate policy, hostility toward environmentalists predicted opposition to climate policy (Hoffarth and Hodson, 2016). In other words, the relational dimension that gave context to climate communication shaped how climate messages were received.

When attitudes toward the messenger of an issue drive attitudes toward an issue itself, opposition to an issue can become part of the social norms of the social identity group (Huddy *et al.*, 2015). This is because in-groupers are motivated to maintain their positive distinctiveness from out-groupers (Haslam, 2000), and so disparaging something out-groupers care about can enhance in-group esteem. For example, social identity groups that are motivated to maintain positive distinctiveness from the identity groups that are perceived to 'own' the issue of climate change may adopt anti-climate norms not necessarily because of anything inherent to the issue of climate change, but in order to express their group belonging and that they are 'not like those other people' (Kurz *et al.*, 2020).

The social identity approach highlights that the relational dimension of communication is far from static. Over time, people will increase and decrease the extent to which a particular social identity is central to their sense of self, and over time the norms associated with a social identity will change (Haslam, 2000). In-group leaders can be influential in changing social norms by adopting a new norm that cues similar norm change among other in-groupers (Kousser and Tranter, 2018). However, in-group leaders run the risk of sanction from the group if they change from the established norm too much or too fast. A leader who goes against the norm may fail to change the identity group norm and instead end up de-legitimized and ostracized. Leaders are most likely to be able to act successfully as identity group 'norm entrepreneurs' when there is early adoption of, or at least openness to, the new norm among the identity group (Centola, 2021). In that way, successful norm change can stem from in-groupers at the periphery of social identity groups, particularly if there is strong social reinforcement for the change (Goldberg *et al.*, 2019; Centola, 2021).

A longer-term view from the relational approach to climate communication is that building bridges across identity group divides can serve to reduce intergroup animosity, opening new opportunities for effective communication (Fielding and Hornsey, 2016). Divides between in-groups and out-groups have power as far as the group divides are salient in a given context.

Communicators seeking effective ways to engage people on climate science, impacts, adaptation strategies and policy options will do well to adopt a relational approach to communication, informed by social identity and intergroup relations. This requires understanding the full range of perspectives on climate change, and how these perspectives are informed by experiences, interests, identities, values and ideologies. Effective communication is not a matter of getting the 'script' right. Rather, it is about recognizing communication as an interaction that is shaped by, and in turn shapes, social context.

Communicators need to consider how the way they communicate will signal to other people whether they are one of 'us or them', and how their position – real or perceived – in some social identity groups but not others affects their ability to communicate to different people. Recognizing when climate norms are responsive to identity drivers, rather than substantive matters to do with climate change, can help communicators to decide whether and how to invest their time in communicating with different people. People who span social identity divides are particularly well-placed to communicate effectively with their in-groups that may be uninspired by out-group communicators. If in-group messengers are messaging against an established social norm, it will be important to recognize and mitigate the risk of losing legitimacy with the social identity group. For this, understanding the relational social context that shapes the identity and norm is key.

References

Abrams, D. and Hogg, M.A. (2004) Metatheory: Lessons from social identity research. *Personality and Social Psychology Review* 8, 98–106. DOI: 10.1207/s15327957pspr0802_2

Badullovich, N. (2022) From influencing to engagement: a framing model for climate communication in polarised settings. *Environmental Politics* 32, 207–226. DOI: 10.1080/09644016.2022.2052648

Brown, R. (2019) The social identity approach: Appraising the Tajfellian legacy. *British Journal of Social Psychology* 59, 5–25. DOI: 10.1111/bjso.12349

Centola, D. (2021) *Change: How to Make Big Things Happen*. Little, Brown and Company, New York.

Colvin, R.M. (2016) *Lessons from the TasWind Proposal in King Island: A Descriptive Report*. The University of Queensland, Brisbane, Australia.

Colvin, R.M. (2020) Social identity in the energy transition: an analysis of the 'Stop Adani Convoy' to explore social-political conflict in Australia. *Energy Research & Social Science* 66, 101492. DOI: 10.1016/j.erss.2020.101492

Colvin, R.M., Witt, G.B. and Lacey, J. (2015) The social identity approach to understanding socio-political conflict in environmental and natural resources management. *Global Environmental Change* 34, 237–246. DOI: 10.1016/j.gloenvcha.2015.07.011

Colvin, R.M., Kemp, L., Talberg, A., De Castella, C., Downie, C. *et al.* (2020) Learning from the climate change debate to avoid polarisation on negative emissions. *Environmental Communication* 14, 23–35. DOI: 10.1080/17524032.2019.1630463

Cox, R. (2013) *Environmental Communication and the Public Sphere*. Sage, Thousand Oaks, California.

Cruwys, T., Platow, M.J., Rieger, E., Byrne, D.G. and Haslam, S.A. (2016) The social psychology of disordered eating: The Situated Identity Enactment model. *European Review of Social Psychology* 27, 160–195. DOI: 10.1080/10463283.2016.1229891

Fielding, K.S. and Hornsey, M.J. (2016) A Social Identity analysis of climate change and environmental attitudes and behaviors: Insights and opportunities. *Frontiers in Psychology* 7, 121. DOI: 10.3389/fpsyg.2016.00121

Fielding, K.S., Hornsey, M.J., Thai, H.A. and Toh, L.L. (2020) Using ingroup messengers and ingroup values to promote climate change policy. *Climatic Change* 158, 181–199. DOI: 10.1007/s10584-019-02561-z

Goldberg, M.H., van der Linden, S., Maibach, E. and Leiserowitz, A. (2019) Discussing global warming leads to greater acceptance of climate science. *Proceedings of the National Academy of Sciences (USA)* 30, 14804–14805. DOI: 10.1073/pnas.1906589116

Haslam, S.A. (2000) *Psychology in Organizations: The Social Identity Approach*. Sage, Thousand Oaks, California.

Hoffarth, M.R. and Hodson, G. (2016) Green on the outside, red on the inside: Perceived environmentalist threat as a factor explaining political polarization of climate change. *Journal of Environmental Psychology* 45, 40–49. DOI: 10.1016/j.jenvp.2015.11.002

Huddy, L., Mason, L. and Aarøe, L. (2015) Expressive partisanship: Campaign involvement, political emotion, and partisan identity. *American Political Science Review* 109, 1–17. DOI: 10.1017/S0003055414000604

Kousser, T. and Tranter, B. (2018) The influence of political leaders on climate change attitudes. *Global Environmental Change* 50, 100–109. DOI: 10.1016/j.gloenvcha.2018.03.005

Kurz, T., Prosser, A.M.B., Rabinovich, A. and O'Neill, S (2020) Could Vegans and Lycra Cyclists be Bad for the Planet? Theorizing the Role of Moralized Minority Practice Identities in Processes of Societal-Level Change. *Journal of Social Issues* 76, 86–100. DOI: 10.1111/josi.12366

Mason, C.M., Paxton, G., Parsons, R., Parr, J.M. and Moffat, K. (2014) 'For the benefit of Australians': Exploring national expectations of the mining industry. *Resources Policy* 41, 1–8. DOI: 10.1016/j.resourpol.2014.01.008

Nyborg, K. (2018) Social norms and the environment. *Annual Review of Resource Economics* 10, 405–423. DOI: 10.1146/annurev-resource-100517-023232

31 Health, Climate and Challenges in Africa: 2024–2100

Colin D. Butler,[1] A. Kofi Amegah,[2] Caradee Y. Wright[3,4] and Devin C. Bowles[5]

[1]National Centre for Epidemiology and Population Health, The Australian National University, Canberra, Australia; [2]Department of Biomedical Sciences, University of Cape Coast, Cape Coast, Ghana; [3]South African Medical Research Council, South Africa; [4]Department of Geography, Geoinformatics and Meteorology, University of Pretoria, Pretoria, South Africa; [5]Faculty of Health, University of Canberra, Canberra, Australia

Abstract

This chapter presents snapshots of existing and likely impacts of climate change in Africa, focusing on linkages, feedbacks and vulnerabilities. While potentially catastrophic risks exist, particularly in sub-Saharan Africa, adaptation steps are also explored, especially the helpful, development-promoting prospects that may emerge if discussing family planning and limits becomes more acceptable, not only to policy makers in the global North, but also within the continent. Population health in Africa may improve, despite the formidable challenges that are faced, if the speed of the nascent 'sustainability transition' in the North is greatly accelerated (avoiding worst-case climate scenarios) and if this is mirrored by the taking up of opportunities that exist in Africa. Overstating the role of calamities in Africa in climate change is problematic; global warming is largely beyond African control; however, many of the co-factors that shape the health of the people of Africa can be improved by local action.

> The continent is too large to describe. It is a veritable ocean, a separate planet, a varied, immensely rich cosmos. Only with the greatest simplification, for the sake of convenience, can we say 'Africa'. In reality, except as a geographical appellation, Africa does not exist.
>
> (Kapuściński, 1998)

31.1 The African Continent

Africa, the 'cradle of humanity', is the home of numerous peoples, languages, customs and cultures. Its size is substantial – almost ten times that of India. Like India, it has a deep history of imperialism and colonialism, still ongoing (Hickel *et al.*, 2021). At a time when the people of the continent had already been raided for centuries, by both Europeans and Arabs, often in collusion with and assistance from people born in Africa, the Berlin conference in the mid-1880s formally 'divided' most of the continent among a few European powers; principally the UK, France, Germany, Portugal, Italy and Belgium, the brutality of whose King, Leopold, spawned the first global human rights campaign of the 20th century (Weisbord, 2003). As a result, Africa is today home to almost 60 separate nations (including eight islands), many of them cartographically captured by straight lines drawn in the 19th century by Europeans, disregarding local cultures, languages and rivalries (Griffiths, 1986). Almost all of Africa was colonized, the exceptions being Ethiopia and Liberia, but the former was briefly occupied by Italy, while the latter was founded as a colony for freed American slaves.

DOI: 10.1079/9781800620025.0031

Today, Africa's population exceeds 1.4 billion. In the almost 40 countries that comprise mainland sub-Saharan Africa (SSA) (plus the island of Madagascar), the population is still growing rapidly. The forecast population for SSA by 2100 is for it to triple to over 3 billion (95% uncertainty interval 2.4–3.8 billion) (Ezeh *et al.*, 2020). This forecast is based on the optimistic assumptions developed by the Institute for Health Metrics and Evaluation, and is substantially lower than that of the median UN Population Division projection (3·8 billion). While this enormous expansion of population may be called a human success, as it means more people will be living in SSA than ever before, it is also recognized by some, including the leading African demographer, Alex Ezeh, as daunting. Ezeh and his colleagues pointed out that such a rise of the SSA population over the next 80 years requires the tripling of 'all existing resources and infrastructure (health, education, housing, energy, and so on) over that period just to maintain existing inadequate levels of basic services and amenities'. They also identified an annual shortfall of 52–64% in financing for infrastructure needs in this region of Africa.

Zulu and colleagues reported that, in 2015, 42% of Africa's population was younger than 15 years. For the world as a whole, the comparable figure in 2021 was 25% (World Bank, nd). Zulu *et al.* (2015) stated that 'such high child dependency burden is widely recognized as a major barrier to socioeconomic development in Africa'. They called for a slowing of population growth in Africa, in order to harness the demographic dividend – the economic and social benefits arising from a significant increase in the ratio of working-age adults relative to young dependents that results from rapid fertility decline. They pointed out that this will increase productivity and savings that can be directed to investments for further economic growth. They also noted that achieving this dividend requires sustained investments in education, skills development, health, job creation and improved governance.

The prospect of this population expansion in SSA is worsened because the consequences of climate change in Africa, and elsewhere in the world, are increasingly menacing. Although the ideas expressed by the term 'demographic entrapment', chiefly popularized by Maurice King (1990), have in recent years become highly unfashionable, we are not so sanguine. André and Platteau (1998) applied an even less popular variant of this term (the 'Malthusian trap') to partly explain the catastrophe of the Rwandan genocide in 1994. These authors analysed data, collected in a densely populated region of Rwanda prior to the genocide (1988–1993) to support the argument that resource scarcity, fuelled by long-standing ethnic tension, a growing scarcity of arable land and a lack of modern agricultural techniques (including fertilizer) combined, with other factors, to spawn genocide. Although catastrophes of that scale have not yet been repeated in other parts of SSA, some areas, including the Democratic Republic of the Congo (DRC), Sudan, South Sudan and, more recently, Ethiopia remain hotspots of intense conflict, in part driven by competition over scarce resources. In turn, conflicts in many parts of Africa reduce local food security, thus creating the potential for reinforcing feedbacks that may intensify poverty and vulnerability, including to climate change (Anderson *et al.*, 2021).

31.2 Climate Change in Africa

The fraction of the world's total emissions of carbon dioxide (CO_2) that originate from African nations for electricity generation and transport is trivial cumulatively, and still minor today, at less than 4% of the world's total in 2020 (Ritchie and Roser, 2022). On a per capita basis, CO_2 emissions from these sources in Africa are also very low; for example, 0.07 tonnes per annum in Niger, compared with over 14 tonnes per person in the USA. However, other forms and sources of greenhouse gas (GHG) emissions are not as low; for example, deforestation in the Congo Basin, the world's second largest tropical rainforest, is rapidly increasing (Kleinschroth *et al.*, 2019), while methane emissions from Africa's large herd of cattle remain disproportionately high, per unit of food supplied. These represent significant African GHG sources (Kiggundu *et al.*, 2019).

All regions and populations of the world are vulnerable to climate change, but many authors, including some based in Africa, agree that the continent is especially at risk, both in its north (Schilling *et al.*, 2020) and in SSA (Connolly-Boutin and Smit, 2016). Climate change heightens many pre-existing vulnerabilities, including of women (see Chapter 31.ii, this volume), fishing communities, rural subsistence farmers and those living in informal settlements. Some of these individuals bear a double burden, for example a single mother living in an unplanned town, or a cultivator relying on the rain to grow her maize or potatoes.

The reasons for Africa's vulnerability are generally attributed to the continent's widespread poverty (Collier *et al.*, 2008). But there are deeper causes and co-factors, including colonialism and unfair exchange rates (Hickel *et al.*, 2021). We suggest that Africa's relative poverty and underdevelopment are also contributed to by its relatively high birth rate in many parts, and by the cluster of factors that determine this. These include: patriarchy, early marriage, inadequate education, literacy and numeracy. These are compounded by the difficulty in accessing family planning services (Koffi *et al.*, 2018). We agree with many analysts, including some based in Africa, that discussion about and reduction of these barriers will reduce poverty (e.g. Mutunga *et al.*, 2012; Zulu *et al.*, 2015; Agyepong *et al.*, 2017; Ezeh *et al.*, 2020). In so doing, we reject assertions that such discussion is to encourage forms of 'eco-fascism' (Tilley and Ajl, 2022), even though some individuals and groups may seek to distort these arguments for their own purposes. All men and women have the right to an education, and women should have control of their reproductive choices, regardless of culture.

Analysts agree that the impacts of climate change on Africa are already apparent (Nangombe *et al.*, 2018). While studies examining heat and health in Africa are comparatively uncommon, they are increasing (Ncongwane *et al.*, 2021), as are average temperatures and heatwaves (Russo *et al.*, 2016). In West Africa, agriculture has been harmed by changes in rainfall patterns, characterized by strong inter-annual rainfall fluctuations, an increased frequency of rainfall extremes and prolonged droughts (Carr *et al.*, 2022).

The desert locust outbreak

From the end of 2019 to early 2020, a desert locust (*Schistocerca gregaria*) outbreak across several East African countries posed a serious risk to livelihoods and food security. The outbreak has been linked to a highly unusual sequence of cyclones. The first (Mekunu) created desert lakes over the 'Empty Quarter' in Saudi Arabia, allowing locusts to hatch and breed. Normally, the dry desert conditions would limit locust survival, but a second cyclone (Luban) followed, five months later. The outbreak spread to Yemen, beset then (as at the time we write) by civil war, and lacking the capacity to apply sufficient pesticides to reduce locust numbers. In turn, winds from a third cyclone (Pawan) carried locusts, in large numbers, to Kenya and other parts of East Africa. Salih *et al.* (2020) attributed the unusual sequence of cyclones to the warming western part of the Indian Ocean. They also described how a weakened response by some East African governments, including from the effects of long-standing conflict in Somalia, hampered access to locust control methods. Lockdowns related to COVID-19 also hindered these efforts, as did lack of funds. Scholars who have studied food security in Ethiopia since the 1980s then attributed the devastating conflict in Tigray, in part, to the locust outbreak (Deckers *et al.*, 2020).

Climate change in the Southern African Development Community region

The region called the Southern African Development Community (SADC) has 15 shared, transboundary river basins (Fig. 31.1). These resources are seen by advocates as an opportunity to achieve short- and long-term benefits through an integrated and coordinated operation of existing and planned hydropower facilities, cooperative flood management and irrigation development. Cooperation among riparian countries in the Zambezi Basin is said to have the potential to balance hydropower and irrigation, enabling stable energy generation of 30,000 Gigawatt hours (GWh)/year and 774,000 ha of irrigated land. Another opportunity for an apparent 'win-win' is the possibility that South Africa is considering: of harnessing water from the DRC in exchange for energy (Nhamo *et al.*, 2018).

On the other hand, Voortman (2013) argued that the irrigation potential in most of Africa is very limited, as there is a lack of extensive tracts

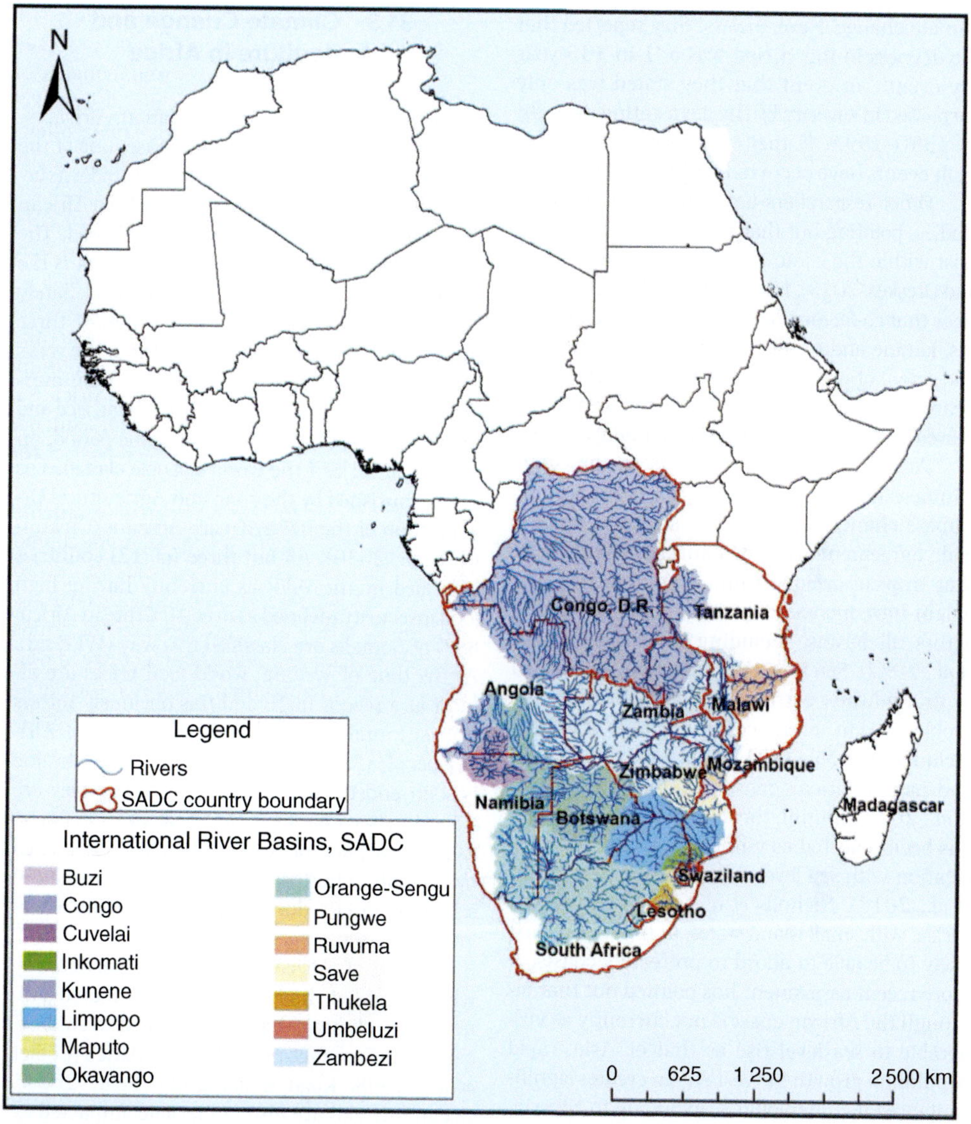

Fig. 31.1. Location of SADC countries and their transboundary river basins. Source Nhamo *et al.*, 2018. (CC BY 4.0 DEED).

of flat alluvial deposits under easy command of large rivers. Higginbottom *et al.* (2021) pointed out that recent irrigation projects in Africa had been disappointing, due to flawed political and management frameworks, including optimism bias.

Unfortunately, many urban areas of the SADC have already been 'ravaged' by climate change, in the form of floods, droughts and cyclones (Mupedziswa and Kubanga, 2017). Madagascar, an island member of the SADC and one of the world's poorest countries, has recently experienced devastating famine, suggested by the World Food Programme (WFP) as 'possibly the world's first climate-induced famine' (Schlein, 2021). However, an attribution study undertaken by Harrington *et al.* (2022) disagreed, finding instead that the occurrence of recent poor rains in Southern Madagascar in the two years following June 2019 was not due to human-caused climate change. Nevertheless, they reported that the dryness in this period was a '1-in-135-year dry event', an event that they stated was only surpassed in severity by the devastating drought of 1990–1992. If they are correct, then two such events have occurred within 31 years.

Other researchers are critical of attribution studies, pointing out that all extreme events now exist within the context of climate change (Lloyd and Oreskes, 2018). Irrespective of this debate, it is clear that co-factors for events such as health status, famine and conflict exist. Although the pace and extent of climate change is largely beyond African control, many of these co-factors can be improved by local as well as by international action.

Attribution of events such as droughts, famines, floods and conflict to anthropogenic climate change remains contested, but there is wide agreement, even by cautious researchers, that tropical cyclone rainfall and storm surge height have increased for individual events and across all basins, including in Africa (Clarke *et al.*, 2022). Sea level rise, sometimes worsened by groundwater extraction, will be increasingly problematic in several coastal regions of Africa, including the Nile delta (Mabrouk *et al.*, 2018) and Lagos, Africa's most populous city (Shiru *et al.*, 2020). Banjul, the capital of The Gambia, has been reported as vulnerable to its entire relocation with sea level rise of only 1 m (Amuzu *et al.*, 2018). Nicholls *et al.* (2011) considered Africa, with small island states, as the region least likely to be able to afford to protect its coasts. A more recent assessment has pointed out that although the African coast is not currently as vulnerable to sea level rise as that of Asia, rapid population growth in West Africa creates significant coastal vulnerabilities, including in Nigeria, Benin, Côte d'Ivoire and Senegal (Neumann *et al.*, 2015).

Development in many sectors (particularly agriculture, water and health) is a vital counterweight to the actual and potential harms of climate change. Before sketching out some of the likely health effects of climate change, we describe some aspects of climate change and agriculture in Africa, as the capacity to grow and protect more food (e.g. store, keep free from mycotoxins) is crucial to positive African development scenarios. Without enough food for all, many other problems will be harder to solve, including those that are health related.

31.3 Climate Change and Agriculture in Africa

Home to 18% of the world's population, Africa, especially its sub-Saharan region, houses one of the largest and greatest concentrations of undernourished children and adults in the world. No African nation is a significant net exporter of food. The continent is a major grain importer; Egypt is the second largest wheat importer in the world, barely below Indonesia, a country that is almost three times as populous (IndexMundi, nd). For the years 2010–2019 inclusive, Africa imported, on average, 38% of its three main cereals (wheat, rice and maize) (FAOSTAT, 2022). In the same period, an average of 32% of the world's people classified as undernourished by the Food and Agricultural Organization of the UN (FAO) are Africans (Otekunrin *et al.*, 2020). All but three (of 12) countries identified by the WFP as currently having high food insecurity (defined as over 40%) are in Africa; 89% of Somalis are classified this way (WFP, nd). At the time of writing, world food prices are almost at a record high, and this obviously harms nutrient intake for hundreds of millions of African people (Fig. 31.2). Even if the war in Ukraine is soon ended, the prospects for food prices are worrying (see Chapter 22). Wheat yields in Egypt, as in many parts of North Africa, are threatened by climate change, but ambitious irrigation schemes could reduce its dependency on wheat imports (Asseng *et al.*, 2018).

Much of Africa's agriculture, particularly where rain-fed, has long been considered vulnerable to climate change, from west to east, including the conflict-prone Horn of Africa (Collier *et al.*, 2008; Kogo *et al.*, 2021; Seife, 2021). Ocean fisheries are also vulnerable, impacted by overfishing, including their illegal plunder by wealthier nations (Merem *et al.*, 2019), while the fish catch in many lakes in Africa, including Tanganyika and Victoria, both important dietary sources, has also been in long-term decline, in part from climate change, but also from the rising human population (Njiru *et al.*, 2018).

Mycotoxins

Both drought stress and floods increase the risk of mould and fungi in staple crops, including

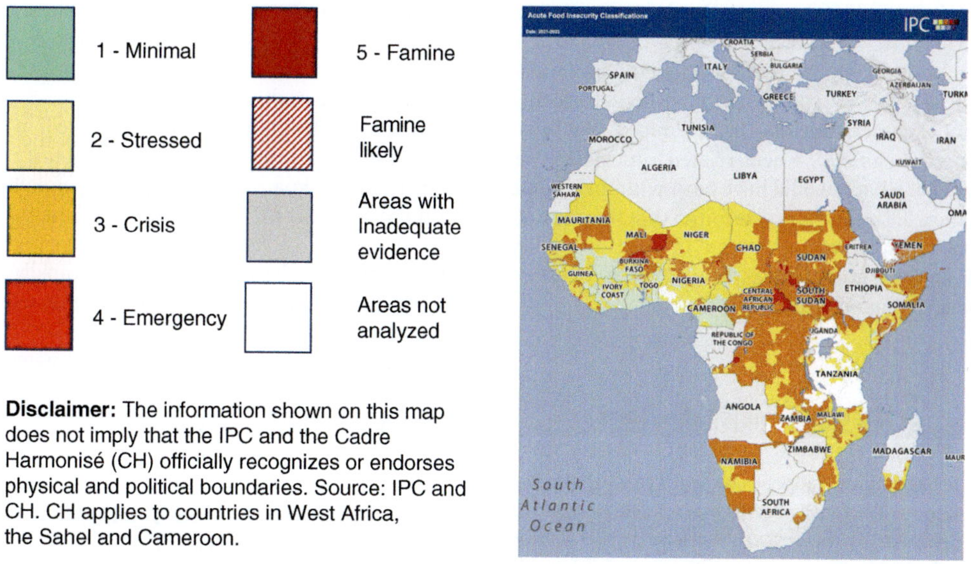

Fig. 31.2. Screenshot estimating insufficient food consumption in African nations. Source: Integrated Food Security Phase Classification (IPC) used with permission https://www.ipcinfo.org/ipc-country-analysis/ipc-mapping-tool/

stored grain. Such food can be contaminated by mycotoxins. People ingesting such contaminated food will usually not see or taste anything abnormal, and can develop liver failure and hepatocellular carcinoma. Though a simple fluorescence test can usually determine aflatoxin contamination, it is rarely used in low- and middle-income countries. Climate change may increase the burden of disease from mycotoxins, including through increased rainfall events during or immediately after harvest (Tefera, 2012).

31.4 Primary Health Effects of Climate Change in Africa

Consistent with the general framework of this book, this chapter next discusses three main categories of health effects in Africa, arising from: primary events (e.g. heatwaves and perhaps reduced cold waves in highland areas), flooding and disasters; secondary events, especially changes to vector-borne diseases and other infectious diseases; and tertiary events, including famine, conflict and large-scale migration. All of these phenomena have important health manifestations and consequences. Note that health systems in Africa are discussed in Chapter 31.i, this volume.

Heatwaves

The number of African megacities is growing (e.g. Cairo, Lagos, Nairobi and Johannesburg) and these are likely to be vulnerable to heatwaves, including via the interaction of extreme heat with air pollution (see Chapter 31.iii, this volume). The forecast temperatures for North Africa beyond 2050, under the worst-case scenarios ('business as usual'), are extremely troubling, with temperatures up to and exceeding 56°C persisting for several weeks. Forecasts show or suggest that by 2100 up to half of the North African population, most of whom will be urbanized, could be exposed to annually recurring super- and ultra-extreme heatwaves (Zittis *et al.*, 2021).

Models and observations have identified SSA as a region where heatwaves are of particular concern. Yet, to date, the reported direct health impacts from heatwaves in Africa appear to be minimal, a paradox that has been interpreted as due to under-reporting (Harrington and Otto, 2020). However, a model-based approach has recently concluded that, among children under the age five (who are highly vulnerable during hot weather, due to their reduced ability to thermoregulate), heat-related mortality is double what it would have been without climate change. The authors also found that this mortality outweighed reductions in heat mortality from improvements associated with development (Chapman *et al.*, 2022).

Flooding-related disasters

Climate change appears to be increasing the frequency of intense precipitation in most parts of Africa, as in the rest of the world (Blenkinsop *et al.*, 2021). The most recent report of the Intergovernmental Panel on Climate Change (IPCC) stated that in Africa, northern and south-western Africa are the likely exceptions (Trisos *et al.*, 2022). The vulnerability of the poor to flooding is amplified by urbanization and inadequate regulation, including of building sites. Those who are poor are frequently forced to build low-quality shelters on flood plains, steep hillsides (at the risk of erosion and landslides) and other locations considered risky. These circumstances are compounded by a lack of insurance and limited or no chance of compensation in the event of disaster. Due to flooding, the poor experience higher rates of trauma and property loss and some flood-related infectious diseases, such as gastroenteritis and leptospirosis. They are also likely to have increased rates of mental stress from their experience of such events in comparison with those who are better off and have agency to cope.

Africa is too vast, and our space in this chapter too limited, to survey flooding events across the entire continent. However, in 2019 Cyclone Ida made landfall in East Africa, generating one of the most devastating extreme weather events to have yet occurred in the Southern Hemisphere. It affected at least 3 million people in Mozambique, Zimbabwe and Malawi (Devi, 2019). Damage to the infrastructure corridor connecting the port of Beira (Mozambique) with three landlocked countries to its west disrupted regional trade and supplies of fuel and wheat, adding to the crisis. Crop damage was also extensive; flooding in some fields persisted for over a month (Bofana *et al.*, 2022).

Increased flooding has been reported in northern Ghana, again impacting food security (Atanga and Tankpa, 2021). Increased rainfall is also projected over the Congo basin (Blenkinsop *et al.*, 2021). In April 2020, flash flooding impacted the inhabitants of Uvira, a rapidly growing city (about 600,000 people) in the eastern part of the DRC, destroying over 5000 homes (Bachinyaga *et al.*, 2022). The damage was worsened by the largely unplanned expansion of the city onto floodplains.

31.5 Secondary Health Effects of Climate Change in Africa

Secondary effects are less direct than primary ones, but more causally apparent, particularly to conservative analysts, than tertiary health effects. An important mechanism for this group is an altered distribution of arthropod vectors, intermediate hosts and pathogens, due to a shift in environmental parameters that make areas more or less welcoming to hosts and pathogens. This in turn will produce changes in the distribution of many infectious diseases. Some diseases may expand their range, but others are predicted to decrease.

Malaria and arboviruses

Many parts of Africa, especially away from the highlands, have very high rates of infectious diseases. West Africa was long called the 'fever coast' by European invaders; mysterious febrile illnesses delayed European incursions and conquest. While most of these fevers were malarial, many illnesses mimic malaria. According to the historians McNeill (1976) and Crosby (1972), both malaria and yellow fever were introduced to the 'New World' from Africa, via the trans-Atlantic slave trade. Malaria has shaped the genome of many Africans, with red cell variants including sickle cell disease, thalassaemia and G6PD deficiency (Eridani, 2011).

The incidence of malaria has declined in recent decades in Africa, due to public health measures, especially the prompt treatment of malarial fevers with the artemisinin combination therapies (ACTs), which also inhibit transmission, and the use of insecticides (particularly pyrethroids in treated bed-nets) (Feachem *et al.*, 2019). However, in 2018, malaria still killed an estimated 400,000 Africans (over 90% of the global burden) and caused a staggering 228 million cases (Mordecai *et al.*, 2020). Although numerous non-climatic factors influence the distribution of malaria, it is implausible that changes in temperature, rainfall and humidity will not also play a role. These factors not only influence the distribution, longevity and activity of arthropod vectors, but in some cases accelerate the life cycle of the parasite within the vector.

There has long been evidence that climate change is contributing to the increased survival of *Anopheles* vectors at higher altitudes (Tulu, 1996; Chen *et al.*, 2009) and thus of malaria transmission (Alonso *et al.*, 2011). While in some cases the changes in the patterns of malaria resulting from climate change may be beneficial (Meyrowitsch *et al.*, 2011), on the whole they are likely to be harmful. It is also plausible that resistance to ACTs and pyrethroids will spread in Africa, again expanding the malarial disease burden (Hemingway *et al.*, 2022).

Lindsay *et al.* (2021) argued that excessive focus on insecticide resistance misses the bigger picture. They suggested that improving the effects of insecticide-treated nets needs increasing funding, improved strategies to increase net use, and enhancing the longevity of the active ingredients and the physical integrity of nets, while simultaneously accelerating the development and evaluation of novel vector control tools. Early warning systems to reduce the burden of disease of malaria in SSA have long been used, with much promise and some success (Thomson *et al.*, 2014; Tompkins *et al.*, 2019). However, the endurance and success of such programmes is vulnerable to a lack of sufficient high-quality health data and the withdrawal of external funding.

Many other vector-borne pathogens, including Rift Valley fever, dengue, chikungunyah, yellow fever, Zika, o'nyong'nyong, West Nile fever, leishmaniasis, onchocerciasis and African trypanosomiasis, circulate regularly in humans, wildlife and livestock in SSA. Mordecai *et al.* (2020) proposed that the collective burden of these, especially those that are arboviruses, may exceed that of malaria, as climate change advances. This argument is based on evidence that shows that malaria transmission by its main vector in Africa (*Anopheles gambiae*) peaks at 25°C, whereas dengue transmission (by *Aedes aegypti*) peaks at 29°C. Although these authors pointed out (probably correctly) that the burden of disease of arboviruses in Africa is substantially underestimated, it is still likely to be low compared with the burden from malaria. Furthermore, there is potential for a very large number of new malaria cases among relatively non-immune populations living in densely populated African highlands, including in Ethiopia (Endo and Eltahir, 2020). Adding to this risk, Sinka *et al.* (2020) discussed the prospect that the vector *Anopheles stephensi*, until recently restricted to Asia, may soon become well established in Africa, following its recent identification in Ethiopia and Sudan. Like *Aedes* (which does not transmit malaria), this species thrives in urban environments. These authors suggested that over 126 million people in cities across Africa could be at risk as a result of this introduction, and suggested support for the call by the World Health Organization (WHO) for targeted control and prioritized surveillance of this vector.

Meningococcal meningitis and other dust-related infections

Epidemics of this vaccine-preventable infection occur worldwide, but the highest incidence is observed in the dusty 'meningitis belt' of SSA, from Senegal on the Atlantic to Ethiopia on the Red Sea (Greenwood, 2006). Populations in northern Nigeria are particularly vulnerable. Other areas in the Rift Valley, the Great Lakes and southern Africa are also affected. Dust storms appear to be increasing in frequency, due to changing land use and climate change. The frequent inhalation of dust predisposes the elderly, the young and individuals with chronic cardiopulmonary diseases to increased respiratory hazards, including influenza A virus, pulmonary coccidioidomycosis and bacterial pneumonia, in addition to meningococcal meningitis (Schweitzer *et al.*, 2018).

Other infectious diseases in Africa

HIV/AIDS could worsen in future in Africa, as a consequence of falling morale and failing public health, for example in a 'four-degree world' with large-scale undernutrition, conflict and declining governance (McMichael *et al.*, 2008). In turn, poorly controlled HIV/AIDs would increase the risk of deepening systemic failure, due to high mortality and consequent lack of cultural transmission and diminished expertise.

The pattern of many other infectious diseases is potentially altered in Africa by climate change. These include schistosomiasis, African trypanosomiasis, cholera and zoonoses carried by bats and rodents. Stress from deforestation and climate change could alter food resources, and thus possibly lower immunity and/or alter behaviour in bats and rats, potentially enhancing diseases such as Ebola and (from rodents) Lassa fever (Tewogbola and Aung, 2020). An increase in palm oil plantations, together with other manifestations of 'neoliberal agriculture', (including large-scale land leases to non-African companies), has been linked with the largest epidemic of Ebola to date in West Africa (Wallace *et al.*, 2014).

Many studies have confirmed and quantified the effects of high temperatures on common forms of food poisoning, such as salmonellosis, including in Africa (Thindwa *et al.*, 2019). Contact between food and pest species, especially flies, rodents and cockroaches, is also temperature sensitive. Fly activity is driven largely by temperature rather than by biotic factors (Goulson *et al.*, 2005).

Climate change-related alterations in rainfall, surface water availability and water quality could also affect the burden of water-related diseases, including cholera. Outbreaks can be associated with droughts, as well as floods (Rieckmann *et al.*, 2018). Cholera outbreaks in many parts of Nigeria, Africa's most populous nation, are comparatively common. For example, by week 40 in 2015, cholera had been reported from 18 Nigerian states. More encouragingly, the reported case fatality rate of cholera in Nigeria in that year was under 4% (Nnaji *et al.*, 2016). The changing climate has also been linked in myriad ways with outbreaks of food- and water-borne diseases in the Republic of South Africa (Chersich *et al.*, 2019).

31.6 Tertiary Health Effects of Climate Change in Africa

One way to think of the tertiary health effects of climate change is as a 'threat multiplier'. Famine, large-scale migration and conflict (all 'tertiary' effects, as defined in this volume) can all occur without climate change-related effects such as warming, sea level rise, and changes in droughts, floods and insect populations. However, in the right circumstances, climate change *can* increase the risk of each of these phenomena. While climate change is neither necessary nor sufficient for these effects, it is increasingly likely to be a factor in them, and in some cases already is, including in Africa. For example, Chapter 24 (on conflict) reviews some of the evidence linking the war in Darfur, Sudan, to the changed pattern of rainfall there.

In the long run, if rapid action is not taken to slow the rate of GHG accumulation, then the health effects of these tertiary events may exceed the primary and secondary effects, perhaps by orders of magnitude. For this reason, all possible steps to minimize their risk should be taken. A 'four-degree world' should remain theoretical, not something that the next generation has to endure. Even a 'two-degree world' may be too dangerous.

Climate change, food security, Africa and the FAO

Climate change and agriculture in Africa has been discussed above, but food security also depends critically on governance (local, regional and global) and purchasing power. Numerous high-level food agencies, especially the FAO, were slow to comprehend the risk to food security represented by climate change (Butler, 2009). Some of this lag seems likely to reflect sources of funding and thinking that are biased towards people and institutions with pre-existing power, few of whom have been based in Africa and few of which lobby on behalf of her people. However, this situation has changed markedly; recognition is 'better late than never'.

Climate change, food security and famine

Africa has experienced several severe famines in recent decades, particularly in the Sahel, including parts of Ethiopia, Somalia and Niger. Contributing factors have been drought, poor governance and conflict. In the 2011 Somali famine, Al Shabaab (Islamist militants) attacked famine relief workers, contributing to critical food supply bottlenecks (Menkhaus, 2012). On the positive side, global governance measures, especially famine relief via the WFP, have been able to limit the scale of overt starvation. However, chronic undernutrition is widespread in many parts of Africa, even without overt famine. Being chronically underfed reflects and creates vulnerabilities, including the risk of being exploited by those who are better fed and stronger.

Climate change was not a factor in the Rwandan genocides of the early 1960s and again in 1994, but food insecurity and the anticipation that this would worsen, due to growing land scarcity, in the absence of sufficient other sources of income, was such a factor (André and Platteau, 1998). Today, the population in Rwanda is roughly double that of 1994 (13.7 million in 2022, according to Worldometer (nd)). The birth rate of this substantially Catholic country (Catholics provide 40% of the nation's health facilities) is declining, and contraceptive use is reported to have risen between 2005 and 2015 from 17% to 53% (Schwandt *et al.*, 2018). These authors identified that one factor for this was the creation by the Rwandan government of secondary health posts located adjacent to or near the faith-based health facilities, which provide only 'traditional" contraceptive methods. Climate change modellers have warned that the yields of several staple crops grown in Rwanda are likely to decline (Austin *et al.*, 2020).

Climate change, migration and conflict in Africa

Africa faces intense geographical exposure to the primary and secondary health effects of climate change and a non-trivial risk of diminished food production, thus further exacerbating food insecurity that is already widespread. This, on the background of Africa's history of colonization by Europeans, substantially heightens the risk of climate change-associated migration and conflict. Africa's social history has deprived her people of financial capital which it could use to adapt to climate change. Even within countries of the global North, colonial histories often impede Indigenous peoples' adaptive capacity relative to others. They may lack economic means for adaptation at the household and community level. At the same time, they lack a fair share of political power, meaning that they are at risk of reduced government investment in adaptation. Most fundamentally, the struggle to adapt to climate change can reinforce cultural assimilation by pressuring people to participate in Western economic structures in order to acquire resources to adapt (Bowles, 2015).

Similar dynamics apply in Africa, where many countries experience paralysing levels of indebtedness and have relatively little influence upon international processes that could foster global adaptation. Colonial histories have often led to poor governance and heightened social divisions in Africa, further inhibiting effective adaptation that could forestall climate change-associated conflict and migration. Conflict and migration each have a number of health consequences, some of which are not immediately obvious (detailed in Chapters 23 and 24).

Climate change and other contributors to 'regional overload' (Butler, 2016) are likely to multiply the risk of conflict, especially by diminishing resource availability. This effect could be substantial in Africa over coming decades, as resources such as water, food, productive cropland and animal pasture are already scarce.

Two socio-political characteristics make conflict a particularly likely consequence of resource scarcity in many African countries. One of these is low levels of institutional inclusivity, which impede other routes to redress grievances (Kahl, 2006). The degree to which a society is divided, whether along religious, ethnic, cultural or class lines, also influences the risk that scarcity will lead to violence. The processes of colonization often intentionally created or heightened such divisions as a method of controlling populations. Additionally, the creation of national borders often ignored local affiliations, creating divisions both within and between modern countries. In societies with high levels of such cross- and sub-national 'groupness', identity and security are associated strongly with membership to a single group. At times of stress, people turn to these groups alone for support, rather than the collective state or a network of groups, in so doing often excluding other groups, who become competitors. Such fault lines are also prone to manipulation by leaders to increase support and distract from more fundamental grievances (Kahl, 2006).

Weak states, of which Africa has a disproportionate number (another legacy of colonialism), heighten the potential for violent spillover, including cross-border diffusion of military skills and grievances. War in Darfur (Sudan) spread into Chad and the Central African Republic. Inconsistency between international and cultural borders and the presence of failing states can both create safe havens for insurgents and internationalize civil conflicts. As mentioned, Tigray (a part of Ethiopia) and the Central African Republic are sources of internally displaced people (IDPs), in part because of recurring conflict. It may be as difficult to define a 'climate war' as it is to define a climate refugee, or to allocate responsibility to each of many causal elements. However, it is not tenable to deny the connection in all cases. The many conflicts between pastoralists and crop growers in the Sahel, from Kenya to Cameroon and Ghana, have been worsened by the shift in rainfall to the south, itself likely contributed to by anthropogenic factors – a pattern likely to continue (Ramanathan, 2020; Mahmoud *et al.*, 2022; Nidumolu *et al.*, 2022; Rosenfeld *et al.*, 2022).

Epidemiologists have long pursued the goal of causal attribution, for example the extent to which a population burden of lung cancer has been caused by smoking, or by air pollution. Even in epidemiology, such studies are problematic and ultimately (often) uncertain, as new datasets become available, and as new interactions are revealed, theorized and, in some cases, understood. Ascribing the precise causal role of climate change (and other important causal factors) to most 'tertiary' events is not our aspiration. However, we stress that the contribution is already substantial, and likely to increase.

Conflict, whether or not it is associated with climate change, often leads to forced migration. The website for the UN High Commission for Refugees (UNHCR, nd) reported that about 30 million IDPs, refugees and asylum seekers live in Africa. This is roughly a third of the world's total (UNHCR, 2022). The largest numbers of recent African IDPs (i.e. in 2021) are in Ethiopia, where 'several' million fled the war in Tigray, and the Central African Republic (roughly half a million). IDPs and refugees who cross national boundaries have long been numerous in and from the DRC and Somalia. Uganda, which hosts about 1.5 million refugees (many from South Sudan), has the largest such number for an African country.

The UNHCR (nd) has an extensive discussion of global warming, stressing the point (with which we strongly agree) that it is extremely difficult to disentangle climate change from many other drivers, including war. The UNHCR also makes clear that it does not support the term 'climate refugee', a term that does not yet exist in international law.

Climate change-associated migration has drivers other than conflict, including the need to adapt to changing resource distribution. When agricultural land becomes less productive, or when there are more people than the land can easily support, a common response is for families to send one or more members (often the most able) to cities in search of paid employment, temporarily or permanently. This is often an effective adaptation at the family level, but comes with trade-offs including disruption to traditional familial and community bonds. It can also lead to exploitation and unsafe work (including sex work) and living conditions for those who have migrated. At a social level, it can increase tension between groups and lead to concentrations of young males, both of which are risk factors for violence.

Most displacement in Africa is internal. Only a small fraction of those who are displaced seek to flee to Europe or succeed in doing so.

Some of these people risk dying of thirst in the Sahara, enslavement in Libya, and drowning in the Mediterranean (Barnes, 2022).

Occasionally, European commentators with long experience of Africa have identified the lack of discussion of factors apparently ignored by most UN agencies. King and Elliot (1997) listed many taboos that limit discussion of human population numbers as a driver of peril. Logie and Benatar (1997), in their paper calling for hope for Africa in the 21st century, also decried the policies of the World Bank and the extent of indebtedness; but they did not discuss the rapid population growth in SSA.

31.7 Conclusion

Perhaps more than anywhere except small island states, the health of Africans is hostage to the increasingly accepted myth, in richer 'developed' countries, that the pursuit of 'net zero' (Dyke *et al.*, 2021) and adaptation are sufficient solutions to climate change. Africa is responsible for only a small part of the annual global emission of heat-trapping gases. Climate change mitigation in Africa may therefore seem of little importance. However, some evidence for an accelerated technological transition exists in Africa. One is the rapid spread of mobile telephony and the internet, which continues to be particularly strong in Kenya (Krell *et al.*, 2021). Another may be the widespread replacement of polluting and increasingly expensive kerosene lamps by highly efficient light-emitting diodes.

Nearly half of the people of Africa lack electricity in their own home (IRENA, 2021). Photovoltaic technology, the cost of which is declining rapidly, may offer the promise of providing electricity in remote locations, where it is currently lacking. Allied with access to the internet, this could accelerate literacy and other determinants of human development (such as greater gender equality and more democratic political environments) and reduce market asymmetries, in ways that reduce poverty and vulnerability. Infrastructure such as railways and roads, designed and built by the Chinese principally to allow the export of raw materials, may also be used by Africans for trade and travel, further reducing isolation.

The International Renewable Energy Agency (IRENA) claims that Africa's renewable energy generation potential (onshore only) is more than 1000 times its projected power demand in 2040, while a report from Climate Analytics (2022) claimed that 'renewable energy technologies are already cheaper than coal and gas power plants in most African countries'. Despite this, fossil fuel subsidies and a range of other factors impede energy transition in Africa.

Improved governance and slower population growth in many parts of Africa are also essential if the numerous challenges of climate change are to be met. President Buhari of Nigeria, Africa's most populous nation, is reported as actively endorsing slower population growth for Nigeria, stating that the purpose of the 'revised national policy on population for sustainable development' seeks to improve the quality of life and standards of living of all Nigerians (Daka, 2022). On the other side of the continent, President Musevini of Uganda, in power since 1986, is reported to have hinted, in 2014, at becoming less pro-natalist (May, 2017). Commenting on this change in 2017, the United Nations Population Fund Uganda Country Representative, Alain Sibenaler, observed that 'Uganda's leadership has come from an attitude of considering family planning taboo to embracing it' (Ndugga, 2019).

Somehow, Africa's population of more than 1.4 billion people needs to find ways to reduce their poverty and to improve their food security and nutrition, and also to find ways to collaborate effectively with people of goodwill everywhere, whether, for example, from China, Europe, the Middle East or South America. Religious and ethnic differences must become subsidiary to the recognition that we are all human. But comparatively affluent populations, represented by almost all who will read this chapter, must reduce their own dependency on coal and other forms of ecological destruction (for example, by discontinuing their high-consumption lifestyles, and by electing governments that prioritize genuinely sustainable and inclusive development over defence spending). This may give not only Africans a chance to flourish but also the whole world.

References

Agyepong, I.A., Sewankambo, N., Binagwaho, A., Coll-Seck, A.M., Corrah, T. *et al.* (2017) The path to longer and healthier lives for all Africans by 2030: the *Lancet* Commission on the future of health in sub-Saharan Africa. *The Lancet* 390, 2803–2859. DOI: 10.1016/S0140-6736(17)31509-X

Alonso, D., Bouma, M.J. and Pascual, M. (2011) Epidemic malaria and warmer temperatures in recent decades in an East African highland. *Proceedings of the Royal Society B* 278, 1661–1669. DOI: 10.1098/rspb.2010.2020

Amuzu, J., Jallow, B.P., Kabo-Bah, A.T. and Yaffa, S. (2018) The socio-economic impact of climate change on the coastal zone of the Gambia. *Natural Resource Modelling* 6, 13–26. DOI: 10.13189/nrc.2018.060102

Anderson, W., Taylor, C., McDermid, S., Ilboudo-Nébié, E., Seager, R. *et al.* (2021) Violent conflict exacerbated drought-related food insecurity between 2009 and 2019 in sub-Saharan Africa. *Nature Food* 2, 603–615. DOI: 10.1038/s43016-021-00327-4

André, C. and Platteau, J.-P. (1998) Land relations under unbearable stress: Rwanda caught in the Malthusian trap. *Journal of Economic Behavior and Organization* 34, 1–47. DOI: 10.1016/S0167-2681(97)00045-0

Asseng, S., Kheir, A.M., Kassie, B.T., Hoogenboom, G., Abdelaal, A.I. *et al.* (2018) Can Egypt become self-sufficient in wheat? *Environmental Research Letters* 13, 094012. DOI: 10.1088/1748-9326/aada50

Atanga, R.A. and Tankpa, V. (2021) Climate change, flood disaster risk and food security nexus in Northern Ghana. *Frontiers in Sustainable Food Systems* 5, 706721. DOI: 10.3389/fsufs.2021.706721

Austin, K.G., Beach, R.H., Lapidus, D., Salem, M.E., Taylor, N.J. *et al.* (2020) Impacts of climate change on the potential productivity of eleven staple crops in Rwanda. *Sustainability* 12, 4116. DOI: 10.3390/su12104116

Bachinyaga, J.M., Deijns, A., Mawe, G.I., Kervyn, F., Michellier, C. *et al.* (2022) *The flash floods of April 2020 in Uvira (DR Congo): story of an event with extreme impacts*. Presented at the EGU General Assembly 2022, Vienna, Austria. DOI: 10.5194/egusphere-egu22-6223

Barnes, J. (2022) Torturous journeys: Cruelty, international law, and pushbacks and pullbacks over the Mediterranean Sea. *Review of International Studies* 48, 441–460. DOI: 10.1017/S0260210522000110

Blenkinsop, S., Alves, L.M. and Smith, A.J.P. (2021) Climate change increases extreme rainfall and the chance of floods. In: Quéré, C.L., Liss, P. and Forster, P. (eds) *Critical Issues in Climate Change Science*. Available at: https://zenodo.org/record/4779119#.YxCnNCFByw4 (accessed 1 September 2022).

Bofana, J., Zhang, M., Wu, B., Zeng, H., Nabil, M. *et al.* (2022) How long did crops survive from floods caused by Cyclone Idai in Mozambique detected with multi-satellite data. *Remote Sensing of Environment* 269, 112808. DOI: 10.1016/j.rse.2021.112808

Bowles, D.C. (2015) Climate change and health adaptation: consequences for Indigenous physical and mental health. *Annals of Global Health* 81, 427–431. DOI: 10.1016/j.aogh.2015.06.004

Butler, C.D. (2009) Food security in the Asia-Pacific: Malthus, limits and environmental challenges. *Asia Pacific Journal of Clinical Nutrition* 18, 577–584. Available at: https://apjcn.nhri.org.tw/server/APJCN/18/4/577.pdf (accessed 6 September, 2022).

Butler, C.D. (2016) Sounding the alarm: health in the Anthropocene. *International Journal of Environmental Research and Public Health* 13, 665. DOI: 10.3390/ijerph13070665

Carr, T.W., Mkuhlani, S., Segnon, A.C., Ali, Z., Zougmoré, R. *et al.* (2022) Climate change impacts and adaptation strategies for crops in West Africa: a systematic review. *Environmental Research Letters* 17, 053001. DOI: 10.1088/1748-9326/ac61c8

Chapman, S., Birch, C.E., Marsham, J.H., Part, C., Hajat, S. *et al.* (2022) Past and projected climate change impacts on heat-related child mortality in Africa. *Environmental Research Letters* 17, 074028. DOI: 10.1088/1748-9326/ac7ac5

Chen, H., Githeko, A.K., Zhou, G., Githure, J.I. and Yan, G. (2009) New records of *Anopheles arabiensis* breeding on the Mount Kenya highlands indicate indigenous malaria transmission. *Malaria Journal* 5, 17. DOI: 10.1186/1475-2875-5-17.

Chersich, M.F., Wright, C.Y., Venter, F., Rees, H., Scorgie, F. *et al.* (2019) Impacts of climate change on health and wellbeing in South Africa. *South African Medical Journal* 109. DOI: 10.3390/ijerph15091884

Clarke, B., Otto, F., Stuart-Smith, R. and Harrington, L. (2022) Extreme weather impacts of climate change: an attribution perspective. *Environmental Research: Climate* 1, 012001. DOI: 10.1088/2752-5295/ac6e7d

Climate Analytics (2022) *Renewable Energy Transition in Sub-Saharan Africa*. Climate Analytics, Berlin. Available at: https://climateanalytics.org/media/renewable_energy_transition_in_sub-saharan_africa.pdf (accessed 2 August 2023).

Collier, P., Conway, G. and Venables, T. (2008) Climate change and Africa. *Oxford Review of Economic Policy* 24, 337–353. DOI: 10.1093/oxrep/grn019

Connolly-Boutin, L. and Smit, B. (2016) Climate change, food security, and livelihoods in sub-Saharan Africa. *Regional Environmental Change* 16, 385–399. DOI: 10.1007/s10113-015-0761-x

Crosby, A.W. (1972) *The Columbian Exchange. Biological and Cultural Consequences of 1492*. Greenwood, Westport, Connecticut.

Daka, T. (2022) Buhari moves against high fertility rate, launches revised population policy. *The Guardian*. Available at: https://guardian.ng/news/buhari-moves-against-high-fertility-rate-launches-revised-population-policy/ (accessed 11 September 2022).

Deckers, S., Nyssen, J. and Lanckriet, S. (2020) Ethiopia's Tigray region has seen famine before: why it could happen again. Available at: https://theconversation.com/ethiopias-tigray-region-has-seen-famine-before-why-it-could-happen-again-150181 (accessed 20 August 2022).

Devi, S. (2019) Cyclone Idai: 1 month later, devastation persists. *The Lancet* 393, 1585. DOI: 10.1016/S0140-6736(19)30892-X

Dyke, J., Watson, R. and Knorr, W. (2021) Climate scientists: concept of net zero is a dangerous trap. *The Conversation*. Available at: https://theconversation.com/climate-scientists-concept-of-net-zero-is-a-dangerous-trap-157368 (accessed 10 September 2022).

Endo, N. and Eltahir, E.A.B. (2020) Increased risk of malaria transmission with warming temperature in the Ethiopian Highlands. *Environmental Research Letters* 15, 054006. DOI: 10.1088/1748-9326/ab7520

Eridani, S. (2011) Sickle cell protection from malaria. *Hematology Reports* 19, e24. DOI: 10.4081/hr.2011.e24

Ezeh, A., Kissling, F. and Singer, P. (2020) Why sub-Saharan Africa might exceed its projected population size by 2100. *The Lancet* 396, 1131–1133. DOI: 10.1016/S0140-6736(20)31522-1.

FAOSTAT (2022) Statistical Database of the Food and Agriculture Organization of the United Nations, Rome. Available at: http://www.fao.org/faostat (accessed 10 September, 2022).

Feachem, R.G.A., Chen, I., Akbari, O., Bertozzi-Villa, A., Bhatt, S. *et al.* (2019) Malaria eradication within a generation: ambitious, achievable, and necessary. *The Lancet* 394, 1056–1112. DOI: 10.1016/S0140-6736(19)31139-0

Goulson, D., Derwent, L.C., Hanley, M., Dunn, D. and Abolins, S. (2005) Predicting calyptrate fly populations from the weather, and the likely consequences of climate change. *Journal of Applied Ecology* 42, 784–794. DOI: 10.1111/j.1365-2664.2005.01078.x

Greenwood, B. (2006) Editorial: 100 years of epidemic meningitis in West Africa – has anything changed? *Tropical Medicine and International Health* 11, 773–780. DOI: 10.1111/j.1365-3156.2006.01639.x

Griffiths, I. (1986) The scramble for Africa: inherited political boundaries. *The Geographical Journal* 152, 204–216. DOI: 10.2307/634762

Harrington, L.J. and Otto, F.E.L. (2020) Reconciling theory with the reality of African heatwaves. *Nature Climate Change* 10, 796–798. DOI: 10.1038/s41558-020-0851-8

Harrington, L.J., Wolski, P., Pinto, I., Ramarosandratana, A.M., Barimalala, R. *et al.* (2022) Limited role of climate change in extreme low rainfall associated with southern Madagascar food insecurity, 2019–2021. *Environmental Research: Climate* 1, 021003. DOI: 10.1088/2752-5295/aca695

Hemingway, J., Field, L. and Vontas, J. (2022) An overview of insecticide resistance. *Science* 298, 96–97. DOI: 10.1126/science.1078052

Hickel, J., Sullivan, D. and Zoomkawala, H. (2021) Plunder in the post-colonial era: Quantifying drain from the global South through unequal exchange, 1960–2018. *New Political Economy* 26, 1030–1047. DOI: 10.1080/13563467.2021.1899153

Higginbottom, T.P., Adhikari, R., Dimova, R., Redicker, S. and Foster, T. (2021) Performance of large-scale irrigation projects in sub-Saharan Africa. *Nature Sustainability* 4, 501–508. DOI: 10.1038/s41893-020-00670-7

IndexMundi (nd) Wheat imports by country. Available at: https://www.indexmundi.com/agriculture/?commodity=wheat&graph=imports 6 (accessed September, 2022)

IRENA (2021) *The Renewable Energy Transition in Africa Powering Access, Resilience and Prosperity*. International Renewable Energy Agency, Abu Dhabi, UAE. Available at: https://www.irena.org/-/media/Files/IRENA/Agency/Publication/2021/March/Renewable_Energy_Transition_Africa_2021.pdf (accessed 2 August, 2023).

Kahl, C. (2006) *States, Scarcity, and Civil Strife in the Developing World*. Princeton University Press, Princeton, New Jersey.

Kapuściński, R. (1998) *The Shadow of the Sun*. Penguin, Melbourne, Australia.

Kiggundu, N., Ddungu, S.P., Wanyama, J., Cherotich, S., Mpairwe, D. *et al.* (2019) Greenhouse gas emissions from Uganda's cattle corridor farming systems. *Agricultural Systems* 176, 102649. DOI: 10.1016/j.agsy.2019.102649

Kleinschroth, F., Laporte, N., Laurance, W.F., Goetz, S.J. and Ghazoul, J. (2019) Road expansion and persistence in forests of the Congo Basin. *Nature Sustainability* 2, 628–634. DOI: 10.1038/s41893-019-0310-6

King, M. (1990) Health is a sustainable state. *The Lancet* 336, 664–667. DOI: 10.1016/0140-6736(90)92156-C

King, M. and Elliott, C. (1997) To the point of farce: a Martian view of the Hardinian taboo – the silence that surrounds population control. *BMJ* 315, 1441–1443. DOI: 10.1136/bmj.315.7120.1441

Koffi, T.B., Weidert, K., Bitasse, E.O., Mensah, M.A.E., Emina, J. *et al.* (2018) Engaging men in family planning: perspectives from married men in Lomé, Togo. *Global Health: Science and Practice* 6, 317–329. DOI: 10.9745/GHSP-D-17-00471

Kogo, B.K., Kumar, L. and Koech, R. (2021) Climate change and variability in Kenya: a review of impacts on agriculture and food security. *Environment, Development and Sustainability* 23, 23–43. DOI: 10.1007/s10668-020-00589-1

Krell, N.T., Giroux, S.A., Guido, Z., Hannah, C., Lopus, S.E. *et al.* (2021) Smallholder farmers' use of mobile phone services in central Kenya. *Climate and Development* 13, 215–227. DOI: 10.1080/17565529.2020.1748847

Lindsay, S.W., Thomas, M.B. and Kleinschmidt, I. (2021) Threats to the effectiveness of insecticide-treated bednets for malaria control: thinking beyond insecticide resistance. *The Lancet Global Health* 9, e1325–e1331. DOI: 10.1016/S2214-109X(21)00216-3

Lloyd, E.A. and Oreskes, N. (2018) Climate change attribution: When is it appropriate to accept new methods? *Earth's Future* 6, 311–325. DOI: 10.1002/2017EF000665

Logie, D.E. and Benatar, S.R. (1997) Africa in the 21st century: can despair be turned to hope? *BMJ* 315, 1444–1446. DOI: 10.1136/bmj.315.7120.1444

Mabrouk, M., Jonoski, A., Oude Essink, G.H.P. and Uhlenbrook, S. (2018) Impacts of sea level rise and groundwater extraction scenarios on fresh groundwater resources in the Nile delta governorates, Egypt. *Water* 10, 1690. DOI: 10.3390/w10111690

Mahmoud, S.H., Gan, T.Y., Allan, R.P., Li, J. and Funk, C. (2022) Worsening drought of Nile basin under shift in atmospheric circulation, stronger ENSO and Indian Ocean dipole. *Scientific Reports* 22, 8049. DOI: 10.1038/s41598-022-12008-8

May, J.F. (2017) The politics of family planning policies and programs in sub-Saharan Africa. *Population and Development Review* 43, 308–329. DOI: 10.1111/j.1728-4457.2016.00165.x

McMichael, A.J., Butler, C.D. and Weaver, H. (2008) *Climate change and HIV AIDS: a joint working position paper*. United Nations Environment Programme, UNAIDS, Nairobi, Kenya. Available at: https://data.unaids.org/pub/basedocument/2008/20081223_unep_unaids_joint_working_paper_on_cca_en.pdf (accessed 6 September 2022).

McNeill, W.H. (1976) *Plagues and Peoples*. Anchor Press, Garden City, New York.

Menkhaus, K. (2012) No access: Critical bottlenecks in the 2011 Somali famine. *Global Food Security* 1, 29–35. DOI: 10.1016/j.gfs.2012.07.004

Merem, E.C., Twumasi, Y., Wesley, J., Alsarari, M., Fageir, S. *et al.* (2019) Analyzing the tragedy of illegal fishing on the West African coastal region. *International Journal of Food Science and Nutrition Engineering* 9, 1–15. DOI: 10.5923/j.food.20190901.01

Meyrowitsch, D.W., Pedersen, E.M., Alifrangis, M., Scheike, T.H., Malecela, M.N., Magesa, S.M. *et al.* (2011) Is the current decline in malaria burden in sub-Saharan Africa due to a decrease in vector population? *Malaria Journal* 10, 188, DOI:10.1186/1475-2875-10-188.

Mordecai, E.A., Ryan, S.J., Caldwell, J.M., Shah, M.M. and LaBeaud, A.D. (2020) Climate change could shift disease burden from malaria to arboviruses in Africa. *Lancet Planetary Health* 4, e416–e423. DOI: 10.1016/S2542-5196(20)30178-9

Mupedziswa, R. and Kubanga, K.P. (2017) Climate change, urban settlements and quality of life: The case of the Southern African Development Community region. *Development Southern Africa* 34, 196–209. DOI: 10.1080/0376835X.2016.1231057

Mutunga, C., Zulu, E.M. and Souza, R.-M.D. (2012) *Population dynamics, climate change and sustainable development in Africa*. African Institute for Development Policy, Nairobi. Available at: https://www.africaportal.org/publications/population-dynamics-climate-change-and-sustainable-development-in-africa/ (accessed 21 August 2022).

Nangombe, S., Zhou, T., Zhang, W., Wu, B., Hu, S. *et al.* (2018) Record-breaking climate extremes in Africa under stabilized 1.5°C and 2°C global warming scenarios. *Nature Climate Change* 8, 375–380. DOI: 10.1038/s41558-018-0145-6.

Ncongwane, K.P., Botai, J.O., Sivakumar, V. and Botai, C.M. (2021) A literature review of the impacts of heat stress on human health across Africa. *Sustainability* 13, 5312. DOI: 10.3390/su13095312

Ndugga, P. (2019) Examining the Mismatch between Fertility Desire and Contraceptive Behaviour in the Extended Postpartum Period in Uganda. PhD thesis, University of Southampton, UK. Available at: https://eprints.soton.ac.uk/444059/1/PATRICIA_NDUGGA_FINAL_THESIS_13th_JUNE.pdf (accessed 15 November 2022).

Neumann, B., Vafeidis, A.T., Zimmermann, J. and Nicholls, R.J. (2015) Future coastal population growth and exposure to sea-level rise and coastal flooding – a global assessment. *PLoS ONE*, 10. DOI: 10.1371/journal.pone.0118571

Nhamo, L., Ndlela, B., Nhemachena, C., Mabhaudhi, T., Mpandeli, S. *et al.* (2018) The water-energy-food nexus: Climate risks and opportunities in Southern Africa. *Water* 10, 567. DOI: 10.3390/w10050567

Nicholls, R.J., Marinova, N., Lowe, J.A., Brown, S., Vellinga, P., de Gusmão, D. *et al.* (2011) Sea-level rise and its possible impacts given a 'beyond 4°C world' in the twenty-first century. *Philosophical Transactions of the Royal Society A* 369, 161–181. DOI: 10.1098/rsta.2010.0291

Nidumolu, U., Gobbett, D., Hayman, P., Howden, M., Dixon, J. *et al.* (2022) Climate change shifts agropastoral-pastoral margins in Africa putting food security and livelihoods at risk. *Environmental Research Letters* 17, 095003. DOI: 10.1088/1748-9326/ac87c

Njiru, J., van der Knaap, M., Kundu, R. and Nyamweya, C. (2018) Lake Victoria fisheries: Outlook and management. *Lakes & Reservoirs: Science, Policy and Management for Sustainable Use* 23, 152–162. DOI: 10.1111/lre.12220

Nnaji, R.N., Ajumobi, O., Bala, U., Oladimeji, A., Sarki, M. *et al.* (2016) Cholera outbreak investigation, Gajala community, Birnin Kudu Local Government Area (LGA), Jigawa State, Nigeria, September 2015. *International Journal of Infectious Diseases* 45S, 144–145. DOI: 10.1016/j.ijid.2016.02.351

Otekunrin, O.A., Otekunrin, O.A., Sawicka, B. and Ayinde, I.A. (2020) Three decades of fighting against hunger in Africa: Progress, challenges and opportunities. *World Nutrition* 11, 86–111. DOI: 10.26596/wn.202011386-111

Ramanathan, V. (2020) Climate change, air pollution, and health: common sources, similar impacts, and common solutions. In*:* Al-Delaimy, W.K., Ramanathan, V. and Sorondo, M.S. (eds) *Health of People, Health of Planet and Our Responsibility.* Springer Open, Cham, Switzerland, pp. 49–59.

Rieckmann, A., Tamason, C.C., Gurley, E.S., Rod, N.H. and Jensen, P.K.M. (2018) Exploring droughts and floods and their association with cholera outbreaks in Sub-Saharan Africa: A register-based ecological study from 1990 to 2010. *The American Journal of Tropical Medicine and Hygiene* 98, 1269–1274. DOI: 10.4269/ajtmh.17-0778

Ritchie, H. and Roser, M. (2022) CO_2 emissions. Available at: https://ourworldindata.org/co2-emissions#global-co2-emissions-from-fossil-fuels-global-co2-emissions-from-fossil-fuels (accessed 20 August 2022).

Rosenfeld, D., Rudich, Y. and Lahav, R. (2022) Desert dust suppressing precipitation: A possible desertification feedback loop. *Proceedings of the National Academy of Science (USA)* 98, 5975–5980. DOI: 10.1073ypnas.101122798

Russo, S., Marchese, A.F., Sillmann, J. and Immé, G. (2016) When will unusual heat waves become normal in a warming Africa? *Environmental Research Letters* 11, 054016. DOI: 10.1088/1748-9326/11/5/054016

Salih, A.A.M., Baraibar, M., Mwangi, K.K. and Artan, G. (2020) Climate change and locust outbreak in East Africa. *Nature Climate Change* 10, 584–585. DOI: 10.1038/s41558-020-0835-8

Schilling, J., Hertig, E., Tramblay, Y. and Scheffran, J. (2020) Climate change vulnerability, water resources and social implications in North Africa. *Regional Environmental Change* 20, 15. DOI: 10.1007/s10113-020-01597-7

Schlein, L. (2021) WFP: Madagascar may be suffering world's first climate-induced famine. *VOA News*. Available at: https://www.voanews.com/a/wfp-madagascar-may-be-suffering-world-s-first-climate-induced-famine/6296676.html (accessed 21 August 2022).

Schwandt, H.M., Feinberg, S., Akotiah, A., Douville, T.Y., Gardner, E.V. *et al.* (2018) 'Family planning in Rwanda is not seen as population control, but rather as a way to empower the people': examining Rwanda's success in family planning from the perspective of public and private stakeholders. *Contraception and Reproductive Medicine* 3, 18, DOI: 10.1186/s40834-018-0072-y

Schweitzer, M.D., Calzadilla, A.S., Salamo, O., Sharifi, A., Kumar, N. *et al.* (2018) Lung health in era of climate change and dust storms. *Environmental Research* 163, 36–42. DOI: 10.1016/j.envres.2018.02.001

Seife, T.K. (2021) The impact of climate change on agriculture and food security in the Greater Horn of Africa. *Politikon* 48, 98–114. DOI: 10.1080/02589346.2020.1861509

Shiru, M.S., Shahid, S., Shiru, S., Chung, E.S., Alias, N. *et al.* (2020) Challenges in water resources of Lagos mega city of Nigeria in the context of climate change. *Journal of Water and Climate Change* 11, 1067–1083. DOI: 10.2166/wcc.2019.047

Sinka, M.E., Massey, N.C., Longbottom, J., Hemingway, J., Moyes, C.L. *et al.* (2020) A new malaria vector in Africa: Predicting the expansion range of *Anopheles stephensi* and identifying the urban populations at risk. *Proceedings of the National Academy of Science (USA)* 117, 24900–24908. DOI: 10.1073/pnas.2003976117

Tefera, T. (2012) Post-harvest losses in African maize in the face of increasing food shortage. *Food Security* 4, 267–277. DOI: 10.1007/s12571-012-0182-3

Tewogbola, P. and Aung, N. (2020) Lassa fever: History, causes, effects, and reduction strategies. *International Journal of One Health* 6, 95–98. DOI: 10.14202/IJOH.2020.95-98

Thindwa, D., Chipeta, M.G., Henrion, M.Y.R. and Gordon, M.A. (2019) Distinct climate influences on the risk of typhoid compared to invasive non-typhoid Salmonella disease in Blantyre, Malawi. *Scientific Reports* 9, 20310. DOI: 10.1038/s41598-019-56688-1

Thomson, M.C., Mason, S., Platzer, B., Mihretie, A., Omumbo, J. *et al.* (2014) Climate and health in Africa. *Earth Perspectives* 1, 17. DOI: 10.1186/2194-6434-1-17

Tilley, L. and Ajl, M. (2022) Eco-socialism will be anti-eugenic or it will be nothing: Towards equal exchange and the end of population. *Politics* 43, 1–18. DOI: 10.1177/02633957221075323

Tompkins, A.M., Colón-González, F.J., Di Giuseppe, F. and Namanya, D.B. (2019) Dynamical malaria forecasts are skillful at regional and local scales in Uganda up to 4 months ahead. *GeoHealth* 3, 58–66. DOI: 10.1029/2018GH000157

Trisos, C.H., Adelekan, I.O., Totin, E., Ayanlade, A., Efitre, J. *et al.* (2022) Africa. In*:* Pörtner, H.-O., Roberts, D.C., Tignor, M., Poloczanska, E.S., Mintenbeck, K. *et al.* (eds) *Climate Change 2022: Impacts, Adaptation and Vulnerability. Contribution of Working Group II to the Sixth Assessment Report of the Intergovernmental Panel on Climate Change.* Cambridge University Press, Cambridge, UK and New York, pp. 1285–1455.

Tulu, A.N. (1996) Determinants of Malaria Transmission in the Highlands of Ethiopia: The Impact of Global Warming on Morbidity and Mortality Ascribed to Malaria. PhD thesis, London School of Hygiene and Tropical Medicine, London. DOI: 10.17037/PUBS.00682286

UNHCR (nd) *Africa*. UN High Commissioner for Refugees, Geneva. Available at: https://www.unhcr.org/en-au/africa.html, accessed 7 September, 2022.

UNHCR (2022) *Global Trends. Forced Displacement in 2021*. UN High Commissioner for Refugees, Geneva. Available at: https://www.unhcr.org/62a9d1494/global-trends-report-2021 (accessed 7 September 2022).

Voortman, R.L. (2013) Why the Green Revolution failed in sub-Saharan Africa. *Rural 21 – The International Journal for Rural Development* 47, 32–33. Available at: https://www.rural21.com/fileadmin/downloads/2013/en-03/rural2013_03-S32-33.pdf (accessed 29 August 2022).

Wallace, R.G., Gilbert, M., Wallace, R., Pittiglio, C., Mattioli, R. *et al.* (2014) Did Ebola emerge in West Africa by a policy-driven phase change in agroecology? *Environment and Planning A* 46, 2533–2542. DOI: 10.1068/a4712com

Weisbord, R.G. (2003) The King, the Cardinal and the Pope: Leopold II's genocide in the Congo and the Vatican. *Journal of Genocide Research* 5, 35–45. DOI: 10.1080/14623520305651

WFP (nd) World Food Programme. Available at: https://dataviz.vam.wfp.org/version2/ (accessed 7 September 2022).

World Bank (nd) *Population ages*. Available at: https://data.worldbank.org/indicator/SP.POP.0014.TO.ZS (accessed 24 September 2022).

Worldometer (nd) *World population*. Available at: https://www.worldometers.info/population/world/ (accessed 10 September, 2022).

Zittis, G., Hadjinicolaou, P., Almazroui, M., Bucchignani, E., Driouech, F. *et al.* (2021) Business-as-usual will lead to super and ultra-extreme heatwaves in the Middle East and North Africa. *npj Climate and Atmospheric Science* 4, 20. DOI: 10.1038/s41612-021-00178-7

Zulu, E.M., Onyango, B., Williams, E.M. and Biaye, M. (2015) *Synthesis Report on the Demographic Dividend in Africa*. UN Population Fund (UNFPA), African Institute for Development Policy (AFIDEP), Johannesburg, South Africa. Available at: https://www.africaportal.org/publications/synthesis-report-on-the-demographic-dividend-in-africa-2/ (accessed 23 September 2022).

31.i Health Systems in Africa

Juliet Nabyonga-Orem
World Health Organization, Africa Regional Office, Brazzaville, Congo; North-West University, Potchefstroom, South Africa

Abstract

This mini-chapter explores persistent challenges facing health systems in African countries that undermine progress towards universal health coverage (UHC) and building health security. Underfunding, weak governance, inadequate and inequitably distributed human resource and suboptimal infrastructure, coupled with limited diagnostic capacity, have hampered access to quality health services and slowed the detection and response to disease outbreaks. Previous efforts to strengthen health systems were plagued with fragmented approaches that resulted from varied interpretation and implementation of primary healthcare as well as the Millennium Development Goals agenda that emphasized disease-focused investments. Africa needs to embrace a comprehensive approach to strengthening health systems that foster people-centred care, mitigate barriers to accessing services, sustain provision of essential health services in the face of shocks and efficiently allocate and utilize available resources to achieve UHC.

31.i.1 Introduction to Health Systems

The World Health Organization (WHO) defines health systems as 'activities whose primary purpose is to promote, restore and maintain health' (WHO, 2000). WHO identifies six elements: (i) service delivery; (ii) health workforce; (iii) information; (iv) medical products, vaccines and technologies; (v) financing; and (vi) leadership and governance (stewardship). The collective, interacting performance of these determines the strength and capacity of the entire health system. Relatedly, health-system strengthening refers to investments and activities that improve the performance of these six components, as well as their interactions, in order to promote greater equity and improved health outcomes (WHO, 2007).

Weak health systems undermine progress towards universal health coverage (UHC). They aggravate excess mortality when disease outbreaks occur, for example the severe outbreak of Ebola in West Africa (Moon *et al*., 2015). Poor health outcomes and weak health systems remain common in Africa (Ibeneme *et al*., 2020). In Africa, underfunding for health is a persistent challenge; 35 countries spend less than US$112 per capita, which is insufficient to deliver an essential package of health services. The minimum for low-income countries is estimated as US$112 per capita; US$146 per capita for lower-middle-income countries (WHO, nd). In response, arguments for increased health investments are made at both the national and continental scale, coupled with calls for improved delivery and spending efficiency. Reliance on donor funding is common for the majority of sub-Saharan African

 Climate Change and Global Health: Primary, Secondary and Tertiary Effects, 2nd Edition (eds C.D. Butler and K. Higgs)
DOI: 10.1079/9781800620025.0031i

countries; donors contributed on average slightly over 20% of health expenditure over the past decade (WHO, nd). Inevitably, conditions are attached; these hamper government priorities (Mwisongo and Nabyonga-Orem, 2016).

Governance is an important determinant of health systems performance. Good governance implies equitable participatory processes, evidence-based decision making, transparency and accountability (Simfukwe *et al.*, 2021). In Africa, it has been noted that weak governance will stall progress in reducing maternal mortality (Olafsdottir *et al.*, 2011). Further, corruption in the health sector has been partially attributed to weak governance; leading mechanisms include: health worker absenteeism; informal payments in public facilities; theft of drugs and supplies; diversion of patients to private facilities; and inappropriate procurement (Onwujekwe *et al.*, 2019). In their review of foreign aid and health governance in Africa, Asongu and Nwachukwu (2016) noted that donor aid negatively impacted institutional governance.

Inadequate and inequitably distributed human resources for health is another challenge in the African region. Only 14.5% of required physicians and 36.7% of required nurses and midwives are available, based on the WHO target of 2.3 doctors, nurses and midwives per 1000 people (Asamani *et al.*, 2019). Inadequate health infrastructure has hampered access to health services and slowed the detection and response to disease outbreaks. Diagnostic capacity is limited; in May 2020 (admittedly early in the COVID-19 outbreak), the number of laboratories that could test for this newly emerged coronavirus was only five in Ghana, three in Malawi and one in Zimbabwe. Although there were 25 in Nigeria, this was to serve a population of over 260 million (WHO, 2020). Weaknesses in health systems notwithstanding, other complexities include poverty, conflict, the epidemiological and demographic transitions, climate change and rapid and unplanned urbanization.

Since soon after the birth of primary health care (PHC) as a concept, efforts made to strengthen health systems have been fragmented. As the global economic system shifted towards neoliberalism, PHC was diluted to a 'selective' weaker version (Werner and Sanders, 1997). The Millennium Development Goals continued this trend, stressing disease-based approaches, rather than comprehensive strengthening of health systems. Instead, most funding focused on achieving agreed outputs, often negotiated from positions of significant power asymmetry. Comprehensive health-system strengthening requires long-term sustained investment; this aligns poorly with the funding timeframes of global health initiatives (GHIs) and, in general, of the Bill and Melinda Gates Foundation (Storeng, 2014). Biesma *et al.* (2009) highlighted the distortion of country processes and policies, distraction of governments from coordinated and comprehensive health-system strengthening, and verticalization of planning and monitoring and evaluation systems as negative GHI effects.

31.i.2 Conclusion

Building resilient health systems has been increasingly called for since the 2014 West African Ebola outbreak. The COVID-19 pandemic reinforces this need. An assessment of health-system resilience for the WHO African Region revealed suboptimal performance with scores ranging from 22.1% (Mauritius) to 87.6% (South Africa). Almost half of the 22 countries assessed scored under 50% (WHO Africa, 2018).

Africa needs to build resilient health systems that sustain the provision of essential health services in the face of acute and chronic shocks. A comprehensive approach to health-system strengthening should be fostered to build systems that provide people-centred care, mitigate barriers to accessing services and efficiently allocate and utilize available resources to achieve UHC.

References

Asamani, J.A., Akogun, O.B., Nyoni, J., Ahmat, A., Nabyonga-Orem, J. *et al.* (2019) Towards a regional strategy for resolving the human resources for health challenges in Africa. *BMJ Global Health* 4, e001533. DOI: 10.1136/bmjgh-2019-001533

Asongu, S.A. and Nwachukwu, J.C. (2016) Foreign aid and governance in Africa. *International Review of Applied Economics* 30, 69–88. DOI: 10.1080/02692171.2015.1074164

Biesma, R.G., Brugha, R., Harmer, A., Walsh, A., Spicer, N. *et al.* (2009) The effects of global health initiatives on country health systems: a review of the evidence from HIV/AIDS control. *Health Policy and Planning* 24, 239–252. DOI: 10.1093/heapol/czp025

Ibeneme, S., Ongom, M., Ukor, N. and Okeibunor, J. (2020) Realigning health systems strategies and approaches; what should African countries do to strengthen health systems for the Sustainable Development Goals? *Frontiers in Public Health* 8, 372. DOI: 0.3389/fpubh.2020.00372

Moon, S., Sridhar, D., Pate, M.A., Jha, A.K., Clinton, C. *et al.* (2015) Will Ebola change the game? Ten essential reforms before the next pandemic. The report of the Harvard-LSHTM Independent Panel on the Global Response to Ebola. *The Lancet* 386, 2204–2221. DOI: 10.1016/S0140-6736(15)00946-0

Mwisongo, A. and Nabyonga-Orem, J. (2016) Global health initiatives in Africa – governance, priorities, harmonisation and alignment. *BMC Health Services Research* 16, 245–254. DOI: 10.1186/s12913-016-1448-9

Olafsdottir, A.E., Reidpath, D.D., Pokhrel, S. and Allotey, P. (2011) Health systems performance in sub-Saharan Africa: governance, outcome and equity. *BMC Public Health* 11, 2017–2033. DOI: 10.1186/1471-2458-11-237

Onwujekwe, O., Agwu, P., Orjiakor, C., McKee, M., Hutchinson, E. *et al.* (2019) Corruption in Anglophone West Africa health systems: a systematic review of its different variants and the factors that sustain them. *Health Policy and Planning* 34, 529–543. DOI: 10.1093/heapol/czz070

Simfukwe, K., Adebisi, Y.A., Oladunni, A.A., Mohammed, E.S.E. and Lucero-Prisno, D.E. III (2021) The role of health service delivery networks in achieving universal health coverage in Africa. *South Eastern European Journal of Public Health*, Special Vol. 2. DOI: 10.11576/seejph-4470

Storeng, K.T. (2014) The GAVI Alliance and the 'Gates approach' to health system strengthening. *Global Public Health* 9, 865–879. DOI: 10.1080/17441692.2014.940362

Werner, D. and Sanders, D. (1997) *Questioning the Solution: The Politics of Primary Health Care and Child Survival*. HealthWrights, Palo Alto, California.

WHO (2000) *The World Health Report 2000: Health Systems: Improving Performance*. World Health Organization, Geneva. Available at: http://www.who.int/whr/2000/en/whr00_en.pdf (accessed 16 July 2021).

WHO (2007) *Everybody's Business – Strengthening Health Systems to Improve Health Outcomes: WHO's Framework for Action*. World Health Organization, Geneva. Available at: http://www.who.int/whr/2000/en/whr00_en.pdf (accessed 16 July 2021).

WHO (2017) *World Health Statistics 2017: Monitoring Health for the SDGs. Sustainable Development Goals*. World Health Organization, Geneva. Available at: http://www.who.int/whr/2000/en/whr00_en.pdf (accessed 16 July 2021).

WHO (2020) *WHO coronavirus disease (COVID-19) dashboard*. World Health Organization, Geneva. Available at: https://covid19.who.int/ (accessed 25 July 2020).

WHO (nd) *Global Health expenditure database*. World Health Organization, Geneva. Available at: https://apps.who.int/nha/database (accessed 24 September 2022).

WHO Africa (2018) *The State of Health in the WHO African Region: An Analysis of the Status of Health, Health Services and Health Systems in the Context of the Sustainable Development Goals*. World Health Organization Regional Office for Africa, Brazzaville, Republic of Congo. Available at: https://apps.who.int/iris/handle/10665/275292 (accessed 24 September 2022).

31.ii Climate Change, Gender and Health in Africa

Caradee Wright
Environment and Health Research Unit, South African Medical Research Council, South Africa; Department of Geography, Geoinformatics and Meteorology, University of Pretoria, Pretoria, South Africa

Abstract
This mini-chapter explores relations between climate change and health in Africa through the lens of gender disparities and inequalities. Power inequalities and social vulnerabilities lead to certain groups bearing the brunt of climate change impacts. Such groups include girls and women, children and young people, and the economically and socially marginalized. It is essential that these groups are included when seeking solutions for solving gender–health–climate-change challenges.

31.ii.1 Introduction: Power and Vulnerability

Climate change is a threat multiplier that increases existing health inequalities (van Daalen *et al.*, 2020). Gender-based effects of climate change adversely impact and exacerbate existing disparities, including inequality, women's empowerment and, of particular concern, human health and well-being.

These gendered impacts of climate change are pervasive and disproportionate. Gender inequality is particularly severe for women and girls. At the same time, climate change is likely to worsen the poverty cycle and this will affect both men and women, especially those living in poverty (Eastin, 2018).

The intersections between gender, ethnicity, age, (dis)ability, power dynamics, socio-economic structures, and societal norms and expectations result in climate impacts being experienced more severely by women and other vulnerable individuals in society (Andrijevic *et al.*, 2020). Power inequalities and social vulnerabilities mean that, in many settings, women and children are more likely than men to die during and in the aftermath of disasters, and women's socio-economic marginalization increases their vulnerability after such disasters (Thurston *et al.*, 2021).

Child abuse occurs during and after natural disasters and conflicts. Sexual violence, including rape against girls and physical violence against boys, are among the most common atrocities (Seddighi *et al.*, 2021). Adolescents and young people are also more susceptible to climate change impacts that lead to socio-economic stress, lack of food, inter-group conflict, economic dislocation, loss of education and forced migration (Sanson *et al.*, 2019).

31.ii.2 Risks in Africa, Including from Climate Change

African women and girls are generally more vulnerable and exposed to climate change effects since they are poorer, receive less education and may be not involved in political and household

DOI: 10.1079/9781800620025.0031ii

decision-making, compared with men (UNDP, 2012). African rural women are even more vulnerable to the impacts of climate change, as they are often disadvantaged by traditional, patriarchal decision-making processes and lack of access to land. These issues are compounded by influences including migrant status, lack of financial services and poor social capital (Ahmed *et al.*, 2016).

Women and children (particularly girls) bear the brunt of health risks associated with climate change, especially those that relate to air pollution, food insecurity and maternal health. Household-associated air pollution is a major environmental health threat in Africa (Bickton *et al.*, 2020). Indoor burning of 'dirty fuels' for heating and cooking exposes women and children to particulate air pollution and diseases such as pneumonia, cataract and respiratory tract infections, among others. As temperatures change, so too do indoor burning patterns. Outdoor air pollution (including from the burning of fuels indoors) also impacts health, especially for families living close to coal-fired power stations. Together, household and ambient air pollution threatens morbidity and mortality, causing premature death – pneumonia in children under 5 years of age is a leading cause of death on the continent.

Among the most concerning gender-based risks relate to food and water insecurity that affect African women and girls more than men and boys. Women account for nearly 80% of the agricultural sector in sub-Saharan Africa; most are vulnerable and poor (Nyasimi *et al.*, 2018). Drought, extreme temperatures and floods threaten crop supplies and amplify risks of food shortages that affect household food hierarchies. Oftentimes, women and children are therefore left undernourished. Women suffer stress in trying to sustain supplies of food and water for their families. During periods of water scarcity, women source water from unsafe sites (e.g. contaminated streams) and store it without proper cleaning or container coverage. Such instances can lead to cross-contamination of diarrhoeal disease from drinking from the same cup used in the stored water container. Less personal hygiene behaviours such as hand and body washing may lead to 'water-washed' diseases such as scabies.[1] 'Risk-sharing networks' or pooling of community assets can protect people from climate-related stressors; however, women and other vulnerable groups may face barriers to accessing these networks (Smucker and Wangui, 2016).

Another concern is women's ability to seek reproductive and maternity health services in general and during extreme climatic events. Box 31.ii.1 describes the urgent need to give women their reproductive rights and information regarding family planning, so as to secure sustainable livelihoods.

Pregnancy-related outcomes can be affected by changes in temperature and nutritional status depending on access to food (e.g. during drought conditions). Maternal health and ambient heat exposure during pregnancy have been associated with congenital birth defects, stillbirths, prematurity and low birthweight. Moreover, mothers face the risk of heat-sensitive diseases, especially vector- and water-borne diseases, during pregnancy and labour. This can jeopardize maternal survival and health, as well as the health of the newborn.

While the gender inequalities affecting women's health in a changing climate are numerous and dire, health threats from climate

Box 31.ii.1. Climate resilience in the Sahel needs radical action to prevent a humanitarian crisis (Graves *et al.*, 2019)

Food security is now and will grow to become an even larger concern in Africa's Sahel. By 2100 (if not before), with an estimated 928 million people living in the region, increased drought and warming of ambient temperatures by more than 3°C, livelihoods will become unsustainable. Famine, conflict and migration will not only disrupt life in the region and neighbouring regions but likely result in destruction and deaths.

Two of the critical actions being called for to help avert a crisis are people's access to family planning information and services and investment in education for girls. The issues are complex and diverse: marriage and childbirth at a young age, stunting and wasting among children under 5 years of age, and more students needing schooling than current facilities and resources can support.

Studies have shown that transitioning girls from primary to secondary school can not only delay marriage (with help from laws that limit minimum marriage age, 18 years, as stated in the African Charter on the Rights and Welfare of the Child) but also stop, delay or limit childbearing and lead to fewer malnourished children in the Sahel. (Also see Chapter 10, this volume.)

change exist for men too. Occupational climate change-related risks such as exposure to extreme heat impacts both men and women working outdoors. For men, mental stress and gender-based violence may be associated with loss of livelihood, substance abuse and other factors; men's loss of life during disasters also has repercussions for families and communities.

The African Union Gender Equality and Women's Empowerment Strategy (2018–2028) (African Union, 2022) aims to be transformational: not only to eliminate threats but also to elevate women and girls to participate in all aspects of society, to strengthen their agency and address challenges (including those related to climate change), equipping women with effective legislation and financing. The Women, Gender and Development Directorate of the African Union has tackled the status of women's rights in refugee and internal displacement settings in Africa, among other issues.

In August 2019, the Future Africa Call to Action on Sexual and Reproductive Health and Rights (SRHR), Gender and Climate Change Resilience symposium took place at the University of Pretoria and its report promised to promote cross-sectoral approaches and encourage donor funding in supporting climate adaptation for all, leaving no one behind (Plant-for-the-Planet Foundation, 2019; UNFPA, 2020). Despite such calls, there remain many unknowns about how climate change will impact, sustain or enhance gender inequalities in Africa.

There is rising intolerance for same-sex relationships across states in Africa (Amusan *et al.*, 2019). There is very little information about lesbian/gay/bisexual and transgender/non-binary people and climate change. This group, often underground in Africa, have different socially and culturally defined gender roles that result in different responsibilities, risk profiles and health needs. Gender-disaggregated health data are under-represented or non-existent in most African countries, making it impossible to use local evidence to lobby for improved conditions. While localized studies are useful and in some cases abundant, their usefulness is not always transferrable to other villages, countries or regions in Africa, given the differing cultural and social contexts. Therefore, regional and global studies are also needed.

31.ii.3 Conclusion

Including the vulnerable and marginalized in understanding and solving gender–health–climate-change challenges is necessary. For example, women have unique knowledge and skills that are important to include in effective and sustainable decision-making for climate change adaptation. Empowering women as agents of change in their communities will require radical shifts in thinking but holds promise for community resilience against climate change threats to health and livelihoods.

Note

[1] Scabies is more common where there is a lack of water, thus reducing the capacity of people to wash.

References

African Union (2022) *AU Strategy for Gender Equality and Women's Empowerment*. African Union, Addis Ababa, Ethiopia. Available at: https://au.int/en/articles/au-strategy-gender-equality-and-womens-empowerment (accessed 12 September, 2022).

Ahmed, A., Lawson, E.T., Mensah, A., Gordon, C. and Padgham, J. (2016) Adaptation to climate change or non-climatic stressors in semi-arid regions? Evidence of gender differentiation in three agrarian districts of Ghana. *Environmental Development* 20, 45–58. DOI: 10.1016/j.envdev.2016.08.002

Amusan, L., Saka, L. and Muinat, O.A. (2019) Gay rights and the politics of anti-homosexual legislation in Africa: Insights from Uganda and Nigeria. *Journal of African Union Studies* 8, 45–66. DOI: 10.31920/2050-4306/2019/8n2a3

Andrijevic, M., Crespo Cuaresma, J., Lissner, T., Thomas, A. and Schleussner, C.-F. (2020) Overcoming gender inequality for climate resilient development. *Nature Communications* 11, 6261. DOI: 10.1038/s41467-020-19856-w

Bickton, F.M., Ndeketa, L., Sibande, G.T., Nkeramahame, J., Payesa, C. *et al.* (2020) Household air pollution and under-five mortality in sub-Saharan Africa: An analysis of 14 demographic and health surveys. *Environmental Health and Preventive Medicine* 25, 67. DOI: 10.1186/s12199-020-00902-4

Eastin, J. (2018) Climate change and gender equality in developing states. *World Development* 107, 289–305. DOI: 10.1016/j.worlddev.2018.02.021

Graves, A., Rosa, L., Mouhou, A.M., Malna, F. and Adoum, D. (2019) Avert catastrophe now in Africa's Sahel. *Nature* 575, 282–286. DOI: 10.1038/d41586-019-03445-z

Nyasimi, M., Ayanlade, A., Mungai, C., Derkyi, M. and Jegede, M.O. (2018) Inclusion of gender in Africa's climate change policies and strategies. In: Filho, W.L., Manolas, E., Azul, A.M., Azeiteiro, U.M. and Mcghie, H. (eds) *Handbook of Climate Change Communication,* Vol. 1. Springer, Cham, Switzerland, pp. 171–185.

Plant-for-the-Planet Foundation (2019) *Future Africa Call to Action, Nairobi Commitment version 31 October 2019*. Available at: https://www.nairobisummiticpd.org/commitment/future-africa-call-action-nairobi-commitment-version (accessed 9 October, 2022).

Sanson, A.V., Van Hoorn, J. and Burke, S.E.L. (2019) Responding to the impacts of the climate crisis on children and youth. *Child Development Perspectives* 13, 201–207. DOI: 10.1111/cdep.12342

Seddighi, H., Salmani, I., Javadi, M.H. and Seddighi, S. (2021) Child abuse in natural disasters and conflicts: A systematic review. *Trauma, Violence, & Abuse* 22, 176–185. DOI: 10.1177/1524838019835973

Smucker, T.A. and Wangui, E.E. (2016) Gendered knowledge and adaptive practices: Differentiation and change in Mwanga District, Tanzania. *Ambio* 45, 276–286. DOI: 10.1007/s13280-016-0828-z

Thurston, A.M., Stöckl, H. and Ranganathan, M. (2021) Natural hazards, disasters and violence against women and girls: a global mixed-methods systematic review. *BMJ Global Health* 6, e004377. DOI: 10.1136/bmjgh-2020-004377

UNDP (2012) *Overview of linkages between gender and climate change. Gender and Climate Change Africa Policy Brief*. UN Development Programme, New York. Available at: https://wrd.unwomen.org/sites/default/files/2021-11/Policy-Brief-Overview-of-linkages-between-gender-and-climate-change.pdf (accessed 6 August, 2023).

UNFPA (2020) *SRHR, Gender and Climate Change Resilience Symposium, 26–28 August 2019*. Report, UN Population Fund, East and Southern Africa, Johannesburg. Available at: https://www.esaro.unfpa.org/sites/default/files/pub-pdf/unfpa_climate_change_symposium_report.pdf (accessed 6 February 2024)

van Daalen, K., Jung, L., Dhatt, R. and Phelan, A.L. (2020) Climate change and gender-based health disparities. *The Lancet Planetary Health* 4, e44–e45. DOI: 10.1016/S2542-5196(20)30001-2

31.iii Ethical Dimensions of Air Pollution and Climate Change in Africa

A. Kofi Amegah
Public Health Research Group, Department of Biomedical Sciences, University of Cape Coast, Cape Coast, Ghana

Abstract
This mini-chapter highlights the ethical dimensions of air pollution and climate change in Africa. The huge air quality data gap in Africa is an ethical issue. This is because inadequate data has implications for community advocacy to minimize exposure among vulnerable groups, and influencing policy action to combat air pollution and avert deaths and illness due to population-level air pollution exposure. The poverty problem at the heart of household air pollution exposure experiences of women and their young children deepens the ethical issues. The ethical issues of the growing climate injustice in Africa relates to the limited capacity of governments and development partners to develop adaptive and response plans. This is as a result of the limited data relating to the past occurrence of extreme weather events, which confounds accurate estimation of the future risk of these events in Africa.

31.iii.1 Air Pollution

Air quality in African cities has deteriorated; this can be partly attributed to rapid population growth and industrial expansion (Amegah and Agyei-Mensah, 2017; Fisher *et al*., 2021). There is, however, a huge air quality data gap in African countries, associated with limited monitoring capacity on the continent. Furthermore, accessing the limited data that do exist, from the responsible agencies for health research, is very challenging due to the huge cost and punishing bureaucratic processes. This air quality data gap in Africa is an ethical issue. Inadequate data has implications for: (i) community advocacy to minimize exposure among vulnerable groups such as street vendors, e-waste recyclers, waste pickers, traffic policemen and wardens, and tollbooth attendants; and (ii) influencing policy action to combat air pollution and avert deaths and illness due to population-level air pollution exposure.

Poverty is at the heart of the air pollution exposure experiences of these vulnerable populations. Many are migrants from rural areas in search of better livelihoods in the city. Although adequate data would not guarantee improved air and other conditions for them, its lack (perhaps itself a consequence of inequality, as well as a cause) limits the ability of advocates to create awareness among and about affected populations. Street vendors, for instance, are ubiquitous in the urban landscape of African countries and are a major driver of the informal economy (Amegah and Jaakkola, 2016). The data gap also creates a vacuum of local evidence with which to analyse the adverse effects of air pollution on health. This thus reduces the capacity to convince governments and development partners

DOI: 10.1079/9781800620025.0031iii

to invest in air pollution control. Despite this data gap, and the associated paucity of air pollution epidemiological studies, the burden of disease attributable to ambient air pollution in Africa is growing, yet underestimated.

According to the latest report by the Health Effects Institute (2022), air pollution is the second leading risk factor for death in Africa, only surpassed by undernutrition. The report finds that the continent is home to five of the top ten most heavily polluted countries globally in terms of ambient fine particulate matter ($PM_{2.5}$). Estimates provided by Roy (2016) suggest that in Africa air pollution kills more people prematurely than unsafe water, undernutrition and poor sanitation. According to Akumu (2016), the cost of air pollution in African cities could be as high as 2.7% of the continent's gross domestic product (GDP). In 2019, air pollution contributed to an estimated 1.1 million deaths in Africa, with 63% linked to exposure to household air pollution (HAP), which mostly impacts women and their young children. Poverty and liquified petroleum gas supply chain issues in Africa mean that over 75% of the population continue to rely on biomass fuels for cooking and heating, generating a wide range of household air pollutants. The poverty problem at the heart of the HAP exposure experiences of women and their young children deepens the ethical issues.

31.iii.2 Climate Change

Climate injustice exists in Africa. Despite Africa contributing relatively little to the anthropogenic causes of climate change, populations in its nations face severe impacts. This can be deduced, despite the relative paucity of studies. One aspect is that Africa has a high burden of climate-sensitive diseases, particularly malaria.

Knowledge gaps also steepen the challenge and bedevil efforts in Africa to prepare for and respond to aspects of climate change unrelated to particulate air pollution. For example, data relating to the past occurrence of extreme weather events are limited. This confounds accurate estimation of the future risk of these events in Africa. This is also an ethical issue, because African governments and development partners are limited in their capacity to develop adaptive and response plans. There is also limited availability of historical data in Africa for daily temperature and precipitation. Observations have either never been recorded, or as yet exist only on paper, with limited accessibility. Encouragingly, several 'data rescue' projects to reduce this gap are under way. A related issue is the limited exchange of recorded data across many African countries, due to limited communication and data management systems, together with variations in data policy.

Unsurprisingly, insufficient epidemiological studies exist that link climate change with changing population health in Africa. As a result, the burden of disease attributable to climate change on the continent remains poorly quantified – a problem globally (Carlson *et al.*, 2023) but especially challenging in Africa. Governments therefore lack local evidence to develop evidence-based climate action plans. Compounding this is a lack of institutional and community-based capacity to develop and implement climate needs assessments, action plans and strategies in many African countries. There is a general lack of awareness of the health risks of climate change. This problematic situation also illustrates ethical concerns.

31.iii.3 Proposed Solutions

There is a growing number of air quality monitoring networks in Africa, including in Ghana, Uganda and across seven Francophone countries of West and Central Africa.[1] Ongoing support for these, by governments and development partners, is essential to bridge the Africa air quality data gap. Funds and determination are needed to strengthen the capacity of the World Health Organization and international health NGOs to further implement and develop programmes designed to build the capacity of health workers to survey for air pollution-related diseases in Africa. Governments should resource their environmental protection agencies to develop air quality management plans and air quality standards, and ensure their implementation and compliance, respectively. International funding organizations should institute funding mechanisms targeted towards African researchers in order to revolutionize air pollution and climate change epidemiological research on the continent and help generate adequate local evidence to inform policy and public health practice.

Note

[1] Ghana Urban Air Quality Project (GHAir) (Ghana) (Sewor *et al.*, 2021), AirQo project (Uganda), International Network to study Deposition and Atmospheric composition in Africa (INDAAF) (https://indaaf.obs-mip.fr/) and AfriqAir Network (https://www.cmu.edu/epp/afriqair/pages/about-us.html)

References

Akumu J. (2016) *Improving Air Quality in African Cities*. United Nations Environment Programme (UNEP), Nairobi. Available at: https://www.unep.org/resources/report/improving-air-quality-african-cities (accessed 25 May 2023).

Amegah, A.K. and Agyei-Mensah, S. (2017) Urban air pollution in Sub-Saharan Africa: Time for action. *Environmental Pollution* 220, 738–743. DOI: 10.1016/j.envpol.2016.09.042

Amegah, A.K. and Jaakkola, J.J.K. (2016) Street vending and waste picking in developing countries: a long-standing hazardous occupational activity of the urban poor. *International Journal of Occupational and Environmental Health* 22, 187–192. DOI: 10.1080/10773525.2016.1209621

Carlson, C.J., Alam, M.S., North, M.A., Onyango, E. and Stewart-Ibarral, A.M. (2023) The health burden of climate change: A call for global scientific action. *PLoS Climate* 2, e0000126. DOI: 10.1371/journal.pclm.0000126

Fisher, S., Bellinger, D.C., Cropper, M.L., Kumar, P., Binagwaho, A. *et al.* (2021) Air pollution and development in Africa: impacts on health, the economy, and human capital. *The Lancet Planetary Health* 5, e681–e688. DOI: 10.1016/S2542-5196(21)00201-1

Health Effects Institute (2022) *The State of Air Quality and Health Impacts in Africa*. Health Effects Institute, Boston, Massachusetts. Available at: https://www.healthdata.org/policy-report/state-air-quality-and-health-impacts-africa (accessed 25 May 2023).

Roy, R. (2016) *The cost of air pollution in Africa*. OECD Development Centre Working Papers, No. 333. OECD Publishing, Paris. DOI: 10.1787/5jlqzq77x6f8-en

Sewor, C., Obeng, A.A. and Amegah, A.K. (2021) The Ghana Urban Air Quality Project (GHAir): Bridging air pollution data gaps in Ghana. *Clean Air Journal* 31. DOI: 10.17159/caj/2021/31/1.11172

32 Climate Change and Health in South Asia

Colin D. Butler and Md Rezanur Rahaman
National Centre for Epidemiology and Population Health, Australian National University, Canberra, Australia

Abstract

This chapter uses the primary, secondary and tertiary health effect framework to explore some of the current and future impacts of climate change on the almost two billion people who currently live in South Asia (chiefly, India, Pakistan, Nepal and Bangladesh). The effects of anthropogenic climate change on health and other aspects of human well-being in this region are likely to be profound; many of its people are already vulnerable, physically and socially. Such effects in this nuclear-armed region may have severe global repercussions.

32.1 South Asia: Vulnerable Physically and Socially

This chapter provides an overview of some of the most important health effects of current and future climate change in South Asia,[1] excluding Afghanistan and the Maldives. Consistent with the framework of this book we discuss 'primary', 'secondary' and 'tertiary' events and their relevant health effects. Mental health impacts, which we designate as 'cross-cutting' (as they can arise via many pathways), are also discussed.

The five most populous countries of South Asia, in 2022, were home to over 1.85 billion people, almost a quarter of the world's population (UNFPA, 2022). Of these five, over 75% live in India. By 2050, the United Nations (UN) estimates that this population will increase by another 20%, to over 2.2 billion.[2] As the impact of anthropogenic climate change worsens, the effects on health in South Asia are likely to be profound, possibly with severe global repercussions in this nuclear-armed region.

Key physical impacts of global climate change include changes to temperature, precipitation, storms[3] and sea level. The most recent findings of the Intergovernmental Panel on Climate Change (IPCC) reported strong confidence for increased heatwaves and heavy rainfall events in South Asia (IPCC, 2021). There are additional regional issues. Changes in the monsoon, on which so much agriculture in South Asia depends, is one such factor. The particulate air pollution in South Asia, which has well-documented and extremely harmful health effects (Balakrishnan *et al.*, 2019), also has complex regional climatic consequences, including to the monsoon (Pandey *et al.*, 2017). Warming, leading to glacial melt, will reduce the reliability and irrigation benefits of the great rivers that flow from the Himalayas, the so-called 'Third Pole'[4] (Immerzeel *et al.*, 2020). China will likely increase its damming and diversion of the Tsangpo-Brahmaputra river system, with potentially adverse effects on Bangladesh and India.

 Climate Change and Global Health: Primary, Secondary and Tertiary Effects, 2nd Edition (eds C.D. Butler and K. Higgs)
DOI: 10.1079/9781800620025.0032

In addition to these physical risks, South Asia has many social factors that increase the vulnerability of its people to a changing climate. These include high population densities combined with extreme poverty, affecting megacities and many rural areas that are vulnerable to flooding and supply chain failures. Fragile health systems, pervasive inequalities and poor housing increase ill health and limit individual and communal adaptive capacity. This parlous social landscape is contributed to by inadequate institutional capacity and a lack of good governance, compounded by political systems that give limited protection to human rights, especially of marginalized populations, including lower castes and scheduled tribes. South Asian democracies are also compelled by global trends to focus on economic 'growth' (as conventionally defined) rather than on addressing climate change[5] and other environmental and social justice issues that are perceived as blocking 'progress'. A free press, which is essential for institutionalizing democracy and for promoting health literacy, is increasingly suppressed in all major South Asian nations (Stone, 2020). These problems are intensified by high regional fertility rates, which, although declining, remain high, especially in poor areas of India's rural north[6] (Samir *et al.*, 2018). The population growth rate in Pakistan is even higher, with a national total fertility rate of 3.6 births per woman in 2017/18 (Goujon *et al.*, 2020). In contrast, desired fertility in Bangladesh is near to replacement (Akram *et al.*, 2020). Throughout South Asia, urban areas are growing rapidly in population, not primarily from high urban fertility but from rural to urban inflow. Population expansion has exceeded the capacity of government to keep pace, creating substantial pressure on the physical and infrastructural capacity of most South Asian cities (Rao *et al.*, 2011).

In the face of climate change, the distribution of some infectious diseases, such as dengue fever, will continue to change in South Asia. If poverty cannot be alleviated, the burden of infectious diseases, including diarrhoeal diseases, particularly among children, may worsen. The cumulative effects of climate change on crop production in South Asia will be substantial, including from effects of heat on agricultural workers (Im *et al.*, 2017), changes in rainfall, and sea level rise. In addition to these better understood (i.e. primary and secondary) links between climate change and health, larger-scale 'tertiary' events and effects may occur, within and beyond South Asia, including famine, mass migration and large-scale conflict.

Adapting to climate change and minimizing its potential adverse health impacts thus requires urgent large-scale and coordinated global effort. Encouragingly, evidence on the specific links between climate change and human health in South Asia is increasing (Akhtar, 2016; Sen *et al.*, 2017). Such evidence may stimulate government, industry and other health stakeholders to collaborate in order to avert potentially disastrous tertiary effects of climate change on human health in South Asia.

32.2 Changing Temperature, Water, Glaciers and Rivers

Mean temperatures across India rose by more than 0.5°C, between 1960 and 2009, with a statistically significant increase in heatwaves (Mazdiyasni *et al.*, 2017). Although historical data on heat morbidity are scarce, these authors reported a 146% increase in the probability of heat-related mortality events for more than 100 people for this period. That time period is now well in the past; heatwaves and their consequences are likely to have worsened (see Chapter 32.i). Already, with barely 1.2°C warming (globally, compared with pre-industrial temperatures), labourers and other outdoor workers in parts of India experience conditions near the upper limits of survival (Voosen, 2021). Saeed *et al.* (2021) suggested that earlier estimates of future exposure to wet-bulb temperatures (Tw)[7] are too conservative. Physical exertion is highly dangerous if Tw > 32°C and human survival is impossible for more than a few hours if Tw is only 3°C greater (see Chapter 12).

Heavy precipitation events in South Asia are expected to increase in frequency, with reduced light rain events (Hijioka *et al.*, 2014). South Asia has a long experience with severe droughts, sometimes with resultant famine (see Chapter 22). However, the most recent IPCC report is undecided about the trend of agricultural droughts in this region. It defines these as periods of abnormal soil moisture deficit, which result from combined shortage of precipitation and excess evapotranspiration, and which, during the growing season, reduce crop production (IPCC, 2021).

Over South Asia, IPCC (2021) also found, with high confidence, that increases in monsoon precipitation, due to warming from greenhouse gas emissions, were offset by decreases in monsoon precipitation due to cooling from human-caused aerosol emissions over the 20th century. If so and if, as forecast, aerosols decline, then the monsoon in South Asia may strengthen (Lelieveld *et al.*, 2019). The availability of water, including groundwater, per person varies considerably in South Asia, with Pakistan and India having the least.[8] Extraction of groundwater, especially in India and Pakistan, is a major supplement to riverine water, canals and rainfall. These supplies are constrained by contamination (with excess salinity and arsenic) and, in some areas, by depletion (MacDonald *et al.*, 2016).

Mountains are highly sensitive to climate change and are warming faster than low-lying regions at similar latitudes. Glaciers in the high mountains of the Hindu Kush and Himalaya act as a buffer, providing a relatively constant flow of water to the plains, even during the hot and dry season via the Indus and Ganges–Brahmaputra 'water towers' (Immerzeel *et al.*, 2020). This water is vital not only for agriculture and industry, but also for generating electricity in certain places.

The Indus river basin supplies water to about 250 million people across most of Pakistan and parts of northern India (Vinca *et al.*, 2021). Already experiencing water stress, the Pakistani population is forecast to increase by 50% by 2050 (UNDESA, 2019). Immerzeel *et al.* (2020) warned that the combination of rising temperature in the Indus headwaters, already existing water stress in the basin and limited government efforts make it unlikely that the Indus water can sustain this pressure. This is considered probable, even assuming a slight annual increase in rain and snow to the catchment.

The steady flow of water along the Ganges and Brahmaputra rivers is also highly vulnerable to glacial melt. Diversion of water from the Tsangpo-Brahmaputra river system by China (which as yet shares no water treaty with either India or Bangladesh) may also be problematic for South Asia. If this diversion proceeds it will reduce the occurrence of seasonal riverine flooding but, by reducing silt delivery to its enormous delta (the world's largest), it could make parts of Bangladesh and India more vulnerable to sea level rise in the cyclone-prone Bay of Bengal (Reddy *et al.*, 2021). Loss of silt may also reduce soil fertility in different regions of India and Bangladesh (Mahapatra and Ratha, 2016). The threat of flooding from overflowing glacial lakes in the Himalaya is another regional issue (Zheng *et al.*, 2021). Such flooding is thought to have contributed to the record flooding in Pakistan in 2022, though the dominant cause is considered to have been an unusually intense monsoon (Hasan, 2022). Poorly maintained irrigation canals, which have accumulated considerable silt, worsened the flooding, particularly in the southern province of Sindh (Dilawar *et al.*, 2022).

32.3 Sea Level Rise

Climate change is causing the global sea level to rise through thermal expansion of seawater and melting of land-based ice and glaciers, including in mountains. Its rate of rise (currently 4.5 mm per annum) is increasing (WMO, 2022). The extent of sea level rise through the rest of the 21st century remains controversial, especially from 2060. An IPCC projection from 2019 is of 0.29–1.10 m by 2100, relative to 1986–2005 levels (McMichael *et al.*, 2020). However, IPCC projections are inherently conservative (Oppenheimer *et al.*, 2019). IPCC projections currently discount, for example, the possibility of rapid collapse of marine ice cliffs, especially in Antarctica. If these (and other) IPCC assumptions are wrong, and if, as now seems highly likely, global average temperature rise exceeds 2°C (compared with pre-industrial levels), then average global sea level rise is likely to approach or exceed 2 m by 2100. By then it will be rising rapidly and will continue to do so at an accelerating rate for the following centuries (Horton *et al.*, 2020; DeConto *et al.*, 2021). A 1 m rise in sea level has been estimated to directly affect up to 6.4 million people in South Asia, in the absence of expensive and effective coastline protection (Dasgupta *et al.*, 2008). Such a rise may also impact agriculture and indirectly cause population displacement. An increase of 2 m or more will be highly problematic for low-lying parts of South Asia, particularly in Bangladesh (Mitchell *et al.*, 2022).

32.4 Forests and Biodiversity

India is one of the world's 17 'megadiverse' countries, accounting for 7–8% of global biodiversity on less than 2.5% of the world's land area. Forest cover in South Asia since 1930 has declined by about 30%, although the rate of deforestation has stabilized since 1995 (Reddy *et al.*, 2018). The value of these forests has also fallen, with much loss of primary forest. Forest fragmentation increases the risk of fires, human penetrations and drying around the forest edge. Biodiversity has declined steeply, including in the Sundarbans, a mangrove forest in the Ganges delta, where human and animal inhabitants are also threatened by sea level rise (Mukul *et al.*, 2019). The retreat, and in some cases extinction, of so many charismatic species in the Indian subcontinent is a source of grief for nature lovers and others with high biophilia (Wilson, 1984).

When India ratified the United Nations Framework Convention on Climate Change Paris Agreement in 2016, it committed itself to sequestering an additional 2.5–3 billion tonnes of CO_2, to be achieved by increasing forest cover (Bawa *et al.*, 2021). However, carbon sequestration in forests has been criticized for harming biodiversity, and in any case is vulnerable to forest fire, including in South Asia.

32.5 Primary Effects and Their Health Impacts

Heat effects

The effects of heatwaves, focusing on India, are discussed in Chapter 32.i. Of course, the linked issue of heat and humidity within South Asia is not confined to India. Average temperatures have also risen in Pakistan, Nepal and Bangladesh. A particularly severe series of heatwaves in May and June 2015 were attributed as causing over 3400 deaths, combining India and Pakistan (UNESCAP, 2016). Such attributions are generally conservative; for example, the Russian heatwave of 2010 caused almost 11,000 non-accidental deaths (Shaposhnikov *et al.*, 2014) but this estimate appears to have excluded more than 1600 deaths by drowning (Matsueda, 2011). In addition, excessive heat is associated with accidents (Umar and Egbu, 2020). In South Asia, record keeping is likely to be less complete than in Russia, particularly for disadvantaged populations, including those working in industries such as brick making (see Chapters 32.i and 32.ii).

The heat island effect has been reported as significant in most South Asian countries (Kotharkar *et al.*, 2018; Dewan *et al.*, 2021). During the daytime in the dry pre-monsoon summer months a reverse urban effect is reported to occur, at least in India. This has been attributed to dwindling vegetation in non-urban regions, leading to reduced evapotranspiration (Shastri *et al.*, 2017). During the pre-monsoon summer nights, the normal heat island effect returns. Nocturnal heat is a major cause of poor sleep and has also been shown to be associated with an increased risk of death (Murage *et al.*, 2017), including in India (Singh *et al.*, 2019).

Natural disasters, trauma and health

South Asia is prone to natural disasters. Ancient (naturally occurring) climate change has been suggested as an underlying determinant in the rise and fall of civilizations. Changes in the Indian summer monsoon have been linked with the demise of the Harappans on the Indo-Gangetic plain about 4000 years ago (Kathayat *et al.*, 2017). Mike Davis (2001) invoked the term 'holocaust' in the title of his book, which documented a series of 19th century calamities in South Asia, often worsened or provoked by the indifference of the colonial power at the time: Britain. Towards the end of British rule in South Asia the Bengal famine caused over 3 million deaths, shaping Amartya Sen's concept of food 'entitlement' (see Chapter 22).

There is growing consensus that climate change will increase the severity of extreme weather events, including storms, droughts and floods (IPCC, 2021). The position, once commonly expressed, that anthropogenic climate change cannot be linked to any single extreme event has substantially shifted. One way to consider this is the view that all extreme weather events occur in an anthropogenically changed climate, interacting with climate variability and human vulnerability (Trenberth, 2011). Even more recently, improvements in attribution science have led to claims of increased precision in determining the chance that any single extreme event is related to anthropogenic climate change. In South Asia to date there are limited case studies to which these methods have been applied (van Oldenborgh *et al.*, 2021). However, it is undeniable that rising sea levels increase the effects of storm surges and flooding in low-lying areas.

Cyclone Bhola, a category 4 storm, which struck Bangladesh (then East Pakistan) in 1970, caused an estimated 250,000–500,000 deaths. The lack of relief by the central government (based in what was then West Pakistan) contributed to the bitter civil war that created modern Bangladesh (Hossain, 2018). Since 1970, despite the consensus that the contribution of climate change to many forms of disaster is worsening, the death toll from such events in South Asia has greatly declined, due, for example, to improved early warning systems, cyclone shelters and more effective relief. Recent evidence suggests that there has been an increased intensity of tropical cyclones in both the Arabian Sea (Murakami *et al.*, 2017) and the Bay of Bengal (Reddy *et al.*, 2021). Storm surges, rather than wind and rain intensity, may be a more important factor in storm-related mortality in South Asia; these are aggravated where coastal geometry can funnel water, as is the case in the northern Bay of Bengal (Mitchell *et al.*, 2022).

Flooding in South Asia can be amplified by glacial melt, changes in monsoon patterns and increased precipitation. These effects are often worsened by inadequate or blocked drainage, sealed surfaces that limit water penetration and poor urban planning. Flooding has primary (direct) impacts on health through drowning and physical and mental trauma, including post-traumatic stress disorder. Additionally, damage to water and sanitation infrastructure may occur via flooding, thus increasing the incidence of diarrhoeal diseases. Other flood-associated health problems include infectious diseases, discussed in the following section. Flooding may also result in infrastructural damage and population displacement. Many South Asian coastal and inland cities are prone to severe flooding (Varughese and Purushothaman, 2021). Rapid urbanization, poverty and population growth throughout the region have led to the proliferation of low-quality human shelter and inferior water and sanitation infrastructure, thus increasing human vulnerability to the adverse impacts of flooding.

A particularly vulnerable megacity in South Asia is Mumbai, which in 2020 was home to an estimated 20 million people. Over 40% of this population lives in squatter communities with poor water and sanitation infrastructure. In 2005, monsoon-related flooding, associated with a high tide, was credited with contributing to the deaths of over 1000 people and was also associated with an increased transmission of infectious diseases, including leptospirosis,[9] the incidence of which rose eightfold (Kovats and Akhtar, 2008). The southern Indian state of Kerala also experienced severe flooding in 2018, when inundation was worsened by release of water from reservoirs (Varughese and Purushothaman, 2021). The effects of flooding on infectious disease epidemiology are discussed in the following section.

32.6 Secondary Effects and Their Health Impacts

Warmer temperatures and flooding, in combination with poverty, are likely to alter the epidemiological characteristics of many infectious and some chronic diseases. In coastal Bangladesh, sea level rise and ground subsidence (Steckler *et al.*, 2022) has already increased salinity in soil and drinking water, with likely adverse health implications for the exposed populations including elevated blood pressure (Khan *et al.*, 2020). Most authors consider these as ‘indirect’ health effects.

Vector-borne diseases in South Asia

Six major vector-borne diseases remain endemic in South Asia: malaria, dengue, chikungunya, filariasis, Japanese encephalitis and visceral leishmaniasis. Zika is regarded as an emerging disease with widespread geographical distribution, including in Nepal. Yellow fever has never been reported in South Asia, for reasons that remain intriguing (Kuno, 2020). The epidemiology of each of these conditions is influenced by factors including climate,[10] poverty, vectors, drug and insecticide resistance, and disease control efforts, such as insecticide-treated bed-nets for malaria. Thus, although climate change will make the control or elimination of these conditions more difficult, there are many co-determinants.

Of these six conditions, malaria still causes the greatest morbidity and mortality in South Asia. Despite intense debate (Hay *et al.*, 2010), there is consensus that the disease burden of malaria in South Asia has recently declined (Dhiman, 2016). There is even hope that malaria will be eliminated from India by 2027, despite the emergence of artemisinin-resistant malaria in the north-east, near its border with Myanmar (Narain and Nath, 2018).

Warmer temperatures at higher altitudes have expanded the habitat for the vectors of malaria as well as for arboviruses, including dengue (Sen *et al.*, 2017; Pandey and Costello, 2019). Higher temperatures can also prolong the transmission period for vector-borne diseases, and shorten the time required for replication of malarial parasites and viruses in an infected mosquito.

The transmission of dengue fever by *Aedes* mosquitoes is highly sensitive to changes in temperature and humidity (see Chapter 16). Although current transmission of dengue is greatest in southern India, its transmission windows are predicted to widen in north India. Dengue is already reported to be an annual epidemic in Nepal (Pandey and Costello, 2019).

Chikungunya, a mosquito-borne viral illness, once thought to have been eliminated from the subcontinent, has re-emerged in southern India, and has also been reported in Nepal and Bangladesh (Anwar *et al.*, 2020). Again, as temperatures rise, transmission periods are likely to increase in northern India. In contrast, the incidence of visceral leishmaniasis (kala azar), a climate-sensitive disease, has declined due to concerted control efforts across Bangladesh, India and Nepal, as was the case for lymphatic filariasis (Yajima *et al.*, 2023). Despite these successes, climate change together with poor governance may jeopardize these programmes.

Diarrhoeal disease

Climate change will likely increase the incidence of infective diarrhoea, due to flooding, rising temperatures and reduced access to clean drinking water. A study among children under five years old in Kathmandu, Nepal, showed an increase in the incidence of diarrhoea in association with rising temperature and rainfall (Bhandari *et al.*, 2020). An identified causal factor for diarrhoea in this area is the use of pathogen-contaminated wastewater for the irrigation of crops, a problem that is widespread in South Asia, especially where fresh water is scarce, such as in Pakistan (Natasha *et al.*, 2020).

32.6 Tertiary Health Impacts

Introduction

This book conceptualizes the 'tertiary' health impacts of climate change as arising at the intersection of people, politics and ecology. Although such effects (large-scale famine, migration, conflict and social collapse) are often overlooked in the health literature, in the long run (perhaps within only a few decades) these effects may be the most damaging (Butler, 2018). Worst-case scenarios (for example, runaway climate change with more than 4°C warming compared with pre-industrial levels) are still avoidable, but avoiding them requires a much greater sense of urgency than currently exists. Experts on climate change have warned that the concept of 'net zero', which underpins much contemporary complacency about climate change, relies on the invention and deployment of affordable technologies that will extract carbon from the atmosphere (Dyke *et al.*, 2021). Such technologies currently consume large amounts of energy. Unless the energy demand of these technologies can be dramatically lowered, the world could find that it is using a significant fraction of its energy to try to reduce carbon levels. Even if such energy is produced from the sun and wind, it is a problematic scenario. Thus, worst-case (or even 'moderately catastrophic') climate change forecasts should not be dismissed.

South Asia is, obviously, part of the Earth system and the world economy. There is a non-trivial risk of 'hothouse' Earth scenarios, meaning that civilization could be crippled by 2100, or even well before, if critical 'eco-social' tipping points are exceeded (see Chapter 1) (Butler, 2016). This could occur even if there are rapid and concerted attempts to slow greenhouse gas emissions (Steffen *et al.*, 2018). The chaos and suffering wrought by the COVID-19 pandemic in South Asia could look like a benign prelude, in comparison. In such a case, good population health in South Asia will be as impossible to achieve as it is anywhere. This chapter is written in the hope that such pessimistic scenarios will be avoided (Torres, 2021).

Even if climate change is restricted to 2°C of warming, profound health and social problems for South Asia may lie ahead. For example, by the year 2100 at Representative Concentration Pathway (RCP) 4.5,[11] there are projections that up to 70% of the population in South Asia will experience wet-bulb temperatures reaching or exceeding 32°C, conditions that will be very difficult to inhabit and will complicate crop cultivation. Even without hothouse Earth scenarios, parts of South Asia will approach the survivability threshold of a wet-bulb temperature of no more than 35°C (Im *et al.*, 2017; Vecellio *et al.*, 2022).

Food security in South Asia

Child stunting is prevalent in most South Asian countries, particularly in India (Kim *et al.*, 2021). The prevalence of 'zero food' days for infants aged 6–23 months[12] is reported to have improved slightly, but still approaches 20% in India (Subramanian *et al.*, 2023). Micronutrient deficiency is also very high in this region (Harding *et al.*, 2017). Importantly, growth failure (and associated suboptimal cognitive development) has multiple causes, including inadequate dietary intake, exposure to infectious diseases and socioeconomic factors. These factors are themselves all vulnerable to climate change. Continuing undernutrition, at high levels, seems inevitable, and the risk of severe famines in the future cannot be dismissed. Food production is endangered by cascading crop failures due to drought and heat in many countries, including in South Asia.

The health consequences of undernutrition are well established. Acute undernutrition causes ill health through starvation, and chronic undernutrition increases susceptibility to other illnesses, including infectious diseases such as pneumonia (Rao and Beckingham, 2013).

Migration

Intra-country, inter-country, regional and overseas migration are all common in South Asia. Migration of Rohingya into Bangladesh (where they are widely known as "forcibly displaced Myanmar nationals"), especially since 2017 has had multi-level consequences, including food insecurity, spread of infectious diseases, violence and an overall economic downturn.[13] India, in recent years, has erected a wall encircling Bangladesh, in order to reduce undocumented migrants (Shahriar *et al.*, 2020). India has already designated at least 4 million people in its north-east as stateless, and who thus risk deportation to Bangladesh. Paulose (2019) claimed that there are 'warning signs' for genocide in Assam.

These pressures already exist; it would be naïve to think that the pressures that are building in South Asia, especially in locations with significant population growth, will not be magnified by future climate change. Health risks of forced migration include anguish, trauma, undernutrition, exposure to emerging and re-emerging infectious diseases and impaired mental health. At sufficient scale, forced migration dehumanizes the other, and opens the door to neglect and to mass killing.

Conflict

The evidence linking climate change and conflict, while generally contested in the academic literature, is not controversial among military thinkers (Bowles *et al.*, 2015). In South Asia, possible pathways between climate change and international conflict include disputes involving three nuclear powers. One concerns the Indus water treaty between India and Pakistan (Bagla, 2010).

Another risk concerns China and India, who have already been involved with two wars since their independence; border tensions remain high. China claims Arunachal Pradesh as its territory. Although this state constitutes only 2.5% of India, it is comparatively well-watered and has a very low population density. Indian possession of this state remains a possible cause for future Indo-Sino war. Similarly, the mooted diversion of waters of the Brahmaputra to the north-east by China could initiate conflict. This risk in turn depends on the severity of climate change in China (Klare, 2020). Bangladesh (though not nuclear armed) is also vulnerable to the withholding of water by China and India.[14]

Any significant decline in countries' agricultural productivity arising via climate change, such as through widespread crop failures, is likely to increase disagreement and tension. All countries of South Asia (including Bhutan[15]) are heterogeneous, having histories of discrimination against religion, language, caste and gender. Therefore, it could be hypothesized that, under the most optimistic scenarios for climate change in South Asia, internal rifts will deepen.

The multilateral global governance structures set up after World War II have played a significant role in stabilizing international relations. However, the international system has, to date, failed to address the complexities and compromise demanded by climate change, as demonstrated by the inability to agree to international carbon reduction targets (Haldén, 2007). This failure challenges the legitimacy of the international system, and leaves it less able to cope with political instability and potential conflict (Foresight International Dimensions of Climate Change, 2011). If multilateral institutions are perceived as ineffective and unable to cope with the climate change challenge, a potential risk may be that countries will take unilateral action to protect their resources (Haldén, 2007), or even become predatory in a quest to secure essential supplies as their own become scarce.

At a national level, natural disasters, internal migration and social disruption associated with climate change challenge governance structures, not only in developing nations, but also in the developed world. An example happened following Hurricane Katrina in New Orleans in August 2005. The inability of national government institutions to plan and implement adequate responses to protect populations from disasters may fuel public unrest and create the conditions for radical political groups to enhance their presence and influence. This is a particular concern in South Asia, where there is some evidence that such changes are already taking place. During the 2010 floods in Pakistan, militant groups were among the first organizations to respond in the worst-hit areas, providing aid and assistance (Foresight International Dimensions of Climate Change, 2011).

32.7 Mental Health

Climate change is associated with poor mental health through many pathways. Recognized mental health effects of extreme weather events include post-traumatic stress disorder, depression, anxiety, aggression, sleep disruption, interpersonal difficulties, worsened psychotic symptoms, substance abuse and suicide (Cianconi *et al.*, 2020). People with existing poor mental health are at particular risk of adverse health outcomes during significant societal disruption. Displaced populations are also at increased risk of mental health problems due to social exclusion, socio-economic disadvantage and exposure to racial or cultural discrimination in receiving communities (Fritze *et al.*, 2008). In India, farmer suicide is documented at over 130,000 deaths per year, disrupting and saddening many more. These have been linked with crop-damaging high temperatures (Carleton, 2017) as well as failed harvests, including from drought (Padhy *et al.*, 2015).

South Asia is at high risk of natural disasters, conflict and involuntary migration related to climate change. As a result, mental health impacts are likely to be substantial across the region. Given that mental illness is already significantly underdiagnosed in South Asia, and that hospital and health resources are likely to be even more stretched, there is a strong likelihood that mental ill health will represent a large, unmet need as a result of climate change.

32.8 Impacts on Women

Women are especially vulnerable to the health impacts of climate change, due to skewed power relations, financial poverty and disparities in decision-making processes, at both the family and policy level (Nellemann *et al.*, 2011). Women are also more at risk of morbidity at all stages of a natural disaster. During natural calamities, household and childcare responsibilities could potentially make it harder for women to follow an evacuation plan, as was documented during the 1991 flood in Bangladesh. A study showed that the mortality of flood-affected females aged 20–44 years was four times higher than that of males of the same age group (Nellemann *et al.*, 2011). Social inequalities and power disparities also make females susceptible to malnutrition, particularly in the post-disaster period. Rape and other forms of gender-based violence also present a significant concern in association with natural disasters (Neumayer and Plümper, 2007).

The flooding in Pakistan in 2022, which is estimated to have harmed more than 33 million people, inundating a third of the nation, particularly hurt women and children, including pregnant women. Some people are thought likely to have been displaced for life (Sarkar, 2022).

32.9 Conclusion

Climate change, in conjunction with poverty and limited resources, is already increasing mortality and morbidity in South Asia. Such impacts are likely to increase as climate change intensifies. The tertiary effects of climate change on health are the hardest to quantify, but, in the long term, are likely to generate the greatest health impact via interlinked, synergistic disruption of the supporting natural, economic and social systems that underpin health.

Overall, marginalized people are the most vulnerable to the health impact of climate change. Poverty is associated with low-quality shelter, an increased risk of trauma from natural disasters, poor sanitation infrastructure and an increased risk of communicable disease. Furthermore, affected people are most likely to migrate from areas adversely affected by disasters secondary to climate change and co-factors. Climate change is likely to exacerbate existing inequalities. Whilst farmers will be affected by declining crop production, the general public will bear rising food prices.

If the nations of South Asia are to fulfil their potential, then an attitudinal transformation – even a revolution – is needed. This peaceful revolution must go much further than high-level acceptance that climate change is real, and that, as a result, South Asia must, like all countries, rapidly transition from fossil fuels to renewable energy. That is necessary, but, even more fundamentally, the people of South Asia need to work in common towards reducing discrimination, easing nuclear weapons tensions and promoting good education and other determinants of the demographic transition.

Climate change will harm every aspect of South Asian development over the next century and beyond. Few people and governments, including those in South Asia, appear to be appropriately prepared for the likely impact. Discussion of the potentially disastrous tertiary consequences may accelerate international efforts to mitigate carbon emissions. Appropriate early planning will help to reduce the health impact of climate change in South Asia; it also represents an opportunity for stakeholders in this region to develop areas of global expertise in innovative approaches to research, mitigation and adaptation.

Notes

[1] Most geographical definitions of South Asia include these countries; here, we exclude them, due to lack of space. Indonesia, a part of the World Health Organization (WHO) region associated mostly with South Asia, is discussed in Chapter 34.

[2] Population forecasts from the Institute for Health Metrics and Evaluation are lower, particularly in the second half of this century (Vollset *et al.*, 2020).

[3] Including cyclones, hurricanes and typhoons; the name varies by location. In the Indian Ocean the word 'cyclone' is generally used.

[4] An alternative name for these Himalayan glaciers is 'water towers'.

[5] The scale of solar and wind-driven electricity, especially in India, is rapidly increasing.

[6] Fertility is highest in the big states of north and north-western India – above three children per woman in Bihar, Uttar Pradesh and Madhya Pradesh, with Rajasthan and Jharkhand being very close to that level. At the other end of the spectrum are eight small states and union territories with fertility levels of less than 1.6.

[7] Wet-bulb temperatures give an indication of heat and humidity, and thus better reflect heat stress (see Chapter 12).

[8] In 2021 each country had less than 1500 m^3/year, with Pakistan less than 1200 m^3/year (http://www.fao.org/aquastat).

[9] A disease transmitted by the penetration of spirochaetes through skin abrasions in contact with water contaminated by the urine of infected rodents.

[10] The extrinsic incubation period, the time for the pathogen's life cycle in the cold-blooded vector, is shortened by warmer temperatures, for both malaria and arboviruses (see Chapter 16).
[11] Equivalent to approximately 4.5°C warming compared with pre-industrial levels.
[12] Presumably, this is possible because such infants are breastfed. However, exclusive breastfeeding after 6 months is harmful (Lutter *et al*., 2021)
[13] Almost 100,000 of these refugees are to be relocated to Jaliar Char, recently renamed as Bhasan (or Vasan) Char, meaning 'floating island', 60 km from the coast (Bremner, 2020). Healthcare on this island will obviously be problematic (Samarasekera, 2021).
[14] 'Bangladesh, at the confluence of three major global rivers, has only 7% of the catchment area of these basins. Its current national development plan recognizes that Bangladesh cannot undertake meaningful water resource development programmes on its own; it aims to strike new transboundary agreements. So far, however, a water-sharing agreement is in place for only one of its 57 transboundary rivers – the Ganges, with India' (UN Water, 2020).
[15] Bhutan has long practised state-sponsored discrimination against its Nepali-speaking minority, most of whom now live in exile (de Varennes, 2008).

References

Akhtar, R. (ed.) (2016) *Climate Change and Human Health Scenario in South and Southeast Asia*. Springer, Cham, Switzerland.

Akram, R., Sarker, A.R., Sheikh, N., Ali, N., Mozumder, M.G.N. *et al.* (2020) Factors associated with unmet fertility desire and perceptions of ideal family size among women in Bangladesh: Insights from a nationwide Demographic and Health Survey. *PLoS ONE* 15, e0233634. DOI: 10.1371/journal.pone.0233634

Anwar, S., Taslem Mourosi, J., Khan, M.F., Ullah, M.O., Vanakker, O.M. *et al.* (2020) Chikungunya outbreak in Bangladesh (2017): Clinical and hematological findings. *PLoS Neglected Tropical Diseases* 14, e0007466. DOI: 10.1371/journal.pntd.0007466

Bagla, P. (2010) Along the Indus river, saber rattling over water security. *Science* 328, 1226–1227. DOI: 10.1126/science.328.5983.122

Balakrishnan, K., Dey, S., Gupta, T., Dhaliwal, R.S., Brauer, M. *et al.* (2019) The impact of air pollution on deaths, disease burden, and life expectancy across the states of India: the Global Burden of Disease Study 2017. *The Lancet Planetary Health* 3, e26–e39. DOI: 10.1016/S2542-5196(18)30261-4

Bawa, K.S., Sengupta, A., Chavan, V., Chellam, R., Ganesan, R. *et al.* (2021) Securing biodiversity, securing our future: A national mission on biodiversity and human well-being for India. *Biological Conservation* 253, 108867. DOI: 10.1016/j.biocon.2020.108867

Bhandari, D., Bi, P., Sherchand, J.B., Dhimal, M. and Hanson-Easey, S. (2020) Assessing the effect of climate factors on childhood diarrhoea burden in Kathmandu, Nepal. *International Journal of Hygiene and Environmental Health* 223, 199–206. DOI: 10.1016/j.ijheh.2019.09.002

Bowles, D.C., Butler, C.D. and Morisetti, N. (2015) Climate change, conflict, and health. *Journal of the Royal Society of Medicine* 108, 390–395. DOI: 10.1177/0141076815603234

Bremner, L. (2020) Sedimentary logics and the Rohingya refugee camps in Bangladesh. *Political Geography* 77, 102109. DOI: 10.1016/j.polgeo.2019.102109

Butler, C.D. (2016) Sounding the alarm: health in the Anthropocene. *International Journal of Environmental Research and Public Health* 13, 665. DOI: 10.3390/ijerph13070665

Butler, C.D. (2018) Climate change, health and existential risks to civilization: A comprehensive review (1989–2013). *International Journal of Environmental Research and Public Health* 15, 2266. DOI: 10.3390/ijerph15102266

Carleton, T.A. (2017) Crop-damaging temperatures increase suicide rates in India. *Proceedings of the National Academy of Sciences (USA)* 114, 8746–8751. DOI: 10.1073/pnas.1701354114

Cianconi, P., Betrò, S. and Janiri, L. (2020) The impact of climate change on mental health: a systematic descriptive review. *Frontiers in Psychiatry* 11, 74. DOI: 10.3389/fpsyt.2020.00074

Dasgupta, S., Laplante, B., Meisner, C., Wheeler, D. and Yan, J. (2008) The impact of sea level rise on developing countries: a comparative analysis. *Climatic Change* 93, 378–388. DOI: 10.1007/s10584-008-9499-5

Davis, M. (2001) *Late Victorian Holocausts: El Niño Famines and the Making of the Third World*. Verso, London.

DeConto, R.M., Pollard, D., Alley, R.B., Velicogna, I., Gasson, E. *et al.* (2021) The Paris Climate Agreement and future sea-level rise from Antarctica. *Nature* 593, 83–89.
de Varennes, F. (2008) Constitutionalising discrimination in Bhutan: The emasculation of human rights in the land of the dragon. *Asia-Pacific Journal on Human Rights and the Law* 2, 47–76. DOI: 0.1163/157181509789025219
Dewan, A., Kiselev, G., Botje, D., Mahmud, G.I., Bhuian, M.H. *et al.* (2021) Surface urban heat island intensity in five major cities of Bangladesh: Patterns, drivers and trends. *Sustainable Cities and Society* 71, 102926. DOI: 10.1016/j.scs.2021.102926
Dhiman, R.C. (2016) Scenario of malaria and dengue in India: Way forward. In: Akhtar, R. (ed.) *Climate Change and Human Health Scenario in South and Southeast Asia*. Springer, Cham, Switzerland, pp. 91–97.
Dilawar, I., Chaudhary, A. and Schultz, K. (2022) *How Pakistan's catastrophic floods spiraled into a nightmare*. BNN Bloomberg News, Toronto. Available at: https://www.bloomberg.com/news/features/2022-09-08/how-pakistan-s-catastrophic-floods-spiraled-into-a-nightmare?leadSource=uverify%20wall (accessed 21 November 2022).
Dyke, J., Watson, R. and Knorr, W. (2021) Climate scientists: concept of net zero is a dangerous trap. The Conversation. Available at: https://theconversation.com/climate-scientists-concept-of-net-zero-is-a-dangerous-trap-157368 (accessed 24 May 2021).
Foresight International Dimensions of Climate Change (2011) *Final Project Report*. The Government Office for Science, London.
Fritze, J.G., Blashki, G.A., Burke, S. and Wiseman, J. (2008) Hope, despair and transformation: climate change and the promotion of mental health and wellbeing. *International Journal of Mental Health Systems* 2, 13. DOI: 10.1186/1752-4458-2-13
Goujon, A., Wazir, A. and Gailey, N. (2020) Pakistan: A population giant falling behind in its demographic transition. *Population & Societies* 576, 1–4. Available at: http://pure.iiasa.ac.at/id/eprint/16409/ (accessed 19 May 2022).
Haldén, P. (2007) *The Geopolitics of Climate Change*. Challenges to the International System. FOI (Swedish Defence Research Agency), Stockholm, Sweden.
Harding, K.L., Aguayo, V.M. and Webb, P. (2017) Hidden hunger in South Asia: a review of recent trends and persistent challenges. *Public Health Nutrition* 21, 785–795. DOI:10.1017/S1368980017003202
Hasan, S. (2022) Blaming melting glaciers for Pakistan floods is far-fetched: experts. *TRT World*. Available at: https://www.trtworld.com/magazine/blaming-melting-glaciers-for-pakistan-floods-is-far-fetched-experts-60520 (accessed 1 October 2022).
Hay, S.I., Gething, P.W. and Snow, R.W. (2010) India's invisible malaria burden. *The Lancet* 367, 1716–1717. DOI: 10.1016/S0140-6736(10)61084-7
Hijioka, Y., Lin, E., Pereira, J.J., Corlett, R.T., Cui, X. *et al.* (2014) Asia. In: Barros, V.R., Field, C.B., Dokken, D.J., Mastrandrea, M.D., Mach, K.J. *et al.* (eds) *Climate Change 2014: Impacts, Adaptation, and Vulnerability*. Cambridge University Press, Cambridge, UK, pp. 1327–1370.
Horton, B.P., Khan, N.S., Cahill, N., Lee, J.S.H., Shaw, T.A. *et al.* (2020) Estimating global mean sea-level rise and its uncertainties by 2100 and 2300 from an expert survey. *npj Climate and Atmospheric Science* 3, 18. DOI: 10.1038/s41612-020-0121-5
Hossain, N. (2018) The 1970 Bhola cyclone, nationalist politics, and the subsistence crisis contract in Bangladesh. *Disasters* 42, 187–203. DOI: 10.1111/disa.12235
Im, E.-S., Pal, J.S. and Eltahir, E.A.B. (2017) Deadly heat waves projected in the densely populated agricultural regions of South Asia. *Science Advances* 3, e1603322. DOI: 10.1126/sciadv.1603322
Immerzeel, W.W., Lutz, A.F., Andrade, M., Bahl, A., Biemans, H. *et al.* (2020) Importance and vulnerability of the world's water towers. *Nature* 577, 364–369. DOI: 10.1038/s41586-019-1822-y
IPCC (2021) Summary for Policymakers. In: Masson-Delmotte, V., Zhai, P., Pirani, A., Connors, S.L., Péan, C. *et al.* (eds) *Climate Change 2021: The Physical Science Basis. Contribution of Working Group I to the Sixth Assessment Report of the Intergovernmental Panel on Climate Change*. Cambridge University Press, Cambridge, UK. DOI: 10.1017/9781009157896.
Kathayat, G., Cheng, H., Sinha, A., Yi, L., Li, X. *et al.* (2017) The Indian monsoon variability and civilization changes in the Indian subcontinent. *Science Advances* 3, e1701296. DOI: 10.1126/sciadv.1701296
Khan, J.R., Awan, N., Archie, R.J., Sultana, N. and Muurlink, O. (2020) The association between drinking water salinity and hypertension in coastal Bangladesh. *Global Health Journal* 4, 153–158. DOI: 10.1016/j.glohj.2020.11.001

Kim, R., Bijral, A.S., Xu, Y., Zhang, X., Blossom, J.C. *et al.* (2021) Precision mapping child undernutrition for nearly 600,000 inhabited census villages in India. *Proceedings of the National Academy of Sciences (USA)* 118, e2025865118. DOI: 10.1073/pnas.2025865118

Klare, M.T. (2020) Climate change, water scarcity, and the potential for interstate conflict in South Asia. *Journal of Strategic Security* 13, 109–122. DOI: 10.5038/1944-0472.13.4.1826

Kotharkar, R., Ramesh, A. and Bagade, A. (2018) Urban heat island studies in South Asia: A critical review. *Urban Climate* 24, 1011–1026. DOI: 10.1016/j.uclim.2017.12.006

Kovats, S. and Akhtar, R. (2008) Climate, climate change and human health in Asian cities. *Environment and Urbanization* 20, 165–175. DOI: 10.1177/0956247808089154

Kuno, G. (2020) The absence of yellow fever in Asia: history, hypotheses, vector dispersal, possibility of YF in Asia, and other enigmas. *Viruses* 12, 1349. DOI: 10.3390/v12121349

Lelieveld, J., Klingmüller, K., Pozzer, A., Burnett, R.T., Haines, A. *et al.* (2019) Effects of fossil fuel and total anthropogenic emission removal on public health and climate. *Proceedings of the National Academy of Sciences (USA)* 116, 7192–7197. DOI: 10.1073/pnas.1819989116

Lutter, C.K., Grummer-Strawn, L. and Rogers, L. (2021) Complementary feeding of infants and young children 6 to 23 months of age. *Nutrition Reviews* 79, 825–846. DOI: 10.1093/nutrit/nuaa143

MacDonald, A.M., Bonsor, H.C., Ahmed, K.M., Burgess, W.G., Basharat, M. *et al.* (2016) Groundwater quality and depletion in the Indo-Gangetic Basin mapped from in situ observations. *Nature Geoscience* 9, 762–766. DOI: 10.1038/ngeo2791

Mahapatra, S.K. and Ratha, K.C. (2016) Brahmaputra River: A bone of contention between India and China. *Water Utility Journal* 13, 91–99. Available at: http://www.ewra.net/wuj/issue_13.htm (accessed 19 May 2022).

Matsueda, M. (2011) Predictability of Euro-Russian blocking in summer of 2010. *Geophysical Research Letters* 38, L06801. DOI: 10.1029/2010GL046557

Mazdiyasni, O., AghaKouchak, A., Davis, S.J., Madadgar, S., Mehran, A. *et al.* (2017) Increasing probability of mortality during Indian heat waves. *Science Advances* 3, e1700066. DOI: 10.1126/sciadv.1700066

McMichael, C., Dasgupta, S., Ayeb-Karlsson, S. and Kelman, I. (2020) A review of estimating population exposure to sea-level rise and the relevance for migration. *Environmental Research Letters* 15, 123005. DOI: 10.1088/1748-9326/abb398

Mitchell, D., Hawker, L., Savage, J., Bingham, R., Lord, N.S. *et al.* (2022) Increased population exposure to Amphan-scale cyclones under future climates. *Climate Resilience and Sustainability* 1, 36. DOI: 10.1002/cli2.36

Mukul, S.A., Alamgir, M., Sohel, M.S.I., Pert, P.L., Herbohn, J. *et al.* (2019) Combined effects of climate change and sea-level rise project dramatic habitat loss of the globally endangered Bengal tiger in the Bangladesh Sundarbans. *Science of The Total Environment* 663, 830–840. DOI: 10.1016/j.scitotenv.2019.01.383

Murage, P., Hajat, S. and Kovats, R.S. (2017) Effect of night-time temperatures on cause and age-specific mortality in London. *Environmental Epidemiology* 1, e005. DOI: 10.1097/ee9.0000000000000005

Murakami, H., Vecchi, G.A. and Underwood, S. (2017) Increasing frequency of extremely severe cyclonic storms over the Arabian Sea. *Nature Climate Change* 7, 885–889. DOI: 10.1038/s41558-017-0008-6

Narain, J.P. and Nath, L.M. (2018) Eliminating malaria in India by 2027: The countdown begins! *Indian Journal of Medical Research* 148, 123–126. DOI: 10.4103/ijmr.IJMR_1175_18

Natasha, Shahid, M., Khalid, S., Murtaza, B., Anwar, H. *et al.* (2020) A critical analysis of wastewater use in agriculture and associated health risks in Pakistan. *Environmental Geochemistry and Health* 45, 5592–5618. DOI: 10.1007/s10653-020-00702-3

Nellemann, C., Verma, R. and Hislop, L. (2011) *Women at the Frontline of Climate Change: Gender Risks and Hopes: A Rapid Response Assessment*. United Nations Environment Programme, Birkeland, Norway.

Neumayer, E. and Plümper, T. (2007) The gendered nature of natural disasters: the impact of catastrophic events on the gender gap in life expectancy, 1981–2002. *Annals of the Association of American Geographers* 97, 551–566. DOI: 10.1111/j.1467-8306.2007.00563.x

Oppenheimer, M., Oreskes, N., Jamieson, D., Brysse, K., O'Reilly, J. *et al.* (2019) *Discerning Experts*. The Practices of Scientific Assessment for Environmental Policy. The University of Chicago Press, Chicago, Illinois.

Padhy, S.K., Sarkar, S., Panigrahi, M. and Paul, S. (2015) Mental health effects of climate change. *Indian Journal of Occupational and Environmental Medicine* 19, 3–7. DOI: 10.4103/0019-5278.156997

Pandey, B.D. and Costello, A. (2019) The dengue epidemic and climate change in Nepal. *The Lancet* 394, 2150–2151. DOI: 10.1016/S0140-6736(19)32689-3

Pandey, S.K., Vinoj, V., Landu, K. and Babu, S.S. (2017) Declining pre-monsoon dust loading over South Asia: Signature of a changing regional climate. *Scientific Reports* 7, 16062. DOI: 10.1038/s41598-017-16338-w

Paulose, R.M. (2019) A new dawn? Statelessness and Assam. *Groningen Journal of International Law & Literature* 7, 99–111. DOI: 10.4103/0019-5278.156997

Rao, M. and Beckingham, A. (2013) Food, nutrition and public health. In: Pielke Sr., R.A. (ed.) *Climate Vulnerability: Understanding and Addressing Threats to Essential Resources*. Elsevier, Amsterdam, The Netherlands, pp. 87–94.

Rao, M., Lakshmi, J. and Samarth, A. (2011) *Addressing Health of the Urban Poor in South-East Asia Region: Challenges and Opportunities*. WHO South East Asia Regional Office, New Delhi, India.

Reddy, C.S., Saranya, K.R.L., Pasha, S.V., Satish, K.V., Jha, C.S. *et al.* (2018) Assessment and monitoring of deforestation and forest fragmentation in South Asia since the 1930s. *Global and Planetary Change* 161, 132–148. DOI: 10.1016/j.gloplacha.2017.10.007

Reddy, P.J., Sriram, D., Gunthe, S.S. and Balaji, C. (2021) Impact of climate change on intense Bay of Bengal tropical cyclones of the post-monsoon season: a pseudo global warming approach. *Climate Dynamics* 56, 2855–2879. DOI: 10.1007/s00382-020-05618-3

Saeed, F., Schleussner, C.-F. and Ashfaq, M. (2021) Deadly heat stress to become commonplace across South Asia already at 1.5°C of global warming. *Geophysical Research Letters* 48, e2020GL091191. DOI: 10.1029/2020GL091191

Samarasekera, U. (2021) Relocated Rohingya face uncertain health-care provision. *The Lancet* 397, 460. DOI: 10.1016/S0140-6736(21)00296-8

Samir, K.C., Wurzer, M., Speringer, M. and Lutz, W. (2018) Future population and human capital in heterogeneous India. *Proceedings of the National Academy of Sciences (USA)* 115, 8328–8333. DOI: 10.1073/pnas.1722359115

Sarkar, S. (2022) Pakistan floods pose serious health challenges. *BMJ* 378, o2141. DOI: 10.1136/bmj.o2141

Sen, B., Dhimal, M., Latheef, A.T. and Ghosh, U. (2017) Climate change: health effects and response in South Asia. *BMJ* 359, j5117. DOI: 10.1136/bmj.j5117

Shahriar, S., Qian, L. and Kea, S. (2020) Anatomy of human rights violations at the Indo-Bangladesh borderlands. *Territory, Politics, Governance* 8, 553–578. DOI: 10.1080/21622671.2019.1582359

Shaposhnikov, D., Revich, B., Bellander, T., Bedada, G.B., Bottai, M. *et al.* (2014) Mortality related to air pollution with the Moscow heatwave and wildfire of 2010. *Epidemiology* 25, 359–364. DOI: 10.1097/EDE.0000000000000090

Shastri, H., Barik, B., Ghosh, S., Venkataraman, C. and Sadavarte, P. (2017) Flip flop of day-night and summer-winter surface urban heat island intensity in India. *Scientific Reports* 7, 40178. DOI: 10.1038/srep40178

Singh, N., Mhawish, A., Ghosh, S., Banerjee, T. and Mall, R.K. (2019) Attributing mortality from temperature extremes: A time series analysis in Varanasi, India. *Science of The Total Environment* 665, 453–464. DOI: 10.1016/j.scitotenv.2019.02.074

Steckler, M.S., Oryan, B., Wilson, C.A., Grall, C., Nooner, S.L. *et al.* (2022) Synthesis of the distribution of subsidence of the lower Ganges-Brahmaputra delta, Bangladesh. *Earth-Science Reviews* 224, 103887. DOI: 10.1016/j.earscirev.2021.103887

Steffen, W., Rockström, J., Richardson, K., Lenton, T.M., Folke, C. *et al.* (2018) Trajectories of the Earth System in the Anthropocene. *Proceedings of the National Academy of Sciences (USA)* 115, 8252–8259. DOI: 10.1126/science.1259855

Stone, R. (2020) COVID-19 in South Asia: Mirror and catalyst. *Asian Affairs* 51, 542–568. DOI: 10.1080/03068374.2020.1814078

Subramanian, S.V., Ambade, M., Sharma, S., Kumar, A. and Kim, R. (2023) Prevalence of zero-food among infants and young children in India: patterns of change across the States and Union Territories of India, 1993–2021. *eClinicalMedicine* 58, 101890. DOI: 10.1016/j.eclinm.2023.101890

Torres, P. (2021) Scared straight: How prophets of doom might save the world. Bulletin of the Atomic Scientists 27 May 2021. Available at: https://thebulletin.org/2021/05/scared-straight-how-prophets-of-doom-might-save-the-world/ (accessed 19 May 2022).

Trenberth, K.E. (2011) Attribution of climate variations and trends to human influences and natural variability. *WIREs Climate Change* 2, 925–930. DOI: 10.1002/wcc.142

Umar, T. and Egbu, C. (2020) Heat stress, a hidden cause of accidents in construction. *Proceedings of the Institution of Civil Engineers – Municipal Engineer* 173, 49–60. DOI: 10.1680/jmuen.18.00004

UNDESA (2019) *World Population Prospects 2019*. United Nations Department of Economic and Social Affairs, New York.

UNESCAP (2016) *Disasters in Asia and the Pacific: 2015 Year in Review*. UN Economic and Social Commission for Asia and the Pacific, Bangkok. Available at: www.unescap.org/sites/default/files/2015_Year%20in%20Review_final_PDF_1.pdf (accessed 9 October 2022).

UNFPA (2022) *World Population Dashboard*. UN Population Fund, New York. Available at: https://www.unfpa.org/data/world-population-dashboard (accessed 25 May 2022).

UN Water (2020) *UN World Water Development Report 2020: Water and Climate Change*. United Nations Water (UN-Water), Geneva. Available at: https://www.unwater.org/publications/world-water-development-report-2020/ (accessed 25 May 2022).

van Oldenborgh, G.J., van der Wiel, K., Kew, S., Philip, S., Otto, F. *et al.* (2021) Pathways and pitfalls in extreme event attribution. *Climatic Change* 166, 13. DOI: 10.1007/s10584-021-03071-7

Vecellio, D.J., Wolf, S.T., Cottle, R.M. and Kenney, W.L. (2022) Evaluating the 35°C wet-bulb temperature adaptability threshold for young, healthy subjects (PSU HEAT Project). *Journal of Applied Physiology* 132, 340–345. DOI: 10.1152/japplphysiol.00738.2021

Varughese, A. and Purushothaman, C. (2021) Climate change and public health in India: The 2018 Kerala floods. *World Medical and Health Policy* 13, 16–35. DOI: 10.1002/wmh3.429

Vinca, A., Parkinson, S., Riahi, K., Byers, E., Siddiqi, A. *et al.* (2021) Transboundary cooperation a potential route to sustainable development in the Indus basin. *Nature Sustainability*, 4, 331–339. DOI: 10.1038/s41893-020-00654-7

Vollset, S.E., Goren, E., Yuan, C.-W., Cao, J., Smith, A.E., *et al.* (2020) Fertility, mortality, migration, and population scenarios for 195 countries and territories from 2017 to 2100: a forecasting analysis for the Global Burden of Disease Study. *The Lancet* 396, 1285–1306. DOI: 10.1016/S0140-6736(20)30677-2

Voosen, P. (2021) Global temperatures in 2020 tied record highs. *Science* 371, 334–335. DOI: 10.1126/science.371.6527.334

Wilson, E.O. (1984) *Biophilia: The Human Bond with Other Species*. Harvard University Press, Cambridge, Massachusetts.

WMO (2022) *The State of the Global Climate 2021*. World Meteorological Organization, Geneva.

Yajima, A., Lin, Z., Mohamed, A.J., Dash, A.P. and Rijala, S. (2023) Finishing the task of eliminating neglected tropical diseases (NTDs) in WHO South-East Asia Region: promises kept, challenges, and the way forward. *The Lancet Regional Health - Southeast Asia* 18, 100302. DOI: 10.1016/j.lansea.2023.10030

Zheng, G., Allen, S.K., Bao, A., Ballesteros-Cánovas, J.A., Huss, M. *et al.* (2021) Increasing risk of glacial lake outburst floods from future Third Pole deglaciation. *Nature Climate Change* 11, 411–417. DOI: 10.1038/s41558-021-01028-3

32.i Heatwaves and Health in South Asia, Focusing on India

Gulrez Shah Azhar
Rand Corporation, Santa Monica, California, USA

Abstract
This mini-chapter warns that the current trajectory of greenhouse gas emissions threatens a combination of heat and humidity that could literally threaten human survival in central India (at least) unless people have access to reliable air-conditioning. Yet, for hundreds of millions of vulnerable people on the subcontinent this is a fantasy. Although the deaths attributed to heat stress in India are relatively modest this is probably a gross underestimate, due to the inadequate keeping of vital statistics. Some adaptation is possible but strategies are also inhibited by cultural practices, which are particularly harmful to the poor, especially to women.

32.i.1 A Vulnerable Population

Historically, humans have remained concentrated in a global climate niche – a small range of optimal temperatures and tolerable humidity. However, in the next few decades, one to three billion people are likely to have to survive in climatic conditions that exceed this favourable space (Xu *et al.*, 2020). In India, later this century, unless the current trajectory of global greenhouse gas accumulation can be dramatically altered, increasing wet-bulb temperatures in densely populated areas are likely to exceed human survival thresholds (Im *et al.*, 2017). These temperatures will lead to an increased probability of heat-related morbidity and mortality (Mazdiyasni *et al.*, 2017).

In recent years, in many parts of South Asia, summer heatwaves have occurred almost annually while heatwave-associated deaths have also shown a modest increase. However, deaths reported from heatwaves are likely to account for only a small fraction of excess mortalities that occur during heatwaves. This is likely due to the way deaths during heatwaves are recorded. Although the causes of some deaths that occur during heatwaves are directly attributed to heat, many are not. It is reasonable to argue that many, if not all, of the *excess* deaths that occur during a heatwave have heat as a contributing cause, but that this is not currently captured by official data (Azhar *et al.*, 2014).

For many of the poor in India and other parts of South Asia, medical certification of the cause of death is rudimentary or completely lacking. Even for the well-off, many deaths may be attributed to a 'proximal' cause such as ischaemic heart disease, without written or even verbal acknowledgement that heat-related dehydration (for example) may also have contributed to the cardiac event. Adding to uncertainty in estimating heat-attributable health impacts, there is variation in the definition of heatwaves (Nori-Sarma *et al.*, 2019).

However, some of the deaths that occur during heatwaves, especially among the elderly,

 Climate Change and Global Health: Primary, Secondary and Tertiary Effects, 2nd Edition (eds C.D. Butler and K. Higgs)
DOI: 10.1079/9781800620025.0032i

likely result in only a modest shortening of life, sometimes called a 'harvesting' effect. This means that while deaths in a geographical region are more concentrated in time (i.e. during heatwaves), the effect on total mortality over a longer period is not as strong. In addition, at least in some parts of South Asia, milder winters due to global heating partially or even fully compensate for heat-related deaths, by lowering the number of people dying during 'cold waves'.

Vulnerability mapping reveals that central India bears the highest risk of heat-related mortality and morbidity (Azhar *et al.*, 2017) (Fig. 32. i.1). The World Bank's South Asia hotspot report also projects a decline in living standards due to unbearable heat and humidity in this region (Mani *et al.*, 2018). While mapping exercises show relative heat vulnerability, absolute heat vulnerability in the country is likely to be higher due to a variety of factors, as follows.

- India has the largest number of poor people of any nation. Many factors related to poverty and inequality increase people's vulnerability to heat. For example, over 40% of the labour force is involved in agriculture, many of whom will undertake manual work, directly exposed to the sun.
- Home to almost a fifth of the planet's population, India's share of the global fresh-water

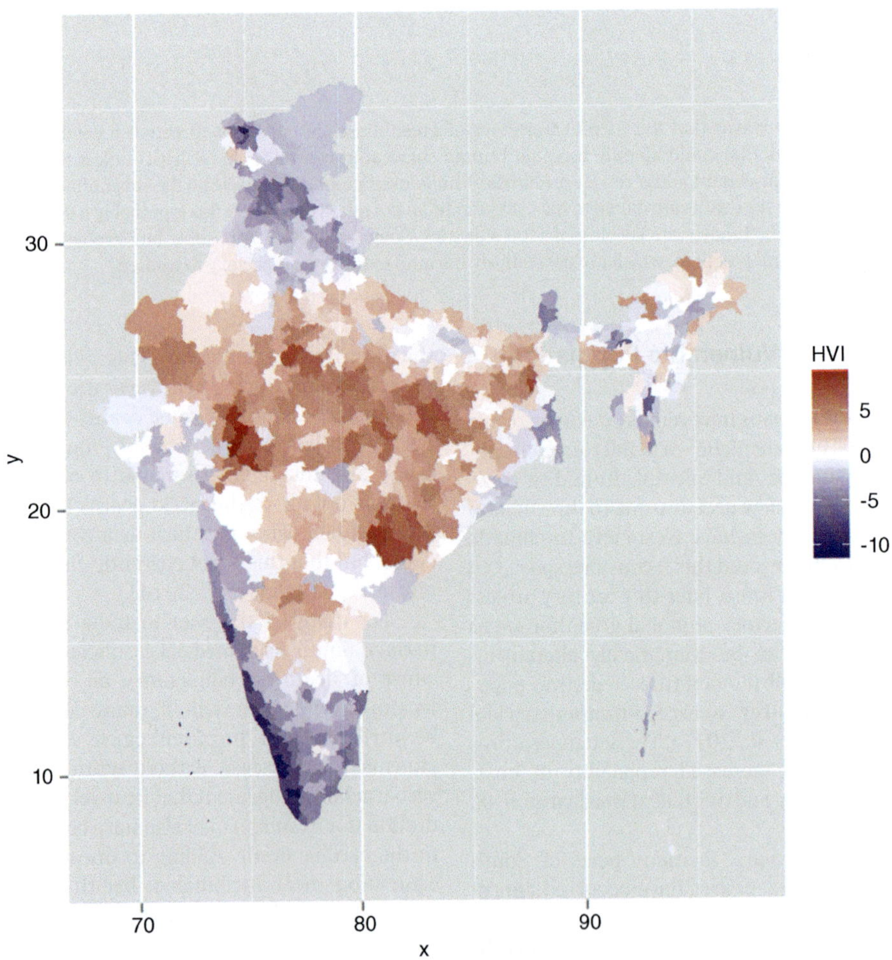

Fig. 32.i.1. Heat Vulnerability Index for India. (Source: Azhar *et al.*, 2017) (http://creativecommons.org/licenses/by/4.0/).

supply is only 4%. With low groundwater levels, especially in Punjab, widespread surface water contamination and projected Himalayan glacier melting, existing water scarcity will likely worsen in the foreseeable future.

- Hundreds of millions of people in South Asia lack access to an electrical connection, regular and affordable electrical power, and potentially life-saving air-conditioning (AC). AC usage, credited with reducing heat deaths in the global North, is still less than 10% in India, Pakistan, Nepal and Bangladesh.
- Heat resilient housing is scarce (even though 'veranda' is a Bengali word). A significant proportion of the urban poor live in slums (informal settlements) lacking an indoor bathroom or toilet. Lack of a separate kitchen and reliance, even in densely populated urban settings, on smoky cooking fuels such as wood shavings and cow dung cakes exacerbate indoor heat. Water supply-related issues are common in low-income urban localities, especially during droughts and heatwaves.
- Economic inequality is compounded by social (including caste-based) and gender disparity – problems that disproportionately affect marginalized populations. Within the country, interstate migrant labourers are often at the mercy of contractors. Workers in labour-intensive small manufacturing industries and the booming construction sectors have little access to heat-protective equipment and limited capacity even to ask for it.
- Environmental health is poorly recognized by the government. India ranks 177 out of 180 in the global Environmental Performance Index, and runs a clear last in environmental health. The healthcare budget is among the lowest in the world, barely 1% of the country's GDP.

Despite this highly challenging starting point, significant opportunities to reduce excess heat-related deaths exist in South Asia. The climate change adaptation literature suggests some traditional behavioural and lifestyle heat adaptation practices that are affordable and could be implemented reasonably easily. These adaptations include wearing suitable light clothing, eating seasonally appropriate food, staying well hydrated, altered shift timing, taking short breaks to reduce heat stress, and an afternoon siesta. Studies in at-risk groups, such as slum dwellers and among construction workers, also provide promising leads. Even though many of these groups are exposed to extreme heat conditions, already exceeding international safety standards (Dutta *et al.*, 2015) simple measures such as seeking and acting on information about heat-related illnesses significantly reduces the odds of falling sick from heat (Tran *et al.*, 2012).

But some of these strategies rely on a rational and humane employer, able to ensure the health and safety of employees and contractors. For most parts of South Asia, deeply rooted culturally based forms of behaviour inhibit the many changes that are needed. These extend to many aspects of gender inequality, including with regard to women's clothing and sanitary needs, such as access to toilets and at times of menstruation.

While laws can and should be strengthened to reduce human exploitation, these still need to be obeyed and enforced. In many cases, such as workers in brick kilns, wages rely solely on output. For such workers, heat and health-protective strategies may reduce that output, endangering health via another pathway.

Current Indian government strategy includes the implementation of municipal heat adaptation plans in some cities. These plans typically include daily weather alerts, public awareness campaigns, and health-worker training (Knowlton *et al.*, 2014). The National Disaster Management Authority of India convenes a technical expert group and provides guidelines. In the future, with increasing income, more people are likely to have access to and be able to afford life-saving air-conditioning. This will require vastly increased amounts of electricity, which will exacerbate climate change unless this new energy is generated using carbon-neutral methods. Further research is required in heat-related morbidity and mortality to inform public health policy.

References

Azhar, G.S., Mavalankar, D., Nori-Sarma, A., Rajiva, A., Dutta, P. *et al.* (2014) Heat-related mortality in India: excess all-cause mortality associated with the 2010 Ahmedabad heat wave. *PLoS ONE* 9(3), e91831. DOI: 10.1371/journal.pone.0091831

Azhar, G., Saha, S., Ganguly, P., Mavalankar, D. and Madrigano, J. (2017) Heat wave vulnerability mapping for India. *International Journal of Environmental Research and Public Health* 14(4), 357. DOI: 10.3390/ijerph14040357

Dutta, P., Rajiva, A., Andhare, D., Azhar, G.S., Tiwari, A. *et al.* (2015) Perceived heat stress and health effects on construction workers. *Indian Journal of Occupational and Environmental Medicine* 19(3), 151–158. DOI: 10.4103/0019-5278.174002

Im, E.-S., Pal, J.S. and Eltahir, E.A.B. (2017) Deadly heat waves projected in the densely populated agricultural regions of South Asia. *Science Advances* 3, e1603322. DOI: 10.1126/sciadv.1603322

Knowlton, K., Kulkarni, S.P., Azhar, G.S., Mavalankar, D., Jaiswal, A. *et al.* (2014) Development and implementation of South Asia's first heat-health action plan in Ahmedabad (Gujarat, India). *International Journal of Environmental Research and Public Health* 11(4), 3473–3492. DOI: 10.3390/ijerph110403473

Mani, M., Bandyopadhyay, S., Chonabayashi, S., Markandya, A. and Mosier, T. (2018) *South Asia's Hotspots: The Impact of Temperature and Precipitation Changes on Living Standards*. The World Bank, Washington, DC. Available at: https://elibrary.worldbank.org/doi/abs/10.1596/978-1-4648-1155-5 (accessed 19 May 2022).

Mazdiyasni, O., AghaKouchak, A., Davis, S.J., Madadgar, S., Mehran, A. *et al.* (2017) Increasing probability of mortality during Indian heat waves. *Science Advances* 3(6), e1700066. DOI: 10.1126/sciadv.1700066

Nori-Sarma, A., Benmarhnia, T., Rajiva, A., Azhar, G.S., Gupta, P. *et al.* (2019). Advancing our understanding of heat wave criteria and associated health impacts to improve heat wave alerts in developing country settings. *International Journal of Environmental Research and Public Health* 16(12), 2089.

Tran, K.V., Azhar, G.S., Nair, R., Knowlton, K., Jaiswal, A. *et al.* (2012) A cross-sectional, randomized cluster sample survey of household vulnerability to extreme heat among slum dwellers in Ahmedabad, India. *International Journal of Environmental Research and Public Health* 9, 3298–3316. DOI: 10.3390/ijerph10062515

Xu, C., Kohler, T.A., Lenton, T.M., Svenning, J.C. and Scheffer, M. (2020) Future of the human climate niche. *Proceedings of the National Academy of Sciences (USA)* 117(21), 11350–11355. DOI: 10.1073/pnas.1910114117

32.ii Health Hazards in the Unorganized (Informal) Sectors of India

Subhashis Sahu and Moumita Sett Chatterjee
Ergonomics and Occupational Physiology Laboratory, Department of Physiology, University of Kalyani, West Bengal, India

Abstract

India, a developing nation at the juncture of economic transition, faces a health and security crisis among its informal (unorganized) sector, a group that constitute about 93% of the total Indian work force, and which contributed half of the gross domestic product in 2008. Internal migrants and those who are seasonally employed are especially at risk. Such workers are exposed to many different occupational hazards, without proper training and protective measures. Females are particularly vulnerable, as they perform a dual role both at paid work and in the household, with little or no assistance from their male counterparts; conditions that are worsened by unhygienic sanitary conditions and lack of potable water. Some recent initiatives such as worker enrolment (e-Shram card) and extension of healthcare facilities may help improve health.

32.ii.1 Introduction

A report by the International Labour Organization (ILO, 2018) showed that 1.3 billion people in the Asia-Pacific region, comprising about 68% of the employed population, work in the unorganized (informal) sectors Most of these workers lack social protection, rights at work and knowledge regarding occupational hazards. Most acquire their skills as children and young adults, through exposure and behavioural modelling on other family members who are already involved. They receive little, if any, professional training. Their low level of education and poor socio-economic status are key factors in understanding their vulnerability; as manual labourers they are poorly paid and can easily be exploited. Informal employment is overwhelmingly dominant in the agricultural workforce of rural areas in South Asia (over 99%). In urban settings, most informal employment is in the construction sector, often involving migrants from nearby villages or less urbanized states. The informal sector accounts for about 72% of the workforce in Pakistan, 87% in Bangladesh and 73% in Nepal.

According to a survey undertaken in 2010 (Indian National Sample Survey Organization, nd), about 437 million workers are employed in the unorganized sectors of India. Of these, about 246 million workers are in the agriculture sector, 44 million in the construction sector and the remainder in manufacturing and service. The Government of India's Ministry of Labour has categorized these workers into four groups, depending on type of occupation, nature of employment, and categories called 'especially distressed' and 'service'. Although the Indian Government enacted the Unorganized Workers' Social Security Act in 2008, most other countries of South Asia lack rules and laws governing this workforce. However, implementation of the Act by Indian States presents a dismal picture of overall delay

 Climate Change and Global Health: Primary, Secondary and Tertiary Effects, 2nd Edition (eds C.D. Butler and K. Higgs)
DOI: 10.1079/9781800620025.0032ii

in extending social benefits, especially for interstate migrants. Apart from poverty, informal workers experience physical, physiological, psychological and environmental stress. These stresses, combined with occupational health hazards, can reduce their work efficiency, in extreme cases leading to lost working days, deepening their poverty and vulnerability. This stress is even higher for female workers in unorganized sectors, who also have domestic responsibilities. Many live in mud dwellings near to the work site, experiencing unhygienic sanitary facilities, including a lack of potable water. Migrant workers (i.e. from other Indian states) are especially vulnerable. Men are generally larger than women and this may grant some protection against heat stress; however, there is no current evidence to suggest that heat stress vulnerability is independently affected by gender (Foster *et al.*, 2020).

32.ii.2 Occupational Hazards in Agriculture and Construction

Physical hazards exist in every construction site. These hazards include noise, heat and cold, radiation, vibration and excessive barometric pressure. These may seriously affect different organ systems, with irreversible consequences. The occupational health hazards in the construction industry can be broken down into six categories: asbestos exposure; silica exposure; excess noise-induced hearing loss; hand-arm vibration syndrome; musculoskeletal disorders; and dermatitis. Construction labourers work long hours and can be compelled to sustain awkward positions (both static and dynamic) for long periods, increasing the risk of musculoskeletal disorders (Chatterjee and Sahu, 2018). Moreover, strains and sprains arising due to slips, trips and falls are very common injuries among construction workers. Labourers perform repetitive, stressful work for long periods of time in a single work–rest cycle, generally with loads in excess of those recommended by the US National Institute for Occupational Safety and Health (Sahu *et al.*, 2010). Apart from the aforesaid causes, a study on brick-field workers (Fig. 32.ii.1) showed that they have biochemical evidence of oxidative stress, probably contributed to by less nutritious food, multiple forms of pollution (including to the air) and excess ambient temperature (Sett and Sahu, 2016). The number of occupationally related deaths in India was estimated in 2016 to be about 48,000, of which at least 24% occurred in the construction sector (Patel and Jha, 2016).

Among Indian farm workers, work-related health risks include to the lungs, hearing and skin, and various forms of cancer due to exposure to toxins, irritants and carcinogens. The burden of accidental deaths among agricultural workers in India is increasing due to the rising use of motorized machinery and electric power. Farmworkers are often barefoot, exposed to unhygienic conditions and biological hazards, including parasitic diseases and even snake bites. Workers are also vulnerable to injuries from sharp tools and machinery. Musculoskeletal disorders can arise due to poor postures adopted at work or from carrying, lifting, pulling or pushing heavy loads, as well as from repetitive work.

32.ii.3 Occupational Hazards: Other Unorganized Sectors

Workers in the informal sector are also employed in numerous other occupations, such as handloom weavers, chikon embroiderers, goldsmiths, jewellery workers, fishermen, rickshaw pullers, potters, sculptors and sawmill workers. All of these occupations carry distinctive work-related risks, such as respiratory conditions from wood processing (Adhikari *et al.*, 2015) and chronic alterations in posture among potters and sculptors, who are often required to adopt very awkward postures for prolonged periods (Sahu *et al.*, 2013a). Handloom weavers also experience increased rates of lower back pain (Durlov *et al.*, 2014). Rickshaws remain a popular mode of transporting people and merchandise in rural areas of South Asia. Pulling rickshaws places a heavy demand on the cardiovascular system (Sahu *et al.*, 2013b). Workplace fumes and gases are responsible for deteriorating lung function status in many goldsmiths (Sahu *et al.*, 2013c). Jewellery workers also face an increased risk of premature respiratory decline, thought to be induced by their occupational exposure to cadmium

Fig. 32.ii.1. Photograph of a typical unorganized brick-field in the construction sector in India. Photo: Moumita Sett Chatterjee.

(Moitra *et al.*, 2013). An estimated 20 million people are also engaged in fishing-related occupations in India. Exposure to adulterated fuel exhaust is common among deep-sea fishermen, who have a significantly higher chance of chronic cough, phlegm and wheeze, together with abnormal spirometry (Moitra *et al.*, 2015).

32.ii.4 Occupational Heat Exposure

Heat stress occurs when people cannot dissipate sufficient heat to their surroundings. Core body temperature must be maintained within 1°C (1.8°F) of normal (about 37°C or 98.6°F) in order to continue to function safely. High environmental temperatures, low wind speed, very high/low humidity and long working hours in open fields under sunlight, along with workload and restrictive clothing, increase the risk of heat stress. A shortage of potable water or shade greatly increases the risk of adverse health effects, as does poverty-related vulnerability. For example, the poor, paid by their production, often have to sacrifice either income or their health.

Heat-related illnesses include heatstroke, heat exhaustion, fainting, heat cramps and heat rashes (see Chapter 12). These conditions can be life threatening. In India, for every degree C rise in temperature, productivity has been estimated to decline by about 2%, in the construction sector (Sett and Sahu, 2014). Moreover, workers in protective clothes (including head gear) are at additional risk due to their generation of metabolic heat from hard physical labour, but with a reduced capacity to cool. An ILO report from July 2019 found that lost productivity due to heat stress in India, brought on by rapidly rising temperatures, will be equivalent to 34 million full-time jobs within 10 years, particularly affecting agriculture and construction workers (ILO, 2019).

References

Adhikari, A., Sahu, S., Bandopadhyay, A., Blanc, P. and Moitra, S. (2015) Fungal contamination of the respiratory tract and associated respiratory impairment among sawmill workers in India. *ERJ Open Research* 6, 00023-2015. DOI: 10.1183/23120541.00023-2015

Chatterjee, A. and Sahu, S. (2018) A physiological exploration on operational stance and occupational musculoskeletal problem manifestations amongst construction labourers of West Bengal, India. *Journal of Back and Musculoskeletal Rehabilitation* 31, 775–783. DOI: 10.3233/BMR-170935

Durlov, S., Chakrabarty, S., Chatterjee, A., Das, T., Dev, S. *et al.* (2014) Prevalence of low back pain among handloom weavers in West Bengal, India. *International Journal of Occupational and Environmental Health* 20, 333–339. DOI: 10.1179/2049396714Y.0000000082

Foster, J., Hodder, S.G., Lloyd, A.B. and Havenith, G. (2020) Individual responses to heat stress: Implications for hyperthermia and physical work capacity. *Frontiers in Physiology* 11, 541483. DOI: 10.3389/fphys.2020.541483

ILO (2018) https://www.ilo.org/asia/media-centre/news/WCMS_627585/lang--en/index.htm accessed 3 July, 2022

ILO (2019) *Working on a Warmer Planet.The impact of heat stress on labour productivity and decent work.* International Labour Organization, Geneva. Available at: https://www.ilo.org/wcmsp5/groups/public/---dgreports/---dcomm/---publ/documents/publications/wcms_711919.pdf (accessed 26 January 2024).

Indian National Sample Survey Organization (nd) *Unorganised Sector*. Available at: https://vikaspedia.in/social-welfare/unorganised-sector-1/unorganized-sector-informal-sector (accessed 24 June, 2021).

Moitra, S., Blanc, P. and Sahu, S. (2013) Adverse respiratory effects associated with cadmium exposure in small-scale jewellery workshops in India. *Thorax* 21, 565–570. DOI: 10.1136/thoraxjnl-2012-203029

Moitra, S., Maity, S., Haldar, P., Pandit, A. and Sahu., S. (2015) Trawler fuel exhaust and respiratory impairments: a cross-sectional pilot study among Indian fishermen working in informal sectors. *International Journal of Occupational and Environmental Health* 21, 185–191. DOI: 10.1179/2049396714Y.0000000057

Patel, D.A. and Jha, K.N. (2016) An estimate of fatal accidents in Indian construction. In: Chan, P.W. and Neilson, C.J. (eds) *Proceedings of the 32nd Annual ARCOM Conference, 5–7 September, 2016.* Association of Researchers in Construction Management, Manchester, UK, pp. 539–548.

Sahu, S., Chattopadhyay, S., Basu, K. and Paul, G. (2010) The ergonomic evaluation of work-related musculoskeletal disorders among construction labourers working in unorganized sectors in West Bengal, India. *Journal of Human Ergology* 39, 99–109. DOI: 10.11183/jhe.39.99

Sahu, S., Moitra, S., Maity, S., Pandit, A. and Roy, B. (2013a) A comparative ergonomics postural assessment of potters and sculptors in the unorganized sector in West Bengal, India. *International Journal of Occupational Safety and Ergonomics* 19, 455–462. DOI: 10.1080/10803548.2013.11077001

Sahu, S., Maity, S., Moitra, S., Sett, M. and Haldar, P. (2013b) Cardiovascular load during summer work of two age groups of van-rickshaw pullers in West Bengal, India. *International Journal of Occupational Safety and Ergonomics* 19, 657–665. DOI: 10.1080/10803548.2013.11077019

Sahu, S., Roy, B. and Moitra, S. (2013c) Assessment of the lung function status of the goldsmiths working in an unorganized sector of India. *Lung India* 30, 33–37. DOI: 10.4103/0970-2113.106131

Sett, M. and Sahu, S. (2014) Effects of occupational heat exposure on female brick workers in West Bengal, India. *Global Health Action* 7, 21923. DOI: 10.3402/gha.v7.21923

Sett, M. and Sahu, S. (2016) Anthropometric characteristics and evaluation of nutritional status amongst female brick field workers of the unorganized sectors of West Bengal, India. *Homo: Journal of Comparative Human Biology* 67, 235–244. DOI: 10.1016/j.jchb.2015.10.003

33 Climate Change and Health in China

Yuming Guo, Ke Ju, Pei Yu, Yao Wu, Bo Wen and Shanshan Li
Climate, Air Quality Research Unit, School of Public Health and Preventive Medicine, Monash University, Melbourne, Australia

Abstract

In recent decades, China has achieved significant advancements in healthcare, driven by rapid economic and technological growth, as well as increased investments in the health sector. However, the challenges posed by climate change will undoubtedly impede China's ability to sustain high-quality healthcare services and further improve the population's health. The consequences of climate change will manifest in diverse ways, affecting temperature patterns, water resources, food production and the spread of infectious diseases. These impacts will directly or indirectly harm the health of the population. This chapter aims to summarize the relationship between climate change and health in China, exploring various aspects of this association in dedicated sections.

33.1 Introduction

The footprint of anthropogenic climate change can already be seen in every corner of the planet. Global warming may bring some regional benefits, such as fewer deaths in wintertime in lands with temperate climates, and increased food yields in areas with currently low agricultural productivity. However, the overall health effects of a changing climate are overwhelmingly negative. Extreme and unpredictable weather events (e.g. droughts, floods, heatwaves and bushfires), water and food insecurity, and an altered distribution of infectious diseases, in part due to climate change, are threatening the health and well-being of populations across the globe. As in other regions, China has recently experienced noticeable climatic changes. These may be a serious challenge for individuals, communities, governments, businesses and industry, in addition to the environment. China has many characteristics that reduce (but will not eliminate) the adverse impacts on health from climate change. These include its rapid social and economic development, its increased investments in healthcare, and a basic medical security system that now covers the entire Chinese population, including minorities and the 'floating' population of workers from rural areas, who do much low-skilled work in cities.

The major health concerns arising from climate change in China include: (i) increased morbidity and mortality from higher ambient heat exposure and intense heatwaves; (ii) higher risk from extreme precipitation and floods; (iii) reduced water sources and impaired food production in certain regions; and (iv) higher risks of food-, water- and vector-borne diseases.

The most recent edition of China's *Blue Book on Climate Change*, published by the China Meteorological Administration (CMA), found that, between 1961 and 1990, extreme precipitation events in China increased, as did extreme heat, from the mid-1990s[1] (CMA Climate Change

 Climate Change and Global Health: Primary, Secondary and Tertiary Effects, 2nd Edition (eds C.D. Butler and K. Higgs)
DOI: 10.1079/9781800620025.0033

Centre, 2021). The complexity of social and climatic systems in China will generate impacts with considerable spatial heterogeneity, as is well summarized by Ding *et al.* (2021) and Feng and Chao (2020).

33.2 Climate Change in China

Comprehensive observations show that the global warming trend, due to human causes, has continued in recent decades. The Fifth Assessment Report of the Intergovernmental Panel on Climate Change (IPCC) Working Group I predicted a further increase in global extreme warm events, with more frequent and longer-lasting heatwaves (Stocker *et al.*, 2014). According to a joint assessment released by the US National Aeronautics and Space Administration (NASA), Berkeley Earth, the UK Meteorology Office and the National Oceanic and Atmospheric Administration (NOAA) (Voosen, 2021), Earth's global average surface temperature in 2020 tied with 2016 as the hottest year on record, consistent with the ongoing dramatic heating trend. The global average temperature in 2020 was about 1.25°C above pre-industrial levels and is on the verge of breaching climate targets set by the Paris Climate Agreement to hold warming to 1.5°C by 2035 and 2°C by 2065.

According to results released in China's *Blue Book on Climate Change 2020*, the previous 5 years (2015–2019) were the five warmest years since complete meteorological observations were first recorded globally. The average annual surface temperature in China has increased by 0.24°C per decade, exceeding the global average surface warming rate during the same period. Since the mid-1990s, extreme heat events in China have increased significantly, and in 2019, the daily maximum temperature reached or exceeded historical extremes at 64 stations, including Yuanjiang (43.1°C) in Yunnan. There are obvious differences in temperature changes between regions in China, with the rate of warming significantly greater in the north than in the south, and greater in the west than in the east. The greatest rate of warming was recorded in Qinghai–Tibet.

33.3 Climate Change, Environment and Health

Temperature

Heatwave hazards, characterized by intense and persistent high temperatures during summertime, are occurring with increasing frequency, and have been known to produce significant effects on human health, regional economies and the natural world. People have differing capacities to withstand high temperature in different regions or climate zones and, because of people's long-term adaptability to the local climate, an objective and universal heatwave definition does not exist. It is believed that heatwaves will have stronger intensities, higher frequencies and longer durations along with background global warming. In summer 2013, the CMA reported that abnormally high temperatures were observed in 19 provinces in China, breaking weather records and generating an unprecedented heatwave across east-central China (Sun *et al.*, 2014). Dian-Xiu and colleagues analysed the temporal and spatial characteristics of heatwaves in China from 1961 to 2010 using 753 *in situ* stations, and summarized that the heatwave frequency, duration and intensity showed a general increasing trend (Dian-Xiu *et al.*, 2014). A recent study modelled the changes of heatwave and population exposure in China at 1.5°C and 2.0°C warming under Representative Concentration Pathway (RCP) 4.5 and RCP 8.5 scenarios to reach the conclusion that heatwaves are projected to be more extreme and more frequent as well as longer-lasting with significant spatial variation (Li *et al.*, 2019). Population exposure will be 1.07–1.10 billion and 1.153 billion for 1.5°C and 2.0°C warming climates, respectively (Li *et al.*, 2019). Similarly, another projection found that the annual mean intensity of high temperature during 1961–2005 showed an increasing trend of 0.07°C/10 years. The intensity of high temperature was projected to increase continuously under RCP 4.5 by 1.9°C at a global warming of 2.0°C compared with reference period (1961–2005) (Fig. 33.1) (Wang *et al.*, 2019).

Heatwaves can directly affect human health by dehydration, overheating, heat exhaustion and heatstroke. Moreover, heatwaves can place additional burdens on the cardiovascular system,

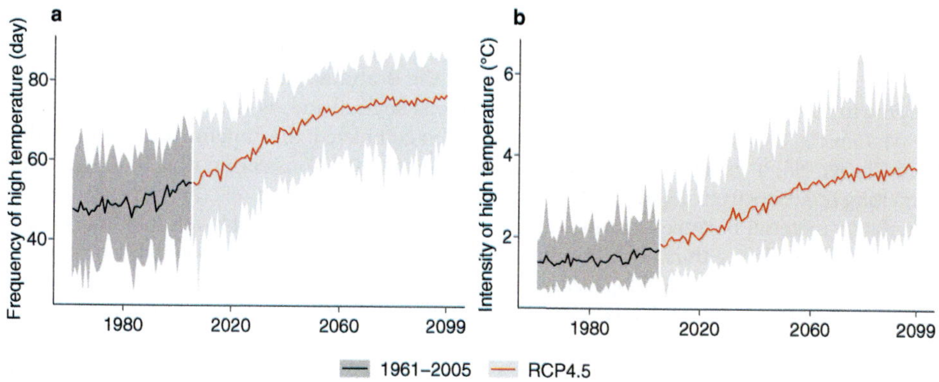

Fig. 33.1. Frequency and intensity of high temperature in Chinese metropolises for 1961–2099. Curves and shadows denote ensemble mean and range, respectively. Frequency of high temperature (days) (a) and intensity of high temperature (°C) (b) in Chinese metropolises for the reference period 1961–2005 (darker grey) and for the future period 2005–2099 with RCP 4.5 (light grey). Adapted from Wang *et al*., 2019, reproduced under a Creative Commons Attribution 4.0 International License (CC BY 4.0), https://creativecommons.org/licenses/by/4.0/

leading to cardiovascular stress and other diseases. Heatwaves have been linked with excess morbidity around the world, including cardiovascular diseases, respiratory diseases and diabetes (Moon, 2021). The increased risk of mortality was also associated with heat exposure. Specifically, higher thresholds (e.g. 95th, 97.5th, 99th) and a longer duration of heatwave days could lead to a greater health burden (Guo *et al*., 2017). It was recognized that about 26,800 deaths were associated with heatwaves in 2019, which has risen by four times as compared with 1990 (Cai *et al*., 2021). Heatwaves are likely to exacerbate potential medical conditions in patients, mainly affecting the elderly, young children and those who work outdoors. Numerous studies have analysed the effects of heatwaves on the health of susceptible populations. One study found that the effects of heatwaves increase with the intensity and duration of the heatwave and are stronger in some patients with diseases that are more sensitive to heatwaves, such as ischaemic heart disease, stroke and respiratory disease (Yang *et al*., 2019). Heatwaves have the greatest impact on women, the elderly and those with lower education levels. Cities with higher particulate matter with a diameter of 2.5 μm or less ($PM_{2.5}$) concentrations have a higher risk of death during heatwaves (Zhou *et al*., 2023). Studies assessing heat-related excess mortality in China under different climate change scenarios show that heat-related excess mortality is projected to increase from 1.9% in the 2010s to 2.4% in 2030 and 5.5% in 2090. Population ageing will increase future heat-related excess mortality by a factor of 2.3 to 5.8, especially in north-eastern China, where the ageing trend is particularly severe (Yang *et al*., 2021). Importantly, many deaths associated with heatwaves usually occur several days after the condition has worsened and might not be properly identified as such. Therefore, the true number of deaths associated with heatwaves is most likely to be underestimated.

To protect human health from heatwaves, an effective heatwave warning system and tips for self-protection should be promoted by governments. During the period of heatwaves, individuals should reduce outdoor activities, stay indoors with air-conditioners, and keep hydrated by drinking fluids. Children and the elderly should be given special care, as they are particularly vulnerable during heatwaves.

Glaciers and fresh water

China has 46,377 glaciers covering an area of 59,425 km^2 (Piao *et al*., 2010). More than 80% of the glaciers in western China are currently in retreat due to the current significant warming.

The evidence indicates that the past 50 years have been the warmest period on the Tibetan Plateau in at least 500 years (Ge *et al.*, 2010). According to the results released by China's *Blue Book on Climate Change 2020*, Urumqi River Source Glacier No. 1 experienced two accelerated melting processes from 1960 to 2017, with a cumulative material balance of minus 19.77 m (i.e. an average glacier thickness thinning of 19.77 m water equivalent, assuming a constant area). Global average sea level has risen 8–9 inches (20–23 cm) since 1880, and the rate of global sea level rise is accelerating: it has more than doubled from 0.06 inches (0.15 cm) per year throughout most of the 20th century to 0.14 inches (0.36 cm) per year from 2006 to 2015 (Lindsey, 2021). Considering that 10% of the global population lives in coastal areas below 10 m above sea level, rising sea level will undoubtedly pose a great threat to human life and health. If nothing is done, 0.5–5% of the world's population will be flooded each year after 2100. Even under RCP 2.6, half of Shanghai, one of the largest cities in China, and 46% of its seawalls and levees are expected to be inundated (Wang *et al.*, 2012).

In the context of climate change, fresh water is a rare and precious resource. China is reportedly planning to build 17 hydropower plants on the main stream of the Yarlung Tsangpo (or Yarlung Zangbo) river (the upper stream of the Brahmaputra river, in the Tibet Autonomous Region) and, as of 2020, two have already been completed: the Zangmu and Gacha hydropower plants. Considering that India and Bangladesh downstream of the Yarlung Zangbo river may be considerably affected, one cannot help but worry about the possibility of a 'water war', including the sharing of hydrological data such as water level and discharge between the countries through which the Yarlung Zangbo river flows, which may affect the preparedness of the countries in the region for flood control.

Precipitation and flood

As the most frequent natural disaster in China, floods are posing an increasing threat to the economy, society and human health. It was estimated that the area of land damaged by flooding per year has shown an increasing trend from 1950 to 2018, which may imply that floods are affecting China in a more profound way (Fig. 33.2) (Guo *et al.*, 2020). Since the beginning of the 21st century, average annual precipitation fluctuations have increased in northern, southern

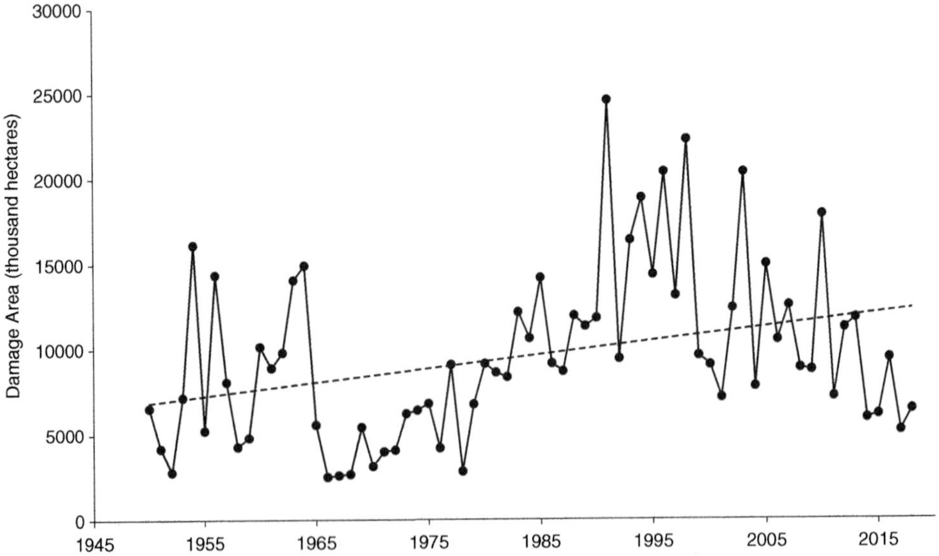

Fig. 33.2. Damage area caused by floods in China from 1950 to 2016. Data source: *The Bulletin of Floods and Droughts in China in 2018* (available at: http://www.mwr.gov.cn/sj/tjgb/zgshzhgb/201912/t20191210_1374387.html, accessed 12 August 2023).

and north-western China, while interannual fluctuations in precipitation have increased in north-eastern and eastern China. In 2020, southern China experienced the highest national average precipitation intensity and amount and the longest duration since 1961. The resulting floods destroyed nearly 400,000 houses, damaged 5 million hectares of farmland, killed more than 200 people and affected more than 60 million people (Guo *et al.*, 2020). On the other hand, the estimated loss in the sustainable shared socio-economic pathways (SSP1, a scenario of projected socio-economic global changes up to 2100 to study the future impacts of a changing climate) increases tenfold in comparison with the reference period 1986–2005 and nearly threefold relative to the interval 2006–2015 (Su *et al.*, 2018). China's Henan Province was attacked by a series of devastating floods in July 2021, which were believed to be caused by unusually heavy rainfall and to be the sign that extreme weather caused by climate change is increasing in China. It is generally accepted that extreme precipitation was responsible for most floods in China, and the unusual increase in precipitation caused by climate change may severely alter the risk of floods. Apart from this, severe drought affected by precipitation and temperature may also change drainage basin conditions and the soil status, which could eventually lead to floods. Moreover, irregular periodic variation in winds and sea surface temperatures, such as El Niño-Southern Oscillation, Pacific Decadal Oscillation and abnormal cyclones, may also influence precipitation and cause severe floods.

Floods have the potential to cause substantial health losses, including mortality, injuries, mental health problems, non-communicable diseases, vector-borne diseases and water-borne diseases. Floods can lead to direct deaths from drowning and injuries from secondary disaster events, such as building collapse. Moreover, floods can also be related to communicable disease outbreaks and infections. For regions with poor hygiene and public health conditions, bacteria in floodwater may lead to infection and contaminate drinking water. For instance, diarrhoea is a health outcome sensitive to floods, especially in areas close to water bodies and among vulnerable populations. Flooding in China's Yangtze River basin from 2013 to 2017 left 274,621 infectious diarrhoeal cases in Anhui Province, China (Zhang *et al.*, 2019). Healthcare professionals and medical researchers identify the problem that the health service or social support systems required by those patients may be interrupted (Ryan *et al.*, 2015). Thus, effective measures must be taken to mitigate and prevent the adverse impacts of floods. Generally, flood control levees, reservoirs and floodways should be constructed to balance the flow of river and discharge excess flood water. For individuals and governments, the early warning system and adequate preparations are necessary to prevent flood hazards.

33.4 Communicable Disease

Another kind of threatened disease affected by climate change is infectious disease. Infectious diseases are one of the major challenges facing human society, and a range of consequences of climate change, such as temperature changes, water pollution and air pollution, have contributed to many infectious diseases, such as dengue fever and malaria. Climate and environmental factors, such as temperature, humidity and air pollution, may influence the spread, transmission and severity of the infection (Fattorini and Regoli, 2020).

Arboviruses are on the rise; one reason is the changing climatic environment. In China, the number of dengue fever cases has increased in recent years, with 59,334 dengue cases reported from 2005 to 2015 (Sun *et al.*, 2017). According to the extent of climate change, the dengue fever risk area in China will continue to expand until 2050 (Jing-Chun and Qi-Yong, 2019). Some regions, like some cities in Hainan Province, may have the temperature conditions that allow dengue fever to be endemic all year round in the future (Li *et al.*, 2018). A high percentage of increases in malaria is also projected in some counties of the north-west, north and north-east, including the northern tip of north-east China if there is no successful global policy to mitigate climate change (Hundessa *et al.*, 2018).

Epidemics following floods are also of high concern, as the immunity of populations would decrease during floods. People are more sensitive to respiratory infectious diseases after floods, and intestinal infectious disease is common after floods as vulnerable people may drink infected water and food; possible infections include infectious diarrhoea, bacillary dysentery, and even hand-foot-and-mouth disease in children (Lin *et al.*, 2019). During the floods in Henan, Hunan and Guangxi provinces, a significantly increased risk of dysentery was observed (Ni *et al.*, 2014; Liu *et al.*, 2016; Liu *et al.*, 2017; Hu *et al.*, 2018), while higher risk of typhoid fever was observed in Yongzhou city of Hunan province (Liu *et al.*, 2018).

33.5 Food

Today, in the 21st century, the global estimate suggested that approximately 720–811 million people continued to endure hunger in the year 2020. Food production systems are particularly vulnerable to adverse impacts of climate change, which is one of the key factors affecting food security in China. Climate change affects food security in a variety of ways, including by influencing the frequency and severity of droughts, causing sea level rise to inundate farmland, and releasing more greenhouse gases that harm crops. Drought is one of the most significant causes of reduced food production. Climate change-driven seasonal drought could, by 2030, lead to substantial losses of nearly 8% in yields of three main crops in China, with a projected drop of nearly 20% of total maize production, followed by wheat with a 4% decline, and rice at 1.5%; this was the finding of an equilibrium analysis under alternative impact scenarios and adaptation scenarios which assessed the impact of climate change on grain production and total land area sown in China (Zhou *et al.*, 2017). Since 2001, the 42 highest-ranked agricultural countries have lost an estimated US$932 billion due to drought (Kogan *et al.*, 2019), which, in Africa, affected millions of people and killed 500,000 between 1981 and 2010 (Masih *et al.*, 2014). The estimated loss under the SSP1 at the 1.5°C warming level increases tenfold in comparison with the reference period 1986–2005 and nearly threefold relative to the interval 2006–2015. In addition, extreme rainfall can pose a major threat to food production. In China, it is projected that extreme rainfall has caused a reduction in rice yields of approximately 7.6 ± 0.9%. Furthermore, by the end of the century, an additional yield reduction of about 8% is anticipated (Fu *et al.*, 2023).

33.6 Future Development

The 2020 *Lancet* Countdown China report gave comprehensive policy recommendations, including five key points (Cai *et al.*, 2021). The suggestion that was the most closely related to public health was to support health and climate change-related research, raise public awareness, motivate media and academia and also speed the Government of China's update to the Healthy China Action Plan (2019–2030) (Cai *et al.*, 2021).

Since even moderate change in global warming could make profound changes in extreme weather and disasters, the mitigation of climate change by cutting carbon emissions could yield much health benefit. To achieve the expected goal, efforts from professionals, policy makers and people in general are needed. Professionals are responsible for educating and encouraging the public on how much we could gain from personal actions. They can also provide evidence for policy makers to help to take regulatory actions and green community building. The policy makers are expected to adopt achievable strategies, recognize the urgency of bold climate action and support environmentally friendly technology improvements. Some of the responsibility rests with each person themselves. However, steps can be taken by governments, such as fostering public transportation, renewable energy and the wise use of energy. These actions would not only reduce the health impacts of climate change, but also yield great reduction of pollution-related health risks.

33.7 Conclusion

Overall, there is now ample evidence that climate change has affected and will continue to affect the socio-economic development and health of populations in countries around the world, including in China. China has taken a series of measures to address the climate crisis, but action cannot stop there. China has huge areas with diverse geographical and climatic variations, with different socio-economical levels; therefore, strategies of adaptation and solutions for mitigating climate change should be tailored to meet the specific requirements of each region. Cross-disciplinary and international collaboration are also an effective way to exchange scientific ideas and to achieve mutual benefits. We need to take more actions to greatly reduce greenhouse gas emissions and significantly increase the adaptive capacity of the global population to the risks of climate change, to ensure that we and our future generations can avoid bearing the worst consequences of the climate crisis.

Note

[1] As summarized by the *Global Times*, available at: https://www.globaltimes.cn/page/202108/1230607.shtml (accessed 28 July 2023).

References

Cai, W., Zhang, C., Suen, H.P., Ai, S., Bai, Y. *et al.* (2021) The 2020 China report of the Lancet Countdown on health and climate change. *Lancet Public Health* 6, e64–e81. DOI: 10.1016/s2468-2667(20)30256-5

CMA Climate Change Centre (2021) *Blue Book on Climate Change in China*. Science Press, Beijing, China.

Dian-Xiu, Y., Ji-Fu, Y., Zheng-Hong, C., You-Fei, Z. and Rong-Jun, W. (2014) Spatial and temporal variations of heat waves in China from 1961 to 2010. *Advances in Climate Change Research* 5, 66–73. DOI: 10.3724/SP.J.1248.2014.066

Ding, Y.-J., Li, C.-Y., Wang, X., Wang, Y., Wang, S.-X. *et al.* (2021) An overview of climate change impacts on the society in China. *Advances in Climate Change Research* 12, 210–223. DOI: 10.1016/j.accre.2021.03.002

Fattorini, D. and Regoli, F. (2020) Role of the chronic air pollution levels in the Covid-19 outbreak risk in Italy. *Environmental Pollution* 264, 114732. DOI: 10.1016/j.envpol.2020.114732

Feng, A. and Chao, Q. (2020) An overview of assessment methods and analysis for climate change risk in China. *Physics and Chemistry of the Earth, Parts A/B/C* 117, 102861. DOI: 10.1016/j.pce.2020.102861

Fu, J., Jian, Y., Wang, X., Li, L., Ciais, P., Zscheischler, J. *et al.* (2023) Extreme rainfall reduces one-twelfth of China's rice yield over the last two decades. *Nature Food* 4, 416–426. DOI: 10.1038/s43016-023-00753-6

Ge, Q.S., Zheng, J.Y., Hao, Z.X., Shao, X.M., Wang, W.C. *et al.* (2010) Temperature variation through 2000 years in China: An uncertainty analysis of reconstruction and regional difference. *Geophysical Research Letters* 37, L03703. DOI: 10.1029/2009GL041281

Guo, Y., Gasparrini, A., Armstrong, B.G., Tawatsupa, B., Tobias, A. *et al.* (2017) Heat wave and mortality: a multicountry, multicommunity study. *Environmental Health Perspectives* 125, 087006. DOI: 10.1289/EHP1026

Guo, Y., Wu, Y., Wen, B., Huang, W., Ju, K. *et al.* (2020) Floods in China, COVID-19, and climate change. *Lancet Planetary Health* 4, e443–e444. DOI: 10.1016/S2542-5196(20)30203-5

Hu, X., Ding, G., Zhang, Y., Liu, Q., Jiang, B. *et al.* (2018) Assessment on the burden of bacillary dysentery associated with floods during 2005–2009 in Zhengzhou City, China, using a time-series analysis. *Journal of Infection and Public Health* 11, 500–506. DOI:

Hundessa, S., Williams, G., Li, S., Liu, L., Cao, W. *et al.* (2018) Projecting potential spatial and temporal changes in the distribution of *Plasmodium vivax* and *Plasmodium falciparum* malaria in China with climate change. *Science of the Total Environment* 627, 1285–1293. DOI: 10.1016/j.scitotenv.2018.01.300

Jing-Chun, F. and Qi-Yong, L. (2019) Potential impacts of climate change on dengue fever distribution using RCP scenarios in China. *Advances in Climate Change Research* 10, 1–8. DOI: 10.1016/j.accre.2019.03.006

Kogan, F., Guo, W. and Yang, W. (2019) Drought and food security prediction from NOAA new generation of operational satellites. *Geomatics, Natural Hazards and Risk* 10, 651–666. DOI: 10.1080/19475705.2018.1541257

Li, C., Lu, Y., Liu, J. and Wu, X. (2018) Climate change and dengue fever transmission in China: Evidences and challenges. *Science of the Total Environment* 622–623, 493–501. DOI: 10.1016/j.scitotenv.2017.11.326

Li, Z., Guo, X., Yang, Y., Hong, Y., Wang, Z. *et al.* (2019) Heatwave trends and the population exposure over China in the 21st century as well as under 1.5°C and 2.0°C global warmer future scenarios. *Sustainability* 11, 3318. DOI: 10.3390/su11123318

Lin, H., Ma, W. and Liu, Q. (2019) *Ambient Temperature and Health in China*. Springer, Singapore.

Lindsey, R. (2021) *Climate change: Global sea level*. National Oceanic and Atmospheric Administration (NOAA) Climate.gov website, US Department of Commerce. Available at: https://www.climate.gov/news-features/understanding-climate/climate-change-global-sea-level (accessed 14 August 2020).

Liu, X., Liu, Z., Zhang, Y. and Jiang, B. (2017) The effects of floods on the incidence of bacillary dysentery in Baise (Guangxi Province, China) from 2004 to 2012. *International Journal of Environmental Research and Public Health* 14, 179. DOI: 10.3390/ijerph14020179

Liu, Z.-D., Li, J., Zhang, Y., Ding, G.-Y., Xu, X. *et al.* (2016) Distributed lag effects and vulnerable groups of floods on bacillary dysentery in Huaihua, China. *Scientific Reports* 6, 1–8. DOI: 10.1038/srep29456

Liu, Z., Lao, J., Zhang, Y., Liu, Y., Zhang, J. *et al.* (2018) Association between floods and typhoid fever in Yongzhou, China: effects and vulnerable groups. *Environmental Research* 167, 718–724. DOI: 10.1016/j.envres.2018.08.030

Masih, I., Maskey, S., Mussá, F. and Trambauer, P. (2014) A review of droughts on the African continent: a geospatial and long-term perspective. *Hydrology and Earth System Sciences* 18, 3635–3649. DOI: 10.5194/hess-18-3635-2014

Moon, J. (2021) The effect of the heatwave on the morbidity and mortality of diabetes patients; a meta-analysis for the era of the climate crisis. *Environmental Research* 195, 110762. DOI: 10.1016/j.envres.2021.110762

Ni, W., Ding, G., Li, Y., Li, H., Liu, Q. *et al.* (2014) Effects of the floods on dysentery in north central region of Henan Province, China from 2004 to 2009. *Journal of Infection* 69, 430–439. DOI: 10.1016/j.jinf.2014.05.016

Piao, S., Ciais, P., Huang, Y., Shen, Z., Peng, S. *et al.* (2010) The impacts of climate change on water resources and agriculture in China. *Nature* 467, 43–51. DOI: 10.1038/nature09364

Ryan, B., Franklin, R.C., Burkle, F.M. Jr, Aitken, P., Smith, E. *et al.* (2015) Identifying and describing the impact of cyclone, storm and flood related disasters on treatment management, care and exacerbations of non-communicable diseases and the implications for Public Health. *PLoS Currents Disasters* 1. DOI: 10.1371/currents.dis.62e9286d152de04799644dcca47d9288

Stocker, T.F., Qin, D., Plattner, G.-K., Tignor, M., Allen, S.K. *et al.* (eds) (2014) *Climate Change 2013: The Physical Science Basis: Working Group I Contribution to the Fifth Assessment Report of the Intergovernmental Panel on Climate Change*. Cambridge University Press, Cambridge, UK.

Su, B., Huang, J., Fischer, T., Wang, Y., Kundzewicz, Z.W. *et al.* (2018) Drought losses in China might double between the 1.5 C and 2.0 C warming. *Proceedings of the National Academy of Sciences (USA)* 115, 10600–10605. DOI: 10.1073/pnas.1802129115

Sun, J., Lu, L., Wu, H., Yang, J., Xu, L. *et al.* (2017) Epidemiological trends of dengue in mainland China, 2005–2015. *International Journal of Infectious Diseases* 57, 86–91. DOI: 10.1016/j.ijid.2017.02.007

Sun, Y., Zhang, X., Zwiers, F.W., Song, L., Wan, H. *et al.* (2014) Rapid increase in the risk of extreme summer heat in Eastern China. *Nature Climate Change* 4, 1082–1085. DOI: 10.1038/nclimate2410

Voosen, P. (2021) Global temperatures in 2020 tied record highs. *Science* 371, 334–335. DOI: 10.1126/science.371.6527.334

Wang, J., Gao, W., Xu, S. and Yu, L. (2012) Evaluation of the combined risk of sea level rise, land subsidence, and storm surges on the coastal areas of Shanghai, China. *Climatic Change* 115, 537–558. DOI: 10.1007/s10584-012-0468-7

Wang, Y., Wang, A., Zhai, J., Tao, H., Jiang, T. *et al.* (2019) Tens of thousands additional deaths annually in cities of China between 1.5 degrees C and 2.0 degrees C warming. *Nature Communications* 10, 3376. DOI: 10.1038/s41467-019-11283-w

Yang, J., Yin, P., Sun, J., Wang, B., Zhou, M. *et al.* (2019) Heatwave and mortality in 31 major Chinese cities: definition, vulnerability and implications. *Science of the Total Environment* 649, 695–702. DOI: 0.1016/j.scitotenv.2018.08.332

Yang, J., Zhou, M., Ren, Z., Li, M., Wang, B. *et al.* (2021) Projecting heat-related excess mortality under climate change scenarios in China. *Nature Communications* 12, 1–11. DOI: 10.1038/s41467-021-21305-1

Zhang, N., Song, D., Zhang, J., Liao, W., Miao, K. *et al.* (2019) The impact of the 2016 flood event in Anhui Province, China on infectious diarrhea disease: An interrupted time-series study. *Environment International* 127, 801–809. DOI: 10.1016/j.envint.2019.03.063

Zhou, L., Wang, Y., Wang, Q., Ding, Z., Jin, H. *et al.* (2023) The interactive effects of extreme temperatures and PM2.5 pollution on mortalities in Jiangsu Province, China. *Scientific Reports* 13, 9479. DOI: 10.1038/s41598-023-36635-x

Zhou, S., Zhou, W., Lin, G., Chen, J., Jiang, T. *et al.* (2017) Adapting to climate change: scenario analysis of grain production in China. *China Agricultural Economic Review* 9, 643–659. DOI: 10.1108/CAER-10-2016-0173

34 Climate Change and Health in Indonesia

Colin D. Butler[1] **and Budi Haryanto**[2,3]
[1]*National Centre for Epidemiology and Population Health, The Australian National University, Canberra, Australia;* [2]*Research Centre for Climate Change, Jakarta, Universitas Indonesia*

Abstract
The population health status of Indonesia, the world's fourth most populous nation, has improved in recent decades. However, these improvements are vulnerable to climate change. Factors that increase this vulnerability include its poverty, its densely populated (still rapidly growing) urban areas, population growth, and vector-borne and other climate-sensitive diseases. Indonesia is also the world's largest coal exporter and a hotspot of carbon-storing tropical forests and biodiversity, much of which is considered vulnerable. It is essential, for its population's future well-being, that the global North decreases its emissions as fast as possible.

34.1 Indonesia: Geography, History, Demography and a Gradual Ascent from Extreme Poverty

Indonesia, the world's fourth most populous country, is by far the largest nation in South-East Asia. Its population, estimated as about 280 million in 2022, is still growing by about 1% per annum, equivalent to a total fertility rate of 2.3, a little over replacement (UNFPA, 2022).[1] Indonesia is the world's largest archipelago, comprising almost 18,000 islands, many of them volcanic. It has the longest coastline of any nation, and about 62% of its total area is marine (Haryanto, 2016). Its 300 ethnic groups speak about 750 languages, a diversity that poses a profound challenge for health systems (Agustina *et al.*, 2019). More than half of the nation's population and even more of its poor (57%) are crowded into the immensely fertile island of Java, the most densely populated large island in the world (BPS, 2016).

The total land area of Indonesia is about 1.8 million km^2, giving it a population density of 155 people per km^2. Java comprises only about 7% of this land area; its population density is almost 1200 people per km^2. Population density of the nation's current capital city, Jakarta, is even higher, at over 4000 people per km^2. About 25% of Indonesians are younger than 15; 6.5% are over 65. About 56% of its population are urbanized, very close to the global average. However, this fraction has increased from 42% in 2000, i.e. at a rate slightly higher than the world as a whole (World Bank, 2022).

Indonesia fought for and achieved independence from the Netherlands in 1945, at the end of World War II.[2] Its founding president, Sukarno, was charismatic and a leading figure in the then emerging alliance of formerly colonized nations, which hosted the Asian–African conference in Bandung, Java, in 1955 (Berger, 2004). At the time, most members of this informal 'Third World' alliance were courted by capitalist

 Climate Change and Global Health: Primary, Secondary and Tertiary Effects, 2nd Edition (eds C.D. Butler and K. Higgs)
DOI: 10.1079/9781800620025.0034

and communist forces during the Cold War. The poverty and inequality in Indonesia, together with support from the international communist movement, led to a strong domestic communist influence. In 1967 Sukarno was replaced by an authoritarian, strongly anti-communist, pro-Western government, led by General Suharto, who remained in power until 1998 (Berger, 2004). As many as 500,000 Indonesians, suspected as communist sympathizers, were killed in the early years of this regime (McGregor, 2013). However, following street protests, with a background of a declining national economy, the Asian financial crisis and the strong El Niño event of 1997/98, Indonesia's government has been democratic (Slater, 2023).

Literacy, education and development indices

For several decades a series of Indonesian governments have invested heavily in promoting literacy. Adult literacy is now reported as almost 93%, slightly higher in males (95.7% compared with 89.9% for women and girls). Many Indonesians have access to social media, especially Facebook. However, a study of eighth-grade students in Indonesia (published in 2012) was reported to have found that only 43% had 'some' understanding of whole numbers, decimals, operations and basic graphs. In contrast, 99% of Singaporean eighth-grade students have this knowledge (Kurniawati *et al.*, 2018). This disturbing finding about the effectiveness of education in Indonesia is consistent with that from many other low-income settings, such as in India (Pritchett, 2013).

The Human Development Index (HDI) of the United Nations Development Programme (UNDP) incorporates data for life expectancy, education (quantity but not quality) and income. UNDP (2020) reported Indonesia's HDI as having improved greatly since 1990. In 2019 its score was 72%, thus classified as a country of 'high human development' with a score over 70%. This is similar to the Philippines and Bolivia; it is much higher than India (65%) but far lower than its former colonizer, the Netherlands (94%).

The Global Multidimension Poverty Index (UNDP, 2022) recently scored Indonesia at 0.014. This is much better than some countries in the global South (e.g. Ethiopia at 0.367) but weaker than Thailand (0.002) and Vietnam (0.008). Years of education constitute one-third of this indicator but, as discussed above, this does not necessarily reflect the quality of education, a caveat explicitly stated in this report.

Based on survey data from 2017, the UNDP (2020) classified 3.6% of Indonesia's population (almost 10 million people) as 'multidimensionally' poor, with an additional 4.7% (12.7 million) considered as 'vulnerable' to that status. Only 0.4% were in 'severe' multidimensional poverty (compared with 4.2% of the population of India). The same report found that 2.2% of the population lived on less than US$1.90 per day, adjusted for purchasing power (UNDP, 2020).

While this indicator of income poverty is widely used, the critical scholar Hickel (2016) and others (e.g. Butler, 2015) pointed out that the goalposts used to measure progress on the reduction of poverty (and hunger) have been moved, in ways that (critics argue) create a greater impression of progress than is justified. Nonetheless, compared with many countries in the global South, including in South Asia and sub-Saharan Africa, Indonesians appear to be moderately prosperous, with grounds for optimism if this recent progress can continue.

As has long been recognized, poverty decreases any population's adaptive capacity to climate change (Carpenter and Brock, 2008). Indonesia's high population density, its large number of poor and its many low-lying coastal regions (Zikria and Suntoya, 2015) create a nation vulnerable to the impacts of climate change. Indonesia also has an extraordinary level of biodiversity and carbon-rich forests and peatlands; its management of these is critical for global climate change. In turn, climate change will have adverse impacts on biodiversity (Arneth *et al.*, 2020).

34.2 Indonesia's Population Health

In recent decades Indonesia has taken significant steps towards improving its population health status. The nation established a centralized healthcare system in 1960 (when its life expectancy was 48 years), promoting the establishment of primary health centres. By 2001 there were almost 21,000 such centres, and life expectancy had increased to 69 years (Agustina *et al.*, 2019). In 2001, a new decentralized approach was tried in response to criticism that the previous approach did not sufficiently address the country's complexity and diversity. This diversity is not only of ethnicity, location and language, but also of diet, disease, local living styles, health beliefs, human development and community participation. Critics argued that this led to stark disparities in health outcomes (Agustina *et al.*, 2019).

However, this reform also faced criticism. In 2014 a third major reform was undertaken: the introduction of universal health coverage (UHC). UHC is described as focusing 'on accommodating diversity with flexible and adaptive implementation features and quick evidence-driven decisions based on changing needs'. By 2019 the Indonesian UHC system had become the largest single-payer scheme in the world, serving over 200 million people. Its supporters argue that this has improved health equity and service access (Agustina *et al.*, 2019).

According to the UNFPA (2022), life expectancy at birth (in 2022) for women was 75 years; for men 70. Other health indicators have also improved. The infant mortality rate (IMR) fell from 46 per 1000 live births in 1997 to 16 in 2020. The maternal mortality ratio (MMR) has recorded a significant decline from 334 per 100,000 live births in 1997 to 228 in 2007 and 177 in 2015. Disparities in this indicator persist, but are not as marked.[3]

The mortality rate in children under 5 years of age is reported to have declined from 58 per 1000 live births in 1997 to 17 per 1000 live births in 2022. This rate is over four times that of its former colonizer, the Netherlands (4), and is also high compared with nearby South-East Asian countries, such as Malaysia (5) and Thailand (8). The principal causes for perinatal mortality are respiratory infections (36%) and premature births (32%); for neonatal mortality the causes are neonatal sepsis (20.5%) and congenital malformations (18%).

The principal causes for both infant and child mortality are diarrhoea and pneumonia (25% and 15.5%, respectively, for children under the age of 5) (Ministry of Health, 2019). There is a wide regional variation in IMR; for example, the rate per 1000 live births in the province of West Nusa Tenggara (an island 1100 km east of Jakarta) is almost four times the national average. A study among Indigenous people in West Papua found life expectancy in Teluk Bintuni was 66 (averaged for male and females) in 2018; however, even here it was increasing (Warami and Marit, 2022).

The public health sector in Indonesia, similarly to many other countries in the global South, faces the twin problems of the continued presence of a high burden of infectious diseases with an increasing trend towards degenerative, non-infectious (chronic) diseases. Such chronic conditions are due to changes in lifestyle, including diet. The emergence of chronic diseases is also accelerated by undernutrition in infancy and childhood (Perng *et al.*, 2016). Obesity is increasing (Arif *et al.*, 2020). Among important chronic, non-infectious diseases are hypertension, cardiovascular disease, type II diabetes mellitus and stroke.

Important infectious diseases in Indonesia still include acute respiratory infections, diarrhoea, malaria, typhoid, dengue haemorrhagic fever, tuberculosis and HIV/AIDS. In addition, newly emerging strains of avian influenza are a perennial threat to human health,[4] as well as to livelihoods.

34.3 Urban Communities, Climate Change and Health

Migration from villages to cities is part of a worldwide trend in the global South (McDonald *et al.*, 2013). Indonesia is no exception. Migrants are lured by greater economic activities and the hope of employment. However, although average incomes are generally higher in cities and some in-migrants obtain work in the formal sector (while others are students), many remain in the informal sector. Such workers are vulnerable to low pay, unsafe conditions and other forms of exploitation. Recently arrived migrants to Indonesian cities are likely to endure worse conditions than those who are better established. Having loosened ties with their ancestral community they are loath to lose face by returning and highly motivated to succeed.

Air pollution is problematic. In 2015, in Palangkaraya (the capital city of Central Kalimantan

Province) daily average particulate matter (PM_{10}) concentrations often reached as high as 3000 μg/m^3, amongst the worst sustained air quality ever recorded worldwide (Gaveau *et al.*, 2021). A study of ambient exposure to inhaled PM and impaired lung function of migrants in South Tangerang City (an inland urban area adjoining Jakarta) and Makassar (a coastal city in Sulawesi) found levels of $PM_{2.5}$ exposure that were 6–8 times the World Health Organization (WHO) standard[5] (Haryanto, 2016).

In Indonesia the number of urban poor living in slums is rising in all its large cities. Urban slum dwellers are at high risk from the impacts of extreme climatic events, including heatwaves (Setiawati *et al.*, 2022) and flooding. Infrastructure and urban services are often inadequate where the urban poor live; housing can be located in precarious settings, such as steep slopes, flood plains, or hazardous industrial areas. Homes are often self-constructed and unable to withstand extreme climate events. Lack of access to early warning systems heightens the risk for these communities.

Nutrition is often inadequate, health is poor, and the capacity of children to learn is harmed, even where there are schools and teachers. This combination of interacting, compounding factors erodes resilience and can result in the loss of basic services, damage or destruction to homes, lost livelihoods, despair and displacement.

A study undertaken in 2016 by the South Tangerang City government reported a drastic shortage of clean surface water, leading to a shortfall met mainly by the extraction of groundwater. This shortage was chiefly attributed to rapid population increase (3.6% per annum for the period 2010/11) outstripping the pace of infrastructure provision. This study (Nastiti *et al.*, 2018) also warned that climate change would increase the frequency and severity of floods and droughts. While the study is silent on the issue of land subsidence worsened by groundwater extraction, leading to vulnerability to sea level rise, this is a well-known problem in much of Jakarta, and also likely to be a factor in Tangerang (Suroso and Firman, 2018). People who rely on community wells, but with inadequate infrastructure to dispose of domestic and industrial solid wastes, face a high risk of contaminated water, exacerbated by floods and droughts. These problems are compounded by high unemployment and poor health literacy.

34.4 Climate Change and Health in Indonesia

Food security

The nutritional status of Indonesians is still low by international standards, with regional variation reported as 'huge' (Arif *et al.*, 2020). Reaching the poor, especially children and women, and providing them with adequate and nutritious food at an affordable price is a significant challenge. A consortium led by the Food and Agricultural Organization (FAO) reported in 2022 that the prevalence of 'undernourishment' among Indonesians had declined by two-thirds between 2005 and 2020, from 19.2% to 6.5% (FAO *et al.*, 2022); but other data (reported below) indicate more caution. We also note that indicators of human physical growth reflect not only nutrient intake but also absorption and loss, for example due to infectious diseases, especially if accompanied by fever.

Riskesdas (Riset Kesehatan Dasar – Indonesia Basic Health Research) data published in 2018 found that 10.2% of Indonesian children under 5 years were wasted (low weight for height) while almost 31% were stunted (low height for age) (Arif *et al.*, 2020). The same dataset reported that undernutrition was more common in children under 5 in rural areas than in urban areas (34.9% compared with 27.3%). A 2013 report also found that undernutrition was a problem even among comparatively well-off Indonesians. For example, stunting was reported as occurring among 25% of children in the highest income quintile (Arif *et al.*, 2020).

The prevalence of low birth weight (< 2500 g) was reported in the 2017 demographic and health survey for Indonesia as 6.7% (Safitri *et al.*, 2022). These data appear to reflect a modest improvement compared with earlier surveys but the geographical patterns are not uniform; some areas show a decline in nutritional indicators.

Partly in response to this nutritional insecurity, Indonesia has, for decades, been pursuing policies to increase its staple food self-sufficiency (Rizal and Anna, 2019). There has been substantial success; today Indonesia reports a food self-sufficiency ratio of over 90% for key staples, including rice, maize, fruit, vegetables and meat

(Lin *et al.*, 2022). As a result, in the face of climate change, Indonesia is now in a much more secure nutritional position than many low-income countries, especially in sub-Saharan Africa. Nonetheless, Indonesia is still a net food importer.

Future food security – for all nations – relies on a functioning global food system, which is likely to be increasingly strained by climate change and other aspects of limits to growth (Butler and McFarlane, 2018). Climate change will affect evaporation, precipitation, extreme heat, the productivity of agricultural labourers and soil moisture. In turn, agriculture and food security will be impacted. Periodic droughts, potentially strengthening due to an intensification of the El Niño/La Niña cycle, place many crops at risk, including Indonesia's most important staple food, rice (Murniati and Mutolib, 2020). The most recent report from the Intergovernmental Panel on Climate Change (IPCC) reported 'high confidence' that climate change has already harmed agriculture, forestry, fisheries and aquaculture, particularly in mid and low latitudes – such as in Indonesia (IPCC, 2022).

In 1997, El Niño-associated droughts reduced the area planted with rice by approximately 700,000 hectares, mostly in Java (Naylor *et al.*, 2001), compared with approximately 15 million hectares in total, as of 2018. This had a marked effect on total rice production, reducing it from 253 kg per person (1996) to 239 kg on average for 1997/8 (Fig. 34.1). The effect of other El Niño events was not as strong.

Infectious diseases

Many infectious diseases are climate sensitive, including leptospirosis, malaria, dengue fever, yellow fever, schistosomiasis, filariasis and plague (Haryanto, 2016). Although not all of these occur in Indonesia, malaria and dengue fever remain especially important. Climate change (with other factors) is likely to make their control even more difficult (see Chapters 15–16). Floods, droughts and warming temperatures also influence the distribution and risk of diarrhoeal diseases, including hepatitis and cholera. Rising sea levels, precipitation changes and increased flooding may also degrade the quality of fresh water and potentially contaminate drinking water. Many of Indonesia's poor live in coastal areas, and most of the small farmers and fishermen lack the financial means to access adequate health services. Thus, the poor lack a safety net to protect them against the threats that climate change causes.

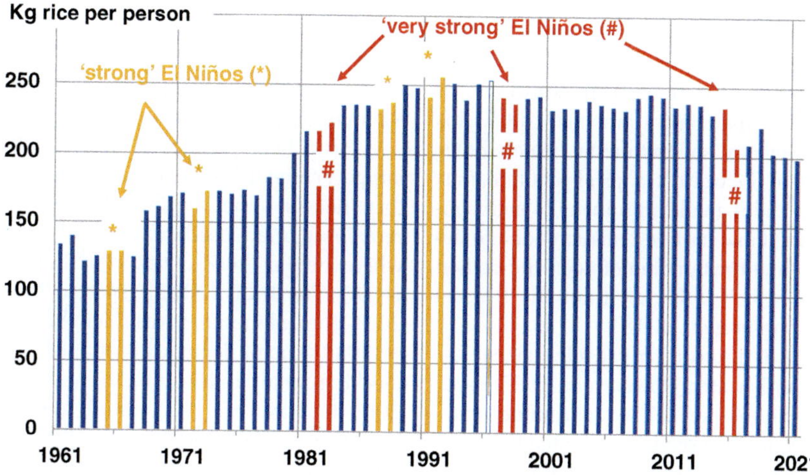

Fig. 34.1. Per capita rice production in Indonesia rose strongly from 1961 until the late 1990s before stabilizing. Also shown are 'very strong' (red) and 'strong' El Niño events. The decline in recent years may be because of a switch to higher-value crops, with any rice shortfall being met by imports. El Niños classed as 'very strong' noticeably reduce the rice harvest. Raw data: Null (2023) and FAOSTAT (nd).

Sea level rise and riverine flooding

Globally, Indonesia has one of the largest numbers and percentages of population in its low-elevation coastal zone, defined as the contiguous and hydrologically connected area of land along the coast and below 10 m of elevation. The damage to infrastructure (and crops) in this zone is compounded by land subsidence and increasing wave heights during storms (WBG and ADB, 2021).

Much of the land on which Jakarta is built has a high rate of subsidence, largely from groundwater extraction. The city faces many other problems, and is widely recognized as excessively inefficient for a national capital. Apart from periodic flooding, these problems include overwhelming traffic congestion and air pollution. As a result, in 2019 the Indonesian government announced its decision to relocate the national capital from coastal Java to coastal Borneo (van der Vuurst and Escobar, 2020).

The population in Indonesia exposed to riverine flooding is also predicted to increase, but not on the scale forecast from coastal flooding due to storm surges (WBG and ADB, 2021).

34.5 Fires, Peat and Coal: Important Indonesian Carbon Sources

Although there is a common perception that the countries of the global North are almost entirely responsible for historic carbon emissions, at least on a per capita basis (Patz *et al.*, 2007) this is not universally so. Indonesia, which holds about 45% of the world's tropical peatland, has, in recent decades, made an unusually high contribution to global carbon emissions due to deforestation and the burning of peat, the carbon content of which is higher than the tropical average (Dohong *et al.*, 2018).

Carbon in peat soil is relatively stable if it is saturated with water. However, in 1995 President Suharto, as part of a programme to increase national food security, initiated the Mega Rice Project in central Kalimantan, on the island of Borneo. This involved the clearance, drainage and conversion of more than a million hectares of forested peat swamps to an ecology planned to enable intensive, Javanese-style wet rice production (Aldhous, 2004). When such swamps are drained, the soil undergoes rapid biogeochemical oxidation, emitting large amounts of carbon dioxide (CO_2) into the atmosphere. For this reason, this project has been called a 'carbon bomb' (Goldstein, 2016). As part of this project, approximately 40,000 rice farmers from Java and Bali were encouraged to migrate; many were given 2.5 ha of land (Goldstein, 2016). Few of these transmigrants harvested much rice; many fled Borneo when the proposal was abandoned, due largely to the acidic soil.[6]

In 1997, fires, mostly deliberately lit and worsened by the drought associated with a strong El Niño event, were estimated to have led to the release of stored carbon (from peat and vegetation) equivalent to 13–40% of the mean annual global emissions from the burning of fossil fuels in that year: 0.81–2.57 Gt of carbon (Page *et al.*, 2002). Some of these fires were unintended; during prolonged dry conditions fires can ignite the desiccated surface of peat, leading to smouldering and even underground burning of drained peatlands. Burning can also spread into relatively undisturbed areas of swamp forest (Huijnen *et al.*, 2016).

In 2015, again during an El Niño event, even more carbon (by some estimates) may have been emitted from fires in Indonesia than in 1997, even though the area burned was smaller than from the earlier event. This paradox has been suggested as explicable because about 50% more peatland was burned in 2015 than in 1997 (Gaveau *et al.*, 2021). Even in non-El Niño years significant carbon is released from annual fires, mainly in southern Sumatra and southern Kalimantan (Huijnen *et al.*, 2016).

Partly in response to criticism, the Indonesian government has enacted numerous regulatory measures that are intended to protect the remaining intact peatland, particularly in areas where the peat is deeper than 3 m (Dohong *et al.*, 2018). The IPCC (2022) reported that, since 2016, the Indonesian government has re-wetted ~380,000 ha of degraded peatlands, mainly by blocking canals and flooding. However, less than 2000 ha have been successfully restored to a condition in which they can sustain native plant species common to peat swamp forests. Replanting native trees has had relatively little success because such trees have low tolerance to prolonged inundation and no fire adaptation strategies (Parmesan *et al.*, 2022).

Despite rules intended to slow or prevent the conversion of remaining peatland for other purposes, such practices are reported to have continued, especially large-scale palm oil plantation. Dohong *et al.* (2018) estimated that over 40% of palm oil plantations located in the former Ex-Mega Rice Project area are situated in areas of deep peat. They estimated that if recent trends continue then between 0.9 and 2 Gt of CO_2 equivalent will be released in the next 25 years; this is substantial, but far less than the amount released in 1997/8.

Compounding this are numerous allegations of human rights abuses in association with palm oil companies (Human Rights Watch, 2021). Palm oil plantations are also a major cause of biodiversity loss in Indonesia, a 'megadiverse' country that harbours the largest expanses of tropical rainforest in South-East Asia, with large numbers of critically endangered species (Laurance *et al.*, 2023). Bornean orangutans (*Pongo pygmaeus*), an extremely charismatic species, have been lost in large numbers (as many as 100,000), due to habitat loss and hunting (Voigt *et al.*, 2018).

Indonesia is also a significant source of coal mining, and is the world's largest coal exporter, with plans to keep expanding this sector (Ordonez *et al.*, 2021). Coal mining is seen as a key economic activity delivering royalties that are vital to local and national public budgets. Ordonez *et al.* (2021) claimed that environmental and climate protection in Indonesia is often framed narrowly as a forestry and land-use issue. These authors also argued that there is limited public awareness of the substantial negative externalities that accompany coal use, not only with regard to climate change, but also in terms of local pollution and health. They suggested that empowering communities affected by local coal mining and burning could help shift political momentum toward cleaner energy sources. But they also admitted that, because political elites derive sizeable rents from coal extraction, climate change mitigation measures will face severe political resistance.

34.6 Conclusion

Indonesia illustrates many dilemmas faced by countries in the global South. These include the confluence of infectious and non-communicable disease. This exists on a background of substantial undernutrition, growing obesity and other factors that have led to and sustained widespread markers of poor physical growth, itself a key determinant of poor school performance, vulnerability and reduced human 'capital'. Though a democracy, Indonesia is a huge and hugely diverse nation, with patterns of corruption familiar to many countries. It has enormous natural resources, but their exploitation is at the cost of carbon emissions, biodiversity loss and human rights abuses to some of its many ethnic minorities.

While slowing (and much slower than in most of sub-Saharan Africa), population growth momentum in Indonesia is contributing to an elusive prosperity for the masses. The impact of future intense El Niño events on Indonesian food security will be adverse but it is unlikely to be critical, provided that the pattern of the El Niño/La Niña oscillation does not itself fundamentally shift. Although climate change-related changes in this cycle have long been postulated (Trenberth and Hoar, 1997), such changes do not, as yet, appear sufficient to lead to nutritional catastrophe. There should be a sufficient buffer in the global food system for Indonesia to import any temporary domestic food shortfall caused by a brief and intense El Niño.

However, if multiple regional 'breadbasket' failures were to occur, due to climate change (Kornhuber *et al.*, 2023), then the poor in Indonesia appear highly vulnerable. While many other aspects of climate change in Indonesia are also likely to be harmful (as briefly outlined above), some of its other major problems – such as the sinking and congestion of Jakarta – are unrelated.

Indonesia's hope to become a hub of electric vehicle manufacture (Veza *et al.*, 2022) is an encouraging sign that it can graduate from mining and deforestation towards cleaner, value-dense exports. It also suggests that its government and policy makers may be starting to appreciate the opportunities and risks that climate change offers. While some Indonesian entrepreneurs have profited disproportionately from palm oil plantations, the burning of peat and the mining and export of coal, the average Indonesian farmer or labourer has barely contributed to global climate change. It is crucial for the future health and stability of Indonesia's people that global carbon emissions are slowed, led by the countries in the global North.

Notes

[1] A recent survey of desired family size in Indonesia reported that 80% of those surveyed felt that the ideal number of children, for a woman in her 40s, was fewer than five. Almost half felt this number was one to two (Assaf and Davis, 2020).
[2] The western half of the island of New Guinea, which has over three times the area of Java, was itself colonized by Indonesia in 1962, after the Dutch finally abandoned it (Pouwer, 1999).
[3] These data should not be viewed as precise, but reasonable approximations.
[4] Particularly if such strains develop human-to-human transmission, something not yet proven to occur on a sustained basis.
[5] In 2021 these levels were 15 μg/m^3 for $PM_{2.5}$ and 45 μg/m^3 for PM_{10} (see Chapter 20).
[6] President Suharto had been warned of this 'by the few local experts who were prepared to speak out' (Aldhous, 2004).

References

Agustina, R., Dartanto, T., Sitompul, R., Suparmi, K.A.S., Achadi, E.L. *et al.* (2019) Universal health coverage in Indonesia: concept, progress, and challenges. *The Lancet* 93, 75–102. DOI: 10.1016/S0140-6736(18)31647-7

Aldhous, P. (2004) Borneo is burning. *Nature* 432, 144–146. DOI: 10.1038/432144a

Arif, S., Isdijoso, W., Fatah, A.R. and Tamyis, A.R. (2020) *Strategic Review of Food Security and Nutrition in Indonesia: 2019–2020 Update*. World Food Programme, The SMERU Research Institute, Jakarta. Available at: https://www.wfp.org/publications/strategic-review-food-security-and-nutrition-indonesia-2019-2020-update (accessed 11 July 2023).

Arneth, A., Shin, Y.-J., Leadley, P., Rondinini, C., Bukvareva, E. *et al.* (2020) Post-2020 biodiversity targets need to embrace climate change. *Proceedings of the National Academy of Science (USA)* 117, 30882–30891. DOI: 10.1073/pnas.2009584117

Assaf, S. and Davis, L.M. (2020) *Comparing Ideal and Completed Family Size: A Focus on Women in Low- and Middle-income Countries with Unrealized Fertility*. ICF (Inner City Fund) DHS (Demographic and Health Surveys) Analytical Studies No. 78, Rockville, Maryland. Available at: https://dhsprogram.com/pubs/pdf/AS78/AS78.pdf (accessed 9 April 2023)

Berger, M.T. (2004) After the Third World? History, destiny and the fate of Third Worldism. *Third World Quarterly* 25, 9–39. DOI: 10.1080/0143659042000185318

BPS (2016) *Statistical data*. Badan Pusat Statistik (Statistics Indonesia), Jakarta. Available at: https://www.bps.go.id/linkTableDinamis/view/id/1119 (accessed 20 November 2020).

Butler, C.D. (2015) Revised hunger estimates accelerate apparent progress towards the MDG hunger target. *Global Food Security* 5, 19–24. DOI: 10.1016/j.gfs.2015.04.002

Butler, C.D. and McFarlane, R.A. (2018) Climate change, food security and population health in the Anthropocene. In: Dellasala, D.A. (ed.) *Encyclopaedia of the Anthropocene.* Elsevier, Amsterdam, pp. 453–459. DOI: 10.1016/B978-0-12-409548-9.09745-1

Carpenter, S.R. and Brock, W.A. (2008) Adaptive capacity and traps. *Ecology and Society* 13, 40. Available at: https://www.ecologyandsociety.org/vol13/iss2/art40/ (accessed 5 June 2022)

Dohong, A., Aziz, A.A. and Dargusch, P. (2018) Carbon emissions from oil palm development on deep peat soil in Central Kalimantan Indonesia. *Anthropocene* 22, 31–39. DOI: 10.1016/j.ancene.2018.04.004

FAO, IFAD (International Fund for Agricultural Development), UNICEF (United Nations Children's Fund), WFP (World Food Programme) and WHO (World Health Organization) (2022) *The State of Food Security and Nutrition in the World 2022.* FAO, Rome.

FAOSTAT (nd) *Food and Agriculture data*. Food and Agriculture Organization, Rome. Available at: https://www.fao.org/faostat/en/#data/QCL (accessed July 16 2023).

Gaveau, D.L.A., Descals, A., Salim, M.A., Sheil, D. and Sloan, S. (2021) Refined burned-area mapping protocol using Sentinel-2 data increases estimate of 2019 Indonesian burning. *Earth System Science Data* 13, 5353–5368. DOI: 10.5194/essd-13-5353-2021

Goldstein, J. (2016) *Carbon Bomb: Indonesia's failed mega rice project*. Environment & Society Portal, Arcadia, no. 6. Rachel Carson Center for Environment and Society, Munich. DOI: 10.5282/rcc/7474

Haryanto, B. (2016) Health adaptation scenario and dengue fever vulnerability assessment in Indonesia. In: Akhtar, R. (ed.) *Climate Change and Human Health Scenario in South and Southeast Asia*. Springer, Cham, Switzerland, pp. 221–236.

Hickel, J. (2016) The true extent of global poverty and hunger: questioning the good news narrative of the Millennium Development Goals. *Third World Quarterly* 37, 749–767. DOI: 10.1080/01436597.2015.1109439

Huijnen, V., Wooster, M.J., Kaiser, J.W., Gaveau, D.L.A., Flemming, J. *et al.* (2016) Fire carbon emissions over maritime Southeast Asia in 2015 largest since 1997. *Scientific Reports* 6, 26886. DOI: 10.1038/srep26886

Human Rights Watch (2021) *Why Our Land? Oil Palm Expansion in Indonesia Risks Peatlands and Livelihoods.* Human Rights Watch, New York. Available at: https://www.hrw.org/sites/default/files/media_2021/06/indonesia0621_web.pdf (accessed 5 June 2022)

IPCC (2022) *Fact Sheet – Food and Water*. Intergovernmental Panel on Climate Change, Geneva. Available at: https://www.ipcc.ch/report/ar6/wg2/downloads/outreach/IPCC_AR6_WGII_FactSheet_FoodAndWater.pdf (accessed 26 February 2023).

Kornhuber, K., Lesk, C., Schleussner, C.F., Jägermeyr, J., Pfleiderer, P. *et al.* (2023) Risks of synchronized low yields are underestimated in climate and crop model projections. *Nature Communications* 14, 3528. DOI: 10.1038/s41467-023-38906-7

Kurniawati, S., Suryadarma, D., Bima, L. and Yusrina, A. (2018) Education in Indonesia: a white elephant? *Journal of Southeast Asian Economies* 35, 185–199. DOI: 10.1355/ae35-2e

Laurance, W.F., Mudhoffir, A.M., Pusparini, W., Meijaard, E. and Engert, J.E. (2023) In Indonesia and beyond: nature conservation needs independent science. *Current Biology* 33, R706-R707. DOI: 10.1016/j.cub.2023.04.068

Lin, H.-I., Yu, Y.-Y., Wen, F.-I. and Liu, P.-T. (2022) Status of food security in East and Southeast Asia and challenges of climate change. *Climate* 10, 40. DOI: 10.3390/cli10030040

McDonald, P., Utomo, I.D., Utomo, A., Reimondos, A. and Hull, T. (2013) Migration and transition to adulthood. *Asian Population Studies* 9, 4–27. DOI: 10.1080/17441730.2012.736700

McGregor, K. (2013) Mass violence in the Indonesian transition from Sukarno to Suharto. *Global Dialogue* 15, 134–145. Available at: https://www.proquest.com/openview/ff5b7b32bc7ba1a390a09edbe95e6e01/1?cbl=426362&pq-origsite=gscholar&parentSessionId=OAUBd%2B7h5XvE78JQTtZwhZyBNYAmLnDrKdXEN43qawg%3D (accessed 16 July 2023)

Ministry of Health (2019) *Country Health Profile – Indonesia.* Pusat Data dan Informasi 2019. Available at: https://www.kemkes.go.id/downloads/resources/download/pusdatin/profil-kesehatan-indonesia/Profil-Kesehatan-Indonesia-2019.pdf (accessed 20 July, 2023).

Murniati, K. and Mutolib, A. (2020) The impact of climate change on the household food security of upland rice farmers in Sidomulyo, Lampung Province, Indonesia. *Biodiversitas Journal of Biological Diversity* 21, 3487–3493. DOI: 10.13057/biodiv/d210809

Nastiti, S., Kusnoputranto, H., Boer, R. and Utomo, S. (2018) A climate risk assessment of clean water supply in an urban area: a case study of South Tangerang city, Indonesia. *IOP Conference Series: Earth and Environmental Science* 129, 012027. DOI: 10.1088/1755-1315/129/1/012027

Naylor, R.L., Falcon, W.P., Rochberg, D. and Wada, N. (2001) Using El Niño/Southern Oscillation climate data to predict rice production in Indonesia. *Climatic Change* 50, 255–265. DOI: 10.1023/A:1010662115348

Null, J. (2023) *El Niño and La Niña years and intensities, based on Oceanic Niño Index (ONI).* Golden Gate Weather Services, Half Moon Bay, California. Available at: https://ggweather.com/enso/oni.htm (accessed 16 July, 2023).

Ordonez, J.A., Jakob, M., Steckel, C.J. and Fünfgeld, A. (2021) Coal, power and coal-powered politics in Indonesia. *Environmental Science & Policy* 123, 44–57. DOI: 10.1016/j.envsci.2021.05.007

Page, S.E., Siegert, F., Rieley, J.O., Boehm, H.-D.V., Jaya, A. *et al.* (2002) The amount of carbon released from peat and forest fires in Indonesia during 1997. *Nature* 420, 61–65. DOI: 10.1038/nature01131

Parmesan, C., Morecroft, M.D., Trisurat, Y., Adrian, R., Anshari, G.Z. *et al.* (2022) Terrestrial and freshwater ecosystems and their services. In: Pörtner, H.O., Roberts, D.C., Poloczanska, E.S., Mintenbeck, K., Tignor, M. *et al.* (eds) *Climate Change 2022: Impacts, Adaptation and Vulnerability.* Cambridge University Press, Cambridge, UK, pp. 197–377.

Patz, J.A., Gibbs, H.K., Foley, J.A., Rogers, J.V. and Smith, K.R. (2007) Climate change and global health: quantifying a growing ethical crisis. *Ecohealth* 4, 397–405. DOI: 10.1007/s10393-007-0141-1

Perng, W., Hajj, H., Belfort, M.B., Rifas-Shiman, S.L., Kramer, M.S. *et al.* (2016) Birth size, early life weight gain, and midchildhood cardiometabolic health. *The Journal of Pediatrics* 173, 122–130.e1. DOI: 10.1016/j.jpeds.2016.02.053

Pouwer, J. (1999) The colonisation, decolonisation and recolonisation of West New Guinea. *The Journal of Pacific History* 34, 157–179. DOI: 10.1080/00223349908572900

Pritchett, L. (2013) *The Rebirth of Education: Schooling ain't Learning*. Center for Global Development, Washington, DC. Available at: https://www.cgdev.org/sites/default/files/rebirth-education-introduction_0.pdf (accessed 4 June 2022).

Rizal, A. and Anna, Z. (2019) Climate change and its possible food security implications toward Indonesian marine and fisheries. *World News of Natural Sciences* 22, 119–128.

Safitri, H.O., Fauziningtyas, R., Indarwati, R., Efendi, F. and McKenna, L. (2022) Determinant factors of low birth weight in Indonesia: Findings from the 2017 Indonesian demographic and health survey. *Journal of Pediatric Nursing* 63, e102–e106. DOI: 10.1016/j.pedn.2021.10.005

Setiawati, M.D., Jarzebski, M.P. and Fukushi, K. (2022) Extreme heat vulnerability assessment in Indonesia at the provincial level. *IOP Conference Series: Earth and Environmental Science* 1095, 012021. DOI: 10.1088/1755-1315/1095/1/012021

Slater, D. (2023) What Indonesian democracy can teach the world. *Journal of Democracy* 34, 95–109. DOI: 10.1353/jod.2023.0006.

Suroso, D.S.A. and Firman, T. (2018) The role of spatial planning in reducing exposure towards impacts of global sea level rise case study: northern coast of Java, Indonesia. *Ocean and Coastal Management* 153, 84–97. DOI: 10.1016/j.ocecoaman.2017.12.007

Trenberth, K.E. and Hoar, T.J. (1997) El Niño and climate change. *Geophysical Research Letters* 24, 3057–3060. DOI: 10.1029/97GL03092

UNDP (2020) Indonesia. In: *Human Development Report 2020. The Next Frontier: Human Development and the Anthropocene. Briefing note for countries on the 2020 Human Development Report*. United Nations Development Programme, New York. Available at: https://www.hdr.undp.org/sites/all/themes/hdr_theme/country-notes/IDN.pdf (accessed 6 June 2022)

UNDP (2022) *Global Multidimensional Poverty Index 2022*. United Nations Development Programme, New York, and Oxford Poverty and Human Development Initiative (OPHDI), Oxford University, UK. Available at: https://hdr.undp.org/system/files/documents/hdp-document/2022mpireportenpdf.pdf (accessed 9 April 2023).

UNFPA (2022) *World Population Dashboard*. United Nations Population Fund, New York. Available at: https://www.unfpa.org/data/world-population-dashboard (accessed 25 February 2023).

van de Vuurst, P. and Escobar, L.E. (2020) Perspective: climate change and the relocation of Indonesia's capital to Borneo. *Frontiers in Earth Science* 8, 5. DOI: 10.3389/feart.2020.00005

Veza, I., Abas, M.A., Djamari, D.W., Tamaldin, N., Endrasari, F. *et al.* (2022) Electric vehicles in Malaysia and Indonesia: Opportunities and challenges. *Energies* 15, 2564. DOI: 10.3390/en15072564

Voigt, M., Wich, S.A., Ancrenaz, M., Meijaard, E., Abram, N. *et al.* (2018) Global demand for natural resources eliminated more than 100,000 Bornean orangutans. *Current Biology* 28, 761–769. DOI: 10.1016/j.cub.2018.01.05

Warami, H. and Marit, E.L. (2022) Characteristics, fertility, and mortality of the population in Teluk Bintuni regency, West Papua Province. *International Journal of Educational Research & Social Sciences* 3, 728–746. DOI: 10.51601/ijersc.v3i2

WBG and ADB (2021) *Climate Risk Country Profile: Indonesia*. World Bank Group, Washington, DC, and Asian Development Bank, Manila, Philippines. Available at: https://climateknowledgeportal.worldbank.org/sites/default/files/2021-05/15504-Indonesia%20Country%20Profile-WEB_0.pdf (accessed 25 February 2023).

World Bank (2022) *Urban population (% of total population)*. United Nations Population Division: World Urbanization Prospects, 2018 Revision. Available at: https://data.worldbank.org/indicator/SP.URB.TOTL.IN.ZS (accessed 16 July 2023).

Zikra, M. and Suntoyo, L. (2015) Climate change impacts on Indonesian coastal areas. *Procedia Earth and Planetary Science* 14, 57–63. DOI: 10.1016/j.proeps.2015.07.085

35 The Health Impacts of the Climate Crisis in the Cook Islands, Niue and Tokelau

Collin Tukuitonga
School of Population Health, University of Auckland, New Zealand

Abstract

The climate crisis is expected to cause significant additional negative impact on the lives and livelihoods of people living in the Cook Islands, Niue and Tokelau. Severe weather events, rising sea levels, ocean warming and acidification will damage crops, fisheries and water supply. Saltwater intrusion, marine pollution and habitat destruction will aggravate the impact of the climate crisis on the fragile ecosystems of these islands. The combined effects of the climate crisis and high prevalence of diseases such as diabetes and heart disease have already affected the health and well-being of people living in these islands. Leaders of the Cook Islands, Niue and Tokelau have been strong advocates for reducing harmful greenhouse gases as part of the global push on the climate crisis. Cook Islands, Niue and Tokelauan people have been active in implementing mitigation and adaptation measures at community level but more needs to be done urgently.

35.1 Introduction

The Cook Islands, Niue and Tokelau are small island states, voluntarily associated with New Zealand. Sometimes called 'realm nations', they share the same head of state (King Charles III). Each retains its own legal system and laws (Quentin-Baxter, 2021). The larger two (Cook Islands and Niue) are self-governing; Tokelau is administered by New Zealand officials.

The climate change crisis is considered to be an existential threat to all the people who inhabit Small Island Developing States (SIDS), including these low-lying islands, which are among the smallest of the SIDS, with extremely fragile natural environments. The communities living in these islands are not only vulnerable to the negative impacts of the climate crisis but also susceptible to the effects of marine pollution and habitat destruction. As a result, their food and water security is threatened; rising sea levels may force them to abandon their home islands. The health impacts of the climate crisis are potentially catastrophic, compounding underlying health and social problems, fragile infrastructure and other resource constraints. More effective actions are needed by all nations in line with the Paris Agreement (UN, 2015) to reduce these threats. All three realm nations are yet to develop an evidence-based health response to the climate crisis; it is now urgent that they do so.

35.2 Cook Islands

The Cook Islands consist of 15 small islands scattered across more than 2 million km^2 of the Pacific Ocean. Rarotonga, with a land area of

 Climate Change and Global Health: Primary, Secondary and Tertiary Effects, 2nd Edition (eds C.D. Butler and K. Higgs)
DOI: 10.1079/9781800620025.0035

67 km^2, is the largest. The total land area of the Cook Islands state is 240 km^2 but its Exclusive Economic Zone (EEZ) covers almost 2 million km^2. The Cook Islands are in two distinct groups. The southern islands tend to be high volcanic islands, whereas the northern group of islands are low-lying coral atolls, thus increasingly vulnerable to rising sea levels and coastal flooding. Tropical cyclones (TCs) have had a major impact on the small islands of the Cook Islands, for example TC Martin that devastated Manihiki in 1997.

The total population of the Cook Islands in 2021 was 17,581, but out-migration to New Zealand has seen a gradual decline in the population over several decades. Migration is an important 'safety' valve for small islands with limited land area, where fertility rates are comparatively high. Border closures and a decline in tourist numbers, caused by the COVID-19 pandemic, have escalated the migration of young Cook Islanders seeking work in New Zealand. The Cook Islands state is self-governing in free association with New Zealand and Cook Islanders are New Zealand citizens. There is open access to New Zealand for Cook Islanders, which often escalates in response to downturns in the local economy.

Tourism accounts for 80% of the Cook Islands' economy, which was initially devastated by the pandemic. Two-way quarantine-free travel with New Zealand, which commenced in May 2021, has revived tourism – an important economic basis.

The climate crisis has already had profound impacts on the lives and livelihoods of the Cook Islands people. Air temperatures and rainfall are projected to increase under both low and high greenhouse gas-emission scenarios by 2035 and 2100. Sea-surface temperatures are also expected to increase; satellite data show that sea level has risen near the Cook Islands by about 4 mm per year since 1993. This is slightly larger than the global average (CIMS *et al.*, 2011). Data show that, since the 18th century, the level of ocean acidification has been slowly increasing in the Cook Islands waters. Ocean acidification threatens the entire marine food chain. The climate crisis is expected to add to the existing local threats to coral reefs, resulting in declines in the percentage of coral cover (Pacific Community, 2017).

People of the Cook Islands are heavily dependent on the sea for their food and income. The average national fish consumption is estimated as 35 kg per person per year. Consumption is higher in rural areas, where it averages > 60 kg per person per year. Coastal fisheries resources are projected to show progressive declines in productivity due to both the direct effects of the climate crisis (coral bleaching) and indirect effects from habitat destruction and marine pollution. Projections also show that the nearshore pelagic fishery component of coastal fisheries is projected to increase in productivity, due to the redistribution of tuna to the east from central and western Pacific waters (Hanich *et al.*, 2018).

The Cook Islands is among the group of Pacific Islands Countries and Territories (PICTs) where the estimated sustainable production of fish and invertebrates from coastal habitats is considered sufficient to supply the national population with 35 kg of fish per person per year (recommended for good nutrition), provided that demand for fish and other forms of seafood does not increase.

Health impacts

In 2016, the World Health Organization Regional Office for the Western Pacific (WPRO) issued a report on human health and climate crisis in PICTs (McIver *et al.*, 2016). The report analysed the impact of the climate crisis on the health of people in 13 Pacific island countries, including the Cook Islands and Niue. The majority of the climate crisis health priorities identified in PICTs were an increasing incidence of foodborne, waterborne and vector-borne diseases and more serious health impacts from heat and natural disasters. In the Cook Islands, dengue fever has become endemic; outbreaks are reported on a regular basis. Given improving cyclone warnings and preparedness, injuries and deaths from storms are now infrequent. The full extent of the impact of the climate crisis on human health in the Cook Islands remains unclear.

Future climate

Projections for all emission scenarios indicate that the annual average air temperature and sea-surface temperature will continue to increase in the future in the Cook Islands. This will generate a rise in the number of hot days and warm nights. There is uncertainty around rainfall projections; model results are inconsistent. However, average annual and seasonal rainfall is generally projected to increase over the course of the 21st century. For the southern group of islands, average rainfall during the wet season is expected to increase due to projected intensification of the South Pacific Convergence Zone (SPCZ). Droughts are projected to become less frequent. Projections show that extreme rainfall days are likely to occur more often. TCs are projected to decrease in frequency but to be more intense when they do occur.

35.3 Niue Island

Niue is an isolated raised coral atoll averaging 69 m above sea level. It is one of the largest coral atolls in the world: its land area is 259 km^2; its EEZ is approximately 450,000 km^2. Niue is a self-governing state in free association with New Zealand. Niueans are New Zealand citizens. Niue had a population (in 2018) of 1620 residents living in 14 village communities, distributed along its coast. The island has experienced continuous population decline for several decades; the threat of depopulation is constant. Niue is situated within the Pacific cyclone belt; the island has experienced many severe TCs, some causing major economic harm (de Scally, 2008; Asian Development Bank, nd).

Surface air temperature and sea-surface temperature are projected to continue to increase over the course of the 21st century. There is very high confidence in this direction of change, because warming is physically consistent with rising greenhouse gas concentrations. Annual mean and wet-season rainfall are also projected to increase, consistent with a projected likely increase in the intensity of the SPCZ. While changes in TCs are not known with high confidence, there is some evidence from models that their frequency in the South Pacific will decrease, but that the proportion of very intense TCs will increase. Mean sea level and ocean acidification are projected to continue to rise over the course of the 21st century.

Niue's marine and coastal ecosystems have cultural value, are sources of nutritious food, are important for recreation and tourism, and collectively comprise a significant share of the island's biodiversity. Increased storm surge damage may harm fisheries development. Niuean agriculture is sensitive to drought and cyclones. Cyclones pose major dangers to human health, causing mortality and injury, and increasing the risk of post-event diseases through disruptions to water, sanitation and power services. Changes in the water supply system, including resetting the depth of pumps, ongoing investigations and monitoring of groundwater, the use of renewable energy to power pumps, and greater investments in rainwater harvesting can help reduce vulnerability (Hanna *et al.*, 2011).

Coastal fisheries are projected to show progressive declines in productivity due to the effects of the climate. Like the Cook Islands, the nearshore pelagic fishery component of coastal fisheries is projected to increase in productivity, due to the redistribution of tuna to the east.

Like the Cook Islands, Niue is among the group of PICTs where the estimated sustainable production of fish and invertebrates from coastal habitats is forecast to easily supply its population with 35 kg of fish per person per year. However, average national fish consumption in Niue is estimated to be 79 kg per person per year; very few other animal-sourced foods are available. Even with this increased per capita consumption, the very large area of coral reefs relative to population size in Niue is expected to enable a sufficient supply of coastal fish to give future food security (i.e. for fish), even if climate change reduces marine productivity in its coastal area.

Health impacts

The majority of the climate crisis health priorities identified in PICTs by the WPRO report were an increasing incidence of foodborne, waterborne and vector-borne diseases and more serious health impacts of heat and natural disasters (McIver *et al.*, 2016). Dengue occurs from time to time on Niue; other viral illnesses have also been reported. The health system capacity and capability on Niue is small and severely constrained. Given predictions on the climate crisis impact on health and health services, there is an urgent need to develop a policy response for the health sector.

Future climate

Niue is very similar to the Cook Islands; projections for all emission scenarios indicate that the annual average air temperature and sea-surface temperature will continue to increase in the future in Niue. Projections show extreme rainfall days are likely to occur more often. Projections tend to show a decrease in the frequency of TCs, but an increase in the proportion of the most intense storms.

35.4 Tokelau

Tokelau consists of three atolls: Atafu, Nukunonu and Fakaofo. Unlike the other two realm nations, Tokelau is directly governed from New Zealand. Although its population and land area are tiny (fewer than 1400 people on a land area of about 10 km^2), its EEZ is enormous: almost 320,000 km^2. These atolls are remote; they can only be accessed by boat from Samoa, a distance of about 500 km that takes about 24 hours. It currently has no airfield; if it did it would be over 3300 km from Auckland, New Zealand's largest city. Many families from Tokelau have at least one member in New Zealand. Tokelau is among the most physically vulnerable low-lying atolls in the world; its altitude ranges from 3 m to 5 m above sea level. It is highly exposed to adverse effects from the climate crisis and natural hazards. Tokelau has invested significantly in renewable energy and it has been described as the first country in the world with 100% renewable energy. Tokelau struggles with access to clean water and droughts are common. The climate crisis is expected to aggravate vulnerabilities in access to clean water and compound environmental and habitat destruction.

The climate crisis is projected to show similar trends in Tokelau as in the Cook Islands and Niue. Air and water temperatures are expected to continue to rise but rainfall projections are uncertain. Higher global temperatures have increased the already hot climate of Tokelau to the point where human health is being impacted. The hotter climate also exacerbates drought periods, when fresh water, already a scarce supply, is increasingly hard to find.

The abundance of reef fish in Tokelau is predicted to decrease but the abundance of pelagic fish is expected to increase within Tokelau's waters due to climate change. The present catch per unit effort analysis, combined with the forecasted effects of climate change, suggests that the domestic fisheries in Tokelau may be on an unsustainable path, highlighting food security concerns, despite the potential for growth in offshore (pelagic) fisheries, such as for tuna (White *et al.*, 2018). There is no higher ground where the people of Tokelau can retreat as the sea rises. The soil is a mixture of coral and sand which can only support limited crops such as breadfruit, coconut, pandanus, giant swamp taro and banana. The climate crisis is likely to drive people towards increased consumption of imported food items.

The people of Tokelau suffer from high rates of non-communicable diseases (NCDs) and related risk factors (Mafile'o *et al.*, 2019). The populations of the atolls have taken considerable steps to prevent and control NCDs but there is limited impact to date. The climate crisis is likely to further drive the shift towards rising consumption of imported highly processed food and sugar-sweetened beverages, which will further increase NCD incidence. On the positive side, Tokelau banned the importation of these drinks in 2013; however, data are lacking to assess the consequences. The climate crisis is likely to further adversely impact the fragile Tokelau environment, including compromising the islanders' ability to grow local staples and to catch reef fish (White *et al.*, 2018).

35.5 Discussion

The people of the Cook Islands, Niue and Tokelau will benefit from additional climate mitigation and adaptation; risks and costs due to impacts of the climate crisis are expected to accelerate over time. Mainstreaming climate crisis actions in development planning is crucial for reducing vulnerability to extreme climate and promoting sustainable development. All three realm island nations are extremely vulnerable to the impacts of the climate crisis.

While all three realm nations have been active in climate action as part of the regional and global activities, more needs to be done. The climate crisis in small island communities threatens to undermine past gains in development and health. The direct effects of the climate crisis include increased heat stress, floods, drought and an increased frequency of intense storms, with the indirect threatening of population health through the spread of disease vectors, food insecurity and undernutrition, displacement and mental ill-health. The climate crisis impacts on the entire socio-economic determinants of health. People in low-lying small islands may be forced to relocate due to sea level rise and saltwater inundation.

Pacific SIDS have benefited from the support of the Pacific Community, Secretariat of the Regional Environmental Programme and the World Health Organization (WHO) on the health impacts of the climate crisis. Between 2010 and 2012, the WHO led a regional climate change and health vulnerability assessment and adaptation planning project, in collaboration with health sector partners, in 13 Pacific island countries, including the Cook Islands and Niue (McIver *et al.*, 2016). Regional and global organizations have been collaborating with the Pacific island countries to develop climate change and health adaption plans that address the health impacts of climate change in the Pacific context.

Nearly all Pacific island countries and areas have completed a National Climate Change and Health Action Plan (NCCHAP) or equivalent, including the Cook Islands, Niue and Tokelau. From a sub-regional perspective, common priorities identified in the national action plans of Pacific island countries and areas include the following.

- Perform vulnerability assessment and adaptation planning processes periodically; updating NCCHAPs to reflect improved knowledge and information regarding climate-sensitive health risks and appropriate management strategies.
- Align climate change and health with activities and ongoing improvements in disaster risk management, food security and water, sanitation and hygiene.
- Mainstream consideration of climate change and health into existing public health systems, while ensuring that such mainstreaming does not lead to dilution, diversion or diminution of the urgency of the targeted activities.
- Improve understanding of the relationship and diverse causal pathways between climate variability and climate-sensitive diseases.
- Explore the previously neglected areas of climate change and health research in the Pacific, including climate change impacts on heat-related illnesses, mental health and noncommunicable diseases, as well as on communicable diseases.
- Develop climate-based early-warning systems.

The effects of the climate crisis on health are already tangible in the Pacific. They have emerged as a huge challenge for the population and health systems of the Cook Islands, Niue and Tokelau.

References

Asian Development Bank (nd) *The economics of climate change in the Pacific*. Asian Development Bank, Manila, Philippines. Available at: https://www.adb.org/sites/default/files/publication/31136/economics-climate-change-pacific-brochure.pdf (accessed 10 August 2023).

CIMS (Cook Islands Meteorological Service), ABM (Australian Bureau of Meteorology) and CSIRO (Commonwealth Scientific and Industrial Research Organisation) (2011) *Current and Future Climate of the Cook Islands.* Australian Government. Available at: https://www.pacificclimatechangescience.org/wp-content/uploads/2013/06/9_PCCSP_Cook_Islands_8pp.pdf (accessed 17 February, 2022).

de Scally, F.A. (2008) Historical tropical cyclone activity and impacts in the Cook Islands. *Pacific Science* 62, 443–459. DOI: 10.2984/1534-6188

Hanich, Q., Wabnitz, C.C.C., Ota, Y., Amos, M., Donato-Hunt, C. *et al.* (2018) Small-scale fisheries under climate change in the Pacific Islands region. *Marine Policy* 88, 279–284. DOI: 10.1016/j.marpol.2017.11.011

Hanna, E.G., Harley, D., Xu, C. and McMichael, A.J. (2011) *Overview of climate change impacts on human health in the Pacific region*. Report to Commonwealth of Australia Department of Climate Change and Energy Efficiency. Australian National University, Canberra, Australia. Available at: https://www.dcceew.

gov.au/climate-change/policy/adaptation/publications/climate-change-impacts-human-health-pacific (accessed 10 August 2023).

Mafile'o, T., Foliaki, S., Koro, T., Leslie, H., Redman-MacLaren, M. *et al.* (2019) *Review Report on Tokelau's Clinical Health Services and Patient Referrals Scheme*. Ministry of Foreign Affairs and Trade, Wellington, New Zealand. Available at: https://www.mfat.govt.nz/assets/Aid-Prog-docs/Evaluations/2019/Tokelau-Clinical-Health-Services-and-Patient-Referrals-Scheme-Evaluation-report.pdf (accessed 10 August, 2023).

McIver, L., Kim. R., Woodward, A., Hales, S., Spickett, J. *et al.* (2016) Health impacts of climate change in Pacific island countries: a regional assessment of vulnerabilities and adaptation priorities. *Environmental Health Perspectives* 124, 1707–1714. DOI: 10.1289/ehp.1509756

Pacific Community (nd) *2.2 Cook Islands*. Pacific Community, Nouméa, New Caledonia. Available at: https://www.spc.int/sites/default/files/wordpresscontent/wp-content/uploads/2017/01/3-cook.pdf (accessed 17 February 2022).

Quentin-Baxter, A. (2021) The Cook Islands, Niue and Tokelau as parts of the realm of New Zealand. *Victoria University of Wellington Law Review* 52, 579–605.

UN (2015) *The Paris Agreement*. United Nations, Rome. Available at: https://www.un.org/en/climatechange/paris-agreement (accessed 17 February 2022).

White, R., Coghlan, A.R., Coulter, A., Palomares, M.L.D., Pauly, D. *et al.* (2018) Future of fishing for a vulnerable atoll: trends in catch and catch-per-unit-effort in Tokelau's domestic marine fisheries 1950–2016. *Frontiers in Marine Science* 4, 76. DOI: 10.3389/fmars.2018.00476

36 Europe and Climate Change: Impacts, Risks and Opportunities

May C.I. van Schalkwyk and Martin McKee
London School of Hygiene and Tropical Medicine, London, UK

Abstract

This chapter explores climate change from a European perspective. It begins with an overview of the impact climate change is already having on countries across Europe and then explores the types of threats that are likely to emerge or continue to occur. These include loss of land due to sea level changes, changes in the distribution of vector-borne diseases, and mounting threats to the availability of food and water. It considers the challenges Europe faces in transitioning away from fossil fuel dependence and its role as a destination for people fleeing other areas of the world that are becoming uninhabitable. The final sections of the chapter explore shifts in public attitudes in Europe, the role of vested interests as barriers to change, and the developments in the aftermath of recent events such as COP26 and the COVID-19 pandemic.

36.1 Introduction

Any lingering doubt among Europeans about the reality of climate change should have been abolished by the scenes that they confronted in the summer of 2021. Torrential rain causing widespread flooding and loss of life in Belgium and Germany was followed by forest fires in Greece, Italy and Turkey. For weeks, newspaper front pages featured apocalyptic scenes. As rescue teams were struggling to save lives and property, the sixth report of the Intergovernmental Panel on Climate Change (IPCC) was published. It was an unequivocal statement that climate change is upon us and is the consequence of human actions. The report was written to inform the 26th Conference of the Parties (COP26), held in Europe – in Glasgow, Scotland – later that year. This conference, hosted by Italy and the UK, was a last chance for the governments of the world to act decisively on what is now indisputably an existential threat to human survival (Butler, 2018). In this chapter we review some of the many consequences of climate change for Europe. We reflect on European responses to the growing realization of the threats posed by climate change. Of necessity, this review is selective; there are many issues that space does not allow us to include but most are covered in other chapters.

We begin by considering the consequences for Europe of the direct effects of climate change on Europe itself and the consequences for Europe flowing from climate change's impact on other parts of the world. The former include the consequences of extreme weather events, sea level rise and changes in ambient temperature, with their many implications for health ranging from the distribution of vector-borne disease to effects on agriculture and fisheries. The latter, characterized in this book as 'tertiary effects',

 Climate Change and Global Health: Primary, Secondary and Tertiary Effects, 2nd Edition (eds C.D. Butler and K. Higgs)
DOI: 10.1079/9781800620025.0036

include the potential for mass migration from parts of the world that are no longer habitable, for example because of high ambient temperatures or drought, threatening lives directly and indirectly through destruction of food systems and fuelling of violence and conflict. We then reflect on the changing policy context in Europe. We examine the European Union's proposal for a European Green Deal, committing to net zero emissions of greenhouse gases by 2050, the decoupling of economic growth from resource use, and that no person or place should be left behind. We place this in the context of growing political support for parties and politicians promoting a green agenda but also the countervailing force from powerful vested interests, many linked to the fossil fuel industry, that seek to undermine this initiative. Finally, we look at developments in Europe in the wake of the COVID-19 pandemic which have implications for how European countries respond to threats to health, including via climate change.

Unfortunately, there are few grounds for optimism, but this should not drown out hope. Europe has repeatedly shown that it can emerge from a crisis, whether in the Renaissance that followed the conflicts of the late Middle Ages, the recovery from the devastation of war after 1945 and the creation of the European Union (EU), or the transition to democracy after the collapse of the Soviet bloc. Young Europeans, many inspired initially by climate activist Greta Thunberg, have been a source of inspiration the world over. However, progress will only be possible if we can overcome inaction, delay, denial and fatalism and we are running out of time to act. As the IPCC has stated clearly, even if emissions of greenhouse gases could be reduced immediately, the planet almost certainly faces the inevitability of further warming, although some models suggest that this could be avoided if they could be reduced to zero or, preferably, to negative levels (MacDougall *et al.*, 2020). Rapid and decisive action is therefore needed both to bring emissions down and to mitigate the otherwise inevitable consequences of failing to achieve adequate progress.

Worryingly, one lesson from the COVID-19 pandemic is that European governments struggle to take rapid and decisive collective action and have sometimes proven susceptible to vested interests that seek to exploit crises for personal and financial gain. Another is that they have been reluctant to invest the necessary resources in the preparedness needed to face a crisis, although this may now be changing following the pandemic, which exposed serious weaknesses in preparedness (Jit *et al.*, 2021). Consequently, the final part of the chapter looks at the lessons that can be learned from the pandemic response and how they might apply to the inevitable consequences of further climate change.

36.2 The Consequences of Rising Temperatures for Europe, as Seen by the IPCC

The 2021 Intergovernmental Panel on Climate Change (IPCC) report divides Europe into four regions: Northern Europe, taking in the British Isles and Nordic countries; the Mediterranean countries, stretching from Portugal to Turkey; Eastern Europe, covering European Russia, Ukraine, Belarus and the Caucasus; and Western and Central Europe, comprising the remaining nations (IPCC, 2021).

The report described some changes that will affect the continent overall. Regardless of future levels of global warming, it projected with high confidence that temperatures will rise in all parts of Europe, at a rate that exceeds global mean temperature changes. It expected that the frequency and intensity of hot extremes, including marine heatwaves, will continue to increase regardless of what happens with greenhouse gas emissions. In contrast, the frequency of cold spells will reduce. It expected an increase in winter rainfall in northern Europe, but a decline in precipitation in summers in the Mediterranean, extending towards northern parts of the continent. This will be accompanied by increased episodes of extreme rainfall and associated flooding in all regions except the Mediterranean. There will be a rise in sea level in all parts of Europe, except for some areas neighbouring the Baltic Sea where land rebound has been occurring since the end of the last Ice Age 12,000 years ago, at a rate close to or exceeding changes in global mean sea level. This is expected to cause increased coastal flooding and the retreat of shorelines on sandy coasts. The melting of glaciers and permafrost seen in recent years is expected to continue.

Turning to the different regions, and depending on the overall increase in global temperature, it projected greater flooding and severe windstorms in Northern, Eastern, and Western and Central Europe. It projected increased droughts in the Mediterranean and in Western and Central Europe, and increased fires in the Mediterranean and Eastern Europe.

36.3 The Impact of Temperature Rise in Europe

The European continent includes many different ecosystems, all of which are vulnerable to climate change in various ways. The EU has distinguished five categories of impact (European Commission, 2021a).

The first relates to increased temperature. This will give rise to more frequent heatwaves, such as that which affected several European countries in 2003, when a combination of high temperatures, especially at nighttime in areas where these were unusual and buildings were poorly adapted, coupled with a lack of social support, led to an estimated 70,000 excess deaths (Robine *et al.*, 2008). The summer of 2022 was the hottest season in Europe on record until that date, with over 60,000 heat-related deaths estimated to have occurred (Ballester *et al.*, 2023) and 2023 saw the highest temperatures ever recorded, reaching into the high 40s (degrees Celsius), while other measures, such as sea temperatures, rose to levels beyond even the most pessimistic predictions, raising questions about whether the climate was reaching a tipping point.

Increased temperature will also lead to changes in the behaviour of animal and plant species, potentially leading to increased numbers of pests and invasion by non-native species, as well as changes in the distribution of vector-borne disease. Changing climate has already been invoked to explain the emergence of West Nile virus in south-east Europe (Hernández-Triana *et al.*, 2014) while the expanding range of the mosquito *Aedes albopictus* into southern Europe has created the conditions for transmission of dengue and chikungunya infection (Jourdain *et al.*, 2020). There are also likely to be profound consequences for agriculture and fishing. Sea warming has already displaced some species from traditional fishing grounds (Baudron *et al.*, 2020) and, in 2023, was accompanied by mass die-offs of fish in some places.

The second set of impacts relate to droughts and wildfires. Droughts have been increasing in frequency in southern Europe but decreasing in northern Europe (Stagge *et al.*, 2017), with the droughts contributing to wildfires (Turco *et al.*, 2017). There has been a substantial increase in fatalities from wildfires in southern Europe since the 1970s (Molina-Terrén *et al.*, 2019) and a recent review of studies modelling the future risk of such fires found most estimating an increase in burnt areas of 12–15% per decade (Dupuy *et al.*, 2020).

The third impact is on the availability of fresh water. Forty per cent of the fresh water in the EU comes from the Alps, which have been experiencing the retreat of glaciers over the past decade (Sommer *et al.*, 2020). Warming of the Alps has many other consequences. Many communities are economically dependent on winter sports and tourism, and may not be viable in a warmer climate. The water that runs from the Alps into rivers such as the Rhine, Danube and Elbe supports a large volume of trade on barges. Low water levels in recent heat waves have disrupted this trade. This poses a major problem to the shipping industry and potentially undermines efforts to reduce emissions from the transport sector by increasing the waterborne movement of goods (Posaner and Cokelaere, 2023). Meanwhile, water shortages threaten the viability of agriculture in some areas, such as the Iberian Peninsula, and consequently the employment in that sector and the local economies it supports.

While most of the consequences for water supply are due to reduced rainfall, torrential downpours can lead to overflow of sewage into the water supply, with obvious risks for health. This takes us to the fourth impact, from flooding. Already, Europe is experiencing increased amounts of flooding from rivers, likely to increase further even if there is some uncertainty about the impacts. A detailed analysis of flow into the Rhine basin suggests that changes in nival (snow melt) and pluvial (rainfall) water may balance out overall but with differential impacts at certain places (Rottler *et al.*, 2021), while another analysis of the Danube basin projects an increasing problem with water scarcity along much of its length (Bisselink *et al.*, 2018). All these models are sensitive to changes in water use and population, among other factors.

Finally, some parts of Europe are relatively low lying, most obviously the Netherlands, which means 'low countries', where large areas are below sea level and some communities have already been abandoned to the sea (O'Leary, 2019).

36.4 The Consequences for Europe of Temperature Rises Elsewhere

Europe has long been a destination for migrants from within the continent and from other parts of the world. People move for a multitude of reasons, often divided into push factors, such as conflict and economic insecurity in the countries of origin, and pull factors, in the form of safety and opportunity in the countries of destination, while certain flows are also influenced by language or historical connections. Climate change is already impacting on the former and is likely to exacerbate many of the conditions that cause people to move. This has important implications for the countries of Europe, as already seen with the large-scale movement of people driven from their homes by conflict in the Middle East since the mid-2000s. The toll of human suffering has been immense, with tens of thousands of people perishing en route, many drowning in the Mediterranean, and others facing horrendous conditions in the countries through which they transit, including what is in effect modern slavery, only to face atrocious living conditions in the refugee camps situated on Europe's southern borders. Those that make it to safety often face discrimination and hardship.

Sadly, not everyone is sympathetic to their plight, and in a number of European countries, populist politicians have exaggerated and exploited their presence to achieve electoral advantage. One country, Belarus, has sought to weaponize them, encouraging attempted migrants to move to the EU's borders to engineer confrontation with Polish and Lithuanian authorities. Migration can also have geopolitical consequences. For example, in Europe, opposition to migration played a major role in the vote by the UK to leave the EU, although, ironically, as the country faces severe labour shortages, recognition of the contributions of migrants is slowly dawning on some of those who supported Brexit.

There are several ways in which climate change can increase the pressure on people to move. First, sea water levels are rising and will continue to do so, a result of both greater volume of warmer water and melting of ice sheets. This will have profound consequences for low-lying areas, including Pacific Island nations, some of which are at risk of being completely overwhelmed (Kumar *et al.*, 2020), and coastal areas such as much of Bangladesh. Even before this happens, low-lying areas face ecological pressure from salination of the water table (Chen and Mueller, 2018). A second is the spread of deserts. Over the past 50 years, the Sahara Desert has expanded markedly. A third threat is from the direct effect of high atmospheric temperatures, especially when they coincide with high humidity. Again, this is already happening, with temperatures reaching almost 50°C in some places (Clinton, 2023). There are also threats posed by shortages of food and, especially, water. Temperature change has consequences for what can be grown, especially when accompanied by soil degradation and water shortages. Increasing temperature and levels of carbon dioxide (CO_2) also reduce the micronutrient content of some crops (Myers *et al.*, 2014). Warmer waters influence the distribution and numbers of fish, with projections pointing to large reductions in waters fished by people from some of the poorest parts of the world (Cheung *et al.*, 2010).

Climate change can also increase migration through its contribution to conflict (Abel *et al.*, 2019). Invariably, there are many factors involved in any particular conflict and climate change is unlikely ever to be the sole cause; but one study has presented a plausible narrative as to how droughts in Syria since the 1990s, coupled with political developments in the country and the broader region, contributed to the civil war that began in 2011. A meta-analysis of studies since 1950 concluded that one standard deviation increase in temperature or rainfall was associated with a 4% increase in interpersonal violence and 14% increase in intergroup conflict (Hsiang *et al.*, 2013).

The relationship between climate change, migration and conflict is extremely complex (Koubi, 2019). Although numerous studies have linked climate-related events to the movement of people, these are rarely the only reasons in a particular case. This reflects the complexity of natural systems and echoes the frequent argument that it is difficult to assign a particular natural disaster to climate change, even though it is clear that there is a pattern. In this way, scholars have proposed referring to climate change as a 'risk multiplier, influencer or co-factor' in relation to these types of complex events (Butler and Kefford, 2018).

It is, however, unclear what this will mean for the overall scale of migration. Several studies have sought to estimate the numbers of people who may be displaced by climate change. Most are in the range of 50–200 million worldwide by 2050 (Burrows and Kinney, 2016). Given Europe's proximity to many of the areas worst affected by climate change, it seems likely that a considerable proportion will be seeking refuge in European countries.

36.5 A European Green Deal

Fossil fuels have played a crucial role in the development of modern Europe. The European Union grew out of the European Coal and Steel Community. Proposed by the French Foreign Minister Robert Schuman, this agreement came into force in 1951, in the aftermath of World War II. Schuman's aim was to 'make war not only unthinkable but materially impossible'. This was to be achieved by a process of regional integration (European Union, 2020). At the same time, the US Marshall Plan, which contributed greatly to the post-war recovery of Western Europe, was heavily influenced by the goal of increasing markets for American oil companies and had a lasting influence on European energy use. Scholars have argued that, in contrast, the inability of the Soviet Union to use oil as a form of soft power in the way that the USA did has limited its ability to exert influence beyond the countries it occupied after 1945 (Painter, 2009). More recently, Russia became dominant in the gas market in Europe, though the situation has changed dramatically following the imposition of sanctions resulting from its illegal invasion of Ukraine.

There is now momentum to take a different path and transition away from a Europe dependent on fossil fuels to 'become the world's first climate-neutral continent' by 2050, as outlined in the EU's Green Deal (European Commission, 2021c). The EU's Green Deal contains many ambitious measures, including an aim to enshrine in EU law the 2050 climate-neutrality target (European Commission, 2021b). Whether these goals are ambitious enough is open to debate and anyway the targets must actually be met – if not, the situation will rapidly get out of control. The governance systems overseeing the implementation of initiatives like the EU's Green Deal will need to be robust. Efforts to delay or weaken the proposed measures by, for example, lobbyists acting on behalf of fossil fuel producers will need to be countered.

36.6 Changing Public Opinion

The very visible impact of climate change, for example in the flooding in Germany in 2021 and the record-breaking heatwaves in 2023, is at last contributing to political change. The 2019 wave of the Eurobarometer survey, which regularly assesses public opinion in the EU, even then reported that 94% of respondents said that protecting the environment is important to them and 91% identified climate change as a serious problem (European Commission, 2020). Eighty-three per cent supported European legislation to achieve progress. In some countries this is translating into greater support for Green parties, most notably in Germany. In other countries, the impact of a greener consciousness is being seen in the changing agenda of existing or new political parties, as with En Marche in France (de la Baume and Oroschakoff, 2021).

European cities have been, and continue to be, hubs of transformation, with many adopting a people-centred approach to how streets and public spaces are designed, a reversal of longstanding policies to prioritize the movement and parking of private cars (Buehler *et al.*, 2017). From Oslo and Copenhagen to Dijon, Paris and London, the reclaiming of city spaces and streets for people and more active modes of transport are delivering important co-benefits for people and the environment (Doran, 2020). The strength of opposition to such initiatives should not be underestimated; they have often been scaled down relative to the original transformative visions.

There is also evidence of change among those living in European cities, 61% of whom supported phasing out of petrol and diesel cars by 2030, according to one poll (Doran, 2020). However, a simple switch to privately owned electric vehicles (EVs) should not be seen as unproblematic. There are many practical challenges, and especially creation of the necessary infrastructure. In many places this will require large-scale investment to provide not only renewably sourced electrical charging points but also active transport systems. Increased 'electronic commuting', stimulated by the pandemic, may lead to long-lasting urban transport transformation, accelerating this transition, but the overall impact on energy consumption is not yet clear.

From a health perspective, there are other considerations, including reduced air pollution from exhaust fumes and less noise (from EVs), as well as lower tyre and brake wear and other co-benefits, such as increased physical exercise associated with use of public transport. On the other hand, some of the costs of change in Europe are likely to continue to be outsourced to other parts of the world, such as the environmental impact of lithium mining, which is needed to produce the rechargeable batteries that power electric vehicles (Balch, 2020).

36.7 The Role of Vested Interests

The scale and nature of the changes that are needed to address the climate crisis and build sustainable communities and ways of life should not be underestimated. This is not to suggest that such a transformation is not possible, only that, at the time of writing, the challenges ahead are formidable. Yet there is no time to lose; too little has been done for too long. The impacts of anthropogenic global warming are already a daily reality for many around the world. It is therefore imperative that the ways in which the problems are being defined and the solutions that are being proposed are scrutinized and challenged, and that barriers to taking decisive action are exposed, considered and overcome. The fossil fuel and associated industries, and the governments that benefit financially from them, have worked hard to avoid being seen as the problem (van Schalkwyk *et al.*, 2022). As in so many areas of public health, the most effective responses are those that recognize and confront the political and commercial determinants of health.

This is important, because there has been a tendency to depoliticize policy proposals. This can be seen in EU climate adaptation policy, which encourages partnerships and multistakeholder approaches. These imply, contrary to all the evidence, the fossil fuel industry and its supporters on the one hand and the environmental and health communities on the other have shared goals and can easily find 'win-win' solutions (Garsten and Jacobsson, 2013; Remling, 2018). This deflects attention from fundamental conflicts of interests and major power imbalances.

Saving the planet, rethinking our relationship with nature, achieving progress and prosperity and creating sustainable economies will only happen if we acknowledge that there are profound ideological and other differences among the various actors, not all of which are explicit. Thus, those seeking to tackle climate change must decide how they will interact with those who profit from the status quo and promote denial and delay.

Increasingly, the fossil fuel industry is being seen as the new tobacco industry, which has implications for how it should be dealt with (Supran and Oreskes, 2020; Marteau *et al.*, 2021). This recognizes how it works hard to manufacture and spread doubt about the science on climate change and tries to block, delay, or weaken policies needed to protect the environment and the climate. We now know that these strategies have been employed for decades, often with support from sections of academia, governments, legal firms, lobbying groups, think tanks and others (Oreskes and Conway, 2011; Bonneuil *et al.*, 2021).

Viewing the fossil fuel industry in this way has profound implications for policymakers, governments, the public health and advocacy communities, and institutions like the World Health Organization (WHO) and the EU. The different sectors that make up the fossil fuel industry have shaped European societies socially, politically, culturally and economically. They have long had a privileged position at the policy table, working closely with governments, legislatures and EU institutions, to the detriment of the climate, public health and the environment (Nelsen, 2016; Corporate Europe Observatory, 2019).

An understanding of how the tobacco industry behaved in ways that were anathema to public health led to agreement on a WHO Treaty, the Framework Convention on Tobacco Control. This specifies, in article 5.3, that the policymaking process should be protected from tobacco industry influence. Given how the fossil fuel industry has adopted similar tactics, there is a strong case for treating it in the same way, with scepticism as the starting point when examining the evidence it draws on (Marteau *et al.*, 2021). As with the tobacco industry, a series of legal challenges is being mounted in the USA, arguing that the fossil fuel industry has supressed evidence and misled the public and policy makers (McGreal, 2021). The outcomes of these trials are likely to have important repercussions for the industry and its political influence and public legitimacy.

36.8 Europe at COP26

Events preceding and during COP26 generated concern that global leaders would fail to commit to taking action commensurate with the scale of change needed to address the climate crisis; leaked documents revealed the extent to which some nations were lobbying the IPCC to tone down or change its recommendations. One analysis concluded that actors representing the fossil fuel industry made up the largest delegation at the conference, and some commentators described COP26 as the most exclusionary of the climate COPs from the perspective of civil society (Global Witness, 2021).

While it was apparent that COP26 needed to be a significant turning point in the global community's approach to combating the future threat from climate change, the near-final agreement, supported by the EU, was watered down at the insistence of India to remove a commitment to phase out coal production. According to media reports, pressure from both India and China led to a shift from the final pledge being 'phase out' to 'phase down' (Ellis-Petersen, 2021). These events are reminiscent of the reported deliberations that occurred during the negotiation of the text in the first IPCC report where the wording favoured by the scientists, that the contribution of human activity to climate change is 'appreciable', was changed to 'discernible' based on protest from one country delegate against the former wording (Mann, 2012). This demonstrates the level of scrutiny being applied from all countries to policy proposals in relation to climate change and energy transitions, and the challenges to transformational change. Even before these developments, COP26 lacked the bold pledges that, according to the science, are needed to meet the target of limiting global heating to 1.5°C above pre-industrial temperature. Moreover, neither the EU nor the USA supported a proposal to create a financial facility to compensate low- and middle-income countries for 'loss and damage', as proposed by the G77 countries (Weise and Mathiesen, 2021). Disappointingly, much of the more challenging negotiations in Glasgow were postponed until COP27, held in Egypt in 2022. The verdict on COP26 has been mixed but governments of nations widely perceived as most at risk, international environmental bodies and civil society have each expressed considerable disappointment. While recognizing that agreement was always going to be difficult, given the interests of fossil fuel-exporting nations, it was also notable that the head of the host government, UK Prime Minister Boris Johnson, was far less engaged in the process running up to the COP than many of his predecessors, such as French President Hollande at COP21 in 2015, which concluded with the Paris agreement. As one recent biography reveals, Johnson, who was a late convert to environmentalism, was beset by infighting among his teams. Always indecisive about appointments, he left the appointment of Alok Sharma as President of COP26 until very late. Although Sharma is widely regarded as having done a good job, he was handicapped by how the UK team 'were frustrated with Johnson and found it difficult to engage him till the event was imminent' (Seldon and Newell, 2023).

36.9 COVID-19 and Climate Change

Europe has been hit especially hard by the COVID-19 pandemic. This came as a shock because, according to assessments undertaken prior to the virus reaching Europe, many European countries were considered to be well prepared technically. Yet the pandemic has shone a light on long-standing vulnerabilities, including the erosion of social safety nets and failures in systems of governance, now the subject of numerous national inquiries.

One manifestation of this reassessment was the Pan-European Commission on Health and Sustainable Development, convened by the WHO Regional Director for Europe (Pan European Commission on Health and Sustainable Development, 2021). Chaired by Mario Monti, a former Italian Prime Minister and European Commissioner, it brought together people from wide-ranging backgrounds to look at what needed to change and be strengthened. It developed a conceptual framework for thinking about health and the 21st century (McKee, 2021). This had the health of humans, animals and the natural environment (One Health) (see Chapter 5, this volume) at its centre. This has much in common with what has been termed planetary health, which emphasizes the interconnections between the health of the planet and the health of human populations and, in practice, represents some of its main elements. This signified a recognition of how climate change, and its myriad effects, would have enormous implications for health in the future.

The Commission's recommendations are wide-ranging, from national to global levels. Within countries, the Commission calls for measures to increase resilience, both in health systems and in the social safety nets needed to protect the most vulnerable. They also include a proposal to establish a European Health Threats Council. This would bring together heads of government, informed by the results of horizon-scanning exercises, to identify ways to prevent or mitigate the many challenges ahead, including the consequences of climate change and other forms of 'planetary overload' (McMichael, 1993). These recommendations are gaining traction, especially in the G20 under the 2021 Italian presidency. However, as has been the case after previous crises, memories often fade fast.

36.10 Conclusion

While 2021 has seen the advent of space tourism and the announcement of further missions to the moon, we must not lose sight of our one and precious planet. The effects of climate change are real and already experienced by millions, with the greatest toll placed on those least responsible and with the least resilience. The impacts of climate change are diverse and devastating and will worsen if transformational action is not taken. Europe has an important role to play in taking such action. Identifying and addressing barriers to change will be essential, including confronting the vested interests of those in the petrochemical and related industries and the governments that derive revenues from them who benefit from a delay in shifting away from a fossil fuel-dependent world. Much can be learned from the COVID-19 pandemic, drawing especially on the evidence assembled to inform the Pan-European Commission, as described above, and on similar reports to inform action at a global level that can help navigate this change. But lessons can only be learned if past mistakes are acknowledged and broken governance systems fixed. There are many challenges ahead, but a starting point has been brought into stark relief by the events of 2020 and 2021: people and the delicate planetary systems upon which life on earth depends must be valued above profit and pipelines.

References

Abel, G.J., Brottrager, M., Crespo Cuaresma, J. and Muttarak, R. (2019) Climate, conflict and forced migration. *Global Environmental Change* 54, 239–249. DOI: 10.1016/j.gloenvcha.2018.12.003

Balch, O. (2020) The curse of 'white oil': electric vehicles' dirty secret. *The Guardian*. Available at: https://www.theguardian.com/news/2020/dec/08/the-curse-of-white-oil-electric-vehicles-dirty-secret-lithium (accessed 7 October 2021).

Ballester, J., Quijal-Zamorano, M., Méndez Turrubiates, R.F., Pegenaute, F., Herrmann, F.R. *et al.* (2023) Heat-related mortality in Europe during the summer of 2022. *Nature Medicine* 29, 1857–1866. DOI: 10.1038/s41591-023-02419-z

Baudron, A.R., Brunel, T., Blanchet, M.-A., Hidalgo, M., Chust, G. *et al.* (2020) Changing fish distributions challenge the effective management of European fisheries. *Ecography* 43, 494–505. DOI: 10.1111/ecog.04864

Bisselink, B., de Roo, A., Bernhard, J. and Gelati, E. (2018) Future projections of water scarcity in the Danube river basin due to land use, water demand and climate change. *Journal of Environmental Geography* 11(3-4), 25–36. DOI: 10.2478/jengeo-2018-0010

Bonneuil, C., Choquet, P.-L. and Franta, B. (2021) Early warnings and emerging accountability: Total's responses to global warming, 1971–2021. *Global Environmental Change* 71, 102386. DOI: 10.1016/j.gloenvcha.2021.102386

Buehler, R., Pucher, J., Gerike, R., and Götschi, T. (2017) Reducing car dependence in the heart of Europe: lessons from Germany, Austria, and Switzerland. *Transport Reviews* 37, 4–28. DOI: 10.1080/01441647.2016.1177799

Burrows, K. and Kinney, P.L. (2016) Exploring the climate change, migration and conflict nexus. *International Journal of Environmental Research and Public Health* 13, 443–443. DOI: 10.3390/ijerph13040443

Butler, C.D. (2018) Climate change, health and existential risks to civilization: a comprehensive review (1989–2013). *International Journal of Environmental Research and Public Health* 15, 2266. DOI: 10.3390/ijerph15102266

Butler, C.D. and Kefford, B.J. (2018) Climate change as a contributor to human conflict. *Nature* 555, 587. DOI: 10.1038/d41586-018-03795-0

Chen, J. and Mueller, V. (2018) Coastal climate change, soil salinity and human migration in Bangladesh. *Nature Climate Change* 8, 981–985. DOI: 10.1038/s41558-018-0313-8

Cheung, W.W., Lam, V.W., Sarmiento, J.L., Kearney, K., Watson, R. *et al.* (2010) Large-scale redistribution of maximum fisheries catch potential in the global ocean under climate change. *Global Change Biology* 16, 24–35. DOI: 10.1111/j.1365-2486.2009.01995.x

Clinton, J. (2023) European heatwave: red alerts issued for 16 cities in Italy. *The Guardian*. Available at: https://www.theguardian.com/world/2023/jul/15/europe-heatwave-red-alerts-italy-cities-weather (accessed 20 July 2023).

Corporate Europe Observatory (2019) *Big oil and gas buying influence in Brussels*. Corporate Europe Observatory, Brussels. Available at: https://corporateeurope.org/en/2019/10/big-oil-and-gas-buying-influence-brussels (accessed 18 November 2021).

de la Baume, M. and Oroschakoff, K. (2021) Green parties struggle to convert climate concern into votes. *Politico*. Available at: https://www.politico.eu/article/green-parties-struggle-gain-electoral-climate-traction/ (accessed 11 August, 2021).

Doran, D. (2020) This is the year a major European city will ban cars from its centre. *Wired UK*. Available at: https://www.wired.co.uk/article/car-city-ban (accessed 1 August 2023).

Dupuy, J.-l., Fargeon, H., Martin, N., Pimont, F., Ruffault, J. *et al.* (2020) Climate change impact on future wildfire danger and activity in southern Europe: a review. *Annals of Forest Science* 77, 35. DOI: 10.1007/s13595-020-00933-5

Ellis-Petersen, H. (2021) India criticised over coal at Cop26 – but real villain was climate injustice. *The Guardian*. Available at: https://www.theguardian.com/environment/2021/nov/14/india-criticised-over-coal-at-cop26-but-real-villain-was-climate-injustice (accessed 1 August 2023).

European Commission (2020) New Eurobarometer survey: Protecting the environment and climate is important for over 90% of European citizens. *Politico.* Available at: https://ec.europa.eu/commission/presscorner/detail/en/IP_20_331 (accessed 11 August, 2021).

European Commission (2021a) *Climate action: how will we be affected?* Available at: https://ec.europa.eu/clima/eu-action/adaptation-climate-change/how-will-we-be-affected_en (accessed 7 October 2021).

European Commission (2021b) *European Climate Law.* Available at: https://ec.europa.eu/clima/eu-action/european-green-deal/european-climate-law_en (accessed 7 October 2021).

European Commission (2021c) *European Green Deal.* Available at: https://ec.europa.eu/clima/eu-action/european-green-deal_en (accessed 7 October 2021).

European Union (2020) *The Schuman Declaration*. Available at: https://europa.eu/european-union/about-eu/symbols/europe-day/schuman-declaration_en (accessed 11 August 2021).

Garsten, C. and Jacobsson, K. (2013) Post-political regulation: soft power and post-political visions in global governance. *Critical Sociology* 39, 421–437. DOI: 10.1177/0896920511413942

Global Witness (2021) Hundreds of fossil fuel lobbyists flooding COP26 climate talks. Available at: https://www.globalwitness.org/en/press-releases/hundreds-fossil-fuel-lobbyists-flooding-cop26-climate-talks/ (accessed 15 November 2021).

Hernández-Triana, L.M., Jeffries, C.L., Mansfield, K.L., Carnell, G., Fooks, A.R. *et al.* (2014) Emergence of West Nile virus lineage 2 in Europe: a review on the introduction and spread of a mosquito-borne disease. *Frontiers in Public Health* 2, 271. DOI: 10.3389/fpubh.2014.00271

Hsiang, S.M., Burke, M. and Miguel, E. (2013) Quantifying the influence of climate on human conflict. *Science* 341, 1235367. DOI:10.1126/science.1235367

IPCC (2021) *AR6 Climate Change 2021: The Physical Science Basis*. Intergovernmental Panel on Climate Change, Geneva. Available at: https://www.ipcc.ch/report/ar6/wg1/ (accessed 7 October 2021).

Jit, M., Ananthakrishnan, A., McKee, M., Wouters, O.J., Beutels, P. *et al.* (2021) Multi-country collaboration in responding to global infectious disease threats: lessons for Europe from the COVID-19 pandemic. *The Lancet Regional Health – Europe* 9. DOI: 10.1016/j.lanepe.2021.100221

Jourdain, F., Roiz, D., de Valk, H., Noël, H., L'Ambert, G. *et al.* (2020) From importation to autochthonous transmission: Drivers of chikungunya and dengue emergence in a temperate area. *PLoS Neglected Tropical Diseases* 14, e0008320. DOI: 10.1371/journal.pntd.0008320

Koubi, V. (2019) Climate change and conflict. *Annual Review of Political Science* 22, 343–360. DOI: 10.1146/annurev-polisci-050317-070830

Kumar, L., Jayasinghe, S., Gopalakrishnan, T. and Nunn, P.D. (2020) Climate change and the Pacific Islands. In: Kumar, L. (ed.) *Climate Change and Impacts in the Pacific*. Springer, Cham, Switzerland, pp. 1–31.

MacDougall, A.H., Frölicher, T.L., Jones, C.D., Rogelj, J., Matthews, H.D. *et al.* (2020) Is there warming in the pipeline? A multi-model analysis of the Zero Emissions Commitment from CO_2. *Biogeosciences* 17, 2987–3016. DOI: 10.5194/bg-17-2987-2020

Mann, M.E. (2012) *The Hockey Stick and the Climate Wars: Dispatches from the Front Lines*. Columbia University Press, New York.

Marteau, T.M., Chater, N. and Garnett, E.E. (2021) Changing behaviour for net zero 2050. *BMJ* 375, n2293. DOI: 10.1136/bmj.n2293

McGreal, C. (2021) Big oil and gas kept a dirty secret for decades. Now they may pay the price. *The Guardian*. Available at: https://www.theguardian.com/environment/2021/jun/30/climate-crimes-oil-and-gas-environment (accessed 30 June, 2021).

McKee, M. (2021). *Evidence Review. Drawing light from the pandemic: a new strategy for health and sustainable development*. World Health Organization, Geneva. Available at: https://www.euro.who.int/en/health-topics/health-policy/european-programme-of-work/pan-european-commission-on-health-and-sustainable-development/publications/evidence-review.-drawing-light-from-the-pandemic-a-new-strategy-for-health-and-sustainable-development.-2021 (accessed 18 November 2021).

McMichael A.J. (1993) *Planetary Overload. Global Environmental Change and the Health of the Human Species*. Cambridge University Press, Cambridge, UK.

Molina-Terrén, D.M., Xanthopoulos, G., Diakakis, M., Ribeiro, L., Caballero, D. *et al.* (2019) Analysis of forest fire fatalities in Southern Europe: Spain, Portugal, Greece and Sardinia (Italy). *International Journal of Wildland Fire* 28, 85–98. DOI: 10.1071/WF18004

Myers, S.S., Zanobetti, A., Kloog, I., Huybers, P., Leakey, A.D. *et al.* (2014) Increasing CO_2 threatens human nutrition. *Nature* 510, 139–142. DOI: 10.1038/nature13179

Nelsen, A. (2016) EU dropped climate policies after BP threat of oil industry 'exodus'. *The Guardian*. Available at: https://www.theguardian.com/environment/2016/apr/20/eu-dropped-climate-policies-after-bp-threat-oil-industry-exodus (accessed 7 October 2021).

O'Leary, N. (2019) When will the Netherlands disappear? *Politico*. Available at: https://www.politico.eu/article/when-will-the-netherlands-disappear-climate-change/ (accessed 11 August 2021).

Oreskes, N. and Conway, E.M. (2011) *Merchants of Doubt: How a Handful of Scientists Obscured the Truth on Issues from Tobacco Smoke to Global Warming*. Bloomsbury Publishing, New York.

Painter, D.S. (2009) The Marshall Plan and oil. *Cold War History* 9, 159–175. DOI: 10.1080/14682740902871851

Pan European Commission on Health and Sustainable Development (2021) *Drawing light from the pandemic: A new strategy for health and sustainable development*. World Health Organization, Geneva. Available at: https://www.euro.who.int/en/health-topics/health-policy/european-programme-of-work/pan-european-commission-on-health-and-sustainable-development/publications/drawing-light-from-the-pandemic-a-new-strategy-for-health-and-sustainable-development-2021 (accessed 8 October 2021).

Posaner, J. and Cokelaere, H. (2023) Deep trouble for EU shipping push as Rhine River runs dry. *Politico*. Available at: https://www.politico.eu/article/germany-rhine-river-climate-change-heath-shallow-rhine-undermines-europes-waterways-push/ (accessed 20 July 2023).

Remling, E. (2018) Depoliticizing adaptation: a critical analysis of EU climate adaptation policy. *Environmental Politics* 27, 477–497. DOI:10.1080/09644016.2018.1429207

Robine, J.-M., Cheung, S.L.K., Roy, S.L., Oyen, H.V., Griffiths, C. *et al.* (2008) Death toll exceeded 70,000 in Europe during the summer of 2003. *Comptes Rendus Biologies* 331, 171–178. DOI: 10.1016/j.crvi.2007.12.001

Rottler, E., Bronstert, A., Bürger, G. and Rakovec, O. (2021) Projected changes in Rhine River flood seasonality under global warming. *Hydrology and Earth System Sciences* 25, 2353–2371. DOI: 10.5194/hess-25-2353-2021

Seldon, A. and Newell, R. (2023) *Johnson at 10*. Atlantic Books, London.
Sommer, C., Malz, P., Seehaus, T. C., Lippl, S., Zemp, M. *et al.* (2020) Rapid glacier retreat and downwasting throughout the European Alps in the early 21st century. *Nature Communications* 11, 3209. DOI: 10.1038/s41467-020-16818-0
Stagge, J.H., Kingston, D.G., Tallaksen, L.M., and Hannah, D.M. (2017) Observed drought indices show increasing divergence across Europe. *Scientific Reports* 7, 14045. DOI: 10.1038/s41598-017-14283-2
Supran, G. and Oreskes, N. (2020) Big oil is the new big tobacco. Congress must use its power to investigate. *The Guardian*. Available at: https://www.theguardian.com/commentisfree/2020/jan/20/big-oil-congress-climate-change (accessed 1 August 2023).
Turco, M., von Hardenberg, J., AghaKouchak, A., Llasat, M.C., Provenzale, A. *et al.* (2017) On the key role of droughts in the dynamics of summer fires in Mediterranean Europe. *Scientific Reports* 7(1), 81. DOI: 10.1038/s41598-017-00116-9
van Schalkwyk, M.C.I., Maani, N. and Petticrew, M. (2022) The fossil fuel industry: Fueling doubt and navigating contradiction. In: Maani, N., Petticrew, M. and Galea, S. (eds) *The Commercial Determinants of Health*. Oxford University Press. Oxford, UK, pp. 111–119. DOI: 10.1093/oso/9780197578742.003.0012
Weise, Z., and Mathiesen, K. (2021) EU, US block effort for climate disaster funding at COP26. *Politico*. Available at: https://www.politico.eu/article/eu-us-block-financial-support-climate-change-cop26/ (accessed 15 November 2021).

37 Climate Change and Health in the Arctic

Shilpa Rao,[1] Kristine Bjerve Gutzkow,[1] Susanne Hyllestad,[1] Solveig Jore,[1] Cathrine Thomsen,[1] Gro Dehli Andersen,[1] Khaled Abass,[2,3] Jim Berner,[4] Anne Lise Brantsæter,[1] Hubert Dirven,[1] Marit Låg,[1] Helle Margrete Meltzer,[1] Arja Rautio,[3] Ågot Aakra,[5] Johan Øvrevik[1,6] and Christine Instanes[1]

[1]Norwegian Institute of Public Health, Oslo, Norway; [2]Environmental Health Sciences, University of Sharjah, United Arab Emirates; [3]University of Oulo, Oulo, Finland; [4]formerly Alaska Native Tribal Health Consortium, Alaska, USA; [5]SINTEF Industry, Oslo, Norway; [6]University of Oslo, Oslo, Norway

Abstract

In this chapter we explore the main issues related to climate change and its impacts on the health of the Arctic population. We summarize evidence related to primary, secondary and tertiary health effects in the Arctic. We explore a broad range of effects on health including increased exposure to extreme weather, impacts from changes in water quality, air quality, contaminant exposure, food systems and patterns of infectious diseases. We also briefly discuss mental health effects in the Arctic.

37.1 Introduction to Arctic Health

Adverse human-driven environmental change, including to the climate, is having an extraordinary and increasing impact on the Arctic environment and its ecosystems. Inevitably, this implies significant effects on human health (Parkinson and Evengård, 2009). A recent scoping review has systematically reviewed the health impacts related to climate change in the Arctic, identifying and ranking 20 key indicators relevant to Indigenous peoples in Nunavat, Canada (Akearok *et al.*, 2019). That study was based on a sixfold conceptualization of the health effects of climate change, outlined by Watts *et al.* (2017). In this chapter, however, consistent with most of this book, we use an older and simpler classification, with only three broad categories of health effects.

'Direct' (in this book also called 'primary') health threats from climate change in the Arctic include morbidity and mortality resulting from an increased severity of extreme events (storms, floods, increased heat and in some cases cold), including an increased incidence of injury and mortality associated with unpredictable ice and storm conditions. Indirect ('secondary') health effects include those from wildfires, potential changes in bacterial and viral diseases, and decreased access to quality water sources. Other health effects that can be conceptualized as secondary include illnesses from failing sanitation infrastructure (resulting from changes in permafrost and storm surges) and dietary alteration arising via changes in the distribution and accessibility of species used for subsistence (Parkinson and Evengård, 2009). Projected warming will also affect the transport, distribution and

 Climate Change and Global Health: Primary, Secondary and Tertiary Effects, 2nd Edition (eds C.D. Butler and K. Higgs)
DOI: 10.1079/9781800620025.0037

behaviour of environmental contaminants, further threatening the safety of the traditional food supply for Arctic peoples, and potentially increasing human exposure (Miner *et al.*, 2021).

We recognize, but do not discuss in detail, that 'tertiary' effects, such as large-scale migration, hunger and even conflict may occur as consequences of climate change. Each of these phenomena is likely to be associated with significant adverse population health effects. Even if these events do not occur in the Arctic, their long-range repercussions may still be relevant.

Finally, in this chapter we note that in addition to physiological health effects, climate change may also adversely affect the mental health of some people and could contribute to social stress.

37.2 Primary Health Effects in the Arctic

Warming of the Arctic is three- to fourfold that of the other areas, due to Arctic amplification (Rantanen *et al.*, 2022). When one part of the Arctic warms, the strength and direction of wind and water currents can alter, resulting in a decrease in temperature elsewhere in the Arctic. Thus, climate impacts may be neither linear nor smooth.

Direct health impacts of climate change in the Arctic occur through rising temperatures and an increased frequency of storms, droughts, floods and even heatwaves. In the Arctic, the rise in heat extremes (Sui *et al.*, 2017; Dobricic *et al.*, 2020) and decrease in cold extremes (Sui *et al.*, 2017) is especially pronounced, in association with intensification of the hydrological cycle and other changes (Box *et al.*, 2019). Death rates in the Arctic are higher in winter than in summer and are expected to stay that way for the next few decades. Globally, low temperatures (mostly in winter) rather than high contribute most temperature-related mortality (Gasparrini *et al.*, 2015). This implies that the gradual warming of the Arctic and the fewer days with low temperatures have generated some positive impacts, such as reduced morbidity and mortality and fewer cold-related injuries. On the other hand, increased variability and unexpected weather conditions can also contribute to accidents in the Arctic, including in winter. For example, a higher occurrence of icy conditions increases the occurrence of slipping injuries. Mortality rates from unintentional injuries are already disproportionately high in Arctic regions, especially for Indigenous peoples. Rates of drowning in Inuit populations have been reported as six times higher than for non-Indigenous populations, and eight times higher for snowmobile-related drownings. Hospitalization rates for unintentional injuries from land transportation in areas with high Inuit populations are greater for adults, children and youth (Durkalec *et al.*, 2014). Climate change is expected to significantly impact weather conditions in the future and thus increase this form of burden for Arctic populations (Clarke *et al.*, 2022).

The Arctic has long been an area at high risk from ultraviolet (UV) radiation damage (Bernhard *et al.*, 2020). Highly reflective snow surfaces result in damaging levels of UV to unprotected eyes and vertical surfaces including faces, trees and shrubs. High UV levels in the Arctic are known to cause sunburn (erythema) and snow blindness (photokeratitis) under normal conditions. UV is an identified immune suppressant, with long-term health consequences such as cataracts and skin diseases, including cancer. Normally, ozone (O_3) in the stratosphere shields the Earth from much harmful UV radiation. However, recent measurements in the Arctic show that climate change is contributing to long-term decreases in the amount of ozone high overhead. In addition, scientists have observed more frequent episodes of extremely low ozone levels, particularly during the springtime, due to climate change (Friedel *et al.*, 2022). The fragile ecosystems of the Arctic, with its peoples and human-constructed surfaces, may thus face additional challenges due to ozone depletion associated modifications to UV-B and UV-A radiation.

37.3 Secondary Health Effects in the Arctic

Impacts on water

Climate change affects water availability, quantity and quality in the Arctic region in different ways. Impacts on water quantity and availability embrace changes to melting glaciers, altered seasonal precipitation and increasing river flow, including increasing contributions by groundwater. Thawing permafrost changes groundwater flow into streams and rivers and increases evapotranspiration (AMAP, 2017; Harper *et al.*, 2020). In many places, increased precipitation results in a transition from a snow- to rain-dominated area (McCrystall *et al.*, 2021).

Impacts on water quality are complex. Lowland areas are increasingly flooded with salty ocean water during storms (made worse by sea level rise). Flooding can also release chemical contaminants stored in the environment. Permafrost thaw and erosion are increasing water turbidity in some places, while runoff has decreased in non-glacial watersheds. Winter river discharge has increased in other regions, as have water temperatures and the occurrence and/or emergence of microbial contaminants (Guzman Herrador *et al.*, 2016; Harper *et al.*, 2020). In remote areas and in dwellings with basic or no drinking-water treatment, allied with inadequate sanitation facilities, such changes will negatively affect northern residents' water safety and availability. In contrast, urban areas whose residents rely on advanced drinking-water infrastructure face higher risks of microbiological contamination due to extreme rainfall events that may overwhelm water distribution systems and drinking-water treatment capacity (Hyllestad *et al.*, 2020; Widerström *et al.*, 2014).

Climate change is anticipated to exacerbate the already high burden of waterborne disease in the circumpolar North (Harper *et al.*, 2020). Strong evidence points, for example, to an association between climatic factors, such as heavy rainfall, and food- and waterborne diseases, such as *Salmonella* and *Campylobacter* in the sub-Arctic region (Hedlund *et al.*, 2014). An association has also been found between heavy precipitation events and waterborne disease outbreaks in four Nordic countries where populations rely on untreated groundwater during summer (Guzman Herrador *et al.*, 2016). In addition to an increased microbial risk from changes in the water cycle, climate change is expected to exacerbate existing chemical contaminant risks in the circumpolar North. Persistent contaminants are already distributed throughout northern ecosystems as a result of local sources (e.g. mining, pesticide use) as well as transport from lower latitudes through air, water and terrestrial routes (Harper *et al.*, 2020).

The impact of climate change on waterborne diseases is of major concern but more research is needed to gain additional insight for the full range of these effects, particularly in urban areas. The breadth of infectious waterborne pathogens has been insufficiently investigated in cold climates in general, and even less so in the context of climate change.

Impacts on air quality

Air pollution is a major environmental challenge at the global level, with adverse effects on human health, well-being and the environment. Air pollution is intrinsically linked to climate change because combustion of fossil fuel is the main source of both carbon dioxide (CO_2) and air pollutants such as particulate matter, nitrogen and sulfur dioxides, and low-level (tropospheric) ozone. Exposure to air pollutants from various sources increases the risk of diseases, including stroke, ischaemic heart disease, chronic obstructive pulmonary disease, lung cancer, pneumonia, type 2 diabetes and reduced birth weight (Health Effects Institute, 2020). Evidence also suggests an association between air pollution and neurological disorders, including Alzheimer's disease (Tsai *et al.*, 2019). Air pollution contaminants can irritate the mucous membranes and aggravate asthma and allergy symptoms induced by pollen exposure (D'Amato *et al.*, 2014).

Permafrost thawing accelerates climate change due to release of CO_2 and methane to the atmosphere, thus potentially impacting health (Ramage *et al.*, 2024). Local Arctic air emissions affect atmospheric composition at both regional and wider scales in the Arctic (Law *et al.*, 2017; Schmale *et al.*, 2018) while a significant number of recent studies have also demonstrated long-range transport of air pollution to the Arctic from lower latitudes. Local emissions typically arise from combustion sources such as fuel for vehicles, and wood, oil, coal, or natural gas for domestic heating, in addition to burning of waste and biomass, and burning of diesel, coal, or natural gas for power generation (AMAP, 2015, 2021a; Schmale *et al.*, 2018). Industrial activities in the Arctic that generate emissions are linked to oil and gas extraction, metal smelting and mineral extraction (AMAP, 2015). Melting of the Arctic Sea ice will open new shipping routes for international trade, possibly as early as the mid-century. As a consequence, increased shipping activity along the coast and inland rivers could be expected, adding to the total regional emission of combustion pollutants (Eckhardt *et al.*, 2013; Aliabadi *et al.*, 2015). Furthermore, the region's rapid warming increases the potential for local emissions of black carbon and other pollutants from boreal forest fires, which are increasing in some parts of the Arctic (Kharuk *et al.*, 2022). Sources of human exposure to pollution in the Arctic vary seasonally and during wintertime people tend to spend a significant amount of time inside, making exposure to indoor air pollution (i.e. from heating by wood) especially relevant as well.

The monitoring of air pollution in the Arctic and its impacts on health is still ongoing. How atmospheric chemical reactions affect air pollutants in the unique Arctic conditions, with extremely cold, dry air and little winter solar radiation, is still poorly understood. There is also a need for improved modelling and forecasting, strengthened efforts to monitor and if possible lower emissions, and intensified monitoring of their associated health effects. These efforts are warranted in all Arctic countries.

Impacts on environmental contaminants

Even though the Arctic has only few local/regional sources of contamination, environmental contaminants from anthropocentric activities reach it from mid-latitudes via atmospheric long-range transport, ocean currents and rivers. Environmental contaminants in the Arctic are divided into three main categories: (i) persistent organic pollutants (POPs) as defined by, and listed under, the Stockholm Convention (e.g. DDTs, PBDEs, PCBs) and some per- and poly-fluoroalkyl substances (PFASs)[1]; (ii) heavy metals (e.g. lead, mercury and cadmium); and (iii) chemicals of emerging Arctic concern (e.g. organophosphate esters, current-use pesticides, and PFAS not listed under the Stockholm convention) (AMAP, 2020). Micro- and nano-plastics are also increasingly found in the Arctic (Carlsson *et al.*, 2021). Through the processes of bioaccumulation in organisms and biomagnification in food webs, persistent contaminants progressively increase in concentration, especially in marine food webs, resulting in high concentration in top predator species, such as predatory fish, seabirds, seals, toothed whales and polar bears (AMAP, 2018). Ultimately this also affects human exposure. Although monitoring for the past decades shows decreasing trends for several POPs and mercury in Arctic air, biota and humans, some POPs and chemicals of emerging Arctic concern are either stable or increasing, with variability between areas and season (AMAP, 2021a, 2021b; Wong *et al.*, 2021).Traditional subsistence hunting of wildlife species forms a large part of the diet of Indigenous populations of northern Canada, Alaska, Greenland and the northern Russian Federation, thus exposing them to significant levels of these contaminants (AMAP, 2021a, 2021b). For metals and POPs, studies performed in Arctic populations report adverse health impacts associated with elevated exposures, including reproductive, immunological, cardiovascular and carcinogenic effects (AMAP, 2021b, 2021c). Maternal placental and breastmilk transfer of contaminants is of considerable concern, as the developing foetus and child are disproportionately vulnerable. Studies of Arctic populations also report that early-life exposure to these contaminants is negatively associated with foetal and child growth, childhood obesity and neurocognitive and behavioural outcomes in children (AMAP, 2021b).

Climate change alters the physical processes (e.g. temperature, precipitation, winds, ocean currents, snow and ice cover, and their seasonality) that drive the transport to and re-distribution of contaminants in the Arctic environment. Stored contaminants in natural reservoirs, including in ice, sediments, vegetation, soil and permafrost, may be released due to increased temperature and precipitation. Release may also be triggered by a higher frequency of extreme weather events and more land erosion, both in and outside the Arctic. Ongoing and even increasing use of pesticides and other agricultural emissions will continue to mean that contaminants are available for long-range transport to and redistribution within the Arctic (de Wit *et al.*, 2022) and uptake in food webs (Borgå *et al.*, 2022).

In addition, climate change is likely to further increase human activities and thus emissions within the Arctic area (such as via tourism, shipping transport and altered land use, including for crops and forests). Climate change is also linked to ecological changes, including shifting predator–prey relationships and new habitats for species. Of special concern is the rapid decline in Arctic sea ice, which threatens the existence of ice-dependent Arctic species (e.g. polar bears and seals) as well as altering food-web structures and thus pathways of contaminant exposures in wildlife and humans. Altogether, these complex and dynamic physical and interlinked ecological factors directly or indirectly caused by climate change will increasingly impact biodiversity and contaminant bioaccumulation within fish and wildlife.

Impacts on food systems

The traditional diet varies among the Arctic nations. In Canada, Greenland, Alaska and the far east of the Russian Federation, there is significant marine mammal consumption by Indigenous peoples. Furthermore, reindeer herding is an important livelihood among the Saami people in Norway, Sweden, Finland and Russia and thus reindeer meat constitutes a significant amount of the diet there. Fish consumption is also a significant part of the diet among both Indigenous and non-Indigenous peoples in many Arctic countries. However, Arctic populations have been moving towards store-bought 'Western' foods for decades, due to greater accessibility, affordability and also concerns about the health impacts from contaminants in traditional and local foods. Higher reliance on imported foods and higher cost of healthy foods contribute to unhealthy diets. The details and pace of these nutritional transitions vary across and within the Arctic countries, but in general the shifts have been greatest in urban areas and towns. Such dietary changes may have negative and positive impacts; their balance is often difficult to ascertain. The benefits of diminished consumption of traditional and local foods include significant reductions of contaminants measured in the blood of pregnant women (Adamou *et al.*, 2020). Increased consumption of vegetables and fruit can also occur in some cases. At the same time, studies have noted troubling trends in deterioration of health status, including a rising prevalence of obesity, diabetes and other metabolic diseases. These harms are triggered not only by increased intake of calorie-dense, nutrient-poor foods, but also by reduced physical activity as the incentive to hunt declines (Little *et al.*, 2021). A deterioration in dental health has also been observed, related to higher consumption of sugary drinks, while the decline in cardiometabolic health is related to reduced consumption of fish and other traditional foods and consequently lower intakes of omega-3 fatty acids, vitamin D and iodine (Andersen *et al.*, 2005; Tvermosegaard *et al.*, 2015; Calder *et al.*, 2019; Nath *et al.*, 2021).

The consequences of climate change pose several threats to food systems and food security in the Arctic region. Rising sea levels, changes in wind and precipitation patterns, and more unpredictable and severe weather events have been documented to result in a decline in fish stocks, birds and marine mammals – species that are traditionally and culturally important, both as food and for human identity in the Arctic region (Little *et al.*, 2021). Climate change-induced increases in biting pests, such as mosquitoes, may also have negative consequences for reindeer and caribou populations, and their role as a local resource (Koltz and Culler, 2021). Other climate change-related phenomena included unpredictable temperature fluctuation and heavier winds and rainfall; these may impact game mating and hunting times and alter fish migration patterns. The overall consequences of this remains unclear. Arctic communities are already burdened by significant disparities in food security and health, and climate change enhances this through changes in local food systems, food security and affordability (Arctic Council, 2021).

Impact on infectious diseases

Infectious disease rates in humans across the Arctic are highly variable, depending on country, ethnicity, disease, age and gender. Current trends in infectious diseases include the re-emergence of tuberculosis (TB), high prevalence of sexually transmitted diseases (STIs), introduction of new pathogens, such as COVID-19, changes in the emergence of food- and waterborne diseases and the spread of vector-borne and zoonotic diseases, partly as a result of climate change (Petrov *et al.*, 2020; Ma *et al.*, 2021). TB and STIs are a significant public health burden in the Arctic populations, with higher incidence rates compared with their respective country's general populations. Environmental and climatic threats include increases in ocean and atmospheric temperature, with resulting changes in biota, habitat, foraging grounds, breeding areas and migration routes, and changes in disease-vector populations, ranges and life cycles (Ma *et al.*, 2021).

Increased human interest and activity in the Arctic bring additional challenges that may impact infectious diseases among wildlife and have the potential to introduce new pathogens and to establish new hosts. Resulting change, such as reduced winter grazing, can stress animals, potentially making them more vulnerable to infectious diseases (Bradley *et al.*, 2005). Many Arctic pathogens may have environmental hosts that are still unknown; for example, *Coxiella burnetti* (causing human Q fever) has recently been identified in Northern fur seals and *Franciscella tularensis* (causing human tularaemia) in polar bears (Atwood *et al.*, 2017). Environmental reservoirs for potentially new Arctic infections are unknown. The ecology of Arctic pathogens requires collaborative investigation, collating local and indigenous knowledge, in partnership with science. There is an increasing infectious disease threat to Arctic wildlife, which includes shifts in disease emergence, rising potential for new and old re-emerging pathogens, and increased infection risk for the human residents.

Warmer ocean and terrestrial temperatures can increase the growth rates of pathogens and animals, including insect vectors and biting pests affecting human health and wellbeing (Koltz and Culler, 2021; Ma *et al.*, 2021). Ticks and tick-borne diseases, such as the *Borrelia* bacteria causing Lyme disease, are gradually moving towards polar regions as temperatures rise. Extreme precipitation may result in flooding and disruption of water/sanitation infrastructure, elevating the risk for waterborne disease outbreaks. Indirectly, climatic factors also affect infectious disease transmission by altering human behaviour. Changes in climatic factors can expand a disease vector's geographical range, enlarge its population and enable more vector species and individuals to survive though winter (Yastrebov *et al.*, 2016). Melting permafrost and warming climate also cause release of previously buried infective agents and contaminants. In 2016 an anthrax outbreak in north-west Siberia killed thousands of reindeer and also affected humans. The outbreak was ascribed to activation of anthrax spores due to permafrost thaw, accelerated by an accompanying heatwave (Ezhova *et al.*, 2021). The warming Arctic climate and permafrost thaw may also influence contaminant exposure and facilitate the spread of zoonotic infectious diseases in the Arctic.

There is still a lack of knowledge about the health impacts and risks of active zoonotic infections in wildlife and the consequent risk to human health. Thus, the development of monitoring networks is needed, especially for new and zoonotic diseases in all Arctic countries, together with the development of useful epidemiological zoonotic and human disease models. Such approaches will better integrate existing data and improve disease forecasts. These models will also facilitate better estimation of the risk and magnitude of human and wildlife health impacts and changes in disease and contaminant exposure in responses to different scenarios of climate change.

37.4 Mental Health Effects in the Arctic

New studies indicate that changing weather patterns also significantly impact mental health in Arctic communities. Three key pathways have been described: (i) shaping daily lived experiences, including connection to place and other determinants of well-being; (ii) altering mood and emotion on a transient basis; and (iii) seasonally influencing individual and community health and well-being (Middleton *et al.*, 2020). Resident Indigenous populations of the Arctic with traditional lifestyles may be especially vulnerable to the adverse mental health effects of climate change because of their close relationship with, and dependence on, the land, sea and natural resources (Akearok *et al.*, 2019). Grief and anxiety may arise as a result of primary, secondary and tertiary effects; we thus group these phenomena as 'cross-cutting'.

For example, unstable ice may cause someone to drown but may also trigger anxiety among survivors. An altered pattern of an infectious disease, or a mysterious condition possibly attributable to an environmental contaminant, may engender feelings of despair and helplessness. Relocating, when and if this becomes necessary in the Arctic due to factors such as sea level rise and infrastructure damage (Garland *et al.*, 2022), is bound to have adverse mental health effects.

Reduced opportunities for subsistence hunting, fishing, herding and gathering are likely to cause psychological stresses due to the loss of important cultural activities. The warming climate has already influenced the 'reindeer culture' of the Saami, the only Indigenous people remaining in the European Union (Jaakkola *et al.*, 2018). Harm to mental health may arise from loss of traditional lifestyles, and also via 'solastalgia' – the long-lasting sadness experienced by some, especially with high biophilia, when familiar landscapes deteriorate within living memory (Albrecht, 2019).

37.5 Tertiary Health Effects in the Arctic

Changes in human and animal migratory patterns and other factors may result in reduced access to traditional food, which is likely to force some Indigenous communities to increasingly rely on non-traditional foods, which can also be less healthy. Flooding, erosion and permafrost thawing related to climate change can negatively affect village habitability and infrastructure and may also contribute to population dislocation and community disruption. Although it may currently seem far-fetched, 'runaway' climate change (Lenton *et al.*, 2019) may in future force people to flee to extreme northerly latitudes, simply for survival.[2]

37.6 Conclusion

The Arctic is strongly affected by climate change. Yet, the likely human health effects remain poorly understood compared with those at lower latitudes. Health effects may arise from direct effects of climate changes, such as via extreme weather events. Secondary effects include via changes in air, water and food quality, and by alterations in the pattern of some infectious diseases. Tertiary effects may arise via alterations in migration and lifestyles. The extent and magnitude of health impacts from these effects are largely unknown, underscoring the need for health research in the Arctic.

Several knowledge gaps must be addressed to project future effects of the climate crisis on environmental contaminant levels in the Arctic. These include contaminant transport to and within the Arctic, and the exposure pathways and levels in food webs, the biota and – ultimately – among humans. Circumpolar temporal monitoring and modelling of contaminant levels in air, water, cryosphere and biota is therefore needed. Such datasets can be integrated with source inventory and emission data and modelling of transport to and within the Arctic, as well as the interlinked ecological changes. Biobanks, storing biological samples from both human and other species, should be established to monitor time trends and to detect new and newly recognized chemicals.

The consequences of climate change are a serious threat to food security of Indigenous populations in the Arctic. Loss of fish, birds, marine mammals and large terrestrial mammals such as reindeer and caribou will reduce the ability of people in the Arctic to harvest their traditional foods and may affect both lifestyle and life quality. This may also contribute to further accelerate the ongoing nutrition transition towards less sustainable dietary patterns more dominated by unhealthy Western diets, with a higher share of unhealthy ultra-processed foods.

Climate change is also anticipated to exacerbate an already high burden of waterborne disease in the circumpolar North. Strong evidence points, for example, to an association between climatic factors, such as heavy rainfall, and food- and waterborne diseases, including *Salmonella* and *Campylobacter* in the sub-Arctic. There are increasing infectious disease threats to Arctic wildlife and hence to the distribution of zoonoses, which include shifts in disease emergence, rising potential for new and old re-emerging pathogens, and thus the possibility for increased infection risk for the human residents. Melting permafrost could release previously buried infective agents and contaminants.

Preventing release of climate drivers, such as CO_2, methane and black carbon, should be the primary goal to reduce the impact of climate change in the Arctic and elsewhere. Considerable extraction of fossil fuels already occurs in several parts of the Arctic. This extraction is projected to increase, creating tension between two major competing interests: nations and communities that profit from such activities; and Indigenous and other people and places who will shoulder most of the burden (Hanaček *et al.*, 2022).

Current climate predictions suggest that further marked warming of the Arctic is inevitable. Adaptation to these changes is therefore crucial, either to prevent or at least reduce the adverse health impacts of climate change in the Arctic. These challenges call for an interdisciplinary research effort, standardization of methodologies and increased availability of environmental and ecological data across Arctic areas for both researchers and policy makers.

Notes

[1] Dichlorodiphenyltrichloroethane (DDT), polychlorinated biphenyls (PCBs) and polybrominated diphenyl ethers (PBDEs).

[2] A prediction made by the pioneering Earth system scientist, James Lovelock, almost 20 years ago (*Nunatsiaq News*, 2006).

References

Adamou, T.Y., Riva, M., Muckle, G., Laouan Sidi, E.A., Lemire, M. *et al.* (2020) Blood mercury and plasma polychlorinated biphenyls concentrations in pregnant Inuit women from Nunavik: Temporal trends, 1992–2017. *Science of the Total Environment* 743, 140495. DOI: 10.1016/j.scitotenv.2020.140495

Akearok, G.H., Holzman, S., Kunnuk, J., Kuppaq, N., Martos, Z. *et al.* (2019) Identifying and achieving consensus on health-related indicators of climate change in Nunavut. *Arctic* 72, 289–299. DOI: 10.14430/arctic68719

Albrecht, G.A. (2019) *Earth Emotions*. Cornell University Press, Ithaca, New York.

Aliabadi, A.A., Staebler, R.M. and Sharma, S. (2015) Air quality monitoring in communities of the Canadian Arctic during the high shipping season with a focus on local and marine pollution. *Atmospheric Chemistry and Physics* 15, 2651–2673. DOI: 10.5194/acp-15-2651-2015

AMAP (2015) *AMAP Assessment 2015: Human Health in the Arctic*. Arctic Monitoring and Assessment Programme, Oslo.

AMAP (2017) *Snow, Water, Ice and Permafrost in the Arctic*. Arctic Monitoring and Assessment Programme, Oslo.

AMAP (2018) *AMAP Assessment 2018: Biological Effects of Contaminants on Arctic Wildlife and Fish*. Arctic Monitoring and Assessment Programme, Oslo.

AMAP (2020) *AMAP Assessment 2020: POPs and Chemicals of Emerging Arctic Concern: Influence of Climate Change*. Arctic Monitoring and Assessment Programme, Tromsø, Norway.

AMAP (2021a) *POPs and Chemicals of Emerging Arctic Concern: Influence of Climate Change*. Arctic Monitoring and Assessment Programme, Tromsø, Norway.

AMAP (2021b) *AMAP Assessment 2021: Human Health in the Arctic*. Arctic Monitoring and Assessment Programme, Tromsø, Norway.

AMAP (2021c) *Human Health in the Arctic 2021, Summary for Policy Makers*. Arctic Monitoring and Assessment Programme, Tromsø, Norway.

Andersen, S., Hvingel, B., Kleinschmidt, K., Jørgensen, T. and Laurberg, P. (2005) Changes in iodine excretion in 50–69-y-old denizens of an Arctic society in transition and iodine excretion as a biomarker of the frequency of consumption of traditional Inuit foods. *American Journal of Clinical Nutrition* 81, 656–663. DOI: 10.1093/ajcn/81.3.656

Arctic Council (2021) *Indigenous Food Security in the Arctic. Implications of a Changing Ocean*. Arctic Council, Tromsø, Norway. Available at: https://oaarchive.arctic-council.org/bitstream/handle/11374/2662/MMIS12_2021_REYKJAVIK_PAME_Indigenous-Food-Security.pdf?sequence=1&isAllowed=y (accessed 1 July 2023)

Atwood, T.C., Duncan, C., Patyk, K.A., Nol, P., Rhyan, J. *et al.* (2017) Environmental and behavioral changes may influence the exposure of and Arctic apex predator to pathogens and contaminants. *Scientific Reports* 7, 13193. DOI: 10.1038/s41598-017-13496-9

Bernhard, G.H., Fioletov, V.E., Grooß, J.-U., Ialongo, I., Johnsen, B. *et al.* (2020) Record-breaking increases in Arctic solar ultraviolet radiation caused by exceptionally large ozone depletion in 2020. *Geophysical Research Letters* 47, e2020GL090844. DOI: 10.1029/2020GL090844

Borgå, K., Mckinney, M.A., Routti, H., Fernie, K.J., Giebichenstein, J. *et al.* (2022) The influence of global climate change on accumulation and toxicity of persistent organic pollutants and chemicals of emerging concern in Arctic food webs. *Environmental Science: Processes & Impacts* 24, 1544–1576. DOI: 10.1039/d1em00469g

Box, J.E., Colgan, W.T., Christensen, T.R., Schmidt, N.M., Lund, M. *et al.* (2019) Key indicators of Arctic climate change: 1971–2017. *Environmental Research Letters* 14, 14 045010. DOI: 10.1088/1748-9326/aafc1b

Bradley, M.J., Kutz, S.J., Jenkins, E. and O'Hara, T.M. (2005) The potential impact of climate change on infectious diseases of Arctic fauna. *International Journal of Circumpolar Health* 64, 468–477. DOI: 10.3402/ijch.v64i5.18028

Calder, R.S.D., Bromage, S. and Sunderland, E.M. (2019) Risk tradeoffs associated with traditional food advisories for Labrador Inuit. *Environmental Research* 168, 496–506. DOI: 10.1016/j.envres.2018.09.005

Carlsson, P., Singdahl-Larsen, C. and Lusher, A.L. (2021) Understanding the occurrence and fate of microplastics in coastal Arctic ecosystems: The case of surface waters, sediments and walrus (*Odobenus rosmarus*). *Science of The Total Environment* 792, 148308. DOI: 10.1016/j.scitotenv.2021.148308

Clarke, B., Otto, F., Stuart-Smith, R. and Harrington, L. (2022) Extreme weather impacts of climate change: an attribution perspective. *Environmental Research Climate* 1, 012001. DOI: 10.1088/2752-5295/ac6e7d

D'Amato, G., Bergmann, K.C., Cecchi, L., Annesi-Maesano, I., Sanduzzi, A. *et al.* (2014) Climate change and air pollution. Effects on pollen allergy and other allergic respiratory diseases. *Allergo Journal* 23, 17–23. DOI: 10.1007/s40629-014-0003-7

de Wit, C.A., Vorkamp, K. and Muir, D. (2022) Influence of climate change on persistent organic pollutants and chemicals of emerging concern in the Arctic: state of knowledge and recommendations for future research. *Environmental Science Processes & Impacts* 24, 1530–1543. DOI: 10.1039/D1EM00531F

Dobricic, S., Russo, S., Pozzoli, L., Wilson, J. and Vignati, E. (2020) Increasing occurrence of heat waves in the terrestrial Arctic. *Environmental Research Letters* 15, 024022. DOI: 10.1088/1748-9326/ab6398

Durkalec, A., Furgal, C., Skinner, M.W. and Sheldon, T. (2014) Investigating environmental determinants of injury and trauma in the Canadian north. *International Journal of Environmental Research and Public Health* 11, 1536–1548. DOI: 10.3390/ijerph110201536

Eckhardt, S., Hermansen, O., Grythe, H., Fiebig, M., Stebel, K. *et al.* (2013) The influence of cruise ship emissions on air pollution in Svalbard – a harbinger of a more polluted Arctic? *Atmospheric Chemistry and Physics* 13, 8401–8409. DOI: 10.5194/acp-13-8401-2013

Ezhova, E., Orlov, D., Suhonen, E., Kaverin, D., Mahura, A. *et al.* (2021) Climatic factors influencing the anthrax outbreak of 2016 in Siberia, Russia. *Ecohealth* 18, 217–228. DOI: 10.1007/s10393-021-01549-5

Friedel, M., Chiodo, G., Stenke, A., Domeisen, D.I.V., Fueglistaler, S. *et al.* (2022) Springtime arctic ozone depletion forces northern hemisphere climate anomalies. *Nature Geoscience* 15, 541–547. DOI: 10.1038/s41561-022-00974-7

Garland, A., Bukvic, A. and Maton-Mosurska, A. (2022) Capturing complexity: Environmental change and relocation in the North Slope Borough, Alaska. *Climate Risk Management* 38, 100460. DOI: 10.1016/j.crm.2022.100460

Gasparrini, A., Guo, Y., Hashizume, M., Lavigne, E., Zanobetti, A. *et al.* (2015) Mortality risk attributable to high and low ambient temperature: a multicountry observational study. *The Lancet* 386, 369–375. DOI: 10.1016/S0140-6736(14)62114-0

Guzman Herrador, B., De Blasio, B.F., Carlander, A., Ethelberg, S., Hygen, H.O. *et al.* (2016) Association between heavy precipitation events and waterborne outbreaks in four Nordic countries, 1992–2012. *Journal of Water and Health* 14, 1019–1027. DOI: 10.2166/wh.2016.071

Hanaček, K., Kröger, M., Scheidel, A., Rojas, F. and Martinez-Alier, J. (2022) On thin ice – The Arctic commodity extraction frontier and environmental conflicts. *Ecological Economics* 191, 107247. DOI: 10.1016/j.ecolecon.2021.107247

Harper, S.L., Wright, C., Masina, S. and Coggins, S. (2020) Climate change, water, and human health research in the Arctic. *Water Security* 10, 100062. DOI: 10.1016/j.wasec.2020.100062

Health Effects Institute (2020) *State of Global Air 2020. Special Report*. Health Effects Institute, Boston, Massachusetts.

Hedlund, C., Blomstedt, Y. and Schumann, B. (2014) Association of climatic factors with infectious diseases in the Arctic and subarctic region – a systematic review. *Global Health Action* 7, 24161. DOI: 10.3402/gha.v7.24161

Hyllestad, S., Iversen, A., Macdonald, E., Amato, E., Borge, B. *et al.* (2020) Large waterborne Campylobacter outbreak: use of multiple approaches to investigate contamination of the drinking water supply system, Norway, June 2019. *Eurosurveillance* 25, 2000011. DOI: 10.2807/1560-7917.ES.2020.25.35.2000011

Jaakkola, J.J.K., Juntunen, S. and Näkkäläjärvi, K. (2018) The holistic effects of climate change on the culture, well-being, and health of the Saami, the only indigenous people in the European Union. *Current Environmental Health Reports* 5, 401–417. DOI: 10.1007/s40572-018-0211-2

Kharuk, V.I., Dvinskaya, M.L., Im, S.T., Golyukov, A.S. and Smith, K.T. (2022) Wildfires in the Siberian Arctic. *Fire* 5, 106. DOI: 10.3390/fire5040106

Koltz, A.M. and Culler, L.E. (2021) Biting insects in a rapidly changing Arctic. *Current Opinion in Insect Science* 47, 75–81. DOI: 10.1016/j.cois.2021.04.009

Law, K.S., Roiger, A., Thomas, J.L., Marelle, L., Raut, J.-C. *et al.* (2017) Local Arctic air pollution: Sources and impacts. *Ambio* 46, 453–463. DOI: 10.1007/s13280-017-0962-2

Lenton, T.M., Rockström, J., Gaffney, O., Rahmstorf, S., Richardson, K. *et al.* (2019) Climate tipping points – too risky to bet against. *Nature* 575, 592–595. DOI: 10.1038/d41586-019-03595-0

Little, M., Hagar, H., Zivot, C., Dodd, W., Skinner, K. *et al.* (2021) Drivers and health implications of the dietary transition among Inuit in the Canadian Arctic: a scoping review. *Public Health Nutrition* 24, 2650–2668. DOI: 10.1017/S1368980020002402

Ma, Y., Destouni, G., Kalantari, Z., Omazic, A., Evengård, B. *et al.* (2021) Linking climate and infectious disease trends in the Northern/Arctic region. *Scientific Reports* 11, 20678. DOI: 10.1038/s41598-021-00167-z

McCrystall, M.R., Stroeve, J., Serreze, M., Forbes, B.C. and Screen, J.A. (2021) New climate models reveal faster and larger increases in Arctic precipitation than previously projected. *Nature Communications* 12, 6765. DOI: 10.1038/s41467-021-27031-y

Middleton, J., Cunsolo, A., Jones-Bitton, A., Shiwak, I., Wood, M. *et al.* (2020) 'We're people of the snow': Weather, climate change, and Inuit mental wellness. *Social Science & Medicine* 262, 113137. DOI: 10.1016/j.socscimed.2020.113137

Miner, K.R., D'Andrilli, J., Mackelprang, R., Edwards, A., Malaska, M.J. *et al.* (2021) Emergent biogeochemical risks from Arctic permafrost degradation. *Nature Climate Change* 11, 809–819. DOI: 10.1038/s41558-021-01162-y

Nath, S., Poirier, B.F., Ju, X., Kapellas, K., Haag, D.G. *et al.* (2021) Dental health inequalities among Indigenous populations: a systematic review and meta-analysis. *Caries Research* 55, 268–287. DOI: 10.1159/000516137

Nunatsiaq News (2006) Arctic – the last refuge from climate change. *Nunatsiaq News*, Iqaluit, Nunavut, Canada. Available at: https://nunatsiaq.com/stories/article/arctic_-_the_last_refuge_from_climate_change/ (accessed 2 April 2023).

Parkinson, A.J. and Evengård, B. (2009) Climate change, its impact on human health in the Arctic and the public health response to threats of emerging infectious diseases. *Global Health Action* 2. DOI: 10.3402/gha.v2i0.2075

Petrov, A., Welford, M., Golosov, N., DeGroote, J., Degai, T. *et al.* (2020) Spatiotemporal dynamics of the COVID-19 pandemic in the Arctic: early data and emerging trends. *International Journal of Circumpolar Health* 79, 1835251. DOI: 10.1080/22423982.2020.1835251

Ramage, J., Kuhn, M., Virkkala, A.-M., Voigt, C., Marushchak, M.E. *et al.* (2024) The net GHG balance and budget of the permafrost region (2000–2020). *Global Biogeochemical Cycles* 38, e2023GB007953. DOI: 10.1029/2023GB007953

Rantanen, M., Karpechko, A.Y., Lipponen, A., Nordling, K., Hyvärinen, O. *et al.* (2022) The Arctic has warmed nearly four times faster than the globe since 1979. *Communications Earth & Environment* 3, 168. DOI: 10.1038/s43247-022-00498-3

Schmale, J., Arnold, S.R., Law, K.S., Thorp, T., Anenberg, S. *et al.* (2018) Local Arctic air pollution: a neglected but serious problem. *Earth's Future* 6, 1385–1412. DOI: 10.1029/2018EF000952

Sui, C., Zhang, Z., Yu, L., Li, Y. and Song, M. (2017) Investigation of Arctic air temperature extremes at north of 60°N in winter. *Acta Oceanologica Sinica* 36, 51–60. DOI: 10.1007/s13131-017-1137-5

Tsai, T.-L., Lin, Y.-T., Hwang, B.-F., Nakayama, S.F., Tsai, C.-H. *et al.* (2019) Fine particulate matter is a potential determinant of Alzheimer's disease: A systemic review and meta-analysis. *Environmental Research* 177, 108638. DOI: 10.1016/j.envres.2019.108638

Tvermosegaard, M., Dahl-Petersen, I.K., Nielsen, N.O., Bjerregaard, P. and Jørgensen, M.E. (2015) Cardiovascular disease susceptibility and resistance in circumpolar Inuit populations. *Canadian Journal of Cardiology* 31, 1116–1123. DOI: 10.1016/j.cjca.2015.06.002

Watts, N., Adger, W.N., Ayeb-Karlesson, S., Bai, Y., Byass, P. *et al.* (2017) The Lancet countdown: Tracking progress on health and climate change. *The Lancet* 389, 1151–1164. DOI: 10.1016/s0140-6736(16)32124-9

Widerström, M., Schönning, C., Lilja, M., Lebbad, M., Ljung, T. *et al.* (2014) Large outbreak of Cryptosporidium hominis infection transmitted through the public water supply, Sweden. *Emerging Infectious Diseases* 20, 4905–4922. DOI: 10.3201/eid2004.121415

Wong, F., Hung, H., Dryfhout-Clark, H., Aas, W., Bohlin-Nizzetto, P. *et al.* (2021) Time trends of persistent organic pollutants (POPs) and Chemicals of Emerging Arctic Concern (CEAC) in Arctic air from 25 years of monitoring. *Science of The Total Environment* 775, 145109. DOI: 10.1016/j.scitotenv.2021.145109

Yastrebov, V.K., Rydakov, N.V. and Rydakova, S.A. (2016) Epidemiology of the transmissible tick-borne infections in Russia. *Public Health and Life Environment (Zdorov'e Naseleniya i Sreda Obitaniya – ZNiSO)* 11, 8–12.

38 How Can Public Health Add Value to Climate Policy?

Alistair Woodward,[1] Kirsty Wild[1] and Jonathan M. Samet[2]
[1]*Epidemiology and Biostatistics, School of Population Health, University of Auckland, New Zealand;* [2]*Colorado School of Public Health, Anschutz, Colorado, USA*

Abstract

Global greenhouse gas emissions continue to rise, despite the growth in our knowledge of climate change and its effects. This simple and rather dismal observation prompts the question: how can public health contribute most effectively to climate policy? Here are four value-adding actions by which public health researchers and practitioners may accelerate effective action on climate change. These are: to provide information on health effects, including both observed impacts and projected risks; to demonstrate how health protection adds teeth to climate policy; to make better use of openings and opportunities to influence policy makers; and to engage more strongly with advocacy and activism.

38.1 Introduction

Health belongs 'at the heart of the climate change agenda', according to the World Health Organization (WHO, 2009). It is no longer adequate to understand climate change as an 'environmental' issue or a 'development' challenge, WHO argued more than ten years ago. According to WHO, climate change now threatens the 'protection and improvement of human health and well-being' everywhere, in every community.

But what weight does health carry in decision making about climate change? If health is indeed 'at the heart of the climate change agenda', one might expect health professionals to be at the centre of decision making. One might expect health outcomes to be explicitly considered in the trade-offs and balances that are part and parcel of establishing priorities. It would be natural for climate change to be framed as a health issue at least as frequently as it is considered an economic conundrum, or a security concern. In fact, in our view, health has played a relatively modest role so far in guiding policy on reducing greenhouse emissions.

There are good reasons for that to change – not just because technical understanding of the impacts on human health is highly relevant to debates about mitigation and adaptation but also because public health brings valuable professional and theoretical resources to the table. These include: (i) commitments to broad, multidisciplinary coalitions that enable us to identify 'co-benefits' or multiple, overlapping gains from climate action in different spheres of community life; (ii) a deep engagement with the 'upstream' socio-economic factors that underpin complex or 'wicked' social problems – including the particularly difficult work of identifying and tackling corporate influence in public life; (iii) skills and knowledge relevant to elaborating and responding to the inequities caused by environmental

 Climate Change and Global Health: Primary, Secondary and Tertiary Effects, 2nd Edition (eds C.D. Butler and K. Higgs)
DOI: 10.1079/9781800620025.0038

hazards; and (iv) experience navigating the challenges and opportunities of 'activist science'. Health considerations are important also when making strategies for adaptation, particularly in response to temperature extremes, air pollution and disrupted food supplies.

In this chapter we explore how public health can make a stronger contribution to climate policy. We draw on a narrative review of early engagements of public health academics with climate policy, writing as three public health researchers – two environmental epidemiologists [AW and JS] and one environmental sociologist [KW]. We have experience in both researcher and clinical/practitioner roles, (AW and JS are public health doctors, and KW has worked extensively in the community sector), and have been closely involved, over many years, in public health 'advocacy', including engagement with politicians and policy makers, and public-good legal cases. We have foundational commitments to a scientific discipline (epidemiology and sociology) as well as ethical and professional commitments to the practice of a multidisciplinary, equity-centred 'public health'.

38.2 Methods

We have used a narrative review to identify public health research that is useful, usable and easily applied. Narrative reviews are particularly well suited for exploratory investigations into relatively new topics without a well-established literature (Sovacool *et al.*, 2018). While a systematic review involves a narrow set of precise questions and summarizes primary studies on a topic, a narrative literature review explores a broader range of studies seeking to identify connections and synergies across a more diverse evidence base (Greenhalgh *et al.*, 2018).

In undertaking the review, we aim to do more than describe themes; our goal is to identify insights that add weight to public health inquiry. In this endeavour we are influenced by calls for more 'problem-driven research ... that confronts social or environmental issues, rather than merely describing them' (Sovacool *et al.*, 2018). These climate and energy researchers implore researchers to work more in the 'boundary space between research and practice' in ways that help us to produce research that is 'immediately useful at addressing a pressing energy- or climate-related problem'. In this vein we utilize our combined 100 years of public health practice to attempt to answer the question: 'What could be useful here?'

We searched the English language literature for papers published between 2000 and 2020, using Google Scholar and the search term: ("climate" OR "climate change") AND ("health" OR "public health") AND "policy". From this search we extracted what appeared in our judgement to be the 50 most relevant papers, and then identified common themes. These are shown in Table 38.1.

38.3 Understanding Health Impacts: Putting Health on the Agenda

The consequences of an overheated global climate system are evident in long-term trends (in temperature especially), and in the frequency and severity of extreme weather events. For example, in June 2021 parts of the north-west USA and Canada experienced temperatures that exceeded any recorded previously by more than 20°C (Bartusek *et al.*, 2022) and resulted in about a thousand deaths (the precise number is not yet known). A rapid attribution analysis found that it was 'virtually impossible' that such extreme heat could have occurred without human-induced climate change (Phillip *et al.*, 2022).

The ultimate aim of the UN Framework Convention on Climate Change is to prevent 'dangerous' human interference in the climate system, and while 'dangerous' is not defined precisely, the Convention refers specifically to adverse effects on 'human health and welfare'. On the other hand, how the world deals with climate change matters hugely to the future of human health and welfare: the main message of the first Lancet Commission on Health and Climate Change was 'tackling climate change could be the greatest global health opportunity of the 21st century' (Watts *et al.*, 2018).

However, health is seldom at the heart of efforts to reduce climate change. The Nationally Determined Contributions submitted by governments signed up to the 2015 Paris Agreement were intended to be the foundation for transformative climate policy globally. About half made only a general reference to health, or said nothing at all. Those that included no mention

Table 38.1. How public health can contribute to climate policy.

	Value-adding actions	Examples of references
	Provide information on health effects, both observed impacts and projected risks	
1.	Ensure there is full accounting of the effects of climate change on health and health equity	Walpole *et al.*, 2009 Bressler, 2021
2.	Identify, test and promote health co-benefits of climate policy	Haines *et al.*, 2009; Whitmee *et al.*, 2021 Chang *et al.*, 2017 Woodward, 2019
	Health protection adds teeth to climate policy	
3.	Control present-day climate-sensitive diseases	Campbell-Lendrum *et al.*, 2015
4.	Apply well-established public health measures to aid mitigation and improve climate resilience	Frumkin *et al.*, 2008 Keim, 2008 Dodson *et al.*, 2020
	Make better use of 'openings' and opportunities	
5.	Seek synergies between the COVID-19 response and climate policy	Semenza, 2021 Pelling *et al.*, 2021
6.	Re-orient research and practice to more closely match the needs of policy makers	Hosking and Campbell-Lendrum, 2012 Browne and Rutherfurd, 2017
7.	Model the changes required within the health system	Pencheon, 2018
	Engage more strongly with advocacy and activism	
8.	Shift the focus to climate action – remember public health is change-making at heart	Szreter, 2003 Rossa-Roccor *et al.*, 2021
9.	Public health perspectives may help reduce political polarization and sectarianism in times of disruption and change	Corner *et al.*, 2014 Finkel *et al.*, 2020
10.	Use health to frame communications about climate change	Petrovic *et al.*, 2014 Bernauer and McGrath, 2016

of health came predominantly from high-income countries, those with the greatest responsibility for historic emissions, such as the EU, the USA, Australia, New Zealand and Japan (Dasandi *et al.*, 2021).

Full accounting of the effects of climate change on health and health equity

International Physicians for the Prevention of Nuclear War (IPPNW) received the 1985 Nobel Peace Prize for their efforts to reduce the numbers of nuclear weapons, and to persuade major powers to step away from risky security policies that relied on nuclear deterrence and rapid response. In making this award, the Nobel Committee emphasized the work of IPPNW in 'spreading authoritative information' and 'creating an awareness of the catastrophic consequences of atomic warfare'. There was no way, IPPNW pointed out, repeatedly and vividly, that medical services could possibly cope with the aftermath of nuclear war. In a similar vein, public health might promote step changes in climate policy by broadcasting a full accounting of the effects of climate change on health, and in particular, by pointing out where effects may be so severe that health services are overwhelmed.

There are examples, certainly, of how fuller accounting of the consequences for health might shift climate policy. The 'mortality cost of carbon' is a concept introduced by Bressler (2021) and based on estimates of the number of temperature-related deaths caused by the emission of an extra tonne of CO_2. Adding this variable to existing integrated assessment models, and including an estimate in dollars of the value of a statistical life, increases what has been called the 'social cost of carbon' sevenfold. (According to Rose (2012) the social cost is a measure of the global marginal monetary burden of an additional unit of carbon dioxide.) A sevenfold increase is likely to make a very big difference to the results of economic models, but whether the increment in costs will matter for decision making is uncertain. Some have argued that it should. For instance, Bressler (2021) argued that the 'optimal path' shifts from gradual emission reductions beginning in 2050 (the remarkably passive course of action indicated by the old social cost value), to achieving full decarbonization by the same date. Whether governments do in fact change course, in line with new approaches to costing impacts, is not known.

It is often said that the move to a low-carbon future must be a 'just transition', but what does this mean? There has been a good deal written about climate justice (Gardiner, 2011), but by and large considerations of health ethics and health equity have not engaged closely with climate change. There is certainly a great need to better understand trade-offs, clarify values and think more critically about choices, especially those relating to equity. A bioethics perspective could, according to Macpherson (2014), 'draw attention to individual and collective responsibilities ... and generate deeper understanding about what is at stake and what it means to live a good life in today's world'.

The bioethics perspective could be applied not just to climate change, but also to the response to climate change. Ideally, investments in climate action take account of the full range of harms and benefits. Renewable energy is not risk-free, for instance: reduced water flows downstream of hydro-electricity plants, and occupational exposures to biomass products, are two examples of health hazards that are relevant to mitigation policies (Levenda *et al.*, 2021).

More fine-grained information on health impacts, new metrics to report the extent and spread of health losses, and a greater awareness of health equity: these are all good things, but do they speed action on climate change? The record so far is not encouraging. Global greenhouse emissions have, with few exceptions, increased year-on-year over the past 70 years, a trend that bears no relation to what is needed to limit warming to 1.5°C above pre-industrial levels (Fig. 38.1).

Identify, test and promote health co-benefits of climate policy

The connection between public health and climate change policy that is most commonly made in the public health literature is 'co-benefits'. We refer here to the health gains achieved by well-chosen policies, over and above any of the improvements due to avoided or less harmful climate change. Mostly this refers to mitigation actions, but the concept applies to adaptation policies also. Here is a compelling reason to move faster on climate, so the argument runs, since the benefits to health may be substantial and will likely be evident long before the impacts on climate.

There are certainly many examples of broad advances due to wise climate policy. In New Zealand the climate is mild, but many houses are draughty, cold and damp, as a result of poor design and unsatisfactory construction. Community trials of retrofitted insulation and clean, efficient heating have shown that it is possible to simultaneously raise temperatures indoors, improve health outcomes (fewer days off school for children with asthma, for instance) and cut greenhouse emissions, due to reduced use of electricity (Howden-Chapman *et al.*, 2007). These trials of housing improvements had substantial influence on policy and legislation in New Zealand, leading for instance to the Healthy Homes standards that have applied since 2019 to rental properties.

The bicycle, originally known as the 'pedestrian accelerator', is a marvellous machine. It shrinks cities and improves access to desired destinations. It is available to nearly everyone and rewards physical effort. And the bicycle is frugal; its carbon footprint is tiny. A study of investments in cycling infrastructure in Auckland, New Zealand, found that in addition to reductions in greenhouse gas emissions, hundreds of millions of NZ dollars would be saved due to improvements in air pollution, car crash injuries, physical inactivity, and fuel costs to households (Macmillan *et al.*, 2014).

However, there is no necessary alignment of policies with such positive outcomes. Climate priorities and needs may coincide with what is required to protect and promote health, or they may not. For instance, the move away from heavy reliance on coal as a primary fuel has simultaneously saved greenhouse emissions and protected human health. But the promotion of diesel as an efficient motor fuel was good for the climate (due to less tailpipe CO_2/km than is emitted from a similar-sized petrol engine), while increasing air pollution deaths and morbidity

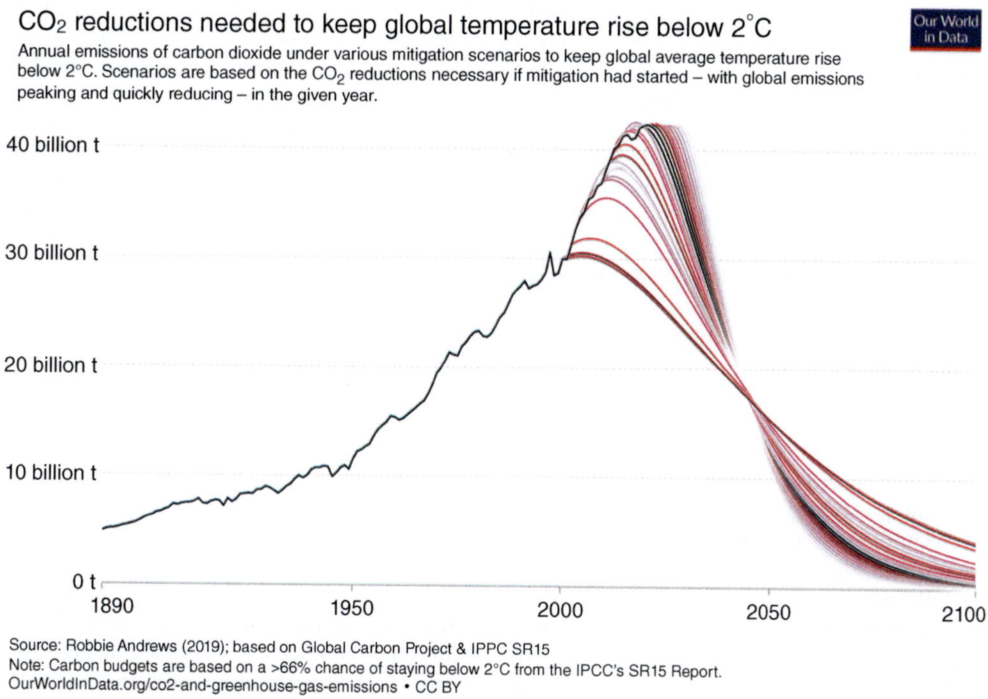

Fig. 38.1. Historic rise in global CO_2 emissions, and future trajectories that hold worldwide average warming to less than 2°C above pre-industrial levels. Source: Our World in Data 2021, available at: https://ourworldindata.org/grapher/co2-mitigation-2c?time=1869..2100 (accessed 19 June 2023). CC BT 4.0 LEGAL CODE.

due to greater quantities of toxic emissions, nitrogen oxides (NO_x) in particular.

Co-benefits may add weight to the case for climate policy, we agree. But they cannot be assumed; climate interventions may be health-neutral, at best. Electric cars are an example, since their use lowers emissions but makes no difference to the transport problems and health risks associated with high levels of motor vehicle traffic. If the focus on electrification of transport was broader, if 'e-vehicles' included battery-powered bicycles, buses and trucks, then environmental and health purposes would be better aligned. But if effective climate policies (such as sharp reductions in travel by private motor vehicles) are presently unacceptable, will health co-benefits make a difference? Can we point to examples? There may be exceptions, such as re-allocation of road space in some cities during the COVID-19 pandemic (Marsden and Docherty, 2021), but on its own, evidence of multiple benefits seldom persuades governments to change what they are doing (McHugh *et al.*, 2021).

Health protection adds teeth to climate policy

Climate policy includes two elements: mitigation (primary prevention); and adaptation (prevention of loss and damage if climate change does occur). Public health practice is particularly relevant to the latter, and may contribute by gauging the magnitude of the problem and establishing causal pathways as the basis for reducing the harm if exposure does occur. Disaster preparedness and emergency responses provide many examples of the contributions that can be made by public health. The steps that must be taken to protect populations and promote recovery are well-described and less contentious than many other elements of climate action. Early warning systems, sanitation and hygiene, protection of drinking-water supplies, control of pests and disease vectors – these are home ground for public health. Funding for basic services has diminished recently in many places, but many hope the experience of COVID-19 will inspire a renaissance of public health.

The health chapter in the IPCC 5th Assessment Report concluded that 'the most effective measures to reduce vulnerability [to climate change] in the near-term are programs that implement basic public health measures such as provision of clean water and sanitation' (Woodward *et al.*, 2014). The reasoning is that climate change acts to multiply risks. Vector-borne diseases such as dengue fever, for instance, are present already in many parts of the world: climate change is projected to increase the intensity of outbreaks, and the range of dengue, as a result of warming and changes in rainfall patterns. It is expected that health losses due to climate change-induced undernutrition will occur mainly in areas that are already food insecure.

Features of public health that make it well-suited to resilience-building include connections with local communities, commitment to reducing pre-existing burdens of disease, and recognition of the importance of social capital. Preparedness is a core principle of public health. Unfortunately, public health does not always live up to its promise, but the basic concepts and tools provide a sound basis for responding to climate change (Whitmee *et al.*, 2021). Public health planners are accustomed to long-term thinking; interventions are commonly framed in systems terms; the importance of effective public communications is acutely understood; and there are precedents for health-sector leadership on big, difficult, important issues, such as global pandemics (Pelling *et al.*, 2021).

Make better use of openings and opportunities

On the whole, public health research does not start with the questions posed by policy makers; inquiry tends to be initiated by the investigators, often motivated by changes in public health statistics. However, if governments reflect on what they need to know in order to respond to threats from climate change, and are explicit about research priorities, then the relevance of public health research and action will be explicit. In 2008, the World Health Assembly identified five priority areas for knowledge generation: the causes of vulnerability to health losses from climate change; health protection strategies; health effects of mitigation and adaptation policies; decision support tools; and costs of measures necessary to protect health. A review of the alignment between these policy priorities and what had been published on climate change and health found there were gaps. In particular, there was evidently a need for more research on the health impact of adaptation measures, as well as more research in general on the needs of the most vulnerable regions (Hosking and Campbell-Lendrum, 2012).

Public health may be in a unique position to help translate aspects of the COVID-19 response to climate action (Pelling *et al.*, 2021). The Nationally Determined Contributions (NDCs) agreed in Paris in 2015 could, in theory, be the blueprint for a joint and therefore more powerful response to the pandemic and climate change. There are COVID-19 relevant themes already in the NDCs from some countries – examples include responses to gendered impacts and risks, the potential for nature-based solutions, and strengthened processes for science-based targets and decision making. As noted before in this chapter, health was not included in many of the initial NDCs, those that were presented in Paris. But there is a stronger representation in the enhanced (second round) NDCs. As of July 2021, 87 countries (about half the total) had submitted updated documents, and 79 (91%) referred to health, although often fleetingly (Romanello *et al.*, 2021).

Finally, climate policy may pick up valuable lessons from the 'health in all policies' approach developed in public health. This recognizes that health is strongly socially determined, so all areas of policy making, not just health policy, need to assess health impacts. 'Environment in all policies' or 'climate in all policies' approaches could help to identify opportunities to integrate climate impact assessment and planning into all areas of policy making, rather than allowing it to be siloed in specialist 'climate departments'.

Engage more strongly with advocacy and activism

The lack of progress on climate action leads some to argue that better public health science is not enough on its own, and the focus must shift to how science, and knowledge more generally, is converted to action. This means re-thinking assumptions about policy that scientists commonly adhere to. According to Rossa-Roccor *et al.* (2021), rather than an orderly 'evidence-based or evidence-informed' policy world, we must recognize that the climate change environment is 'political, contested, ideology-driven ... and stricken by complex power dynamics'.

In many ways, this is nothing new (Christoffel, 2000; Campbell and Cornish, 2021). Public health has always been an activist discipline (Berridge, 2007), and has not shied away in the past from politics, contestation and clashing ideologies. The COVID-19 pandemic is one example, but there are other instances in history of social movements and political mobilizations that have been inspired and led by public health professionals. These have sometimes been unsuccessful, but on occasion have contributed substantially to the health of populations. We touch here on improvements in urban environments in the 19th century; and achievements of tobacco control about a hundred years later.

The arrival of the steam age brought great advances but there was also collateral damage. In the first half of the 19th century mortality was exceptionally high in many industrial cities – it was not until the last decade of the century that life expectancy in Liverpool, UK, exceeded 30 years, for instance. The public health response to perils of the industrial revolution may tell us something about how to approach disruption due to climate change – success drew on science (which was patchy: knowledge of the causes of common urban diseases was partial and contested), but it was politics and institutions and activism that made the biggest difference.

The historian Simon Szreter (1988) argued that effective public health in Britain in the Victorian period was distinguished by focused and definitive direction from the centre, energetic local agency, and accountability. The report of the Royal Sanitary Commission of 1869–1871 and legislation that came into force shortly afterwards provided direction; and action on the ground by local authorities (councils and health boards, for instance) was essential to improve urban sanitation and transform living conditions (though frequently opposed by local property owners and business interests). Greatly improved health statistics, routinely collected and regularly published, ensured that local authorities were accountable.

The same success factors can be applied to climate change, although the scale of action is global rather than national. In this instance, central direction depends on international agreement, and the 2015 Paris Climate accord is a model in many ways, although it lacks sanctions and enforcement powers. Cities were engine houses for environmental and public health action in the transition to coal and oil, and are prominent once more in the move away from fossil fuels. The C40 Network, for example, leads the way in planning, mitigating and adapting. In partnership, these cities are seeking local action that will bend the emissions curve far enough to meet targets agreed in Paris. Priorities include energy reforms, low-carbon buildings, mobility policies and waste management.

One of the best examples of effective public health action in the 20th century is tobacco control. Tobacco use remains one of the biggest causes of preventable death and illness worldwide, but large improvements have been made, particularly in high-income countries where the problem was first identified. (For instance, in New Zealand, adult daily smoking, which was more than 30% in the 1980s, fell to 9% in 2020/21 (and only 1% among youth) (Smokefree New Zealand, 2022). As with the clean-up of cities in the 19th century, progress was based on science, and in particular, the epidemiological studies that linked smoking, both active and passive, with cancer and other serious health outcomes as well as the recognition that nicotine is highly addictive. But epidemiological findings were not enough, on their own, to reduce the consumption of tobacco products and to prevent an epidemic of tobacco diseases. The harm caused by smoking was evident in the 1950s, but smoking prevalence in countries like the USA and the UK continued to rise until the 1970s.

Simon Chapman (2008) has examined what turned the prevalence curve, what it was that 'made smoking history' (in some countries), and his conclusions are relevant to climate policy, in our view. Elements of effective public health action, according to Chapman, included getting close to politicians (at least to understand the problem through their eyes), influencing the framing and reporting of tobacco control (news-making is key), challenging social norms (such as smoking is cool, smoking is free choice, anti-smoking is anti-freedom), and what Chapman called 'vector control'. He was referring here to the tobacco industry, and how important it was that public health challenged the industry and helped to undermine its influence and standing. The list is not complete, of course: other lessons that have been drawn include the value of multi-level interventions that were required to halt the promotion and use of tobacco, and the importance of an international framework to coordinate and focus national policies.

Will public health professionals lose their credibility if they follow the tobacco example, and are more directly involved with advocacy and political agitation on climate? A survey of citizens in the USA and Germany found that this was not the case: the majority of respondents thought that scientists should advocate for climate policies and should work closely with policy makers (Cologna *et al.*, 2021).

Use health to frame communications about climate

The framing of climate issues is important. Adding health and other co-benefits to public messaging about climate change may help people to engage with what otherwise seems to be an overwhelming and hopeless problem. It seems sensible to link warnings about future climate change with positive signals about what can be done immediately, and almost everyone wishes to protect their health.

One reason may be the connections between health and climate change are not obvious to many people, including those in the media whose preconceptions and perspectives influence the world view of the general public. A study was undertaken of media reports on the three components of the 5th Assessment Report of the IPCC, published in 2014 (O'Neill *et al.*, 2015). The ways in which the IPCC reports were framed varied between print, TV and social media, but across the board the consequences for health were seldom mentioned. The frames applied most commonly were settled science, uncertain science, disaster and political or ideological struggle. The conclusion: climate change, as reported by the IPCC, was hardly ever treated in the media as first and foremost a health issue.

We have argued already that framing climate as a health issue does not necessarily lead to action. Some researchers have reported that co-benefits increase motivation to act on climate change, although positive effects on economic and scientific development may be stronger influences than the prospect of health gains (Bain *et al.*, 2016). But others have found no evidence that re-framing climate policy in this way increases public support (Bernauer and McGrath, 2016).

38.4 Conclusion

We acknowledge that health is not necessarily the paramount consideration in decision making on matters affecting public health. Economic consequences are often emphasized, though these are not so helpful, in our view, when based on incremental cost–benefit approaches that are poorly suited to large-scale system problems requiring transformative change. Nevertheless, it is possible for public health to make significant contributions to climate policy. These extend beyond characterizing the human health impacts of this ecological crisis. Public health resources include: considerable and successful experience with big picture, multi-disciplinary systems thinking that aims to break through institutional policy silos; a history of concern with identifying and getting ahead of compounding risks and 'risk multipliers'; connections to communities that will be affected; and a comfort with and commitment to advocacy and the practice of an 'activist science'. As public health researchers and practitioners begin to engage more deeply with climate policy, they bring a wealth of valuable change-making history and experience with them. Those success factors include an appreciation of the contested and socially dynamic nature of policy-making, the influence on that process of interests and perspectives other than those of public health and science, the strength of coalitions, and the importance of narratives and story-making in capturing attention and persuading decision takers.

References

Bain, P.G., Milfont, T.L., Kashima, Y., Bilewicz, M., Doron, G. *et al.* (2016) Co-benefits of addressing climate change can motivate action around the world. *Nature Climate Change* 6, 154–157. DOI: 10.1038/nclimate2814

Bartusek, S., Kornhuber, K. and Ting, M. (2022) North American heatwave amplified by climate change-driven nonlinear interactions. *Nature Climate Change* 12, 1143–1150. DOI: 10.1038/2021

Bernauer, T. and McGrath, L.F. (2016) Simple reframing unlikely to boost public support for climate policy. *Nature Climate Change* 6, 680–683. DOI: 10.1038/nclimate2948

Berridge, V. (2007) Public health activism: lessons from history. *BMJ* 335, 1310–1312. DOI: 10.1136/bmj.39399.612454.AD

Bressler, R. (2021) The mortality cost of carbon. *Nature Communications* 12, 4467. DOI: 10.1038/s41467-021-24487-w

Browne, G. and Rutherfurd, I. (2017) The case for 'Environment in All Policies': Lessons from the 'Health in All Policies' approach in public health. *Environmental Health Perspectives* 125, 1–6 DOI: 10.1289/ehp294

Campbell, C. and Cornish, F. (2021) Public health activism in changing times: re-locating collective agency. *Critical Public Health* 31, 125–33. DOI: 10.1080/09581596.2021.1878110

Campbell-Lendrum, D., Manga, L., Bagayoko, M., Manga, L. and Sommerfeld, J. (2015) Climate change and vector-borne diseases: what are the implications for public health research and policy? *Philosophical Transactions of the Royal Society B Biological Sciences* 370, 20130552. DOI: 10.1098/rstb.2013.0552.

Chang, K.M., Hess, J.J., Balbus, J.M., Buonocore, J.J., Cleveland, D.A. *et al.* (2017) Ancillary health effects of climate mitigation scenarios as drivers of policy uptake: a review of air quality, transportation and diet co-benefits modeling studies. *Environmental Research Letters* 12, 113001. DOI: 10.1088/1748-9326/aa8f7b

Chapman, S. (2008) *Public Health Advocacy and Tobacco Control: Making Smoking History*. Wiley-Blackwell, Hoboken, New Jersey.

Christoffel, K.K. (2000) Public health advocacy: process and product. *American Journal of Public Health* 90, 722–726. DOI: 10.2105/ajph.90.5.722

Cologna, V., Knutti, R., Oreskes, N. and Siegrist, M. (2021) Majority of German citizens, US citizens and climate scientists support policy advocacy by climate researchers and expect greater political engagement. *Environmental Research Letters* 16, 024011. DOI: 10.1088/1748-9326/abd4ac

Corner, A., Markowitz, E. and Pidgeon, N. (2014) Public engagement with climate change: the role of human values. *WIREs Climate Change* 5, 411–422. DOI: 10.1002/wcc.269

Dasandi, N., Graham, H., Lampard, P. and Mikhaylov, S.J. (2021) Engagement with health in national climate change commitments under the Paris Agreement: a global mixed-methods analysis of the nationally determined contributions. *The Lancet Planetary Health* 5, e93–e101. DOI: 10.1016/S2542-5196(20)30302-8

Dodson, J., Dérer, P., Cafaro, P. and Götmark, F. (2020) Population growth and climate change: Addressing the overlooked threat multiplier. *Science of the Total Environment* 748, 141346. DOI: 10.1016/j.scitotenv.2020.141346

Finkel, E., Bail, C., Cikara, M., Ditto, P.H., Iyengar, S. *et al.* (2020) Political sectarianism in America. *Science* 370, 533–536. DOI: 10.1126/science.abe1715

Frumkin, H., Hess, J., Luber, G., Malilay, J. and McGeehin, M. (2008) Climate change: the public health response. *American Journal of Public Health* 98, 435–445. DOI: 10.2105/AJPH.2007.119362

Gardiner, S. (2011) *A Perfect Moral Storm. The Ethical Tragedy of Climate Change*. Oxford University Press, Oxford, UK.

Greenhalgh, T., Thorne, S. and Malterud, K. (2018) Time to challenge the spurious hierarchy of systematic over narrative reviews? *European Journal of Clinical Investigation* 48, e12931. DOI: 10.1111/eci.12931

Haines, A., Wilkinson, P., Tonne, C. and Roberts, I. (2009) Aligning climate change and public health policies. *The Lancet* 374, 2035–2038. DOI: 10.1016/s0140-6736(09)61667-6

Hosking, J. and Campbell-Lendrum, D. (2012) How well does climate change and human health research match the demands of policymakers? A scoping review. *Environmental Health Perspectives* 120, 1076–1082. DOI: 10.1289/ehp.1104093

Howden-Chapman, P., Matheson, A., Crane, J., Viggers, H., Cunningham, M. *et al.* (2007) Effect of insulating existing houses on health inequality: cluster randomised study in the community. *BMJ* 334, 460. DOI: 10.1136/bmj.39070.573032.80

Keim, M.E. (2008) Building human resilience: the role of public health preparedness and response as an adaptation to climate change. *American Journal of Preventive Medicine* 35, 508–516. DOI: 10.1016/j.amepre.2008.08.022

Levenda, A., Behrsin, I. and Disano, F. (2021) Renewable energy for whom? A global systematic review of the environmental justice implications of renewable energy technologies. *Energy Research and Social Science* 71, 101837. DOI: 10.1016/j.erss.2020.101837

Macmillan, A., Connor, J., Witten, K., Kearns, R., Rees, D. *et al.* (2014) The societal costs and benefits of commuter bicycling: simulating the effects of specific policies using system dynamics modeling. *Environmental Health Perspectives* 122, 335–344. DOI: 10.1289/ehp.1307250

Macpherson, C.C. (2014) Climate change matters. *Journal of Medical Ethics* 40, 288–290. DOI: 10.1136/medethics-2012-101084

McHugh, L.M., Lemos, M.C. and Morrison, T.H. (2021) Risk? Crisis? Emergency? Implications of the new climate emergency framing for governance and policy. *WIREs Climate Change* 12, e736. DOI: 10.1002/wcc.736

Marsden, G. and Docherty, I. (2021) Mega-disruptions and policy change: lessons from the mobility sector in response to the Covid-19 pandemic in the UK. *Transport Policy* 110, 86–97. DOI: 10.1016/j.tranpol.2021.05.015

O'Neill, S., Williams, H.T.P., Kurz, T., Wiersma, B. and Boykoff, M. (2015) Dominant frames in legacy and social media coverage of the IPCC Fifth Assessment Report. *Nature Climate Change* 5, 380–385. DOI: 10.1038/nclimate2535

Pelling, M., Kerr, R.B., Biesbroek, R., Caretta, M.A., Cissé, G. *et al.* (2021) Synergies between COVID-19 and climate change impacts and responses. *Journal of Extreme Events* 8, 2131002. DOI: 10.1142/S2345737621310023

Pencheon, D. (2018) Developing a sustainable health care system: the United Kingdom experience *Medical Journal of Australia* 208, 284–285. DOI: 10.5694/mja17.01134

Petrovic, N., Madrigano, J. and Zaval, L. (2014) Motivating mitigation: when health matters more than climate change. *Climatic Change* 126, 245–254. DOI: 10.1007/s10584-014-1192-2

Phillip, S.Y., Kew, S.F., van Oldenborgh, G.J., Anslow, F.S., Seneviratne, S.I. *et al.* (2022) Rapid attribution analysis of the extraordinary heatwave on the Pacific Coast of the US and Canada June 2021. *Earth System Dynamics* 13, 1689–1713. DOI: 10.5194/esd-13-1689-2022

Romanello, M., McGushin, A., Di Napoli, C., Drummond, P., Hughes, N. *et al.* (2021) The 2021 report of the Lancet Countdown on health and climate change: code red for a healthy future. *The Lancet* 398, 1619–1662. DOI: 10.1016/S0140-6736(21)01787-6

Rose, S. (2012) The role of the social cost of carbon in policy. *WIREs Climate Change* 3, 195–212. DOI: 10.1002/wcc.163

Rossa-Roccor, V., Giang, A. and Kershaw, P. (2021) Framing climate change as a human health issue: enough to tip the scale in climate policy? *The Lancet Planetary Health* 5, e553–e559. DOI: 10.1016/s2542-5196(21)00113-3

Semenza, J.C. (2021) Lateral public health: Advancing systemic resilience to climate change. *The Lancet Regional Health – Europe* 9, 100231. DOI: 10.1016/j.lanepe.2021.100231

Smokefree New Zealand (2022) *Smoking and its effects. Facts & figures*. Available at: https://www.smokefree.org.nz/smoking-its-effects/facts-figures (accessed 4 March 2022).

Sovacool, B., Axsen, J. and Sorrell, S. (2018) Promoting novelty, rigor, and style in energy social science: Towards codes of practice for appropriate methods and research design. *Energy Research & Social Science* 45, 12–42. DOI: 10.1016/j.erss.2018.07.007

Szreter, S. (1988) The importance of social intervention in Britain's mortality decline c.1850–1914: a re-interpretation of the role of public health. *Social History of Medicine* 1, 1–37. DOI: 10.1093/SHM/1.1.1

Szreter, S. (2003) The population health approach in historical perspective. *American Journal of Public Health* 93, 421–431. DOI: 10.2105/ajph.93.3.421

Walpole, S., Rasanathan, K. and Campbell-Lendrum, D. (2009) Natural and unnatural synergies: climate change policy and health equity. *Bulletin of the World Health Organization* 87, 799–801. DOI: 10.2471/blt.09.067116

Watts, N., Amann, M., Arnell, N., Ayeb-Karlsson, S., Belesova, K. *et al.* (2018) The 2018 report of the *Lancet* Countdown on health and climate change: shaping the health of nations for centuries to come. *The Lancet* 392, 2479–2514. DOI: 10.1016/S0140-6736(18)32594-7

Whitmee, S., Green, R., Phumaphi, J., Clark, H. and Haines, A. (2021) Bridging the evidence gap to achieve a healthy, net zero future. *The Lancet* 398, 1551–1553. DOI: 10.1016/S0140-6736(21)02278-9

Woodward, A. (2019) Climate change: Disruption, risk and opportunity. *Global Transitions* 1, 44–49. DOI: 10.1016/j.glt.2019.02.001

Woodward, A., Smith, K.R., Campbell-Lendrum, D., Chadee, D.D., Honda, Y. *et al.* (2014) Climate change and health: on the latest IPCC report. *The Lancet* 383, 1185–1189.

WHO (2009) *Protecting Health from Climate Change: Connecting Science, Policy and People*. WHO, World Health Organization, Geneva.

39 Health Activism and the Challenge of Planetary Change, including to the Climate

Colin D. Butler,[1] Sue Wareham[2] and Andrew Harmer[3]

[1]National Centre for Epidemiology and Population Health, The Australian National University, Canberra, Australia; [2]Medical Association for Prevention of War (Australia), Canberra, Australia; [3]Centre for Primary Care and Public Health, Queen Mary University of London, UK

Abstract

This chapter argues that health practitioners have a duty of care to embrace and to support non-violent social activism, to the extent to which they feel comfortable, but perhaps including the risk of arrest. We believe this is justifiable, given the increasing peril in which civilization finds itself. However, it is also clear that the 'milieu' matters; the social environment in which protests occur. Health practitioners, in some countries, have contributed to significant social reform; such reform may happen again; the challenges are formidable, yet we must try.

> All health professionals have a duty and obligation to engage in all kinds of non-violent social protest to address the climate emergency.
>
> Richard Horton (2019)

> I think it is insane that people are gathered here to talk about the climate and they arrive here in private jets.
>
> Greta Thunberg (2019)

39.1 Introduction

The direct relationship between patient and doctor (or other healthcare professional) has traditionally been the essence of clinical practice. Most healthcare practitioners train to do this; it is how they spend most of their working lives. However, it is a role that is increasingly recognized as having limits, for it can often be one of patching up or simply reacting to the damage wrought by a host of factors that affect human health, whether social, economic or environmental, or a combination of these. Because addressing these deeper causes effectively is rarely possible in either the clinical or research setting, activism has been introduced as a subject in some health courses. While inculcating an awareness of these issues (sometimes called social medicine) is valuable, it also seems important to sensitize students to the arduous path they can face when trying to blend awareness of social justice with the pressures of traditional health settings such as hospitals (Finnegan *et al.*, 2017).[1] We believe that the issues discussed in this chapter (and many parts of this book) are vital for health workers, including medical students, to be exposed to and to understand. Yet, their visible application is likely to arise in less predictable ways, interspersed through a lifetime. Consider the all-too-common counterfactual of zero *formal* exposure to the relationship between environmental and social justice. In contrast, we

DOI: 10.1079/9781800620025.0039

insist that these issues are vital for modern health care students to be exposed to.

Many important advances in human health and welfare have occurred through better understanding and, where possible, amelioration of the links between health and these external factors, i.e. the social, commercial (Kickbusch *et al.*, 2016), economic and environmental determinants of health. Such advances have occurred in association with improved living conditions for people. They hold the promise that, in the future, good health could be a basic human right, i.e. one that is realized, rather than simply asserted.

Historically, activism has been associated with public health practitioners more than their medical clinical counterparts (Berridge, 2007) although this association is less evident today with the creation – in some countries – of medical activist groups such as Doctors for Extinction Rebellion (XR) (Fig. 39.1). Many health workers with little formal public health training have

Fig. 39.1. Dr Scott Bell, a retired medical practitioner (named with permission), in attire inspired by an ancient Greek sybil (a prophetess associated with oracles) bearing witness for Extinction Rebellion at a rally to warn of the risk of climate change. Scott had several companions, all standing mute during the speeches. Launceston, Tasmania, Australia, May 2022. (Photo: anonymous, with permission).

played important reforming roles, often working with others such as family members and friends of ill people.[2] They have strived to improve health issues, from those differentially besetting Indigenous communities to road trauma, alcohol abuse, domestic violence, asbestos exposure, gun control legislation and the advertising of tobacco, junk food and other harmful substances. Unsurprisingly, success has varied with time, cultural context and location. Sometimes, health reformers have led public opinion; at other times, they have been swept along by unstoppable cultural change. Health activism by trained health workers can lag far behind that of affected communities, such as when concerns about HIV arose among gay men in the USA in the 1980s. In that case, activism was led mainly by the gay community (Hilliard *et al.*, 2007).

Health activists can be particularly visible in contexts where they are desperately needed. The life imprisonment sentence of the award-winning Indian community health activist, Dr Binayak Sen (Kennedy and King, 2011; Sen, 2011), for allegedly providing care to people involved on both sides of a vicious conflict, is rendered all the more terrible because of the atrocious health conditions of millions of rural poor in the northern states of India.[3] On the other hand, in China, the work of individual activists is rarely recognized (and sometimes suppressed by the state); yet Chinese society has mobilized successfully on a vast scale to lower the burden of many diseases, such as schistosomiasis. Mao Zedong (founder of the People's Republic of China), despite numerous other actions which proved problematic to public health in China, was personally inspirational for and supportive of the long campaign to reduce the effects of this parasitic infection (Utzinger *et al.*, 2005).

Overall, average health in China is now far superior to that in India. However, some health activists in China have been oppressed, such as Zhao Lianhai, the father of an infant whose kidneys were damaged by the inadvertent consumption of the toxic substance melamine, used to adulterate many processed foods, including infant formula, in order to increase profits. Zhao organized a website ('Kidney Stone Babies') for parents of affected children (as many as 300,000), but was found guilty of 'inciting social disorder' and in 2010 was sentenced to 2.5 years in prison, a year after his arrest (Reuters Health, 2010). Even in 2018, deep distrust of the safety of Chinese-made formula was reported as persisting; this nurtured a flourishing import industry, providing foreign-sourced milk for those who could afford it (Huang, 2018). Dr Li Wenliang, the Chinese ophthalmologist who recognized the risk of COVID-19 in very early 2020, was vilified and humiliated, before dying of the disease he tried to warn about (Nie and Elliott, 2020).

In some countries, health activists have had remarkable success in limiting cigarette smoking (Chapman, 2008; Thomson *et al.*, 2015). An activist campaign, involving some with medical training, graffitied cigarette billboards and helped to stigmatize smoking in Australia (Chapman, 1996). In 2004, Ireland became the first nation to implement a comprehensive smoking ban in indoor workplaces, including restaurants and bars. Contrary to expectations, the ban was both popular and enforceable. Pleasingly, other countries followed (Koh *et al.*, 2007). The right to clean air was being affirmed. However, in China – the world's largest producer and consumer of tobacco products – attempts to reduce that nation's once pervasive tobacco addiction, especially prevalent among males, have faltered, due in part to the Chinese government's dependence on tobacco revenue (Leng and Mu, 2020).

39.2 The Great London Smog

As our natural environment has come under greater strain, its links with human health have been increasingly recognized. For centuries, people complained of poor air quality in London, the original 'Big Smoke' (McNeill, 2000). In 1950, air pollution in most British cities – due largely to the combustion of coal, then considered essential for heating and industry – was so pervasive that, in combination with high rates of smoking, virtually every man had a chronic wheeze by the age of 50 (Fig. 39.2). In December 1952, London was again[4] afflicted by a heavy, unshifting smog, reducing visibility to metres and bringing the city to a standstill. Policemen had to guide buses through the haze. Hospital admissions for respiratory illnesses rose, as did mortality, the latter by an estimated 3500–4000 excess deaths for the month. At first, there was a lack of recognition of the severity of these health

Fig. 39.2. A trace of London's great smog remains in its profusion of chimneys, still plentiful more than 70 years since their last use (Photo: C.D. Butler).

impacts, which was perhaps surprising (Davis *et al.*, 2002). However, four years later, the Clean Air Act followed growing agitation, which included leadership from the National Smoke Abatement Society, founded in Manchester and chaired (at least initially) by a barrister (Heys, 2012). This was passed even though parliamentarians remained fearful of the reception of new restrictions on behaviour (Brimblecombe, 2006). Other regional smoke abatement committees had also been founded in the late 19th century, including in Leeds and London. National organizations to reduce smoke were founded in the early 20th century (Mosley, 2017).

39.3 Other Forms of Health Activism in and Since the 19th Century

While the doctors of London in 1952 did not seem to recognize and react rapidly to the adverse impact of smoke on health, many health pioneers before and since have seen far beyond their individual patients, to explore deeper causes. Some of these are social and environmental conditions (Table 39.1). About a century before the great London smog, the German physician and pathologist, Rudolf Virchow (1821–1902) broke ground with his revolutionary understanding of the social and physical bases of ill health (Virchow, 1879). Alongside his discoveries that cells were the fundamental unit of life (as well as pathfinding work on the pathology of thrombosis, leukaemia and other conditions), Virchow was equally innovative in social medicine. He advocated for social changes as a means to fundamentally improve many health conditions, sometimes at personal risk. He offended the German politician Otto von Bismarck (partly by criticizing the expansion of military spending) so much that von Bismarck challenged him to a duel (Eisenberg, 1984).

Table 39.1. Selected health activists since 1790, in approximate chronological order. Health workers can be proud of this record, yet much urgent work remains unfinished.

Name	Nationality	Period	Major contribution
Philippe Pinel	French	Late 18th century	Psychiatric reform
James Kay	British	Early 19th century	Public health reform, Manchester
Thomas Wakley[b]	British	Early 19th century	Founder of *The Lancet*, London
Edwin Chadwick	British	Mid-19th century	Sanitation
Florence Nightingale	British	Mid-19th century	Nursing reform
David Livingstone	British	19th century	Anti-slavery
William Farr[b]	British	Mid-19th century	Statistics, social activism
Lemuel Shattuck[b]	American	Mid-19th century	Sanitary reform
John H. Griscom[b]	American	Mid-19th century	Sanitary reform
Rudolf Virchow[b]	German	Late 19th century	A founder of social medicine
Oswaldo Cruz	Brazilian	Early 20th century	Public health
Natan Vigdorchik	Russian	Early 20th century	Social security, public health
Ronald Ross	British-Indian	Early 20th century	Malaria control
John Boyd Orr	British	Mid-20th century	Founding Director General, FAO
René Dubos	French-American	Mid-20th century	Ecology and health
Frantz Fanon	French Caribbean	Mid-20th century	'Third World' rights, decolonization
Albert Schweitzer[b]	Swiss	Mid-20th century	Human rights, international health
Jonathan Mann	American	Late 20th century	HIV/AIDS, human rights
John Goldsmith	American	Late 20th century	Environmental epidemiology
Fred Hollows	New Zealander	Late 20th century	Global ophthalmology
Frank Fenner	Australian	Late 20th century	Ecology and health
Paul Epstein	American	20th–21st century	Climate change, sustainability, justice
Catherine and Reginald Hamlin	Australian	20th–21st century	Obstetric fistula repair and prevention
Maurice King	British	20th–21st century	Primary health care, sustainability
Binayak Sen[b]	Indian	Contemporary	Human and health rights for the poor
Rani and Abhay Bang	Indian	Contemporary	Rural health care
Tushar Joshi	Indian	Contemporary	Asbestos reduction
Wan Yanhai, Tian Xi[a]	Chinese	Contemporary	HIV/AIDS activism
Zhao Lianhai[a,b]	Chinese	Contemporary	Health promotion (melamine danger)
Bob Brown[b]	Australian	Contemporary	Environmental health
Geoff Koffman	British	Contemporary	Illegal organ trading
Gro Harlem Brundtland	Norwegian	Contemporary	Sustainable development
May Haddad	Lebanese	Contemporary	Public health
Mechai Viravaidya[a]	Thai	Contemporary	HIV/AIDS, family planning ('Mr Condom')
Zackie Achmat[a]	South African	Contemporary	HIV/AIDS
Waris Dirie[a]	Somali	Contemporary	Female genital mutilation

[a]Not a trained health worker; [b]mentioned or discussed in this chapter.

Another 19th century advocate for the role of primary prevention and public health was New York doctor, John H. Griscom. Griscom advocated measures such as public hygiene, medical care for the destitute and prison and tenant reforms (Bloom, 2001). He was an early proponent of public health as a speciality, and advocated the collection of vital statistics, similar to his British contemporary, William Farr (Hamlin, 1995). Griscom also played a significant role in the formation of the American Medical Association. His driving force was ethical and moral concern, combined with a strong commitment to scientific principles.

For some physicians in more recent times, directly witnessing large-scale environmental destruction has led to a life in politics. Dr Bob Brown, the first leader of the Australian Greens Party, swapped his career as a general (family) practitioner for parliamentarian in response to massive deforestation of pristine wilderness in his adopted Australian state of Tasmania, along with other assaults on the health of our planet. During and following his long political career, Brown remained engaged with the health community, including with medical students.

Reform attempts continue. A growing number of health workers have been arrested, drawing attention to the diverse health risks that global health faces. A broad-based and powerful voice, comparatively recent, is Extinction Rebellion (Box 39.1), whose actions in 2018/19 precipitated national-level declarations of climate emergency (Kinniburgh, 2020) and facilitated UK Prime Minister May's 'net zero' carbon dioxide emissions targets.

Box 39.1. Extinction Rebellion

My name is Andrew Harmer. I am a parent of two young children and, in my professional life, a Senior Lecturer in Global Health at a university in London. I have been a global health scholar for two decades and have specific expertise in the health impact of climate change on the global population.

I have supported the social movement Extinction Rebellion since 2018 and participated in a number of direct actions organised by the movement. In April 2019, I participated in two weeks of direct action in London. On the 21st April, I walked onto Waterloo Bridge around midday and sat next to protestors at the south end of the bridge. A police officer approached me, informed me that a condition had been imposed on protest on the bridge and explained the consequences of failing to comply with that condition. I refused to comply with the condition, was arrested and subsequently found guilty of violating section 14 of the 1986 Public Order Act.

With knowledge comes responsibility, as the adage goes. This is particularly applicable to academics, who spend much of their time trying to understand how the world works. If you know from your reading, teaching and research that the world is warming and that, if it warms too much, significant harm will result; if you know that the targets your government has set for greenhouse gas emissions reductions are insufficient to prevent the world from warming beyond 1.5C; and if you know that the negative health effects of exceeding 1.5C will impact future generations, including your own children, then you have to ask yourself a fundamental question: what do I do?

Noam Chomsky's 1967 essay *The Responsibility of Intellectuals* argued: 'It is the responsibility of intellectuals to speak the truth and to expose lies'. I don't think that that is enough. Faced with imminent climate breakdown, it is not sufficient to speak the truth and expose lies; it is incumbent on those who understand the gravity of the situation facing not just humanity but potentially all life on Earth to intervene physically. From an early age I was inspired by the fictional character Arthur Dent, the pyjama wearing, unassuming individual who was prepared to lie down in front of a bulldozer in order to protect his home from destruction.[5] This is the situation we now face, writ large.

It is all too easy to pretend that speaking the truth and exposing lies is enough – it's not. Climate scientists have known the truth since the 1980s and have repeatedly reported it year-on-year to politicians who either don't care, are too busy, or who simply do not believe the facts. Some politicians are obviously corrupted by fossil fuel suppliers. And the lies peddled by the fossil fuel industry have been exposed; yet lies continue even today. So, what then? You have told the truth, you have exposed the lies, and yet we are on the brink of irreversible climate change – change that may play out over the centuries whether we want it to or not.

Given the vast expanse of time in which life on Earth has evolved, it is hard to comprehend that what we do in the next decade will determine the fate of literally millions of living organisms, including our own species. That's on me, that's on you, that's on us. I know what's coming if I don't act. That's why I sat down in the road on Waterloo bridge, knowing that I would be arrested. I am resolved to do it again. As a parent, as an academic who knows what is at stake, I have a responsibility to act – now – and so do you.[6]

39.4 Medical Journals

Some medical journals have been important vehicles for promoting new, reforming and humanitarian ideas. Prominent among them has been *The Lancet*, founded by the activist surgeon, Thomas Wakley, in 1823. In its early days, this journal highlighted issues such as the rights of prisoners to medical care and the conditions under which the poor worked (Kandela, 1998). In more recent times, the journal has continued to address causes considered by some as too 'political' for professional involvement, such as opposing apartheid in South Africa and highlighting the plight of Palestinians. A former *Lancet* editor, Ian Munro, became founding president of Physicians for Human Rights (UK) in 1989 (Kandela, 1998). *The Lancet* has led its peers in recognizing environmental issues including climate change and many aspects of sustainable development (King, 1990; Anonymous, 1992). It was crucial in the campaign against its own parent, the conglomerate Reed Elsevier, which for a time conducted a lucrative sideline in the arms trade, marketing cluster bombs and torture equipment (Anonymous, 2005).

Similar to *The Lancet* in the UK, the US-based *New England Journal of Medicine* played a key role in alerting the profession and the wider community to the impossibility of providing medical care in the event of a nuclear war, publishing four articles on this topic in one issue (1962). It has reinforced this role several times. Perhaps coincidentally, this famous journal's headquarters are in Shattuck Street (Boston), named for another pioneering public health figure, legislator Lemuel Shattuck (Fair, 1952; Cassedy, 1975).

Numerous other journals have played roles in promoting research and opinion on the social and environmental factors that improve health. Some journals have failed; some have even expressed bias towards interests inimical to health, including the tobacco industry. There are also increasing concerns that some publishers are overly influenced by commercial and political actors, yet fail to recognize or declare the conflicts of interest that thus arise (Butler *et al.*, 2007). While the most obvious facet of this is 'predatory publishing', the problem may be broader (Bishop, 2020).[7]

39.5 The Medical Campaign to Reduce the Threat of Nuclear Weapons

In the latter part of the 20th century, an increasing suite of ecological threats joined the known dangers to health: poverty, dirty water and undernutrition. These new concerns included the deterioration and contamination of land, river systems, biodiversity and fish stocks in many parts of the globe (Ceballos *et al.*, 2020). The past 80 years have also seen the emergence of a qualitatively new risk to global health: the development and possible use of nuclear weapons (see also Chapter 3).

It is worth examining the medical profession's response to the threat of nuclear weapons for clues that might be enlisted for advocacy for climate change action. This is done recognizing that, first, the nuclear weapons problem is still potent, and, second, the medical movement for nuclear weapons abolition is not held up as a model that other similar movements should uncritically emulate.

The challenges to healthcare posed by nuclear weapons were obvious from the very start of the nuclear weapons age (Rhodes, 1986). In 1951, British medical practitioners formed the Medical Association for Prevention of War (facilitated by correspondence published in *The Lancet*). Its focus was on the destructive capacity of modern warfare generally, but in particular the newly developed weapons of mass destruction. Physicians for Social Responsibility was formed 10 years later in the USA (Sidel and Levy, 2009). Both organizations were later to become part of the global movement: the International Physicians for the Prevention of Nuclear War (IPPNW).

In the early days of the campaign against nuclear weapons, the primary goal was to end their atmospheric testing; this was recognized as an important public health measure, due to the health effects of the radioactive fallout such tests generated. An early champion of this movement was the distinguished physician, theologian, humanitarian and international health advocate, Albert Schweitzer, the 1953 Nobel Peace Prize laureate. In April 1957, urging an end to nuclear testing, Schweitzer read his *Declaration of Conscience* over Radio Oslo. Broadcast in approximately 50 nations and reported in countless newspapers, his appeal had a significant impact on global opinion (Wittner, 1993).

The finding that radioactive strontium, released by atmospheric tests, had been incorporated into the deciduous teeth of US children proved a powerful argument in the medical campaign, and hastened the banning of atmospheric testing, formalized in the 1963 Partial Test Ban Treaty (Humphrey, 2009). While a critical milestone, this treaty generated a certain degree of complacency and quiescence, which lasted until the late 1970s. At that time, several factors combined to strengthen the opposition to nuclear weapons; these developments included a deterioration in East–West relations, new nuclear missile deployments and official talk of actually fighting (and supposedly 'winning') a nuclear war. These factors stimulated the formation in 1980 of IPPNW. Australian paediatrician Dr Helen Caldicott played a powerful and central role in the USA, Australia and globally, in the renewed struggle to prevent nuclear war, dubbed at the time 'the final epidemic' (Abrams *et al.*, 1981).

IPPNW had a pivotal function in raising awareness of these dangers. The medical message was stark and simple: in the event of nuclear war, doctors would have very little to offer survivors. Medical evidence to back up the message was plentiful, including the report, *The Medical Effects of Nuclear War*, which concluded that the UK's health services could not cope with a single Hiroshima-sized bomb attack (BMA, 1983). Articles in the general science literature warned of 'nuclear winter', a global haze triggered by multiple nuclear explosions that might envelop the atmosphere for at least a year and kill almost all life (Turco *et al.*, 1983). In May 1983, the World Health Organization stated that 'nuclear weapons constitute the greatest immediate threat to the health and welfare of mankind, and that physicians have both the right and the duty to draw attention in the strongest possible terms to the catastrophic results that would follow from any use of nuclear weapons'.

A further element added to the group's success in capturing the attention of both the public and governments: its unity across the enormous political divide of the Cold War. IPPNW's co-founders were eminent cardiologists, an American (Bernard Lown) and a Russian (Eugene Chazov) who was physician to the Soviet political leadership. Cardiology brought them together, but a mutual concern to avert nuclear catastrophe drove their ground-breaking alliance. The movement that they co-founded proclaimed loudly and clearly that health trumps politics. We are all in this together. Five years after its formation, IPPNW was awarded the Nobel Prize for Peace for its role in educating the world on the health effects of nuclear weapons.

The medical message did not, of course, penetrate every corridor of power, but it did have a profound impact on Mikhail Gorbachev, Soviet president in the late 1980s, who credited IPPNW with influencing his decision that nuclear weapons must be abolished. In his book, *Perestroika*, he wrote:

> It is impossible to ignore what these people are saying. What they are doing commands great respect. For what they say and what they do is prompted by accurate knowledge and a passionate desire to warn humanity about the danger looming over it ... In the light of their arguments and the strictly scientific data which they possess, there seems to be no room left for politicking. And no serious politician has the right to disregard their conclusions or neglect the ideas by which they take world public opinion a stage ahead.
>
> (Gorbachev, 1987)

Even before Gorbachev's presidency, the IPPNW message spread within the Soviet Union. In June 1982, an hour-long unscripted discussion between three US and three Russian members of IPPNW was broadcast – twice – on Moscow television, reaching an estimated 100 million people.

In contrast to that, Lown described in his book *Prescription for Survival*, on his time at the helm of IPPNW, a different experience in the US:

> I became aware of an astonishing fact. IPPNW could penetrate the Iron Curtain far more readily than it could enter the free halls of power of a democratic society. Leaders in the highest echelons of the Soviet Union were ready to meet and converse as well as listen. Such consistent access was denied to us in the West. Not only were we ignored by the political establishment in the United States and other NATO countries, but the western media shut us out as well.
>
> (Lown, 2008)

Lown observed that advances on the political front must be 'wrested by an unrelenting, well-organized struggle ... by a public clamouring for change', which remains true today.

As noted in Chapter 3 (this volume), medical activism to abolish nuclear weapons took a huge leap forward with the establishment in 2007 of ICAN, the International Campaign to Abolish Nuclear Weapons. The campaign was the brainchild of Malaysian obstetrician and former IPPNW co-president Dr Ron McCoy, and was initiated by IPPNW's affiliate, the Medical Association for Prevention of War (Australia). ICAN was instrumental, with supportive governments and other peak organizations, in achieving the 2017 Treaty on the Prohibition of Nuclear Weapons (TPNW). There is evidence that the widespread revulsion against weapons with such horrific effects goes a long way to explaining why they have not been used since 1945 (Wittner, 1993). By rendering the weapons illegal, the TPNW greatly reinforces the stigma associated with not only their use but even their possession. That stigma must grow until the weapons are completely abolished.

39.6 Barriers and Solutions

Climate change, like nuclear war, threatens the health of not only individuals and communities but also that of whole nations. Some argue that civilization, as we know it, is imperilled. In the face of such challenges, health professionals can, like the rest of society, succumb to feelings of fear and hopelessness.

For health professionals who undertake advocacy on these issues, another problem awaits. Coming from a world of medicine and science, where evidence is taken seriously and acted upon, they enter a sphere in which scientific evidence may be treated with contempt. Few leaders are like Gorbachev. Instead of decisions based on knowledge painstakingly collected to promote and to protect public health and well-being, policies may aim to preserve or enhance corporate profits and political careers. Respect for scientific knowledge is rarely a prerequisite for holding political office.

Another major problem we face is that human behaviour – which must change at individual, governmental and systemic levels (Mann, 2021) – is based not so much on what we know but on what we feel; that is, our emotions. Many people avoid confronting unpleasant information, dismiss evidence that makes them feel uncomfortable, and retreat into denial. Hence, dissemination of knowledge of the health impacts of climate change is insufficient, though a core task. We must change behaviours not just by presenting information but also by inspiring hope and a desire for change.

To bring about change, it is helpful to start with a statement of values with which most people agree. This is convenient for health professionals, because good health is highly valued by practically all the public. Our advocacy must reinforce health as a fundamental value that must be upheld. This need not and should not apply only to current generations but also to future generations, who will have to live with the effects of today's climate policies. Most people would agree with the principle of intergenerational equity, or, more simply, not leaving a profound mess of a planet for our children and grandchildren.

Ed Maibach, of the Centre for Climate Change Communication in Virginia, USA, is a forceful advocate for the use of simple, clear messages, repeated often, by a variety of trusted and caring messengers (Maibach *et al.*, 2023). But is messaging enough? Our view is that protest and a degree of sacrifice by opinion leaders are also necessary, if not sufficient.

While the role of health professionals in bringing about climate change action is unique and compelling, it is vital to recognize the importance of other disciplines and perspectives. As with nuclear weapons, there are many experts in many fields with whom health activists could join forces, and an even greater number who are not experts but are worried nonetheless and prepared to take action. Climate change can be framed not only as a health issue but also as an environmental, human rights, development, security and parenting issue, among others.

Ultimately, we cannot know whether our efforts to prevent catastrophic climate change will succeed. The poet and environmental activist Wendell Berry said: 'We don't have a right to ask whether we are going to succeed or not. The only question we have a right to ask is "what's the right thing to do?" What does this Earth require of us if we want to continue to live on it?' (Berry and Moyers, 2013).

Protests that are disruptive or mildly damaging, such as throwing tomato soup on the glass cover of a famous photo, do not appear to damage public support for the underlying cause (Davis, 2022).[8] Support for environmental protest remains strong in many countries, including the UK (Carrington, 2022). Perhaps in response in many places, today, activists face increasingly heavy jail sentences and, in some places, even the death penalty. The 'milieu' in which protests occur is important. Gorbachev no longer heads the Soviet Union. Activists are rare and likely to remain so. The protest theorist Erica Chenoweth estimated that regimes can topple if only 3.5% of a population is sufficiently motivated to risk arrest (Kinniburgh, 2020). Perhaps this percentage is perceived as within reach, generating another reason some governments are legislating harsher penalties for peaceful protest.

The only certainty is failure if we do not try. Our hope lies in each one of us, and in the global movement of organizations and individuals working for this common cause. Our collective wisdom, learning, advocacy skills and integrity are vast, but these skills must be used to the full. Humanity's future is not pre-ordained, but will depend on the choices we make.

In relation to nuclear weapons, Bernard Lown once said that if we are to prevail, 'we must never delegate in the presence of challenge, and never whisper in the presence of wrong' (Lown, 1993). Similarly with climate change, raising the alarm about its dire health impacts cannot be delegated. While there are many partners in other fields who are staunch campaigners on other costs of the climate crisis, it is up to us, as health professionals, to educate, advocate and agitate for a healthier planet. Our message must be loud and clear, and combined with a determination to challenge the political inertia, the vested interests, the deniers of science and every other obstacle that prevents meaningful action.

Notes

[1] One graduate of a US-led course on social medicine, partly undertaken in Uganda, confessed during her internship that 'there is literally no time for any "social medicine" stuff at all and I'm finding it extremely frustrating. I have been repeatedly told by seniors and attendings that unfortunately with the high patient volume there is not time to do any social history or worry about anything that will not affect the patient "in the next 3 hours"' (Finnegan *et al.*, 2017).

[2] An early example was The Health of Towns Association (1844–1849), formed in Britain.

[3] Sen was released on appeal in April 2011. As of June 2023, he was free, yet still facing life imprisonment. We hope he still is free. A 2019 interview with Sen and his wife is available at: https://www.facebook.com/watch/?v=2046792618735385 (accessed 25 June 2023).

[4] In 1873, some Londoners accidentally walked into the Thames, unable to see the river.

[5] Arthur Dent is a character in Douglas Adams' book (1979) *The Hitchhiker's Guide to the Galaxy* (Harmony Books, New York).

[6] More on Dr Harmer's story, including his defence, can be found at: https://andrewharmer.org/2019/12/11/the-imminence-of-climate-change-and-the-peril-it-presents-to-life-or-serious-injury-a-defence/.

[7] Springer Nature, in 2020, announced its intention to introduce an article processing charge of €9500 (£8290 or US$11,390), with a subsidiary pilot scheme whereby authors pay €2190 in exchange for peer review, an editorial assessment and a recommendation of which journal in the *Nature* hierarchy their article is suitable for (if any) (Bishop, 2020). Such high fees clearly reduce the opportunity for disadvantaged authors, including from low-income settings, to publish in this prestigious family of journals.

[8] This may not be the case for violent protests.

References

Abrams, H.L., Caldicott, H.M., Chazov, E.I., Constable, J.D., Feld, B.T. *et al.* (eds) (1981) *The Final Epidemic: Physicians and Scientists on Nuclear War*. University of Chicago Press, Chicago, Illinois.

Anonymous (1992) Horse manure after Rio. *The Lancet* 339, 1515–1516. DOI: 10.1016/0140-6736(92)91272-A

Anonymous (2005) Reed Elsevier and the arms trade. *The Lancet* 366, 868. DOI: 10.1016/S0140-6736(05)67292-3

Berridge, V. (2007) Public health activism: lessons from history? *BMJ* 335, 1310–1312. DOI: 10.1136/bmj.39399.612454.AD

Berry, W. and Moyers, B. (2013) Renowned Farmer and Writer Wendell Berry: Confronting the consequences of runaway capitalism. *AlterNet*, Washington, DC. Available at: https://www.alternet.org/2013/10/renowned-farmer-and-writer-wendell-berry-confronting-consequences-runaway-capitalism/ (accessed 29 December 2021).

Bishop, D. (2020) Nature's OA fee seems outrageously high – but many will pay it. *Times Higher Education*. Available at: https://www.timeshighereducation.com/blog/natures-oa-fee-seems-outrageously-high-many-will-pay-it (accessed 17 May 2021).

Bloom, M. (2001) Primary prevention and public health: an historical note on Dr. John Hoskins Griscom. *The Journal of Primary Prevention* 21, 305–308. Available at: https://www.proquest.com/openview/4024e3804aa76c98817f271f92aaad4a/1?cbl=48283&pq-origsite=gscholar&parentSessionId=1rqNsPrq-jUl1%2BtG6Z3xEfYtb2CkuKMCJaJsrq8AaSkM%3D (accessed 26 June 2023).

BMA (1983) *The Medical Effects of Nuclear War.The Report of the British Medical Association's Board of Science and Education*. J. Wiley, London.

Brimblecombe, P. (2006) The clean air act after 50 years. *Weather* 61, 311–314. DOI: 10.1256/wea.127.06

Butler, C.D., Castleden, W., Ruff, T.A., Westberg, N.G. and Corra, L. (2007) A call for publishers to declare their conflicts of interest. *Journal of the Royal Society of Medicine* 100, 355. DOI: 10.1177/014107680710000805

Carrington, D. (2022) Huge UK public support for direct action to protect environment – poll. *The Guardian*. Available at: https://www.theguardian.com/environment/2022/oct/24/huge-uk-public-support-for-direct-action-to-protect-environment-poll (accessed 29 June 2023).

Cassedy, J. (1975) The roots of American sanitary reform 1843–47: seven letters from John H. Griscom to Lemuel Shattuck. *Journal of the History of Medicine and Allied Sciences* 30, 136–147. DOI: 10.1093/jhmas/xxx.2.136

Ceballos, G., Ehrlich, P.R. and Raven, P.H. (2020) Vertebrates on the brink as indicators of biological annihilation and the sixth mass extinction, *Proceedings of the National Academy of Sciences (USA)* 117, 13596–13602. DOI: 10.1073/pnas.1922686117

Chapman, S. (1996) Civil disobedience and tobacco control: The case of BUGA UP. *Tobacco Control* 5, 179–185. DOI: 10.1136/tc.5.3.179

Chapman, S. (2008) *Public Health Advocacy and Tobacco Control: Making Smoking History*. Wiley-Blackwell, Hoboken, New Jersey.

Chomsky, N. (1967) A special supplement: the responsibility of intellectuals. *The New York Review*, 23 February 1967. Available at: https://www.nybooks.com/articles/1967/02/23/a-special-supplement-the-responsibility-of-intelle/ (accessed 17 May 2021).

Davis, C. (2022) Just Stop Oil: do radical protests turn the public away from a cause? Here's the evidence. *The Conversation*. Available at: https://theconversation.com/just-stop-oil-do-radical-protests-turn-the-public-away-from-a-cause-heres-the-evidence-192901 (accessed 29 June 2023).

Davis, D., Bell, M. and Fletcher, T. (2002) A look back at the London smog of 1952 and the half century since. *Environmental Health Perspectives* 110, A734–A735. DOI: 10.1289/ehp.110-a734

Eisenberg, L. (1984) Rudolf Ludwig Karl Virchow, where are you now that we need you? *The American Journal of Medicine* 77, 524–532. DOI: 10.1016/0002-9343(84)90114-1

Fair, G.M. (1952) Shattuck, Chadwick, and the engineer in public health. *Public Health Reports* 67, 443–449. Available at: https://www.ncbi.nlm.nih.gov/pmc/articles/PMC2030776/ (accessed 19 February 2023).

Finnegan, A., Morse, M., Nadas, M. and Westerhaus, M. (2017) Where we fall down: Tensions in teaching social medicine and global health. *Annals of Global Health* 83, 347–355. DOI: 10.1016/j.aogh.2017.03.515

Gorbachev, M. (1987) *Perestroika*. Collins, London, p. 154.

Hamlin, C. (1995) Could you starve to death in England in 1839? The Chadwick-Farr controversy and the loss of the 'social' in public health. *American Journal of Public Health* 85, 856–866. DOI: 10.2105/ajph.85.6.856

Heys, R. (2012) The Clean Air Act 1956. *BMJ* 345, e5751. DOI: 10.1136/bmj.e5751

Hilliard, N., Jenkins, R., Pashayan, N. and Powles, J. (2007) Informal knowledge transfer in the period before formal health education programmes: case studies of mass media coverage of HIV and SIDS in England and Wales. *BMC Public Health* 7, 293. DOI: 10.1186/1471-2458-7-293.

Horton, R. (2019) Health and climate. Available at: https://www.youtube.com/watch?v=YEVGNeneYug (accessed 24 June 2021).

Huang, E. (2018) Ten years after China's infant milk tragedy, parents still won't trust their babies to local formula. *Quartz*. Available at: https://qz.com/1323471/ten-years-after-chinas-melamine-laced-infant-milk-tragedy-deep-distrust-remains (accessed 26 June 2023).

Humphrey, J. (2009) The development of the physicians' peace movements. *Medicine, Conflict and Survival* 25, 87–89. DOI: 10.1080/07488008508408620

Kandela, P. (1998) Medical journals and human rights. *The Lancet* 352, 7–12. DOI: 10.1016/S0140-6736(98)90292-6

Kennedy, J.J. and King, L.P. (2011) Understanding the conviction of Binayak Sen: Neocolonialism, political violence and the political economy of health in the central Indian tribal belt. *Social Science & Medicine* 72, 1639–1642. DOI: 10.1016/j.socscimed.2011.03.040

Kickbusch, I., Allen, L. and Franz, C. (2016) The commercial determinants of health. *The Lancet Global Health* 4, e895–e896. DOI: 10.1016/S2214-109X(16)30217-0

King, M. (1990) Health is a sustainable state. *The Lancet* 336, 664–667. DOI: 10.1016/0140-6736(90)92156-c

Kinniburgh, C. (2020) Can Extinction Rebellion survive? *Dissent* 67, 125–133. DOI: 10.1353/dss.2020.0007

Koh, H.K., Joossens, L.X. and Connolly, G.N. (2007) Making smoking history worldwide. *New England Journal of Medicine* 356, 1496–1498. DOI: 10.1056/NEJMp068279

Leng, W. and Mu, R. (2020) Barriers to tobacco control in China: A narrative review. *Societies* 10, 101. DOI: 10.3390/soc10040101

Lown, B. (1993) *Never Whisper in the Presence of Wrong. Selections from Speeches on Nuclear War and Global Survival*. International Physicians for the Prevention of War, Cambridge, Massachusetts, p.26.

Lown, B. (2008) *Prescription for Survival: A Doctor's Journey to End Nuclear Madness*. Berrett-Koehler Publishers, San Francisco, California, pp. 382–383.

Maibach, E.W., Uppalapati, S.S., Orr, M. and Thaker, J. (2023) Harnessing the power of communication and behavior science to enhance society's response to climate change. *Annual Review of Earth and Planetary Sciences* 51, 53–77. DOI: 10.1146/annurev-earth-031621-114417

Mann, M.E. (2021) *The New Climate War*. Scribe, London.

McNeill, J.R. (2000) *Something New Under the Sun: An Environmental History of the Twentieth-Century World*. W.W. Norton, New York.

Mosley, S. (2017) *Clearing the air: can the 1956 Clean Air Act inform new legislation?* History & Policy, Institute of Historical Research, London. Available at: https://www.historyandpolicy.org/policy-papers/papers/clearing-the-air-can-the-1956-clean-air-act-inform-new-legislation (accessed 17 March 2023).

Nie, J.-B. and Elliott, C. (2020) Humiliating whistle-blowers: Li Wenliang, the response to Covid-19, and the call for a decent society. *Journal of Bioethical Inquiry* 17, 543–547. DOI: 10.1007/s11673-020-09990-x

Reuters Health (2010) China court sentences melamine milk activist to jail. Available at: https://www.reuters.com/article/instant-article/idAFTRE6A914Y20101110 (accessed 17 May 2021).

Rhodes, R. (1986) *The Making of the Atomic Bomb*. Simon and Schuster, New York.

Sen, B. (2011) Securing the right to health for all in India. *The Lancet* 377, 532–533. DOI: 10.1016/S0140-6736(10)62182-4

Sidel, V.W. and Levy, B.S. (2009) Homage to Janus: looking backward and forward. *Medicine, Conflict and Survival* 25, 271–274. DOI: 10.1080/13623690903417143

Thomson, I., Dey, C. and Russell, S. (2015) Activism, arenas and accounts in conflicts over tobacco control. *Accounting, Auditing & Accountability Journal* 28, 809–845. DOI: 10.1108/AAAJ-08-2013-1439

Thunberg, G. (2019) World Economic Forum. Available at: https://libquotes.com/greta-thunberg/quote/lbw-4m8o (accessed 27 January 2024)

Turco, R.P., Toon, O.B., Ackerman, T.P., Pollack, J.B. and Sagan, C. (1983) Nuclear winter: global consequences of multiple nuclear explosions. *Science* 222, 1283–1292. DOI: 10.1126/science.222.4630.1283

Utzinger, J., Xiao-Nong, Z., Ming-Gang, C. and Bergquist, R. (2005) Conquering schistosomiasis in China: the long march. *Acta Tropica* 96, 69–96. DOI: 10.1016/j.actatropica.2005.08.004

Virchow, R. (1879) *Gesammelte Abhandlungen aus dem Gebiet der Öffentlichen Medicin und der Seuchenlehre*. Vol 1. Hirschwald, Berlin. [Cited by: Waitzkin, H. (2006) One and a half centuries of forgetting and rediscovering: Virchow's lasting contributions to social medicine. *Social Medicine* 1, 5–10.] Available at: https://www.socialmedicine.info/index.php/socialmedicine/article/view/6/6 (accessed 8 March 2022).

Wittner, L.S. (1993) *The Struggle Against the Bomb. Vol. 2 Resisting the Bomb*. Stanford University Press, Redwood City, California, p. 31.

40 Climate Change and Global Health: Developing A Social Vaccine to Motivate Transformation

Colin D. Butler,[1] Andrew Harmer[2] and Devin C. Bowles[3]
[1]*National Centre for Epidemiology and Population Health, The Australian National University, Canberra, Australia;* [2]*Centre for Primary Care and Public Health, Queen Mary University of London, UK;* [3]*Faculty of Health, University of Canberra, Canberra, Australia*

Abstract
Global civilization is imperilled by the combination of climate change, ancient and emerging rivalries at varying scales, and the existence of weapons of mass destruction and their threatened use. There is excessive hope, approaching a quasi-religious faith, that technologies, many of them either not yet invented, or unaffordable, will rescue us. Workers in public health, on the whole, have badly underestimated the scope and urgency of the emergency that humanity faces and, thus, the risks to global public health. Such workers, as a generalization, continue to do so. We apply and expand the 'social vaccine' metaphor as an instrument by which to appeal to readers, including within the health community, to help engage with this crisis and to help develop viable solutions.

> The main threats to sustained human existence now come from people, not from nature.
>
> (Rees, 2013)

40.1 Constrained Choices

In this final chapter, we recap some of the impending and existent sub-crises and some responses to them. It is clear that a catastrophe arising from decades of disregard of the warnings of climate and ecological activists is now with us. We conclude by identifying some hopeful and constructive actions that are not yet out of reach, including the concept of a 'social vaccine'.

Earlier chapters of this book have surveyed evidence for the primary, secondary and tertiary effects on health from climate change, observable now and foreseeable in coming decades. These health effects interact with broader evolutionary and socially influenced health determinants, including ideological, political and corporate. In July 2023 the World Meteorological Organization (WMO) reported that the world had probably experienced its hottest known month, with global mean temperature temporarily exceeding 1.5°C above its pre-industrial level (WMO, 2023). United Nations (UN) Secretary General Antonio Guterres warned that 'the era of global boiling has arrived' (Anonymous, 2023). A nuclear disaster appears to loom in Ukraine, and in any case the invasion of that country by Russia shows that the fantasies of a new era of global peace, as forecast by Steven Pinker (2011) and seen by some as widely credible, are unlikely. In the global South, underpublicized conflicts smoulder or rage, from Manipur in north-eastern India (Ganguly, 2023) to the Central African Republic (Médecins Sans Frontières, 2023).

 Climate Change and Global Health: Primary, Secondary and Tertiary Effects, 2nd Edition (eds C.D. Butler and K. Higgs)
DOI: 10.1079/9781800620025.0040

Also in July 2023, Volker Turk, the UN High Commissioner for Human Rights, stated that climate change threatens to deliver a 'truly terrifying' dystopian future of hunger and suffering (Associated Press, 2023). A new report involving the UK Institute and Faculty of Actuaries has concluded that the forecasts of many financial institutions are excessively optimistic, and severely underestimate the risk of damage should the Earth pass 'tipping points' – self-reinforcing and irreversible negative planetary changes. The report also found that financial institutions often failed to understand the assumptions and limitations of their models (Trust *et al.*, 2023).

In June 2023, a submersible vessel ruptured while attempting to observe the remains of the *Titanic*, deep in the Atlantic Ocean. Five individuals perished, two of whom were ultra-wealthy Pakistanis. Their dismal plight attracted widespread publicity, at least in the global North. In the same week, a severely overcrowded boat that had sailed from Libya towards Italy sank in the Aegean Sea, drowning up to 500 people who were seeking a better future in Europe. Prior to sinking, an estimated six passengers had died, possibly from thirst. Up to 400 of the people aboard were also from Pakistan, many of whom were allegedly forced below deck, along with women and children, where they had a markedly diminished chance of surviving the capsize (Smith *et al.*, 2023). Although this tragedy was noted in some media, it received far less attention.

This book has attempted to trace some of the relationships between these and many other developments. It has argued that events such as this latest calamity in the Mediterranean are neither random nor completely unpredictable. Of course, the actual time and location of this boat's demise could not be foreseen, but the reported indifference of the European authorities towards them is consistent with the 'fend signal' (i.e. deterrence) (see Chapter 23, this volume) evident to would-be migrants, in an attempt to prevent them from entering Europe, or other parts of the global North. The Greek coastguard was alleged to have ignored the migrants' pleas for help (though they, in turn, deny this). Another example of such a fend signal is the determination by the current British government that asylum seekers be sent to Rwanda (Roy, 2023).

Human rights and health: the case of Rwanda

Human rights and freedoms are of great relevance to global health and human well-being (Butler *et al.*, 2005). This edition was originally to have a chapter focusing on human rights; alas, this proved impossible. However, human rights are implicit in many chapters. To complement those allusions, and because the British government seeks to 'resettle' asylum seekers in Rwanda, we wish to briefly discuss freedom (or its lack) in this central African nation, which, until 1962, was a Belgian colony.

There are persistent claims that Rwandan dissidents are persecuted, journalists hunted and sometimes killed (Sundaram, 2016). We are pleased that the editor of *The Lancet* publicized some of these claims (Horton, 2016) and that the same journal allowed three pages for an extended essay by a Pulitzer prize-winning author who dared to challenge the official narrative about Rwanda (Garret, 2018). We were thus interested that, with two colleagues, a former Rwandan health minister, later founding Vice Chancellor of the University of Global Health Equity (based in Rwanda), criticized *The Lancet* for publishing this essay, comparing it to something by Holocaust denier David Irving (Binagwaho *et al.*, 2019).

In our view, the evidence for significant loss of human freedom in Rwanda is overwhelming, and extends far beyond the evidence cited here. Yet, tangible data indicate a level of access to healthcare, including vaccinations (even against human papilloma virus), in Rwanda that is excellent for a country in the global South, and better than many in the global North (Binagwaho *et al.*, 2014). Life expectancy in Rwanda is also increasing. Although household food insecurity and severe stunting are still important in this small, densely populated and land-locked nation (Agho *et al.*, 2019), the country – a recipient of considerable development assistance – seems universally acknowledged as steadily and effectively improving healthcare. One of the principal drivers of this is the late Paul Farmer, widely recognized as an outstanding promoter of global health. A co-founder of the NGO Partners in Health, Farmer has been acclaimed by friends ranging from the philanthrocapitalist Bill Gates to a Jesuit priest who has called for

Farmer's canonization (Dear, 2022). Some may consider that the lack of freedom in Rwanda is an acceptable, or perhaps even necessary, trade-off for the nation's improving physical health. However, if so, we are disturbed that this shadow side of Rwandan life is so rarely acknowledged, including within global health literature.[1]

Rising energy costs, the economic crisis and unburnable carbon: a seemingly insoluble melange

The rising cost of fossil fuel-derived energy used for transport, electricity and industry is intimately linked to the challenge of climate change. Although the sustained price of oil has declined from its peak of about a decade ago, oil prices again exceeded US$100 per barrel in early 2022 (World Bank, 2023) while the price of coal rose in the same year to its highest level since 2008 (BP, 2022). Gas prices are also historically high, due to the war in Ukraine and associated disruptions to the supply of gas from Russia to Europe. Overall, energy prices are much higher than during the 1990s, the era of the 'cornucopian enchantment' (Butler, 2007). This was a period when hubristic optimism dominated the landscape, leading, for example, to dismissal of warnings about climate change issued by then US Vice President Al Gore (1992). In this time, US President Bill Clinton also revoked the Glass–Steagall Act, which had been introduced during the Great Depression to try to prevent banks gambling with the money of depositors. Robert Reich, Secretary of Labor in the Clinton administration, argued that this reversal was at the request of the 'big Wall Street banks' (Reich, 2015). In summary, the 1990s were a wasted decade: the warnings about 'limits to growth' that US President Jimmy Carter had heard in the 1970s were forgotten; the decade became like a party to be enjoyed, for which the future would pay (Fig. 40.1).

These high energy prices far exceed the modest carbon price that was once widely promoted as a way to accelerate the energy transition. That transition is occurring, yet the annual consumption of fossil fuels has scarcely altered (BP, 2022).[2] The comparatively high cost of energy erodes purchasing power, even for those who do not travel (except by foot). This is because energy is often a hidden component of the price of food and other essentials.

Major exporters of fossil fuels (e.g. Australia, Norway, Russia and Saudi Arabia) benefit from high prices, and have long used their influence to slow the energy transition (Krange *et al.*, 2019; McKie, 2021). The apparently perennial world crises also slow the transition to renewable energy; populations experiencing declining living standards are less willing to pay for the required infrastructure. Funds are found in Western nations to bail out the global banking system, or to fund the defence of Ukraine, but such crises

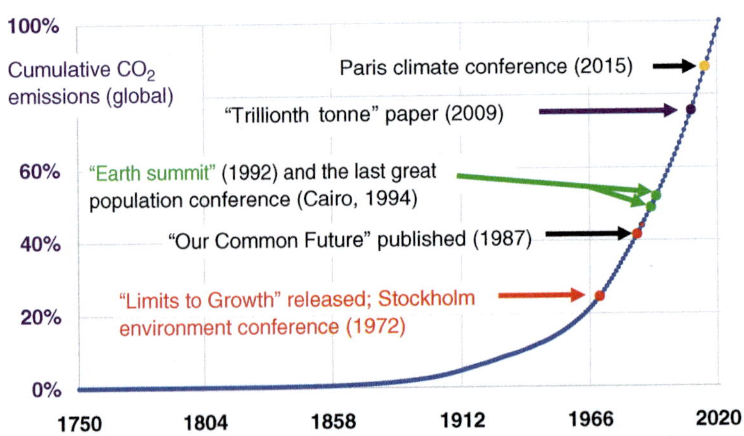

Fig. 40.1. More than half of global CO_2 emissions since the start of the industrial revolution (1750) have occurred in the past 30 years. Numerous warnings and pledges have been, effectively, ignored. Data source: https://ourworldindata.org/ (accessed 27 July 2023).

are also used to justify delaying the global sustainability transition, including the roll-out of less polluting forms of energy.

The rising energy price reflects less accessible fossil fuel sources. Although the concept of 'peak oil' is too simplistic, fossil fuels that remain unburned are of lower quality and costlier to exploit than those available decades ago. Even so, most remaining fossil fuel supplies, other than tar sands and 'heavy oil', have far higher 'net energy' than modern renewables (Hall, 2022). There has been a prolonged rush to locate and extract these fuels, because not only will their accessibility continue to decline, but so too will the 'social licence' permitting their extraction. But, while supplies are claimed as plentiful, their rate of extraction is managed in order to maintain comparatively high prices. Unfortunately, ample fossil fuels remain to raise average global temperature well beyond the 2°C warming mark, once considered the 'guard rail'. This could then trigger runaway warming, threatening civilization as we know it; a point made repeatedly in earlier chapters (Besley and Peters, 2020; Kemp *et al.*, 2022).

Recognizing this, terms such as 'unburnable carbon' and the 'carbon bubble' have been coined (Leggett, 2014). These attempt to generate investor awareness of the possibility of 'stranded assets', i.e. worthless investments such as a railway to a coal mine – wasted financial capital if the coal has no market. The encouragement of such consciousness is a major factor for fossil fuel divestment campaigns. If the world was one in which most people truly understood the risks of exceeding 1.5°C (let alone 2°C) then fossil fuel prospecting would already be in sharp decline. This is far from the case, and it is in the interest of fossil fuel sellers (and the governments and shareholders with whom they collude) to delay such awareness for as long as possible.

Fossil fuel investments may also become stranded if sufficient momentum for renewable energy accrues. A tenfold increase in the 'net energy' of renewables, generating good financial returns to the companies investing in them (some of whom currently profit from fossil fuels), suggests an escape from this dilemma, and some steps in this direction have been taken (Raugei, 2019). But the pace of this energy transition is slowed greatly by the stubborn adherence of governments and financial markets to conventional (non-ecological) economics, by social denial, by technological obstacles and by organized forces who profit from 'business as usual' (e.g. Oreskes and Conway, 2010).

40.2 Climate Change Mitigation

Since climate change emerged as a serious global issue at the first Rio de Janeiro 'Earth Summit' in 1992, which saw the establishment of the UN Framework Convention on Climate Change, the stated strategies to deal with it have focused on mitigation and adaptation. In healthcare, mitigation is analogous to primary prevention (e.g. a vaccine or smoking cessation); adaptation can be likened to health interventions following an adverse health event (for example, an exercise programme to regain cardiovascular health after cardiac surgery).

The philosophy underlying mitigation, sometimes called the precautionary principle, has vernacular analogies such as 'look before you leap'. Yet a counterview is 'he who hesitates is lost'. In other words, some argue that progress requires risk taking, even if this leads to the most distressing consequences outlined in this book, especially via the tertiary pathways.

Climate change mitigation relies on ways to reduce fossil fuel emissions, including energy efficiency, and new carbon-sparing forms of electrical energy capture such as from the wind or sun. Energy use can be made more efficient, such as via improved lighting or the use of the internet for meetings, reducing physical travel. These measures illustrate 'technological leapfrogging', a term originally applied to the way in which economies powered mainly by livestock could transition to those fuelled by the sun, bypassing the fossil fuel intermediary (Reddy and Goldemberg, 1990). However, 'progress' can create unforeseen energy burdens, such as those used to provide electronic data storage, or to 'mine' for bitcoins.

Value changes promoted by education, peer behaviour, environmental stewardship and voluntary simplicity are also vital, as championed by Hanlon and Carlisle (2010). Urbanization contributes to warming in ways additional to the release of greenhouse gases (GHGs), such as by changing transpiration and albedo (Ouyang *et al.*, 2022). Modifying agricultural land use in ways that will ameliorate climate change (e.g. greater carbon sequestration in soils) is also a form of climate change mitigation, as is reduced deforestation. In turn, some of these changes can be encouraged by new economic signals that will shift civilization away from fossil fuels to the 'Green Economy' (Kosoy *et al.*, 2012), based on principles of ecological economics (Daly, 1996).

40.3 Climate Change Adaptation

In the 1990s, the dominant discourse concerning climate change focused on the need to avoid 'dangerous' warming, then largely judged as 2°C or more above the average pre-industrial global temperature. In more recent years, attention shifted away from climate mitigation towards adaptation; too far, in our judgement. Of course, adaptation is essential, not only because climate change has already occurred but also because additional effects are inevitable, even if the world were, henceforth, to practise aggressive mitigation. In this respect, adaptation is also precautionary. However, as the evidence of climate change has strengthened and as carbon-sparing, energy-generating technologies have improved and reduced in cost, there is a strengthening view that civilization need not avoid this risk but can, instead, adapt to it. A reliance on adaptation is likely to magnify inequity, given differential adaptive capacity. For instance, a reliance on adaptation will see most of the world's Indigenous peoples disproportionately disadvantaged, including through the erosion of traditional cultures (Bowles, 2015). Adaptation is necessary, but should not be seen as an alternative to mitigation.

Adaptation could even be a pretence, a justification by rich polluting populations to indefinitely defer mitigation. Such thinking may well have slowed mitigation, in the same way that the chimera of clean coal (carbon sequestration) has. Beyond a critical point of climate change, adaptation is certain to be inadequate.

40.4 Values Shift

Common to many strategies for both mitigation and adaptation is the need for a shift in values. While some groups may express extreme techno-optimism, in which no social changes are required because technology solves everything, such positions are increasingly rare. Collectively, the global shift in cultural, political and behavioural change might be called an eco-noösphere – a global understanding that we are one civilization on one planet (Vernadsky, 1945). Increasingly, leaders in science, development and the media are calling for such changes.

Values can also shift concerning family size. Replacement or below-replacement fertility is now widespread, not only in most of Europe but also in China and much of east Asia, other than the Philippines. Increased access to family planning has been called for as part of the national (climate) adaptation plans for many least-developed countries (Bryant *et al.*, 2009).

Pope Francis (2015), whose name evokes a love of animals, is a powerful advocate for stewardship (rather than exploitation) of 'Our Common Home'. The Pontifical Academy of Science has hosted a series of meetings involving scientists and Christian faith leaders on this theme (Al-Delaimy *et al.*, 2020). Unfortunately, Catholic leaders continue to deny the environmental and social benefits of contraception, other than by use of the 'rhythm' method, which is disproportionately inaccessible to the poorest.

In an interview cited by Hitchens (1995), Catholic nun Mother Teresa (later canonized), was asked, 'So you wouldn't agree with people who say there are too many children in India?' She replied: 'I do not agree, because God always provides. He provides for the flowers and the birds, for everything in the world He has created. And those little children are his life. There can never be enough.'

The position of Evangelical Christians (30% of the US population) towards climate change has recently been posited as becoming less polarizing (Anderson, 2020). In 2008, the Ecumenical Patriarch Bartholomew was named one of *Time Magazine*'s 100 Most Influential People in the World, defining environmentalism as a spiritual responsibility (Anonymous, 2018). Tibetan Buddhists have long sought to protect nature (Butler, 2014). Environmental protection can be interpreted as embedded in Islamic and Hindu scriptures (Sadiqa and Nadeem, 2022). Yet, greenhouse gases rise, while, in the global South, many rivers and canals are choked with rubbish. As with other religious teachings, it may be that those on environmental protection are only partly observed. Nevertheless, their partial effectiveness is often through individuals' self-policing due to the threat of supernatural sanction. This means that religious adherence offers a unique challenge to people operating in their own self-interest at the expense of the common good or ecological sustainability (Rappaport, 1999).

Taylor (2020) hypothesized that the popularity of the 2009 motion picture *Avatar* may herald evidence that Earth is witnessing the emergence of the 'nascent stages of a new global nature religion'. Adherents believe that nature is sacred, entailing an ethical obligation to try to protect other species. Some supporters of the Green movement may fit in this category, as may some Indigenous populations. The traditional religions of many societies played vital roles in managing ecological balance. Former President of the American Anthropological Association, Roy Rappaport, argued that religion, which can be modified to be explicitly ecological, may be humanity's best chance to stay within the Earth's ecological limits (Rappaport, 1999).

Expanding selves

Patterns of environmental awareness and consumption are linked intimately with people's sense of themselves. For over a century (but accelerating in more recent decades) a shift has occurred in the self-identity of many people, especially in the global North, from producers to entitled consumers (Hamilton, 2010; Higgs, 2014). This has been incited by advertisers utilizing public relations techniques, creating messages that products will enhance their purchaser, making them more attractive, better parents and even more fully realized individuals (Wiedmann *et al.*, 2020). For many (at least in the global North) the prospect of reduced consumption is an affront, which dwarfs the amorphous shadow of climate change.

Those with a narrow, egocentric concept of self may act with less regard for the environment than those who conceptualize themselves as linked with others, especially those whose self-identity incorporates aspects of nature (Hamilton, 2010). Hard economic times may prove useful in this regard. As the economy shudders and people find it harder to buy goods that would define them, they may turn to other sources of identity, such as community, family, religion, voluntary activities and connection with nature.

Medical and other health professionals have a potentially powerful role to play in helping people grow their sense of self. As they tend to be near the top of the economic and social hierarchy, their example is potent. The social impact of bike racks, rather than parking places, being reserved for doctors outside of hospitals could be considerable, as the definition of success moves from fast cars toward toned calves.

Expanding selves: awareness of others

Some religious practices benefit physical health, such as taboos on eating pork (at least in areas where parasites in pigs are common). Religious involvement also promotes mental health through greater social connectivity. Prayer and meditation may be of benefit, irrespective of the existence of a deity (Kawachi, 2020). Some religiously enforced edicts, such as the Hindu protection of cows, have been postulated as having ecological benefits (Harris *et al.*, 1966). In addition, all religious traditions encourage an awareness of the value of the other, be they a Samaritan or a stranger. Yet, in practice, tolerance of the other easily transforms to mass *in*tolerance. If we are to survive as a civilization (and perhaps even as a species), we must overcome this.

Contraction and convergence

The concept of contraction and convergence (Meyer, 2000) appeals to those attracted by fairness, rationality and incrementalism, by creating pollution budgets that are divided between over- and under-consumers. In this model, high consumers voluntarily reduce their 'footprint', such as of carbon emissions, while under-consumers increase theirs, so that per capita consumption converges on an equitable level. Thus, this limits total environmental harm, while also protecting the most vulnerable. A group that also warrants consideration is future generations; another is nature herself.

In 2009, investigators concluded that, to stay beneath the 2°C 'guardrail', the world had to restrict cumulative carbon emissions, released by the burning of fossil fuels and deforestation, to approximately one trillion tonnes (Allen *et al.*, 2009). They calculated that approximately half of this 'budget' had already been consumed. Since then, a further quarter of this budget (i.e. half that was remaining) has been used (Fig. 40.1). This suggests a gross disregard for the 'rights' of the future, a paralysed incapacity to act, or perhaps, elements of each.

Without a fundamental value shift the impact of contraction and convergence is greatly constrained. To date, there is little evidence that those who are the most affluent have reduced their carbon or other forms of environmental footprint (Wiedmann *et al.*, 2020). Another form of voluntary redistribution, the Tobin tax, remains theoretical and has been stoutly resisted by business interests. The Pareto Principle holds that scarcity is best solved not by redistribution but by lifting consumption by the poor, i.e. convergence *without* contraction. Pareto's strategy fails in a resource-constrained world; humanity does not yet seem to have gained sufficient selflessness for contraction and convergence to be plausible.

Responses involving the health sector

Nevertheless, one variant of contraction and convergence holds promise. This is where health and other elements of 'co-benefits' may result from ecologically positive measures. This strategy appeals to the sense of justice and fairness held by those who are more fortunate, for example with regard to global meat consumption. People who ingest high levels of red meat (especially from digastric animals such as cattle and sheep) have an enhanced risk of bowel cancer and heart disease, and also contribute disproportionately to GHG emissions. Therefore, reduced ingestion of red meat by some people offers to improve their health, slow climate change and improve animal welfare (McMichael *et al.*, 2007; Bonnet *et al.*, 2020). On the other hand, an increased intake of red meat will benefit more than a billion people who are chronically iron deficient, though other ways to improve their iron status, such as treatment of parasites, will also help (UNEP, 2022).

Two more win–win strategies are increased active transport and better housing. Such strategies increase human fitness (and health), reduce energy use and lessen social isolation, thus potentially improving social capital. Being a climate activist may also deliver co-benefits, such as for mental health (Schwartz *et al.*, 2022).

Many health organizations have been founded by health workers who attempt to promote this cultural shift towards greater environmental protection. These include umbrella organizations such as the Global Climate and Health Alliance and the International Society of Doctors for the Environment (and their affiliates, including Physicians for Social Responsibility (US), and the Canadian Association of Physicians for the Environment, Doctors for the Environment Australia, MedAct (UK), The Planetary Health Alliance, and XR Health Professionals). An important NGO focusing on global health is The People's Health Movement (Musolino *et al.*, 2020).

40.5 Muddling Through with Technology

A comparatively high energy price appears permanent, even if the Ukrainian war ends. Scarcity may drive civilization towards shared global frugality, but an intensifying fortress world, in which anxious populations try to preserve their high levels of material and energy consumption, even though surrounded by want, is plausible. Such prospects may seem blasé to those used to living within the fortress, but conditions within currently prosperous regions could deteriorate. For example, the riots in France in mid-2023 may be seen as a portent, unless racism in France and Europe more broadly can be reduced. Although we think this would be a wise strategy, an intensification of inequality and an even more intrusive surveillance society seems more likely. Already, some may look back wistfully to three decades ago, when low energy prices and comparative prosperity offered greater opportunity to prepare for and to avert problems which are now upon us.

Planetary engineering and risk

Proposals of intentional planetary 'geoengineering', using stratospheric aerosol 'injections', have existed since at least 1991 (Teller *et al.*, 1997).[3] Such proposals may help offset warming, but have not been proven to be successful and will have an uncertain effect on the hydrological cycle (Irvine and Keith, 2020). They have also been criticized for further concentrating power among elites, as well as being vulnerable to militarization (Surprise, 2020; Stephens *et al.*, 2021). An artificial haze is also likely to reduce solar electricity generation and crop growth.

The prospect of geoengineering also contributes to moral hazard (Keith, 2000), as may other putative 'rescue strategies' such as large-scale tree planting (vulnerable to fire), carbon capture (expensive and also not proven at scale) and biomass sequestration (Anderson, 2015) and other forms of direct CO_2 removal (Dyke *et al.*, 2021). The deliberate sinking of massive quantities of kelp to safely sequester significant quantities of carbon has also been proposed, but Filbee-Dexter (2020) reported a 'strong' risk that such carbon could re-enter the atmosphere if unfavourable environmental conditions evolve.

40.6 Balancing Fear and Hope: Devising a Social Vaccine to Protect Civilization

Many trends in population are cyclic, driven by the waxing and waning of social memory and behaviour (Butler, 2005), in a way that can be likened to physical immunity. Following epidemics, surviving populations have sufficient antibodies to inhibit repeat infection until enough immunologically naïve people accrue, due to waning immunity and the maturation of a new generation. An analogous idea is that of a social vaccine. This metaphor has been suggested as a method to promote and maintain popular 'resistance' to unhealthy policies and practices (Baum *et al.*, 2009). Not only do we extend this concept to the causes of climate change and other aspects of 'planetary overload' (McMichael, 1993) but we also argue that an appropriate 'dose' of bad news is required to create an instrument that can effectively protect global society (Fig. 40.2).

Following the horror, terror, loss and suffering of two global conflicts in only 30 years and the intervening Great Depression, strong collective action led to a vigorous social contract seeking to prevent a Third World War. This movement was fortified by numerous institutions, led by the United Nations. Almost 80 years later, this form of 'social immunity' is still fairly powerful. This period of intense suffering, allied with damage wrought to and by European powers, liberated social forces that not only gave birth to the UK National Health Service but also reduced inequality for several decades outside the UK. Decolonization, for example, became unstoppable. However, gradually, many forms of inequality have reappeared, as has the risk of nuclear confrontation.

Today, global civilization is again clearly at risk. Numerous civilizations have collapsed in the past, driven by tight 'eco-socially' connected systems (Redman, 1999). An effective social vaccine could help stimulate civilization to protect itself, due to collective self-interest. Like a biological vaccine, a social vaccine needs to convey a small 'dose' of the disease; that is, a sufficiently compelling

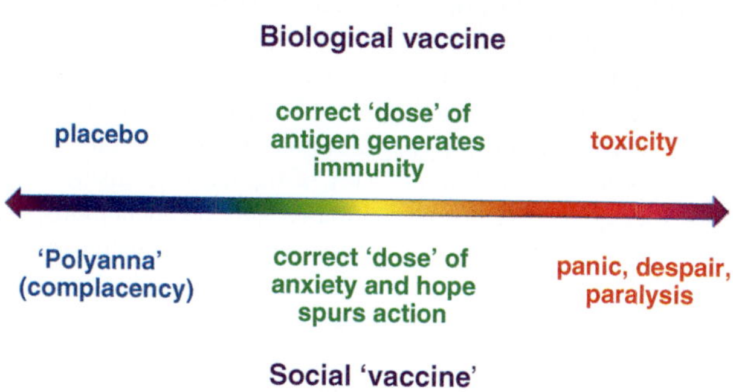

Fig. 40.2. Excess antigen in a biological vaccine causes toxicity, such as serum sickness. Excess bad news harms a social vaccine, but some anxiety is required to stimulate a helpful response. For a mass audience, in which people have varying knowledge and sensitivity, the optimal dose of hope and fear will vary. Although that dose is difficult to determine, policy makers in the global North have long erred towards too much hope.

portrayal of the consequences of inaction to provoke a protective stimulus. Its aim is to 'immunize' enough people to generate sufficient self-corrective action. Something akin to this was employed by the peace movement during the Cold War, for example by describing the effect of a thermonuclear explosion in Boston, Massachusetts (Sidel *et al.*, 1962; Chivian, 2013).[4]

On the other hand, many have pointed out that too blunt a warning has a paralysing, counterproductive effect (Albrecht, 2011). 'Eco-paralysis' has other causes too, such as the indefinite wait for the always elusive perfect deal (Albrecht, 2019). As Torres (2021) stated, 'the key question is how to talk about global catastrophic risks in ways that inspire people to become activists, rather than nudge them into fundamentalist beliefs or an attitude of defeatism'.

Paralysis, despair or other negative 'earth emotions', arising from excessive fear, is analogous to how too much antigen in a physical vaccine can cause harm. But a placebo social vaccine may be more dangerous, by inducing complacency, and fuelling moral hazard – the illusion that all we have to do is wait and the problem will be solved by others. A balance between these two extremes is thus needed. Failure to alter planetary consciousness is likely to lead to an enforced and unpleasant change, following a time of turmoil that may make the Dark Ages seem desirable. Perhaps the most plausible ugly scenario is of a fortress (triage) world in which diminishing populations maintain, at least for a time, reasonable order and well-being amidst an increasing sea of disorder, violence and coercion. Instead of this outcome, it is preferable to nurture a sufficient number of influential people to call for, develop and support the new policies that are required. Such policies include acceptance that resources are limited, green (ecological) economic systems that explicitly price negative externalities, and a revival of fairness of opportunity. But, of course, education is not enough; people need to change behaviour.

The eco-noösphere

No single solution is likely. Former US Vice President Al Gore referred to 'silver-buckshot', rather than a single magic 'silver bullet'.[5] This extends not just to technologies but also to social strategies and subsidies such as feed-in tariffs. The preceding chapters in this book indicate the complexity of the web of interconnections in the eco-noösphere, which links GHG emissions and ill health. This complexity, which resists human understanding, can lend the challenge of climate change an aura of intractability. Yet the high degree of interconnection can aid solutions. Webs have key supporting strands on which they rely. Individual efforts, when well targeted, have the potential to yield disproportionately powerful results.

One such key strand is charismatic individuals who serve as catalysts and sources of inspiration. There are many: from the late Wangari Maathai to Mary Robinson, Jane Goodall and David Attenborough. Robert May, a former president of the Royal Society, once told of a game in which cooperation led to a better group outcome (such as repeating the prisoner's dilemma puzzle). In some settings, groups learned quickly to cooperate. But in other settings, some players seemed to derive perverse satisfaction from seeing others lose, even if at a personal cost (Nowak and May, 1992). If this is the situation now, we need to ask: to what extent can those who seek or at least passively observe the demise of Earth's support systems be restrained by those who seek to protect them?

Following a sufficiently severe global crisis, might a more cooperative global community emerge? For example, could a limited nuclear war or a sharp increase in the rate of sea level rise finally stimulate effective global action? Perhaps, but it would be even better if the collective imagination of these possibilities could avoid such a crisis, before its worst manifestations. Increasingly, this seems unlikely, yet it is vital not to lose all hope.

Possibly the most realistic promise of a solution lies in adequate high-level understanding of the risks of 'business as usual' (including over-reliance on adaptation). Numerous warnings by a series of UN Secretary Generals, including Antonio Guterres, give hope (Anonymous, 2022). The founding director of the UN Environment Programme once asked 'where on Earth are we going?' (Strong, 2001). However, to date, such leadership is largely symbolic. Increasing discussion of the Green Economy seems rhetorical while fossil fuel subsidies remain so high (Coady *et al.*, 2017).

Albrecht (2019) argued that use of the term Anthropocene is fatalistic. If so, this is surely a populist misunderstanding, because Paul Crutzen, widely credited with introducing this word, campaigned vigorously to reverse harmful processes that were associated with the Anthropocene, such as stratospheric ozone depletion, for which he shared the Nobel Prize (Schellnhuber, 2021). Lenton and Latour (2018) called for 'Gaia 2.0' in which humans add a level of self-awareness to Earth's self-regulation.

Few inhabitants of the Anthropocene are either inherently evil or seek to deliberately ruin our only habitable planet; indeed, most of us appear – on an individual scale – to behave in ways more 'civilized' than our Roman or Aztec ancestors. Yet, humans now possess the capacity to destroy not only our own species quickly, such as by the use of nuclear weapons, but also slowly, as we collectively shape a planet hostile to human life. Albrecht has called for a 'Symbiocene' in which humans live together for mutual benefit, recognizing the interconnectedness of life and all living things. Relatedly, the physician Bryan Furnass coined the word 'Sustainocene', in 2012. This was summarized by Faunce (2018) as a period when 'humanity acts as Earth stewards, preferring eco-centric to anthropo-centric (or even corporato-centric) governance structures and scientific endeavours in ways that deliver long-term prosperity'. Albert Einstein and his colleagues famously argued that 'we have to learn to think in a new way' if civilization is to survive (Born *et al.*, 1955).

Greater use of the internet, powered by clean energy, is an obvious way to increase opportunities, knowledge and social interaction and stimulation, especially in rural areas. This will also slow rural to urban migration and could lead to a decline in birth rates in many low-income countries and locations, such as Niger, Somalia and rural India. Greater gender equity, combined with other forms of human rights, will also slow fertility, an important adaptation and a strategy to reduce future climate change. Perhaps the 'Great Transition' (Raskin *et al.*, 2002) foreseen by some optimists will not emerge, but at least a 'muddle-through world' may (Butler, 2010).

William Shatner, commander of the starship *Enterprise* in the television series *Star Trek*, described the 'Overview Effect' – a term he attributed to Frank White – which Shatner (2022) experienced during his trip to space, as 'a sense of the planet's fragility' taking hold in an 'ineffable, instinctive manner'. He noted that 'every day, we are confronted with the knowledge of further destruction of Earth at our hands: the extinction of animal species, of flora and fauna ... things that took five billion years to evolve, and suddenly we will never see them again because of the interference of mankind'. White (1987) described the Overview Effect as 'the foundation of the philosophy necessary to build a planetary civilisation and a planetary system'. Much more of this kind of awareness is required.

Designing and administering an appropriate social vaccine with which to protect civilization is as formidable as it is audacious. History reveals that many leaders and visionaries have issued warnings, even prophecies, that have been ignored. The Roman empire vanished, as did empires in the so-called New World, despite rituals, sacrifices and systems intended to provide protection. Sometimes collapse was staggered over centuries, sometimes it was fast. If global civilization is to survive, then its leaders (apart from Guterres, who seems genuinely concerned) need a booster dose of the social vaccine which we have postulated, but with a far higher dose of alarm.

Acknowledgements

This chapter is adapted from Butler, C.D., Stott, R., Bowen, K., Haigh, M., Smith, B., Braidwoood, M. and Bowles D.C. (2014) Climate change and health: from adaptation towards a solution. In: Butler, C.D. (ed.) *Climate Change and Global Health*. CABI, Wallingford, UK, pp. 284–292.

Notes

[1] Under Rwandan law the use of ethnic identifications has been prohibited since 1994. The official designation of the most recent Rwandan mass killing as 'genocide against the Tutsis' thus seems disingenuous,

not least as tens of thousands of Hutus were also killed. If the Rwandan government was sincere about ending ethnic identification, why would it repeatedly remind its population of which group faced the greatest persecution in 1994?

[2] Between 2019 and 2021, renewable energy increased by over 8 exajoules. However, consumption of fossil fuels was broadly unchanged; accounting for 82% of primary energy in 2021, down from 83% in 2019 and 85% in 2017 (BP, 2022).

[3] Kolbert (2021) traced the concept to a report presented to President L.B. Johnson in the 1960s; however, the method proposed was to reflect sunlight from the ocean.

[4] '... as a result of these concrete stories, I believe we helped make nuclear war more real for people, we made it harder for them to think about such wars in vague, abstract, technical terms, and in the process, I believe, we helped change public opinion and indeed maybe even public policy about the use of these weapons' (Chivian, 2013).

[5] In a speech to the New York University School of Law (18 September 2006), Gore referred to 'what Bill McKibben has called "silver-buckshot" – numerous important solutions, all of which are hard, but no one of which is by itself the full answer for our problem', in contrast to finding a single magic 'silver bullet'.

References

Agho, K.E., Mukabutera, C., Mukazi, M., Ntambara, M., Mbugua, I. *et al.* (2019) Moderate and severe household food insecurity predicts stunting and severe stunting among Rwanda children aged 6–59 months residing in Gicumbi district. *Maternal & Child Nutrition* 15, e12767. DOI: 10.1111/mcn.12767

Albrecht, G.A. (2011) Chronic environmental change: Emerging 'psychoterratic' syndromes. In: Weissbecker, I. (ed.) *Climate Change and Human Well-being. International and Cultural Psychology*. Springer, New York, pp. 43–56. DOI: 10.1007/978-1-4419-9742-5_3

Albrecht, G.A. (2019) *Earth Emotions*. Cornell University Press, Ithaca, New York.

Al-Delaimy, W.K., Ramanathan, V. and Sorondo, M.S. (eds) (2020) *Health of People, Health of Planet and Our Responsibility*. Springer Open, Cham, Switzerland. Available at: https://link.springer.com/book/10.1007%2F978-3-030-31125-4 (accessed 6 July 2023).

Allen, M.R., Frame, D.J., Huntingford, C., Jones, C.D., Lowe, J.A. *et al.* (2009) Warming caused by cumulative carbon emissions towards the trillionth tonne. *Nature* 458, 1163–1166. DOI: 10.1038/nature08019

Anderson, K. (2015) Talks in the city of light generate more heat. *Nature* 528, 437. DOI: 10.1038/528437a

Anderson, L. (2020) Faith in God and the health of people. In: Al-Delaimy, W.K., Ramanathan, V. and Sorondo, M.S. (eds) *Health of People, Health of Planet and Our Responsibility*. Springer Open, Cham, Switzerland, pp. 265–272.

Anonymous (2018) *The Green Patriarch – Ecumenical Patriarch Bartholomew*. Greek Orthodox Archdiocese of America, New York. Available at: https://www.goarch.org/society/greening-the-parish/green-patriarch (accessed 27 July, 2023).

Anonymous (2022) UN chief warns against 'sleepwalking to climate catastrophe'. *UN News*. Available at: https://news.un.org/en/story/2022/03/1114322 (accessed 2 October 2022).

Anonymous (2023) Hottest July ever signals 'era of global boiling has arrived' says UN chief. *UN News*. Available at: https://news.un.org/en/story/2023/07/1139162 (accessed 30 July 2023).

Associated Press (2023) Climate change spells 'terrifying' future: UN rights chief. *France* 24. Available at: https://www.france24.com/en/live-news/20230703-climate-change-spells-terrifying-future-un-rights-chief (accessed 5 July 2023).

Baum, F., Narayan, R., Sanders, D., Patel, V. and Quizhpe, A. (2009) Social vaccines to resist and change unhealthy social and economic structures: a useful metaphor for health promotion. *Health Promotion International* 24, 428–433. DOI: 10.1093/heapro/dap026

Besley, T. and Peters, M.A. (2020) Life and death in the Anthropocene: Educating for survival amid climate and ecosystem changes and potential civilisation collapse. *Educational Philosophy and Theory* 52, 1347–1357. DOI: 10.1080/00131857.2019.1684804

Binagwaho, A., Farmer, P.E., Nsanzimana, S., Karema, C., Gasana, M. *et al.* (2014) Rwanda 20 years on: investing in life. *The Lancet* 384, 371–375. DOI: 10.1016/S0140-6736(14)60574-2

Binagwaho, A., Hinda, R. and Mills, E. (2019) Rwanda and revisionist history. *The Lancet* 320, 318–319. DOI: 10.1016/S0140-6736(19)30121-7

Bonnet, C., Bouamra-Mechemache, Z., Réquillart, V. and Treich, N. (2020) Viewpoint: Regulating meat consumption to improve health, the environment and animal welfare. *Food Policy* 97, 101847. DOI: 10.1016/j.foodpol.2020.101847

Born, M., Bridgman, P.W., Einstein, A., Infeld, L., Joliot-Curie, F. *et al.* (1955) *Russell–Einstein Manifesto*. Atomic Heritage Foundation, Washington, DC. Available at: https://www.atomicheritage.org/key-documents/russell-einstein-manifesto (accessed 8 October 2022).

Bowles, D.C. (2015) Climate change and health adaptation: Consequences for indigenous physical and mental health. *Annals of Global Health* 81, 427–431. DOI: 10.1016/j.aogh.2015.06.004

BP (2022) *Statistical Review of World Energy*, 71st edn. BP, London. Available at: https://www.bp.com/en/global/corporate/energy-economics/statistical-review-of-world-energy.html (accessed 4 October 2022).

Bryant, L., Carver, L., Butler, C.D. and Anage, A. (2009) Climate change and family planning: least developed countries define the agenda. *Bulletin of the World Health Organization* 87, 852–857. DOI: 10.2471/BLT.08.062562

Butler, C.D. (2005) Peering into the fog: ecologic change, human affairs and the future. *EcoHealth* 2, 17–21. DOI: 10.1007/s10393-004-0090-x

Butler, C.D. (2007) Globalisation, population, ecology and conflict. *Health Promotion Journal of Australia* 18, 87–91. Available at: https://search.informit.org/doi/abs/10.3316/ielapa.452266146183185 (accessed 6 July 2023)

Butler, C.D. (2010) The climate crisis, global health, and the medical response. *World Medical Journal* 56, 56–58. Available at: https://www.wma.net/wp-content/uploads/2016/11/wmj26.pdf (accessed 2 October 2022).

Butler, C.D. (2014) Tibet, China, protest immolation and social medicine. In*:* Gislason, M.K. (ed.) *Ecological Health: Society, Ecology and Health*. Emerald Press, Bingley, UK, pp. 67–89.

Butler, C.D., Chambers, R., Chopra, K., Dasgupta, P., Duraiappah, A. *et al.* (2005) How might ecosystems and their services change in the future under various plausible scenarios? In: Millennium Ecosystem Assessment (ed.) *Ecosystems and Human Well-being: Synthesis*. Island Press, Washington, DC, pp. 71–84

Chivian, E. (2013) Global environmental threats: why they are hard to see and how a medical model may contribute to their understanding. *Cardiovascular Diagnosis and Therapy* 3, 93–104. DOI: 10.3978/j.issn.2223-3652.2013.02.09

Coady, D., Parry, I., Sears, L. and Shang, B. (2017) How large are global fossil fuel subsidies? *World Development* 91, 11–27. DOI: 10.1016/j.worlddev.2016.10.004

Daly, H. (1996) *Beyond Growth*. Beacon Press, Boston, Massachusetts.

Dear, J. (2022) Paul Farmer was my friend. He should be made a saint – and a doctor of the church. *America*. Available at: https://www.americamagazine.org/faith/2022/02/23/paul-farmer-john-dear-remembrance-242450 (accessed 5 July 2023).

Dyke, J., Watson, R. and Knorr, W. (2021) Why net zero policies do more harm than good. In*:* Böhm, S. and Sullivan, S. (eds) *Negotiating Climate Change in Crisis*. Open Book Publishers, Cambridge, UK, pp. 39–52. DOI: 10.11647/OBP.0265.04

Faunce, T.A. (2018) Global artificial photosynthesis and renewable energy storage and policy for the Sustainocene. *Advanced Sustainable Systems* 2, 1800035. DOI: 10.1002/adsu.201800035

Filbee-Dexter, K. (2020) Ocean forests hold unique solutions to our current environmental crisis. *One Earth* 2, 398–401. DOI: 10.1016/j.oneear.2020.05.004

Ganguly, M. (2023) *Ethnic Violence Spirals in India's Northeastern Manipur State*. Human Rights Watch, New York. Available at: https://www.hrw.org/news/2023/06/30/ethnic-violence-spirals-indias-northeastern-manipur-state (accessed 6 July 2023).

Garret, L. (2018) Rwanda: not the official narrative. *Lancet* 392, 909–912. DOI: 10.1016/S0140-6736(18)32124-X

Gore, A. (1992) *Earth in the Balance*. Houghton Mifflin, Boston, Massachusetts.

Hall, C.A.S. (2022) The 50th anniversary of The Limits to Growth: Does it have relevance for today's energy issues? *Energies* 15, 4953. DOI: 10.3390/en15144953

Hamilton, C. (2010) *Requiem for a Species. Why We Resist the Truth About Climate Change*. Allen and Unwin, Sydney, Australia.

Hanlon, P. and Carlisle, S. (2010) Re-orienting public health: rhetoric, challenges and possibilities for sustainability. *Critical Public Health* 20, 299–309. DOI: 10.1080/09581596.2010.482581

Harris, M., Bose, N.K., Morton, K., Mencher, J.P., Oberg, K. *et al.* (1966) The cultural ecology of India's sacred cattle. *Current Anthropology* 7, 51–66. https://www.jstor.org/stable/2740230

Higgs, K. (2014) Growth and consumerism. *Collision Course: Endless Growth on a Finite Planet*. MIT Press, Cambridge, Massachusetts, pp. 65–78.

Hitchens, C. (1995) *The Missionary Position: Mother Teresa in Theory and Practice*. Verso, London and New York.

Horton, R. (2016) Offline: The King and his courtiers. *The Lancet* 387, 1800. DOI: 10.1016/S0140-6736(16)30309-9

Irvine, P.J. and Keith, D.W. (2020) Halving warming with stratospheric aerosol geoengineering moderates policy-relevant climate hazards. *Environmental Research Letters* 15, 044011. DOI: 10.1088/1748-9326/ab76de

Kawachi, I. (2020) Religion as a social determinant of health. *American Journal of Epidemiology* 189, 1461–1463. DOI: 10.1093/aje/kwz204

Keith, D.W. (2000) Geoengineering the climate: history and prospect. *Annual Review of Energy and Environment* 25, 245–284. DOI: 10.1146/annurev.energy.25.1.245

Kemp, L., Xu, C., Depledge, J., Ebi, K.L., Gibbins, G. *et al.* (2022) Climate endgame: exploring catastrophic climate change scenarios. *Proceedings of the National Academy of Sciences (USA)* 119, e2108146119. DOI: 10.1073/pnas.2108146119

Kolbert, E. (2021) *Under a White Sky. The Nature of the Future*. Penguin, Random House, New York.

Kosoy, N., Brown, P.G., Bosselmann, K., Duraiappah, A., Mackey, B., Martinez-Alier, J. *et al.* (2012) Pillars for a flourishing Earth: planetary boundaries, economic growth delusion and green economy. *Current Opinion in Environmental Sustainability* 4, 74–79.

Krange, O., Kaltenborn, B.P. and Hultman, M. (2019) Cool dudes in Norway: climate change denial among conservative Norwegian men. *Environmental Sociology* 5, 1–11. DOI: 10.1080/23251042.2018.1488516

Leggett, J. (2014) *The Energy of Nations. Risk Blindness and the Road to Renaissance*. Earthscan, Routledge, London.

Lenton, T.M. and Latour, B. (2018) Gaia 2.0. *Science* 361, 1066–1068. DOI: 10.1126/science.aau0427

McKie, R.E. (2021) Obstruction, delay, and transnationalism: Examining the online climate change counter-movement. *Energy Research & Social Science* 80, 102217. DOI: 10.1016/j.erss.2021.102217

McMichael, A.J. (1993) *Planetary Overload. Global Environmental Change and the Health of the Human Species*. Cambridge University Press, Cambridge, UK.

McMichael, A.J., Powles, J., Butler, C.D. and Uauy, R. (2007) Food, livestock production, energy, climate change and health. *The Lancet* 370, 1253–1263. DOI: 10.1016/S0140-6736(07)61256-2

Médecins Sans Frontières (2023) *Ten years of atrocities and violence in Central African Republic*. Available at: https://www.msf.org/ten-years-atrocities-and-violence-central-african-republic (accessed 6 July 2023)

Meyer, A. (2000) *Contraction and Convergence. The Global Solution to Climate Change*. Green Books for the Schumacher Society, Foxhole, Devon, UK.

Musolino, C., Baum, F., Freeman, T., Labonté, R., Bodini, C. *et al.* (2020) Global health activists' lessons on building social movements for Health for All. *International Journal for Equity in Health* 19, 116. DOI: 10.1186/s12939-020-01232-1

Nowak, M.A. and May, R.M. (1992) Evolutionary games and spatial chaos. *Nature* 359, 826–829. DOI: 10.1038/359826a0

Oreskes, N. and Conway, E.M. (2010) *Merchants of Doubt: How a Handful of Scientists Obscured the Truth on Issues from Tobacco Smoke to Global Warming*. Bloomsbury Press, New York.

Ouyang, Z., Sciusco, P., Jiao, T., Feron, S., Lei, C. *et al.* (2022) Albedo changes caused by future urbanization contribute to global warming. *Nature Communications* 13, 3800. DOI: 10.1038/s41467-022-31558-z

Pinker, S. (2011) *The Better Angels of our Nature*. Viking, New York.

Pope Francis (2015) *Encyclical Letter Laudato Si' of the Holy Father Francis on Care for Our Common Home*. Vatican City. Available at: https//www.vatican.va/contentakistanicesco/pdf/encyclicals/documents/papa-francesco_20150524_enciclica-laudato-si_en.pdf (accessed 22 October 2022).

Rappaport, R. (1999) *Ritual and Religion in the Making of Humanity*. Cambridge University Press, Cambridge, UK.

Raskin, P., Gallopin, G., Gutman, P., Hammond, A., Kates, R. and Swart, R. (2002) *Great Transition: The Promise and Lure of the Times Ahead*. Stockholm Environment Institute, Boston, Massachusetts.

Raugei, M. (2019) Net energy analysis must not compare apples and oranges. *Nature Energy* 4, 86–88. DOI: 10.1038/s41560-019-0327-0

Reddy, A.K.N. and Goldemberg, J. (1990) Energy for the developing world. *Scientific American* 263, 63–72. Available at: https://www.jstor.org/stable/24996935

Redman, C.L. (1999) *Human Impact on Ancient Environments*. University of Arizona Press, Tucson, Arizona.

Rees, M. (2013) Denial of catastrophic risks. *Science* 339, 1123. DOI: 10.1126/science.1236756

Reich, R. (2015) Hillary's Glass-Steagall mistake. *Chicago Tribune*. Available at: https://www.chicagotribune.com/columns/sns-201507211100--tms--amvoicesctnav-b20150721-20150721-column.html (accessed 6 July 2023).

Roy, A. (2023) Suella Braverman's Rwanda dream disappears as judges rule migrant policy illegal. *The Telegraph (India)*. Available at: https://www.telegraphindia.com/world/suella-bravermans-plan-to-deport-illegal-migrants-to-rwanda-is-unlawful-appeal-court-rules/cid/1948770 (accessed 6 July 2023).

Sadiqa, U. and Nadeem, Q. (2022) Deep ecology: A contemporary response of Islam and Hinduism for sustaining man-nature relationship in an era of environmental crisis. *Pakistan Journal of Humanities and Social Sciences* 10, 1211–1220. DOI: 10.52131/pjhss.2022.1004.0278

Schellnhuber, H.J. (2021) Paul Josef Crutzen: Ingeniousness and innocence. *Proceedings of the National Academy of Science (USA)* 118, e2104891118. DOI: 10.1073/pnas.210489111

Schwartz, S.E.O., Benoit, L., Clayton, S., Parnes, M.F., Swenson, L. *et al.* (2022) Climate change anxiety and mental health: Environmental activism as buffer. *Current Psychology* 42, 16708–16721. DOI: 10.1007/s12144-022-02735-6

Shatner, W. and Brandon, J. (2022) *Boldly Go: Reflections on a Life of Awe and Wonder*. Atria Books, New York.

Sidel, V.W., Geiger, H.J. and Lown, B. (1962) The physician's role in the postattack period. *New England Journal of Medicine* 266, 1137–1145. DOI: 10.1056/NEJM196205312662205

Smith, H., Baloch, S.M., Michaelson, R. and Graham-Harrison, E. (2023) Pakistanis 'were forced below deck' on refugee boat in Greece disaster. *The Guardian*. Available at: https://www.theguardian.com/world/2023/jun/18/pakistanis-were-forced-below-deck-on-refugee-boat-in-greece-disaster (accessed 5 July 2023).

Stephens, J.C., Kashwan, P., McLaren, D. and Surprise, K. (2021) The risks of solar geoengineering research. *Science* 372, 1161. DOI: 10.1126/science.abj3679

Strong, M. (2001) *Where On Earth Are We Going?* Alfred A. Knopf, Toronto, Canada.

Sundaram, A. (2016) *Bad News: Last Journalists in a Dictatorship*. Bloomsbury, London.

Surprise, K. (2020) Geopolitical ecology of solar geoengineering: from a 'logic of multilateralism' to logics of militarization. *Journal of Political Ecology* 27, 213–235. DOI: 10.2458/v27i1.23583

Taylor, B. (2020) Dark green religion: A decade later. *Journal for the Study of Religion, Nature and Culture* 14, 496–510. DOI: 10.1558/jsrnc.34630.

Teller, E., Wood, L. and Hyde, R. (1997) *Global warming and ice ages: I. Prospects for physics-based modulation of global change*. Paper submitted to 22nd International Seminar on Planetary Emergencies, Erice (Sicily), Italy. Available at: https://inis.iaea.org/collection/NCLCollectionStore/_Public/29/043/29043613.pdf (accessed 2 October 2022).

Torres, P. (2021) Scared straight: How prophets of doom might save the world. *Bulletin of the Atomic Scientists*. Available at: https://thebulletin.org/2021/05/scared-straight-how-prophets-of-doom-might-save-the-world/ (accessed 6 July 2023)

Trust, S., Joshi, S., Lenton, T. and Oliver, J. (2023) *The Emperor's New Climate Scenarios*. The Institute and Faculty of Actuaries and The University of Exeter. Available at: https://actuaries.org.uk/media/qeydewmk/the-emperor-s-new-climate-scenarios.pdf (accessed 6 July 2023).

UNEP (2022) *COVID-19: A Warning. Addressing Environmental Threats and the Risk of Future Pandemics in Asia and the Pacific*. United Nations Environment Programme, Bangkok, Thailand. Available at: https://wedocs.unep.org/20.500.11822/40871 (accessed 21 October 2022).

Vernadsky, V. (1945) The biosphere and the noösphere. *American Scientist* 33, 1–12. Available at: https://monoskop.org/images/5/59/Vernadsky_WI_1945_The_Biosphere_and_the_Noosphere.pdf (accessed 8 February 2024)

White, F. (1987) *The Overview Effect: Space Exploration and Human Evolution*. Houghton-Mifflin, Boston, Massachusetts.

Wiedmann, T., Lenzen, M., Keyßer, L.T. and Steinberger, J.K. (2020) Scientists' warning on affluence. *Nature Communications* 11, 3107. DOI: 10.1038/s41467-020-16941-y

WMO (2023) *July 2023 is set to be the hottest month on record*. World Meteorological Organization, Geneva. Available at: https://public.wmo.int/en/media/press-release/July-2023-set-to-be-hottest-month-recird (accessed 28 July 2023).

World Bank (2023) *World Bank Commodity Price Data (The Pink Sheet)*. Available at: https://www.worldbank.org/en/research/commodity-markets (accessed 6 July 2023).

Index

Printed and bound by CPI Group (UK) Ltd, Croydon, CR0 4YY

13/09/2026

14961109-0005